MODERN MICROELECTRONIC

CIRCUIT DESIGN, IC APPLICATIONS, FABRICATION TECHNOLOGY

VOL. I

Staff of Research and Education Association,
Dr. M. Fogiel, Director

Research and Education Association
505 Eighth Avenue
New York, N. Y. 10018

MODERN MICROELECTRONIC CIRCUIT DESIGN,
IC APPLICATIONS, FABRICATION TECHNOLOGY

Printed in the United States of America

Library of Congress Catalog Card Number 81-50168

International Standard Book Number 0-87891-520-6

PREFACE

This edition of "Modern Microelectronics" recognizes the importance for the circuit designer to be familiar with the structure and functional operation of microelectronic devices. Without such knowledge it is not possible to obtain optimum design with respect to the number of circuit components, size and weight, cost, and reliability. The special features and operating characteristics must be known to the designer in order to achieve optimum results. The modern designer will find that he cannot achieve such results by treating microelectronic devices as "black boxes" and applying to them conventional circuit analysis techniques.

This edition, therefore, commences with thorough descriptions of the important devices and circuits used in microelectronics. These descriptions are accompanied by numerous illustrations to aid the reader in understanding the subject matter rapidly. The essential findings of theoretical and experimental investigators are presented to give the reader a complete picture of the state of the art.

Following the discussions on the hardware aspects of microelectronics, is a practical section on the design of highly reliable circuits. Reliability of circuit operation is stressed here, since this is of utmost importance in electronic applications. For this reason the circuits that have been proven highly reliable in practice are presented in this section, rather than circuits that may appear interesting in some ways but do not have a record of reliable operation Reliability is also especially important in military applications—major consumers of microelectronics.

Historically, the primary object of the original workers in microelectronics was to develop methods for miniaturizing electronic equipment so as to result in technical and economical advantages through reduction of the large bulk and weight that was often associated with electronic installations. As the development program progressed, however, it was found that the methods used to fabricate the miniaturized electronic devices, also yielded two essential features: a very significant increase in the reliability of operation and a sharp reduction in cost. It was soon realized that these two features are of greater importance than the miniaturization characteristic, in promoting widespread acceptability of electronic devices.

Thus, before the advent of microelectronics, the use of complex electronic equipment was limited because of the large amount of continuous servicing that was required to maintain the equipment in operating condition. At the same time, the initial high cost of the equipment also confined its acceptance to relatively few users. Since microelectronic techniques are steadily removing these two obstacles, both industry and government are increasingly relying upon electronic equipment to perform functions previously accomplished through manual means, as well as to perform entirely new functions not possible before. Although microelectronics may still give the impression, through its name, that it is mainly concerned with miniaturization concepts, reliability improvements and cost reductions have actually become its main assets.

When the advantages of microelectronic techniques were discovered, the industry moved rapidly to transfer laboratory developments to practical applications. As a result, the electronics industry with its untiring appetite for R & D, made rapid advancements toward making microelectronic circuits commercially available in quantity. While considerable progress has been made in this direction, much remains to be accomplished to produce microelectronic devices having the same electrical properties as their conventional counterparts which they are intended to replace. Microelectronics, furthermore, is still very much an art rather than a science, and considerable work remains to be done to provide the understanding necessary for duplicating fabrication processes and products with precision and ease.

To discuss the potential and possible applications of microelectronics, is no longer fruitful. Microelectronics has made its impact, and the market for microelectronics is assured. There is a demand for microelectronics for almost every application where electronics is involved, provided that the technological and price requirements are met. However, to attractively price microelectronics for the large markets available, it is essential to overcome the technical problems that are encountered in fabrication and design.

Consequently, the successful marketing and utilization of micro-electronics today is accomplished essentially through technical know-how and thorough familiarity with the subject.

It is particularly important to be aware that microelectronics is not merely a matter of buying a series of black boxes in IC form, and then interconnecting them. Thus, successful microelectronics utilization cannot be achieved by sub-dividing an electronic equipment design into separate basic or standard IC circuits, obtaining these circuits in individual forms, and then interconnecting the individual circuits. This approach will not yield the advantages of minimum size, high reliability, and low cost available from microelectronics.

To benefit from microelectronics, it is necessary to treat the equipment or design as a whole or composite unit, and not as a combination of separate interconnected circuits, even if such circuits are individually in integrated form. This, in turn, requires familiarity with the various technical processes involved.

In applying microelectronics, it is essential that the system designer be acquainted with the underlying theory as well as their fabrication technology. In contrast to what has prevailed in the past, microelectronics has brought about the inextricable intertwinement of system design with component fabrication. Thus, it is essential that the practitioner in one of these areas have a thorough understanding of the methods and problems of the other. Experience has shown that the system designer is at a disadvantage if he is not schooled in the microelectronic fabrication processes and their capabilities.

The contents of this book are written and organized from the viewpoint that the designer or fabricator may need to become acquainted with design tools outside of his discipline. Also included are separate sections detailing the packaging, testing, reliability, maintenance, and procurement aspects of microelectronics for those who specialize in these areas. Thus, the book is intended to be a helpful source of information for the designer, fabricator, and those involved in supplying the necessary parts and materials, as well as maintaining the equipment in service. The first section of the book, ''Microelectronics at a Glance,'' is arranged in the form of a summary and outline with sufficient detail to provide the knowledge required to deal with microelectronics at the management level.

In writing and organizing the contents of this book, a large amount of available information was analyzed, screened, and reduced into form believed to be most useful to the reader. Available information from the USSR was also analyzed, and led to the conclusion that Soviet workers in microelectronics lacked know-how of the processing refinements that result in the higher precision and higher reliability devices produced in the U.S.A.

To disseminate the information which has been developed by those active in the microelectronics field, REA has given seminars at which leaders from government, industry and universities gathered to learn the techniques of microelectronics and to become acquainted with the latest developments in this field. The active participation of the members of these organizations at the seminars, encouraged the publication of a previous book, ''Microelectronics.'' The present book combines some basic information from the earlier volume with the latest technological advances. These advances have been considerable and have progressed at an increasingly rapid pace. Much gratitude is, therefore, due to the many participants, too numerous to name individually, who aided in the presentation of these seminars and gave the views of their government, industry, and university organizations. Particularly significant contributions to the book, through either seminar participation or published research, are gratefully acknowledged from the staffs of Aerojet-General, Air Force Cambridge Research Labs., Arinc Research Corp., Battelle Memorial Institute, Bell System, Bendix, Boeing, CBS Labs., Components Branch of Bureau of Naval Weapons, Corning Glass, Electronic Design, Electronic Engineer, Fairchild Semiconductor, General Electric, General Instrument, General Telephone and Electronics, Harvard University, Hughes Aircraft, IBM, IEEE, IRC, ITT, Johns Hopkins University, Lockheed Missiles and Space, Massachusetts Institute of Technology, Motorola, NASA, Naval Air Development Center Crane, Naval Research Lab., North American Rockwell, Philco-Ford, Princeton University, RCA, Research Triangle Institute, Rome Air Development Center, Royal Radar Establishment, Sprague Electric, Stanford Research Institute, Stevens Institute of Technology,

Texas Instrument, University of Illinois, U.S. Army Forth Monmouth, Varian Associates, Westinghouse Electric, Wright Patterson Air Force Base. Special thanks are due to H. J. Kaufmann for assistance in the preparation of this book.

Max Fogiel

CONTENTS

MICROELECTRONICS AT A GLANCE

PART I

THIN/THICK FILM TECHNOLOGY

1 FUNDAMENTAL PRINCIPLES OF THIN FILMS 62

2 FABRICATION OF THIN FILM MICROELECTRONIC CIRCUITS 90

3 CHARACTERISTICS OF THIN FILM MICROELECTRONIC CIRCUITS 136

PART II

INTEGRATED CIRCUITRY/ SEMICONDUCTOR DEVICES

5 MATERIALS FOR SEMICONDUCTOR DEVICES

OPTOELECTRONICS 233

8 EPITAXY 361

9 MONOLITHIC CIRCUIT FABRICATION 396

10 CHARACTERISTICS OF SEMICONDUCTOR MICROELECTRONIC CIRCUITS 432

PART III

MICROELECTRONIC CIRCUIT CONFIGURATIONS

PART IV

INTERCONNECTIONS
PACKAGING
CIRCUIT TESTING

12 INTERCONNECTIONS 613

13 PACKAGING 654

14 CIRCUIT TESTING 728

PART V

RELIABILITY
REDUNDANCY
MAINTAINABILITY
IC SPECIFICATION/PROCUREMENT

PROBLEMS AND REVIEW QUESTIONS FOR PARTS I – V 824

PART VI

ANALYTIC CIRCUIT DESIGN FOR RELIABLE EQUIPMENT OPERATION

1 BASIC PRINCIPLES 861

2 VARIABLES AND PARAMETER RELATIONS 881

3 DEVELOPMENT OF INTRINSIC DEVICE THEORY AND RELATED FUNDAMENTAL LIMITATIONS AND THEIR MEASUREMENTS 907

4 CIRCUIT PARAMETER RELATIONS 939

5 DESIGN OF TRANSISTOR *R-C* AMPLIFIERS 954

6 CIRCUIT STABILIZATION 985

7 TRANSFORMER-COUPLED AMPLIFIERS 995

8 RF AND IF AMPLIFIERS 1019

9 NONLINEAR THEORY OF OSCILLATORS 1038

10 PRACTICAL *L-C* OSCILLATORS 1050

11 *R-C* OSCILLATORS AND TIME-DELAY OSCILLATORS 1066

12 DESIGN OF MIXERS AND CONVERTERS 1083

13 TRANSISTOR MULTIVIBRATORS 1096

14 SWITCHING AND SAMPLING CIRCUITS 1113

APPENDIX A

DERIVATION OF DISTORTION EQUATIONS 1135

APPENDIX B

TOPOLOGICAL EQUATION DERIVATION 1136

APPENDIX C

LEGENDRE AND ORTHOGONAL POLYNOMIALS 1147

MICROELECTRONICS AT A GLANCE

INTRODUCTION

In the past, electronic systems were designed around components of known characteristics, and the selected components were then assembled to form the system. A feature of this situation has been the division of the entire electronic equipment production into basically three separate operations: design, component fabrication, and assembly. Initially, this had advantages, but eventually this process was considered undesirable from the viewpoints of unwieldy size, poor reliability, poor producibility, and high cost. The advent of sophisticated electronic systems requiring the use of an increasing number of components, forced a re-evaluation of the fabrication and assembly technology.

The result of this re-evaluation has been the development of two different solutions. One is based upon the use of small high-reliability components combined with new assembly techniques. In this method, the electronic system is still essentially an assembly of components, even though these may have new geometrical form factors, and improvements in size, weight, and reliability. Examples of this method, often designated as ''micropackaging,'' are pellet or dot components, micromodules, and cordwood. The second solution is based upon a totally new approach: the functions of the component producer, assembler, and designer are integrated, so that, ideally, all are performed by one group. In this case, the electronic systems are formed directly from raw materials, using physical and chemical processing techniques. This method has been designated as ''Microelectronics.''

This new method eliminates a great deal of the flexibility inherent in the older assembly methods. Thus, it is now difficult, and sometimes impossible, to change components or trim values. Furthermore, these new processes cannot, at the present time, produce the entire range and variety of components and devices available with the older methods. Because of these limitations, it is of the utmost importance that the system and circuit designer work very closely with the fabricator. This is necessary in order to take the greatest advantage of the available technology and to avoid difficult or impossible specifications.

At the present time, two principal microelectronic processes are available: ''Thin Film Microelectronics'' and ''Semiconductor or Integrated Circuit Microelectronics''. A thin film may be defined as one whose specific electrical properties depend upon the film thickness. At the present time, it is possible to produce resistances, capacitors, and inductances in thin film microelectronics. A variety of methods are available for making these thin films, each with its own set of advantages and disadvantages. Due to the limited range of components which can be made, most thin film circuits require the use of conventional components in addition to the thin film components.

In semiconductor microelectronics, the components that are commonly made are transistors, diodes, pnpn switches, and resistors. Component interconnections are made with thin films. New semiconductor devices with improved tolerances, better isolation, and higher frequency operation, would greatly ease the present constraints on circuit designers.

In tracing the history of microelectronics, thin films and semiconductors were both in use as electrical components and as light detectors at the beginning of this century. Attempts to develop methods of circuit fabrication embodying improvements in reliability, size, weight, and cost, started with screened resistor-capacitor networks on ceramic supports, about 25 years ago. This process is still being used, and has been applied in some of the most reliable systems.

Microelectronics, as known today, began about 1950 when it became possible to make thin film passive components and combine them with transistors. Semiconductors entered the picture near the end of the same decade when the introduction of the multistage thermal diffusion planar technology made it possible to make a complete semiconductor circuit.

There are significant differences between the cost factors of the thin film and semiconductor processes. The preparatory costs are far greater for the semiconductor microelectronic circuits than for the thin film circuit. A new thin film circuit can be made operable and put into production in much less time than required for the corresponding semiconductor circuit. Furthermore, the unit production cost is less for the thin film circuit, for small quantities. However, for long production runs the unit cost is less for the semiconductor circuit.

To take advantage of microelectronic techniques, the electronic designer should first understand something about the fabrication of microelectronic circuits. It is not necessary for the electronic designer to understand the complex physics involved in the fabrication procedures. He should, however, know the general processes involved, the physical and electrical characteristics of all the individual component parts, the expected interaction between these component parts, and the methods of assembling these parts into a circuit or system. He should also understand the characteristics of the various microelectronic approaches. This will enable him to compare the advantages and disadvantages of each approach, and decide which approach is best to solve his design problem.

The successful design of a microelectronic system depends upon the cooperation between the electronic designer and the microelectronic fabricator. They both must have some understanding of each other's problems. The decision as to the type of microelectronic approach that should be used to fabricate an electronic system, is as much the responsibility of the electronic designer as the fabricator.

PHYSICAL CHARACTERISTICS OF MICROELECTRONIC DEVICES

Whereas the physical characteristics of microelectronic devices appear very attractive, it is not always possible to benefit fully from them. For example, an integrated circuit of four transistors, six diodes, and ten resistors can be produced in a silicon chip of only 40 × 40 mils. However, current device-packaging techniques increase the surface-area (and volume) requirement considerably. Thus, a flat pack requires a surface area of at least 125 × 250 mils, excluding leads, while the dual-in-line package (DIP) typically requires 250 × 700 mils of surface area, excluding leads. Leads require additional surface area. When simple circuits are packaged separately, the surface-area and volume efficiency is poor.

Space efficiency can be improved when multiple circuits are contained in a single package. Identical circuits in a single package are frequently used. Circuit density is limited by the number of package leads on standard packages.

The hybrid microcircuit is one technique currently being pursued to increase circuit density. The ultimate approach is that of LSI (Large Scale Integration) in which many thousand circuits could be packaged in a space no larger than 3 × 3 inches.

The surface-area and volume requirements of film and multichip circuits are usually greater than those of the monolithic integrated circuit. The multichips may be inserted into a modified TO-5 package with a surface-area requirement of about 10^5 square mils. Film circuits typically require a minimum of 250 × 250 mils of surface area, excluding leads.

The weight of a small integrated-circuit package in general use may be approximately 0.1 gram. Again, the assembled equipment averages a much higher weight — an estimated 1 gram per unit — than the simple sum of the weights of the integrated-circuit packages. As with volume, the weight of multichip and thin-film microcircuits is greater than that of the monolithic integrated circuit. Nevertheless, microelectronic circuits weigh $\frac{1}{5}$ to $\frac{1}{50}$ as much as conventional circuits when connected into equipments.

The estimates given above are very approximate and are valid only for comparisons at the circuit level. A given equipment's volume or weight may be changed severely or only slightly with microelectronics conversion, depending mainly on the weight and volume of the parts that are not converted. Furthermore, there is a strong trend to increase the functional complexity of the integrated circuit, since increased complexity can often be achieved without materially increasing overall package weight or volume.

An integrated circuit does not always consume less power than an equivalent conventional circuit. While microelectronic equipments are often designed to consume considerably less power than conventional versions, this reduction cannot be attributed to an inherent power efficiency in the individual microelectronic circuits. Any power-speed combination available today in a microelectronic circuit can also be achieved with conventional components. The sharp reductions in power that often accompany conversion of an equipment to microelectronic form are the result of changes in design. There is also a reduction in lead lengths and mechanical joints in the integrated circuit, with a corresponding reduction in power loss. This power difference is negligible in most cases at present, but as the power levels continue to decrease, the difference will become significant. For example, in nanowatt circuits, the microelectronic circuit is indeed significantly more efficient than conventional circuitry.

PERFORMANCE CHARACTERISTICS OF MICROELECTRONIC DEVICES

The integrated-circuit technology has been developed extensively for digital circuitry because such circuitry does not require passive elements with tight tolerances and broad ranges of values, and because there is usually a high degree of repetition of the same circuit in digital equipment. All digital functions are currently available in integrated-circuit form.

Several thousand items are available off-the-shelf, but many of these circuits overlap in function. The circuits are also not generally interchangeable. Thus, most of the major manufacturers have developed their own compatible circuit family to provide most of the required digital functions. Mixing items from different families is generally not feasible. Consequently, in working from the standard inventory, a designer must initially choose the logic scheme and characteristics that best suit his needs, and then design his system around a single family of circuits.

Special requirements, not covered by the stock circuits, can be met by custom-connecting (metalizing) the elements of standard matrix wafers, which are maintained in inventory for this purpose by some manufacturers. To fill even more exacting needs, fully custom-made circuits are offered generally by all manufacturers.

Although monolithic structures are particularly appropriate for digital circuitry, some multichip and hybrid circuits are used – usually for the extremes of the operating frequency range.

Significant progress has been made in the application of microelectronics to analog functions, although these did not receive as much attention, at first, as the digital applications. The most common linear circuit available is the operational amplifier. Other linear circuits include audio amplifiers, oscillators, preamplifiers, If-limiting amplifiers, R-f amplifiers, video amplifiers, and wide-band amplifier-discriminators.

The offerings in the linear area, however, are not as extensive as those in the digital area. The proportion is out of balance with the potential usage of the two types, and because of the later start made on the linear functions. One of the factors that made the digital system so attractive to manufacturers of integrated circuits was its unitized design based on the repetitious use of a few basic circuit forms. This advantageous design feature does not exist to any such degree with linear devices, so that there is a constraint against the establishment of a stock inventory and a tendency to rely more on the custom approach.

COST

The price of the integrated circuit has been steadily decreasing for some time and is now generally competitive with that of conventional circuitry. Much of the cost of an integrated circuit is incurred in the packaging of the crystal. This packaging cost remains essentially constant for a considerable range of circuit complexity. The cost per circuit function is expected to drop well below that of conventional circuitry in the future as the capability for increasing complexity (with constant yield) improves. Prices for the packaged device itself can be expected to decrease, but the added cost of testing may have a significant effect on the price paid by the consumer.

Savings resulting from the widespread use of microelectronics begin at the device-procurement stage but becomes far more significant after the equipments are placed in operation. Both maintenance and logistics costs can be drastically reduced even without standardizing circuit-performance characteristics. These reduced costs for maintenance and logistics result directly from increased reliability of the devices.

Microcircuits are tested extensively to improve equipment reliability, since a reduction in equipment failures will decrease the number of maintenance actions required and thus reduce maintenance costs. If all other costs are equal, the life-cycle cost is also reduced.

MICROELECTRONIC DEVELOPMENTS

A common reaction to one's first encounter with an integrated circuit (properly magnified), is that "the whole production process must be completely automated". This is not the case, however; a major portion of integrated-circuit manufacturing is performed by highly skilled personnel.

The manufacturing steps are broken into two major groupings — wafer processing and device assembly. The first group relies heavily on automation and the other on human labor.

The major role of people in the wafer-processing steps is in the operation of automatic equipments, loading and unloading these equipments, and transporting groups of wafers from place to place. Little increase in automation is expected in this area. However, improved automatic process control and testing by computers can be expected.

Manufacturing steps that involve device assembly, offer promise for automation (device testing is already highly automated in most plants). The chip-mounting and lead-attachment steps are particularly inviting for automation, since they are almost completely dependent on human skill, with the accompanying cost and reliability penalties. Further, errors at these points are very difficult to detect. (There is no practical technique available for the nondestructive testing of minute connections and joints.) The flip-chip technique, in which the chip is turned face downward and the metalized contact areas are joined to similar areas preformed on a substrate, or some variation of it, may provide the solution to the mounting and lead-attachment problems. Such techniques also introduce difficulties such as bond-inspection restrictions, and sharply reduced thermal-transfer capabilities.

In addition to manufacturing steps, major areas of concern in the automation of integrated-circuit production, are the production tooling (particularly the diffusion and intraconnection masks) and circuit design.

The electrical design of current integrated circuits is much more constrained by the topological layout of the circuit elements than that of conventional circuits. The coupling problems are more severe (because of the p-n junction method of isolation); typically a single level of intraconnections is utilized. As the circuits become more complex, the number of variables that must be considered becomes enormous. Considerable effort has, therefore, been expended on development of computer techniques to assist the designer in handling these variables.

The trend in microelectronics packaging is toward higher circuit density. Complex arrays are available to a limited degree; research and development is being performed on extremely complex devices that have the required intraconnections on the silicon wafer. Such an approach is called Large-Scale Integration (LSI) and has as its objective, the intraconnection of hundreds of undiced circuits packaged in a container comparable in size to a silver dollar.

THIN FILMS

Film circuits are usually used where specialized applications such as high power, high frequency, and low component tolerances, cannot be met with silicon monolithic circuits.

Both thin-film circuits and thick-film circuits are available. Each has advantages and disadvantages when compared with either each other or with monolithic circuits. The major disadvantage is the lack of a suitable film-type active device. Thus both film circuits involve the deposition of passive film components on flat substrates and then the attachment of appropriate active devices.

The development of the thin film technology for the fabrication of electronic circuits, somewhat parallels semiconductor technology, for it was not until transistors became available that the potential rewards associated with the development of thin film techniques became apparent. The applications which appeared possible for thin films involved resistors, capacitors and interconnections of electronic circuits. Before the advent of the transistor, these portions of the electronic circuits were relatively small, reliable, and inexpensive. The change in this situation caused a large increase in the attention given to thin film techniques.

Thin-films are deposited in a vacuum chamber at a pressure of 10^{-5} torr. The material being deposited is electrically heated in a vessel in the chamber until it melts and vaporizes. The vaporized material condenses on a glass or ceramic substrate typically one inch square. The deposition pattern may be controlled by a metal mask. The mask is usually 0.002-inch-thick stainless steel, and is placed over the substrate during vapor deposition. An alternative to this approach, is to deposit over the whole substrate and then etch out the desired pattern, using photolithographic techniques and selective etchants.

In simplified form, a thin film resistor is made by depositing a metallic film of thickness, t, length, L, and width, W. Its resistance, R, is given from Fig. 1 as

$$R = \rho L/tW = (\rho/t)(L/W) = \text{sheet resistance} \times \text{aspect ratio}$$

where ρ is the material resistivity. The ratio ρ/t is known as the sheet, or ohms/square resistance and is given in ohms. The ratio L/W is known as the aspect ratio. Conductors are made by depositing thick films of a low resistivity material such as aluminum.

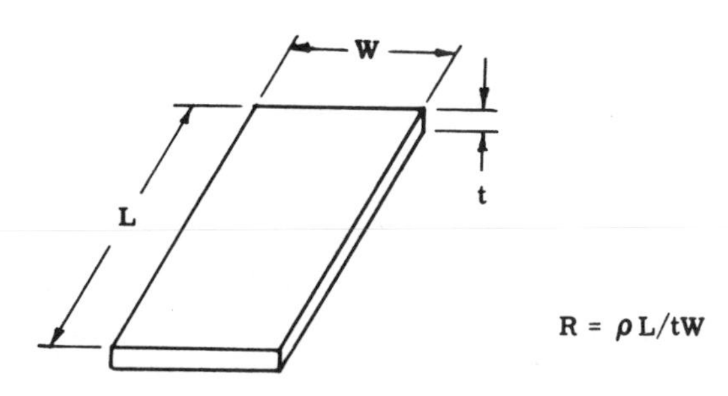

Fig. 1 Thin film resistor

A number of materials have been used for thin-film resistors, including nichrome, SiO-Cr "cermets," tantalum, and titanium, with nichrome being the most widely used. Substrates of finely polished glass, glazed ceramic, and oxidized silicon have been used. Nichrome resistors can be deposited with a sheet resistance of between 1 and 300 ohms/square, and with a temperature coefficient of resistance of ± 50 ppm/deg C. Resistors can be deposited to tolerances of ± 1 percent. Tantalum is also widely used for thin-film resistors. The advantage of tantalum resistors is that they can be adjusted to close tolerances by electrolytic anodizing of the tantalum. Another advantage of tantalum is that the anodic oxide has a high dielectric constant and is suitable for the fabrication of capacitors in the range of 2.5 pf/sq mil. Thus, both resistors and capacitors can be fabricated from a single metal. However, this approach has the serious disadvantage that the tantalum metal reacts with the tantalum oxide at temperatures above 200°C and cannot withstand the temperatures encountered in packaging integrated circuits. A number of metal compounds have been investigated for use in thin-film devices, including oxides and nitrides of such metals as chromium, titanium, and tantalum.

A simplified thin film capacitor is made, as shown in Fig. 2, by sandwiching a dielectric film between two conductive electrodes. The capacity is given by,

$$C = 0.0885KA/t \text{ picofarads (pf)}.$$

Here K is the dielectric constant, A the area in cm^2, and t the dielectric film thickness in cms. An inductor is usually made by forming a conductive spiral. All of these components are formed on a glass or ceramic substrate by many methods. However, the vacuum evaporation process is the most-commonly used one.

The process for producing thin-film capacitors involves three deposition steps: the deposition of a bottom electrode, the deposition of a dielectric material, and the deposition of the top electrode. The electrodes and dielectric are all deposited in succession on the substrate. The dielectric materials commonly used are silicon monoxide and silicon dioxide.

Silicon monoxide has a higher dielectric constant than silicon dioxide. However, evaporated silicon monoxide is subject to pinholes. The resulting loss in yield is usually not acceptable; in addition, the dielectric constant of silicon monoxide is strongly dependent on deposition conditions – making good reproducibility difficult to maintain. Silicon monoxide cannot be etched and therefore must be deposited through masks. This introduces the problems of achieving mask alignment and maintaining it during substrate heating and fabrication.

Silicon dioxide can be deposited by the simple process of reactive sputtering of silicon in a mixture of argon and oxygen at a pressure of a few microns. The use of silicon dioxide gives a high yield at a capacitance of 0.25 pf/sq mil since the sputtered silicon dioxide is relatively pinhole-free. In addition, sputtered silicon dioxide is stable and has the properties of bulk-fused silica. The dielectric constant and the capacitance are thus reproducible. An equally important advantage of using silicon dioxide is the fact that it can be chemically etched with hydrofluoric acid, which is compatible with integrated-circuit processing. This permits the fabrication of the thin-film capacitors and resistors by photo-masking techniques – techniques that all integrated-circuit manufacturers are currently using.

The complete thin film system is designed keeping in mind the capabilities of the fabrication process to be used. Normally, a prototype of conventional components is built to check out the design. The microelectronics components are then designed to be accommodated by the glass substrates. The layout of the components must take account of the problems of power dissipation, conductors, space for add-ons and lands, and the substrate size. Assuming the vacuum evaporation process is used, a precise drawing is then made, many multiples of full size, for each material to be deposited. This drawing is reproduced on a photographic plate at full size. The photograph is then used to make an evaporation mask – a thin sheet of metal with openings through which the thin film is to be deposited. These masks can be made by either an etching process or an electroplating process. A separate mask is required for each material to be deposited.

The basic vacuum evaporation process is illustrated in Fig. 3. Practically all of the air is removed from the chamber. A crucible with a charge of the material to be deposited, is heated to vaporize the charge. The vapor atoms travel in straight lines and deposit on all exposed surfaces. In this fashion, the vapor passes through the mask openings to deposit on the exposed substrate, and form thin films of the required area at the required location. Thickness is controlled by a monitoring process. All the required films are deposited in succession to build up the thin film microelectronic circuit. After removal from the chamber, the required add-ons, principally transistors and diodes, are attached in the spaces left for them. Lead wires can be attached by a number of processes such as soldering, welding, conductive cements, or ultrasonic bonding. Some of the major advantages of thin films are as follows:

1. Thin film circuits do not exhibit spurious interactions any more

than do conventional component circuits. For this reason, a one to one extrapolation can be made from conventional components

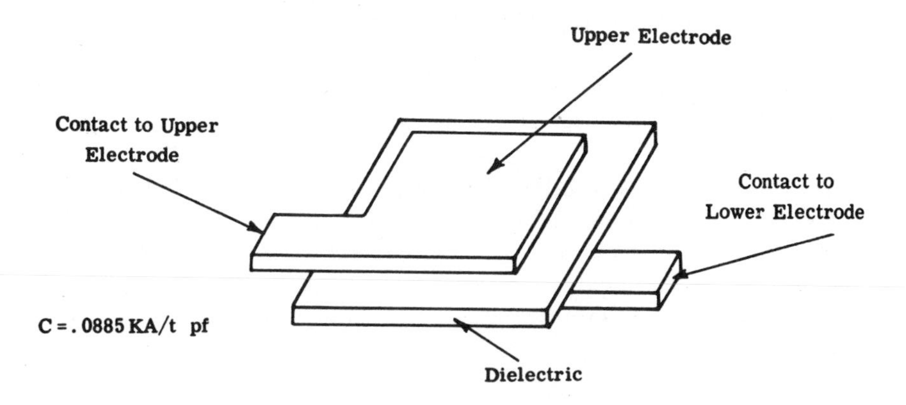

Fig. 2 Thin film capacitor

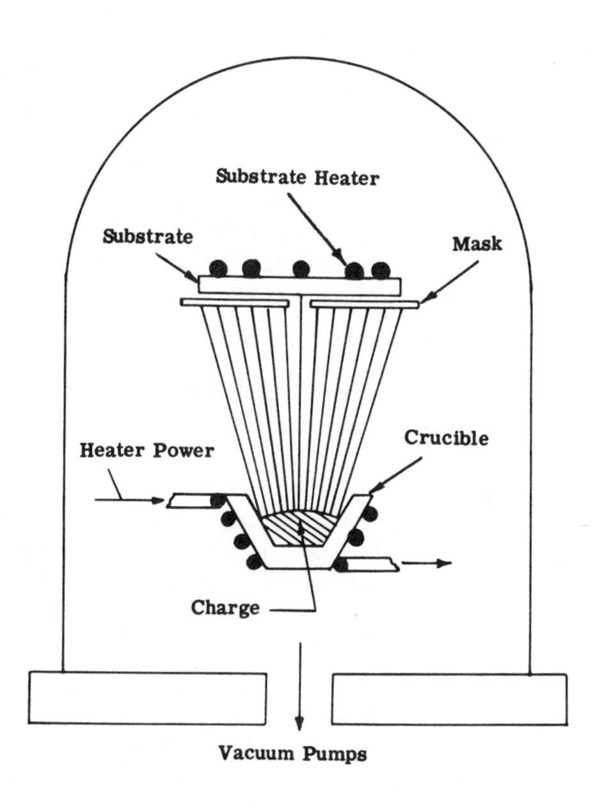

Fig. 3 Vacuum evaporation process

to thin films; the sometimes troublesome interactions characteristic of silicon integrated devices are not present. This results in a bias towards thin film techniques by engineers who are familiar with conventional circuit design, and also offers a high frequency circuit capability which has not yet been demonstrated by the silicon technology.

2. Silicon circuits are inherently small and have limited power handling capabilities. When the size increases in order to accommodate higher power, the cost goes up rapidly. In this higher power end of the circuit spectrum, thin films, for use as passive components and for interconnections, are very practical. High resistance, high power resistors, for example, can be made with thin films but not as part of a silicon integrated device.

3. Thin film circuits are believed to be considerably more radiation resistant than are semiconductor integrated circuits. This apparent ability to operate in a radiation environment for longer periods of time than silicon integrated circuits, has been a prime justification in thin film circuit development.

4. Thin film circuits can consist of a multitude of materials, each chosen to be optimum for its particular application. This allows a more complete assortment of component parameter values and circuit types than is possible with silicon. Counter to this, however, is a theory that the addition of new materials to a given electronic structure may introduce new failure modes and may detract from the reliability.

5. Thin films can be deposited upon a silicon substrate and used to improve the capabilities of silicon integrated devices. This adds additional processing steps to the fabrication, increases cost, and perhaps decreases reliability. The hybrid integrated circuit, however, is currently obtaining increasing popularity because of its capability for obtaining certain extreme parameter values impossible with silicon alone.

6. Another hybrid technology involves the deposition, on a substrate, of thin films for passive components and interconnections, after which silicon integrated circuits are inserted. A favorite method for accomplishing this, is to provide ball type electrical connections to the silicon integrated structure, facing it down on a thin film substrate, and bonding it by soldering or other appropriate techniques. This approach, in which the best attributes of both the silicon and thin film technology can apparently be used, possesses a large potential.

Each of the preceding points for thin films could, perhaps, be made for thick film structures. It is difficult to tell which is better able, in a particular case, to provide the best answer. Thick film structures might be expected to be more reliable due to the high temperature processes employed in their fabrication, but techniques have been developed for many more materials with thin films than with thick films. Also for thick films there is apparently no foreseeable chance to obtain an active device structure; this is not the case with thin films. If size is important, thin films are, at present, made by techniques which allow higher resolution in the structures.

A somewhat detailed outline of the vacuum deposition method is given below in order to provide the reader with a fundamental understanding of this essential process. The other major thin film fabrication process is also briefly described. For a thorough description of the thin film technology, however, the reader should refer to the appropriate chapters on this subject.

VACUUM DEPOSITED THIN FILM CIRCUITS

As mentioned earlier, the equipment for fabricating thin film circuits consists basically of a vacuum system, masks, evaporation sources, and substrate heaters. The complexity of this equipment may vary a great deal ranging from an "oil" pumped, 18-inch bell jar system with mask changers and evaporation sources, to a complex in-line system of multiple vacuum chambers which allow substrate transfer from one processing chamber to the next, without breaking the vacuum.

Regardless of complexity, the actual processes are basically the same. Given a circuit design, suitable masks are made which define the areas of deposition for the various materials. Objectives in mask design are avoidance of crossovers, minimization of interconnection lengths, minimum area for the desired parameter tolerances, graded power dissipation, and a realizable mask geometry. A typical circuit may require four masks: the first for the resistor pattern, the second for bottom capacitor plate, the third for dielectric coating and the final one for top capacitor plates and interconnections. With some processes an additional pattern is needed in order to provide suitable materials for soldering external leads to the circuit. Masks are normally formed by photo-etching openings in thin copper-nickel alloy sheets.

The masks are mounted in a changer. While a variety of changer designs are available, a typical one has a rotary plate on which the masks are mounted and stepped sequentially in front of the substrate. Positioning is often achieved by aligning suitable holes in the masks with locator pins. Alternative systems involve movement and location of the substrate rather than the mask. The exact mechanism is unimportant, as long as the alignment is close enough for the desired resolution in the deposited pattern. Tolerances of several thousandths of an inch between successive masks are possible.

Source heaters are provided for heating the appropriate materials to evaporation temperatures. There are a variety of designs for sources. Aluminum, a very common material in thin film circuits, is most often evaporated from a stranded tungsten coil filament which is directly heated by passage of current. Silicon monoxide, the most common evaporated dielectric, requires the use of carefully designed evaporation sources which prevent line of sight evaporation from the solid silicon monoxide to the substrate. This is to avoid the ejection of particles from the source to the substrate which causes defects in the films. These sources are normally fabricated from sheets of refractory metal and are heated by passage of large currents. Nickel-chrome alloys which are employed for resistors, can be evaporated from coiled tungsten wire or by direct passage of current through the alloy wire. A variety of other materials is also employed, each requiring some attention to the design of the evaporation source.

Both alumina and glass substrates are employed for thin films. These substrates are mounted in the vacuum in a suitable holder. It is usually necessary to heat the substrate at various stages of the deposition process, both for cleaning the substrate and for good adhesion of the deposited material. This is accomplished by a suitable resistance heater mounted behind the substrate. Normally the evaporation sources are mounted at the bottom of the vacuum system, with the substrate at the top. In some of the more advanced systems, very high vacuums are possible—down to the 10^{-10} torr range, and the evaporation sources may be heated by electron beam bombardment.

Normally all depositions are made without breaking the vacuum. This has been found necessary in order to avoid defects in the films brought about by contamination in the air. Dust is a particularly severe problem. In some cases the adherence of the successive depositions will be poor if there has been exposure to the atmosphere.

With the apparatus described, the deposition of the thin film circuits is a relatively straightforward process. In general, substrates are placed in the holders, the vacuum is generated, the substrates are heated for cleaning, and the materials are evaporated through their respective masks in sequence. The substrates are then removed from the vacuum chamber, tested, resistors adjusted if necessary, active components inserted, lead wires attached, and the circuit encapsulated. Particular processes, of course, differ in detail due to the variety of materials which can be handled in a vacuum evaporation scheme. While nickel-chrome alloys are popular for thin film resistors, rhenium, titanium, nickel, and other resistive materials have also been vacuum deposited. For dielectric coatings silicon monoxide has been used most often, but other dielectrics such as magnesium fluoride have also been employed. The silicon monoxide is sometimes oxidized to silicon dioxide during evaporation in order to improve its dielectric properties. Conductor materials used include aluminum, silver, copper, and other good metallic conductors.

TANTALUM THIN FILM TECHNIQUES

Tantalum thin film circuit technology has been the subject of increasing interest. The advantages of using a tantalum film are high annealing temperature, film stability, anodic adjustability of resistors, and high dielectric strength and dielectric constant of its oxide for capacitor use. Unfortunately, it is not possible to fabricate transistors or diodes on tantalum patterns. However, these devices can usually be mounted directly onto the substrate.

The high melting point of tantalum requires deposition by cathodic sputtering rather than by vacuum evaporation (electron beam heating can be used, but is usually less desirable). The sputtered films adhere better than evaporated films, and are more uniform. Sputtering is more economical of material because the deposit is localized. However, the deposition is difficult to monitor and to mask for geometry control.

Sputtering is a vapor deposition process in which an electrical discharge is set up between two plates in the presence of a low pressure inert gas such as argon. The ionized gas atoms are accelerated by the high electric field to the tantalum cathode, and release their kinetic energy, thus knocking off tantalum atoms (a few of which may become ionized).

These are then free to diffuse to the glass substrate on the anode.

In the sputtering process, the vapor deposition chamber is used only to produce thin films. The resistor and capacitor geometry patterns and circuit layout are done externally by high-resolution photoengraving techniques, which permit the fabrication of high density micro-circuits.

For tantalum thin film resistors, a uniform and stable film with a predetermined sheet resistivity must be obtained on the glass or ceramic substrate. Nonuniformity in the resistance of sputtered tantalum films is due to the presence, during deposition, of temperature gradients and electric charge on the substrate.

Resistance adjustment and stabilization are obtained by anodic oxidation of the tantalum at temperatures over 350° C. Resistance values are monitored externally during trimming. Adjusted values to within ±3 percent are typical, but values can be trimmed to better than ±0.02 percent where extreme precision is required.

Stabilization of the resistor is obtained by gold doping and oxide formation. A layer of gold with a thickness of about 7 percent of the tantalum, is diffused into the tantalum film at a temperature of around 400° C for about 30 minutes. The film obtained has the same resistivity as the original undoped film, but is much more stable and has a temperature coefficient close to zero. A heat treatment for 25 hours at 250° C forms a protective oxide coating thicker than any that could grow during the resistor's lifetime. This oxide cover provides effective protection from the atmosphere, and makes further encapsulation unnecessary unless additional mechanical protection is required. Finally, conductive material is deposited over the film, and circuit patterns are produced employing standard photoengraving techniques. Conducting lands are usually obtained by depositing gold. A titanium layer between the gold and tantalum creates a more reliable high-strength bond.

Introducing reactive gases such as nitrogen, oxygen and hydrocarbons into the argon sputtering atmosphere, changes the electrical properties and structure of the resulting tantalum film. The introduction of nitrogen to the sputtering chamber affects the specific resistivity of the sputtered film, and the temperature coefficient of resistance. Thus, the consistent fabrication of precision resistors with reproducible parameters is possible by using nitrogen at various background pressures in the sputtering process.

The precision of thin film capacitors is a function of the uniformity and reproducibility of the thickness of the oxide dielectric and the electrode area. A means to attain the high precision required for microcircuit capacitors is the anodic oxidation of tantalum films. An electrochemical cell consisting of a tantalum anode and a platinum cathode in a heated electrolyte is used to form the oxide dielectric. A potential of 100-150 volts dc applied to the cell for a specified time, converts the tantalum metal to a very uniform pentoxide (Ta_2O_5). The oxide thickness is directly proportional to the applied voltage at room temperature, and forming constants of 25 A/V are typical. The capacitor characteristics are, therefore, predictable.

A method for improving capacitor yields may be used. After the tantalum has been anodized, a layer of aluminum is evaporated over it. This aluminum layer is then removed by etching, with care taken to assure that the etch does not undercut the glass substrate or destroy the mask.

The film is then reanodized to the previous final voltage, and counterelectrodes of aluminum are evaporated to form the capacitor. The area of the counterelectrode determines the final capacitance values. This process is believed to fill in film defects with aluminum oxide.

The rapid increase in the dissipation factor with increasing frequency for tantalum thin film capacitors, limits their use in high frequency logic circuits. Improvements to extend this frequency limitation involve the use of a relatively thick layer of tantalum and the application of a conductive metal underlay, such as aluminum, under the tantalum. Capacitors having a Q of 50 at 5 Mc may be fabricated by underlaying the tantalum film with aluminum, and by using thick counterelectrodes of copper.

THICK-FILMS

Thick films are produced by screening or photo-etching formations of conducting and insulating materials on ceramic substrates. Noble metals are favored for conducting materials, and ceramics are normally used for the insulating portions.

Hundreds of different cermet formulations can be used to obtain a wide range of component parameters. The material used for a 10-ohm/sq resistor is quite different from that used for a 100-kΩ/sq resistor.

The formulations are fired at temperatures above 600°C to form an alloy that is permanently bonded to the insulated substrate. The characteristics of the final material can be controlled to a limited extent by the firing temperature/time profile.

The cermet is composed of metal combined with a glass frit. Inks with resistivity of 500 ohms/sq, 3kΩ/sq, 8kΩ/sq, and 20kΩ/sq are used routinely. Inks in the range of 50, 100, and 200kΩ/sq are available for higher resistance values. Resistance values can be specified to a tolerance of ±10 percent. Closer tolerances can be obtained by adjusting each resistor after fabrication.

Because of resolution limits of the screen and spreading of the edges of the resistor during firing, the minimum dimensions for a screened pattern are about 0.030 in. However, 0.050 in. is a more desirable minimum width for design purposes.

For a finer pattern resolution, a technique has been developed for etching cermet resistors that permits an order-of-magnitude reduction in the minimum line width. The process consists of the following steps: A cermet ink is applied uniformly over the entire area of a ceramic substrate; the substrate is coated with a photoresist; the substrate is exposed through a photomask and the photoresist is developed, leaving a protective coat of photoresist material over those areas that will comprise the resistor pattern.

Capacitors are formed by a sequence of screenings and firings; they consist of a high-temperature conductor (capacitor bottom plate and intraconnections), a dielectric, and a top-plate conductor. For R-C networks a resistor material is also screened and fired. The final step is screening and firing of a glass encapsulant. The temperature profiles are dependent upon the materials used.

Thick-film capacitor sheet capacitance is a function of the material used and typically varies from 5,000 – 100,000 pf/in^2. The temperature and voltage characteristics are dependent upon the materials used. These capacitors have high breakdown voltages and good frequency response into the hundreds of MHz.

SCREEN CIRCUIT PROCESSES

The general thick-film approach consists of deposited components and metallic interconnections, with discrete semiconductor chips attached. It is similar to the thin film method with the advantage that high-quality transistors, of either n-p-n or p-n-p types, may be used. Furthermore, medium-power output transistors may be used in conjunction with this technique. Thick film circuits are flexible in fabrication, low cost, radiation resistant, and have good reliability characteristics.

The most commonly used substrate material for thick film circuits, is a high-alumina porcelain containing 4 to 6% glass. The selection of this material is based on a compromise of thermal, electrical, and mechanical properties, taken in combination with cost considerations. Substrates are generally used in the ''as-fired'' state. They are available in a wide variety of sizes and shapes with dimensional tolerances of 1% being standard. Substrates are often produced with holes for the purpose of simplifying layout, assembly, and interconnection of the circuits. When designing such holes, care must be exercised to consider their location in relation to the substrate thickness, to prevent stress-cracking problems in substrate processing and packaging.

Circuit components are deposited on the substrate through a screening process. A wide variety of machines (hand-operated, semi-automatic and automatic) are available to perform this process. However, they are all based on the same principle involving a squeegee driven across the screen at a constant rate. During this action, the squeegee deforms the screen, to the extent where it contacts the substrate, and forces the paste, to be deposited, through the mask. Since material deposition must be controlled in three dimensions, the machine's control over the squeegee rate and pressure, as well as its ability to maintain the required settings, is an important factor for producing uniform products economically.

The screen can be made of silk or stainless steel—the latter being more durable. The mesh size of the screen varies with the application, but generally falls in the range of 150 to 200 mesh. Finer screen sizes result in incomplete circuit patterns since the paste may not extrude through to the ceramic surface. Coarser screens, on the other hand, may cause voids in the pattern and have lower resolution. The application of the paste is, in principle, the same as that used by artists in the silk screen process for many years.

All pastes, used for the screening process, have general properties in common. This applies whether the pastes are conductive, resistive, or dielectric. The pastes consist of fine particles suspended in organic media. The percentage of solids by weight, is usually within the range between 50 and 80%, depending upon the specific gravity and film thickness desired. Thorough mixing of high viscosity pastes is essential to realize homogeneity. Useful values of viscosity lie within the range from 20,000 to 250,000 centipoises. The higher values of viscosity are used for resistors and conductors, whereas the lower values are applicable to dielectrics.

The organic media must provide lubrication for the squeegee, and should not possess excessive vapor pressures. A moderate degree of thixotropy is often used to provide good screening definition. Highly thixotropic pastes, however, tend to leave the imprint of the screen mesh.

A firing process is applied next, often in a continuous conveyor type of furnace. During the firing process the organic media are oxidized and vaporized. Alloying of the metals, oxidation, and softening of the glass then take place, to produce a complex heterogeneous microstructure. Such kinetic processes as alloying and oxidation are diffusion controlled, and wetting rates are determined by interfacial energies and controlled by the viscosity of the glass. Both the heating rate and time, at peak temperature, influence significantly the sheet resistivity, long-term stability, temperature coefficient, and the current noise characteristics of resistors.

The temperature profiles in the firing furnace depend upon the materials selected. Thus, the resistor furnace profiles are determined by the desired resistivity and related properties. The optimum temperature profiles for dielectrics are determined by the desired dielectric properties and vitrification. One family of dielectric pastes is fired for 30 minutes between 800 to 900° C. To achieve high dielectric coefficients, considerably higher temperatures are required.

A low-melting glaze is subsequently applied to provide a protective seal for the resistors and capacitors, and to serve as an insulating medium for conductors. This glaze is fired at temperatures below 600° C, so that the components will not change values as a result of this heating.

Conductors used in thick film techniques have sheet resistivities which are generally less than .05 ohms/square. Palladium-silver conductors are most often used for solderable leads. Platinum-gold and palladium-gold are also frequently used. Pure gold is most satisfactory for chip and wire bonding.

The adhesion of the conductors to the substrate is of prime importance, if reliable units are to be obtained. In this regard, tests involving peel strength are more meaningful than axial-pull tests. This stems from the condition that when the products are in use, the flexing of the interconnecting leads applies the main stress.

Thick film capacitors permit considerable design versatility by making sheet capacitances available in the 1,250 to 100,000 pfd/in.^{2} range. The temperature and voltage characteristics vary depending upon the dielectric material used. Glasses are applied primarily in the fabrication of crossovers and small capacitors in the 1 to 25 pfd range. The glass-based dielectrics have low temperature coefficients within the −50 to 150° C range, possess Q's greater than 500 at frequencies up to 100 MHz, and exhibit small variation of capacitance over a wide range of frequency.

Resistivities available with thick film resistors lie within the range of 100 to 200,000 ohms/square. After firing, and even with careful control over the processes, resistor values are approximately ±10%. When closer tolerances are required, resistors are often designed for about 80% of the desired magnitude, and then trimmed to value. The trimming process is usually performed by an air-abrasive process in which a stream of abrasive alumina particles impinges on the film resistor, to remove resistor material from the substrate. The material is removed so as to increase the effective number of squares of material, and thereby increase the resistor value. During trimming, the magnitude of the resistor is monitored with an electrical bridge which compares the trimmed value to a preset amount of resistance. When the trimmed resistor attains the desired preset value, the bridge becomes balanced, and a signal is transmitted to the abrasive unit to cease the stream of particles. With this trimming procedure, resistor tolerances of $\pm 0.1\%$ may be realized.

Similar such methods may be adapted for the purpose of trimming capacitors. A laser beam may be used to remove material from the top conductor area, without damaging the dielectric.

SEMICONDUCTOR INTEGRATED DEVICES

The rapid advances in semiconductor-device technology would not have been possible except for the impressive progress made in the purification and preparation of semiconductor materials. Near-perfect crystals of virtually absolute purity are needed if consistent, high-quality devices are to be made. The impurity level must be controlled to less than one impurity atom per million — at times, to less than one atom per billion. This level is not detectable chemically or spectroscopically. Its measurement depends entirely on electrical conductivity. The expression for conductivity is

$$\sigma = q(\mu_p p + \mu_n n)$$

where

μ_p = mobility of holes
μ_n = mobility of electrons
p = number of holes per cubic centimeter
n = number of electrons per cubic centimeter
q = electron charge

For an intrinsic semiconductor n or p can be assumed equal to the donor or acceptor impurity density, respectively. When n is very much greater than p, the semiconductor is n type (electrons are the majority carriers); also, generally, n is much greater than n_i (the intrinsic free-electron density), so that n_i can be ignored when the concentration of donor impurity is being determined.

The design of a semiconductor microelectronic circuit wafer is far more difficult than the design of the same circuit in thin films. Principally, this is due to the fact that the silicon substrate used is an active substrate, and the devices diffused into it are capacitatively coupled to it. This causes serious interactions between the components, and limits operating frequencies. Furthermore, since the production is a multistep high temperature diffusion process, successive diffusions affect previous ones. Attempts to compensate for this effect are generally difficult. Therefore, evolving an adequate design and production process for a semi-conductor microelectronic circuit is usually a long laborious procedure, and relies heavily on past experience.

The following description of the fabrication of a semiconductor microelectronic circuit is a highly simplified version of an actual process. The semiconductor material used is silicon since it is available as doped wafers at a reasonable cost, its technology is well known, and its permissible operating temperature exceeds that of germanium.

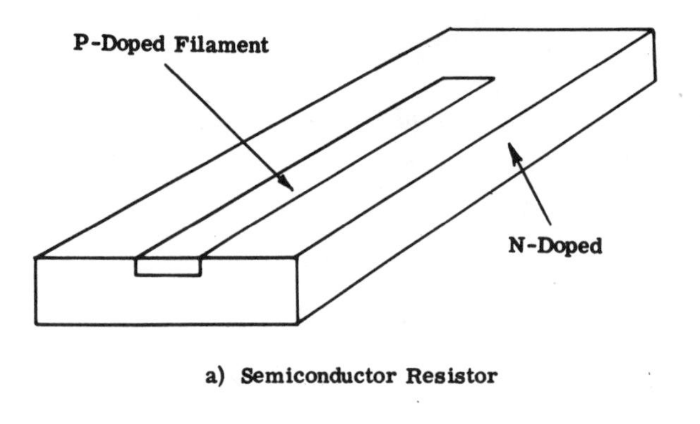

a) Semiconductor Resistor

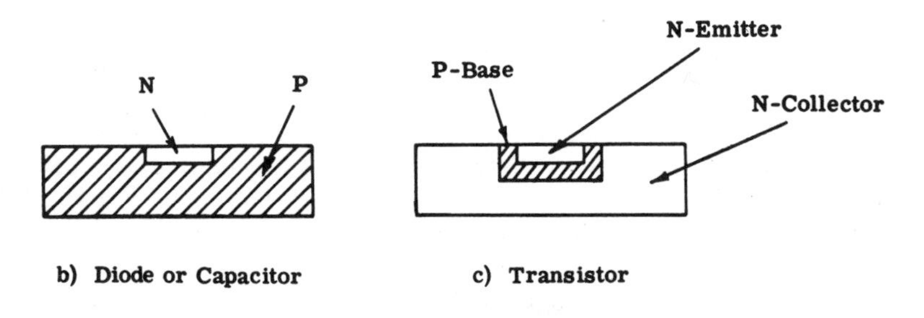

b) Diode or Capacitor

c) Transistor

Fig. 4 Semiconductor microelectronic devices

A semiconductor microelectronic resistor is made by diffusing into the surface a dopant which converts a filamentary volume to the opposite type. Thus, Fig. 4a depicts a resistor of p silicon in an n-region. The resistor is isolated from the n-region by virtue of the p-n junction, provided the resistor is back-biased with respect to the substrate. This isolation, however, is not complete since it is well known that a back-biased p-n junction behaves like a nonlinear capacitor. As in the case of the thin film resistor, the value of the resistance equals the sheet resistance multiplied by the aspect ratio. The sheet resistance depends, in a complicated manner, on the distribution of dopant concentration as a function of depth. The usual values of sheet resistance used are 100 to 200 ohms. The resistor width is usually in the neighborhood of 20 μ(microns).

A diode, as shown in Fig. 4b, is a small button of oppositely doped substrate material, producing a p-n junction. The button diameter is usually in the 20-100 μ range. As previously mentioned, not only is the junction a diode, but in the back-biased state it acts as a capacitor, the value of which depends on the magnitude of the bias used. Thus, the diode may serve two functions. In practice, the p-n junction needed for a diode or capacitor is obtained from a transistor. Depending upon the desired characteristics, the fabricator may use the base-collector or the base-emitter junction, or even both in parallel.

The basic structure of a transistor is shown in Fig. 4c. In this case the silicon substrate is the collector, and it is of n-material. Some p-dopant is diffused in to form the base with a diameter of perhaps 60 μ and a depth of 10 μ. Finally, some n-dopant is diffused in to form the emitter. The emitter diameter may be 20 μ and its depth 9 μ, making the base width 1 μ.

The fabrication of a semiconductor microelectric circuit is begun with an appropriately doped silicon wafer about one inch in diameter and 100 μ thick. The surfaces are lapped flat, polished, and chemically etched to remove broken crystallites on the surfaces. The result is shown in Fig. 5a.

In the next step, a thin surface layer of the silicon is converted to silicon dioxide, as shown in Fig 5b. This dioxide film is about 0.75 μ thick and is intended to serve as a barrier to the diffusion of dopants into the silicon. This dioxide film can be formed in many ways, but it is usually done by exposing the silicon surface to water vapor at about 1100° C for one to two hours. The details of the procedure determine the quality and thickness of the oxide film.

The next step is to apply a photosensitive resist to the oxide, as shown in Fig. 5c. A photosensitive resist is a lacquer-like material which upon exposure to light and development, is converted to a film which adheres to the support and resists chemical action. Where it is not exposed to the light, it washes away in the developer. Such a resist is called a negative resist; a positive resist is one which washes away where it is exposed to light, and is fixed where not exposed.

For each step in the diffusion process, an appropriate photographic mask is required. For this purpose, a large scale drawing, as much as 500 times full size, is made of those areas on the oxide, where holes are to be etched either for diffusion or aluminum deposition. The drawing is then reduced to full size on a photographic plate. This plate comprises the mask through which the resist is exposed and developed. As a result, the resist is removed and the oxide film exposed in those areas of the silicon wafer where the dopant diffusion is to take place. The result of this operation is shown in Fig. 5d. The wafer is then immersed in a chemical etchant which removes the oxide and exposes the silicon, after which the resist is completely removed. The result is shown in Fig. 5e.

The wafer is now placed in a high-temperature furnace, usually about 1000 to 1300° C, and exposed to a gas containing the dopant atom to be diffused into the silicon. The depth of the diffusion and the concentration of doping atoms as a function of depth, is controlled by the details of the high temperature process. In this way, a controlled doping of a desired area on the silicon wafer is obtained. The entire process can be repeated as often as necessary by stripping the old oxide mask, forming a new masking film of oxide, etching holes in the oxide as required for the next diffusion, etc. The entire circuit of components and isolation barriers can be diffused into the wafer in this manner.

To complete the circuit, the components are interconnected by high vacuum evaporation of aluminum conductors. A complete oxide film is produced, and holes etched over the spots where contact is to be made. A thick aluminum film is deposited over the entire wafer by vacuum evaporation, and the photoresist is removed. This aluminum film passes through the holes in the oxide, to contact the appropriate points on the diffused components. The wafer is now baked at 600° C to alloy the aluminum into the silicon.

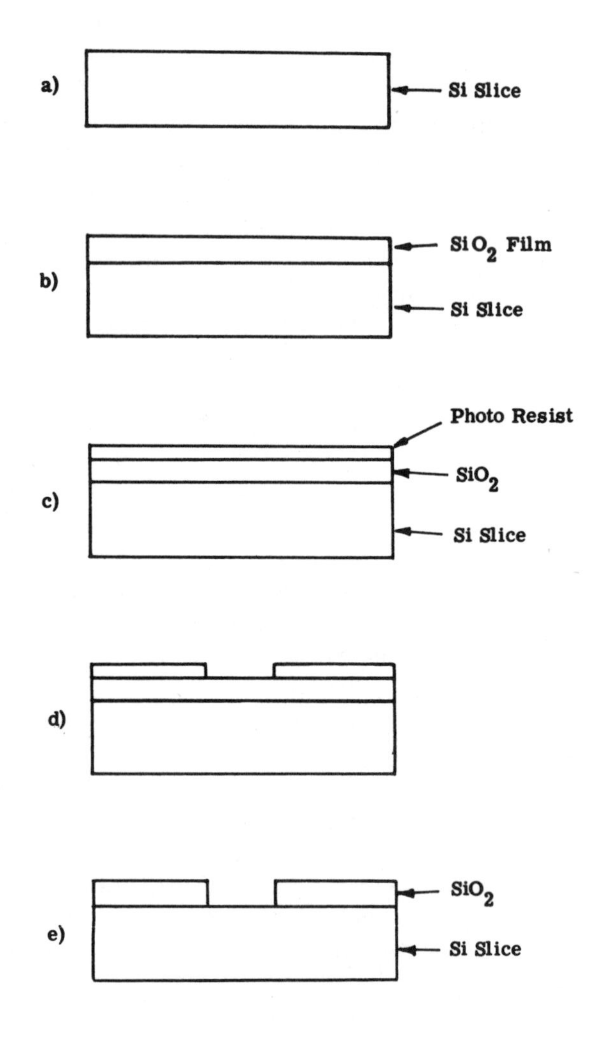

Fig. 5 **Dioxide masking on silicon wafer**

Since a typical semiconductor microelectronic circuit size is less than 0.1 of an inch on a side, and the silicon wafer is at least one inch in diameter, it is possible to make a large number of circuits on one wafer and in one operation. To accomplish this, the single photographic image of the circuit pattern, used in the mask fabrication, is repeated to form a rectangular array of identical images over the wafer. This is known as the step-and-repeat operation. In a typical case 200 circuits may be obtained on a 1¼-inch wafer. As the capability of the silicon crystal producer increases, and larger wafers are made available, as many as 2000 circuits in one sequence of operations should be possible. The resulting rectangular array of semiconductor microelectronic circuits are separated into individual circuits and then packaged.

It is useful to examine the reliability, size and power requirements of silicon integrated devices. With respect to reliability no basic limitations have yet been found. As with any new device many early design and production methods turn out to be inadequate. With these eliminated or corrected, however, the life of silicon integrated devices is quite long. Thus, a silicon integrated 3-input NOR gate used for a space guidance computer, has a failure rate of less than 0.005 percent/1000 hr$=5/10^8$ hr (at 90 percent confidence level) estimated from actual operating use.

In discussing microelectronic devices, their size and weight are often compared to those of conventional circuits performing the same function. Integrated silicon devices do offer a very significant reduction in size and weight; in fact, so much so that the emphasis on their size and weight has, to a large extent, been forgotten. In most commercial, industrial, and military applications, the size and weight of the system is no longer significantly affected by further size reduction of integrated devices; the input, output and interconnections of these systems are the limiting factors. In space research where size and weight are still extremely important, integrated circuits are being used to good advantage. Equally important, if not more so than size and weight, is the low power requirement of the integrated device.

Power dissipation of silicon integrated devices is between one and several hundred milliwatts per device, dependent upon the function of the device and its design. Both ends of this power spectrum are being pushed by further research and development. For space research and other applications in which power is at a premium, devices which operate at microwatt power levels are needed. The high power end of the spectrum is being pushed, at the same time, because it is necessary to obtain enough power from silicon integrated devices to perform useful functions. The upper limit is set by the ability of the package to remove heat from the integrated device. There is a high power level at which the utility of integrated devices is less than that of conventional component circuits. Thin film circuits have the ability to operate at higher temperatures, and thus have more efficient heat transfer for handling more power.

Some of the attributes of silicon integrated devices are listed below. They apply primarily to the monolithic device rather than to the multichip structure which is sometimes also classified as an integrated silicon device.

1. Batch Processing—One important attribute of silicon integrated devices is that numerous devices are processed as a unit up to the stage where leads are attached and they are encapsulated. This allows a high degree of process control and device uniformity, with relatively low unit cost.

2. Processing Simplicity—The number of processes which are involved in the fabrication of a silicon integrated device is very small, when compared to the total number of separate processes required to fabricate the components of the conventional equivalent circuit.

3. Device Diversity—The identical processes can fabricate a variety of integrated devices by variation of the necessary photographic patterns.

4. Materials—In the silicon integrated device a small number of different materials are employed. For example, one class of devices employs silicon, silicon oxide, aluminum and gold; no other materials are necessary. In other types of integrated devices, additional materials may be employed. They may be other conductors or contacting materials or a resistive metal employed for thin film resistors. Even with these, the number of different materials involved in the integrated device is small. This tends to promote high reliability.

5. Area Factor—The surface area of the single crystal silicon die on which the integrated device is fabricated, is very important. It influences yield and thus cost, allowable power dissipation, required power to operate, package size, and functional capability. For a given structure the present lower limit on area may be set by dissipation, current carrying ability, capacitor and resistor parameters, or by resolution limits of the photoengraving process. For low power circuits, the latter is most important since for a fixed resolution, the only tradeoffs are between component tolerances and circuit size.

6. Inverted Economics—Because of the greater area required for capacitors and resistors on silicon integrated devices, these components add more cost to the integrated devices than do transistors or diodes. This is the reverse of the situation for circuits designed with vacuum tubes and separate transistors. The inversion of relative cost of the active and passive circuit components will continue to have significant impact upon the design of integrated circuits.

As for the case of thin films, a somewhat detailed summary of the semiconductor device fabrication technology is presented below. For a complete understanding of this area of microelectronics, however, the reader should refer to the appropriately designated chapters.

SILICON INTEGRATED DEVICE TECHNOLOGY

The processing technology by which silicon integrated devices are realized, varies little from one manufacturer to another. The details are different, however, and these details can be extremely important. The major steps involved are:

1. substrate preparation
2. photoengraving
3. oxidation
4. diffusion
5. epitaxy
6. chemical processing
7. interconnection, lead attachment, and packaging

Substrate Preparation

The substrate on which silicon integrated devices are formed, is a wafer of single crystal silicon which may be between 0.7 and 1.5 inches in diameter. This circular wafer is cut from a single crystal which is grown either by pulling the crystal from a melt or by the float-zone method. The former is characterized by a lower dislocation density while the latter is more free from dissolved impurities—primarily oxygen. Six of the properties which may be specified in silicon wafers are:

Base material type and impurity.—This is a gross specification indicating whether n- or p-type material is required. The particular impurities which are employed to dope the silicon have normally been boron or phosphorous, but other impurity dopants are available, and are employed in particular applications. Although not normally specified, it is desirable that the silicon have a high resistivity before doping in order to minimize compensation effects.

Orientation.—Crystals sliced from a grown ingot normally have a specified crystalline orientation which may be confirmed by X-ray diffraction. The orientation is important for the dicing operation and for wafers on which epitaxial layers are grown. Wafers with a <100> orientation have cleavage planes at a 90° angle. Therefore, when the wafers are scribed with a diamond tool and broken, a high yield is obtained. In a <110> orientation, the cleavage planes are at 60° angles, and the yield is lower. The normal orientation tolerance is 1.5°.

Resistivity.—The resistivity may or may not be an important parameter, depending upon the particular design approach used for the integrated circuit. For epitaxial devices the substrate silicon material is a passive supporting structure, and its resistivity is relatively unimportant. However, if parts of the substrate are used in active structures of the integrated device, the resistivity is very important. The resistivity tolerance is affected not only by the accuracy of the doping and crystal growing procedures, but also by the variation of resistivity along a diameter of a circular wafer, which results from temperature gradients existing in the growing process.

Etch pit count.—When a silicon wafer is etched, in a so-called preferential solution, pits are formed at the sites of crystalline dislocations. The density of these is employed as a measure of the crystal perfection and is sometimes specified. Float-zone crystals are characterized by a large dislocation density (up to 50 000/cm^2) while some suppliers of pulled crystals claim a zero etch pit count. A specification on pulled crystals of less than 1000 etch pit counts/cm^2 is readily obtainable. The dislocation density is important in its relation to device yield, particularly for epitaxial structures wherein the crystalline imperfections are propagated throughout the epitaxial layer.

Surface finish.—Silicon wafers may be purchased either as-cut from the ingot, or with subsequent surface preparation steps already performed. The diamond slicing operation is followed by mechanical lapping and polishing. The final surface employed for the first wafer processing step may be either a mechanically polished surface with a mirror-like finish, or a chemically polished surface. When a mechanically polished surface is used, it is assumed that in subsequent processing, the mechanical damage is removed either by oxidation or by gas etching in the epitaxial furnace. The chemically polished surface is not optically flat, but is characterized by the so-called lemon peel undulation; this is sufficiently small so as not to interfere with the photoengraving process.

Dimensions.—If the wafers are obtained with as-cut surfaces, then the thickness should be approximately twice that needed in the device processing, in order to allow for polishing and lapping one surface of the wafer. An 8-10 mil thick wafer is used most often. This keeps breakage to a minimum and does not waste material. The diameter of the wafers is not critical unless limited by processing equipment. There is a trend toward wafers of greater than 1 inch diameter in order to have more devices processed on a wafer. The upper limit is set by crystal uniformity and allowed parameter spread.

Wafers with polished surfaces are ready for subsequent processing which may take one of several different paths. If an epitaxial layer is to be provided, the wafers are put into the epitaxial process directly. If dielectric isolation is to be provided, then the actual substrate used in device fabrication involves considerably more processing before it is ready for use. If diffusion isolation is to be employed, the wafers are put directly into the photoengraving, oxidation and diffusion steps.

The substrate preparation for dielectric isolation is rather complex, as noted above. The first step is to etch into one side of the wafer a grid pattern several mils deep. After this, the surface is oxidized, and a thick layer of polycrystalline silicon is then deposited over the oxide by pyro-

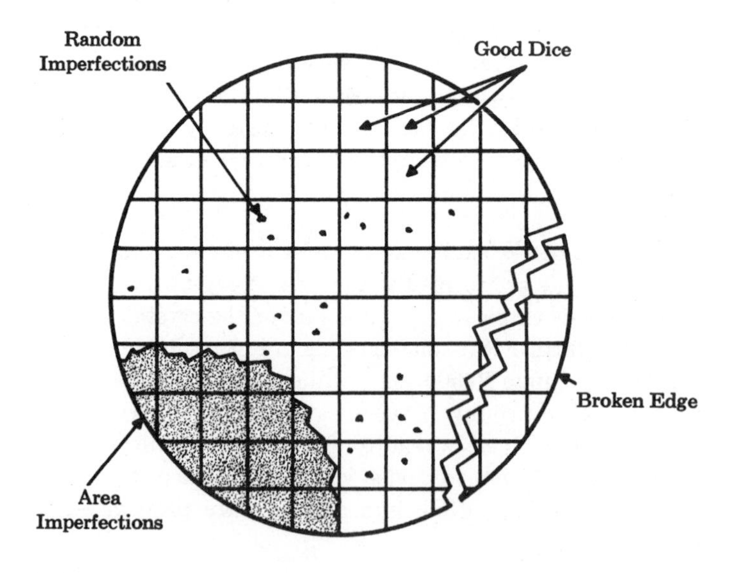

Fig. 6 Imperfections which may occur in a wafer

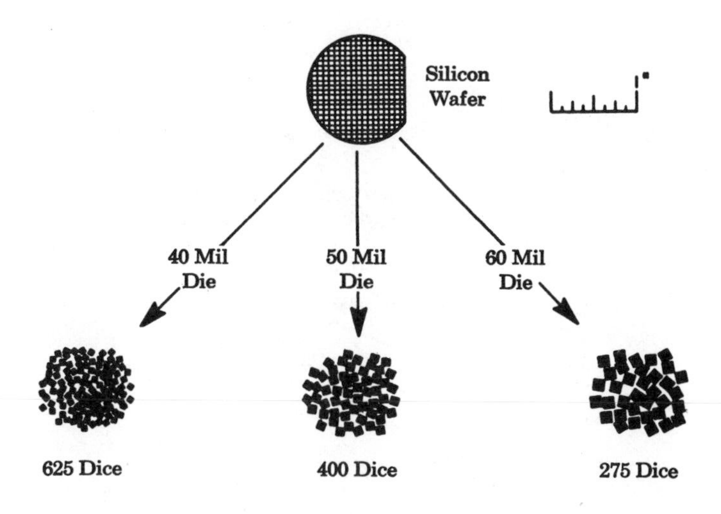

Fig. 7 The effect of die size on number of dice

lytic deposition processing, similar to that used in epitaxy. This layer must have sufficient thickness to provide mechanical support for the final structure. After deposition, part of the original silicon wafer material is removed by a combination of mechanical polishing and etching, so that a grid pattern of oxide is exposed on the surface. This surface of the silicon wafer then consists of a large number of single crystal silicon squares, completely isolated from each other and from the substrate by the silicon oxide layer. The wafer is then ready for further processing.

Photoengraving

The techniques of photoengraving are necessary for planar processes and interconnections in silicon integrated device technology. The photoengraving masks which give the geometry definition on the silicon wafer are made through the use of high resolution photographic and photoreduction equipment, step and repeat camera, and precision drafting equipment. Using these and the circuit topology layout provided by the design engineer, a series of photographic masks are prepared which define the areas for diffusion, contacts, or for interconnection patterns.

Wafer cleanliness is extremely important for the photoengraving process. Normally the photoresist is placed on a wafer by an eye dropper or similar dispensing apparatus, while the wafer is spinning at high speed on a vacuum chuck. After the photoresist is allowed to air dry, it is usually baked at a low temperature before the printing operation. The patterns are aligned under a microscope. All masks, after the first, must align with the existing patterns on the wafer to high degrees of accuracy. Commercial apparatus is available which allows this to be done. All photoresist operations, up to this step, must be performed in yellow red (non-blue) light in order to avoid exposure of the resist. After alignment, the wafer with the mask tightly pressed against its surface is exposed to light with a high

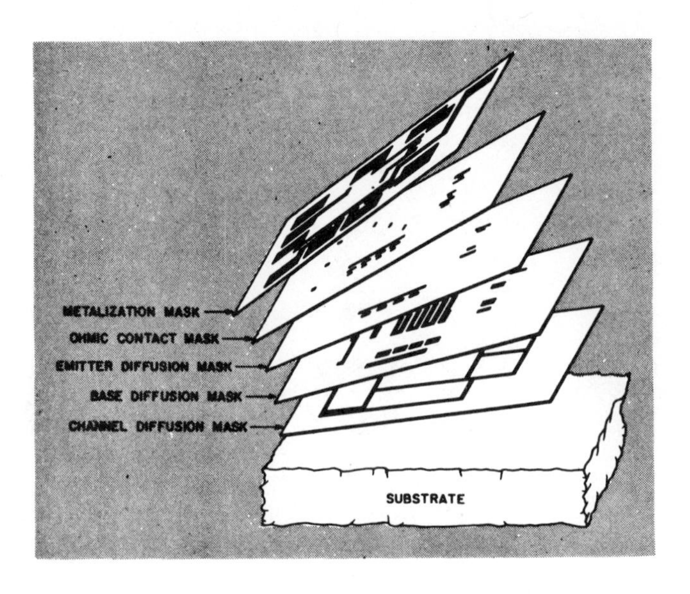

Fig. 8 The use of photolithography to prepare a series of masks

ultraviolet content. This changes the polymerization of the resist. The wafers are developed in proprietary developing solutions or xylene and dried. Then the wafers are baked in a vacuum oven in order to harden the remaining resist and improve its acid resistance. This pattern is then etched through the silicon oxide so that a suitable pattern of openings exists through which diffusing impurities will pass. In forming contact or interconnection patterns of metal, it is most common to use a reverse process wherein the unwanted metal is removed from the surface, leaving behind the desired pattern. After the resist is used for the etching operation, it is removed from the surface before subsequent silicon processing. This may be accomplished by heating the wafer in hot sulfuric acid for a short period of time.

The state-of-the-art of photoengraving is such, that it is an inexact science. There is much to learn about all the steps required in obtaining patterns with the desired qualities. At the same time it is apparent that photoengraving techniques have been very successful in both the integrated silicon device technology and in the fabrication of silicon transistors. Photoengraving has been proven far superior to other techniques for obtaining close geometry control in very intricate patterns. The main problems associated with photoengraving, are that the quality of the resist is variable, the optimum processing steps are poorly defined, better resolutions are desirable, mask making is expensive, and surface preparation and printing techniques need further development.

Diffusion

By diffusing donor and acceptor impurities into semiconductor materials, it is possible to fabricate p-n junction devices that have superior electrical characteristics. The diffusion process has many distinct advantages over other fabrication processes, particularly if silicon is the material used. It is the basic process and in most cases the only process in which superior transistor characteristics can be realized. Junction depths and impurity concentration of the layers can be controlled more precisely than in alloyed structures.

After the introduction of diffusion techniques into semiconductor

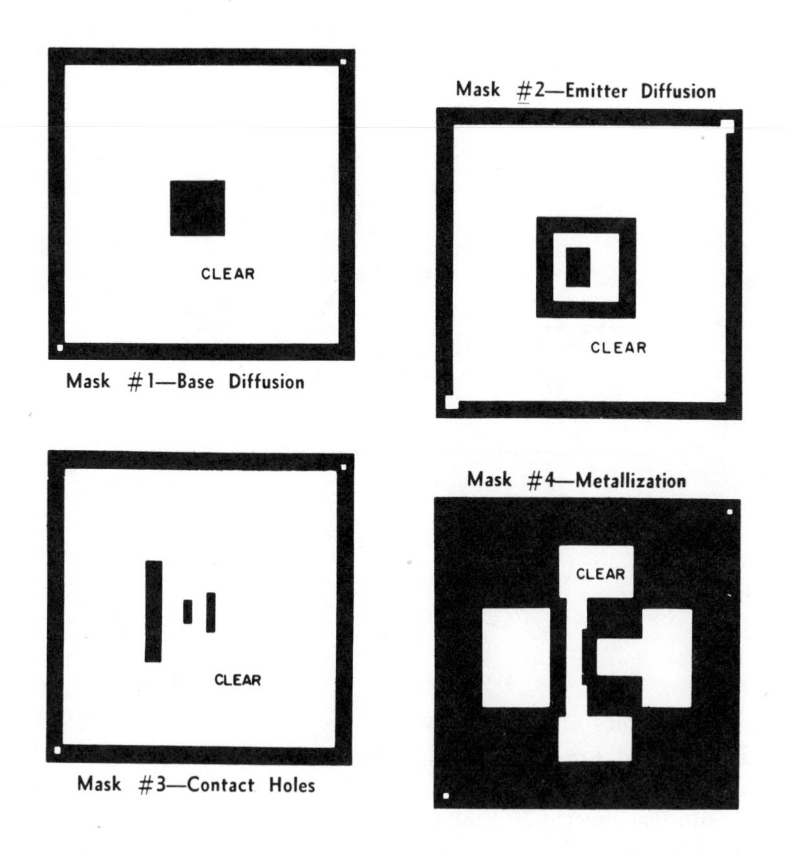

Fig. 9 The lateral dimensions of an IC are set by the photo masks

fabrication, it became possible to achieve base widths, with high accuracy, of less than 1 micron, which improves considerably the transport factor (and therefore α) and the transport time (and therefore f_a). Other means of improving the high-frequency response would be to reduce the collector-base junction area, as has been done with mesa transistors. A typical mesa transistor is shown in Figure 10.

In such a structure, however, the collector-base junction is exposed to the surrounding atmosphere, and surface effects at this point are severe. To minimize surface effects on junctions, the planar diffused structure was introduced. A typical planar structure is shown in Figure 11.

The planar process permits the passivation of the surface by an oxide layer at an early fabrication stage. The silicon-oxide coating is grown on

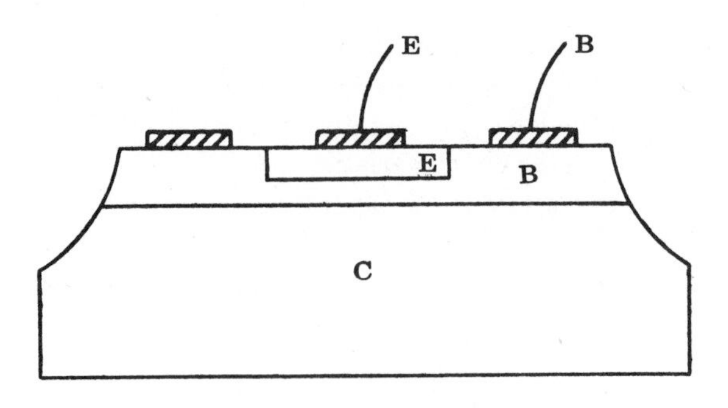

Fig. 10 Mesa transistor

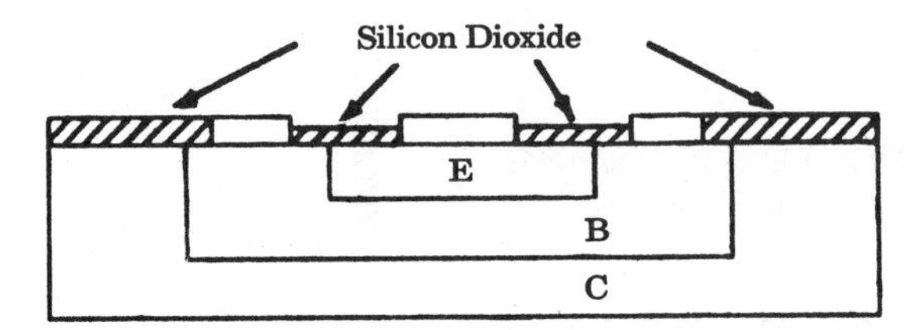

Fig. 11 Planar Structure

the surface before any junctions are diffused. This greatly improves the parameters that are particularly sensitive to surface conditions. These parameters are the reverse leakage currents, breakdown voltages, noise figure, and low-current β. Planar structures, also made it possible to solve many problems in the realization of integrated structures, in which several elements are built simultaneously on the same piece of semiconductor.

Impurity diffusion is another one of the basic tools of silicon integrated device technology. While mathematical theories describing simple kinds of diffusion are known, the diffusions are performed on the basis of empirically determined methods. As with photoengraving, wide variations may be found in the details of the diffusion processes used in practice. These variations include the use of different types of impurity sources, different diffusion procedures, and different degrees of control. The basic process requires a suitable impurity (dopant) source, almost always either a phosphorus or boron compound. The dopant is transported in vapor form by a carrier gas, usually nitrogen, over the silicon wafers, and then discharged. The entire system is contained in a glass or quartz tubing depending on the temperature. The silicon wafers rest on a quartz boat at a thermally flat region of the furnace which is held very precisely at a specified temperature. Diffusion temperatures range from 900° C to 1200° C. Typical impurity sources for phosphorus are phosphorus pentoxide, phosphorus oxychloride, red phosphorus, ammonium phosphate, phosphorus tribromide, and phosphine. For boron, typical impurity sources are boric acid, boron tribromide, methyl borate, boron trichloride, and diborane. The liquid sources, phosphorus oxychloride and boron tribromide, are the most common. There is some evidence that the gas sources, phosphine (PH_3) and diborane can be used to more advantage.

By following any of the prescribed processes, acceptable diffusion results can be obtained. Normally the diffusion parameters are varied by changing the temperature of the furnace. As silicon integrated device technology becomes more sophisticated, a higher degree of control, than is now possible, will probably be required. Some of the inadequacies which exist in diffusion technology are: (1) agreement between theory and practice is difficult to obtain, (2) data describing diffusion processes vary a great deal between laboratories due to the use of different models, and (3) present diffusion operations are largely empirical and many of the second order effects are not understood.

Thus, the present techniques of diffusion are sufficiently well developed so as to present no problem for most contemporary silicon integrated devices. However, better control and understanding is desirable in order to make devices which are more exactly alike, and whose characteristics can be predicted more accurately.

Oxidation

The ability of the thermally grown oxide on the surface of the silicon wafer to mask against impurity diffusions and to protect the junction from the environment, is very important in silicon integrated device technology. The requirements for the oxide in the silicon integrated device structure are based on the following conditions:

1. It is first used as a diffusion mask, and becomes thereby inherently contaminated with impurity (dopant).
2. It protects silicon junctions which have fringing electric fields on the order of 10^4 V/cm (10^6 V/cm in field effect devices).
3. Oxides are used as substrates for metallic conductors.
4. In some designs, the oxide is the dielectric for a capacitor structure.
5. During processing, the oxide surface helps to protect the silicon from mechanical abuse.

In conjunction with this, the oxide is expected to exhibit negligible conductivity, low dielectric loss, and be thermodynamically and metallurgically compatible with the other materials with which it comes into contact.

Oxides on silicon can be prepared by a variety of techniques, but the three most common oxidizing atmospheres are steam, wet oxygen, and dry oxygen—all at high temperatures. Oxides formed by the three processes have different porosity, conductivity growth rates, and other physical properties. The most common process is steam oxidation which has a higher growth rate at a given temperature.

A typical oxidizing procedure is to place clean silicon wafers, with prepared surfaces, onto a quartz boat, and to insert it into the quartz tube of a furnace. The furnace is very similar to those used for diffusion. A flat zone sufficient to accommodate all of the wafers is desirable. The appropriate environment is provided in the tube before the wafers are inserted. Empirical relationships give the approximate length of time required to form an oxide, of the desired thickness, at a given furnace temperature. For most processing, the required thickness is on the order of 7000 Å=0.7 microns. A carrier gas is not necessary with steam oxidation. Instead, a container of very pure water is maintained at, or close to, its boiling point. The vapor is forced to flow through, and out of the open quartz tube.

For most processing purposes, the color of the oxide, which is obtained in an oxidation process, is an accurate enough indication of its thickness. If sufficient care is taken, the oxides, when removed from the furnace, are clear, uniform in color, and free from imperfections.

Oxides can be made which are adequate for realizing good silicon integrated devices. They perform very well for diffusion masking, passivation, etc. Inadequacies stem from the fact that in processing silicon integrated devices, the proper procedures are largely empirical, and at times, these empirical processes fail. When the oxides on the resulting devices degenerate—often for unknown reasons, the production process must be "retuned" until successful structures again result. A failure of a production process may take the form of collector-emitter transistor shorts, leaky junctions, shorts, or similar electrical problems. Some troubles are a result of stored charge, in or on the oxide, which produces layers of different conductivity on the surface of the silicon. If these layers are of opposite conductivity type, i.e., inversion layers, device failures can result. Similarly, control of the charge content in surface field-effect transistor structures is essential.

Epitaxy

Epitaxy is the descriptive word for that process wherein a thin layer of material is grown in single crystal form on a suitable substrate. While the term may apply to many materials or combinations of materials, microelectronics is concerned only with the growth of silicon single crystal films on single crystal silicon substrates. These epitaxial films may be of opposite conductivity type and different resistivity than that of the substrate on which they are prepared. The ability to grow high quality films of this type has provided an important tool in device fabrication. In silicon integrated devices, better transistor parameters and improved isolation techniques are available through the use of epitaxial films. One may note that if the impurity distributions which determine device behavior are formed completely by diffusion, then the impurity densities, as one moves into the silicon from the surface, must be rapidly decreasing, even when the type of impurity changes. Epitaxial techniques allow variation of impurity profiles such that larger concentrations may be below the surface, and lower concentrations near the surface. The most important use of epitaxial films is to provide a very thin active region in which to fabricate silicon integrated devices, and thus to use the substrate silicon only as a mechanical support and "ground" plane.

Epitaxial techniques are diverse. A typical system consists of an induction heater, a quartz tube, a wafer holder or susceptor, and gas handling apparatus. The substrate silicon wafers are heated to approximately 1100° C by heating a susceptor (made of a conducting material) on which the wafers are resting. Different gases flow through the quartz tube in which the substrate wafers and susceptor are placed. First, the silicon wafers may be heated to high temperatures and exposed to hydrogen in order to clean and etch the surface. After cleaning, silicon tetrachloride ($SiCl_4$) is added to the hydrogen carrier gas. At the hot surface of the silicon, a reaction occurs in which the $SiCl_4$ decomposes giving HCl and Si. The silicon deposits on the substrate wafer, and at the temperature of the process, the atoms rearrange themselves into a minimum energy configuration, i.e., a single crystal layer.

The combination of planar diffusion techniques with epitaxial structures to give improved transistor characteristics, is shown in Figure 12. There are several advantages of incorporating an epitaxial layer on the collector substrate. The epitaxial layer has comparatively higher resistivity than the starting material, so that base-collector junction capacitance is low. The series-collector bulk resistance remains low since the epitaxial layer is thin, and the substrate is heavily doped. Thus, the saturation resistance is low, which permits higher current flow for a given dissipation, and shorter collector-saturation time. Also, the collector-base breakdown is high because of the high resistivity of the epitaxial layer.

It is possible to adjust impurity doping quite precisely. A number of methods can be used for this purpose. An impurity compound can be mixed in a small quantity with the silicon tetrachloride; this mixture can then be reduced in the reaction zone, and the impurity atoms will be included in the resultant crystal. Alternatively, separate vessels can be provided for the main silicon tetrachloride supply, for a p-doped supply, and for an n-doped supply. As in the main supply, a carrier gas can be sent through the doped vessels. Also, diffusion can be employed as the delivery mechanism. In the latter, it is possible to control vapor pressure by adjusting temperature. Still another approach employs gaseous impurity compounds diluted in a carrier gas.

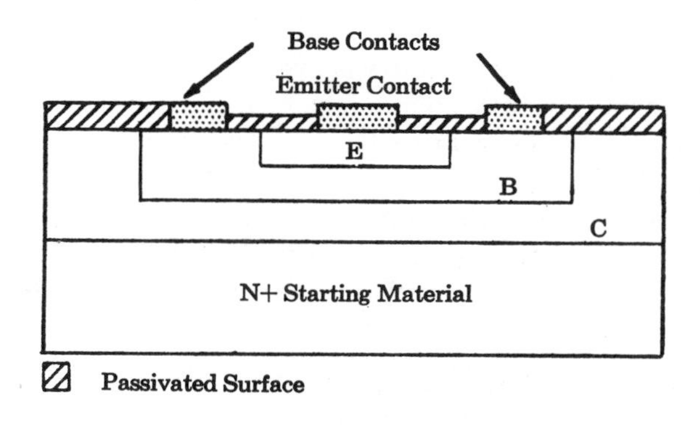

Fig. 12 Planar-expitaxial transistor

ISOLATION

The isolation between components on a monolithic substrate is usually accomplished by reverse-biased p-n junctions. Associated with the p-n junction, are capacitance and normal leakage currents. Therefore, complete isolation is not accomplished, and the effect is less-than-optimum performance.

Using the example indicated in Figure 13, there is capacitive coupling between the collectors of the two transistors because of the capacitance associated with the junctions. At normal doping levels, the value may be about 0.1 pf/sq mil at 0.5 volts of reverse bias. In addition, the leakage current is a function of temperature and may be appreciable at high temperatures. It would be best to isolate the devices from each other by a dielectric, and obtain isolation more closely related to discrete components. A number of techniques have been developed to accomplish this. Some manufacturers are also developing methods for growing single-crystal silicon on an insulating substrate.

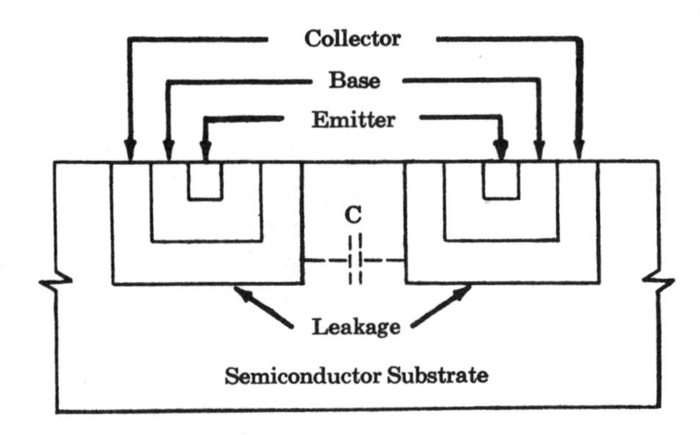

Fig. 13 Capacitance and leakage in P-N junction isolation

A circuit fabrication procedure combining oxide isolation with other techniques – which include localized epitaxial growth, localized etching and backfilling, and localized gold doping – is described below.

(1) An epitaxial substrate wafer is prepared by conventional processing techniques with an n+ layer on one surface. The starting resistivity is chosen to be that required for one or more of the finished circuit elements.

(2) Depressions are etched into the back of this wafer in locations corresponding to regions where the required conductivity type and impurity density will be different from those of the region selected in (1).

(3) Each depression is filled with epitaxial silicon that has the impurity density and type required for additional devices.

(4) Grooves are etched into the silicon around each of the desired regions from the back side of the wafer.

(5) In a two-step process, metal and oxide layers are deposited.
(6) Polycrystalline silicon is then grown over the entire back side of the wafer to a thickness approximating that of the original wafer.
(7) The top surface of the wafer is then lapped or etched down to expose each of the desired isolated regions.
(8) The wafer is then processed conventionally to form each of the desired devices.

Before final metalization, gold is deposited through openings in the oxide in those regions where it is desired to reduce lifetime. A short heat treatment (approximately 1000°C for 15 minutes) is sufficient to distribute the gold throughout the desired regions.

The capacitance associated with oxide isolation is in the range of 0.02 pf per square mil per micron of oxide thickness. Thus, a 5-micron-thick oxide reduces the capacitance coupling by a factor of 25 below that of p-n junctions which have a slight reverse bias.

Chemical Processing

Chemical processes are vital in fabricating silicon integrated devices. Examples are etching of silicon and silicon oxide, cleaning of the silicon surface, cleaning of apparatus used in other processes, and diffusion sources. For silicon etching, hydrochloric acid—nitric acid combinations are employed; for oxide etching, hydrofluoric acid (usually buffered with another chemical) is employed; and a variety of solvents are used for cleaning the silicon surface. Ultrasonic cleaning, vapor degreasing, and other special cleaning and chemical handling techniques are used. The most important aspect of all chemical processing is to obtain chemicals of the highest practical purity. Gases used in diffusion should be dry and free of solid particle content. Deionized water employed throughout chemical processing should have a resistivity in the 10- to 20-megohm-cm range. Suppliers of gases and other chemicals have special high purity grades for use in semiconductor processing.

Some studies have been made of the effect of trace impurities on device performance. One important contaminant of oxides is sodium ions which may be leached from glass containers or from quartz used in the processing. Heavy metal ions in processing chemicals have been found to stick to the silicon surface and modify its properties. For diffusion sources such as phosphorus pentoxide, the chemical impurity as well as the water content can be very important. In many cases clean areas are provided to remove airborne contamination from the device environment, but these clean areas are very expensive and difficult to maintain at the required level of cleanliness. For this reason, there is a trend toward the use of small enclosures to which the devices are confined during critical operations.

INTERCONNECTION AND PACKAGING

After the structure within the silicon is completed, openings are etched in the oxide layer, over the silicon, for the metallic contacts. Aluminum is then evaporated over the entire surface, and removed by photoengraving from all areas except where contact to the silicon is to be made. Then the silicon is heated briefly above the silicon-aluminum eutectic temperature and cooled, to cause alloying between the silicon and aluminum. After this, aluminum is again evaporated over the entire surface and removed by photoengraving, so that the desired pattern of interconnecting conductors is left on the surface. In this step, relatively large area aluminum pads are provided on the surface of the oxide. Using thermo-compression bonding,

Fig. 14 Two silicon "chips" mounted on a memory module. Each contains 128 memory circuits and 46 support circuits

for example, gold wires are attached to the aluminum and to the feed-throughs for the particular encapsulation being employed. The silicon die is mounted in the encapsulation normally by a gold silicon eutectic bond to a metalized region on a ceramic substrate, or by direct bonding to the metal header. Usually TO-5 type transistor headers or especially designed flatpacks are employed for packaging the devices.

Modifications of this basic procedure are becoming more common particularly since the purple plague intermetallic reaction between gold and aluminum has been observed in the lead contact areas. Some manufacturers prefer the all aluminum system wherein aluminum rather than gold wires are bonded to the aluminum pads on the silicon oxide. Others are providing different metallic structures for the interconnections. These may consist of a layer of chromium overcoated with a layer of gold. Flatpacks are Kovar-glass structures or glass-ceramic-Kovar structures both of which can be hermetically sealed and are inexpensive. The package sealing is performed in a clean inert gas atmosphere and hermeticity is checked by helium leak detector techniques.

CIRCUIT DESIGN

In the design of microelectronic circuits it is essential to take into account the manner in which the circuits are fabricated.

The various fabrication methods (i.e., monolithic silicon and thick and thin films) introduce design constraints. These result from the small dimensions of components, parasitics introduced by certain component-isolation techniques, component-tolerance limitations, and various voltage and current limitations.

The monolithic fabrication technology limits performance because all of the components are fabricated *in situ* and different components or parts of different components are fabricated at the same time. Thus, no advantage can be taken of optimum processes for constructing various components. When discrete circuits are fabricated, each component can be tested and optimum choices can be made.

Thin- and thick-film microcircuits provide more design freedom, and approach more closely the flexibility associated with discrete circuits. Components can be adjusted to achieve desired close tolerances when required. In addition, discrete transistors may be tested before attachment in the circuit.

STANDARD FUNCTIONAL BLOCKS

Before the design of a microelectronic system can be commenced, a decision must be made as to which microelectronic approach will be used. A good starting point is to examine the standard lines of microelectronic functional blocks now on the market. These standard blocks are fabricated using mainly semiconductor techniques, and the majority of the blocks are designed for use in digital systems. The costs of these standard blocks are, generally, reasonable since they are mass produced. However, the number of different types available, is limited. If the electronic system being designed is compatible with these semiconductor standard functional blocks, the design problem becomes one of fitting these blocks together. This is the easiest and most appealing approach. However, the rules listed below should be followed.

a) The specifications of the functional block should be carefully studied to ascertain that they meet the specifications of the electronic system. Samples of the standard blocks should be obtained and a sample breadboard assembled and tested.

b) The standard line of compatible functional blocks should be used without exception. This usually means the standard blocks of one manufacturer.

c) The electronic system must be designed such that the need for individual components such as resistors and capacitors, not included within the functional blocks, is limited. If this is not followed, the advantages of using standard blocks over custom designed systems are destroyed.

An important item to remember in the design and fabrication of a microelectronic system, is that the cost of the functional blocks is only approximately one third the cost of the completed system. The other two thirds consist of the interconnection matrices, support members, electrical connectors, testing, etc. It may be possible, by various design techniques, to reduce the cost of the microelectronic functional blocks. However, if the costs and size of the other items are thereby increased, the intended net gain may be a loss. Before a decision to use standard functional blocks is made, therefore, the advantages of custom microelectronic fabrication should be investigated.

CUSTOM DESIGN

The main advantage of custom microelectronics is that it gives the electronic designer complete freedom of design-from the individual circuit component values to the complete system. In most research and development efforts, circuit design freedom is essential. Futhermore, standard functional blocks are not available for most linear circuitry and for special type digital circuitry.

In order to aid the electronic designer in deciding which approach to pursue, the following guide lines for custom design may be used.

1. The thin film approach allows a much faster fabrication time. The semiconductor fabrication procedure is much more complex and timely. Also, since the semiconductor substrate does not electrically isolate all the components, the chances of a successful first fabrication effort is somewhat slim. In the case of the thin film approach, the chances are considerably better.
2. The thin film approach is lower in cost for small quantities of circuits. For high mass production, the semiconductor approach is probably the most economical one.
3. The thin film approach is capable of tighter component tolerances. In the design of digital circuitry this factor may not be important. In linear circuitry, however, tolerance considerations may be extremely important.
4. A wider range of component values is available in the thin film approach. The restrictions imposed on the circuit designer are therefore not as severe as with the semiconductor approach.
5. Semiconductor circuits are generally much smaller and lighter than thin film circuits. The interconnections between the semiconductor circuits are therefore more critical and costly.
6. Circuit design changes can be incorporated in thin film circuitry much easier, faster, and more economical than in semiconductor circuitry.
7. Thin film circuitry can be more easily adapted to existing conventional circuit and system design.
8. Higher power dissipations can be obtained using thin film circuitry.
9. Thin film circuits can be operated at higher voltage ratings.
10. High frequency operation can generally be obtained easier using thin film circuitry. This depends a great deal on the type of circuit and the amount of effort devoted to the fabrication design process.

The design of film microelectronic devices is analogous to conventional circuit design. Silicon integrated devices, however, present unique problems because of the interactions which exist between components within the structure, and because of the deviations of component parameters from the optimum values which are available with separate components. The technology of silicon integrated devices has advanced considerably beyond the ability of present models to explain and predict the phenomena observed. The present models are from transistor theory which was developed to account only for first order phenomena. When applied to integrated silicon structures, the second order phenomena, which have not been treated theoretically, are often important in determining performance.

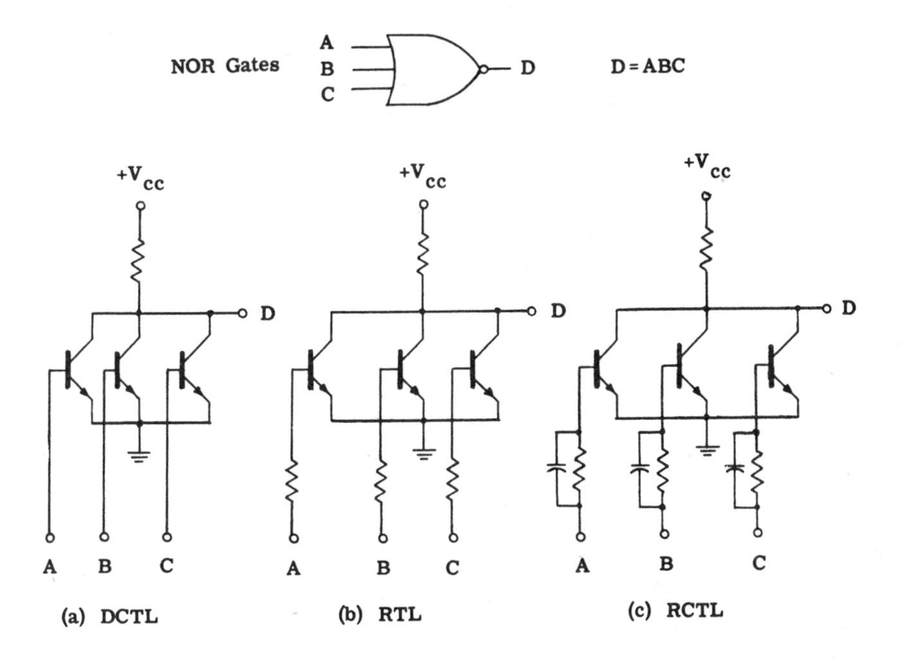

Fig. 15 Direct coupled transistor logic gate circuits

Logic devices are required to provide gating, level restoring, memory, etc. The realization of these functions in microelectronic devices is based upon conventional circuit design methods. Again, in thin film type structures, the conventional circuit design procedures can be carried over on a 1 to 1 basis. Any circuit available in conventional components can be realized in the film technology, as long as inductors and transformers (both are relatively rare in logic circuits) are not required. In the silicon integrated device structures, however, special design considerations must be taken into account. One of these is the inverted economics of the silicon structure; i.e., the transistors and diodes are very inexpensive when com-

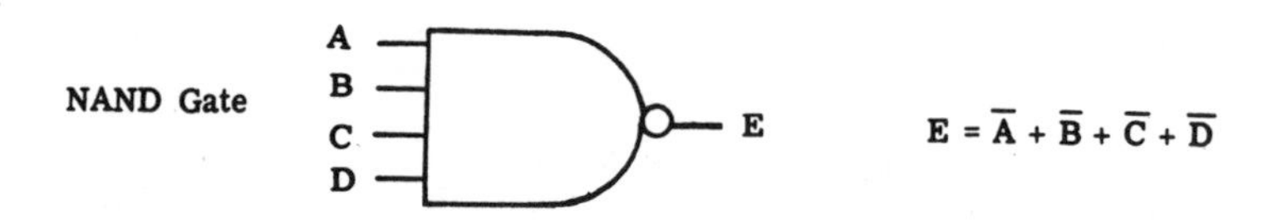

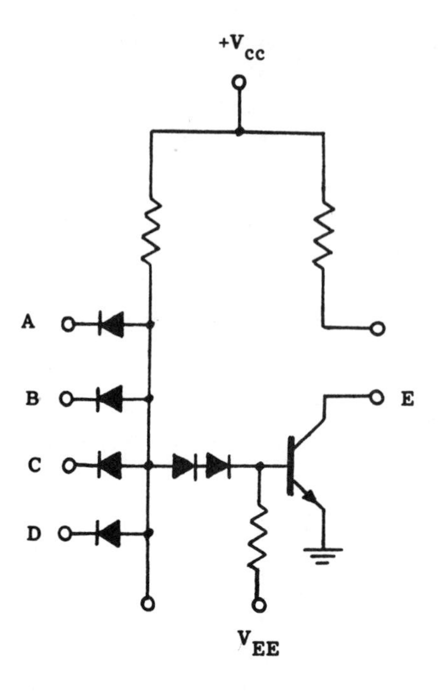

Fig. 16 Diode transistor logic gate

pared to resistors or capacitors.

The simplest form of logic circuit which is available in integrated form, is DCTL or direct coupled transistor logic. The basic circuit configuration of DCTL is shown in Figure 15a. Variations of DCTL include resistor-transistor logic (RTL) and resistance-capacitance transistor logic (RCTL) which are shown in Figures 15b and 15c. A second form of logic which has been applied in silicon integrated devices is diode transistor logic (DTL). One version of this is shown in Figure 16. Other important logic designs are transistor-transistor logic (T^2L) shown in Figure 17, and emitter coupled transistor logic (ECTL) which is shown in Figure 18. Thus, most of the logic devices available, can be classified into three major types, viz, DCTL, DTL, and ECTL. Most devices which are available are of the DTL type.

Because of the smaller market for linear circuits and the more difficult design and fabrication features of these circuits, fewer linear circuits are available in commercial form. Attempts are being made to increase the potential market for particular linear integrated device designs by making them more flexible. Either extra devices are provided in the structures with a variety of possible interconnection patterns, or several components are left out of the silicon structure so that flexibility can be achieved by providing different external components.

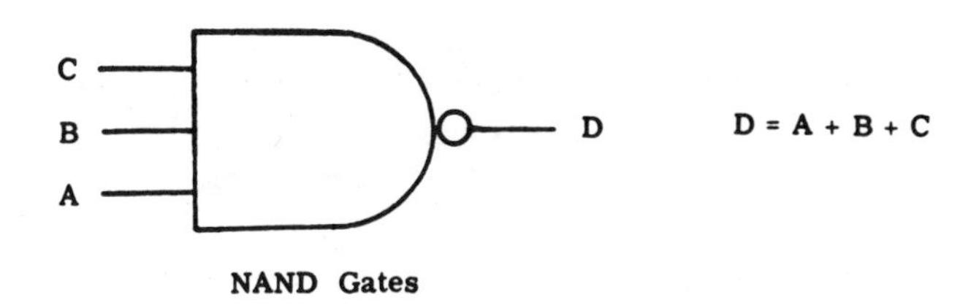

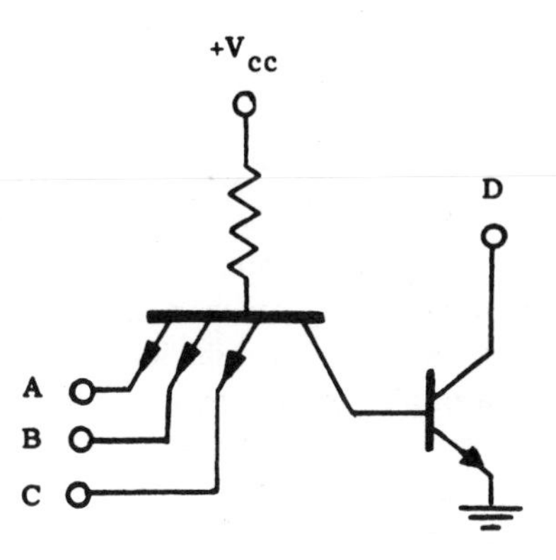

Fig. 17 Transistor-transistor logic gate

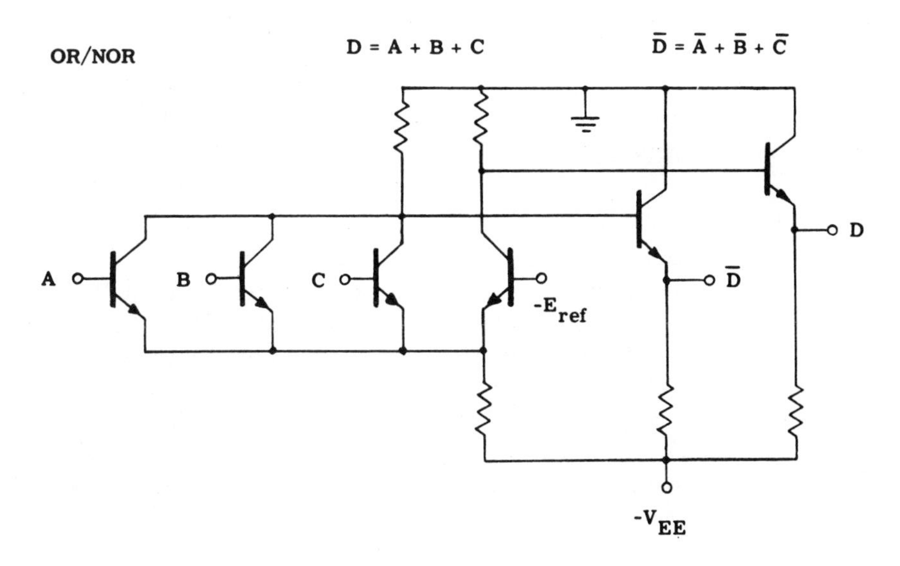

Fig. 18 Emitter coupled transistor logic gate

Because dc coupling is easiest in silicon integrated structures, bias point stabilization has been a problem since silicon resistors and junction parameters are highly temperature-sensitive. Techniques for avoiding this temperature dependence have been developed by using resistance ratios and compensating junctions which help avoid shifts with temperature. DC feedback is often used to eliminate resistors and to provide additional stabilization. Differential amplifier stages are employed to improve the stability of single ended amplifiers. In linear structures wide use has been made of the Darlington amplifier connection to provide moderately high imput impedances. Integrated linear circuits possess the same advantages of reliability, as their digital counterparts.

Attempts to convert electronic systems from conventional circuits to microelectronic integrated devices, without considerable redesign, are of little value. Before redesigning, the technical feasibility of performing the required circuit function and the economics of the integrated circuit design must be determined. Once a decision is made to go to microelectronic devices, other factors must be considered. First, what form of microelectronics—film or integrated structures? Usually the silicon integrated structure is desirable because most readily available commercial microelectronic devices are of this type. When using silicon integrated devices, decisions are necessary on the type of package to be employed, the interconnection methods, and other factors in system design. The cost factor is, of course, not negligible. For a small computing system, the industrial type silicon integrated devices now offer considerable cost savings over standard transistor circuits. In military systems where higher cost structures must often be employed, this saving may not be present, but the added cost may justify the reliability improvement of the equipment. Although size and weight are not important for many applications, they are in others. The value of weight saving has been estimated as $0.05 per pound for stationary electronics, and $20,000 per pound for deep space systems—most applications fall between these extremes.

HYBRID MICROCIRCUITS

A hybrid microcircuit is one that is fabricated by combining two or more circuit types (i.e., film circuits and semiconductor circuits) or a combination of one or more circuit types and discrete elements. The primary advantage of hybrid microcircuits is design flexibility. Hybrid microcircuits find wide application in specialized applications, such as low-volume circuits, and high-frequency circuits.

Several elements and circuits are available for hybrid applications. These include components that are discrete and that are electrically and mechanically compatible with monolithic IC's. They may be used to perform functions supplementary to monolithic IC's, and can be handled, tested, and assembled with essentially the same technology and tools.

Such devices may be packaged individually or in groups, or they may be fabricated as beam-leads devices. Their small size allows their use in both thin- and thick-film circuits. Such devices are easily tested.

Elements are available in a variety of packages (including flip chip) that may be soldered, welded, or ultrasonically welded to an appropriate substrate. Size, cost, and performance are dependent upon the manner in which the elements are packaged. In addition to elements, complete circuits are available in the form of uncased chips (unencapsulated IC dice) and as flip chips. These chips are usually identical to those sold as part of the manufacturer's regular production line. It is necessary that they be handled with extreme care and be properly packaged and connected by the user if a high-quality final assembly is to be obtained.

An interesting application of the hybrid technology is the use of flip-chip circuits in complex arrays. The flip chip has small conducting lumps instead of bonding pads on the die for making electrical connections. The electrical connections are made by ultrasonically bonding or by soldering these lumps to a set of bonding pads on the substrate. Thus, the die is inverted when bonded to the substrate. Such an approach introduces alignment problems during bonding, and also may sharply reduce the heat-transfer capability of the device, because the die is attached to the substrate only through the bonding lumps.

SPECIFYING THE CIRCUIT

The design engineer can follow one of two approaches in microelectronic circuit design. In one approach, he will keep current with and be experienced in the state of the art of monolithic-circuit fabrication. He will then be in a position to give the circuit manufacturer detailed specifications on all components required in his circuit. The other approach is to provide the manufacturer with complete "black box" terminal specifications to which the final circuit must be manufactured and guaranteed.

Basic to both of these approaches is an understanding of the limits or boundary conditions that put constraints on the circuit design. One such constraint involves the fabrication technology itself. For example, what types of devices can be fabricated on the same substrate? It should be understood that the more complicated the circuit the more costly it will be to fabricate.

Microcircuits are fabricated in a variety of ways. Monolithic, hybrid, thin film, and thick film are some of the approaches. Generally, only monolithic silicon integrated circuits and some variation or hybrid of the monolithic form are considered here. The sheet resistance of monolithic circuits is limited to that associated with the base and emitter diffusions. This is necessary from a fabrication standpoint. The passive components are formed at the same time the transistor base or emitter is fabricated.

The hybrid offers a wider variation in sheet resistance than is possible with diffusion only. Thin-film resistors are available for use with monolithic circuits in a wide range of sheet resistances. Diffused resistors above 300-400 ohms per square are not recommended for close tolerances, because of the change in value with the temperature associated with lead-bonding operations.

Capacitors are either silicon-oxide (nonpolar with relatively high Q at high frequencies) for bypass and high-frequency tuning, or junction type (low Q due to effective series resistance of the top contact) for bypass and voltage tuning.

The yield and, therefore, the cost of a circuit is more dependent on the area the circuit requires, than on the number of components. In this respect, it is wise to keep the circuit area to a minimum consistent with other requirements, such as component value and power requirements.

Area efficiency is a key factor with regard to capacitance. Large capacitors (500 to 10,000 pf) become progressively more expensive because of the area requirement. Values above 10,000 pf are not practical.

Though state-of-the-art monolithic transistors impose certain constraints, geometry nevertheless can be varied, and many combinations of transistor characteristics are possible.

Another constraint is imposed upon the designer by the fact that if inductance is required, it is necessary to synthesize or design around it.

MICROWAVE INTEGRATED CIRCUITS

Integrated microwave devices are made with both silicon monolithic and hybrid fabrication techniques. By "integrated" the microwave engineer means a group of components or circuits that are fabricated in a single package. "Microwave IC" represents a grouping of microelectronic components or circuits in a single package. Such groupings might include monolithic elements, thin-film elements, beam-lead elements, or silicon monolithic circuits. Thus, microwave IC's use both monolithic and hybrid techniques applied either together or separately.

Hybrid fabrication is usually performed on an aluminum-ceramic substrate whose dimensions are of the same order of magnitude as that of thin-film circuits. Quartz or other suitable materials may be used as the substrate. The basic requirements are that it be a good insulator (for component isolation), have acceptable thermal properties, have the capability for achieving a micro-smooth surface (because thin-film elements are fabricated on the surface), and have an acceptable dielectric constant.

Microstrip transmission lines are used to transmit the signals. The substrate acts as the dielectric between the ground plane and the top conductor. The characteristic impedance (Z_o) of an unshielded line is a function of the dielectric constant of the substrate (ϵ_r), the width of the top conductor, and the separation between the top conductor and the ground plane. Most applications require a Z_o of 50 ohms or greater. To achieve these values within the constraints of substrate thickness (typically 20 mils) and top-conductor width (about 2 mils minimum), a dielectric constant whose value lies in the range of 9 to 35 appears to be best.

The active elements of a hybrid device are typically separate, unpackaged semiconductor devices. Passive components such as resistors, capacitors, and inductors are formed by thin-film techniques. Ferrite devices are used where no reciprocal functions are required. When low-frequency circuits – such as control circuits – are required, a monolithic silicon circuit may be used and mounted directly on the alumina substrate.

Planar techniques that are used to manufacture low-frequency monolithic silicon IC's, may also be used to fabricate microwave IC's. The substrate for such circuits is high-resistivity (1500 ohms per cm) p-type silicon. For this value of resistivity, silicon is a reasonably good insulator and provides the required isolation between components.

Active components are fabricated in low-resistivity n-type pockets epitaxially grown on the high-resistivity silicon substrate. Inductors and capacitors are fabricated on the substrate by thin-film techniques. The transmission lines are microstrip; typical conductor dimensions are about 6 mils. The silicon substrate is 10 mils thick. The number of circuits that can be fabricated on a single wafer is dependent upon the surface-area requirements of the various components.

Most of the microwave IC's are fabricated as hybrid circuits. These are often simpler to build than monolithic circuits, and may be less costly to build. The cost is very dependent upon yield, and if high yields can be obtained, monolithic circuits could be the least expensive.

TRANSISTORS

Inherent noise properties is a particularly important consideration which must be given to microwave transistors when they are used to amplify low level signals. This is the case because in microwave transistors, the noise level asymptotically increases to a 6-dB per octave slope as the operating frequency approaches the alpha cut-off frequency of the device. The noise property of microwave transistors is the result of thermal and shot noise mechanisms in the base-emitter junction, and in the collector regions. Analytical considerations given to these noise generating sources permit deriving a high frequency noise model of the microwave transistor as well as an expression for the noise figure of the device. This same analytical approach may be extended to derive the noise figure of a field-effect transistor.

Other than noise considerations, ultimate physical limitations place an upper boundary on the frequency response of microwave transistors, and this is of concern because such a limitation places constraints upon the frequency response. This limitation is base width, as the charge-carrier transit time is dependent upon the distance which base carriers must travel.

Power generation at microwave frequencies gives rise to thermal problems, as heat dissipation is confined to a small area. In order to dissipate more power, the device areas involved should be enlarged. Scaling-up must be accomplished while keeping the high frequency parameters intact. One approach to power generation is the use of overlay geometry.

Germanium and silicon remain the major semiconductor materials used in microwave transistors. The advances in diffusion technology have led to the development of new geometries. The field effect transistor is now developed so that designers may utilize the low noise, vacuum tube characteristics of these devices at L-band frequencies. These devices perform well as RF amplifiers or mixers in the VHF region, in stripline configurations, and microstrip techniques are feasible.

SEMICONDUCTOR DEVICES

Varactor diodes have found wide application as frequency multipliers, divider circuits, and as capacitors in tuned networks. The capacitance of the junction may be made to vary as much as 11 to 1 by applying a reverse dc voltage bias.

Two types of varactors are in use: the abrupt junction (Schottky barrier), and the diffused junction varactor. The junction capacitance of either varactor is made to vary by variation of an applied reverse biasing voltage. In the case of the abrupt junction diode, the junction capacitance varies approximately inversely to the square root of the reverse bias voltage, whereas, the junction capacitance of a diffused junction diode varies inversely to the ⅓ power of the reverse bias voltage.

Two factors were responsible for the breakthrough to the long life Gunn effect device. The first was development of a high purity epitaxial GaAs chip. The thickness and doping levels were optimized for Gunn effect performance. The second critical factor is heat dissipation. Gunn effect devices are very inefficient, and therefore great amounts of heat must be removed from the chip. Improved mounting techniques have resulted in proven long term CW operation at increased powers.

In the 2—3-GHz range, the chip thickness becomes relatively large, and resonant cavities become cumbersome. These facts plus low efficiency and poor frequency stability make it undesirable to use Gunn effect devices at 2.5-GHz, especially in view of the fact that microwave transistors will operate reliably at this frequency.

The tunnel diode is a unique device which exhibits a region of negative resistance, i.e., the current decreases with increasing applied voltage. By definition, a positive resistance dissipates energy, so conversely, a negative resistance should be capable of supplying energy, and such is the case.

The tunnel diode is a two terminal device consisting of a single PN junction, but the conductivity of the P and N materials is 1000 times as high as that used in conventional diodes. As a result, the width of the junction (the depletion layer) is very small, of the order of 10^{-6} inch. The very small junction thickness makes it possible for valence electrons, under the influence of a small forward or reverse dc bias, to "tunnel" the junction without overcoming the potential barrier at the junction. As the forward dc bias is increased, the flow of valence electrons across the junction decreases, giving rise to the negative resistance effect. Then, as the bias is further increased, holes and electrons gain the energy to overcome the potential barrier at the junction, and behavior corresponding to that of a conventional diode results. The theoretical frequency limit for this phenomenon is 10^7 MHz. The practical limit is a function of the packaging and connecting techniques. In microwave packaging, frequencies over 10 GHz are possible.

The tunnel diode amplifier operates on the basis that when the source and load conductance are made very nearly equal in magnitude to the negative conductance of the tunnel diode at the chosen bias point, the total conductance of the circuit is near zero (resistance very high). Therefore, a small varying signal introduced into the circuit (e.g. the source), results in much larger signal variations at the individual components of the circuit (e.g., the output). When the tunnel diode is connected in parallel with the source and the load, the voltage gain is unity, and the current gain may be up to 30 dB. If the connection is series, the current gain is unity, and the voltage gain may be up to 30 dB.

The primary advantages of the tunnel diode amplifier are high frequency capability, low power consumption, low noise, and circuit simplicity. The disadvantages are relatively low power output, a large dependence on all conductances (source, diode, and load) remaining constant over the entire operating environment, and the resulting necessity for temperature compensation.

The main application of the tunnel diode amplifiers is the first RF amplifier in microwave receivers.

The tunnel diode oscillator differs from the amplifier only in the value of the total series resistance of the circuit. It has the same advantages and disadvantages as the amplifier, except that impedance matching over the environmental range is not as critical. The circuit must be designed to oscillate at the most critical environment, and temperature compensation is determined by the frequency stability requirements. If these requirements are severe, crystal control may be used.

Two types of oscillations are possible. Sinusoidal oscillations utilize only the relatively linear portion of the negative conductance region. Harmonic content is held to a minimum, but as a consequence, the output power is quite low. Greater power output, at the expense of considerable harmonic content, can be achieved with relaxation oscillations which traverse large portions of the static VI characteristics of the diode.

The tunnel diode circuit may operate as a mixer-converter by simultaneously performing oscillation at the LO frequency, amplification at the RF frequency, mixing due to non-linearity, and amplification at the IF frequency.

Another two-terminal active device is the avalanche diode. With these diodes, microwave oscillation or amplification is obtained by biasing into the avalanche (Zener) breakdown region. The efficiency and power output are superior to Gunn devices, but a very high noise figure and frequency instability overshadow the advantages.

LINEAR INTEGRATED CIRCUITS

Current linear integrated circuits are fabricated using monolithic silicon epitaxial construction, and will perform well in the frequency spectrum from dc to 200 MHz. Within this frequency range, the monolithic linear integrated circuit satisfies many functions required in receiver design. A standard operational amplifier or balanced differential configuration will perform such functions as amplification, limiting, and mixing. Performance may be varied by changing external biasing or feedback configurations. The use of integrated configurations also offers the designer an increase in reliability as the number of interconnections to the active element is reduced by the monolithic diffusion process.

Operation at higher frequencies normally requires the use of thin-film hybrid processes. Although monolithic microwave structures are feasible in theory, this technology is not sufficiently developed. Presently, the semiconductor substrate employed in monolithic fabrication presents a lossy structure at higher frequencies. As microwave frequencies are approached, the conventional passive networks require transmission line construction which present matching and isolation properties. These problems are overcome by use of thin-film processes which have provided circuits that have been successful in microwave applications.

THIN FILMS

As a specialized area of thin film technology, microstrip construction offers many advantages to circuit design. The parameters of microstrip have been determined experimentally with sufficient accuracy to ensure reliable design. The primary advantage of the microstrip is that it is deposited in a single plane and, therefore, can make direct connection to other deposited components. In similar fashion, external connections such as those from a chip device may be made easily and with the shortest possible leads.

The characteristic impedance of a microstrip is determined by the width of the line, the width of the substrate, and the dielectric constant of the substrate. This constant may range from 20 to 100 ohms, with substrate materials such as silicon and ceramic. For a given substrate, the characteristic impedance is directly proportional to the W/h ratio, where W is the width of the conductor, and h is distance from the conductor to the ground plane, i.e., the substrate thickness.

The ratio of a wavelength in free space to a wavelength in the microstrip is, for all practical purposes, constant in the 1—10 GHz range for a given characteristic impedance and dielectric substrate. The ratio decreases as the characteristic impedance increases and/or the dielectric constant decreases.

Another important consideration is the loss per unit length of microstrip. Loss is a function of resistivity and substrate temperature, and it decreases as the resistivity increases. This loss is relatively constant from 0—100°C but increases beyond this range.

In applications where extremely small size is important, it is possible to use very high dielectric material. By increasing the dielectric constant from 10 to 100, the guide wavelength of a 50-ohm microstrip line can be reduced by a factor of 0.35. However, in many cases, the disadvantages will outweigh the advantage of decreased size.

The thermal conductivity of present high-dielectric materials is very poor. High characteristic impedances cannot be obtained because the width of the conductor must be very small, or else the substrate thickness must be relatively large. A thick substrate is not desirable from the standpoint of heat dissipation.

In some cases it is desirable, from the viewpoint of size or performance, to use lumped element components. These components are small enough so that distributed effects play no part in their operation. In the microwave region, inductors may be fabricated by a small thin-film spiral with a diameter on the order of 0.060 inches or smaller. Capacitors and resistors may also be deposited onto the same substrate, making an entire circuit to which the active elements may be chip mounted.

FERRITES

Ferrite materials exhibit ferromagnetic properties which are particularly applicable in the design of such microwave components as circulators and isolators.

The electrical behavior of a ferrite substance is specified by its complex permeability, which is different for opposing circular polarization. Thus, a right circular polarized wave has different transmission constants through a ferrite sample, than a negatively circular polarized wave. This nonreciprocal electrical property principle is employed to construct circulators and isolators. In the case of an isolator, this property allows transmitting a microwave signal with a minimal amount of transmission loss, and prevents reflections from the load from interacting with the generating source. For circulators, this property accounts for the circulation of the applied signal to the respective ports.

For high frequency operation, ferrite dimensions on the order of 0.5cm are required. This size is compatible with integrated circuity. The dc magnetic bias requirements are not excessive due to present high remanent ferrite materials, and X-band integrated circulators are easily feasible. Integrated X-band isolators are also feasible.

Preparation of fine grain ferrite materials have resulted in an order of magnitude increase in the power absorption threshold, P_{cric} of ferrite materials. Thus, ferrite devices may now be devised which are capable of higher peak power handling capabilities. Previously, the small external fields required to achieve resonance would not saturate the ferrite sample, and operation in the VHF range was impractical. Synthesized hexagonal ferrite of polycrystalline compounds have high values of magnetic anisotropy. Thus, smaller external magnetic fields are required for saturation, and operation at lower frequency ranges is more easily obtainable.

FILTERS

Conventional design of microstrip ceramic parallel coupled filters at lower microwave frequencies, results in physical configurations that may lead to cracking and mechanical instability. Attempts to circumvent this problem and to reduce the total length, led to "folded" and vertical folded configurations having the same performance characteristics as the conventional design.

Bandpass filters exhibiting extremely steep rejection skirts, can be realized by means of a composite bandpass, bandstop filter. Employing a pair of band rejection filters in the composite structure, reduces the number of resonators required to achieve a given response, and permits lower mid-band insertion loss, as well as lower Q resonators.

Magnetically tunable filters constructed with such ferromagnetic materials as yttrium iron garnet (YIG) and gallium substituted YIG (GAYIG) have been investigated. These devices have unloaded Q's greater than 1500 in a 600-MHz to 12-GHz frequency range, resulting in narrow bandwidths and low insertion loss. The filters are capable of being electronically tuned by an external magnetic field.

MIXERS

Microwave receivers generally require a mixing operation, in order to translate the RF signal to a lower frequency where gain may be readily achieved. The mixing of the RF signal with a signal produced by a local oscillator, produces a different frequency which effectively converts the incoming signal to a lower frequency. At microwave frequencies, the mixing process is conventionally accomplished by utilizing the nonlinearities of Schottky barrier mixer diodes. The nonlinearity of these diodes produces undesired spurious responses which, to some extent, may be reduced by proper filtering. In the case where two diodes are employed in the mixer configuration, an improvement in spurious deduction is achieved by applying different biasing voltages to the individual diodes. This technique reduces the higher ordered terms appearing in the power expansion expression for diode non-linearity.

The inherent intrinsic properties of the diode contribute a conversion loss to the mixing process, which degrades its noise performance. In receivers where RF gain precedes the mixer stage, the noise figure performance of the mixer is not significant, but when RF gain is lacking, the noise figure of the mixer causes concern. The conversion loss of modern planar junction Schottky barrier diodes is much lower as compared to older crystal point contact diodes. The major contribution to conversion loss is the series resistance of the diode. In modern diodes made of silicon and gallium arsenide, the series resistance is typically much less than 10 ohms as compared to the 50 to 100 ohms for crystal point contact diodes.

Another consideration in mixers is the proper termination of the image frequency. The broadband mixer will respond equally well to an image frequency which increases conversion losses, and degrades noise performance.

At frequencies within the VHF range, the Field-Effect-Transistor, FET, may be used as a mixer to provide conversion gain, low noise figures, and a minimum of cross-modulation. The nearly ideal square law transfer function for the FET, reduces the spurious intermodulation products for this type of mixer (as compared to the diode or bipolar transistor stage).

THERMAL DESIGN

Power dissipation is specified by device manufacturers, but different manufacturers specify under different operating conditions. Usually the maximum power dissipation at a 25°C ambient temperature is specified.

This is typically the static (d-c) power dissipation. It is also necessary to consider the effects of the a-c component of power dissipation on the device.

Operating characteristics that affect the power dissipation are ambient temperature, capacitive loading, operating frequency, and bias voltage. In all cases, the power dissipation of the device increases for increases in these parameters.

A temperature change from —55°C to +25°C will typically increase the dynamic power dissipation by 400 percent; from +25°C to +125°C the dynamic power-dissipation increase is 200 percent. The average power dissipation increases as a function of capacitance loading, and is different for different frequencies of operation. An increase in capacitance from 15 pF to 100 pF will cause an increase of about 48 percent in average power dissipation in the frequency range of 10 to 25 MHz. The power dissipation may increase as much as 35 percent for a 10-percent increase in bias voltage.

The amount of energy required to energize a load is fixed and independent of the circuit bandwidth speed. However, the power dissipated in a circuit will vary according to the frequency of operation. The relationship between power and frequency is illustrated in Figure 19. The minimum power and the maximum speed of the circuit are generally determined by

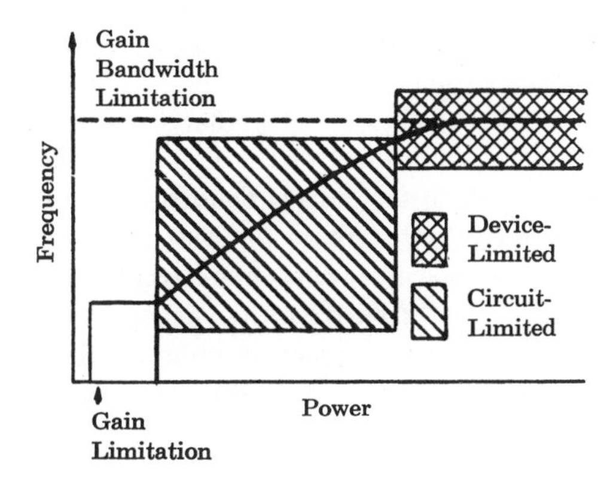

Fig. 19 Speed vs power dissipation

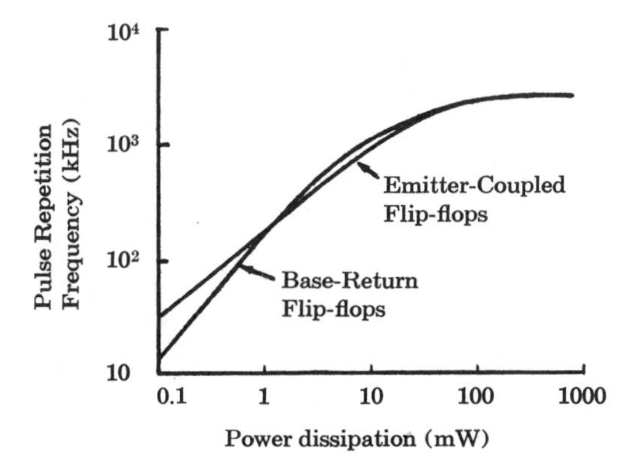

Fig. 20 Flip-flop prf vs power dissipation

component or device properties. Figure 19 shows that power dissipation increases as frequency increases. A specific illustration of this principle is shown in Figure 20. There the PRF of a flip-flop is plotted against power dissipation. For a change from 1 to 25 MHz at 25°C, a typical increase in power dissipation is about 300 percent.

The ability of an IC to operate at a given power dissipation is a function of the power density on the substrate. This is not a problem for simple IC's, but becomes important as complexity increases while substrate size remains the same.

LSI THERMAL CONSIDERATIONS

High component densities in LSI devices will require effective heat-transfer mechanisms. Digital IC's operate at a few mW power dissipation per gate function, and present no serious thermal problem. For a complex array with perhaps as many as 1000 gates, the power dissipation of the array would be tens of watts. The average power density of a 1000-gate LSI with an average power dissipation of 10 mW per gate on a 0.3 by 0.3-inch substrate, is in excess of 100 watts per square inch. This is average power density; component power densities will be much higher. To achieve integrated-circuit gate delays in the nanosecond region, much higher component power densities from 5 to 20 kW/inch^2 are required, based upon the estimate that resistor dimensions approach 0.25-by-0.25 mil.

The requirement of higher power for higher speed and its detrimental effects on failure rate, requires effective thermal management for the LSI device. The pedestal to which the chip is bonded must have a sufficiently high thermal conductivity to distribute the heat flux across the chip and then transfer the heat to the package without an excessive thermal gradient. The heat must then be transferred from the package to a heat exchanger. To keep thermal gradients within acceptance limits it may be necessary to mount the chip on a metal pedestal instead of an insulator. The package may be required to have fins. The use of forced-air cooling may also be required to keep junction temperature within acceptable limits.

FREQUENCY CONSIDERATIONS

The IC operating frequency stated by the manufacturer on the specification sheet, is usually a maximum device frequency under ideal conditions. Only when the effects of temperature, variations in supply voltage, actual physical layout, and capacitive loading from other devices are determined, is it possible to determine the actual operating frequency.

When buffer amplifiers and driver units are required, additional speed may be lost because some of these units are slow. For logic circuits an estimate of the actual maximum operating frequency is about half the ideal repetition rate specified by the manufacturer.

Aside from the frequency constraints associated with the design of the IC circuit, the most important single factor affecting frequency is capacitive loading. This effect can be reduced by understanding where the difficulties occur, and packaging the system to minimize these parasitics.

There is capacitance associated with the input of the device. This is nominal and usually amounts to between 3 and 5 pF. The total capacitance associated with IC inputs depends upon the fan-out. For a high fan-out, it could be appreciable. There is also a capacitance associated with printed-circuit wiring, in which case it is a function of conductor width, conductor length, and separation between conductors. The capacitance associated with a double-sided printed-circuit board, ranges from 0.1 to 1 pF per inch of conductor length.

When multiple-layer printed-circuit boards are used, the capacitance load increases because of the decrease in separation between conductors. Typical values are 0.3 pF per inch between adjacent conductors on the same plane, 3 pF per inch for parallel lines on adjacent layers, and 7.5 pF for lines on layers next to a power plane.

Capacitance is also associated with the printed-circuit board connector. Typical values are from 1 to 3 pF. If these capacitances can be reduced or eliminated, higher operating frequencies can be obtained.

LSI aims at complex circuit arrays with high component density and high speed. Arrays with as many as 1000 gates on a single chip with logic-block delays of only a nanosecond or so per gate are sought. These short delay times are possible because of the close proximity of the circuits, and the reduction of capacitance associated with wiring, printed-circuit boards, packages, and connectors.

As the switching speed increases, serious consideration must be given to the introduction of false signals caused by reflections from unterminated transmission lines. The close spacing of circuits, with their resultant short transmission lines, tend to minimize this problem on the wafer. However, transmission lines between LSI packages may require resistive terminations to reduce signal reflections.

LARGE-SCALE INTEGRATION

There is no clear definition for LSI within the industry. It is agreed, however, that it involves a high degree of complexity on a single silicon substrate, and a component density requiring at least two levels of metalization. The LSI concept implies the fabrication of a complete function block or sub-system on a single substrate.

The LSI concept uses monolithic IC techniques at the system and subsytem levels. With LSI, IC's on the same substrate are intraconnected by metalization techniques. The circuits are arranged in a matrix and interconnected to achieve a desired signal processing.

The matrix elements may be identical IC's or groups of components that can be intraconnected to form multiple gates or various types of flip-flops. The first level of metalization provides the intraconnections for the components that make up the matrix element. It is also possible to place different circuits on the same substrate. For example, flip-flops may be interspersed with gates by combining the masks for each circuit during the step-and-repeat masking preparation. When the circuits are different, the circuit requiring the greater area is usually designed with a surface area that is an even integral number larger than the area of the other circuit.

The LSI device is fabricated in the same manner as silicon monolithic integrated circuits up to the point at which the wafer is diced. The wafer may contain hundreds or thousands of circuits depending upon circuit complexity, component density, or various design considerations. Before separating the dice (scribing), each circuit is electrically tested to determine its performance characteristics. Those not meeting the performance requirements are marked. If the wafer is to be used for IC's, the marked dice are thrown away after scribing.

When the wafer is to be used for LSI, the wafer is not scribed and thus the "bad" circuits remain. They are, however, not intraconnected. The LSI device may not require the use of the complete wafer, but only a matrix of circuits in an array. Ideally, it is desirable to select an array of circuits from the wafer that does not include any bad circuits. If this is possible, a single set of metalization masks can be used for all subsequent identical arrays. When bad circuits are interspersed with good ones within the array, metalization must be done on a custom basis for each array.

The matrix of circuits, once metalized according to signal and bias requirements, is mounted in a package; the inputs and outputs are brought out from the array through the package leads. The result is a complex function in a single package instead of a single circuit (or a few identical independent circuits) in a single package.

LSI will probably find initial application in digital assemblies with a high incidence of repetitive circuits, such as registers. Ultimately, LSI may advance to a micro-computer (or "computer on a slice"). Any application of LSI has the advantage of reducing the back-panel wiring at the expense of increasing device metalizations, because a large amount of circuit interconnecting is done on the silicon substrate.

The proximity of the circuits reduces interconnection parasitics and thus increases frequency response. While substrates are large, component density is high.

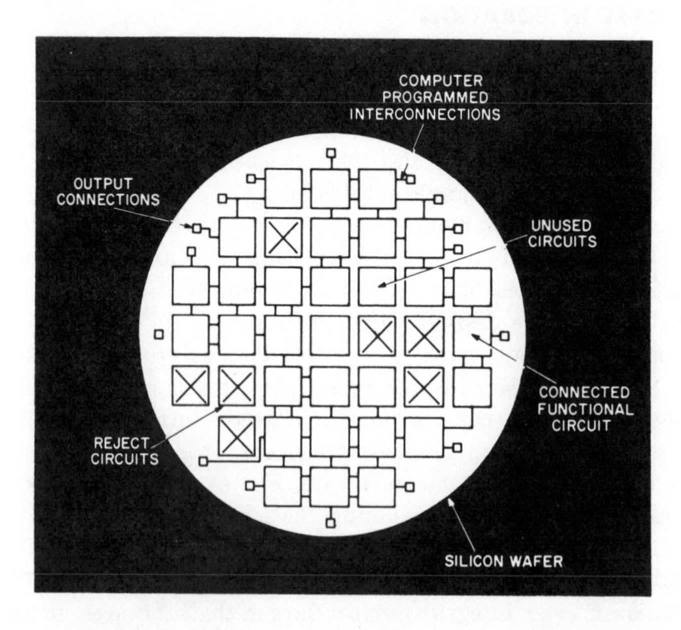

Fig. 21 LSI arrangement. Only good circuits on wafer are connected

The LSI device has the potential for being especially low in cost. In limited volume and special one-time cost, LSI may keep its price high, but its cost advantage can be so great that for highly repetitive digital applications its use may be mandatory.

LSI devices may require custom design. While some applications are similar enough to use the same off-the-shelf LSI, these may be the exception unless system designers adopt a set of procedures for designing at a level higher than the circuit. Thus, the re-use factor of any LSI device may be poor.

Testing the LSI device may be complex and costly, if complete testing at each pin is required. It may be necessary to perform only selective tests or to test by substitution.

Thermal properties of the device package and methods of heat transfer must be carefully analyzed in LSI applications. The high circuit density requires packages with excellent thermal properties. While thermal management is always a serious consideration, it is even more important for LSI.

THE COMPUTER AS A MICROELECTRONIC-DESIGN TOOL

Computers are used to aid in the design and analysis of microelectronic circuits. They can be expected to play an ever-increasing role in future microelectronic developments, aside from their obvious role as a major consumer of the circuits. Their use as an engineering tool is not simply desirable, but could become mandatory for the design and analysis of the complex devices that are currently being developed.

Computer analysis of electronic circuits has became common in recent years. This is due, in large part, to the development of general-purpose circuit-analysis programs which can be used by engineers who do not understand machine programming and may be applied to a range of problems dealing with a-c, d-c, and transient analysis.

Circuit analysis, such as worst-case design, and a-c, d-c, and transient analysis, are applicable across the board. Computers also have more specialized applications in microelectronics, such as component layout to optimize surface-area requirements, design of diffusion and metalization masks (especially the complex metalization patterns required for LSI), partitioning for optimum grouping of circuits into LSI devices, and real-time process control of the fabrication sequence in device manufacturing.

To illustrate the application of a computer in microelectronic design and fabrication, a general discussion of various stages in design and processing of an electronic subassembly to be fabricated as a complex IC array will be considered. It is assumed that the subassembly has been specified at least to the extent of the input-output requirements.

The first step is determining the number and type of circuits that are required. Once the circuit requirements have been specified, preliminary design can be performed by use of a graphics-interfaced general-purpose computer program. D-c, a-c, and transient analysis can be performed and changes made as required to meet the circuit specifications. Performance sensitivity to parameter drift and element tolerances can be analyzed with computer-aided worst-case design studies.

Assuming that the subassembly is reasonably complex, it will be necessary to partition the subassembly into complex arrays that are technologically realizable. This involves determining the type and number of circuits on each array that will optimize some objective function, subject to a set of constraints.

The objective function may represent a figure of merit indicating the effective integration of each possible grouping of circuits. The constraints represent the real-world restrictions imposed upon the fabrication of such arrays. For complex arrays, computer analysis is mandatory if meaningful comparisons of all possible groupings are to be made.

Once the number and circuit type of the required arrays have been determined, computer-aided component and circuit layout may be performed to optimize surface-area requirements. Ultimately, the information generated by the computer may be used to cut the required fabrication masks automatically on digitally controlled equipment.

The arrays are now ready to be fabricated. Computers can be used on a real-time basis to control the various process steps such as diffusion and etching. Continual testing and comparison of the results with a set of standards provides the necessary data for the computer to make decisions, such as whether to increase doping concentrations or alter temperatures or etching times. Errors that cannot be corrected will require the computer to make a decision on whether or not to start the processing over. Such tests and decisions may be made by the computer up to the time the array is ready for shipment.

After the wafer has been processed, each circuit is tested and the data are stored in the computer for subsequent use in determining a feasible metalization routing for the array. This is an iterative process that can easily be performed by the computer, assuming that a solution actually exists.

After the array has been packaged, a computer-controlled automatic tester may be used. Test data accumulated for each device (including results of various screening tests) may be stored to be easily accessible for future reference. (This might include examination of the data in relation to the results of subsequent failure analysis performed on the array.)

This example is intended to illustrate possible uses for computers in microcircuit design and fabrication. The illustration has not exhausted all possible applications. Computers are used in varying degrees in some of the applications mentioned in the example, while other applications are only in the conceptual stage.

TESTING OF MICROELECTRONIC DEVICES

Testing methods for microelectronic devices made by the film technologies are analogous to methods used for conventional component circuits. For integrated silicon devices, however, testing has proved to be a problem area for which new concepts have been developed. While individual component structures are tested by probing during fabrication, the prime testing operation on monolithic silicon devices is performed on the complete circuit both before and after encapsulation. The large number of tests which are necessary in order to categorize fully the integrated circuit operation has resulted in this being one of the more expensive and slow parts of the production process. In order to solve this problem, automatic test equipment capable of performing a large variety of measurements sequentially, has been designed. In the design of this apparatus, advantage has been taken of the integrated circuits themselves, in order to come up with low cost, high speed testing methods.

The fabricator is interested in achieving control over his processes and in meeting certain device specifications. His tests are mostly device-oriented rather than circuit oriented. They include evaluation of the quality of the photographic images, and their alignment; the measurement of sheet resistivities, junction depths, and diffusion profiles; the measurement of component parameters and the determination of thermal characteristics. These tests are usually applied at specific process stages during fabrication.

Certain simple circuit function tests are performed on the completed slices to determine whether or not a slice will be suitable for scribing and dicing into separate units for packaging. Such tests are made with multiprobe instruments which contact the lead extensions of each circuit for testing. By precisely indexing the wafer, each circuit of the matrix may be tested. This technique is one of the most frequently used.

More sophisticated testing can be accomplished with more elaborate equipment. For example, a novel nondestructive method, utilizing electron microscopy for inspection of completed circuits, may be applied. This system uses a very small diameter electron beam which scans the surface of a circuit wafer. The secondary electron emission from the surface is amplified, and the resulting signal is used to modulate the brightness of a cathode-ray tube having an electron beam moving in synchronism with the primary electron scanning beam. When various potentials are applied to the circuit under inspection, the region immediately beneath the surface may be examined for faults, by mixing the generated photo-voltages with the secondary emission video signal.

The scanning electron microscope may be used to inspect production samples; to observe potential gradients across integrated resistors; to inspect surface contours of p-n junctions; to detect surface inversion layers; and to study the conditions of the oxide surface, the evaporated interconnections, and the quality of wire bonds. Many faults not readily observed with conventional microscopes can be found with the scanning electron microscope.

Infrared techniques are also applicable for testing the operability of devices. This stems from the condition that devices may fail from a chemical or electro-chemical viewpoint, due to the application of heat. In infrared testing, the heat dissipated by energized components is measured by its infrared emission. Thus, the infrared energy transmitted by the component, is representative of the current flowing through it.

The thermographic technique is another heat detection method in which a high-resistivity thermographic phosphor is sprayed on top of the component. When irradiated with an ultraviolet source, the bright yellow fluorescence is quenched or darkened by the heated component. The darkening effect is proportional to the temperature.

Devices may also be tested for their noise characteristics. Abnormal noise effects are indicative of defective conditions within the device. Five basic types of noise effects may be detected in semiconductor devices: 1) *thermal noise* which is generated by random thermal motion of charged particles; 2) *shot noise* is due to random passage of discrete carriers across a barrier or discontinuity, such as a semiconductor junction; 3) *excess noise* results from the passage of current through a semiconductor material; 4) *avalanche noise* is produced when carriers in a high voltage gradient acquire sufficient energy to dislodge other carriers through physical impact; and 5) *multistate noise* which is due to erratic switching effects generated within the device, when particular magnitudes of current are applied.

RELIABILITY

One of the goals of microelectronic circuitry is to improve the reliability of the electronic system. One of the reasons for this is that, as increased component densities are made possible through the use of microcircuits, the resulting equipments are essentially non-repairable. Furthermore, reliability is of prime importance in two major areas of application; in space applications and in large systems. Therefore, along with reduced size, weight, cost, and power consumption, increased reliability is one of the chief aims of microelectronics.

If one adopts the qualitative definition of reliability as being a measure of one's confidence that the device will perform as intended, then most microelectronic structures are reliable. This is because system designers do have considerable confidence in the performance of microelectronic structures. If one desires quantitative reliability numbers, however, the situation is more complex. This is because with the excellent reliability of microelectronic devices, statistics are difficult to accumulate which allow the assignment of realistic reliability numbers. At the present time, however, sufficient testing has been performed on certain types of silicon integrated devices, so that meaningful numbers can be given.

The best approach for obtaining reliable devices, is to study them in great detail in order to identify failure causes. If these failure modes are eliminated, and screening techniques are adopted which allow the detection of latent failure modes with a minimum of testing, then high reliability can be achieved. Standardization on a relatively few circuit types can be of great assistance, since the testing and evaluation are time consuming and expensive. In one version of this testing procedure, the performance of devices is studied in detail, and mathematical models are determined to express the performance in terms of those factors which affect it, i.e., equivalent circuits are developed. The equations of the models state the interrelations between the various parameters, and predictions can be made with respect to the accuracy of the equations and over the range in which they are applicable.

By far the largest amount of data has been accumulated on semiconductor digital functional blocks. An average value obtained from approximately 50 million hours of room temperature operation of standard functional semiconductor blocks, is .04 failures per 100,000 hours. This value corresponds to blocks fabricated by a number of different manufacturers. The units are designed to operate from −55° C to +125° C and are hermetically sealed. Reliability data on thin film circuitry is much more scarce although the thin film art is much older. This is due to the large diversity of circuitry fabricated and the lack of mass-produced units. The best data thus far obtained indicates that thin film circuitry has a failure rate between 50% and 70% lower than conventional circuitry. A comparative value for semiconductor blocks is 10%.

Most of the reliability considerations which apply to conventional components are equally valid when applied to microelectronics. Special advantages resulting from the use of microelectronics are:

1. Fewer dissimilar materials have to be connected, resulting in fewer interconnection failures. When connections are made using dissimilar material, certain disadvantageous conditions prevail which do not arise when connections are made using similar materials. Some of these are:
   a) The joining material may be less stable. Thus, solder melts, silver migrates on the surface, and aluminum corrodes.
   b) When a joining operation is made, the area around the joint may be abused, its electrical characteristics degraded, and its strength impaired.
   c) Unwanted chemical reactions may occur which are caused by chemical residues left from the joining operation.
   d) Phase changes in the solid state may occur in the joining operation and lead to new stress situations.
   e) Differences in thermal expansion between the dissimilar materials may cause stresses at various temperature levels.
2. Because of their close proximity to each other, the components within a device will tend to show similar changes in characteristics.
3. The lower cost of microelectronics may be traded for higher reliability through product improvements. Furthermore, the lower cost makes increased reliability, by using redundancy, less costly.
4. In the manufacture there is less handling of the devices, and therefore a more uniform product results.
5. Smaller size allows better protection against hazards.

Some factors which are unfavorable to the reliability of microcircuits, are:

1. Small components are generally more difficult to manufacture than larger ones, and therefore there may be more flaws in the fabrication. This objection, however, vanishes when increased experience promotes greater manufacturing skill.
2. While strong coupling, thermally, electrically, and mechanically is considered an advantage in that the parameters change together, it may also be a disadvantage since a failure in one part of the circuit affects another part. Accordingly, any catastrophic failure may spread.
3. Small size results in small inertia, and the device can be extremely sensitive to dynamic overload.

FAILURE MECHANISMS

In order to assure that the microelectronic devices, being used, are good in the first place, elaborate screening tchniques have been devised which have succeeded in identifying a number of failure modes including:

1. Open bonds due to plague formations caused by poorly controlled bonding techniques.
2. Open bonds due to poor aluminum adhesion to the silicon dioxide passivation.
3. Open circuits in the interconnection pattern due to scratches on the device surface or to corrosion of the aluminum conducting path. Corrosion is caused by inadequate cleaning procedures and is accelerated by heat.
4. Cracks in the silicon chip which result from strains induced in the assembly process.
5. Poor layout resulting in excessive spalling of the silicon chip.
6. Poor arrangement of the internal lead wires that provide contact between the silicon chip and the feedthroughs of the package.

One of the more frustrating failure modes observed in practical electronic systems has been that of poor wiring joints. In the past, most of these connections have been made by soldering. More recently welding methods have been investigated in an attempt to find a more reliable process. The conclusion drawn from such studies is that redundant welded interconnections are more reliable. Redundancy here means that the device lead wire is welded to the circuit board in at least two places. Numbers which have been quoted give the probability of a bad welding connection as 10^{-4}. Redundant welds reduce this probability to 10^{-8} (assuming statistical independence). Other factors which have been found to influence the quality of the connection are the shape and time duration of the welding pulse, and the nature of the materials between which the weld is being made. For example, no copper is recommended for the integrated circuit interconnection board—Kovar and nickel combinations are compatible. Interconnection and assembly methods are not only a reliability problem but are also important in cost and weight considerations.

RELIABILITY DETERMINATION

Drift or degradation of microelectronic devices is difficult to detect because of their complexity and high stability. Device specifications generally do not include information regarding parameter drift but only state performance measures under conditions that are seldom realized in practice. To realize the full inherent reliability of microelectronic devices, it is important that design engineers have information regarding performance in any specific application and knowledge of any degradation in performance that is likely to occur.

One study sponsored by NASA in these specific reliability areas, used a dual NAND-NOR gate as an example. Performance parameter distributions as determined for a sample of these devices, in standard test configurations, are shown in Figure 22 The devices were then placed in logic circuits designed to determine the influence of system parameters on performance, e.g., the loading factor. From these measurements, empirical mathematical models were derived which described these influences. The mathematical models are given in table 1.

In attempting to determine time dependence of the parameter distributions, these devices were stored at 125° C for a total of 56,000 device hours, after which the performances were remeasured. No statistically significant changes could be detected in the parameter distribution. It is expected, however, that accelerated testing would cause variations and that these can then be related to system performance by means of appropriate equations. If the acceleration factors can be evaluated, then it is possible to predict the time dependence of performance (or reliability) for these devices under normal use conditions.

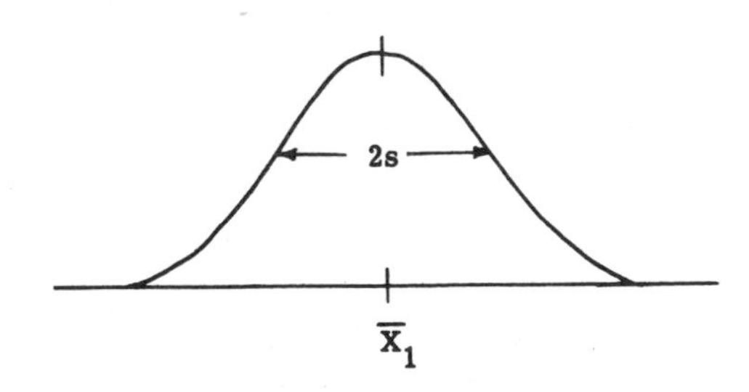

	s	Mean	Inputs
Output Pulse Height, Va (volts)	0. 115	$\bar{X}_1 = 4.98$	all
No Specification			
Delay Time, t_d (nsec)	4. 116	$\bar{X}_1 = 56.4$	all

Specification: typical - 55 max - 80

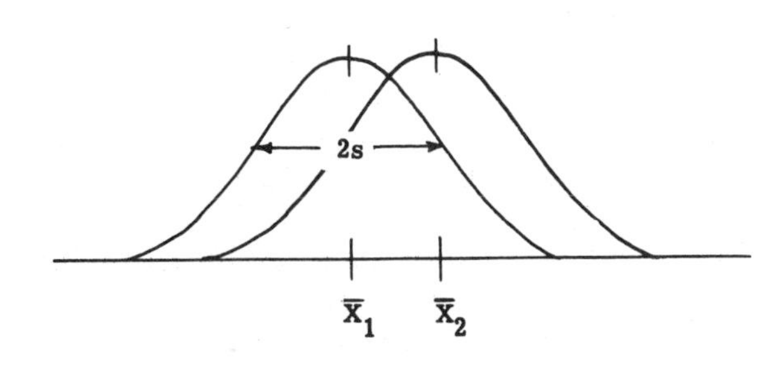

	s	Mean	Inputs
Fall time, t_f (nsec)	52. 79	$\bar{X}_1 = 785.1$	4, 5, 10
		$\bar{X}_2 = 838.5$	1, 2, 6

Specification: typical - 900 max - 1200

Fig. 22 Performance parameter distributions for NAND - NOR gate

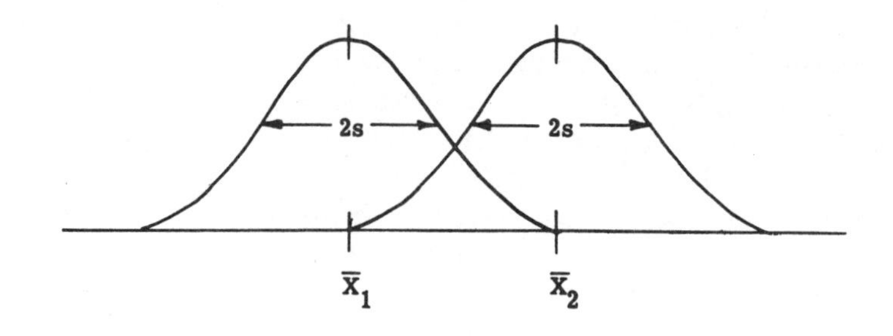

	s	Mean	Inputs
Rise time, t_r (nsec)	4.108	$\overline{X}_1 = 76.5$	1, 2, 6, 4, 5
		$\overline{X}_2 = 87.5$	10

Specification: typical - 75 max - 110

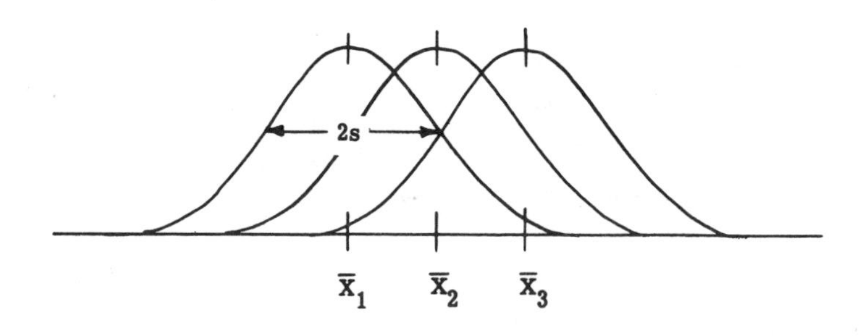

	s	Mean	Inputs
Storage time, t_s (nsec)	3.415	$\overline{X}_1 = 37.5$	4, 5
		$\overline{X}_2 = 40.6$	1, 2, 6
		$\overline{X}_3 = 44.9$	10

Specification: typical - 50 max - 90

Fig. 22 (continued) Performance parameter distributions for NAND-NOR gate

Table 1 Mathematical Models for Performance of Silicon Integrated NAND-NOR Gate

	Units	s	R
$\hat{V}_a = 0.20 - 0.50F_o + 1.11V_a^*$	volts	0.25	0.93
$V_a^* = 5.70 + 0.000192\ h_{FE} - 0.000178\ R_L - 0.00419\ R_{in} + 0.00201\ C_{in} + 0.0751\ C_{oe} + 0.00236\ f_\beta$	volts	0.022	0.984
$\hat{t}_f = -287.2 + 117.2F_o + 1.18t_f^*$	n sec	73.33	0.96
$t_f = 0.1616 + 0.000123\ R_L + 0.0774\ C_{oe} + 0.00285\ f_\beta$	μ sec	0.0293	0.925
$\hat{t}_r = -14.8 + 40.0F'_o = 17.4F_i - 8.2F_o + 1.45t_r^*$	n sec	21.79	0.94
$t_r^* = 326.05 + 0.175\ h_{FE} - 0.00718\ R_L - 0.538\ R_{in} - 1.71\ C_{in} - 34.69\ C_{oe} - 1.78\ f_\beta - 38.21\ l_n\ h_{FE}/C_{in}$	n sec	4.91	0.746
$\hat{t}_d = 46.2 + 29.9F'_o - 15.3F_i + 14.3F'_i + 3.9F_o$	n sec	18.93	0.93
$t_d^* = 153.98 - 3.65\ R_{in} - 23.24\ C_{oe} - 0.575\ f_\beta + 0.0864\ R_{in}$	n sec	2.64	0.753
$\hat{t}_s = 1.4 - 2.3F'_o + 2.8F_i + 3.1F'_i + 0.83t_s^*$	n sec	7.93	0.57
$t_s^* = 46.87 - 0.0338\ h_{FE} + 0.00578\ R_L - 0.284\ R_{in} - 0.200\ C_{in} - 0.202\ f_\beta$	n sec	2.285	0.867

V_a—pulse amplitude
t_r—rise time
t_f—fall time
t_s—storage time
t_d—delay time
s—standard deviation
R—correlation coefficient

$\wedge$—predicted mean value
*—value of quantity in standard test circuit
F_o, F_i—fan-in and fan-out
F'_o, F'_i—fan-in and fan-out of driving stage
h, R_L, R_{in}, C—component parameters in silicon structure

$$1 \leq \frac{F_o}{F'_o} \leq 5 \qquad 1 \leq \frac{F_i}{F'_i} \leq 3$$

PART I

THIN/THICK FILM TECHNOLOGY

CHAPTER 1

FUNDAMENTAL PRINCIPLES OF THIN FILMS

1.1 GROWTH AND STRUCTURE OF THIN FILMS

How thin is thin? Most of the films used in thin-film circuitry are from 50 Å (15 atomic layers) to 5000 Å (1/50 of a mil) thick; the thicker being 1/5 the thickness of the thinest films. "Thin" or "thick" films are more often defined by the fabrication processes rather than the actual depth of the film. Thick films are generally screened on in paste form and subsequently fired to form a ceramic with the desired properties, i.e., resistive, conductive or dielectric. Thin-films, on the other hand, can be vacuum evaporated, sputtered, electrodeposited, vapor plated, anodized, or polymerized.

Thin-films were originally used for decorative purposes, i.e., gold leaf and silver plating. World War II brought about both the development of the hybrid type of circuit in certain fuze applications, and thin-film components used as waveguide terminations. These events along with the discovery of the transistor led to the development of the thin-film technology of today.

The physical properties of thin films are determined by their structure and composition, and these are established by their growth mechanisms. In bulk semiconductor materials it is possible to produce essentially perfect crystals which closely resemble the descriptions of the perfect media used for analytical purposes. In thin films, however, growth mechanisms are not very well understood. Their use is dependent mainly on the experimental results.

Among the properties vital to the behavior of thin film components are: structure and composition, electrical properties, temperature coefficient of resistance, mechanical properties, quality of adhesion, and internal stresses in the films. Most of these properties are very dependent on the method of producing the film, and various parameters associated with a given process.

Thin films are commonly grown, under high vacuum, by thermal evaporation, or sputtering of elements or compounds onto a variety of substrates. There are a number of factors known to influence the characteristics of films; these include rate of evaporation, evaporation temperature, film thickness, structure of the substrate, and the substrate temperature. These factors are important in determining the properties of thin films, of which the most important ones are crystal structure, homogeniety and continuity, imperfection and impurity density, and development of stresses.

Depending on the material to be evaporated, a given minimum temperature is necessary to produce a significant vapor pressure and evaporation rate. In addition, a specific vapor density must be achieved, for a given substrate, in order to condense vapor on the substrate. The adhesion of the impinging atoms is determined by the forces prevailing between them and the substrate. The substrate provides a surface to which the atoms in the vapor, may transfer energy and become attached in an adsorbed state. This energy is dependent on the binding forces between the condensate and the surface, and, therefore, on the condition of the surface. The binding energy may vary from weak electrostatic van der Waals forces, to strong metallic or ionic bonding. Surface contamination of a few atom layers may substantially change these binding forces. Once a deposit is a few monolayers thick, impinging atoms only, interact with these layers and the substrate ceases to have significant effect on the process.

A very important factor in the growth of the film, is the mobility of the condensing atoms, once they are on the substrate. This ability to move about in a limited region is imparted, to the atoms, by the energy of condensation. The thermal energy is important in giving the condensing atoms a degree of surface mobility. Surface chemical contamination can influence the degree of this mobility. While this mobility can affect the degree of rearrangement of the atoms, and, hence, the structure of the film, there is a complex relationship between this effect and the rate of evaporation. If the evaporation rate is extremely high, there is only a limited time for the atoms on the surface to rearrange and combine into clusters before the next monolayer is deposited. Hence, rapidly deposited films contain a large number of small islands, whereas slowly deposited films initially contain a few large islands. As the average film thickness is increased, the distance between islands decreases. Eventually a continuous film is produced in either case, if the growth is allowed to continue for a sufficient period.

The structure of the substrate is often an extremely important factor in determing the characteristics of the film. Often a substrate can have a substantial ordering effect on the film structure, tending to make the film more crystalline. Films deposited on amorphous substrates such as glass, are usually composed of a large number of small polycrystalline grains. Use of single crystal substrates together with the proper deposition parameters, can result in the formation of single crystal films. In order that the film be of this quality, the deposition rate must be slow enough and the substrate temperature high enough, so that the impinging atoms can arrange themselves to a sufficient degree. The oriented growth of films can occur on a variety of substrates, and does not, necessarily, depend on a close crystallographic match between the deposit and the substrate.

Oriented growth of films by such techniques, is often referred to as "epitaxial growth." Mismatches of over 70 percent between normal atomic spacings of the two have been found to be allowable in certain cases. However, the usual technique makes use of substrates of the same material or one with only a slight mismatch in spacing.

1.2 PHYSICAL PROPERTIES OF THIN FILMS

The two methods most prominently used for producing thin films are vacuum evaporation and sputtering. A very significant difference exists in the films produced by these techniques due to the different vacuum conditions which exist. In vacuum evaporation, the films are deposited, at fairly high rates, in vacuums of 10^{-5} to 10^{-6} torr, while sputtered films are produced at pressures of 10^{-1} to 10^{-2} torr, in the presence of various specific gases. Sputtered films generally contain much higher amounts of gaseous impurities which are effective in altering not only their chemical composition, but also their structural and electrical properties.

Films made by either process have similar structural properties while their composition is considerably more variable. The structure of most thin films, is a collection of small grains with the grain size dependent on the material. The material and film thickness also determine the degree of continuity of the film structure. Observations have shown that materials such as gold and silver tend to form an agglomerated structure, i.e., a number of widely separated crystallites of fairly large sizes ($\sim$500Å diameter). These materials become continuous films at thicknesses of more than several hundred angstroms. Most other materials have a grainy film structure, but a fairly continuous one as has been observed in iron, nickel, and chromium. An even more continuous structure is observed in the case of aluminum which becomes continuous at a thickness of a few angstroms and contains few observable grain boundaries. In most cases, the structure of a thin film, after it becomes continuous, is a large number of small grains of random orientation. Common orientations may be obtained by annealing processes or by the epitaxial techniques. In general, films deposited at higher substrate temperatures, will have larger grains with a greater degree of common orientation, due to the higher surface mobility of the deposited atoms at higher temperatures.

Film composition depends on the process used in fabrication, as well as the parameters associated with a given process such as pressure and rate of deposition. Those films produced by vacuum deposition at pressures of about 10^{-6} torr, are basically composed of the deposited element and some amount of the residual gas in the system – probably air or oxygen. The amount of oxygen content depends on the rate of deposition and the affinity of the material for oxygen. High deposition rates will, of course, lead to films with lower percentages of residual gases. Films such as gold and silver are relatively pure, due to their poor affinity for oxygen, even though their agglomerated form provides large areas for possible adsorption. Copper which also exists in an aggregated form and has a high affinity for oxygen, tends to oxidize readily. Continuous films, such as aluminum, undergo a certain amount of surface oxidation, but little oxygen is present below the surface layer.

Sputtered films which are produced at pressures of 10^{-1} to 10^{-2} torr, and at slower rates than evaporated films, are much more mixed in their composition. If a film is sputtered using an argon glow discharge, chemical reaction between the deposited element and the gas can be avoided. However, if reactive sputtering of metals in nitrogen, oxygen or air is carried out, the deposit will be a chemical compound. Such materials as tantalum and titanium together with their oxides and nitrides, have been prepared by this technique. Such films have generally a more homogeneous mixture of their constituents, than those prepared by vacuum evaporation.

The electrical properties of thin films have been found to be influenced by the thickness, structure, and composition of the film. Such properties as electrical conductivity, temperature coefficient of resistivity, and Hall mobility are severely affected by the thickness and continuity of the film. Films, such as aluminum, which are continuous for even very thin specimens, have positive temperature coefficients of resistance, and obey the classical resistivity relations. Conduction in agglomerated films, with gold and silver, is by tunneling between the separate islands of the structure. This leads to a different conductivity relationship, and a negative temperature coefficient of resistance. In films which are generally continuous, however, the electrical resistivity increases as the film thickness decreases, with the greatest rate of increase occurring for thicknesses less than that of the carrier mean free path (typically 500 Å). The Hall mobility decreases significantly for thin films, with the rate of decrease being largest below thickness equal to the mean free path of electrons. Since the electrical conductivity and thermal conductivity are directly related, the thermal conductivity also falls far below the bulk value for films less than the electron mean free path in thickness.

Very thick films reach a value of electrical conductivity which is independent of the thickness. However, this value is not, in general, equal to the conductivity of a bulk specimen. This higher resistivity in film specimens, is generally due to the larger number of impurities and imperfections present in films grown by vacuum evaporation or sputtering. Generally, sputtered films being higher in impurity content, are of higher resistivity than comparable vacuum evaporated films. A few attempts have been successful in growing thin films of resistivity equal to the bulk material. This was achieved through the use of ultra high vacuum to reduce the impurity content, and epitaxial growth on crystalline substrates to reduce the number of structural imperfections.

Another property of interest is the adhesion of the evaporated films to their substrates. It has been observed that materials with a strong affinity for oxygen, appear to adhere best to a glass substrate. Apparently, the formation of an oxide layer between the film and the substrate is required to obtain good adhesion. Materials with a strong affinity for oxygen and a fairly open structure such as magnesium, iron, chromium, nickel, molybdenum, and zirconium have been found to be strongly adherent. The adherence of most of these materials increases with time as oxygen diffuses through the structure to complete the formation of a layer at the film-substrate interface. Gold, silver, and copper have an open structure, but a low affinity for oxygen, and tend to adhere very poorly to glass. Some metals with very high surface mobility such as zinc, cadmium, and tin do not adhere well to the substrate due to the instability of the film structure. Their tendency is to form large aggregates. Despite its great affinity for oxygen, aluminum adheres only moderately well, because it forms a very continuous film and does not permit much oxygen to diffuse to the interfacial region.

Thin films have been observed to have mechanical properties differing from the comparable bulk materials. In particular, a number of experiments have shown that films have considerably greater strength than the parent material. A relationship, between the thickness and strength, has been observed for gold and silver polycrystalline and single-crystal films. The films tend to have the bulk fracture stress down to thicknesses of several thousand angstroms, while, for thicknesses less than this, the fracture stress may reach ten times the bulk value. Some measurements show that films can withstand several times the bulk fracture stress, at thicknesses of several microns. The films can also withstand elastic strains of one percent or more. These properties are apparently associated with the microstructure of the film, the immobility of dislocations in these specimens, and the lack of formation of new dislocations.

The mechanical properties are also affected by the mechanical state of the film, due to its preparation. Substantial residual tensile strains arise in the films during preparation. These strains may amount to as much as 0.3 to 0.4 percent. The corresponding internal stress is partially due to the difference in thermal expansion coefficients between the film and the substrate. The rest arises from the internal structure of the film, and the fact that it is not able to assume a minimum energy situation during growth. Some relief of the latter condition is often possible by resorting to annealing procedures.

1.3 SUBSTRATES FOR THIN FILMS

"Sub" means under and "stratum" means layer and that is what a substrate is. The substrate is the supporting layer that provides a smooth surface for the deposits, provides electrical isolation between the components, and provides the thermal path to remove power dissipated by the circuit elements. Such factors as the substrate strength, thermal properties, and cost must be considered in selecting a material.

The materials generally used for substrates are ceramics and glasses. Glass has a high degree of smoothness—of the order of 300 Å while ceramics, on the other hand, have a surface roughness of the order of 20,000 Å.

Glass can combine most readily the properties of surface smoothness with wide range of thermal expansion coefficients. The thermal conductivity for glass ranges between .002 and .004 cal-cm/cm^2-sec-deg C at room temperature. One method to overcome partially the undesirable thermal conductivity, is to draw the glass into very thin sheets. The degree to which the glass may be drawn in this manner is limited essentially by the residual strength.

Ceramics have a wider range of thermal conductivity than glass. Thus, berylium-oxide ceramics has a conductivity of 0.55 cal-cm/cm^2-sec-deg C at room temperature, as compared to cordierite ceramics which exhibits a corresponding conductivity of .003 cal-cm/cm^2-sec-deg C.

Glazed ceramics can incorporate the desirable properties of the surface of glass and the thermal conductivity of ceramics. The use of glazed ceramics represents an improvement in thermal conductivity over glass of about an order of magnitude.

Surface imperfections are generally analyzed by interference or electron microscopy, and stylus methods. In general, empirical methods must still be resorted to, in determining the most applicable substrate.

1.3.1 SUBSTRATE PROPERTIES

1.3.1.1 SMOOTHNESS

Smoothness refers to the fine surface finish of a substrate and is normally measured in micro inches (1 micro inch = 250 Å). To be suitable for vacuum deposition technology, a substrate should have a peak-to-valley smoothness of less than a micro inch. This smoothness is then on the order of the resistive film thickness and an order of magnitude less than that of dielectric films. Films deposited on substrates of this smoothness have quite uniform characteristics, and dielectrics show no appreciable reduction in working voltage due to high electric field intensities caused by peaks in the substrate. Substrates having a smoothness in this range are usually glasses or glazed ceramics having a fire polished finish. To further enhance the surface finish for vacuum deposited films, an undercoat of SiO may be deposited on the substrate prior to any of the thin-film circuit depositions.

Surfaces with a peak-to-valley smoothness in the range of from 1 to 10 microinches are still usable in many applications such as resistor networks, but thin-film capacitors fabricated on substrates with these surfaces tend to have D factors much higher than those obtained with less than microinch substrates and also show greater leakage and lower breakdown voltage. These surfaces have one advantage. They provide a tooth for conductive films that are to be later plated up for use in microstrip applications. Substrates in this smoothness range are generally ground and polished ceramics.

Surfaces with a roughness greater than 10 microinches are generally not suitable for use in thin-film technology but would be suitable for thick-film technology.

1.3.1.2 CAMBER

The camber of a substrate material is a measure of gross smoothness or waviness of the surface, and is normally expressed in terms of so many mils variation in height per lineal inch. Most of the substrate materials used have a camber of less than 5 and are generally under 3. The flatness is desirable for one main reason in two areas. In thru-mask deposition, intimate contact between the substrate and mask is necessary for good edge definition and in the photolithographic processes, the same contact must be made during the exposure of the resist. Other requirements for flatness are better contact between substrate and heat sink for thermal energy transfer.

1.3.1.3 ELECTRICAL PROPERTIES

The electrical properties of importance for general use are dielectric strength, resistivity, dielectric constant and D factor. All of the common substrate materials meet the requirements when acting as a more or less passive substrate. For microstrip applications, both the dielectric constant and loss tangent should be known as functions of frequency before adequate design in this area can be done.

1.3.1.4 MECHANICAL PROPERTIES

Some of the more important mechanical properties are temperature coefficient of expansion (TCE), mechanical strength, thermal conductivity and resistance to thermal shock. The TCE of the substrate should be compatible with the material to which it is to be attached. Most of the materials have a TCE of about 5 PPM °C which is less than that of the more common metals except tantalum and titanium. The mechanical strength of the common substrates are quite high, but the materials are brittle—especially the glasses. Thermal conductivity is important when one is concerned about power dissipation. Most of the glasses will dissipate on the order of 0.012 watts/cm°C while the alumina ceramics can dissipate 0.35 watts/cm°C. Beryllia has a thermal conductivity of over 2 watts/cm°C. Resistance to thermal shock is best in materials such as quartz and sapphire, the ceramics are next, and the common glasses are the worst. All the materials, though, are usable for most thin film applications.

1.3.1.5 CHEMICAL PROPERTIES

Chemical properties of interest in thin film technology are: (1) stability during the etching processes used in the photolithographic technology and (2) freedom from alkaline metal ion migration under conditions of high humidity and voltage stress. Most of the commercial glazes and glasses sold for micro-electronic applications have this property. Soda-lime glass, commonly used for window glass and for microscope slides, is not, and therefore should not be used.

1.3.2 SUBSTRATE MATERIALS

The substrates in common use are of two types: the glasses and the ceramics. Others such as sapphire and certain plastic films are in use, but they have limited application.

1.3.2.1 GLASSES

Most of the glasses in use for substrate materials are proprietory blends having a very low alkaline content. They are hence pratically free of the ion migration problems found in the soda lime glasses, and the surface smoothness of the glasses is quite good if drawn or polished. One of the more popular types of glasses for thin film hybrids is a barium alumino-silicate glass available in many sizes and thicknesses. This is an excellent material in all respects except one. It is extremely difficult to score and break into small pieces. A boro-silicate glass that is available in very thin sheets is used mainly in applications requiring that particular property. Glass is rarely used for thick-film hybrid circuits because of their low thermal conductivity.

1.3.2.1.1 THERMAL PROPERTIES

Glasses in common use are usually described by the main constituents other than silica, which are added primarily to assist in melting and to make the flow characteristics such that the shaping operation can be done economically. Thus, the common glass of windows, containers etc. are soda-lime glasses (13-18% Na_2O, 5-13% CaO) with these modifiers added in the proportions giving the best compromise between ease of working and a chemical durability or weather resistance appropriate to the intended use. The high alkali content impairs the electrical properties because of electrolytic conduction, and where this is of concern much of the alkali and most of the lime is replaced by lead oxide to form what is commonly called lead glass. These glasses have thermal expansion coefficients of 8-10 ppm/°C and are easily broken by thermal shock. At some sacrifice of ease in melting, the alkali content can be reduced to about 4% by using boric oxide as a flux, to yield a borosilicate glass with thermal expansion about 3.2 ppm/°C and with a three-fold improvement in thermal shock resistance.

In many applications of glass, the sought-after property is a relative one, e.g., the resistivity should be greater than some limit, or the chemical resistance adequate. But in applications where the glass is to adhere to some other material, there is an absolute requirement in that the two materials must be thermally compatible. The most important characteristic is then the thermal expansion coefficient and this parameter provides a convenient means of typifying the glass.

In a comparison of available glasses, these may be arranged in order of their linear expansion coefficients (α) most usually measured over the range 20-300°C. The coefficient is seldom strictly constant over this range of temperature. Thus, one generally finds that a measurement made over the range 20-100° will produce a value about 5% lower. At higher temperatures, the curvature of the elongation vs temperature graph becomes much more pronounced, and is then usually a function of the previous heat treatment of the glass. At somewhat higher temperatures the usual dilatometer record shows a maximum followed by an apparent contraction, where the glass starts to yield under the stress imposed by the apparatus. In the cooling of a glass-to-metal or glass-to-ceramic seal, a differential stress will not begin to arise until the temperature has dropped below some rather indeterminate 'setting' point—a little below the dilatometer softening point, and at room temperature the differential strain will be determined by the differential contraction experienced between the setting temperature and room temperature.

Since the setting temperature depends on process variables such as the cooling rate, one can only crudely define the expansion required for compatibility, and for this purpose it is found that for most glasses the over-all expansion coefficient from the setting to room temperatures is about 10% higher than α, the figure for the range 20-300°C. A second rule of thumb is to assume that the setting point is 15°C below the annealing point (viscosity = 10^{13} poise).

The sintering of glass is usually conducted at a temperature where the viscosity of the glass is between 10^6 poise (for loose powders) and 10^8 poise (for pressed compacts). The Littleton softening point, at which temperature the viscosity is $10^{7.6}$, is therefore a useful guide, and taken in conjunction with the annealing point, gives some idea of the working range of the glass.

For glasses to be used for soldering where fluid flow and wetting are required, the viscosity of the glass should be in the range 10^4 to 10^6 poise. Solder glasses are usually of very short working range so that the soldering temperature is not far removed from the softening point e.g., 50-100°C higher.

Soft glasses tend to have relatively high expansion. The correlation between softening point and expansion is largely a consequence of the approximately additive nature of expansion, and the fact that silica has abnormally low expansion and high refractoriness. Furthermore, the greater its dilution by fluxes, the higher the expansion tends to be. An important exception to the additive rule is provided by B_2O_3. This easily melted oxide, which by itself is of high expansivity (15 ppm/°C), has an anomalous behhavior; when added to the extent of no more than about 15%, its first influence is to negate some of the large contribution of alkalis to expansion, and so enable the formulation of glasses of low expansion which can nevertheless be melted fairly easily. Glasses in the region α = 3-5 ppm/°C, are borosilicate glasses. Glasses centered around α = 4.5 ppm/°C, are mostly glasses of high alumina content which, although of high softening points, become quite fluid at higher temperatures and can therefore be melted readily. It is in this group that stable substrate glasses are likely to be found. The group in the region α = 9-10, are common soda-lime or lead glasses.

1.3.2.1.2 ELECTRICAL PROPERTIES

From the physical rather than the chemical point of view there are a number of generalizations which can be made which lend a measure of validity to a comparison between glasses based only on, for example, the resistivity at one or two temperatures.

Most common glasses contain 4-20% alkali, and these metal ions are relatively mobile in the glass structure and act as current carriers. The electrolytic nature of the current, gives rise to the usual effects such as polarization at the electrodes, and interference with any semi-conducting materials at the surface. The non-additive contribution of carriers of differing species is exemplified by the 'mixed alkali effect', i.e., by the fact that if the alkali content of the glass is in the form of, for example, equal amounts of two different alkalis, the electrical resistivity may be several orders of magnitude greater than if the glass contained the same total amount of one alkali only.

Not all glasses owe their conductivity to alkali ions. In fluoride glasses the current carrier is the fluorine ion. Glasses containing appreciable amounts of Fe_2O_3, V_2O_5 or titanium oxides are normally electronic conductors, and there is evidence to suggest that at low temperature, electronic conduction predominates even in common glasses. In all cases, however, the variation of resistance with temperature, in the range of interest between room temperature and the setting point, follows the relationship Log R = a +b/T where T is the temperature in degrees absolute. As a result, measurement at two temperatures only, is sufficent to obtain the constants a and b and define the behavior of the glass. A further circumstance of interest is that if the value of the slope b is plotted against resistance at a chosen temperature, all glasses with the exception of fluoride and chalcogenide glasses are found to lie near one or the other of two straight lines, depending on whether they are ionic or electronic conductors. Were this strictly true, it is evident that a single measurement of resistance would suffice to define the entire range of resistance behavior. While this is not the case, a single measurement does in fact, provide a guide which is adequate for many purposes. Rather than specify a resistance at a stated temperature, it is often more useful to state the temperature at which the resistivity has a certain value. A frequently used point known as the Tk100 temperature is the temperature at which the resistivity is 10^8 ohm cm.

It has been shown that a relationship exists between DC conductivity and migrational losses in glasses in an AC field. In the empirical form,

For (log f + log R) greater than 11

$$\log \epsilon'' = -1.9 + 40 \times 10^{-0.1(\log fR)}$$

for (log f + R) equal to, or less than 11

$$\log \epsilon'' = 12.25 - \log fR$$

where f is the frequency, R the resistivity of the glass and ϵ'' is the loss factor. Here again, the DC resistivity or the Tk100 temperature, for example, can provide a figure of merit as to the relative dielectric behavior of glasses.

1.3.2.1.3 ION MIGRATION

Electrolytic current in glass is the product of the number of mobile ions and their mobility, and to this extent the electrical resistivity is an inverse measure of the mobility. However, in considering the suitability of glasses as passivation coatings, concern is more specifically with the mobility in isolation, because the source of the ions to which a barrier requires to be presented is external to the glass.

The addition of alkalis to silica is primarily to assist in its melting, and the alkali ions finally reside in holes in the structure while the additional oxygens contribute to the volume of the glass. However, the holes are formed around the intruding ion so that their size affects the volume to the extent that the holes are larger (or in the case of Li, smaller) than those natural to the SiO_2 random structure. The alkali ion is only loosely held, and may be dissociated from its original site by thermal energy to migrate through the structure. The freedom to migrate is reduced by the presence of other ions undissociated from their holes, and which are therefore blocking possible migration channels. This is true also of undissociated ions of the mobile species. Addition of alkali to the glass composition is, therefore, expected to affect the resistivity in three ways: to increase conductivity by providing carriers, to increase mobility to the extent that the skeleton is expanded, and to provide hindrance to the extent that the ion fails to dissociate.

A small ion is less likely to dissociate because of the higher field strength, but once dissociated will migrate more easily. A large ion of the same charge will dissociate more easily but migrate more slowly. It may be seen, in general, that a glass containing a mixture of alkalis is expected to have a higher resistivity than one containing an equivalent amount of one ion only. Consider a glass containing sodium oxide with some of the soda replaced by lithia. The lower dissociation of lithium ions means that some of the channels originally free, are now blocked by lithium so that the mobility of the sodium ions is reduced. Those lithium ions which have dissocated have inherently a much greater mobility, but both the undissociated sodium ions and the slow moving mobile sodium ions hamper the movement so that the traffic is slowed down—to the speed of the slowest mover, if most of the channels are occupied. Thus, the number of current carriers is reduced without a corresponding gain in over-all mobility.

It would seem, therefore, that a measure of the tightness of the structure, such as the permeation rate for helium gas, would supply a more direct indication of the quality of a glass as a passivation layer. It is interesting to note that the helium atom and the sodium ion are approximately equal in size. As a generalization, it may be said that SiO_2, B_2O_3 and P_2O_5 in the glass composition are involved in the formation of the network structure, and that the cations of all other oxides lodge in the interstices of this structure. They therefore constitute blockages irrespective of their chemical nature. On this basis one may expect good correlation between the permeation rate and the mole per cent ($Sio_2+B_2O_3+P_2O_5$) content of the glass. Relatively poor performance may be expected from fused silica and the borosilicates. Several orders of magnitude of impenetrability could be achieved by use of glasses containing large proportions of immobile additives such as calcium, barium, and lead which occupy cavities in the network, and are themselves of low mobility.

1.3.2.1.4 CHEMICAL DURABILITY

In general, the attack on glass by water and acids is different in nature to the attack by alkalis. The network forming oxides which bind the glass together are all acidic, while the modifying oxides for the most part are alkaline. With the exception of HF which rapidly dissolves the network, the acids and water tend to leach out the modifiers to leave a skeleton or gel of silica which may or may not be self supporting. The action of alkalis, on the other hand, is a direct attack on the network, and a general breakdown of the surface follows.

1.3.2.1.5 TRANSPARENCY

The transmission characteristics of common glasses are usually determined by trace impurities, even when these are not deliberately added to develop a color. Thus, ultra-violet transmission is usually limited by the presence of Fe_2O_3, and infra-red by FeO, so that the yellow-green or blue-green tint of common glass depends on the oxidizing state of the furnace atmosphere during melting. 'Decolorizing' agents are often added to neutralize the color. A good quality glass will normally transmit from 0.3 to 2.9 microns with an integrated visible transmission of about 90%.

1.3.2.1.6 OTHER GLASSES

With the realization that many useful properties can be exploited through the partial or controlled devitrification of glass, the boundary regions of glass-formation fields acquire an importance which has largely been neglected in the past. Systems with only a minor glass-forming tendency may now be of interest, provided the tendency is great enough to allow control of the devitrification to be exercised, or the production method is such that a stable quenched glassy material is obtained. Alloys such as silicon and gold or silver can be obtained in glassy form. Vapor deposition onto a cold surface might be regarded as a quenching technique of this kind, although it is a question of definition as to whether amorphous films formed in this way are to be called glasses. With normal glass-making techniques, the possibility of glass formation is determined by the rate of cooling, and therefore by the size of the article. Where this is small — as in the case of microelectronics, a different criterion as to what constitutes a stable glass is required.

1.3.2.1.7 SEMICONDUCTING CHARACTERISTICS

When mono-valent cations are absent, electrolytic conduction in glasses is negligible because of the lack of mobile carriers· Conduction is then generally electronic and dependent on the variable valency of poly-valent ions. In crystalline transition oxides such as iron oxide, electron mobility is very low, and the effective free path so short that long range order is not a factor, and conductivity is insensitive to impurities. In glasses which have short-range order only, a similar situation exists — the electron progressing by hopping from one possible site to the next. The insensitivity to impurities (to which there are exceptions e.g. Ag in As_2S_3) and the absence of polarization effects, combined with the working characteristics of a glass, make these materials valuable, particularly where electrostatic charges are required to be drained away. For example, in image-orthicon tubes the long term performance is greatly improved by their use, and in glass bonded mica, dielectric break-down due to static charge build-up is prevented.

Current-voltage characteristics of some elemental glasses indicate that they can be used as switching and memory devices. The threshold switch is a thin film of amorphous material (e.g., the semi-conducting glass Te 47.7, As 29.9, Si 12.6, Ge 9.76 mole%) that exhibits an avalanche effect when in a field exceeding a certain threshold. The below-threshold resistance may be several megohms and capacitance a few picofarads. For a typical film thickness, the threshold is about 12 volts in both directions, and when conducting may pass 300 milliamps with less than 2 volts drop. The ability to work in both directions enables these switches to do the work of two conventional devices in certain applications. Experimental circuits using commercially available glass devices at slow pulse rates, have switched 1 amp currents millions of times without sign of wear. Typical devices are made by vacuum evaporation of semi-conducting glass on individual electrodes which are fused to leads, while the joint is encapsulated in glass.

Many oxide glass semiconductors have large positive ($80\mu V/°C$) and negative ($—300\ \mu V/°C$) Seebeck coefficients, suggesting their use as temperature sensors. An almost linear dependence of conductivity on pressure in, for example, 80% V_2O_5-20% P_2O_5 glass, suggests its use as a pressure transducer. Thermo-stimulated current in vitreous $Tl_2SeAs_2Te_3$ has been used to explore the local levels in the forbidden band.

Glasses based on selenium tellurium and arsenic are photo-conductors insensitive to impurity content. Phototropic or photo-chromic glasses free from fatigue effects have been developed from silicate glasses containing about ½% silver halide. Although darkening and recovery is slow (2 minutes to hours) the optical resolution is claimed to be 10-20 times that of photographic film.

An optically transparent magnetic glass (Ferro-glass) has been produced in small commercial quantities.

1.3.2.1.8 DEVITRIFIED GLASS

The technique of devitrifying glass may be used to produce partially or completely crystalline materials. At one time in the past, devitrification was regarded as a plague which would embrittle the glass to the extent that its strength was seriously reduced. However, by exercising control over the process, materials having higher strength than the parent glass can be obtained.

Loss of strength was due, in the past, to the expansion incompatibility between the crystals and the host glass. If the crystals are, however, sufficiently small and sufficiently numerous, high strength glass-ceramic bodies can be obtained — because the skeletal structure of the residual glass exhibits the high strength normally associated with glass fibers of similar small section. The success of a glass-ceramic formulation, therefore, depends largely on the ability to devitrify it to yield this fine-grained structure.

Devitrification normally proceeds most readily from a surface, and this leads to an inhomogeneous structure. It is necessary to promote internal devitrification either by nucleation or, in effect, to create an internal surface by first forming a glass-glass phase separation.

Metals, (Pt, Cu, Ag) may be dispersed in the glass to act as nucleation centres, or sparingly soluble oxides such as TiO_2 or SnO_2 may be precipitated or immiscibility promoted by using more than one type of network forming oxide. Cations of high field strength such as Li and Zn are also useful in promoting bulk devitrification. The phase diagram for the system is not always a reliable guide to the course of crystallization; epitaxial considerations and the similarity of the crystal structure to that of the parent glass, are of importance in determining the order of crystallization.

If the glass working character is to be preserved during the shaping of the article, it is necessary that devitrification should not proceed too readily. On the other hand, the maturing must be completed in reasonable time for commercial exploitation. Fortunately, development of nuclei is usually at a maximum at about the annealing temperature of the glass, whereas the rate of crystal growth is a maximum at somewhat higher temperature. The glass can then be formed into the required shape, held at the nucleation temperature for a sufficient time, and then heated at a rate determined by the developing crystals' ability to hold the article's shape.

1.3.2.2 CERAMICS

The two common types of ceramic substrates in common use are beryllia and alumina. Both are available in various densities and with or without a surface glaze. Because it is very expensive and because it has a high toxic hazard rating, beryllia is normally only used where its extremely high thermal conductivity is needed. The most common material used is a glazed or unglazed alumina depending on the hybrid application. For most applications at frequencies below 250 mHz and at moderate power levels, alumina substrates are adequate.

One particular advantage available only in the ceramics is the availability of prescored or grooved ceramic sheets which can be diced after deposition to final substrate size by simply breaking with the hands. If substrate sizes can be standardized, then the scored ceramic substrates, even though they are more expensive per square inch than glass substrates, are usually cheaper per finished substrate than the glasses if the dicing and handling costs are considered.

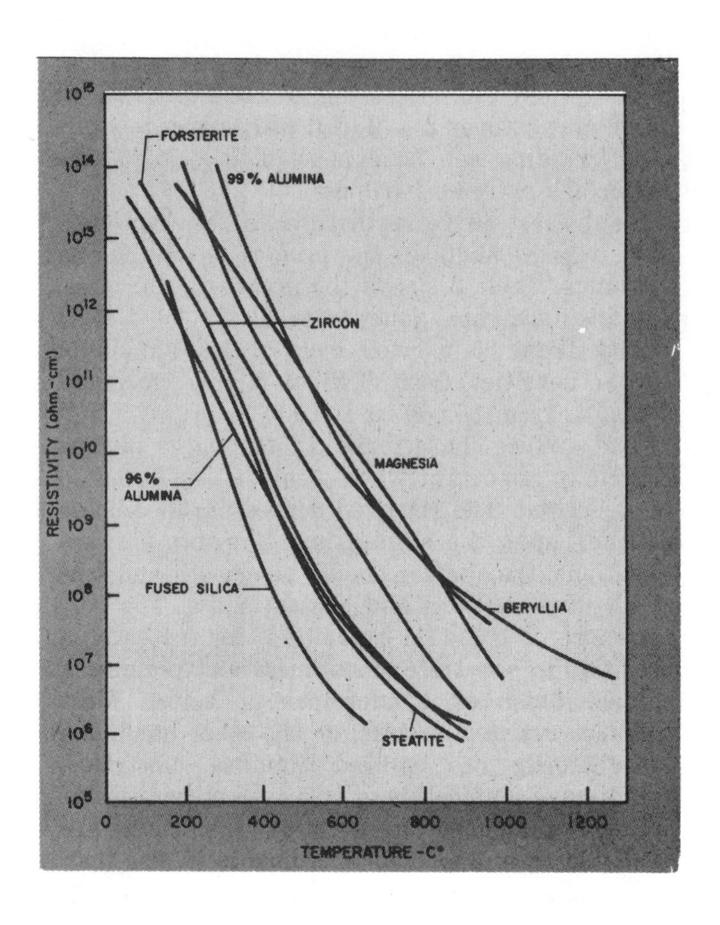

Ceramics have thermal characteristics that come close to some metals, and they also have the high resistivities needed for use as microcircuit substrates, insulators and dielectrics.

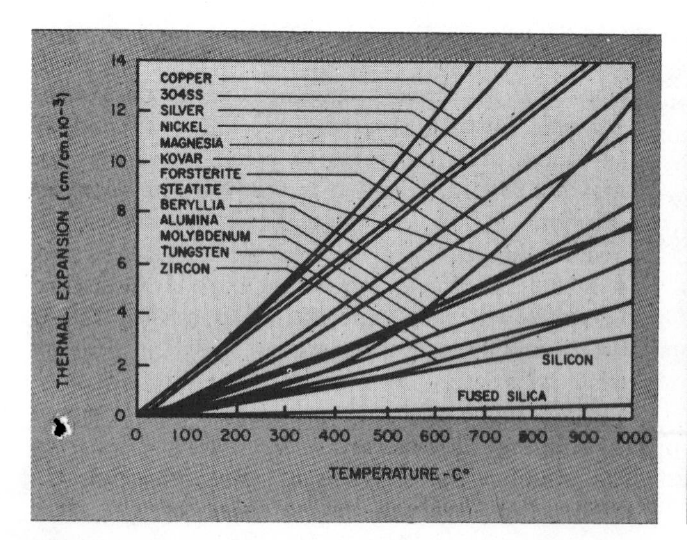

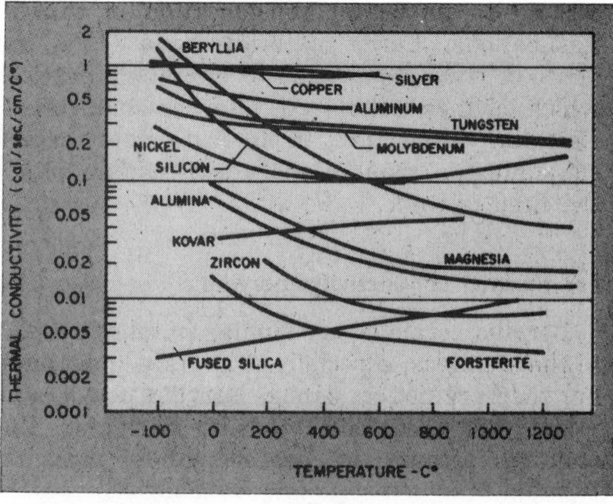

Some representative properties of ceramic substrates

Material property	Unit	Steatite	Forsterite	Zircon	Alumina (99.5% pure)	Beryllia
Water absorption	%	Impervious	Impervious	Impervious	Impervious	Impervious
Hardness	Moh's scale	7.5	7.5	8	9	9
Thermal expansion	25 to 900 C cm/cm/°C	8.5×10^{-6}	11.7×10^{-6}	4.9×10^{-6}	7.7×10^{-6}	8.7×10^{-6}
Dielectric strength (60 Hz ac, 1/4" test discs)	Volts/mil	220	240	220	220	230
Dielectric constant 1 MHz		5.3	6.2	8.8	9.9	6.9
Dissipation factor 1 MHz		0.0026	0.0004	0.0010	0.0001	0.0002
Loss factor 1 MHz		0.014	0.002	0.009	0.0011	0.0014

Typical properties of various alumina substrates

Properties	Units	Per cent Al_2O_3		
		96	99.5	99.7
Alumina's X particle size	microns	2.9	1.5	0.6
Density	cms/cc	3.70	3.89	3.86
Thermal expansion linear coefficient cm/cm × 10^{-6} per °C.	25-300 C. 25-700 C. 25-900 C.	6.4 7.5 7.9	6.6 7.4 7.7	6.5 7.4 7.7
Flexural strength specimen 0.025" T × 0.070" W, 0.500 L	lb/in²	60,000	70,000	85,000
Dielectric constant @ 1 MHz, 25 C.		9.3	10.5	10.3
Dissipation factor @ 1 MHz, 25 C.		0.0003	0.0001	0.0002
Average crystal size	microns	4.0	3.0	1.5
Surface finish	microinches	25	8	4

1.3.2.3 OTHER MATERIALS

Sapphire is used sometimes as a substrate when a high thermal conductivity is desired. The conductivity of sapphire is approximately that of alumina, but the sapphire can be polished to such a smoothness that a glaze is not required. The normal glaze required on a ceramic is one or two mils thick and is, therefore, equal in conductivity to from 30 to 60 mils of ceramic.

Depositions have been made on plastic films such as mylar. Use of these materials has been generally in applications other than hybrid circuits such as thermocouples, strain gauges, and solar cells.

Silicon wafers are used as substrates in thin film compatible technology. Their use is limited, though, because of the high parasitic capacitance between the film components and the substrate.

1.4 RESISTIVE AND CONDUCTIVE FILMS

The properties of thin metallic films depend on the structure of the film, the number of impurities and imperfections, and the thickness of the film. The electrical conductivity of a metallic film is altered, from the conductivity of the same bulk metal, by anything which causes scattering of the carriers and limits thereby their mobility. In thin films which are thick enough to be regarded as continuous, the primary causes of scattering are imperfections and the surfaces of the film. The problem is treated analytically by considering the film to be perfectly flat with an appreciable portion of the electrons scattered at the surfaces, so that the mean free path of the electrons is limited. The simplest assumption which may be made, is that every electron trajectory is terminated by collisions at the surface, and that the scattering is entirely random or diffuse. Solution of the Boltzmann transport equation yields a resistivity-thickness relationship of the type indicated in Fig. 1-1. For relatively thick films, σ, the conductivity of the film, may be expressed as

$$\sigma_0/\sigma = 1 + 3/8K \qquad K >> 1. \qquad (1\text{-}1)$$

where σ_0 is the bulk conductivity, and K is the ratio of film thickness to mean free path. Resistivity is denoted by ρ. For very thin films,

$$\sigma_0/\sigma = 4/[3K\, ln(1/K)] \qquad K << 1. \qquad (1\text{-}2)$$

If it is assumed that a fraction p of the electrons are specularly scattered at the surface, approximate expressions can be developed in the form

$$\sigma_0/\sigma = 1 + 3/8K(1-p) \qquad K >> 1, \qquad (1\text{-}3)$$

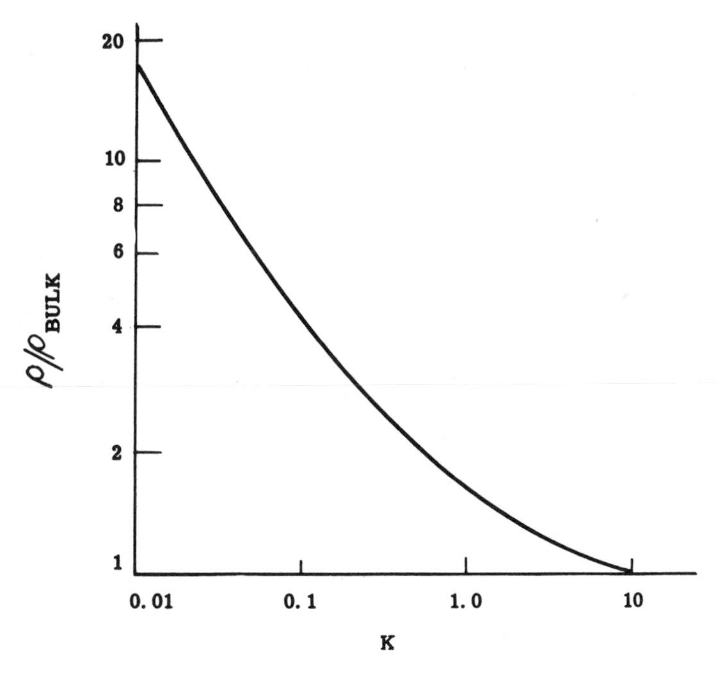

Fig. 1-1 Resistivity vs. Thickness for thin metallic films

and

$$\sigma_0/\sigma = (4/3)\ \frac{1-p}{1+p}\ \frac{1}{K\ ln(1/K)} \qquad K << 1. \tag{1-4}$$

For alkali metals resonable agreement has been found between this theory and experiment. For other metals, agreement between theory and experiment is less exact unless modifications are made to the theory. Most experiments find best agreement with the diffuse scattering expressions.

Another important property of metallic films is their temperature coefficient of resistance. In bulk materials this coefficient is determined by electron-phonon scattering. In the case of thin films, however, it is believed that it is also dependent on the surface scattering. On the basis of this, the temperature coefficient of resistance, α, should be given by,

$$\alpha_0/\alpha = ln\,(1/K) + 0.43 \tag{1-5}$$

where α_0 is the bulk temperature coefficient. There is only limited agreement between this expression and experimentál data.

If the film is so thin that it is composed of islands, the preceding discussion, based on a condition of extremely flat surfaces, does not hold. In this case the conduction mechanism consists of a thermal excitation of charge carriers, followed by electron tunneling across the gaps between islands. If a film is composed of a number of identical islands of radius r, the conductivity σ is given by,

$$\sigma = C \exp\left(\frac{-e^2/\epsilon r}{kT}\right) \tag{1-6}$$

where C is a constant, e is the electronic charge, and ϵ is the dielectric constant of the substrate. Such a film will have a negative temperature coefficient of resistance, since the fractional change in resistivity, ρ, per unit temperature change, is

$$d\,ln\rho/dT = \frac{-e^2/\epsilon r}{KT^2} \qquad (1\text{-}7)$$

This is in agreement with the negative temperature coefficient observed in many thin metallic films.

While the temperature during deposition affects the properties of the films, subsequent temperatures, to which it may be exposed, are also of some importance. Thus, various annealing procedures may cause rearrangement and improvement of the film structure. The reduction of imperfections and boundaries in the film tends to cause an increase in carrier mobility and thereby decrease the observed resistivity. In very thin films this procedure may cause an improvement in the film continuity and temperature coefficients.

1.4.1 CONDUCTIVE FILMS

Conductive films are employed to replace conventional solder and wire connections, and to provide for "lands" or external connections. Conductive films will interconnect resistive and capacitive areas in one layer as well as in a number of layers. The general requirements of conductive films are low sheet resistance, low contact resistance, ability to adhere to the substrate, and resistance to atmospheric attack.

Conductive films may be prepared by chemical and electrochemical deposition, cathodic sputtering, and vacuum evaporation. The oldest example of the chemical method is the electroless process as illustrated by the silver mirror. It consists of depositing the metal constituents upon the substrate through reduction of the soluble salts. Metals that can be readily deposited in this fashion are copper, rhodium, nickel, and cobalt. Ruthenium, palladium, iridium, platinum, iron, chromium, aluminum, magnesium, and manganese are reported to have also been deposited by this method. By maintaining the temperature of the coating solution closely controlled, it is possible to obtain the desired film thickness. Although the process is very adaptable to single-component films, it is not as readily applicable to alloy deposition.

Numerous resinates are commercially available for applying to substrates through brushing, spinning, silk-screening, roller-printing, or dipping at room temperature. When subsequently fired at elevated temperatures, the conductive films are generated.

Vacuum evaporation stands out as a relatively simple process when confined to single-layer films. When applied to multilayer films, however, considerable complications result. The process consists of the volatilization of metals in a vacuum, and may be applied to many metals if heating sources are suitable. Aluminum is an excellent example of a metal for this process. It adheres well to glass, is easy to evaporate, and is resistant to corrosion. Gold films have better corrosion resistance, but do not adhere to glass. By interposing, however, a chromium film between the gold and the glass, a certain degree of adherence may be realized.

In the vapor plating of aluminum, the metal halides are reduced, and decomposed at high temperature. The pyrolysis of organo-metallic compounds in the absence of moisture or air, produces then the desired sought-after metal film. The reducing gas, generally hydrogen, is routed first over a heated metal halide to gather up its vapors, and then into a coating chamber where both reduction and deposition occur at high temperature.

Cathodic sputtering is employed for depositing conductive refractory metal films. In the process the metal becomes the cathode of an electric discharge. Thus, material leaving the cathode is deposited on the nearest available surface. Tantalum which is used for resistive, capacitive, and conductive films, is generally deposited in this manner.

By observing the growth of films under an electron microscope, it has been found that the first step from a supersaturated beam of material is nucleation. Nuclei then grow into a continuous film. Because nucleation involves the association of several atomic species coming into contact, it is presumed that a nucleus is formed when two or more atoms collide on the surface. Atoms that do not collide with others prior to emerging at the surface, migrate to adjoing nuclei or re-evaporate from the surface. Many variables affect the growth process. The principal ones are pressure, substrate temperature, temperature of the evaporation source, vapor concentration, angle of incidence of the vapor stream, and the residual gases in the evaporation chamber as well as the substrate.

1.4.2 RESISTIVE FILMS

The electrical resistance of a rectangular prism is given by

$$R = \rho \frac{L}{w\,b} \qquad (1\text{-}8)$$

where ρ is the resistivity (ohm-cm)
L is the length (cm)
w is the width (cm)
b is the thickness (cm)
R is the resistance (ohms)

The sheet resistance or surface resistivity is defined by

$$R = \rho_S \left(\frac{L}{w}\right) \qquad (1\text{-}9)$$

Where ρ_s is the resistivity (ohms)
The relationship between ρ and ρ_s is therefore given by

$$\rho = b\,\rho_s \qquad (1\text{-}10)$$

The temperature coefficient of resistance, α, is a function of temperature, and for this reason two average values for α are specified within the range of –55 to 150° C, the common operational range for conventional film circuits. The two α's correspond to the temperature ranges of –55 to 20° C, and 20° C to 150° C respectively.

To determine α, three resistance values at two temperatures, T_1 and T_2, are generally taken in the order R_{oT1}, R_{T2}, R_{T1}. The amount by which R_{T1} and R_{oT1} differ indicates the stability of the material with temperature cycling. This deviation, expressed in per cent is given by

$$\frac{R_{oT1} - R_{T1}}{R_{T1}} \times 100 \tag{1-11}$$

The resistance of polycrystalline film materials often varies with voltage. The voltage coefficient in ppm/volt is defined as

$$\frac{R_1 - R_2}{R_2 \, \Delta V} \times 10^6 \tag{1-12}$$

where R_1 and R_2 correspond, respectively, to the resistance values at V_1 and V_2, and ΔV is the difference between the voltages or $V_1 - V_2$. V_2 is generally made one-tenth of the rated voltage V_1.

The conduction processes in metals and semiconductors are effectively analyzed through the technique of the Hall effect. The Hall constant and independent-resistivity measurements combine to yield information on the number of carriers and their mobility in film materials. The Hall constant is given by the ratio $\pm 1/Ne$ where N is the number of carriers per cubic meter, and e is the electronic charge in coulombs.

For metal film thicknesses exceeding 100-150 Å, the film resistivity is believed to be composed of the following factors:

$$\rho_i + \rho_r + \rho_t \tag{1-13}$$

where ρ_i is the effect from electron scattering due to thermal motion of ions in the metal;

ρ_r is the effect from electron scattering due to structural defects, i.e., lattice defects, missing atoms, and impurities;

ρ_t is the effect from electron scattering arising from the boundary conditions of the film. This effect is very much dependent upon film thickness.

Metal oxides have been extensively used in resistive films. The oxides of cadmium, indium, and tin are good conductors and can be hydrolytically or pyrolytically deposited. Metal oxides added to tin oxide affect the sheet resistance of a film to the extent that metal oxides having stable valency less than 4, act as electron acceptors and poison the film. The oxides of cadmium, aluminum, indium, iron, and boron are examples of these. Oxides with stable valency of 5, however, act as electron donors and tend to lower the film's sheet resistance. The oxides of tungsten, arsenic, antimony, tellurium, fluoride and phosphorous are examples of these.

Resistor networks of tin and antimony-oxide films can be chemically deposited on glass or ceramic substrates so as to obtain the pure resistive material. Interconnections may be formed by electroless deposition of copper. Resistance tolerances of $\pm$ 1 per cent of the resistive areas are possible if a tailoring pad is included.

Metal films are often used for the resistive function in both element or alloy form. Films in which the metal is in element form appear to be lower in cost and to require fewer masking operations. Their limitation is that high sheet resistances, at zero temperature coefficient, are only possible through thin films that tend to be very unstable. Alloy films, on the other hand, have the added expense of composition control, but they have high sheet resistances with a temperature coefficient close to zero. Metals deposited by vacuum evaporation and cathode sputtering may exhibit higher sheet resistances with better stability through protective films of dielectrics such as silicon monoxide, magnesium fluoride, or silica.

Nickel-chromium films have been extensively experimented with. The most stable film is obtained by depositing 80Ni-20Cr between 250 and 350° C on glass or ceramics substrates. The temperature coefficient of the films depends upon the substrate, the composition of the film, the rate of deposition, and the gas pressure during evaporation. The coefficient is radically affected by oxidation of the alloy in poor vacuums. A negative shift of the coefficient is detected with decreasing film thickness.

The use of tantalum for resistive films has received much interest due to the successful application of the tantalum oxide capacitor. Stable films may be produced through reactive sputtering in nitrogen. This results in the presence of nitrogen in the film. The composition is dependent upon the concentration of the nitrogen. Stable films may also be produced by providing a gold film between two layers of tantalum. Upon heating the film in vacuum, the gold diffuses into the grain boundaries of the tantalum. Stability of tantalum films depends much on the gases that are present during the period that the film is formed.

Cermit films are combinations of metals and dielectrics. They are used to obtain large sheet resistances at low temperature coefficients. These films may be deposited by a flash-evaporation technique or vacuum evaporation with the simultaneous operation of multiple sources. In the latter technique the composition is controlled through the separate adjustment of the silicon monoxide and chromium sources. In the former technique a stream of silicon monoxide and chromium powder is passed over heated tantalum. The substrate is generally preheated to 400° C. By varying the concentration of chromium in silicon monoxide, the sheet resistance may be varied over a wide range of magnitude. At low chromium concentrations, temperature coefficients are negative, but they will be near zero when the concentration is increased to approximately 40%. A variety of cermet compositions may be applied to ceramics or glass through screening, dipping, roller-printing, or spraying. The technique employed depends largely on the consistency of the mixtures of metals, metal oxide, and glass frit in an organic vehicle.

1.5 DIELECTRIC AND INSULATOR FILMS

A dielectric film is characterized by the same parameters as a bulk dielectric material, with the exception that the structure and composition of a film may differ from those of a bulk dielectric of the same material. The dielectric constant ϵ is given by,

$$\epsilon = D/E = 1 + 4\pi(P/E) \qquad (1\text{-}14)$$

where D is the electric displacement, E is the electric field strength, and

P is the polarization. The polarizability of an atom α is

$$\alpha = p/E \tag{1-15}$$

where p is the dipole moment. Then the polarization is given

$$P = Np \tag{1-16}$$

where N is the number of atoms per unit volume (assuming all identical atoms). It can be shown that the relation between the dielectric constant and the atomic polarization is

$$\frac{\epsilon - 1}{\epsilon + 2} = \frac{4\pi}{3}\frac{\rho}{M} L\alpha \tag{1-17}$$

where ρ is the density, M the molecular weight, and L is Avogadro's number. This can also be written in terms of the optical refractive index, n, since $\epsilon = n^2$.

The polarizability can be separated into four portions: electronic, ionic, orientational, and space charge. Electronic polarizability arises from displacement of electrons relative to the nucleus, while the ionic contribution stems from displacement of a charged ion with respect to other ions. Orientational polarizability exists in materials containing molecules possessing a permanent electric dipole moment which may change orientation in an electric field. The space charge polarizability arises from the trapping of charges, flowing in the dielectric, by either imperfections or interfaces. These effects are dependent upon the frequency of the applied electric field, and therefore, the dielectric constant is a complex function of the frequency. The difference between the low-frequency and high-frequency dielectric constant is often attributed to a damping out or relaxation of the orientational contributions. The relaxation time, τ, is the interval required for restoring the disturbed system to equilibrium. If the polarizability is expressed in terms of the static polarizability α_0 as

$$\alpha = \alpha_0 / (1 + i\omega\tau)\,, \tag{1-18}$$

where ω is the frequency, then the complex dielectric constant is

$$\epsilon = \epsilon_1 - i\epsilon_2 = 1 + \frac{4\pi\alpha_0 N}{1 + \omega^2\tau^2} - i\,\frac{4\pi\alpha_0\omega\tau N}{1 + \omega^2\tau^2} \tag{1-19}$$

The power dissipation per unit volume, W, is

$$W = j_E E\,. \tag{1-20}$$

where j_E is the component of current density in phase with E. Since j =

$\sigma E + \frac{1}{4\pi}\frac{\partial D}{\partial t}$, the sum of conduction and displacement current,

$$j = \left(\frac{\epsilon_2\omega}{4\pi} + \frac{i\epsilon_1\omega}{4\pi}\right) E \qquad (\sigma = 0) \tag{1-21}$$

or

$$W = \frac{E^2}{4\pi}\,\omega\epsilon_2 = \frac{\epsilon_1 E^2}{4\pi}\,\omega \tan \delta \tag{1-22}$$

where the loss tangent is

$$\tan \delta = \epsilon_2/\epsilon_1. \tag{1-23}$$

This loss tangent is also referred to as the dissipation factor.

The properties of dielectric films are influenced by essentially the same parameters as metallic films, although the order of significance is not necessarily identical. Included among the most important are film thickness, imperfections, and chemical impurities. Another factor is the presence of considerable stress in the film.

The thickness of a film of any given material will, normally, determine the voltage which it can support before breakdown occurs. Breakdown is often caused by electron impact ionization. It may also occur as thermal breakdown caused by the heat generated by currents flowing in the dielectric. Premature breakdown may be caused by imperfections or impurities, in the dielectric, which partially bridge the path between electrodes.

Chemical impurities may also cause a change in the composition of the dielectric and, thereby, cause a considerably different dielectric constant. This is particularly true in films which are vacuum evaporated at slow rates, allowing considerable oxidation of the dielectric to occur. For example, SiO with a refractive index of two, may condense as SiO_2 with an index of 1.5, if the rate is slow enough to permit substantial oxidation.

Impurities, imperfections, and certain evaporation parameters may also cause residual stresses of considerable magnitude to be present in the film. These stresses may be sufficient to cause fissures in the film or other structural defects which cause premature failure. Very thin dielectric films may be composed of isolated islands which also make them unsuitable for most applications.

Dielectric films are generally composed of inorganic substances such as the halides and oxides of metals and semiconductors. The material most commonly used is silicon oxide deposited by vacuum evaporation. Compositions of silicon oxide may be obtained ranging from SiO to SiO_2. Variations in composition lead to differences in physical and electrical properties. High electrical strength and mechanical durability is exhibted by SiO. To prevent residual stresses in the SiO film after deposition, the substrate has to be heated during the deposition process. The film may peel or buckle because the material is thermodynamically unstable and will react with O_2 as well as H_2O. The material is also hard and brittle. Its coefficient of thermal expansion is quite low.

When dealing with semiconductor devices made from diffused junctions in silicon and germanium, considerable use is made of dielectric materials of the oxides produced by direct chemical reaction. This may consist of the sputtering of a metal in the presence of a reactive gas, pyrolytic deposition upon a heated substrate, or a reaction between a surface and a gas. The oxide under these conditions, is not only a dielectric, but it also serves as a mask for selective diffusion. The dielectric properties of these oxides are generally quite good, and they are especially useful for high-temperature operation. Oxides produced by thermal evaporation exhibit the likelihood of holes or voids even when unusually thick due to the absence of proper nucleation centers. This undesirable feature is overcome through anodically-formed oxide films such as AL_2O_3, TaO_2 and TiO_2. In anodization, any defective area will be subject to faster reaction and thus keep the thickness of the film uniform.

Organic polymer films are superior to inorganic films in several respects. They have more chemical inertness and low dissipation factors. They also exhibit higher resistance to cracking and chipping. The main problem associated with such use of polymers is that they will not withstand high operational temperatures for extended periods of time. Acrylic esters and styrene have been the principal polymer films for dielectric purposes. Polysilicone oils have been used to a lesser degree.

Polymer films are produced by electrical discharges in monomer vapor, bombardment of adsorbed monomer molecules with low-energy electrons, ultraviolet irradiation of adsorbed monomers, and pyrolytic deposition. Electron bombardment of a variety of adsorbed materials is probably the technique used most for generating polymer films. Polystyrene films produced by this technique, have low dielectric constants. Ultraviolet photolysis produces reliable films with electrical properties similar to those of the bulk material. This technique, however, produces films which are softer and more plastic than those produced by other polymerization techniques.

The anodizing of tantalum is employed for thin film capacitors, since it satisfies the requirement of a dielectric material with properties that do not change over a wide range of thickness.

Thermal expansion coefficients of dielectric films are less than those of semiconductors or metals, but they have been found to vary considerably. This induces stresses, and for this reason polymer films find favorable application, because they undergo plastic deformation and thus maintain residual stresses at lower values than most inorganic dielectric films.

The insulation of a flat surface from a single metallic film has been found easier than the insulation of multilayer films. Breakdown of the insulation appears mainly at the edges where conducting films cross over one another. It is therefore desirable to maintain, at a minimum, the number of such crossings, as well as the superposition of conductive films. The relative thicknesses of conductive and insulating films must also be taken into account, in order to avoid the necessity of having to insulate a sharp, steep edge with a thinner dielectric film.

1.6 SEMICONDUCTING FILMS

Semiconducting films are comprised chiefly of silicon and germanium, and compounds of GaAs, InP, and InAs. Such films are epitaxially grown on oriented single-crystal substrates. By diffusing dopants into the epitaxial films, active devices are formed. Semiconductor films may also be produced by depositing CdS, CdSe, PbS, and ZnS, on an inactive substrate by thermal evaporation. This process generally requires close control of the evaporation parameters as well as heat treatment, in order to realize the desired properties.

Devices using semiconductor films are diffused-junction and field-effect transistors, diodes and injection lasers, photovoltaic radiation detectors and converters, electroluminescent photoconductors, space-charge-limited triodes, and optical transistors. The unipolar field-effect transistor consists of a narrow semiconducting channel between two electrodes with a dielectric film insulating the channel from a metallic electrode. By biasing this electrode, the current flow between source and drain is regulated.

The space-charge-limited triode modulates the space-charge field and current by the use of a fine-line metal grid film between the emitter and collector surfaces. The use of an evaporated thin-film semiconductor in this device has not proven feasible because of the resulting poor transfer coefficients.

Photovoltaic devices in the form of infrared-radiation detectors include evaporated thin films of PbS and PbSe. In this application, evaporation is accomplished upon a cooled surface in a high vacuum. Thin films of CdS, PbS, GaAs, selenium, and doped germanium and silicon have been used also in photovoltaic cells such as solar converters. Such cells require a coating for the active semiconductor to permit passage of solar radiation, and at the same time protect the film.

The small-area electroluminescent cell consists of a thin-film junction of metal-phosphor conductor. It has found application in display panels, direct read-out from data processing systems, and persistent-data display. These devices, however, have found limited use because of the problems associated with the phosphor film in regard to long-term instability and operating frequency.

1.7 SUPERCONDUCTING FILMS

The cryotron and the magnetic flux-trapping sheet film are the two principal devices in the area of superconducting films. The cryotron is a switching element while the other device is a storage element. Lead, tin, and indium have been mainly employed in these metallic films which are generally produced by thermal evaporation in a high vacuum.

The switching speed of a cryotron, for a given geometry, is determined by a time constant which is proportional to the ratio L/R, where L is the inductance between gate and control, and R is the gate resistance. In order to obtain a low time constant, L may be minimized by having a thin insulating film between gate and control. The resistance R may be maximized by alloying the gate metal with other metals, and by making the gate film as thin as possible. The coherence length of a superconducting electron, limits the degree to which the film may be made thin.

In the storage element, the flux is trapped in a superconductive sheet below two insulated superconductive films. Depending on the direction of current flow in the drive films, flux is trapped in one or the other direction. It is this property that makes the device a bi-stable storage element. The packing density of the element is unusually high.

Superconducting films have the disadvantage of requiring expensive equipment in the form of liquid-helium refrigeration. Such equipment is necessary to contain the circuit, because the latter will not function properly above a certain temperature. On the other hand, superconducting films permit high packing densities and reliability. When the circuit is in its cryogenic environments, its physical and chemical processes have slowed down to the extent that it should perform with almost no change in its characteristics.

1.8 MAGNETIC FILMS

The principal application of ferromagnetic films is for switching elements in rapid-access memories. The films are composed of alloying elements such as iron, nickel, cobalt, molybdenum, and vanadium. They are deposited by chemical deposition, sputtering, electroplating, vacuum deposition, and thermochemical decomposition. The films attain a uniaxial anisotropy when deposited in the presence of an orienting magnetic field applied in the plane of the substrate surface. A hysteresis loop is obtained when an alternating magnetic field is applied parallel to the film surface, and along the direction of the orienting field.

When a ferromagnetic film is magnetized to a maximum remanent state along its easy direction, the film can be remagnetized by applying a switching field which is larger than the coercive force, and is parallel to the easy axis of the film and in opposition to the remanent magnetization. When the film is large and sufficiently thin, and is in a maximum remanent state, it consists mostly of a single domain magnetized parallel to the easy axis. If the switching field is applied as in the preceding manner, the small domains of reverse magnetization, that may be present near the polar edges of the film, will grow until they merge and reverse the magnetization state. If the small domains of reverse magnetization do not initially exist, they will become nucleated when the applied switching field equals the nucleating field. Their rate of growth will then be proportional to the ratio of the switching field to the coercive force.

The binary alloy 81Ni-19Fe has been extensively applied in magnetic films because of its small magnetostriction. This alloy has, however, considerable crystalline anisotropy. The ternary alloy 4-79 molybdenum Permalloy is a more desirable film material because it has low magnetostriction as well as low anisotropy. When rolled to the proper thickness and heat-treated, a nearly rectangular hysteresis loop results. However, since molybdenum has a melting point much greater than nickel or iron, difficulties are encountered in alloying these three elements.

Films of 81 Permalloy have been extensively applied as storage elements in high-speed destructive-readout memories. By magnetizing the film elements parallel to one of the two directions along the easy axis, it is possible to store binary information. Readout results upon applying a strong field along the hard axis. Sensing of the information is done in a line parallel to the difficult axis. The information is restored through magnetization steering by either energizing the sense line itself with a pulse, or by energizing a "digit line." For high-speed operation care must be taken to maintain eddy-current losses at a minimum in these lines.

In the construction of memory arrays, the films are often deposited on glass substrates and the wiring, in printed form, overlays the films. With this arrangement nearly twice the field is provided by the word and digit lines for a given current. To maintain demagnetizing effects at a mimimum, film elements, usually in the form of circular or rectangular dots, are made as thin as possible. In such a configuration, however, the film elements in adjoining word registers must be spaced apart sufficiently to prevent magnetization creep. On the other hand, wide spacing may give rise to difficulties in high-speed operation.

A memory array may also be constructed where the film is deposited on a pair of glass substrates backed up by a metal plane. The wiring in two or three layers with proper insulation between them is located between the pair of substrates. Each storage element consists of a pair of oppositely magnetized films. By arranging the spacing between elements so that it does not exceed 10% of the element's length, it is unnecessary to take into account the demagnetizing field. This configuration, therefore, permits line delays that are short enough to make high-speed memories with large capacities possible.

1.9 ADVANTAGES OF THIN FILM CIRCUITS

On the basis of the material presented in this chapter, the advantages of using thin film microelectronic circuits, are summarized as follows:

1. Improved reliability due to the substantial decrease in interconnections, and mutual compatibility of materials used.
2. Thin films minimize the effects of intense radiation fields.
3. Frequency capabilities are higher since components are smaller, distributed effects are smaller, and precisely reproducible.
4. Possibility exists that thin film circuits can be made usable at higher temperatures than are now used.
5. When compared to semiconductor microelectronic circuits, thin film components can be made with a wider range of values, capacitors are linear, tolerances and temperature coefficients are smaller, and substrates are nonactive.

CHAPTER 2

FABRICATION OF THIN FILM MICROELECTRONIC CIRCUITS

Whereas many techniques for fabricating thin films have been proposed, only a few have actual or potential usefulness in the fabrication of thin film circuits. These are:

1. Vacuum deposition
2. Sputtering
3. Anodization
4. Silk screening
5. Epitaxy
6. Vapor plating
7. Plasma decomposition
8. Electron Beam decomposition

The usefulness of any of these processes, in the fabrication of circuits, is primarily dependent upon the fabricator's ability to control the physical and geometrical properties of the film. However, the choice of one process over the other is often determined by specific circuit requirements such as number of circuits to be made, cost, precision, temperature dependence, stability, and circuit complexity.

Vacuum deposition is the most commonly-used process because it possesses the greatest versatility in forming films of a large variety of materials, and appears to have the greatest potentiality for future growth. Therefore, most attention is paid to it, while the others are discussed only briefly.

2.1 VACUUM DEPOSITION

Figure 2-1 illustrates the major steps to be taken in the fabrication of vacuum deposited thin film circuitry. Thus far, this approach to circuit fabrication has been used mainly in the production of passive elements, but in recent years much research has been done to develop an active thin film device. This recent effort has resulted in a thin film field effect transistor. However, most thin film circuits still require either the use of add-ons, i.e., active elements which are added to an inactive substrate

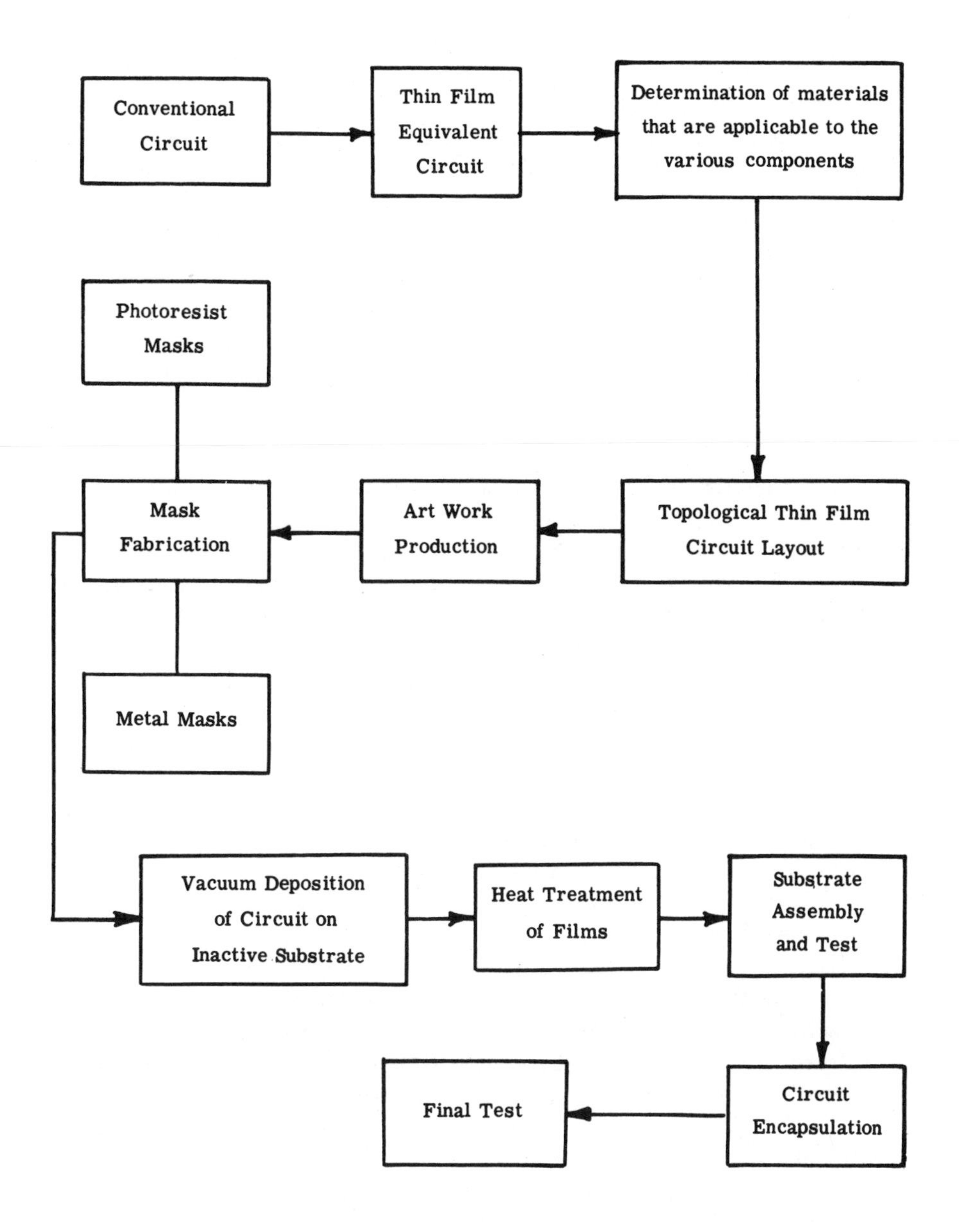

Fig. 2-1 Block diagram illustrating fabrication steps for thin film - vacuum deposited circuitry

containing passive components; or the use of hybridization, i.e., the use of thin film passive elements on an active substrate (silicon) containing transistors and diodes. It is anticipated that complete thin film circuits, containing both passive and active elements in thin film form, will be made available. They have not, as yet, been produced beyond the laboratory level.

The equipment used for the vacuum deposition of thin films can be divided into two parts: (a) the vacuum chamber or bell jar with its associated pumps, valves, base plate and gauges; and (b) the mechanical hardware installed within the chamber to support, and allow manipulation of the sources, masks, substrates, substrate heaters, and film monitoring equipment.

The proper selection of the vacuum equipment is very important since its cost is a large portion of the total installation cost. Furthermore, the specific design affects the size and number of substrates which can be processed per unit of time, the type of film monitoring equipment which can be used, ease of making the many manipulations required to fabricate a thin film circuit, and ease of making modifications to the equipment.

At the present time most depositions are made within the 10^{-6} to 10^{-5} torr region. The value of working in the ultra-high vacuum region has not yet been definitely established. Since the cost of UHV equipment is substantially greater than high vacuum equipment, the pump-down time is much longer, and its usefulness remains doubtful, equipment operating in the high vacuum range is given the most attention.

A commonly used unit is the 18-inch glass bell jar vacuum system available from a number of manufacturers as a standard item. These units have jars 17-inch I.D. x 29 inches high, with the top being a section of a sphere. They are usually equipped with a nominal 6-inch diffusion pump, a mechanical backing pump, and a bypass or roughing line. A minimum of four valves are required: one large high vacuum valve between the diffusion pump and the base plate; a second valve in the roughing line; a third valve between the diffusion pump and the mechanical pump; and a fourth valve to let the jar down to atmospheric pressure. A fifth valve in the foreline to permit connecting a leak detector, when required, is very useful. Refrigerated baffles are universally used. A collar between the base plate and the bell jar which makes available about 12 ports is also very useful. Vacuum gauging is usually performed through an ion gauge for the high vacuum range, and a Pirani or thermocouple gauge for the range above one micron.

High-current low-voltage supplies are often included as standard items, or can be supplied as extras. A wide range of accessories are available for use with this 18-inch system. Examples are: high-voltage and high-current feedthroughs, rotary feedthroughs, multiple terminal headers, quick connect fittings, standard leaks, and stainless steel bell jars as replacements for the glass jars. Many manufacturers make larger standard systems and will make any vacuum system to a customer's specifications.

In normal operation, the gases removed by the pumps originate from small residual leaks, surface desorption, and gases given off by the vapor source. Expressing this mathematically,

$$Q = Q_S + Q_L + Q_W \tag{2-1}$$

where Q is the total quantity of gas in torr-liters/second, Q_S, Q_L, and Q_W are the amounts, respectively, originating in the source, through leaks, and from all the surfaces exposed to the high vacuum. If the pump speed measured at the base plate is S in liters/second at a pressure P, then at equilibrium

$$PS = Q\,. \tag{2-2}$$

The left side of this equation is the "through-put" at pressure P, or the amount of gas removed by the pumps in unit time at pressure P. Since the speed S is substantially constant in the pressure region of interest, the through-put decreases with the pressure.

With modern equipment the pump-down time is short; it depends upon the ratio of volume to speed. The equilibrium pressure is given by

$$(Q_L + Q_W)/S . \qquad (2\text{-}3)$$

When the vapor sources are turned on they give off copious quantities of gas so that

$$Q_S >> Q_L + Q_W \qquad (2\text{-}4)$$

and

$$Q >> Q_L + Q_W . \qquad (2\text{-}5)$$

The equilibrium pressure rises to

$$Q/S$$

which is usually 2 to 10 times as large as

$$(Q_L + Q_W)/S .$$

Since it is desired to maintain low pressure during deposition, the pump speed, *at the chamber,* must be substantially greater than that required to produce a short pump-down time.

The main reason for using a high vacuum (of the order of 10^{-6} torr) is that it greatly enhances the efficiency of deposition, and standardizes the nature of the film. As the vapor stream passes from the source to the substrate, it can be deflected from its path by collisions with residual gas atoms. It is well known that out of N_0 atoms leaving a source and traversing a distance x to the substrate, the number N completing the trip is

$$N = N_o \exp(-x/L) \qquad (2\text{-}6)$$

where L is the mean free path of the atoms. L is approximately one meter at a pressure of 10^{-4} torr and varies inversely with the pressure. If the pressure is low enough and $x << L$,

$$N = N_o(1 - x/L) \qquad (2\text{-}7)$$

where x/L is the fraction of atoms deviated by collisions. For efficient vacuum deposition of thin films, it is necessary that x/L be small. If this ratio is made equal to 0.01, then, for a typical source-to-substrate distance of 20 cm, L must be at least 20 m, corresponding to a pressure of approximately 5×10^{-6} torr.

Another important reason for using a high vacuum is that it tends to prevent contamination of the film, as it is being formed, by residual gases. It has been shown that the number of residual gas atoms, n, striking a unit area of the substrate per unit time is

$$n = P/(2\pi mkT)^{1/2} \qquad (2\text{-}8)$$

where P is the gas pressure, m is the mass of the gas atom, k is the Boltzman constant, and T is the absolute temperature. This expression clearly indicates the necessity for low pressures in keeping the residual gas contamination low. However, even at a pressure of 10^{-6} torr, a monomolecular layer of residual gas atoms is formed within a few seconds. The residual gas contamination may be virtually eliminated by using pressures less than 10^{-9} torr (UHV region), but this is very difficult, from the technological viewpoint.

The vacuum chamber is a glass jar, typically 30-inches high and 18 inches in diameter. Within it is located the multiplicity of evaporation sources containing the materials used to form a thin film circuit, the evaporation masks, and the substrates. The evaporation sources, are heated to temperatures as high as 3000 °C. The atoms emitted from a given source, propagate in straight lines until they strike the substrate through the appropriate mask with which it is in contact. As a result, a film is formed on those areas, of the substrate, which are exposed through the mask openings. By changing mask patterns and source materials, several films may be successfully deposited on the substrate to form the desired circuit. Figure 2-2 illustrates the arrangement used to deposit a single film on a heated substrate.

2.1.1 SUBSTRATES

The selection of a suitable substrate is governed by the following requirements: a) it should be chemically stable so that it doesn't interact with the films that it supports; b) its thermal expansion coefficient should be at least as low as that of any of the materials to be deposited; c) it should have a high enough thermal conductivity to be effective in heat dissipation; d) it should exhibit good electrical insulating properties; and e) its surface should be smooth.

Standard glass microscope slides have been commonly used as substrates for thin films, since these are cheap, readily available, and possess smooth fire-polished surfaces. However, glass substrates containing alkali metals, such as sodium or potassium, have been found to cause considerable trouble. When current is passed through a thin film circuit which has been deposited upon such a substrate, an electric field is established which may draw alkali ions, such as sodium or potassium, to the substrate surface. These ions interact with the thin films and cause them to break down and deteriorate. This condition is common with thin film capacitors having aluminum counter-electrodes. To overcome this problem, it is necessary to first coat the substrate with a thin film of silicon monoxide. This film acts as a buffer between the thin film circuit and the substrate. A glass has been especially developed for use as a thin film substrate. It is an aluminosilicate, alkali-free, glass. Its use has eliminated the need for coating the substrate with a silicon monoxide film. In addition, its thermal coefficient of expansion is lower than that of the ordinary lime glass microscope slide.

The thermal conductivity of glass is rather low (between 0.002 and 0.004 cal.cm/sec/°C), and attempts have been made to replace it with such ceramics as alumina (Al_2O_3), which has 20 times as high a thermal conductivity, or with beryllia (BeO), which has 200 times as high a thermal

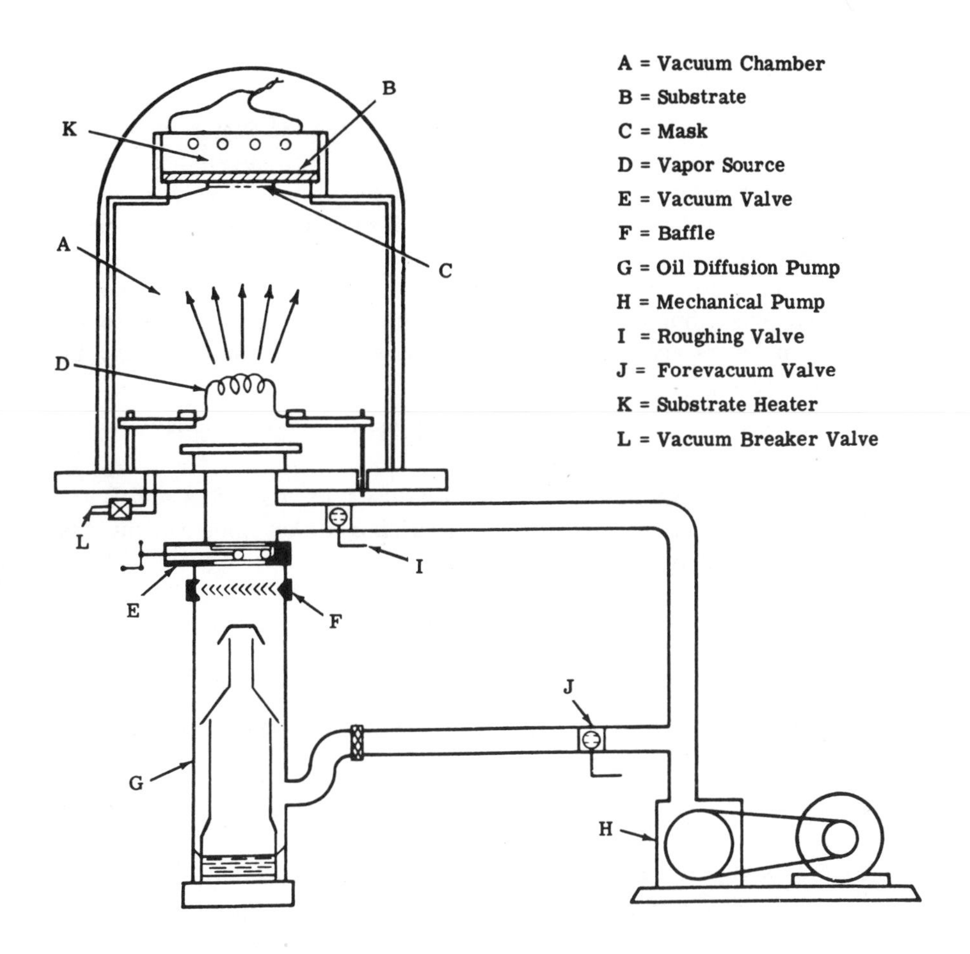

Fig. 2-2 A vacuum evaporator

conductivity. The only serious shortcoming of ceramic substrates is their lack of surface smoothness, which renders them useless to support some thin film components. This is gradually being overcome by the development of suitable glazes for use on the ceramics. Since the glaze is very thin, its effect on the thermal conductivity is small. However, glazing does introduce new problems such as crazing (cracking) of the glaze, departure from flatness, and high cost.

It is desirable to inspect a substrate before it is used. A simple procedure useable for glass is to pass a beam of light through a polished edge of the substrate. Any surface defects will then reveal themselves by scattering a small amount of light through the surface.

The cleaning of the substrate may be performed by many methods. A simple procedure commonly used includes ultrasonic washing in a detergent solution, followed by a thorough rinsing in deionized water and then a final rinsing with double distilled acetone. In all cases, the final cleaning operation should be performed immediately prior to placing the substrate in the vacuum chamber. It is not advisable to store precleaned substrates.

2.1.2 EVAPORATION SOURCES

The selection of a good vapor source is a critical factor in making good thin film circuits. The number of source types available is very large. A good source should provide a particle-free vapor stream/which is free of foreign atoms, and which has adequate area coverage, beam intensity, and charge capacity.

The evaporant may convert to the vapor phase from either the solid or liquid phase, but in either case, the critical factor is the vapor pressure-temperature characteristic. To a first approximation, this may be expressed as

$$\log P = A - B/T \qquad (P < 1\,\text{torr}) \qquad (2\text{-}9)$$

where P is the vapor pressure at temperature T in °K, and A and B are material constants. If the charge is vaporizing so that none returns, the mass evaporated in grams/cm^2/sec is

$$E = 0.0585P(M/T)^{1/2} \qquad (2\text{-}10)$$

where P is in torr and M the molecular weight. In practice, the actual evaporation rate is less than the ideal so that the mass evaporated is given by

$$0.0585aP(M/T)^{1/2} \qquad (2\text{-}11)$$

where a is an empirical factor. Combining Eqs. (2-9) and (2-11) and differentiating the result,

$$\Delta E/E = (2.3B/T - 1/2)\Delta T/T\,. \qquad (2\text{-}12)$$

Since 2.3(B/T) is in the range 10 to 20, a small fractional change in the temperature will result in a large fractional change in the vapor pressure. It therefore follows that in those deposition processes where the source vapor pressure must be controlled, the temperature must be controlled with great accuracy.

The simplest vapor source is a hot wire. It can be used when vaporization is by sublimation and the vapor pressure is adequate at temperatures below the melting point. Tungsten and rhodium, for example, have been evaporated in this way. To avoid burning out the wire and permit operation at the highest temperature, it is necessary to control its temperature carefully. This can be done if the function IV^3 is maintained constant, where I is the current heating the wire and V the voltage drop across the wire. It has been shown that this function is independent of the wire diameter and depends only on the temperature.

If an evaporant can be electroplated onto a wire it can be sublimed off, or melted to form drops which will vaporize. Chromium has been evaporated from tungsten wire in this fashion.

If the evaporant is available as a wire or ribbon, it can be wrapped around a refractory metal wire or hung on it in the form of inverted "U's." As the wire temperature is raised, the evaporant will melt and form drops which will vaporize. Figure 2-3a illustrates this type of source. The liquid evaporant should wet or alloy with the hot wire.

Sources in the form of boats fabricated from the refractory metals are useful if larger amounts of the vapor are needed, or if the charge does not wet the wire. The boat may be a depression in a metal ribbon, a V-shaped trough, or a cup. The latter is shown in Fig. 2-3b. Heating is achieved by current passing through the metal.

Crucibles have been used which are made of refractories or metal. Some of the refractories used are zirconia, aluminum oxide, thoria, beryllia, boron nitride, aluminum nitride, magnesium oxide, fused quartz, and silicon carbide. These crucibles can be heated by embedding a heater in the wall, or by radiation from a nearby heater. If a conductive charge is used, it may be heated by R-F induction.

Conductive crucibles are most often made of graphite. This material is available at low cost and in pure rods which are easy to machine. They can be heated by R-F induction, radiation from adjacent heaters, electron bombardment, or by current passing through the graphite. Graphite

a) Tungsten Filament Source

b) Boat Source

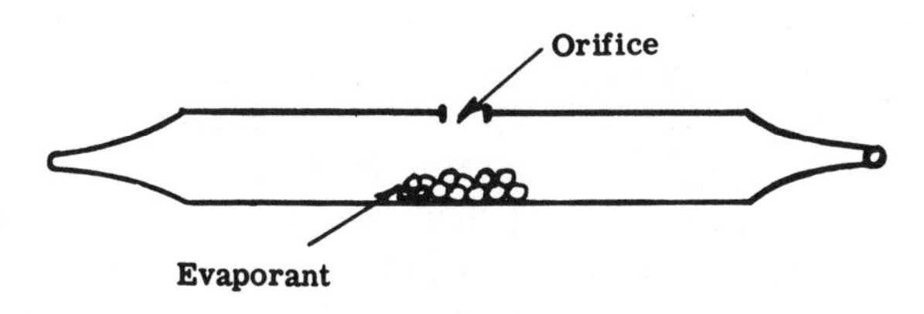

c) Cylindrical Tube Source

Fig. 2-3 Evaporation sources

has good resistance to attack by liquid metals. Metallic crucibles are usually used to vaporize dielectrics such as silicon monoxide. The reactions between liquid metals and hot metal crucibles are usually severe, and great care must be exercised in their use. Metal crucibles can be heated by radiation, electron bombardment, and R-F induction. Current heating involves very large currents because of the low resistivity of the metals used. A metal tube with its ends closed and an effusant hole in the side as shown in Fig. 2-3c, is a simple and effective source.

If the evaporant is available in wire form it can be fed from a spool so that it contacts a hot ribbon and vaporizes. The vaporation rate depends upon the wire feed rate. If the charge is available as a powder or mixture of powders, it can be dropped onto a hot ribbon and flash evaporated.

Electron bombardment is a useful method for generating heat. When electrons are accelerated from an electron source or cathode to an anode at potential V, the heat generated at the anode is IV, where I is the electron current. The anode may be a metal crucible, an evaporant charge resting on a hearth, or the end of a metal rod which vaporizes. The cathode is usually a loop of tungsten wire heated by current from a low voltage transformer. The electron current is controlled by the wire current. Probably the first use of this method, for heating a crucible, involved a helical coil of tungsten surrounding a graphite crucible. Since the accelerating voltage is usually 5 to 10 KV, and currents of 100 to 200 ma are easy to generate, it is easy to produce 1 to 2 KW of power. Electron bombardment is particularly useful in evaporating the refractory metals which must be maintained at approximately 3000° C.

A disadvantage of the simple wire cathode surrounding the anode, is the possibility of contamination by tungsten from the cathode. This can be avoided if the emitter is placed to one side and the electrons are deflected through 180° and focused onto the charge by a permanent magnet. Such equipment is commercially available.

The wire cathode can be replaced by an electron gun placed to one side. It focuses the electron beam onto a small area if the electronic charge will leak off and so avoid the generation of a field which defocuses the beam. In this case, the accelerating voltage is usually 20 KV. The size of the charge to be heated can be made small. This type of system can be used to vaporize insulators. Insulators are characterized by a negative resistance coefficient. If it can be preheated by, for example, radiation from adjacent hot surfaces, the resistance can be lowered to the point where charge leakage is adequately fast to avoid defocusing of the electron beam. In certain cases preheating by auxiliary heat sources is not required. While the insulator is cold, the charge accumulation on the insulator will defocus the electron beam to the extent that the area covered is large. As the evaporation temperature increases, it becomes conductive, and the conductivity is finally sufficient to allow good beam focusing and hence good power concentration.

Because of the importance of silicon monoxide, special sources have been developed to supply intense beams of particle free vapor for long periods of time. One example of such a source has essentially two concentric cylinders made of tantalum. The inner cylinder is the heating element and is perforated with many small holes. The SiO charge is packed between the cylinders. The SiO adjacent to the heater vaporizes, and diffuses through the perforations into the inner cylinder, and from there out through the top.

2.1.3 CIRCUIT LAYOUT

When a suitable schematic for a thin film circuit has been prepared, all the required components may be tabulated, along with their magnitudes and tolerances, as well as the various materials to be used in their fabrication. After this information has been compiled, it is possible to proceed to construct an enlarged layout of the circuit. It is usually necessary to subdivide the circuit into smaller units. Each unit must be small enough to fit onto the substrate accommodated by the high vacuum evaporator. Substrates as small as 1/4 square inch and as large as 20 square inches are being used.

The next recommended step is the construction of paper models of each component, enlarged 10 times. These components can then be easily arranged and maneuvered to determine the best topological layout for each substrate. Upon completion of the scaled paper models of each substrate, detailed layouts of the geometrical patterns of each material, may be constructed. For example, one such layout might consist of the resistor pattern or a conductor pattern for a given substrate. These layouts form later the basis for the evaporation mask fabrication. If all of these layouts were superimposed, the composite layout topology for a given substrate would be formed.

In making the topological layouts for each substrate in the system, it is important to keep interconnection leads, between substrates, as short as possible. The layout of interconnecting leads or conductors should be placed so as to avoid the creation of undesirable feedback loops and crossovers. The locations of input and output terminals, power feeds, and grounds are extremely important. Also, the layout should allow sufficient room for any necessary add-ons. For a good topological layout of any circuit, the designer should keep these factors in mind. A poor layout requiring redesign of the mask, after the circuit fabrication has commenced, may result in considerable delay and much added expense.

From the topological layout for each material, a high precision reproduction is made suitable for photography on a high-resolution photographic plate. This is known as the artwork.

2.1.4 ARTWORK PREPARATION

The material most commonly used in artwork preparation consists of a clear sheet of 5 mil mylar covered by a red film 2 mil thick. The artwork is produced by cutting and peeling the red film where the thin film components are to be located. This is known as the ''cut-and-strip'' method.

An undesirable feature of the cut-and-strip process resides in the condition that the knife, used in the cutting operation, penetrates into the clear mylar, and produces a V-shaped groove as shown in Fig. 2-4. Since the preferred method for illuminating the artwork is by backlighting, the effect of the groove on the light rays is as shown in the figure. Ray A is absorbed in the red film, ray D passes on through the clear film, but rays B and C are reflected from the side walls of the groove. Since the angle of incidence can readily exceed the critical angle, the reflection will be

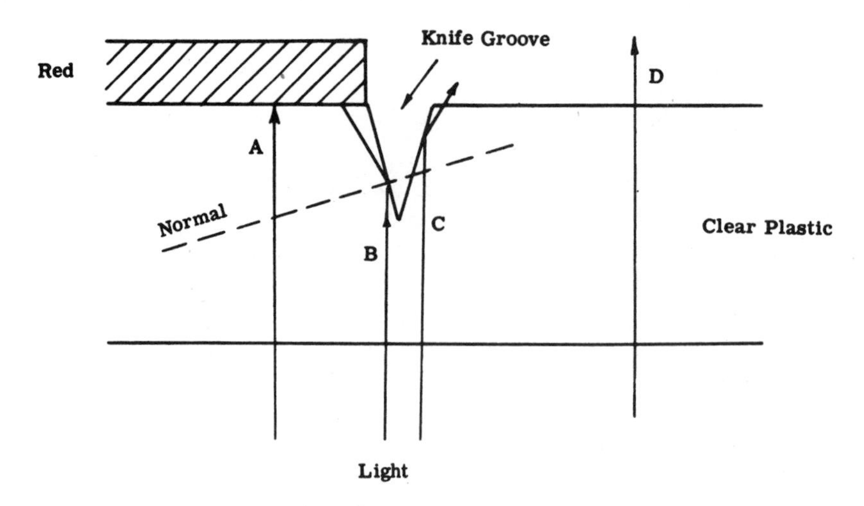

Fig. 2-4 Effect of knife penetration on light transmission through clear plastic sheet

nearly complete. Careful use of the cutting knife is required to produce a reproducible groove of minimum width.

A material useable for preparing thin film artwork is often referred to as ''scribing'' material. This material is a thin opaque film applied to a clear support of either plastic or glass. The opaque film is removed by a wedge whose width equals that of the desired line. The advantages of the scribing process are: one pass of the scribing wedge produces the line; the line width is fixed by the wedge width; and the edges are superior. The disadvantages are cost, the greater care required in the preparation of the artwork, and the necessity for separate wedges for each line width.

It is possible to use cruder methods for preparing the artwork. Examples of these are layouts using masking tape, and black and white layouts made by a draftsman. In these cases, however, it is necessary to use very large-size artwork which substantially increases the difficulties of the subsequent photographic process.

It is customary to use a precision coordinatograph to prepare the artwork. Machines are available with an accuracy of 0.1 mil and they may be motor driven.

If a flexible base material is used, it is undesirable to roll it up, or use tape or tacks to hold it in place. A convenient easel for supporting the flexible artwork during both the preparation and photographic steps, is the vacuum easel shown in Fig. 2-5.

2.1.5 MICROPHOTOGRAPHY

The high-precision photoreduction of the artwork onto photographic plates, requires a high level of skill applied to good photographic equipment and materials, and the plate processing operation. A good camera system, and this includes the light source, should meet the requirements described below. Many commercial camera systems are available, but few are specifically designed for microelectronics use.

Because of very long exposures usually required, the camera system should not be affected by vibrations. The choice of a good location for the camera, often obviates the necessity for special mountings or vibration absorbers.

If the camera reduction factor is variable, then it should be possible to reproduce a specific value with good precision. This is necessary since it is often required to replace one plate of a series.

The image sharpness is a critical function of the lens-to-plate spacing. It is necessary that the plate location within the camera be precise to better than 0.25 mil, and the lens position be adjustable to at least 0.25 mil, and preferably to 0.1 mil if a 2-inch lens is used. If lenses of longer focal length are used, looser tolerances are permissible.

The magnification depends upon the copy-to-lens spacing, given by $S = f(1+m)$ where f is the focal length of the lens, and m is the reduction factor. Through differentiation,

$$\Delta S/S = \Delta m/1+m \cong \Delta m/m.$$

For a typical case in which $f = 5$ inches, $m = 10$ and $S \pm 55$ inches, an error of ½-inch in the copy-to-lens spacing results in only a 1 percent error in m. Since the position of the nodal points of the lens are rarely known, it is not possible to fix the copy-to-lens spacing with precision in a single measurement. A simple and convenient procedure is to set up the camera system by measuring S from the lens center, or any other convenient reference point on the lens, measure the actual value of the reduction factor, calculate Δm and ΔS, and, then, readjust the value of S.

The copy or artwork is best illuminated by transmitted light. For this purpose planar light sources are available up to approximately 40 inches x 40 inches with sufficient uniformity. They use high-voltage, cold-cathode

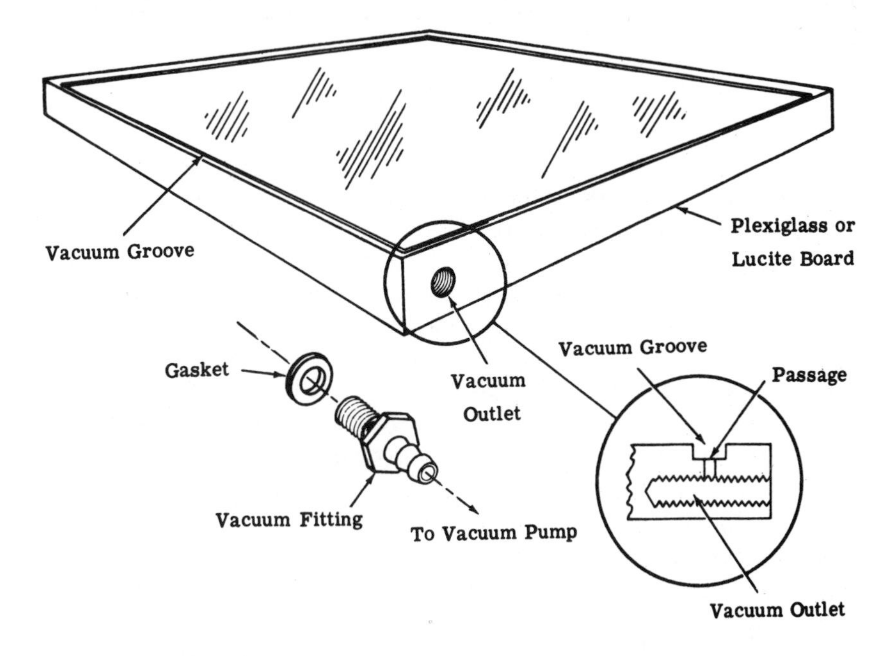

Fig. 2-5 Vacuum copyboard used for supporting flexible artwork during preparation or copying

(neon sign) tubing in a closely-spaced rectangular grid. A sheet of translucent diffusing material is used to improve the brightness uniformity. To improve lens performance, it is desirable to use light of a single color and so minimize the chromatic aberrations. The photographic plates, usually used, are sensitive to the blue-green region of the spectrum. The grain size in the emulsion is comparable to the wavelength of light so that they scatter in accordance with the $1/\lambda^4$ law. For this reason the green scatters less than the blue and will produce a sharper image. Green is, therefore, the color usually selected. Since the luminescent tubing uses a mercury arc and a fluorescent coating rich in the green, it is only necessary to place a sheet of green plastic over the tubing to convert it to a green source. The use of the green light entails an increase in the exposure time,

Fig. 2-6 A computer-controlled drafting machine produces artwork at 100 times actual size.

but this is well worth the improvement in image sharpness. A diffusing sheet should also be used over the lights to smooth out the intense areas immediately over the tubes.

The selection of a lens is critical, and is generally achieved on a cut-and-try basis. The lens should be evaluated for line edge sharpness and image distortion. The usefulness of a lens for microelectronic applications is not guaranteed if it has its full theoretical resolution. A more useful criterion for selecting a lens is the "accutance" which is a measure of the density gradient across an edge on the photographic plate. In using the lens it is not evident that the best quality is obtained at the smallest aperture. In fact, the best quality usually results if a slightly larger aperture is used. Another important factor in selecting a lens is the useful field angle. Again, this must be determined individually for each lens, but a reasonable ratio of focal length to image diameter is five.

There is a need for lenses especially designed for this application. Consider the case of a conventional lens, used at a ratio of focal length to field size of 5, to reduce artwork for a 1-inch x 3-inch substrate by a factor of 10. The focal length should be

$$5(1^2 + 3^2)^{1/2} \cong 16 \text{ inches.}$$

The lens-to-artwork spacing must be

$$16 \times (1 + 10) = 176 \text{ inches.}$$

The entire camera system requires about 20 feet and this much space in a "white room" is expensive. Furthermore, good lenses of 16-inch focal length are expensive, and the camera heads are large and expensive. If the lens permits doubling the field angle, then the overall size of the camera system can be 10 to 11 feet—which is not too unreasonable. If the camera is used for one reduction factor with green light, the lens design is greatly simplified, since chromatic aberration and a varying lens-to-artwork distance are eliminated as sources of quality degeneration. In that case doubling the usable field angle may not be too unreasonable a goal.

The photographic emulsion to be used should meet certain requirements. To minimize the effects of variations in temperature and relative humidity, glass plates should be used. It is also desirable to work in a "clean room" since temperature and humidity are usually well controlled there. It is also of advantage to eliminate dust particles on the plates, and this is another reason for working in a clean room. The emulsion should be capable of being processed to a high density at a high gamma. This means that the emulsion latitude is short and so requires precise control of the exposure. The emulsion should be fine-grained, have high resolution, be available on glass plates of a variety of sizes, and be well controlled in manufacture.

If very short exposure times are used—of the order of a few milliseconds, the high-resolution emulsion shows a degradation of gamma. For this reason, the high intensity flash lamps should be used only where necessary, such as short exposures, to minimize the effects of vibration. Another deficiency of these emulsions is the rapid decay of the latent image, or exposure leakage, after exposure and before development. This decay is especially severe in the first hour. For this reason the plate should be processed as quickly as possible after exposure. Visual focusing, using a microscope, will rarely coincide with actual best focus. Final focusing is normally a cut-and-try process.

Careful plate processing is important. The solutions should be temperature controlled and agitated. The wash water should also be temperature controlled and filtered. Good darkroom housekeeping is essential. Drops of fixing solution on walls or drain boards, for example, when dry, allow hypo dust to pass into the air and fall onto the plates.

2.1.6 PHOTO-RESISTS

Photo-resists are a class of materials widely used both in the thin film and the semiconductor processes for making microelectronic circuits. These resists are lacquer-like materials and are of two types: (a) negative resists where polymerization is caused by exposure to light, and (b)

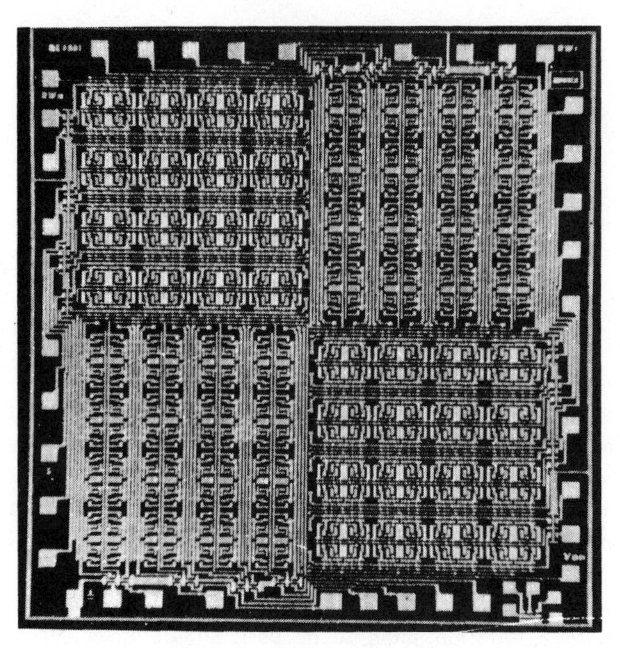

Fig. 2-7 Microphotograph.

positive resists where a polymer is degraded by exposure to light. The polymerized resist adheres to the support while the unpolymerized resist is washed away by the developer and washing solutions. The fixed resist protects the support from attack by the chemical solutions used in processing.

The resists are applied to a thickness of about 2 microns, usually by spinning, and then prebaked. After that they are exposed through a photographic plate upon which the desired pattern has been produced. Upon developing and washing, it is dried and postbaked, and is then ready for use. The relative humidity in the work area should be low. In general, resists are slow-speed emulsions and require long exposures to intense light sources rich in ultraviolet.

2.1.7 MASK FABRICATION

The evaporation masks, which are used to define the location and size of the deposited thin film patterns, are usually made by either etching or electroforming. Of these two techniques, etching is usually the simplest.

An etched mask may be made from a thin metal sheet of stainless steel, copper, or beryllium copper, usually two or three mils thick. A negative photosensitive resist is applied to the metal sheet and exposed through the appropriate reduced photonegative of the artwork. Upon development, the resist is fixed in place where exposed, and is washed away from the unexposed areas. As a result, the unexposed areas, which correspond to the patterns on the photonegative, are susceptible to etching, while the coating of resist on the rest of the metallic sheet protects it from being etched. The metallic sheet is then placed in a spray etching machine and exposed to the etchant, usually ferric chloride, for 2 to 3 minutes at 40°C. The result is a mask with the pattern etched out. After removing the resist and cleaning, the mask is ready for use. Etched masks are generally not as precise as electroformed masks, because of the random etching of line edges and undercutting of the resists.

Electroforming is the reverse of etching. In this process, the metal mask is built up by electroplating nickel upon a metallic sheet, or matrix, having a chemically passivated surface. The function of passivation is to insure that the electroformed nickel films can be separated from the supporting matrix after formation. A positive photosensitive resist is applied to the matrix and exposed to ultraviolet radiation through the reduced photonegative of the artwork. Upon development, the resist remains on the unexposed areas, and is washed away from the exposed areas. As a result, the resist coated areas, where the mask openings are to appear, are impervious to electroforming, while the rest of the metal support, where the mask is to be built up, is receptive to electroplating. The matrix is then immersed in a nickel plating bath, and the mask is built up to a thickness of a few mils, after which it is stripped from its support.

Mounting the above mask in the deposition equipment so that it can be accurately aligned with the circuit substrate presents various problems. First of all, the difference between the thermal expansion coefficients of the mask and its support, in the deposition equipment, upsets the alignment after the entire assembly is heated to the 100 to 400°C temperatures in use. Furthermore, the mask is thin and fragile to the extent that it may be damaged during handling or mounting. Thick masks have been made, but they introduce shadowing and are difficult to make flat.

A better electroforming process produces the mask as an integral part of a precision nickel frame which can be dropped over locating pins in the evaporator to provide good alignment. Since the mask itself is also of nickel, no thermal expansion problems are encountered. This fabricating process makes use of a stainless steel matrix with a high polish on one side. Stainless steel is used because it is self-passivating, due to the oxide on its surface. A positive resist is applied to the polished surface, exposed through the reduced photonegative of the artwork, and then developed, as in the conventional electroforming process. The nickel frame is then bolted to the matrix, stop-off tape is applied to the areas where plating is not to occur, and the whole assembly is finally immersed in a nickel sulfamate solution for electroplating. The plated nickel layer may be as thin as one mil, and it is firmly attached to the inside of the mask frame. When the matrix is removed, a taut mask is left attached to a frame. It can thus be handled with a minimum of risk, and can be easily aligned in the evaporator. It is important that registration marks on the photoplate be superimposed on corresponding marks on the matrix, and that the mask frame also be precisely located with respect to these registration marks. In use, the substrate holder is registered to the frame.

Other techniques may be used for mask fabrication. One simple process is the careful machining of a sheet of mica. A similar process makes use of a tape-controlled high precision milling machine to mill the mask from a thin sheet of graphite. This procedure does not require the use of artwork and photographic plates, but it does require very costly equipment, and the narrowest line which can be milled is about 15 mils. A glass sheet known as ''Fotoform,'' can also be used for masks. This glass, when exposed to ultraviolet radiation and heat treated, can be preferentially etched in its exposed portions.

2.1.8 MASK AND SOURCE CHANGERS

The formation of an entire thin film circuit on a substrate, entails the deposition of a sequence of films each of which requires its own mask and source.

The simplest method is to repeat the entire evacuation and deposition operation for each film, and change the masks and sources between operations. Inasmuch as the vacuum chamber must be evacuated for each deposition and then opened to the atmosphere for mask and source changing, this method is time consuming, but minimizes the amount of equipment required. In addition, the films are exposed to the atmosphere before the usual protective coating of silicon monoxide is deposited on the completed circuit.

These disadvantages may be avoided if the entire sequence of films is deposited in one composite operation without breaking vacuum. Such an approach necessitates the use of a mechanical device within the vacuum chamber, known as a "mask changer," which positions the substrate over the appropriate mask and source for each deposition.

A typical changer containing seven channels for mask-source combinations, is shown in Fig. 2-8. The sources and masks are fixed, while the substrate is mounted on a holder attached to a rotatable arm. The latter contains a heating element to maintain the substrate at the desired temperature. A shutter is positioned directly below each mask. The indexing mechanism provides accurate alignment of the substrate and masks. Since the alignment of one film pattern with respect to another is crucial to the operation of the circuit, the indexing mechanism must fulfill tolerance requirements as low as 0.001 inch.

Mask changers are commercially available. One of these can process six 2-inch x 2-inch substrates in one pump-down, and fits inside an 18-inch vacuum jar 30 inches high. The top plate contains the substrates and substrate heaters, and is rotatable. Below the substrate plate is a mask holder plate which can rotate about a central axis, and move up and down to provide contact between the mask and substrate. Spaced below the mask plate, is a rotatable six-position source ring. Registration between the mask and substrate is within one mil. A multistation mask changer is also available.

While it is possible to construct multistation evaporators in an 18-inch bell jar, due to the limited available volume, they are of limited usefulness with respect to production capability. When high production rates on large substrates are required, it is necessary to use larger chambers. An example is the six-station, in-line machine constructed by the International Business Machines Corporation. One machine is installed at the Naval Avionics Facility in Indianapolis, and another machine is at IBM in Owego, New York. As an example of the capability of this machine, it can process substrates 3-3/4 inches x 5-3/16 inches. These substrates are cut into 16 smaller panels (9/16 inch x 1-3/4 inches) each containing 24 digital circuits. Each circuit is made up of 10 components; one capacitor, four resistors, one transistor, and four diodes. The semiconductor devices are on

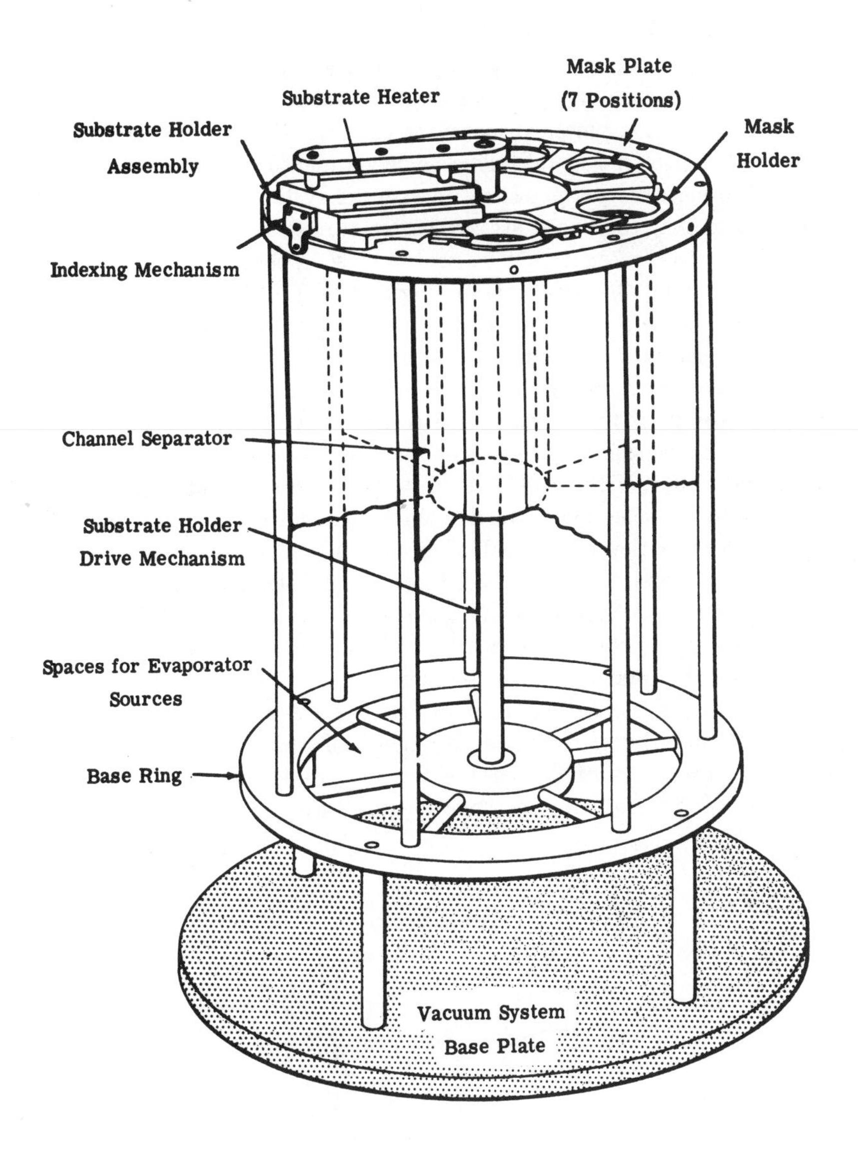

Fig. 2-8 Seven-position mask changing mechanism

three silicon chips bonded to conductive lands. Each substrate has 384 logic circuits, and the production capability is approximately 20,000 logic circuits per 8-hour day.

2.1.9 MONITORING OF FILM THICKNESS DURING DEPOSITION

While conductive films can have any thickness greater than the minimum needed for effective conductivity, resistive and capacitor dielectric films must be formed to specified thickness values. The thickness of these resistive and dielectric films can be controlled or "monitored" in several ways.

In one method the vaporization rate may be controlled by monitoring the power input to the source in order to control the source temperature. This method is generally difficult to apply because of the erratic fluctuations in source temperature during evaporation, and the difficulty in maintaining constant source vaporization characteristics.

A better monitoring method measures the rate of deposition with the triode gauge shown in Fig. 2-9. This device is identical to the ionization gauge used to measure gas pressures in vacuum systems. A helical coil or grid, placed so that the vapor stream passes through it, accelerates electrons from an electron emitter. The grid is biased positively with respect to the latter. Along the grid axis, a single wire, biased negatively with respect to the emitter, acts as a positive ion collector. Electrons are accelerated by the grid into the region through which the vapor stream passes. Some of these electrons collide with and ionize atoms in the vapor stream, forming thereby positive ions which are collected by the ion collector. The resulting ion current is a measure of the instantaneous vapor stream, intensity and, hence, of the rate of deposition. The integral of the ion current is, therefore, a measure of the film thickness, and when the integrated ion current reaches the desired value, the deposition is terminated.

An oscillating quartz crystal placed near the substrate can also be used to monitor the film thickness during deposition. The frequency of oscillation of such a crystal decreases if a face is loaded by a small mass. For example, a typical quartz crystal oscillating at 10 mcs has its frequency decreased 20 cycles/second by the deposition of 1 Å of gold on one of its faces. Thus, the total decrease in the frequency of such a crystal, during a deposition, is a measure of the mass of the deposited film.

The preceding monitoring methods are applicable to all types of films.

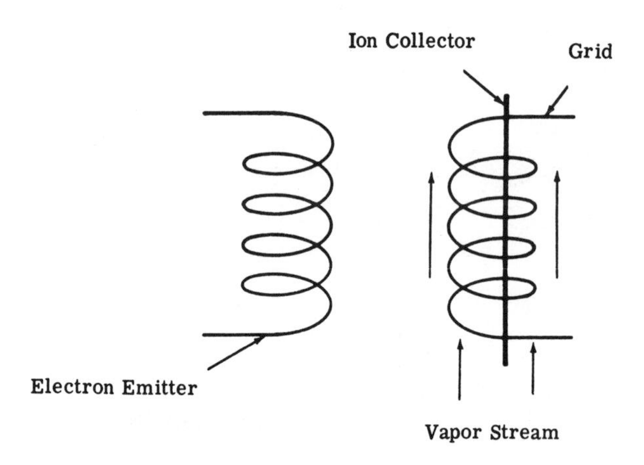

Fig. 2-9 Triode gauge for measuring deposition rate

However, there are simpler methods for monitoring films of a specific type, such as resistive and dielectric films.

The simplest method for monitoring sheet resistance is to measure the resistance of a resistor formed on a monitoring substrate placed near the circuit substrate. If the monitor resistor has a known aspect ratio, it is possible to determine its sheet resistance and hence the sheet resistance of the resistors on the circuit substrate, provided the latter is sufficiently near the monitor and the substrate temperatures do not differ appreciably.

The monitor can be part of an automatic controller which terminates the deposition when the desired monitor resistance value is reached.

Dielectric film thicknesses can be monitored by using a double beam interferometer. During deposition, a beam of monochromatic light, incident normally on the film, is divided into two separate coherent beams; one is reflected from the top face of the film, and the other from the bottom interface. The beams are focused onto a light detector as shown in Fig. 2-10a. The two beams interfere constructively or destructively depending on their phase difference ϕ, which (at normal incidence) may be expressed as

$$\phi = 4\pi nt/\lambda \tag{2-13}$$

where n is the refractive index of the film, t its instantaneous thickness, and λ the wavelength of the incident light. When $\phi = 2N\pi$, i.e., $nt = N\lambda/2$, where N is an integer, the beams interfere constructively and the detector output records a maximum of intensity. Thus, as the film grows in thickness, the detector output passes through a series of maxima and minima, as shown in Fig. 2-10b, with successive maxima (or minima) indicating thickness increments of $\lambda/2n$. It is not necessary to know the value of n since deposition can be terminated when a previously determined number of maxima and minima have been passed through.

2.1.10 MEASUREMENT OF FILM THICKNESS

During circuit fabrication one monitors a function of the film thickness or mass per unit area, as already described. However, it is also often necessary to measure the absolute values of these quantities. Several methods exist for performing such measurements in the laboratory.

For the thinnest of thin films used, it is often difficult to visualize what is actually meant by "thickness." The films are well known to be granular in structure, i.e., small crystallites on a surface separated from neighboring crystallites. Accordingly, mass measurements will result in a form of

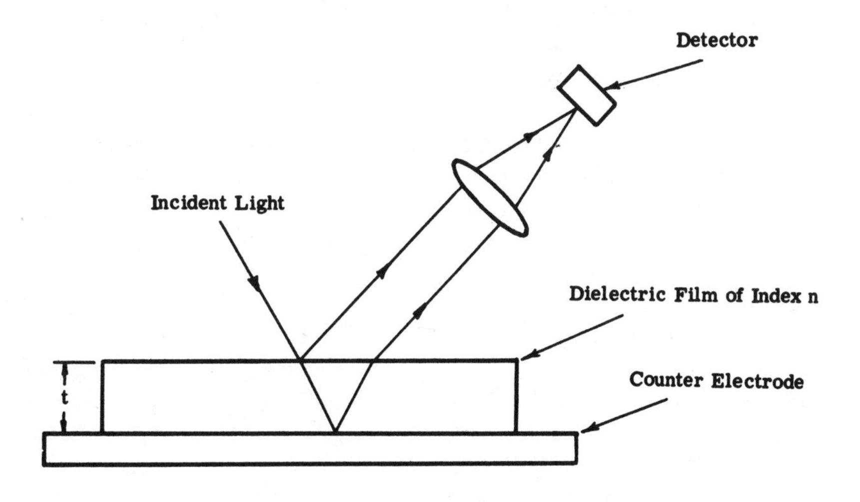

a) Double Beam Interferometer Arrangement

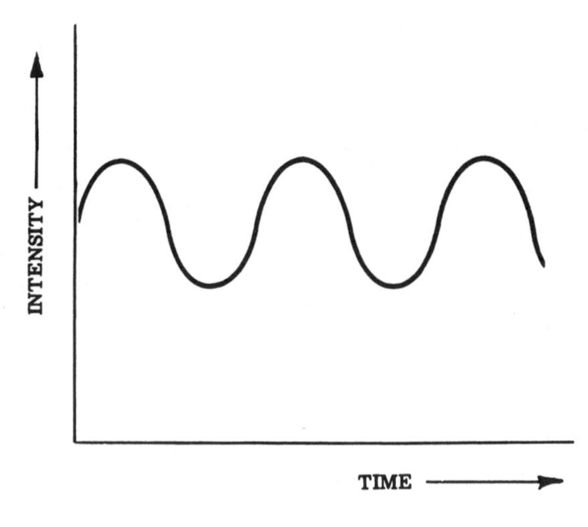

b) Variation of Interferometer Output

Fig. 2-10 Dielectric film thickness monitoring

average value, and thickness measurements will also be affected by the film structure. Many physical properties of thin films differ appreciably from their bulk values. It is, therefore, not generally permissible to divide the mass per unit area by the bulk density to compute the film thickness.

A variety of methods for measuring film thickness or mass are available. The more useful ones are described below:

(a) Two-Beam Interferometry: If two beams of light, of the same wavelength, are caused to combine on a receptor such as a photographic plate or the human eye, the resultant light intensity is found by vectorially adding the electric intensity vectors and squaring the result. If the two are in phase, or differ in phase by an integral number of wavelengths, the intensity is a maximum. This situation is known as constructive interference. If the vectors differ in phase by half a wavelength, or half a wavelength plus an integral number of wavelengths, the resulting intensity is a minimum, and the situation is known as destructive interference. If the vector amplitudes are, at the same time, equal, the destructive interference produces zero intensity. These phase relationships must be of sufficiently long duration to permit the experiment to be carried out. The simplest way to assure this is to generate the two beams from the same source. A detailed description of interference phenomena can generally be found in books on physical optics.

Consider the simple configuration of Fig. 2-11b of the typical arrangement in Fig. 2-11a. Here I is a beam of monochromatic light of wavelength λ incident first on a glass F, optically flat (assumed ideally flat). Part of the light is reflected as beam R, and part passes on to the sample S, where a second part is reflected as R′. For the sake of simplicity, assume normal incidence and ignore other beams generated in the system. The phase difference, in wavelengths, between R and R′ is

$$2tn/\lambda$$

where n is the index of refraction of the space between F and S, (in this case it is usually placed equal to one), and t is the separation of reflecting surfaces between F and S at the point of examination. Bright bands, or fringes, occur if this phase difference equals a whole number of wave lengths or if

$$N = 2t/\lambda, \qquad N = 0, 1, 2, \ldots$$

Since the separation t increases uniformly, from one end to the other of the test sample S, N will also increase progressively across the face of S.

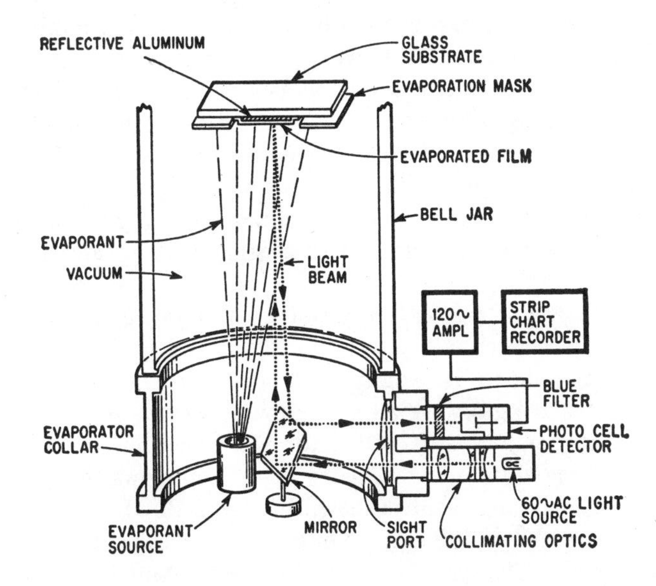

Fig. 2-11a Two-beam interferometer used to measure thickness of a thin-film component

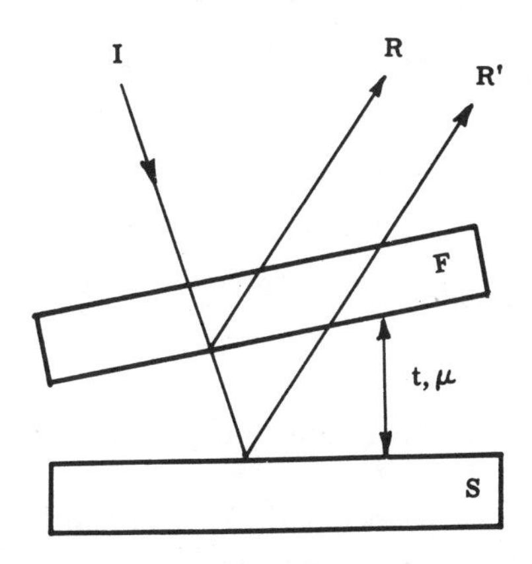

Fig. 2-11b
Basic arrangement for measuring surface defects with two-beam interferometer

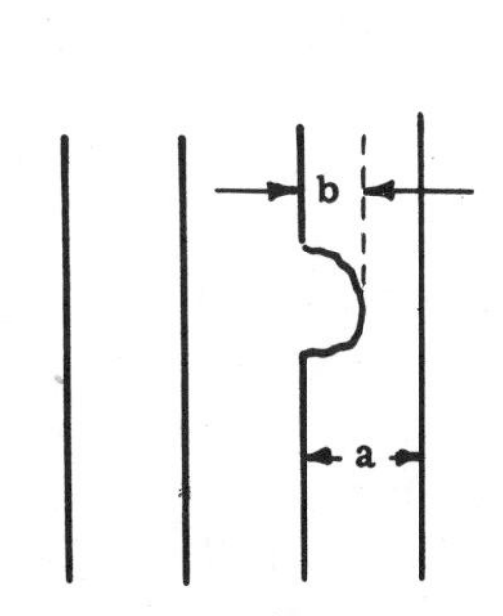

Fig. 2-11c
Appearance of a fringe crossing a surface defect

The result will be a set of parallel bright fringes across S, assuming S to be perfectly flat. The spacing of these fringes depends upon the angle between F and S. For two successive fringes, where N increases by one, t increases by $\lambda/2$. The wavelengths, usually used, are either the sodium yellow line ($\lambda = 5893Å$) or the mercury green line ($\gamma = 5461Å$). The unit of measurement (1 fringe) is therefore 2947 or 2730Å which is 11 to 12 micro inches.

If the surface of S, at a small area, deviates from perfect flatness, the fringe crossing that area will shift. The magnitude of the deviation from perfection is found from Fig. 2-11c as

$$(b/a)(\lambda/2)\,. \tag{2-14}$$

This two-beam interferometer technique suffers from a serious drawback; the fringes change gradually from maximum to minimum intensity, and thus are difficult to locate with accuracy. It is usual to assume that the accuracy of fringe shift measurements is 1/10 fringe or 250 to 300 Å, and this is often not good enough for measurements of film thickness in microelectronics. Equipment, for generating the fringes, is available in the form of an accessory for use with a microscope. If the microscope has a micrometer in the eyepiece, the measurement on the fringes can be made easily. If the microscope is equipped with a camera, they can be photographed.

(b) Multiple Beam Interferometry: A slight modification of the two-beam interferometer will convert it into the multiple beam interferometer. The resulting device is usually good enough for microelectronic applications. If the fringes were very narrow, it would be possible to measure their separation and shift with improved precision, and that is precisely what multiple beam interferometry does–it narrows the fringes.

The basic arrangement is shown in Fig. 2-12. Here a standard optical flat surface is shown at S and the unknown at S′. The supporting structures are not shown. The space between S and S′ is assumed to be air (index of refraction = 1). The test sample is coated with an evaporated film of reflectance R and transmission T,–the same as the flat S. The emerging beams of intensity, R, RT^2, R^3T^2, . . . are collected by a lens and brought to a focus. The phase difference between successive beams is

$$\Delta = (2t/\lambda) \tag{2-15}$$

where normal incidence is again assumed. The intensity, I, is

$$I = \frac{T^2}{(1-R^2)} \times \frac{1}{1+\sin^2\Delta/2[4R/(1-R)^2]} \tag{2-16}$$

If $\sin \Delta/2 = 0$, I is a maximum. If, on the other hand, $\sin \Delta/2 = 1$, it is a minimum.

Figure 2-13 compares the fringe shapes obtained with two and multiple beam interferometers. It is necessary that $R \rightarrow 1$ to get sharp fringes, and this is accomplished by evaporating a film of silver across the film step to be evaluated.

The multiple beam interferometer can be used in two forms. In one, monochromatic light is used exactly as in the two beam case, and the fringes are known as Fizeau fringes. In the second case white light is used and the fringes are contours on which t/λ is constant. As the fringe passes

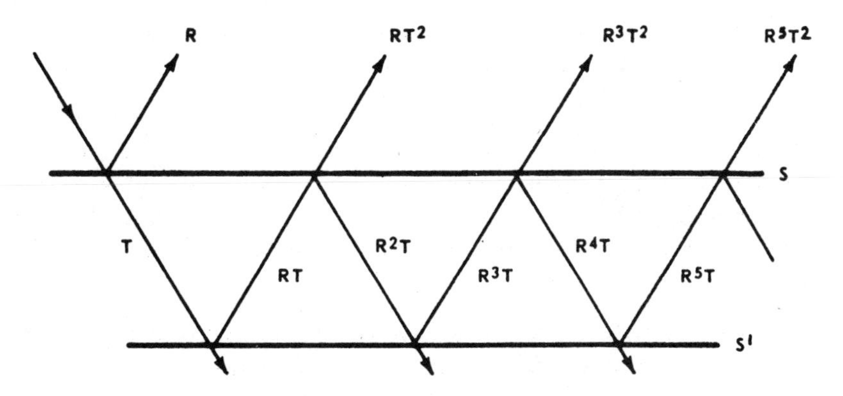

Fig. 2-12 Basic arrangement for measuring surface defects using a multiple-beam interferometer

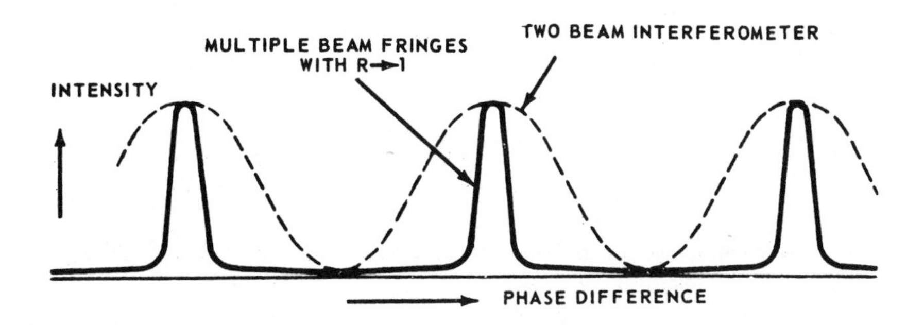

Fig. 2-13 Relative appearance of fringes in two-beam and multiple-beam interferometry

over a step so that t changes the wave length, the light which makes up that specific fringe changes to keep t/λ constant. These fringes are known as "fringes of equal chromatic order" or FECO fringes. In this case, t is calculated by measuring the wavelength of the light which makes up the various parts of the fringe. This is easily done by focusing the beams on the slit of a constant deviation spectrometer, and measuring the wavelengths on the spectrometer drum.

The FECO fringe method is capable of greater accuracy than the Fizeau fringe method. The latter is precise to 1/100 of a fringe or 30 Å, whereas the FECO fringe method can produce results precise to 5 Å, if great care is used.

(c) Weighing: A relatively simple technique consists of weighing

the film. Balances are available which can weigh 20 gram masses to an accuracy of 1 to 2 micrograms, with a sensitivity of 1 μgram. One square cm of gold, 10 Å thick, weighs 2 μgrams.

(d) Stylus Measurements: If a stylus, connected to a sensitive displacement transducer, is dragged across a film and over the step at the edge of the film, the transducer output will record this event. The accuracy attained is 50 to 100 Å. Care must be taken when using the stylus on soft films.

(e) Chemical: If the thin film is caused to react with a predetermined amount of another chemical, the amount of the reactant left, permits calculation of the weight of the film. For example, assume a Ba film is boiled in a known amount of a solution of HCl of known normality. The HCl which remains is determined by titration, and from this data the weight of the Ba film can be computed.

(f) Ellipsometry: This method involves measurements of phase shift and intensity of the two components of reflected polarized light. The method can be very sensitive, but the optical constants n and k must be known—usually they are not.

(g) Electrical Method: A method of determining the thickness of vacuum-deposited gold films using film resistance and Hall voltage measurements, resides on the following basis: for a given thin isothermal film of thickness t in the x-y plane, in the presence of an electric field E_x and magnetic field H_z, the effective electron mean free path, λ_{eff}, is

$$\frac{\lambda_{eff}}{\lambda_0} = \left(\frac{A_H(t)}{A_{HO}}\right)^{2/3} \frac{\sigma(t)}{\sigma_0}, \tag{2-17}$$

where λ_o is the effective mean free path in the bulk material, A_H is the Hall coefficient, σ is the conductivity, and the zero subscripts denote bulk material quantities. The parameters $A_H(t)$ and $\sigma(t)$ are complex functions of the thickness.

If the resistance, R, and the Hall voltage, V_y, for a given current I_x and magnetic flux density B_z, are measured for a section of film of length Y and width X, λ_{eff} is

$$\frac{\lambda_{eff}}{\lambda_0} = \frac{Y}{RX}\frac{1}{\sigma_0}\left(\frac{V_y}{I_x B_z}\frac{1}{A_{HO}}\right)^{2/3} t^{-1/3} \tag{2-18}$$

Thus, given the Hall voltage and resistance, the thickness of a film can be determined by plotting equation (2-18) for various values of t, and noting the intersection of this curve with the curve plotted from equation (2-17) as a function of t. This graphical method is equivalent to equating equations (2-17) and (2-18) and solving for t.

Using this method, the calculated thickness of unannealed gold films agrees with optical interferometer measurements, within 5 percent for films with a thickness greater than 400Å. However, for thinner films, greater differences occur, probably due to variations in film structure. The resistance measurements in the unannealed films are particularly susceptible to errors caused by stresses in the film.

2.1.11 ADHESION OF THIN FILMS

An important characteristic of the film is its adhesion to the glass substrate. While it is possible to work with films which do not adhere to the substrate, it is not practical to do so, since, for example, it is not possible to solder such films. The substrate is usually glass because if alumina is used, it is coated with a glassy phase glaze. If the substrate is improperly cleaned and degassed, then no depositant will adhere well. Accordingly, the substrate surface should be properly prepared for deposition. In that condition no foreign materials are on the substrate surface, except a gas film from the residual gaseous atmosphere in the vacuum chamber.

A theoretical explanation of all of the factors associated with film adhesion is difficult to present for at least two reasons: (a) reliable, self-consistent measurements of the forces of adhesion between films and their substrates are very difficult to make, and (b) the detailed physical nature of the film and glass structure is very complex. For this reason, the production of adherent films remains an empirical art.

Film adhesion is most easily checked by rubbing a finger across the film. If it withstands this test, the adhesive tape test should be used. A strip of pressure-sensitive adhesive tape is applied to the film and firmly rubbed in place. It is then lifted off. If the film withstands this test, more elaborate testing procedures must be used, and usually these are of a quantitative nature.

In one such method a rubber pad, impregnated with an abrasive, is rubbed across the film. The results of this technique are highly variable, even when factors such as the rubber pad, pressure, and speed of dragging the pad across the film are standardized. The best that can be said for the method is that it is useful for comparative monitoring, if the comparisons are made on the same machine.

Another method attempts to measure the adhesion of films on a rotor spinning at very high speeds. The speed at which the film parts from the rotor, measures the adhesion. In this method, films with good adhesion cannot be spun off before the rotor explodes. For poorly adherent films, forces of 10^9 dynes/cm^2 are required. Adhesive forces in the range 10^8 to 300 x 10^8 dynes/cm^2 have been found. In a third method, films are deposited on a piezoelectric crystal. When oscillating, the surface is subjected to high acceleration normal to the surface. A fourth method resides on cementing a flat-headed pin to the film, and then pulling it off with a force normal to the film. In a still further method, a smooth, hard steel, ball is dragged across the film and the force pressing the ball into the film is gradually increased until the ball strips off the film. An analysis of the mechanics of the system, allows the adhesive force between the film and its substrate to be computed.

In general, it has been found that metals, which oxidize readily, adhere well to glass, and that the adhesion improves with aging up to about 500 hours. The improvement observed with aging is assumed to be due to oxidation. The importance of oxygen in the adhesion process can be demonstrated by depositing Al and Fe films under 10^{-3} torr of 0_2, H_2, and air. The films deposited in 0_2 have better adhesion than those deposited in air.

Films deposited in H_2 have poorer adhesion, than those deposited in air.

The films with exceptionally good adhesion do not solder well. To solve this problem it is customary to deposit a bimetallic film. The metal on the glass is selected for good adhesion, and on that metal is deposited a second metallic film to which good solder bonds may be made readily. Examples are chromium plus gold, and manganese plus silver. A combination used for good reflectivity is chromium plus aluminum.

2.1.12 FRACTIONATION OF AN ALLOY

Most of the conventional evaporation techniques used in the vacuum

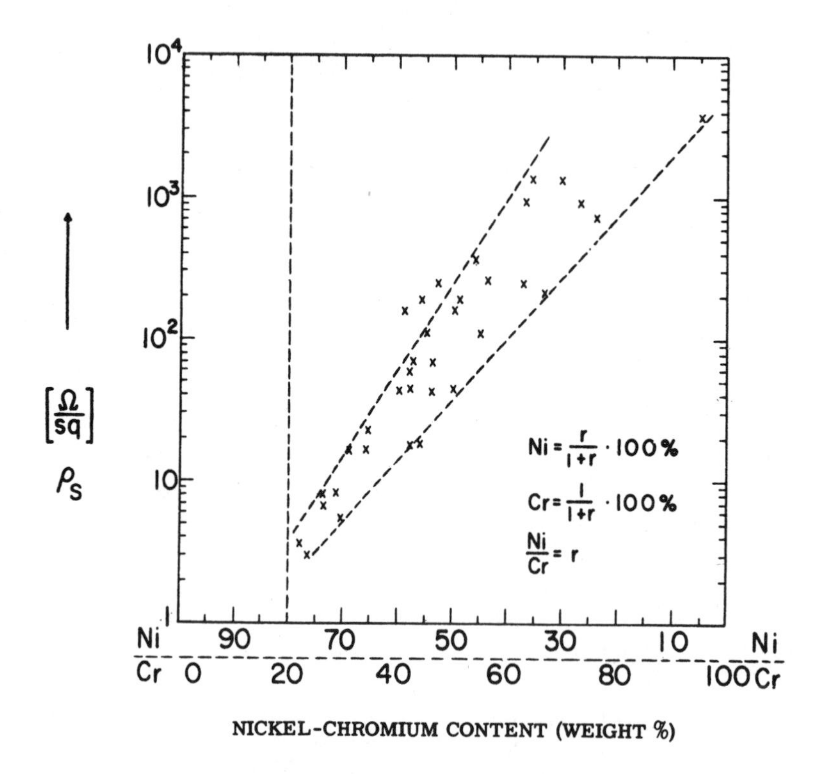

Fig. 2-14 Correlation between nichrome sheet resistivity, ρ_s and % weight composition

deposition of alloys, result in some degree of fractionation of the alloy. The composition of the film resulting from the vacuum evaporation of a binary alloy will differ from that of the source bulk material because of unequal volatilization rates of the two components. For example, fractionation takes place in an 80/20 nickel-chromium evaporant with the vapor being initially rich in chromium, becoming sufficiently rich in nickel as evaporation progresses, and finally reaching near bulk composition for relatively thick films. There is no precise law which relates the partial vapor pressure of the alloy constituents to their vapor pressures in the pure state. However, Raoult's Law, which states that vapor pressure of the solution is lower than that of the pure solvent by an amount which is proportional to the concentration of the solute, can be used to estimate the behavior of an alloy's constituents. Applying this relation to the 80/20 nickel chromium alloy, then, at 1200°C the rate of evaporation for chro-

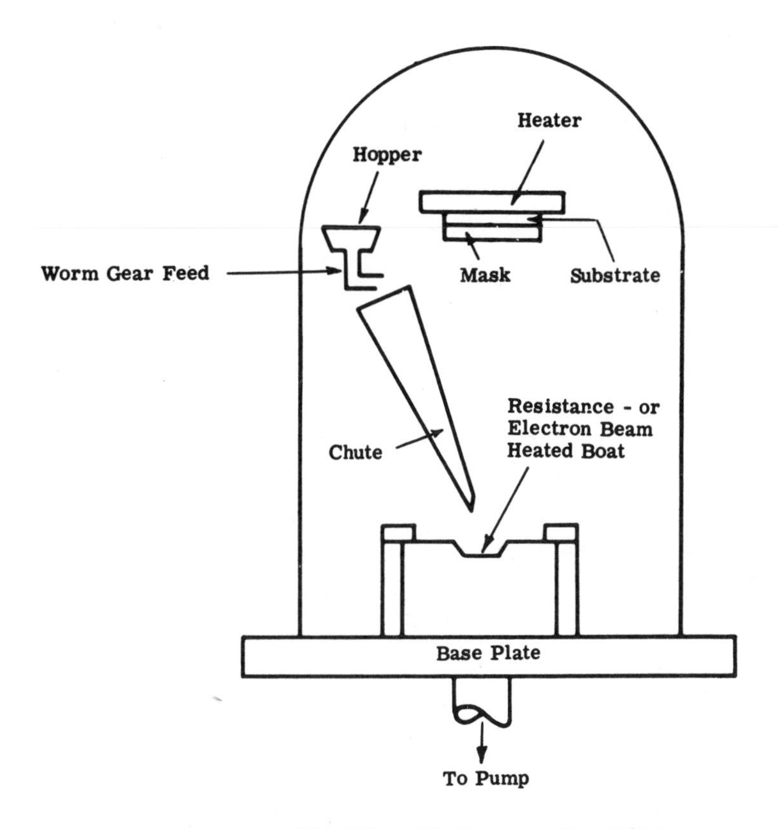

Fig. 2-15 Flash evaporation system

mium will be approximately 100 times as fast as that of nickel. Figure 2-14 illustrates, for an 80/20 nickel-chromium charge, the correlation between the sheet resistivity and the related weight for chromium $1/(1 + r)$ and nickel $r/(1 + r)$ of the evaporated films, where, $r = 4$ for the starting charge. The results indicate that the thinner the film, or the higher its resistivity, the richer is the chromium content.

A means of controlling the composition of the film so that it is very nearly chemically equivalent to the starting compound or mixture, is by flash evaporation (Fig. 2-15). A finely divided powder of the source material is dropped, in measured amounts, onto a refractory strip which is heated to temperatures above the evaporation temperature of the source material by any suitable means. By this process, the fine grains can be vaporized almost instantaneously thereby keeping fractionation of the material to a minimum.

2.1.13 METAL OXIDES

Metal oxides can seldom be deposited in their highest state of oxidation using conventional vacuum evaporation techniques. The dissociation of such oxides at high temperatures has been a major problem in producing capacitor dielectrics by vacuum evaporation procedures. One of the most commonly used thin-film dielectrics is silicon monoxide. The major attributes of silicon monoxide are its relative ease of deposition, and its relative high DC breakdown strength (2×10^6 V/cm). However, as a dielectric, silicon monoxide suffers from a relatively low dielectric constant ($k = 4.5 - 6.8$), which limits its application in thin-film circuitry where larger capacitance per area is needed. Reproducibility of the electrical characteristics is another problem associated with silicon monoxide films. It is now well established that the dielectric properties of SiO capacitors are dependent (to varying degrees) upon both deposition rate and film thickness (Fig. 2-16). The general conclusion is that low deposition rate favors incorporation of oxygen into the film, and shifts the oxygen/silicon ratio somewhere between Si and SiO_2, with the actual composition of the

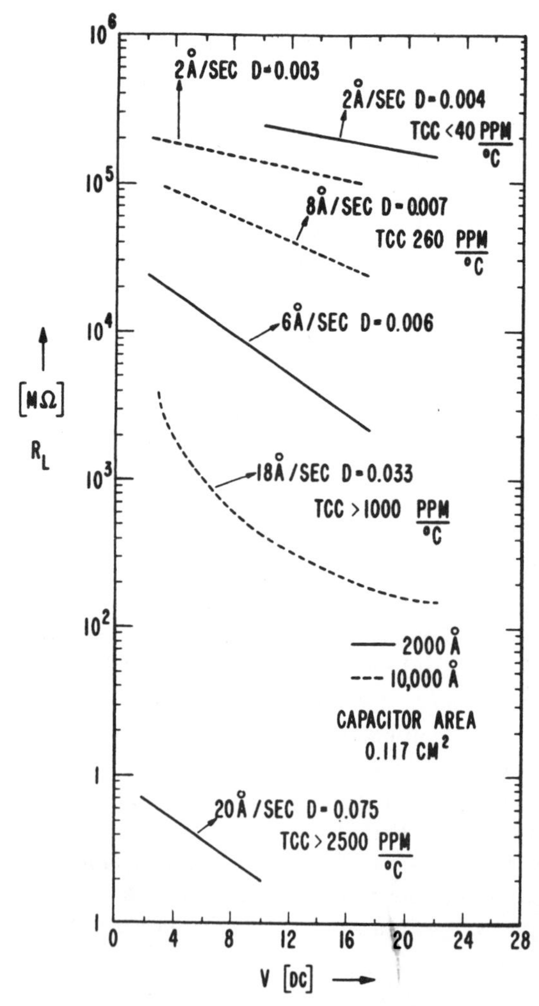

Fig. 2-16 Leakage resistance, R_L, dissipation factor, D, and temperature coefficient of capacitance, TCC, of thin (2000 Å) and thick (10,000 Å) SiO films as a function of deposition rate, Å/sec

film consisting of varying percentages of Si, SiO, S_2O_3, and SiO_2. (The dielectric constant of Si is 11.6 and for SiO_2, 3.81.)

2.2 ANODIZATION PROCESS FOR FABRICATING THIN FILM CIRCUITS

This process applies three different methods for making thin films to

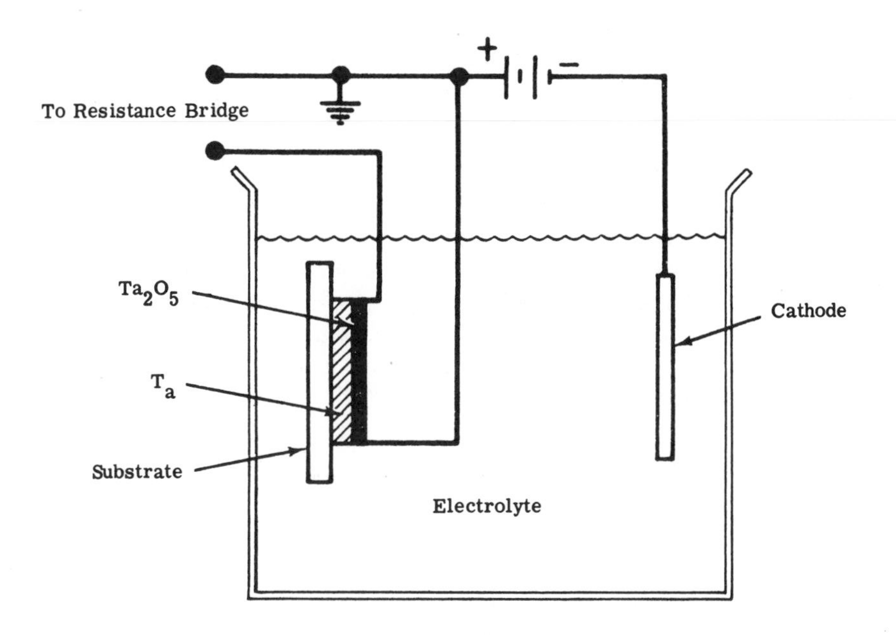

Fig. 2-17 Basic anodization arrangement

produce a completed microelectronic circuit: sputtering, anodization, and vacuum evaporation. The unique feature of the process is the use of anodization, and therefore it is commonly referred to as the anodization process. The anodization can be used to make particularly the passive devices of resistance, capacitance, and inductance.

The process has some advantages over that of vacuum deposition. For example, in the case of the capacitor, the temperature coefficients and loss factors are comparable, but the dielectric constant for the Ta_2O_5 is 26, while the value for SiO (used in evaporation) is 6. Thus, larger valued capacitors are easier to make with the anodization process. Resistors are stable and can be made with an initial tolerance of 0.01 percent.

In practice, two basic materials are being used in the anodization process — tantalum and titanium, with by far the greater emphasis on the first. In order to understand the fabrication of thin film microelectronic circuits by anodization, it is necessary to be acquainted with two technical processes: anodization and sputtering.

2.2.1 ANODIZATION

Anodization is an electrochemical process for converting a metal to its oxide. It has been used for many years, for example, in making aluminum and tantalum electrolytic capacitors, and to form a protective film of aluminum oxide on aluminum.

If, as shown in Fig. 2-17, Ta is made the anode in a solution which produces oxygen ions and does not dissolve the metal oxide, the electric field, in the metal oxide, drives the metal ions through the oxide and to the surface where new metal oxide molecules are formed. As the oxide thickness increases, the field strength decreases and eventually will be too weak to force the ions through the metal oxide film. The current will then be very small, and the oxide film will substantially have stopped growing. The oxide is tantalum pentoxide (Ta_2O_5) having a thickness, t, proportional to the voltage, V, or

$$t = KV. \tag{2-19}$$

One cell which has been applied uses a platinum cathode and an electrolyte made of one part oxalic acid, two parts water, and three parts ethylene glycol. The ratio of thickness to voltage (K in Eq. 2-19), is usually about 20 Å/volt, and depends on the temperature. The resulting anodic films are amorphous, of low porosity, and have high resistance to abrasion and chemical attack.

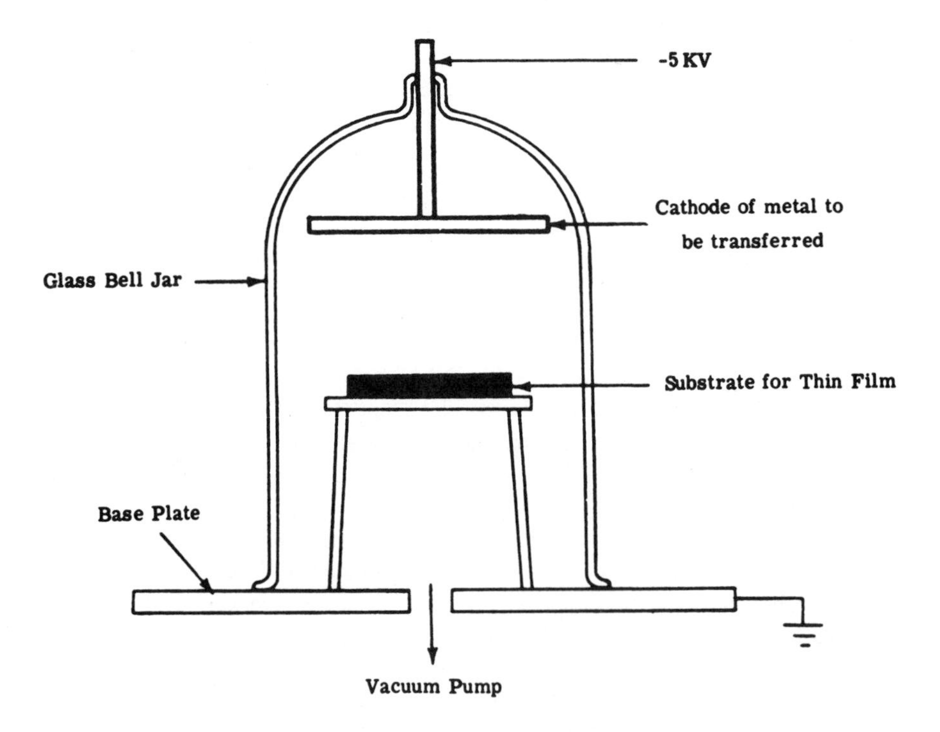

Fig. 2-18 Simplified sputtering apparatus

2.2.2 SPUTTERING

If the positive ions in a gaseous discharge bombard a metallic cathode at sufficiently low gas pressures (10^{-2} to 10^{-1} torr), free atoms leave the cathode and can be deposited on a surface in its vicinity. This transfer of material from the cathode is known as ''sputtering.'' Although the discovery of this phenomenon is quite old, the detailed mechanism by which the cathode atoms are liberated, has yet to be fully understood. There are two principal theories: the ''evaporation theory'' holds that the cathode atoms are thermally evaporated due to local heating by the bombarding ions, while the ''impact theory'' holds that the cathode atoms are liberated by a direct momentum transfer from the bombarding ions. Unfortunately, both theories can claim the support of experimental evidence. A simple experimental arrangement is illustrated in Fig. 2-18. A commercial unit used in practice is shown in Fig. 2-19.

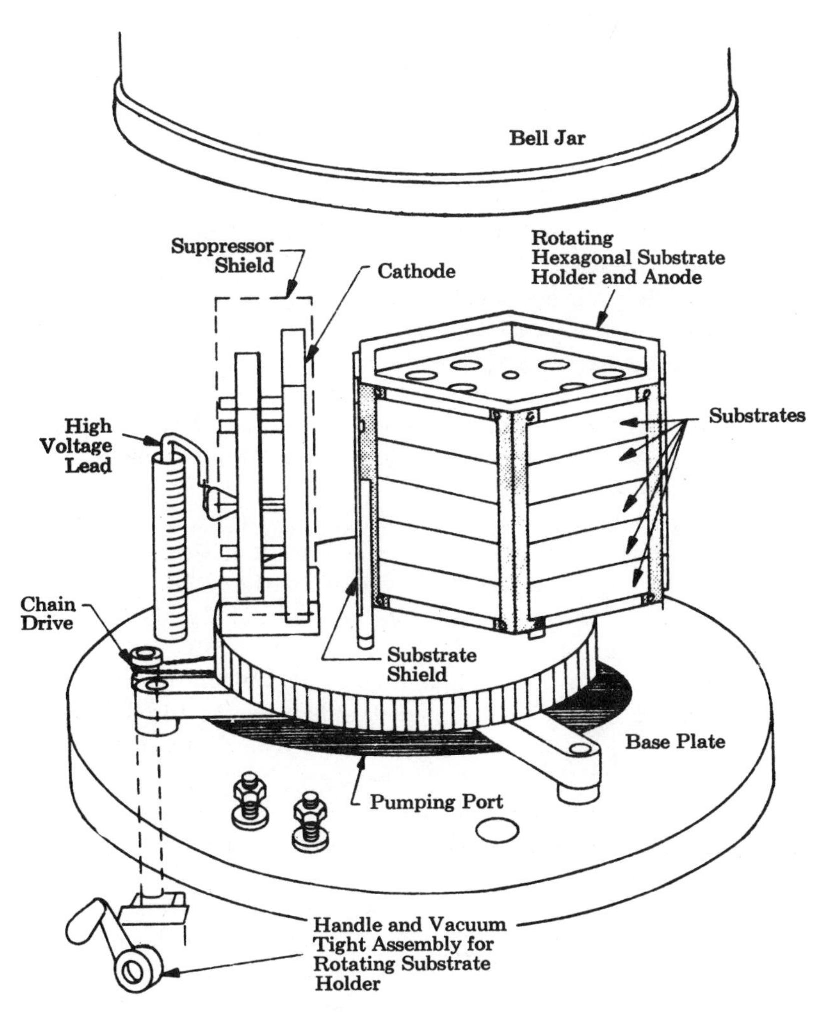

Fig. 2-19 Bell jar assembly of sputtering unit.

The rate at which atoms are liberated from the cathode, which is a measure of the rate of deposition of the sputtered film, depends upon a number of factors including the following: pressure, temperature, and constitution of the discharge gas; the energy of the ions bombarding the cathode; the distance from the cathode to the receiving substrate — usually incorporated as the anode; the angle of incidence of the bombarding ions; and the temperature and constituency of the cathode. Conditions affecting the sputtering rate may be summarized as follows:

a) to a good approximation, it is linearly proportional to the number and energy of the bombarding ions;

b) it increases with the temperature of both the discharge gas and the cathode;

c) it increases with the atomic weight of the discharge gas; and

d) at low-discharge voltages ($\sim$100V) it increases with the angle of incidence of the bombarding ions.

The sputtering process can be used to deposit a wide variety of metals and is particularly useful for metals which are difficult to deposit by thermal vaporization; these include tantalum, platinum, molybdenum, and rhenium. Semiconductor materials have also been deposited by sputtering. Typical conditions for the sputtering of tantalum, include the use of argon as the discharge gas at a pressure of 5×10^{-2} torr, and an applied voltage of 1 to 2 kilovolts.

By adding small amounts of other gases to the argon, the composition can be modified or changed. This is known as ''reactive sputtering''. By using the appropriate cathode material and added gases, it is possible to form films of oxides, nitrides, carbides, etc. The result of adding nitrogen to the argon, in sputtering Ta, is of special interest. The resulting film is a nitride of tantalum which is superior to pure Ta as a thin film resistor material. For a 1000 Å film, the sheet resistance is increased by a factor of 5 to 25 ohms, and the temperature coefficient of resistance changes from over 1000 PPM/°C to –60 PPM/°C.

A large automatic machine has been built for sputtering Ta on wide substrate areas. In this machine, clean substrates are inserted at one end at atmospheric pressure, are degassed, Ta film is sputtered on, and they are then returned to the atmosphere. The machine uses 11 chambers in an in-line configuration. The pressure in each chamber is maintained at the correct process value by an interrelated design of a vacuum pump (one for each chamber), and carefully designed leakage paths through the openings at the ends of each chamber. The substrate is locked in a carrier which is carried on a track through the 11 chambers.

The Ta film, made in this machine, may be of pure Ta or its nitride. The production capability of the machine is 50 square inches per minute of 1200 Å thick Ta film.

Insulated films may also be sputtered on a silicon wafer. Using an RF energy source, an insulator surface is bombarded alternately with ions and electrons. The ions eject molecules from the insulating materials which are then able to diffuse to the substrate where they are deposited as a thin film. The electrons neutralize the positive charge which would otherwise build up on the insulator surface. In one arrangement, a dielectric plate is used as the target for RF sputtering. The RF potential is applied to a metal electrode positioned behind the dielectric plate. A grounded metal shield is placed at the back of the electrode to extinguish the glow on that side and prevent sputtering of the electrode. Insulating films may be made in this manner from quartz, boron nitride, alumina mulite, and a variety of glasses. The method is applicable for depositing a number of layers, each having different properties to suit different purposes. Thus, multi-layers may be deposited taking into account such factors as matching thermal expansion coefficients, dielectric constants, and moisture permeabilities.

2.2.2.1 Hot Cathode Sputtering

Triode, hot cathode, or low energy sputtering differs from cathodic or reactive sputtering, in that the processes of ionization of the gas and bombardment of the material to be sputtered, are separated from each other. A filament serves as a heat furnace which, when energized, emits electrons to ionize the sputtering gas—Fig. 2-20. An ion target or source material is introduced as a third element and, with a negative potential, attracts the positive ions in the argon gas. The use of the thermionic cathode permits a glow discharge to be maintained at pressures as low as 1×10^{-4} torr, or approximately 2 orders of magnitude below that of normal cathodic sputtering. The method permits higher deposition rates when compared with cathodic sputtering, probably due to the larger mean-free paths accounting for less backscattering of the sputtered atoms at the lower pressure, with an increase in the atoms arriving at the substrate. In addition, triode-sputtering may minimize structural and chemical changes induced in films grown in a glow discharge environment.

2.2.3 GENERATING METAL PATTERNS

Due to the available processing steps of etching, film deposition, and photolithography, it is possible to devise numerous processes for generating patterns of Ta films applicable to resistors, capacitors, terminals and conductors.

Mechanical masks can be used during the sputtering process to define the substrate area to be coated with Ta, but this is not a recommended method, since the sputtered metal edges are not well defined. A better procedure consists of coating the entire substrate with an appropriate film of Ta by sputtering, and then coating the Ta with a positive photoresist. Using the artwork and camera techniques described earlier, photographic plates are prepared on which the Ta film devices appear as high-density areas. The resist is exposed through these plates and processed. This leaves the Ta in the device areas protected by the fixed resist. The rest of the Ta film is exposed, and can be etched by an appropriate etchant. Since the preferred etchants use hydrofluoric acid, they will attack the substrate unless great care is used. This can be avoided if a protective film of 100 to 300Å of Ta_2O_5 is formed on the substrate as a first step.

In a third process, a film of easily etchable metal is first deposited by evaporation or sputtering. A good example of this is copper. By using photoresist technology and etching, the copper is removed from the substrate area where the Ta is to be retained, and the Ta film is sputtered over the entire substrate. The substrate is then again exposed to the copper etchant which removes the copper beneath the undesired Ta and allows it to float off.

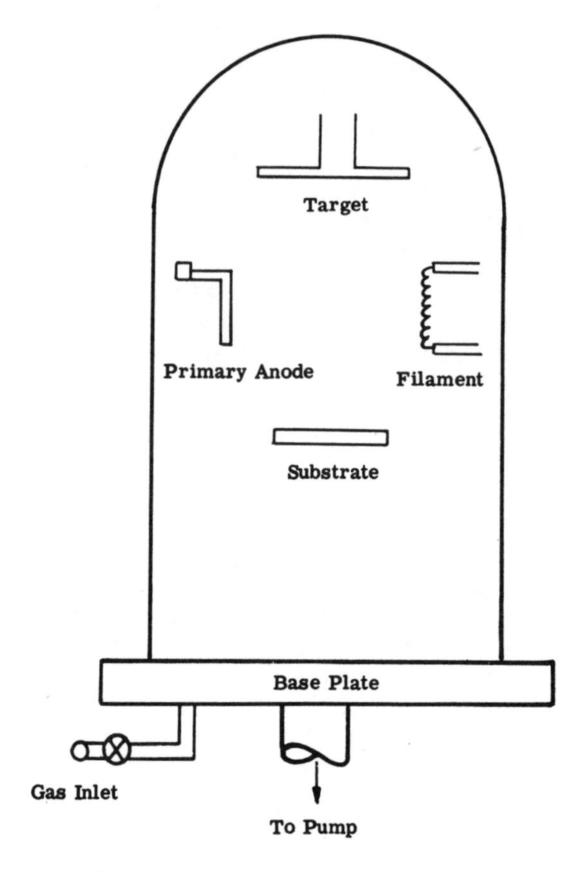

Fig. 2-20 Hot cathode, low pressure sputtering

2.2.4 ANODIZED TANTALUM THIN FILM CIRCUITS

Resistors, capacitors, and inductors made by anodizing a metal film are structurally the same as in the vacuum deposition case. A resistor is a thin metal film of appropriate length, width, and thickness, but it is now covered by a film of its own oxide. A capacitor is a sandwich of dielectric between two conductive electrodes. The dielectric is the oxide of one conductive electrode.

The starting point for a resistor is a Ta film of appropriate length and width but thicker than necessary. Its thickness is reduced by anodization to the point where its sheet resistance equals the design value. The result is a metal film resistor of Ta, or its nitride, covered by a film of Ta_2O_5 which is an excellent protective layer. Because of its superior properties as a resistance material, the nitride is used for this purpose. If high precision is necessary, the resistor can be anodized, baked in air at 250°C for 5 hours to stabilize the films, and finally trim anodized to its final value. Monitoring during anodization is also possible. Since the anodization process uses dc current, monitoring is possible if an ac signal is used.

To make a capacitor, an extra thick film of pure Ta is deposited. The selection of dielectric thickness and area to be used depends upon the operating voltage and capacity. A figure of merit for Ta capacitors is 5 μf-volts/cm^2; the field in the dielectric should not exceed 5 x 10^6 volts/cm to avoid breakdown.

Since the dielectric constant of Ta_2O_5 is 26, the capacity/cm² is given by

$$C = \frac{0.09 \times 26}{t} 10^{-12} = \frac{2.34}{KV\ 10^{-8}} 10^{-12} \text{ farads}$$

$$CV = \frac{2.34 \times 10^{-12}}{20 \times 10^{-8}} = 11.7\ \mu f - \text{volts}$$

where K = 20 Å/volt in Eq. (2-19). In practice, the maximum operating voltage should not exceed one-half the anodizing voltage, so that the CV product should not exceed 11.7/2 = 5.8 μf-volts. Some typical values are:

Ta_2O_5 film thickness = 2000 Å
Capacity = 0.03 μf
Area = 0.3 cm²
Dissipation factor = 0.01
Breakdown voltage = 100 volts = 5×10^6 volts/cm
Leakage current at 75 volts = 1×10^{-9} amperes.

When the required oxide thickness has been determined, the anodizing voltage is calculated using Eq. (2-19), with the value of K being usually about 20 Å/volt. The exact value of K depends upon the details of the process.

After anodization has been completed, the second electrode (often of gold) is deposited by evaporation in a vacuum chamber. To minimize the dissipation factor (maximize the capacitor Q factor), the sheet resistance of the electrodes must be low. For this reason, they are as thick as can be made, and tantalum nitride is not used. It has also been found that the oxides made from tantalum nitride have significantly lower values of dielectric constant and breakdown voltage.

2.2.5 ANODIZED TITANIUM THIN FILM CIRCUITS

The anodization technology has been described from the viewpoint of tantalum oxide since most applications do use Ta, but titanium is also being used. The oxide in this case is TiO_2. The titanium anodization process is, in its essentials, the same as the Ta process, but a few differences do exist in practice.

Commonly, the Ti film is formed on a ceramic substrate using a fused salt process. The substrate is placed in a carbon boat, coated with a layer of alkali metal halide salt, and a sheet of pure Ti is placed on the salt. The combination is placed in a furnace at 600°C, using normal atmosphere. At this temperature the salt melts, and the Ti dissolves in the molten salt. The Ti is deposited out on the ceramic substrate through a disproportionation reaction. The film is of high purity and suitable for anodization. The properties of the film depend upon the details of the reaction such as time, temperature, and salt composition. The sheet resistance can be made to vary from 1 to 50 ohms ±5 percent. Ti films can be formed by vacuum evaporation, sputtering in argon, and reactive sputtering in argon plus nitrogen and argon plus oxygen.

2.3 THIN FILM CIRCUIT FABRICATION BY ETCHING

This fabrication technique for making thin film microelectronic circuits, has advantages over the vacuum deposition method in the form of simplified equipment and processes, and low-cost equipment.

To make resistors, an etching step always forms part of the sequence of steps. In one procedure, a film of copper is deposited over the entire substrate by sputtering or evaporation. By using photolithographic techniques, holes are etched in the copper where the resistors are to be deposited. A film of resistive material is then deposited on the substrate having the appropriate sheet resistance. The resistive material must be such that it is not affected by the etchants used to dissolve the copper. The substrate is next exposed to a copper etchant which dissolves the copper and floats away the resistive material it supports. The resistive material deposited through the openings in the copper film, remains in place and forms the resistors.

In a second procedure of this type, a film of chromium, of the appropriate sheet resistance, acts as the resistor film. A gold film is deposited thereon, thick enough to form good conductors. Using photolithographic methods, the gold is etched away except on those areas where the resistors and conductors are to be located. The exposed chromium is removed by etching. Finally, the gold is etched away from the chromium rectangles which will form the resistors. The gold is allowed to remain, where required, to form conductors.

Capacitors can be deposited using the vacuum deposition or anodization process.

2.4 SILK SCREENING

Silk screening is a technique that has been borrowed from the graphic arts industry. It is used in microelectronic thin film fabrication when close component tolerances are not required.

A screen made of silk or woven metal is tautly mounted on a rigid frame. The screen is coated with a photosensitive emulsion or resist which is exposed through a photographic mask and developed. This leaves the screen clear where the thin film is to be deposited, and blocked elsewhere by the fixed emulsion. The screen is placed on the circuit substrate and carefully aligned. A viscous suspension of the thin film material and the constituents of glass, is squeegeed through the open areas of the screen to deposit a layer of the suspension onto the substrate. The unit is fired in a furnace to drive off liquids, and to form the glass and bond it to the substrate. The resistive metals are entrained in the glass which acts as a matrix. Some of the materials which may be applied by screening and firing are carbon, gold-palladium, silver-palladium, gold-platinum, and tin-antimony. A typical suspension is a mixture of powders made of 24 percent palladium, 52 percent lead oxide, and 24 percent silver, suspended in an organic binder. Sheet resistance values up to 10,000 ohms may readily be obtained thereby.

This process has been used to deposit resistors, capacitors and conductors. The process does not lend itself to accurate control of the sheet resistance or aspect ratio, but the fired part can be trimmed to a particular value. Overall registration for large areas (about 12 inches x 12 inches) is approximately 5 mil. Line edge raggedness is about 1 mil peak-to-valley.

2.5 VAPOR PLATING

The deposition of thin films through thermal degradation (pyrolysis), by reduction of vapors, or by chemical combination of volatile substances is termed vapor plating. The process consists essentially of passing a volatile compound, of the material to be deposited, over a substrate where the compound may be decomposed, reduced, or synthesized. In contrast to

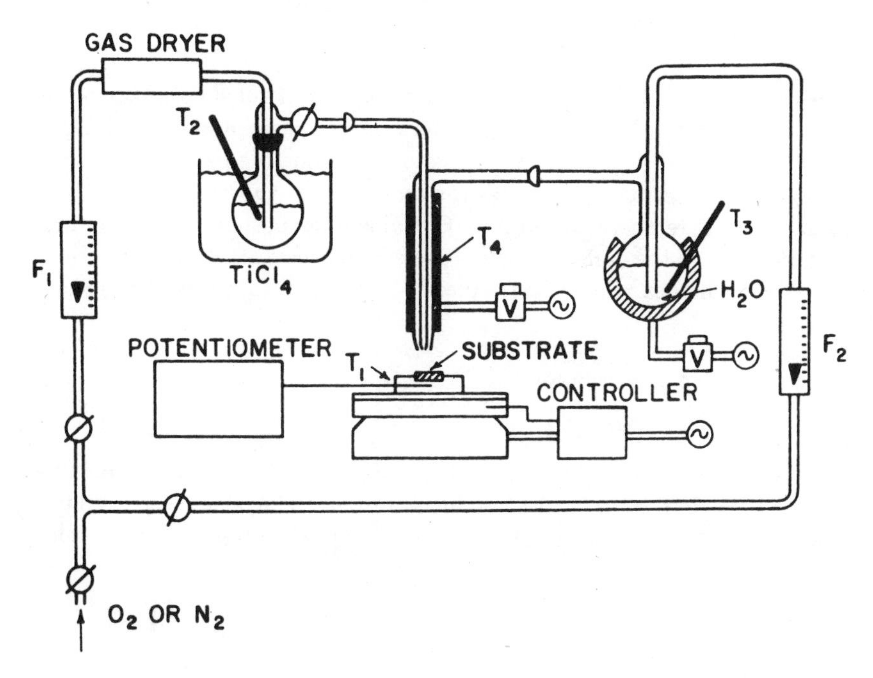

Fig. 2-21 **Titanium dioxide film deposition apparatus. (F = flow meter, T = thermocouple, and V = variable voltage transformer)**

vacuum evaporation, sputtering, and anodization, vapor plating requires a heated substrate to catalyze the reaction.

The nucleation and growth of films from the vapor phase are accomplished by various techniques with the design of the apparatus, type of construction materials, geometry of the reaction chamber, and mode of operation playing a part in affecting the properties of the deposited material. Other parameters which affect the final film in a complex pattern include rate of flow of the carrier gas, the purity of the system, the effect of substrate preparation, temperature gradient across the substrate, and deposition temperature and time.

In general, carrier gases are mixed with reagent vapors. The vapors are then reacted near the substrate to form the film. Reaction rates are basically controlled by carrier gas flow rates and temperatures of the reagents. Substrate temperatures vary with the process, encompassing a range from room temperature up to 1200°C. For thin films, compatible substrate temperatures in the range of 300°C are preferable.

2.5.1 TITANIUM DIOXIDE DEPOSITION

A controlled chemical vapor reaction technique has been developed for depositing titanium dioxide films on both metal and semiconductor substrates—Fig. 2-21. The process involves the synthesis of TiO_2 by the hydrolysis of titanium tetrachloride; the reaction proceeding as follows:

$$TiCl_4 + H_2O \rightarrow TiOCl_2 + 2HCl$$
$$TiOCl_2 + H_2O \rightarrow TiO_2 + 2HCl.$$

The rate at which the reagents are generated is controlled both by monitoring the flow rate of the chemically inert carrier gases, and controlling the temperature of the reagent reservoirs to determine the partial vapor pressures of the reagents. To catalyze the reaction, the substrate is held at the relatively low temperature of 150°C. Capacitors prepared with titanium dioxide films as the dielectric, yield dielectric constants up to 82, with the dielectric constant relatively frequency-insensitive into the gc range.

2.5.2 GLASS DEPOSITION

Vapor plating techniques have been developed for preparation of thin metal oxide glass films for use as capacitor dielectrics, electrical insulators, and encapsulants. A listing of several such types of glasses is given below:

Dielectric	Approximate Formula	Source Materials
Alumina	Al_2O_3	Tri-Ethyl-Aluminum + O_2
Alumina Silica	Al_2O_3 SiO_2	Tri-Ethyl Aluminum + Tetra-Ethyl-Orthosilicate + O_2
Alumina-Boro-Silica	Al_2O_3 SiO_2 B_2O_3	Tri-Ethyl Aluminum + Tetra-Ethyl-Orthosilicate + Tri-Isopropyl-Borate + O_2
Silica	SiO_2	Silane (SiH_4) + O_2

Approximate deposition temperatures at the heated substrates are in the range of 350°C. Typical apparatus for the vapor plating of such glasses is shown in Fig. 2-22. An evaluation of the above dielectrics on both MOS (metal-oxide-silicon) and MOM (metal-oxide-metal) structures indicates best electrical characteristics, for the alumina-boro-silica glass dielectric, based on a figure of merit involving capacitance per area, breakdown dc voltage, and conductance.

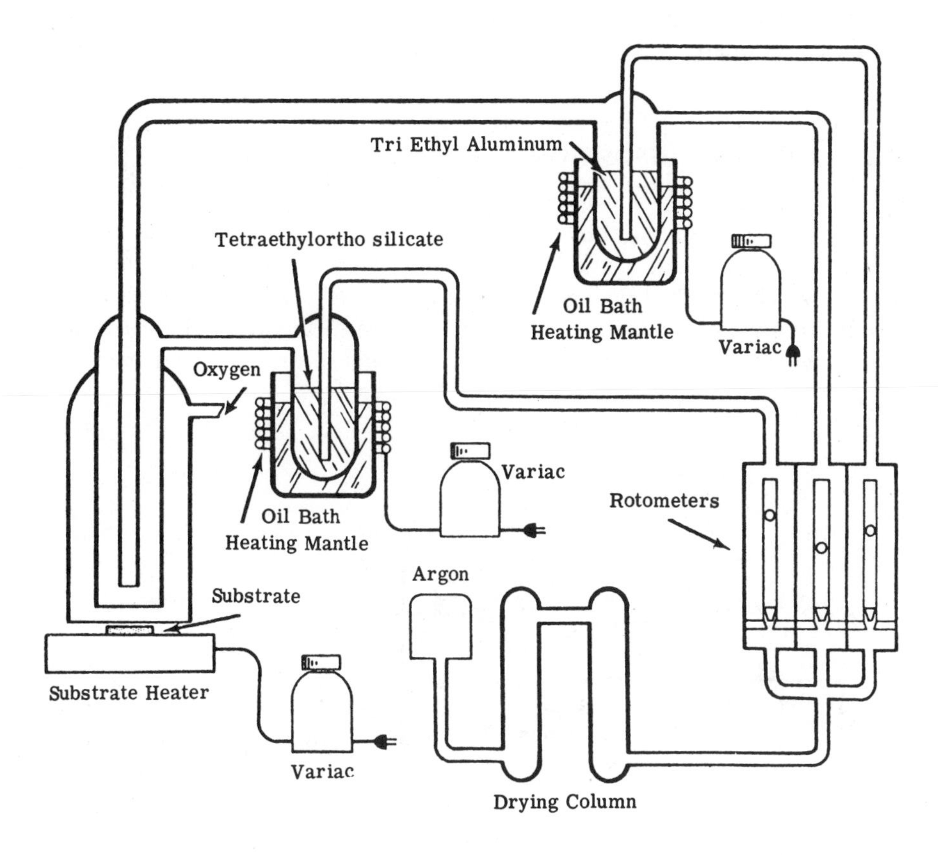

Fig. 2-22 Apparatus for reactive vapor plating of Al_2O_3 - SiO_2 films

2.6 PLASMIONIC ANODIZATION

In this process, the electrolyte is an oxygen plasma maintained by a glow discharge serving as the oxygen reservoir.

By subjecting a metal to the oxygen plasma atmosphere, relatively thin oxide films (25Å) can be grown with a thickness vs time relationship which resembles that for thermal oxide formation, except that the terminal thickness is greater. Application of a positive bias to the metal film permits the growth of an oxide film on the metal over a set period of time-Fig. 2-23. In this case, the mechanism of film growth is essentially similar to electrochemical anodization in that the oxide growth is a function of the anodizing voltage. A growth rate of 22 Å/per volt may be obtained for aluminum oxide prepared by plasmionic anodization. Process variables such as the oxygen pressure (20 to 100 microns) and applied voltage (10-20 volts) have been found to affect the time for the aluminum oxide film formation.

There are apparent advantages to be obtained by this process. In contrast to the electrochemical process, plasma anodization will permit the anodization of metals, which may be readily dissolved by the commonly used electrolytes, and will also permit the heating of the substrate to relatively high temperatures during the oxide growth.

In addition to aluminum oxide, thin-film capacitors using dielectrics of tantalum pentoxide, lanthanum titanate, and hafnium oxide may be prepared by this process.

2.7 ION BEAM DEPOSITION

In this technique the thin films are produced by generating a focused beam of ions, of the depositant, and writing the thin film circuit onto the substrate. This method does not require the use of evaporation masks, it results in films of greater purity than those resulting from the conventional evaporation methods, it allows fabrication of smaller components,

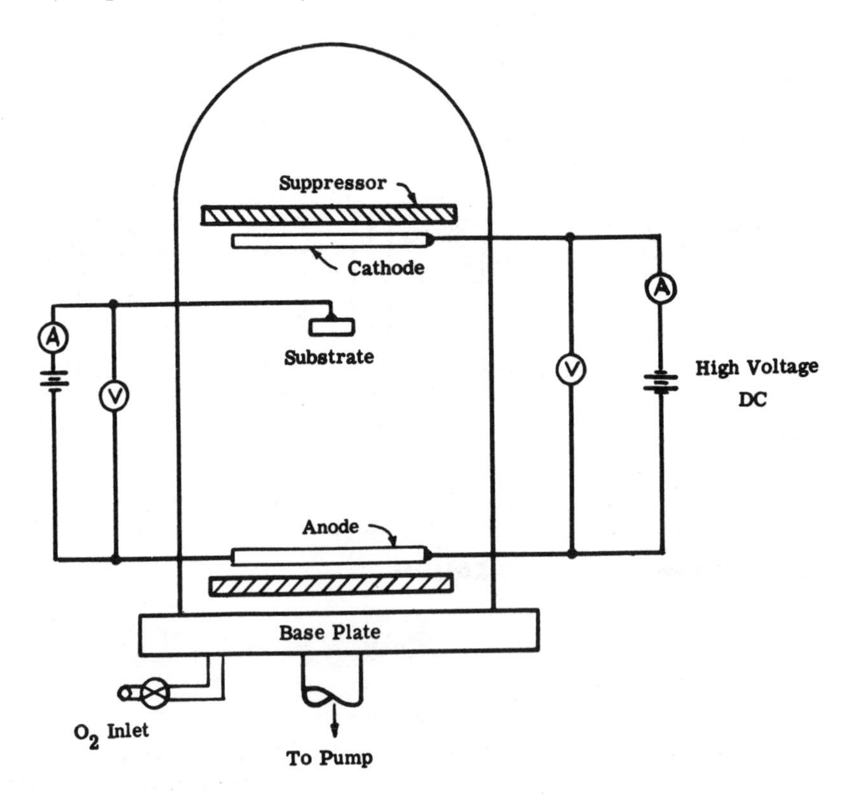

Fig. 2-23 Plasma anodizing system

and finally it is capable of making thin film circuits on an automated basis.

The concept is illustrated in Fig. 2-24. A conventional vapor source generates a beam of neutral atoms, of the depositant, passing into the region labeled "ion plasma". A beam of electrons is accelerated into this same region and ionizes the vapor. A magnetic field produced by an appropriately shaped electromagnet, confines the positive ions within the ion plasma region. A plate with a small aperture and at a large negative potential, is adjacent to the positive ion plasma. The ions are extracted from the plasma, and pass through a mass analyzer which allows only the desired ions to pass through and be focused onto the substrate. Since the substrate is an insulator, the positive charge of the ion beam is neutralized by flooding the substrate with electrons from an adjacent source. The ion beam must be slowed down or decelerated since an energetic beam will cause sputtering of the substrate. This is accomplished by a grid at appropriate potential.

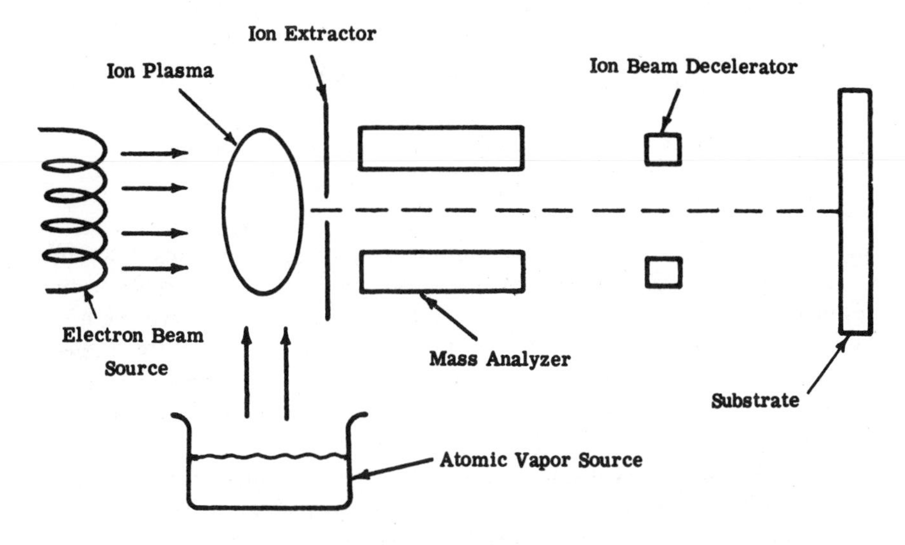

Fig. 2-24 Arrangement for generating a beam of ions and depositing them on a substrate

Beam diameters of one micron are feasible, and accordingly, very small resistors may be made. Since the ion beam can be made to pass through a deflection system which can be a combination of electric and magnetic fields, the thin film stripes can be written on the substrate. By appropriate automatic control of all the factors, the deposition of the thin films can be automated.

2.8 SILICON THIN FILMS

These represent an attempt to combine the good characteristics of both the semiconductor and thin film technologies.

The starting substance is a sapphire (Al_2O_3) crystal with an appropriate face optically polished. Onto this face is formed a film of Si crystal using epitaxial techniques. The latter is described under semiconductor devices. The deposition temperature is 1200°C. Silicon crystals, of the appropriate orientation, may be grown on a number of faces of the sapphire crystal. The Si films may be grown using hydrogen and silicon tetrachloride, and also with silane (SiH_4). The Silane reaction is simple— it decomposes to deposit Si and release H_2. The Si films can be grown to

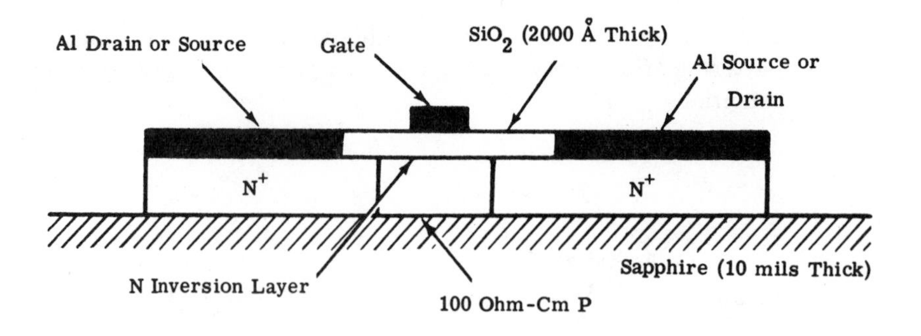

Fig. 2-25 Structure of silicon thin film field effect transistor

any desired thickness from a fraction of a micron up to many microns.

The first step in fabricating the components of the microelectronic circuit, is to remove all excess Si and to leave Si areas which will be used to make the components. Similar to the case of the diffused Si microelectronic circuit, the complete circuit is small, and therefore many circuits can be situated on a single sapphire substrate. Using photoresist as a stop-off medium and a Si etch, the excess Si is removed.

An active device resulting from this process is the insulated gate field effect transistor. Since this is a thin film majority carrier device, it is relatively insensitive to radiation. The structure is shown in Fig. 2-25. The starting point is an epitaxial 100 ohm-cm P-film. N^+ regions are diffused into the ends of the rectangular Si area. It should be noted that the junctions are normal to the substrate plane, and not parallel as in the diffused semiconductor technology. When the oxide is grown after the N^+ diffusions, an inversion layer is formed in the P section. The current path is from the N^+ source, through the N inversion layer, and into the N^+ drain. The voltage potential on the gate produces the field which modulates the surface states in the inversion layer and, therefore, the current flow.

Resistors have been made by diffusion into the Si film. Since the diffusion depth is held to one micron, the sheet resistance is 10^4 to 10^5 ohms.

High quality single crystal films of germanium may be obtained by using an electron beam to produce a small molten region, and moving this repetitively over the film area. Germanium films are prepared by normal evaporation techniques on sapphire substrates, and then heated to approximately 850°C (their melting point is 950°C). A small zone of this film is further heated to above the melting temperature by means of an electron beam. This electron beam is scanned in a regular grid pattern. During this scanning process, small crystallites of germanium which are found at the edge of the molten zone grow in size. Certain preferred crystalline orientations propagate preferentially and, after repeated scans, dominate the morphology of the film. The crystalline structure of the sapphire does not contribute to this growth process. The crystalline nature of the films may be confirmed by both X-ray and electron diffraction studies. Hall effect measurements may be employed to obtain resistivity, carrier concentration, and carrier mobility parameters. A typical film has a resistivity of 0.7 ohm-cm, p-type with a carrier concentration of

$5 \times 10^{15}/cm^3$, and a mobility of 1900 ± 500 $cm^2/Vsec$. The mobility values suggest that the zone-melted film approaches the properties of bulk single crystals.

2.9 LEAD ATTACHMENT

Due to the nature and size of thin films, the attachment of leads to thin film microelectronic circuits presents a considerable problem.

The oldest, and probably still the most widely used method is that of soldering. Soldering directly to thin films without the use of an intermediate layer to improve the quality of the bond, may be accomplished using indium rich solders. The best alloys use small amounts of either silver or lead, and melt in the range 144 to 160° C. Successful joints have been made to thin films including Al, Cr, Cu, Au, Fe, Mg, Ni, Ag, Sn, Ti, and Zn. An important point to be noted is that fluxes were not used. The very low melting points of the indium solders, however, limit their usefulness.

In most cases lead-tin soft solders are used. Occasionally they include a small amount of other metals to improve specific characteristics. Since gold is a common material in microelectronic circuits or wires, the effect of gold on lead-tin solder is of interest. When present in amounts exceeding 5 percent, gold causes brittleness, while smaller amounts may be advantageous. A typical solder might be 60 percent tin, 37 percent lead, and 3 percent silver or gold.

In general, a pad or land must be prepared on the substrate containing enough metal to combine with the solder and adhering sufficiently well to the substrate. A common method uses a mixture of precious metal powder and glass frit suspended in a liquid vehicle. This suspension is applied to the land area by screening, spraying, or even brushing, and is then fired on. The result is a glass matrix in which is suspended the metal powder. The resulting film is about 5 millionth of an inch thick. In another solution to this problem, the pad is produced by evaporating a bilayer. The layer adjacent to the glass substrate is of a material which adheres well to the glass. The second layer, of the combination, is of a metal which combines well with soft solder. While a number of metal combinations have been tried, Mn-Ag appears to be best. An alloy ingot of 20 percent Mn and 80 percent Ag may be made by melting in a vacuum. The ingot is pressed into sheet form from which pieces of convenient size may be cut and evaporated from a Ta boat. The resulting film is pure Mn on the substrate. Standard solder may be used to attach Cu wires.

The equipment used to raise the temperature of the solder to its melting point is important. The most common device, for this purpose, is the soldering iron in miniature form or a modification thereof. The shape of the soldering tip, its size, cleanliness, and temperature control are all critical. The solder temperature and duration, while in the liquid phase, is important. The solder dissolves the noble metal in the land, and if allowed to do so sufficiently long, it may deplete the land to the extent that its adhesion to the substrate will be seriously impaired. This effect can be minimized by using a solder saturated with the whole metal. The temperature of the solder should be only a few degrees above melting, and the duration should be kept as short as possible.

A better method, for transferring heat to the joint assembly, is through the use of infrared radiation or hot gas. The advantages are better control of the heat transfer to the work, ease of automation, short duration to complete a soldering operation, and the possibility for using inert or even reducing gases. It is best to use assembly jigs which hold the substrate and lead wire in place, and to apply the solder as a preform.

A more detailed description of the techniques and apparatus used to attach leads in microelectronic circuits, may be found in Chapter XII.

2.10 SUMMARY

1. The circuit or systems design engineer who desires to utilize thin films in any form of hybrid integrated circuits should be familiar with the thin-film fabrication processes, and be aware that film characteristics are normally controlled to a considerable degree by the specific processing procedure.

2. For vacuum evaporation, various methods have been developed to heat the charge to the point where evaporation is effected. Resistive materials such as nichrome, chromium-silicon monoxide, and one dielectric material—silicon monoxide—have gained relatively wide acceptance as materials suitable for fabricating thin-film passive parts. A problem associated with vacuum evaporation is that of depositing capacitor dielectrics with relatively high dielectric constants. This is due to the limitations of materials with high dielectric constants and, in addition, to the fact that those charges that are suitable in bulk form, dissociate at high temperatures and low pressures in the vacuum chamber.

3. Sputtering has been utilized mainly to deposit refractory materials such as tantalum. The capability of the process has been enhanced by adding reactive gases to the sputtering gas, resulting in the deposition of the metal-oxide mixtures. Stable resistive films of tantalum nitride formed by sputtering tantalum in a nitrogen atmosphere (sheet resistivities of 10-100 ohms per square) may be formed with characteristics equivalent to the best metal film resistors. Likewise, sputtering of tantalum in the presence of oxygen permits the fabrication of thin-film resistors and capacitors. Further refinements and modifications will, no doubt, result in the capability to deposit purer metal films by triode sputtering. In addition, the method of depositing insulating films by RF sputtering may lead to the development of means of depositing dielectric films with relatively high dielectric constants.

4. The anodization method has the advantage of allowing a single metal system to supply all the passive parts of a thin-film circuit. Tantalum nitride resistors can be adjusted to exact values by anodizing. Capacitors are fabricated by anodizing sputtered tantalum and evaporating a metal counterelectrode over the oxide layer. The use of such a process demands that all materials be relatively unaffected when subjected to the anodic solutions.

5. The growth of films by vapor plating is not a new technique in so far as thin-film deposition, per se, is concerned. However, its use specifi-

cally for the preparation of complete thin-film passive circuits has been relatively limited. Vapor plating has found wide usage commercially for the epitaxial growth of both silicon and germanium for transistor and integrated circuit fabrication. The technique appears to have its greatest potential in the thin-film field as a means of preparing glass films for use as dielectrics or as protective coatings.

6. Most films follow the same growth pattern. The exact thickness at which the film's structure becomes continuous depends upon the material itself, the temperature of deposition, and the method of deposition. Likewise, each process used for film deposition imparts impurities and foreign matter which have related effects on the electrical behavior of the deposited film.

CHAPTER 3

CHARACTERISTICS OF THIN FILM MICROELECTRONIC CIRCUITS

3.1 RESISTORS

A resistor is fabricated by depositing a resistive material, as a thin film, in a predetermined geometrical pattern on an electrically insulating substrate. An example of a thin film resistor is shown in Figure 3-1. The

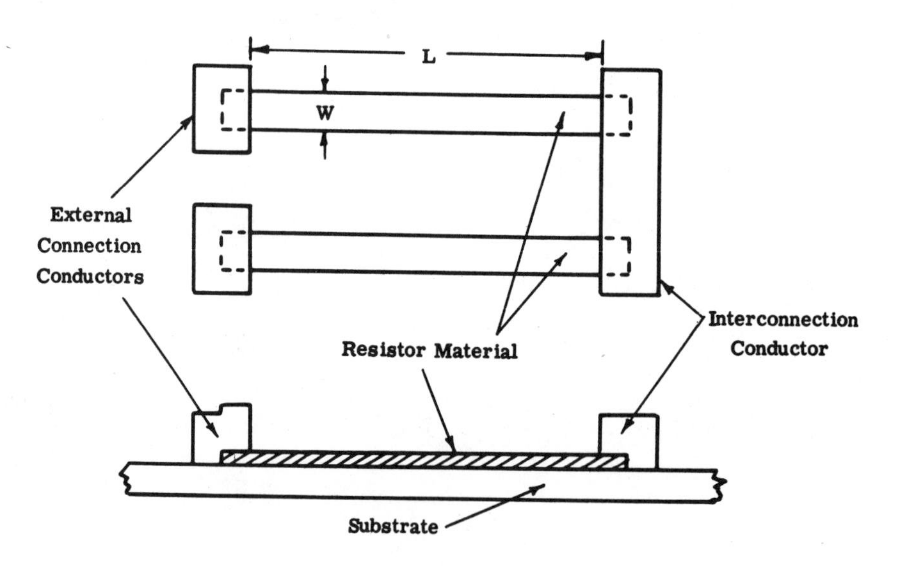

Fig. 3-1 Thin film resistor

resistor may have many legs, or one leg, depending upon its final value. It may, or may not, have any interconnecting conductors. The factor used to measure the resistive property of a thin film, is the ohms per square of the film. The ohms/square of a resistive film is the value obtained when the resistance of one square of the film is measured between opposite sides of the square as indicated in Figure 3-2.

The value of a resistor is determined by the ratio of its length to its width, times the ohms per square factor of the film or

$$R = \frac{L}{W} \times (\text{ohms/square}) \tag{3-1}$$

In order to fabricate a 10k resistor, for example, from a film of 400 ohms/sq., the length to width ratio, or aspect ratio, would be 25 to one.

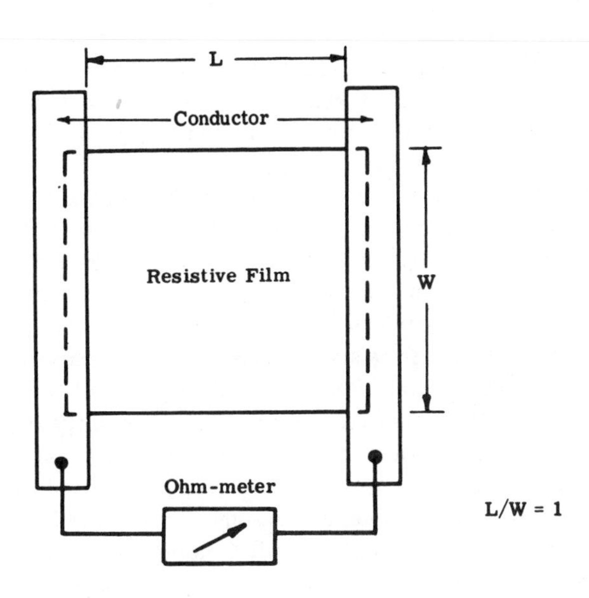

Fig. 3-2 Measurement of ohms/square

3.1.1 EVAPORATED THIN FILM RESISTORS

The fabrication of thin film resistors is a critical process. The properties of the resultant devices depend upon the materials and procedures used, including:

a) The substrate material
b) The cleaning procedure used on the substrate
c) Handling of substrate between cleaning and pump down
d) Substrate processing in the evaporator, high temperature bake-out or glow discharge ion bombardment
e) Type of vapor source, evaporation temperature, deposition rate, and, in the case of alloys, the effect on film composition
f) Substrate temperature and gradients
g) Pressure and constitution of residual gases in the vacuum evaporator
h) Thermal annealing procedures
i) Protective evaporated films.

The resistors used in thin film microelectronic circuits are of two types: films of pure metal or metallic alloys; and a mixture of metal and insulator.

Examples of the first type are chromium, tantalum, nickel-chromium, and of the second, chromium-silicon monoxide (cermet). While it is theoretically possible to produce films having large values of resistivity by making them very thin, such films are, in fact, not useful. Thus, they are difficult to control in production, are unstable during use since they undergo irreversible changes with temperature, and have usually large thermal coefficients. In practice, therefore, the films have sheet resistances of the order of 100-200 ohms/square, while cermets have been used to about 1000 ohms/square. Relatively few materials have found extensive use in making thin film resistors.

Electron conduction through metals, alloys, and combinations of other basic materials, is well documented in the literature. However, the conduction of electrons in a thin film is quite different from that found in the same material in bulk form. This difference, in terms of resistance to electron flow, is most important in the study of thin film resistors for electronic use.

Resistance in a metal conductor exists because of two types of scattering of the conduction electrons. The first is scattering caused by the existence of imperfections created by thermal motions of the ions, and has a reversible dependence on temperature. The resistivity associated with this scattering is known as the "ideal resistivity." The second is scattering caused by lattice defects such as missing atoms, impurity atoms, stacking faults, and interstitials. The resistivity associated with this scattering is known as the "residual resistivity." However, a third type of scattering occurs in the metal when one dimension, the thickness of the conductor, is comparable to or less than, the mean free path of the conduction electron. This thickness limitation exists to some degree in all thin film resistors.

Variations in the structure will change the residual resistivity characteristics, and thickness variations will contribute to changes in the total resistivity. Efforts to control the characteristics of thin film resistors must, therefore, be primarily concerned with these two basic parameters—structure and thickness.

Consider the initial formation of a thin film on a smooth substrate, during a vacuum deposition process. The first atoms to arrive at the condensing surface, experience an energy exchange with the atoms of the surface. As a result, some of them are captured and nucleation is initiated. Subsequent atoms arriving from the vapor stream, strike both the substrate surface and some of the condensed atoms. If these atoms on the surface possess sufficient mobility, they move about on the substrate and group together to form separated island structures. As more atoms condense, the island structures grow larger, and finally they touch each other to form a continuous film. If the substrate surface is not perfectly smooth, the surface mobility is impeded by atoms striking a non-normal plane, and the energy exchange at the surface is affected. The result, in this case, is that a thicker film is required to achieve the initial continuity of the film. This additional thickness depends on the surface roughness of the substrate.

The temperature of the substrate during the film formation will affect both the energy exchange and the surface mobility of atoms. When the substrate temperature is varied, the mechanical structure of the film is changed. Lower temperatures tend to freeze the atoms in place, and usually produce a highly porous film with many imperfections. These films have a higher resistance, for a given thickness, and are usually very unstable. Higher temperatures, however, increase both the surface mobility and the film density. At the same time, such higher temperatures cause some imperfections to disappear. These films have a lower resistance, for a given thickness, and are more stable. If very high temperatures are used on the substrate, the energy exchange can cause the reverse effect whereby the arriving atoms gain enough energy to re-evaporate. The result of this condition is that nucleation will not occur, and a film cannot be formed. An optimum substrate temperature must, therefore, be applied for the particular substrate being used, and for the particular type of vapor atom striking the condensing surface.

When a metal is heated in a vacuum, the atoms in the vapor stream have kinetic energy which depends on the source temperature. If the evaporation source temperature is further increased, the evaporation process is accelerated and the kinetic energy of the vapor atoms is increased. Lowering the pressure in the vacuum chamber has a similar effect. Thus, the energy exchange phenomenon at the condensing surface is a function of both, source temperature and vacuum pressure. These two parameters should therefore be controlled, to optimize certain thin film resistor characteristics.

Impurity atoms contribute to the resistance of thin films. The most obvious source of impurities is the evaporant material, and it may contain impurities by choice, in order to produce a certain characteristic. It is often not possible to get materials without some degree of impurity. Another source of impurity in thin film resistors, is the chemisorption in the film, during formation, from residual gaseous atoms in the vacuum chamber. Carbon dioxide, oxygen, and water vapor are among the most common residual gases found in vacuum systems. These, as well as other gases, are quite active chemically with many of the evaporant materials used in the formation of thin film circuits. Generation of oxides is the most frequently encountered effect during and after film formation. Some control can be exercised over these impurities which affect the resistor characteristics. Increasing the evaporation rate, for example, always reduces the number of residual gas atoms absorbed during film formation. However, very high evaporation rates increase the mechanical faults in a film. This is because successively arriving atoms at the condensing surface cover those already there, before they can move to a minimum stress location or preferred site. It is apparent, therefore, that to achieve a desired characteristic in a resistor, requires a judicious selection of vacuum pressure, rate of evaporation, and substrate temperature.

Some materials used for thin film resistors experience surface oxidation and consequent resistance increase when exposed to the atmosphere, upon removal from the vacuum chamber. It is a general practice to evaporate a protective film of some other material–insulating in nature–on top of the resistance film, prior to removal from the evaporator. Silicon monoxide is a satisfactory material for many resistor materials.

Postdeposition exposure of the resistance film to temperatures greater than those used during film formation, is a desirable method for reducing the density of imperfections in the film structure. During such a heat treatment process, in the absence of an oxidizing atmosphere, the resistance decreases, while the film density increases.

The temperature coefficient of resistance, of a perfectly continuous metal film, is always positive. In thin films, the temperature coefficient of resistance is most frequently negative. When positive, it is at least an order of magnitude less than that of the bulk material. The temperature coefficient of resistance, of a thin film resistor, can be considered as a "figure of merit" for evaluation purposes. For a fixed resistance value, the temperature coefficient of resistance does not always need to be constant. By adjusting or optimizing certain process parameters, a range of temperature coefficients can be associated with a given resistance value. As the thickness decreases, the temperature coefficient of resistance becomes more negative, and both resistor stability and magnitude tolerance degrade.

Only single-element, vacuum-evaporated metal films have been considered above, in order to establish a basic understanding of thin film resistors. Films prepared from metal alloys and metal oxides, however, also exhibit desirable resistance characteristics. The resistance characteristics of bulk metal alloys and metal oxides can usually be achieved in thin film resistors. These characteristics may include: high corrosion resistance; a combination of high optical transmission with low resistance values; or a combination of magnetic and resistance characteristics.

A method which may be used to achieve high resistivity without going to very thin films, which are basically unstable, is to combine a metal and a ceramic. Such a combination may be as chromium and silicon monoxide or silicon dioxide, in a homogeneous amorphous film structure. This film is called a "cermet." Other suitable materials for high value resistors are in the form of nitrides and silicides.

Characteristics of thin film resistors may be placed into categories similar to those applicable to conventional resistors, with a few exceptions. Specification of size, lead length, pull strength, etc., no longer apply. The following list characterizes the thin film resistor parameters which are of primary interest:

a) sheet resistance value for each material
b) tolerance on resistance magnitude
c) temperature coefficients of resistance
d) power rating
e) operating temperature
f) stability
g) noise characteristics
h) substrate material and surface condition.

In order to describe the characteristics of various resistance materials, assume a set of specific geometric conditions. Therefore, let the available substrate area be 0.500 inch x 0.800 inch. For each resistor, let the maximum length L be 0.800 inch, the minimum width W be 0.020 inch, and the spacing between resistors be 0.020 inch. Consequently, the maximum number of resistors of this width and spacing, in a 0.500 inch width of substrate surface, is 12. The aspect ratio for each resistor is 40, and the total number of squares in the pattern is 480. The largest resistor value which can be considered, for the geometry defined above, may then be calculated for various resistance materials, provided the maximum value of sheet resistance is known for each material.

The ability of a thin film resistor to dissipate power is a function of many variables. Two possible situations may arise: the temperature of the resistor may be increased so that the value of the resistor exceeds its permissible tolerance; or, the temperature of neighboring semiconductor devices, such as attached transistors and diodes or deposited thin film triodes, may be raised to the point where they become inoperative.

The factors to be considered in making a thermal analysis of a substrate with thin film components on it are: substrate dimensions and material; heat sinking of potting compounds; substrate assembly structure; container; number of substrates; ambient temperature; resistor dimensions and location on substrate; effect on temperature distribution of other components such as conductors and capacitors.

If a single resistor is located on a substrate with given heat sink characteristics and material, the temperature distribution on the substrate depends on the factors mentioned above. If two resistors are on the substrate, the total substrate temperature is approximately the sum of the distributions produced by each resistor acting alone.

From these general remarks, it is apparent that it is not possible to prescribe rules for the power dissipation of thin film resistors. This accounts for the wide range of power dissipation values which can be found in the current literature. If a resistor power dissipation figure in watts/area is available, it is possible to derive the necessary resistor dimensions.

$$P/A = I^2R/A = (I^2R_S/LW)(L/W) = I^2R_S/W^2 \qquad (3\text{-}2)$$

where

P/A is the power dissipated/unit area.

I is the current through the resistance R of length L, width W and sheet resistance R_S.

Since R_S, I, and P/A are known, W can be computed. L may be finally obtained from

$$R = R_SL/W.$$

The power dissipation capabilities of Ta resistors are greater than those of evaporated resistors. The upper operating temperature limit for many resistor materials can conservatively be rated at 300° C. Under special conditions such materials as chromium and rhenium will perform satisfactorily in air at 500° C.

The noise characteristics of thin film resistors are related to their stability under load, their temperature coefficient of resistance characteristics, and their shelf storage stability. Current noise is greatest in carbon-composition resistors and relatively less in deposited carbon resistors, metal film resistors, and metal oxide resistors, in that order. A discrete noise level can be selected which divides the stability or temperature coefficient of resistance into either normal or abnormal groups. This becomes a useful tool for production quality control, as well as a research evaluation technique, when applied to thin film resistors.

Protective coating materials for various types of resistive films must be chemically compatible with the films. Other important considerations are porosity, continuity, abrasion resistance, compatibility of temperature coefficients of expansion, high temperature limitations and moisture absorption. By these criteria, silicon monoxide is found to be a suitable protective film for most materials. However, there are some exceptions, and tungsten is perhaps the most outstanding. When tungsten film resistors are coated with SiO using a vacuum evaporation technique, an oxygen exchange is believed to take place at the tungsten-silicon monoxide interface, so that the SiO is converted to SiO_2 in molecular layers next to the tungsten. Presumably, any oxygen which combined with the tungsten during deposition is transferred; this is observed as a sharp decrease in resistance for the first few seconds of SiO deposition.

3.1.2 NICKEL-CHROMIUM RESISTORS

Probably the most common metal used for film resistors is a nickel-chromium alloy. The evaporant is generally, an alloy made for use as a heater wire, and commonly called "Nichrome." This name is a trade mark of the Driver-Harris Company. Nickel/chromium alloys have three advantages which account for their wide use. They adhere well to the substrate; they have small thermal coefficients; and they can be used at large values of sheet resistance.

The film is invariably evaporated from a commercial alloy. The film composition depends, therefore, upon the composition of the alloy and the source temperature. The theory of alloy evaporation is discussed in the section on cermet resistors. The substrate temperature during deposition should be about 300° C to assure the formation of good films. This high substrate temperature produces good diffusion of the components on the surface during deposition. In this manner the film is homogeneous with respect to composition.

The simplest Nichrome source is a resistance-heated wire. Unfortunately, the rate of deposition, at the highest permissible operating temperature, is too low to be useful as a routine procedure. The alloy wire can also be wrapped around a tungsten wire and evaporated from the wire in the 1400-1500° C range. The nickel in the alloy, however, dissolves the tungsten so that filament life is short and the film composition is affected. It is necessary to use heavy tungsten wires (1/16 inch) if this source is used.

A method based on an electron bombardment source may also be used. With this device a conductive hearth of Al_2O_3 plus molybdenum is connected as the anode, and bombarded by electrons from a tungsten wire emitter. The Nichrome wire is fed to the hearth until a molten ball is formed. This ball is then evaporated. The evaporation rate as well as the film composition are approximately linear with temperature.

Finally, the alloy can be evaporated from a ceramic crucible using RF induction heating. By using very large sources, the film composition is kept constant. By varying the alloy composition and temperature, the film composition can be varied. Vaporization temperatures are in the range 1425 to 1525° C.

3.1.3 CHROMIUM RESISTORS

This metal is often used as a thin film resistor material. Its principal disadvantage when compared to Ni/Cr alloys is higher values of the temperature coefficient of resistance. It does, however, have the advantage that the problems of variable Ni/Cr ratios, which are a nuisance in the case of Nichrome, are not present.

The vapor source problems are not too serious. Though Cr is not available as a wire it can be electroplated onto tungsten wire and vaporized off the wire. It has been vaporized by electron bombardment, and from lumps held in a tungsten spiral or conical cup of wire—though the vaporization is often erratic because of an oxide film on the surface. At the proper vaporization temperature, Cr sublimes, and therefore interactions with metal crucibles are not as severe as when it is vaporized from the molten state. Resistance-heated tantalum sources are often used. Alumina crucibles lined with a tungsten cup and heated by rf induction may also be used. As commonly known, the adhesion of Cr films to ceramic on glass surfaces is excellent. In fact, it is often used as a base layer with materials of poor adhesion, to assure their adhesion to the substrate.

For 100-300 ohms/square, the film thickness is about 200Å. Temperature coefficients of resistance are in the range of –50 to –100 PPM/° C, depending on the film thickness and its heat treatment.

3.1.4 EVAPORATED CERMET RESISTORS

The deposition of resistive thin films of metals or metallic alloys, has always resulted in films of useful sheet resistance values not over a few hundreds of ohms. An exception to this is when rhenium is used. However, the problem of developing a material, with an order of magnitude higher sheet resistance, predates any work with rhenium. A solution to this problem is to form resistive films which are mixtures of a metal and a dielectric, and to control the resistivity by varying the relative proportions of the metal and dielectric, as well as the film thickness.

While a number of different cermet systems can be deposited by evaporation in high vacuum only, one has found appreciable use in the form of a mixture of chromium and silicon monoxide. Since one of the constituents of the cermet is a compound, the source used is important, because it may decompose if the source temperature is too high. If the source temperature is too low, on the other hand, the rate of deposition will also be low and thus encourage oxidation of the chromium on the substrate.

To assure the formation of good cermet films, the substrate temperature should be about 250° C. The substrate has been found to affect the resistivity, the thermal coefficient, stability, and the change in resistance during the thermal annealing cycle.

Simultaneous deposition from separate sources for the Cr and the SiO may also be applied. This technique has the advantage of avoiding possible interactions between the two components at the high temperatures used. Accordingly, this permits controlling of the metal/dielectric ratio by individual rate control of the two sources.

Another method which may be used resides in the flash evaporation of a premixed powder mixture. This method has the advantage of conceptual simplicity, but incurs problems in the fabrication of a hopper and feed mechanism, for the powder mixture, which will operate reliably at elevated temperatures without causing segregation and rate variations.

The cermet film may also be deposited from a mixture of powdered Cr and SiO, in a single crucible, at the appropriate temperature. In this method the film composition for a given powder mixture is a function of the source temperature.

The formation of a cermet resistor is a two-step process. The film, of desired composition and thickness, is laid down, and the substrate is annealed at a temperature of approximately 400° C. The annealing is done in an air furnace, and the annealing schedule is a function of the desired sheet resistance and the actual value. A third step is sometimes used if

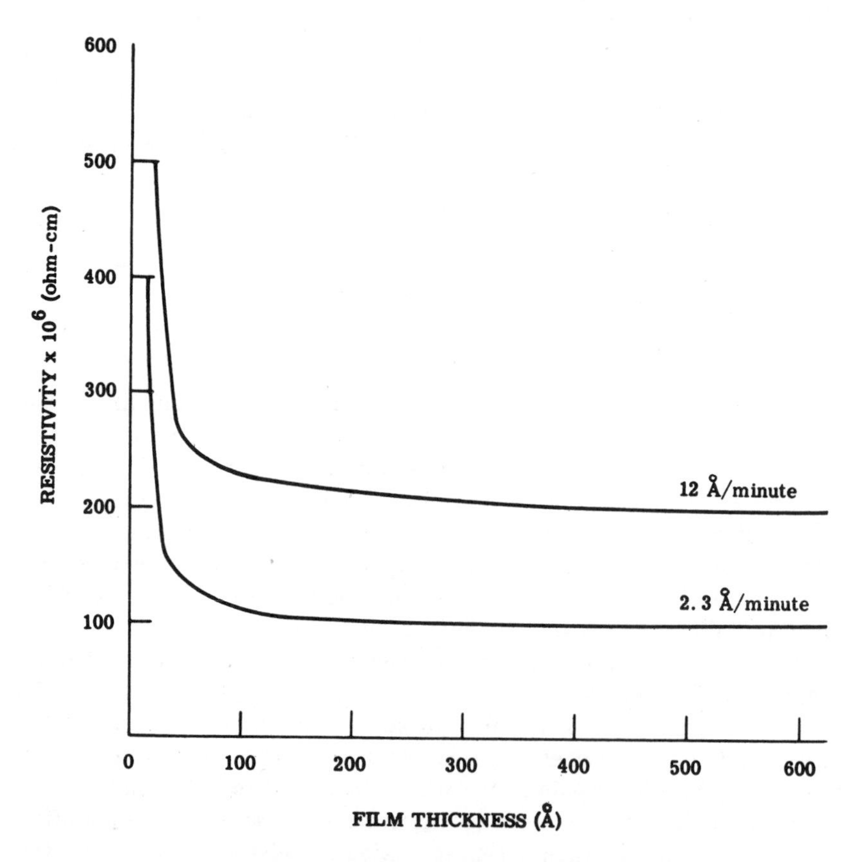

Fig. 3-3 Variation of rhenium resistivity with thickness and deposition rate

tolerances of a few percent are necessary. This consists of trimming the individual resistors by thermal pulsing. In this way the tolerance can be reduced to 1 percent, and tighter tolerances are possible if greater care is used.

By varying the Cr/SiO ratio, the sheet resistance may be varied.

Resistors have been made using very large values of sheet resistance. However, evaporated resistors, most often, use films which do not exceel 1000 ohms/square. For a given composition, the sheet resistance may be varied by varying the film thickness. With the available range of sheet resistance values, it is not too difficult to make one-megohm resistors. Temperature coefficients within the range of ±100 PPM/° C are common.

3.1.5 RHENIUM RESISTORS

The application of Cr and Ni/Cr is limited to the 100-300 ohms/ square range. Cermets, SiO/Cr, should probably not be used at sheet resistance values exceeding 1000 ohms. A disadvantage of cermets is the difficulty of depositing a well-controlled film. There is, therefore, a need for a pure metal resistor material, capable of being used at sheet resistance values up to about 10,000 ohm/square. Rhenium has been investigated, in this connection, with the following results: it is usable at high temperatures and therefore at high power levels; it has a low temperature coefficient of resistance; possesses excellent age stability; and exhibits low variation of of resistivity with film thickness, even for large values of sheet resistance.

Rhenium must be evaporated at about 3200° C to obtain useful rates of deposition. Re is available sufficiently pure to be used in an electron bombardment source, or vaporized with an electron gun. To obtain reliable films, the substrate must be maintained over 275° C during film deposition, and preferably around 450° C. A protective film (SiO) must be deposited immediately without dropping the temperature. Glazed ceramics, glass, and fused quartz substrates may be used.

Figure 3-3 illustrates the variation of sheet resistance with film thickness and rate of deposition. The deposition rate is important. Since the resistivity of the film deposited at the faster rate is higher, it appears that the structure rather than impurity content determines the resistivity. The TCR is zero in the 250-300 ohms range and increases to –350 PPM/° C for 10,000 ohms/square films. These data are for films on glass or quartz substrates maintained between 400 and 500° C, and allowed to anneal for 30 minutes at the deposition temperature.

3.1.6 ALUMINUM RESISTORS

The thin film resistor materials, customarily used, are not adapted to making small valued resistors of the order of 10 ohms. This is because the substrate area that would be required, would be very large. However, it appears that aluminum is a suitable material. It is undoubtedly the best known evaporant. Its adhesion to glass is good, the required substrate temperature during evaporation, about 100° C, is easy to maintain, and it will not affect other films. Furthermore, it is easy to evaporate.

Aluminum films thicker than 300Å have bulk resistivity. Therefore, a 1000Å film has a sheet resistivity of approximately 0.3 ohm. This is a convenient value for making small value resistors. The temperature coefficient is about 1500 PPM/° C.

3.1.7 ANODIZED TANTALUM RESISTORS

An extensive Ta resistor evaluation program has been conducted. The substrates used were glass, glazed alumina, and sapphire. All of the films used were reactively sputtered tantalum nitride. The following conclusions were drawn from the program:

a. Tantalum nitride resistors can be made to an initial tolerance of ±0.1 percent, with a 20-year stability if the value exceeds 100 ohms.
b. The substrates, in order of preference, are sapphire, glazed alumina, and glass.
c. Resistor quality is independent of film thickness, line width and anodizing voltage.
d. Sheet resistivity should be in the range 5 to 75 ohms.
e. The temperature coefficient for these nitride tantalum resistors is in the range of –50 to –100 PPM/° C.

Thus, while the useful sheet resistance value of 75 ohms is considerably less than that used with the evaporated resistive films, it is possible to use narrower lines. This is made possible by the fact that each resistor can easily be trimmed to the desired value through anodization, allowing, thereby, the use of larger values of aspect ratio on a given area.

3.1.8 ANODIZED TITANIUM RESISTORS

Titanium resistors may be made by the anodization process. One process for forming the basic Ti film is that of disproportionation from an alkali halide salt. The range of sheet resistance values in the region of interest, is from 1 to 2000 ohms/square. Control of time and temperature of the film deposition is used to control the sheet resistance in the range 1-50 ohms/square. Above 50 ohms/square anodic conversion is used to increase the sheet resistance. Tolerances of ±1 percent are possible through continuous monitoring during the anodization process. The commonly used value for the sheet resistance is 1000 ohms/square. At this value, the temperature coefficient of resistance is 50 PPM/° C. A voltage coefficient of 425 PPM/V is also present. Operation at 5 watt/in^2 for 1000 hours decreases the resistance by 0.3 percent.

Evaporated Ti films are also used to form Ti resistors by anodization. In this case, the electrolyte used is a mixture of ethylene glycol and a saturated aqueous solution of oxalic acid. The films used are of 100 ohms/square. The temperature coefficient is very small.

3.2 THIN FILM CONDUCTORS

A conductor is fabricated by depositing a conductive material, as a thin film, in a predetermined geometrical pattern on an electrically insulating substrate. The rules for fabricating resistors apply as well to conductors, except that the goal here is to obtain very low ohms/sq. values. Good conductor films should have values of 0.1 ohms/sq. to 0.01 ohms/sq.

Many types of conductive materials are used, the most common being aluminum, copper and gold. Deposition thicknesses from 4 μ inches to 100 μ inches are used. Potentially every conductor is a resistor. The electronic designer must therefore be careful in specifying the high conductance paths necessary in his design.

3.3 THIN FILM CAPACITORS

A capacitor is fabricated by first depositing a thin film conductor material on a suitable substrate. On top of this conducting film, a dielectric material is deposited, and then a conductor material is deposited on top of the dielectric, as the counter electrode. An example of a thin film capacitor is shown in Figure 3-4.

In a lossless capacitor the applied voltage and current are 90° out of phase. The impedance is $1/j\omega C$ or $-j/\omega C$. An actual capacitor does have losses due to the resistance of the leads, the plates, protective and supporting materials, and the finite resistance of the dielectric. A simple equivalent circuit is shown in Fig. 3-5 where R_L is the resistance of the leads and plates, and R_P represents all other losses.

The power dissipated in the capacitor is

$$\frac{v^2 \cos\theta}{(R^2 + X^2)^{1/2}} \tag{3-3}$$

where $X = 1/\omega C$

$R = R_L + R_P$

$\tan \theta = X/R$

$V =$ voltage across the device

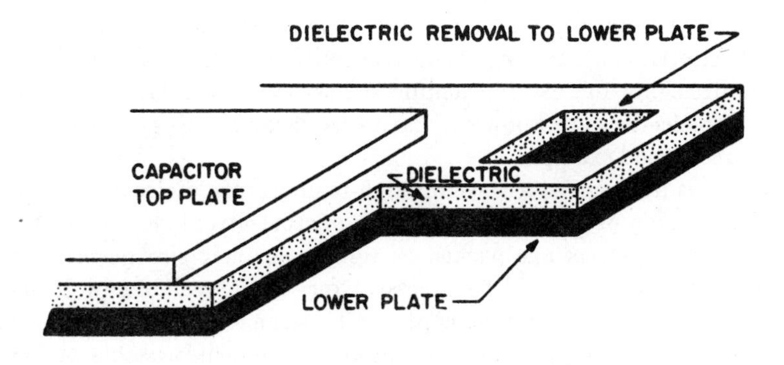

Fig. 3-4 Geometry of thin film capacitors

For a perfect capacitor $\theta = 90°$ so that this power vanishes. For an actual capacitor θ deviates from 90° by δ where $\tan \delta = \omega CR$. Tan δ is usually known as the "loss factor" or dissipation D, and since it is small for a useful device, $D = \delta = \omega CR$. If the usual definition for the quality factor Q is used, then $Q = 1/\delta$. Lead and electrode losses can be calculated from

$$\tan \delta_L = \omega CR_L \tag{3-4}$$

while all other losses can be calculated from

$$\tan \delta_P = \omega CR_P . \tag{3-5}$$

For a thin film capacitor, the capacitance is given by

$$C = 0.08842\ KA/t \text{ picofarads} \tag{3-6}$$

where

K = dielectric constant (air = 1)
A = capacitor area in cm^2
t = dielectric thickness in cm.

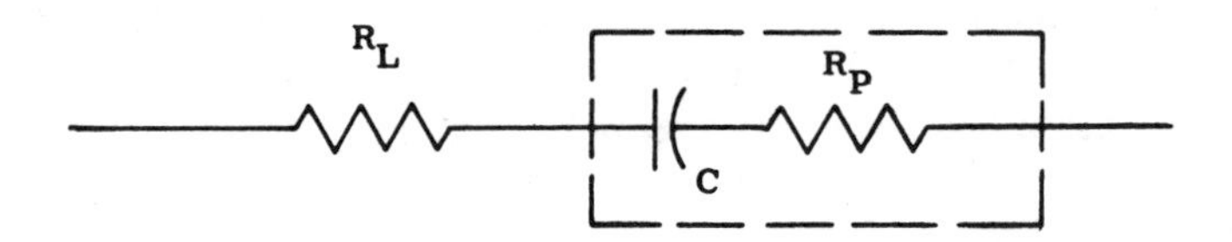

Fig. 3-5 Equivalent circuit for a thin film capacitor

Dielectric films cannot be made too thin for two principal reasons: (a) since the permissible field strength in the dielectric is limited (approximately 10^6 volts/cm), the film thickness for usable voltages cannot be too small; and (b) very thin films are not homogeneous and reliable. It is customary to use films within the range 0.1 to 1 micron in thickness. The available area is limited. To increase the capacity therefore, it is necessary to use either dielectrics having large dielectric constants, or multilayered devices. Dielectrics having large dielectric constants, and meeting also the other requirements for use in vacuum deposition, are relatively scarce. Certain ferroelectrics are known to have very large K values—the barium titanates are a good example.

Research is in progress to develop high-temperature dielectrics. If full advantage is taken of the small size of the microelectronic circuit so that large electronic systems are packed as tightly as their size permits, the operating temperature would be excessive for the materials now in use. The packing density is therefore, kept small, or some means for removing the excess heat must be provided. If, however, materials usable at high temperatures were available, high packing densities would be possible. Research has been performed with mixtures of neodymium and boron oxides, and dysprosium and boron oxide. The results indicate that dielectrics made from these mixtures may be usable up to 500° C.

Two principal techniques are used to make thin film capacitors—anodization and vacuum deposition. Since the properties of the resulting capacitors are different, they are described separately.

3.3.1 VACUUM EVAPORATED CAPACITORS

The dielectric most commonly used in making vacuum deposited capacitors is silicon monoxide (SiO). The nature of the depositant is important, since it affects the properties of the deposited film. SiO is made by reacting Si and SiO_2 in a vacuum. A mixture, of two parts of finely powdered SiO_2 and one part of finely powdered Si, is placed in one end of a quartz tube. The loaded end of the tube is placed in a 1250° C furnace, while the other end protrudes so that the temperature changes gradually from 1250° C to room temperature. The pressure within the tube is maintained at less than 10^{-4} torr. The reaction which occurs is

$$SiO_2 + Si \rightarrow 2\,SiO\,.$$

The monoxide, in vapor form, condenses in the cooler section of the tube. That portion which condenses at less than 800° C, is substantially SiO. Commercially available SiO is in the form of black lumps. Silicon monoxide finds extensive use in thin film circuitry not only as a capacitor dielectric, but also as a buffer film on the glass substrate, as a protective film over the

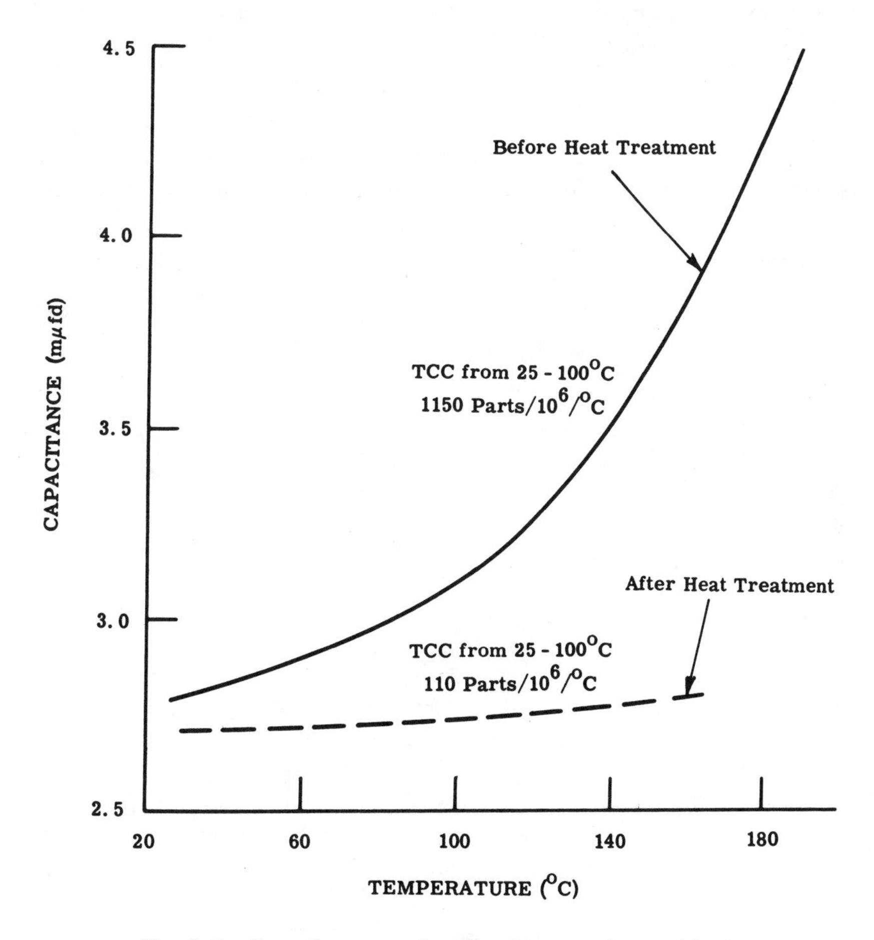

Fig. 3-6 Capacitance as a function of temperature and heat treatment

components, and as an insulator between crossing conductors.

The crucible used to vaporize SiO is usually made of sheet tantalum. At the vaporization temperatures used, the Ta does not contaminate the film. The source temperature should be about 1250° C. At higher temperatures the film tends towards SiO_2, while at lower temperatures the films are porous and of poor quality. The substrate should be held at approximately 250° C during deposition. The pressure in the evaporator should be less than 10^{-5} torr. The rate of film deposition should not be too slow, since the film tends to oxidize to SiO_2 with a lower dielectric constant. The rate should be greater than 10Å/sec, to reduce the total evaporation time.

The counter-electrodes should adhere well to glass and SiO, and also have a low value of sheet resistance to minimize losses. The materials, generally used, are aluminum or chromium plus copper.

Typical characteristics for SiO thin film capacitors are:

capacitance/unit area = 100 $\mu\mu F/mm^2$ or 0.6 $\mu f/in^2$
dielectric constant (SiO) = 6.0
refractive index at 5800Å = 2.1
dissipation factor (tan δ) at 1 kc and 25° C = 0.1 percent
temperature coefficient of capacity = 110 ppm/°C from 25°C to 100°C
breakdown potential at 25° C = 50 volts.

When using silicon monoxide, proper heat treatment of a thin film capacitor is important. Figure 3-6 illustrates capacitance as a function of

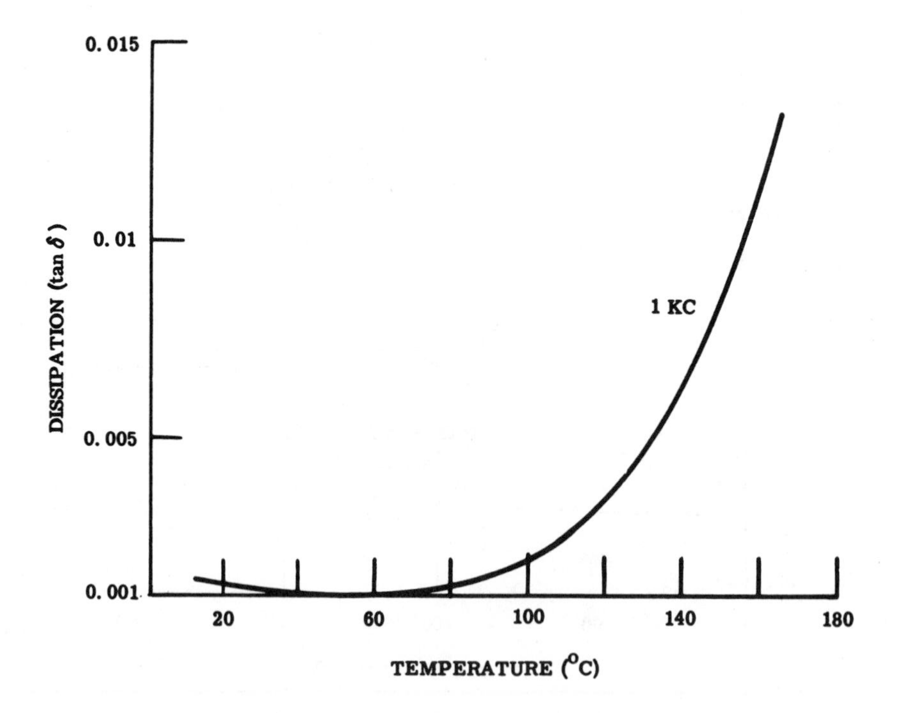

Fig. 3-7 Dissipation factor versus temperature for Al-SiO-Al capacitors

temperature, before and after subjection to a heat treatment cycle. The solid curve indicates that, prior to heat treatment, the temperature coefficient of capacitance (TCC) is very large and increases with temperature. After proper heat treatment, however, the TCC may be an order of magnitude smaller, and fairly constant over a large temperature range, as indicated by the dashed curve. Figure 3-7 shows the capacitor dissipation factor

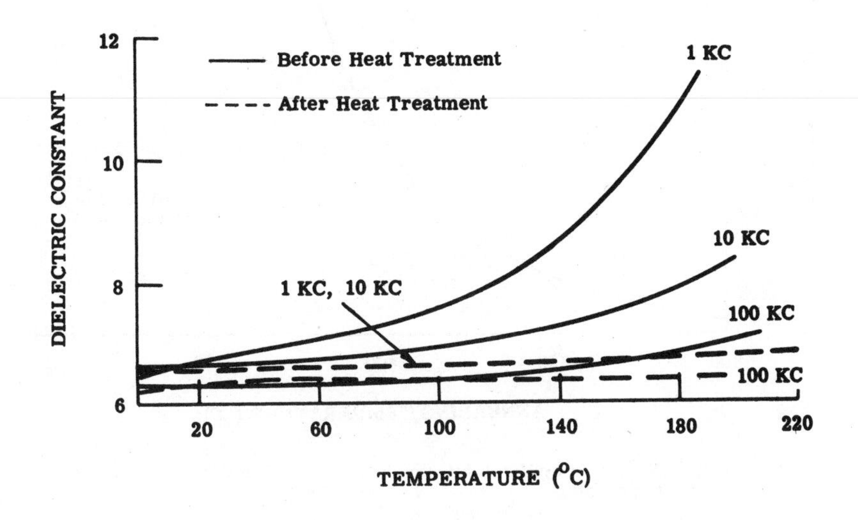

Fig. 3-8 Dielectric constant versus temperature for Al-SiO-Al capacitors as a function of frequency and heat treatment

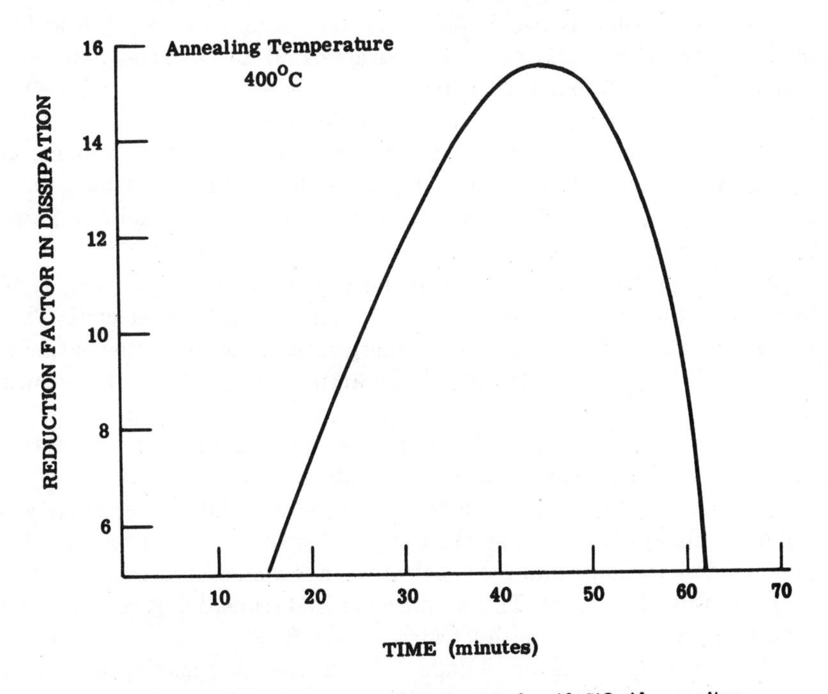

Fig. 3-9 Effect of annealing on dissipation for Al-SiO-Al capacitors

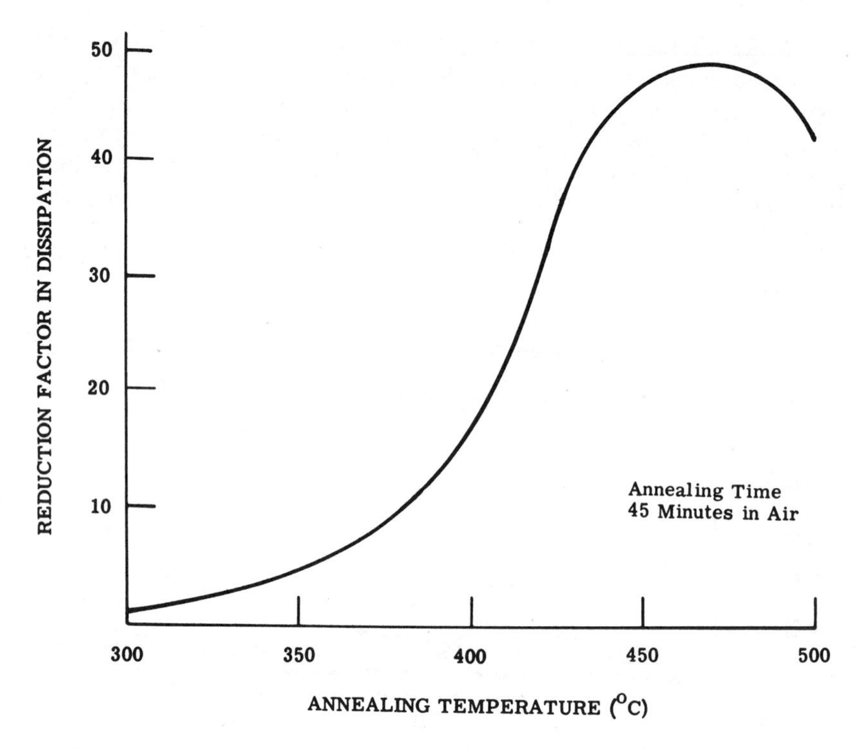

Fig. 3-10 Reduction factor in SiO dissipation as a function of annealing temperature

as a function of operating temperature, after subjection to heat treatment. The variation of the dielectric constant, before (solid lines) and after heat treatment (dashed lines), as a function of operating temperatures, for various frequencies, is depicted in Fig. 3-8. Substances exhibiting high temperature coefficients, are also characterized by a large dissipation factor. Figures 3-9 and 3-10 indicate the time-temperature cycles which yield the greatest reduction in the dissipation factor.

Typical capacitor leakage characteristics, after heat treatment, are shown in Fig. 3-11. It is of interest to note that at an operating point of 20 volts, the leakage current of a random selection of capacitors, is well below one microampere.

A problem often encountered in making thin film capacitors, is the shorting of the counter-electrodes due to pinholes in the dielectric. These shorts can often be burned out by discharging a capacitor connected across the plates. This vaporizes the metal film around the pinhole thus isolating it.

Another weak point in a thin film capacitor can occur if the edges of the films are sharp. The situation is illustrated in Fig. 3-12a. Thus, the contact to the upper counter-electrode may be unreliable, particularly if the lower plate is thicker than the upper. This situation can be corrected by separating the mask and the substrate by a few mils, thereby generating a shadow at the film edges. This allows a gradual transition from the upper plate to the contact lead as illustrated in Fig. 3-12b.

The permissible working voltage depends upon the maximum permissible field strength in the dielectric, and the leakage currents. In the case

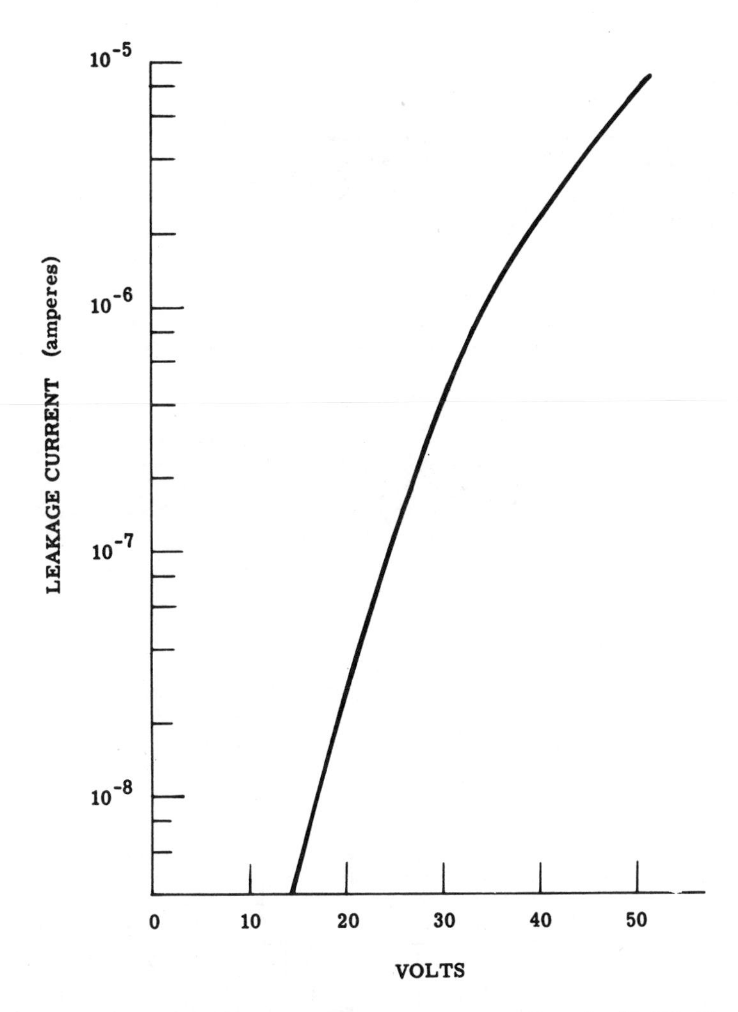

Fig. 3-11 Capacitor leakage current versus voltage for Al-SiO-Al capacitors

of SiO the field should, generally, not exceed 2-4 x 10^5 volts/cm. Thus, for a capacitor of 0.01 μfd/cm^2 the working voltage should not exceed about 20 volts. The interrelationship between capacity and working voltage is $CV = 0.2$. It is best not to use films thinner than 0.4μ, due to pinholes and inhomogeneities, or films thicker than one micron because of the possibility of fracturing in thick films due to mechanical stresses.

The dissipation factor can be unduly large unless counter-electrodes and leads, of high conductivity, are used. The effect of plate resistance on dissipation is illustrated in Fig. 3-13.

Leakage of evaporated capacitors may be reduced by using two evaporated dielectrics, SiO plus a 1000Å film of Al_2O_3, M_gF_2, or SiO_2. If, for example, SiO_2 is used, the leakage at 40 volts and 100° C may be reduced by a factor of 1000. The total capacity of the bilayer is calculated as two capacitors in series.

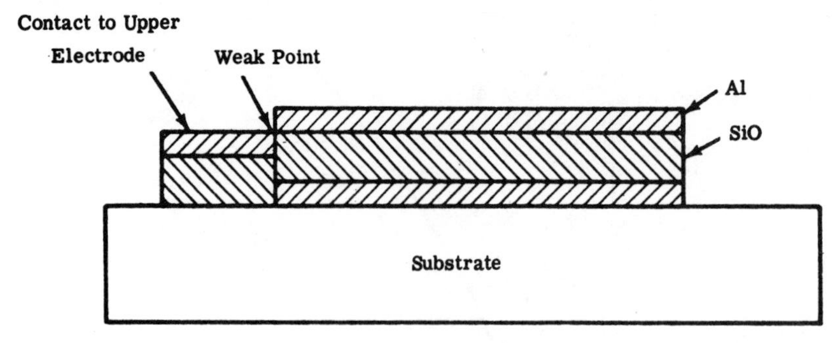

(a) Weak point due to sharp film edges

Al
SiO
Substrate

(b) Weak point eliminated by allowing edges to be shadowed

Fig. 3-12 Effect of sharp film edges on contact to upper capacitor electrodes

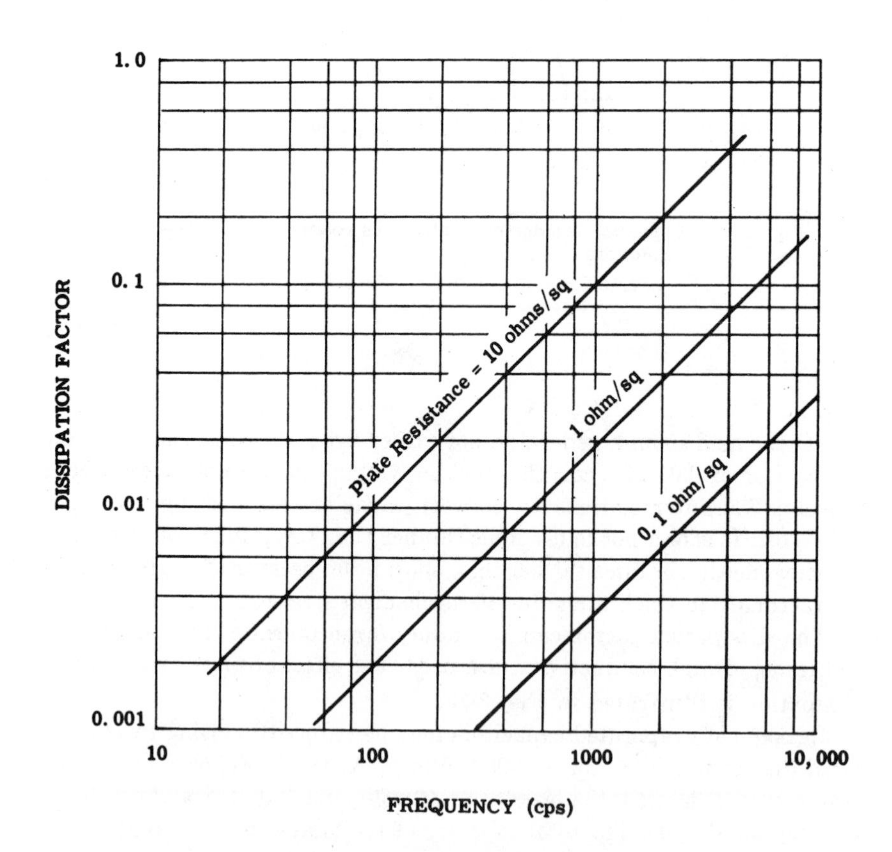

Fig. 3-13 Effect of capacitor plate resistance on dissipation factor

3.3.2 ANODIZED CAPACITORS

Tantalum pentoxide film capacitors have been widely used. These capacitors have a layer of sputtered tantalum, part of which is anodized to tantalum oxide. The oxide layer is then coated with a film of conductive material, such as gold, to form the upper counter-electrode. The fabrication techniques have been described in the preceding chapter. Figure 3-14 illustrates the structure used. Due to the high dielectric constant of tantalum oxide, this method yields capacitors which occupy much smaller areas than the equivalent silicon monoxide capacitors.

One of the limitations of tantalum capacitors, is their poor performance at frequencies above 10 kc. This is due to the high series resistance of the tantalum film, which causes a rapid increase in the dissipation factor. To overcome this limitation, either one of two methods may be applied: use either very thick Ta films of small sheet resistance (5000Å), or use a conductive film of aluminum as an underlay for the sputtered tantalum.

As described in Chapter II, the anodized capacitor can be trimmed to the desired magnitude by monitoring its value during the anodization process. The choice of the second counter-electrode requires a compromise between operating voltage and adhesion. Metals which have good adhesion allow low operating voltages. Examples of these are Ta and Al. Metals with poor adhesion on the other hand, have large values of breakdown voltages. Examples of the latter are gold and palladium. One recommended practice is to use gold or palladium, and to carefully refrain from abrading the surfaces. It is considered that the adherent films set up strains in the oxide and open up fissures through which they can short to the other plate. The metals with poor adhesion, however, tend to agglomerate and are therefore less likely to form conductive paths through fissures or pinholes in the oxide. The effect of operating temperature on the capacity and dissipation factor is shown in Fig. 3-15. A useful design guide for anodized tantalum

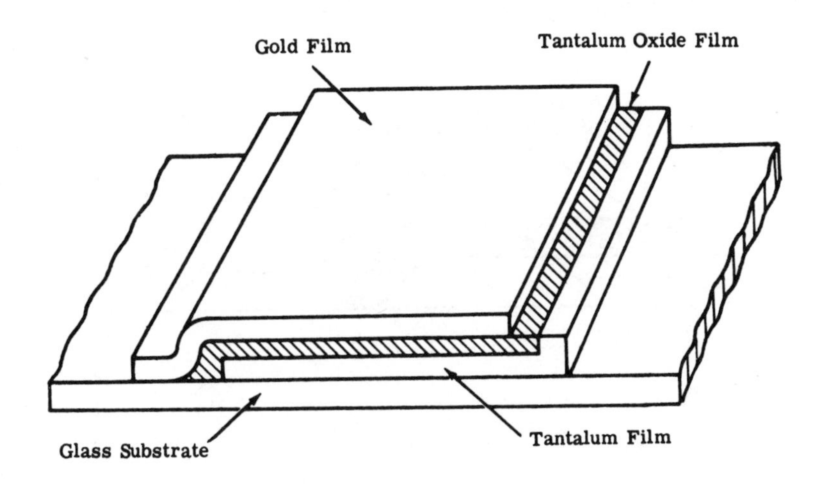

Fig. 3-14 Tantalum capacitor design

capacitors is the relation: maximum operating voltage $\times$ capacity/$cm^2 = 5$.

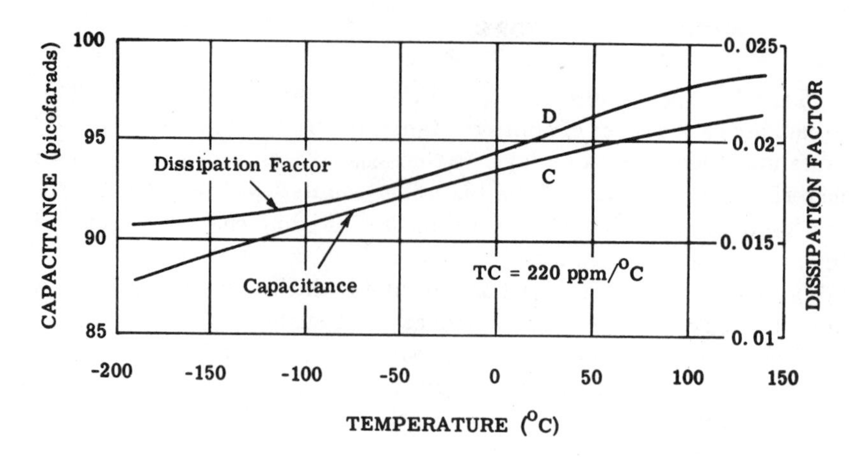

Fig. 3-15 Capacitance and dissipation factor versus temperature for a tantalum capacitor

Bulk Ta capacitors are known to be polarity sensitive, but thin film Ta capacitors have very little such polarization sensitivity. There is, however, some polarization which is found to depend upon the materials used, and the details of the fabrication process. A typical leakage current characteristic, for a Ta capacitor in the forward and reverse directions, is shown in Fig. 3-16.

Anodized aluminum capacitors have been used for high frequency applications. The processing technology used, is essentially the same as that used with Ta. The behavior of these devices as a function of temperature, is shown in Fig. 3-17. Since the dielectric constant of alumina is about 10, twice as much area is needed for anodized Al as for anodized Ta. The temperature coefficients of capacitance are similar, but the dissipation for the aluminum oxide is substantially less than that for the tantalum. Anodized titanium (TiO_2) capacitors are also being made.

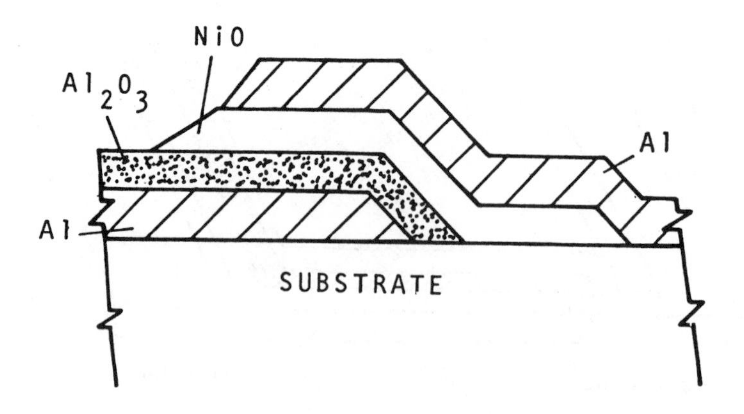

Fig. 3-15.1

Cross Section of 10μF/in^2 Thin-Film Capacitor

The thin film capacitor shown in cross section in Fig.3-15.1 is fabricated in four steps that include evaporation of the aluminum base electrode, thermal oxidation of the aluminum, reactive sputtering of nickel in pure oxygen, and evaporation of the aluminum top electrode. The thermal oxidation step is a critical one, since it forms the capacitor dielectric, which is an Al_2O_3 film less than 100Å thick. The relationship between the Al_2O_3 film thickness and the oxidation rate is a reproducible one which has not been explained. There may be a change in grain size or surface roughness during the aluminum deposition that affects the oxidation rate. The capacitors are usually formed by oxidizing a 1 μm aluminum film in dry oxygen at 500°C for three minutes. The nickel oxide film, which is approximately 1000Å thick, is formed by dc sputtering of nickel for 15 minutes at 50-millitorr oxygen pressure and a cathode potential of 2.5 kV

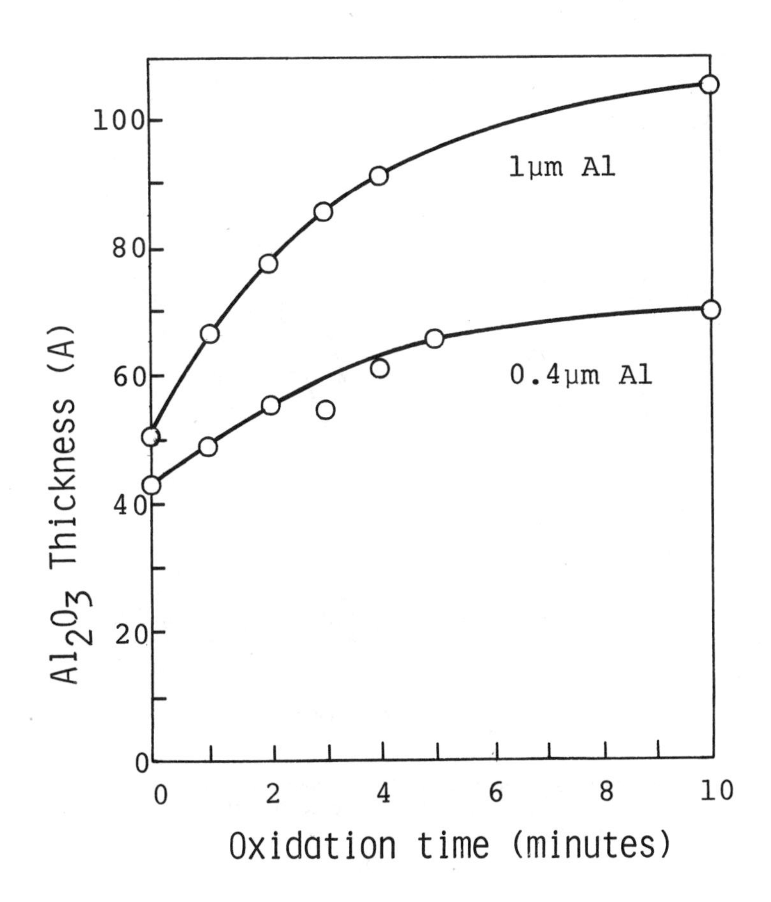

3.4 THIN FILM INDUCTORS

In the thin film art inductors up to about 5 microhenrys in value and with relatively low Q values, may be made by forming a conductive planar spiral. Figure 3-18 illustrates a square spiral inductor. This inductor, however, is an inefficient device. As a result of the high coil resistance and large stray magnetic fields, the Q value is low. Furthermore, the stray fields can give rise to undesirable coupling effects. Generally, the designer avoids the use of inductors in microelectronic circuits operating at frequencies below about 30 mcs.

In the spiral configuration of Fig. 3-18, $D = 5D_1$ for optimum Q. It is usual to make $p = q$, and p is often about one mil. If $D_1 = 0$ the inductance L is

$$L = 8.5 \times 10^{-3} DN^{5/3} \text{ microhenrys} \tag{3-7}$$

where the number of turns $N = D/2(q + p)$. The total spiral length $d = 2DN$. The resonant frequency f_o is given by (assuming the substrate dielectric constant $= 4$)

$$f_o = 7.5 \times 10^9 p/D^2 \tag{3-8}$$

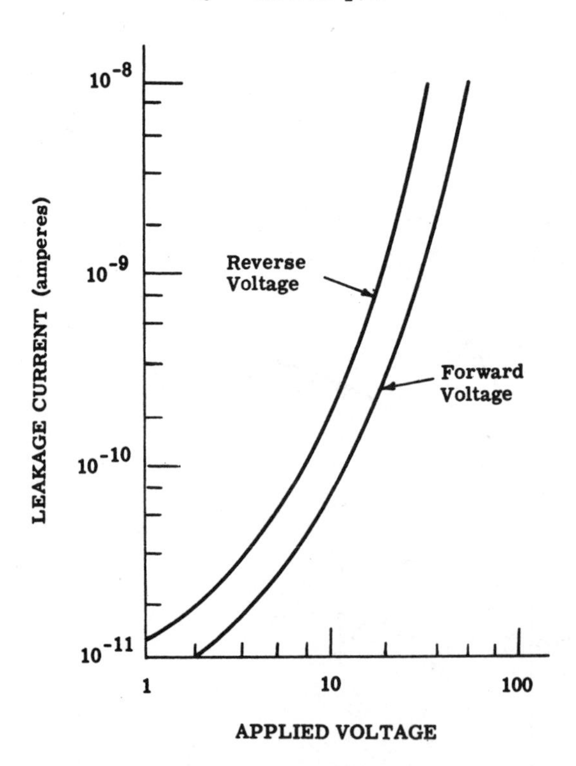

Fig. 3-16 Leakage current at 25°C as a function of applied voltage in forward (anodization) and reverse directions, for a tantalum capacitor

The Q value is given by

$$Q = 4\ (f/f_o)\ (10^4 t)\ Q_o \tag{3-9}$$

where Q_o is the value of Q at $f_o/4$ and is

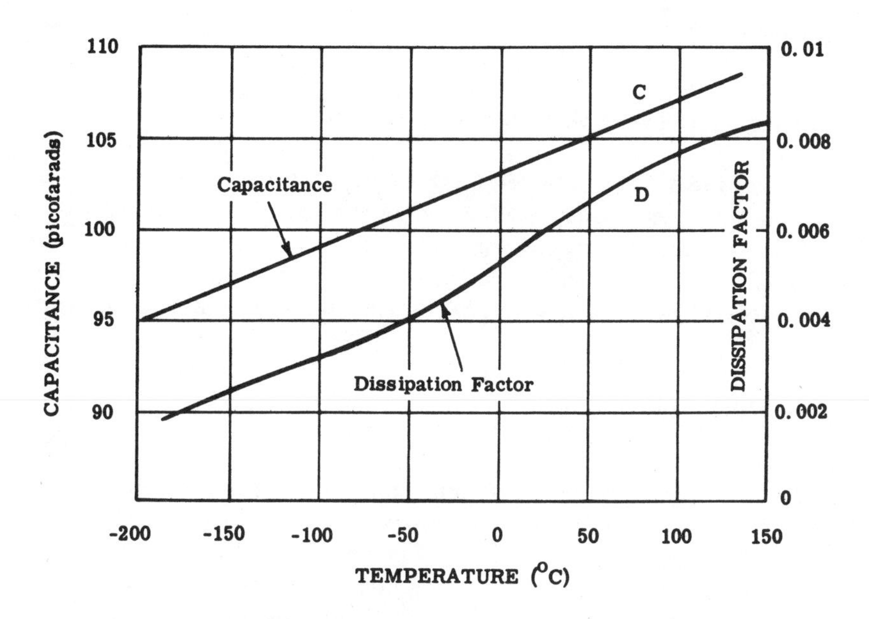

Fig. 3-17 Capacitance and dissipation factor versus temperature for an anodized aluminum capacitor

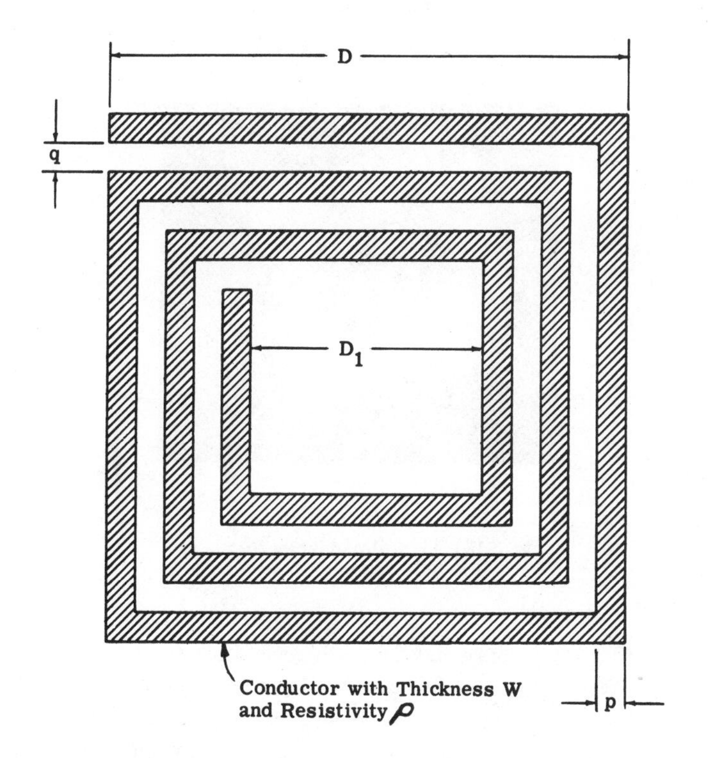

S = Surface Area of Coil ($D_1 = 0$)

N = Number of Turns

r = p/q

Fig. 3-18 Flat, spiral inductor coil

$$Q_o = 20p^{4/3}D^{-4/3}R_S^{-1} \tag{3-10}$$

where R_S is the sheet resistance of the conductive film.

Several ideas have been set forth to improve the characteristics of thin film inductors. Examples of these are stacked turns in a multi-layer construction; spirals situated on both sides of a thin substrate and connected in series; and enclosing the coil in layers of magnetic material. All of these suggestions, however, are difficult to produce with good yield.

3.5 THIN FILM DIODES

The principle of operation of thin film diodes, is the same as that of bulk semiconductor diodes, namely, P-N junction barrier rectification. The essential difference lies in the choice of materials, the fabrication method, and the polycrystalline state of the thin films. CdS and Te may be used as the N and P materials, respectively. Gold and Gold-Indium Oxide may be used to form contacts to the Te and CdS, respectively. Some important features of these diodes remain unexplained, such as the behavior of both the current and capacitance under a forward bias exceeding a given value. However, such behavior does not render the device less useful.

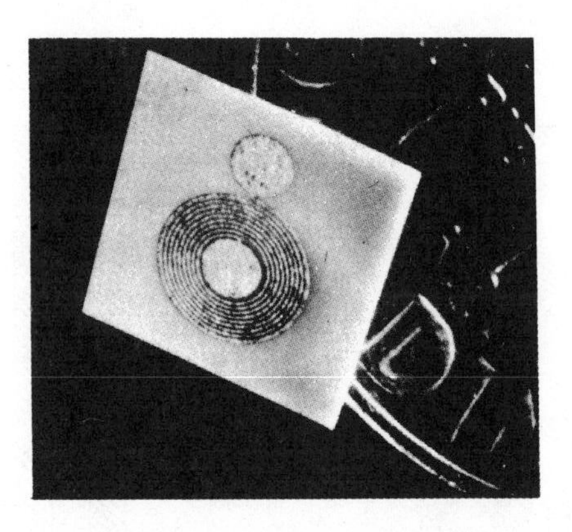

Fig. 3-19 Thin-film inductor for use in uhf and microwave circuits.

A thin film diode may have a high-resistance semiconductor between metal films. When the electrical conductivity of the semiconducting film is very low (i.e., the concentration of equilibrium charge carriers is low), injection currents are observed which are limited by the space charge. The diodes with this structure are often called analog diodes in view of the analogy of their operation with vacuum diodes, with respect to physical phenomena. Sometimes they are called dielectric diodes to emphasize the basic condition of the investigated operating mechanism: the presence of a film with very low electrical conductivity between the electrodes.

In high-impedance semiconducting films which have a low concentration of the equilibrium charge carriers and also a large number of capture centers, deviation from Ohm's law is observed for large electric fields as a result of limitation of the flowing currents by the space charge. In weak fields, the large carrier concentration does not depend on the electric field intensity: the current is proportional to the applied voltage, and the electrical conductivity of the film is constant at the given temperature. In strong electric fields the concentration of the carrier depends on the intensity, and can turn out to be greater than the equilibrium concentration which is observed in weak fields.

The equilibrium concentrations of electrons and holes are created only by thermal motion; they do not depend on the electrodes and intensity of the weak electric fields. The film as a whole is not charged, and there are no space charges in it. The nonequilibrium (or excess) electrons and holes are pulled into, that is, injected into the film from the electrodes under the effect of the electric field.

The phenomenon of electron injection in a high-impedance semiconductor can be studied by analogy with their emission into a vacuum. In the first approximation it is possible to neglect the internal carriers, and consider that all the electrons (or holes) were injected into the film from the electrodes. In this case space charges in excess of the equilibrium charge occur, and this limits the current.

At low voltages, the equilibrium carriers succeed in being redistributed within the film during passage of the injected carriers through its thickness, so that the charge of these carriers is neutralized and does not affect the current. In this case, the current is defined by Ohm's law and the concentration of only the equilibrium carriers. At high voltages, the time of passage of the injected carriers through the film is so small, that neutralization of their charge cannot occur.

The traps contained in the film have an essential effect on the magnitude of the currents limited by the space charge. The traps are created by lattice defects caused by foreign admixtures, and they give foreign levels in the forbidden zone.

In a structure consisting of a film, for example, of cadmium sulfide between two different electrodes, one of which is a barrier electrode and the other, an ohmic electrode, the current in the transmission direction is limited by the space charge. The function of the cathode in this analog diode is assumed by the ohmic contact electrode from which the electrons are injected. Therefore, it is called the injecting electrode.

The motion of the electrons in a high-impedance semiconductor is different from the motion in a vacuum; in the case of the semiconductor, the electrons are scattered on the lattice defects, admixtures and thermal lattice vibrations. Their speed is determined by the product of the mobility times the electric field intensity. At the same time, the speed of the electrons in a vacuum diode increases proportionally to the square root of the applied anode voltage.

The function of the anode is performed by the barrier contact. The electrons can move from the semiconductor film to the barrier electrode, but the reverse transition is difficult; in the semiconductor film, a physical barrier layer is formed near the barrier electrode. Actually, if the work function of the electrons from the semiconductor is less than from metal, the electrons from the layer of semiconducting film next to the electrode layer drift into the metal film of the electrode, charging it negatively. A positive space charge layer depleted with respect to electrons remains in the semiconductor; therefore, the resistance of the layer next to the electrode is high.

If a voltage is applied to the diode in the transmission direction, a field is created in the semiconductor film, and the electrons in it move from the injecting to the barrier electrode under the effects of this field. The transition of the electrons from the injecting electrode is not difficult, in view of the fact that a negative space charge is created in the high-impedance semiconductor film, which somewhat limits the injection current. The effect of the potential barrier of the barrier electrode on the current in the transmission direction is insignificant, since the height of the barrier for transition of the electrons from the film to the barrier electrode decreases in this direction as a result of the approach of the electrons. Consequently, in practice the current is limited only by the space charge in the transmission direction.

When a cutoff voltage is applied to the diode, the flow of electrons from the injecting electrode into the film is almost halted because very high energy is required to overcome the potential barrier. As a result, the current densities in the direction of easy flow and the barrier direction differ sharply. Here, by analogy with the vacuum diode, the direction of easy flow is the direction where the negative voltage is applied to the injecting electrode (on injection of holes the polarity is reversed).

In manufacturing an analog diode component on a glass substrate, the barrier electrode film made of tellurium is deposited in a vacuum; then the cadmium sulfide film is deposited on it followed by the injecting electrode film made of indium or gallium, as shown in Figure 3-20a.

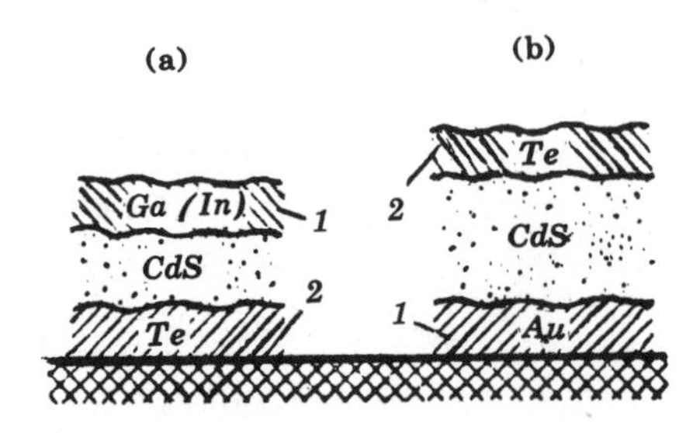

Fig. 3-20 Structure of an analog diode component.
1 – injecting electrode; 2 – barrier electrode.

Another version of an analog diode is obtained by deposition of the injecting electrode made of gold and indium oxide, followed by deposition of a thick layer of cadmium sulfide (approximately 5 microns), and then the tellurium barrier electrode (Figure 3-20b). The direction of easy flow is the direction when a positive voltage is applied to the tellurium film.

3.5.1 TUNNEL DIODE COMPONENTS

Electrons can pass through very thin (up to 100 Å) dielectric films by means of the tunnel effect. For this purpose, the transparency of the potential barrier must be appreciable. This transparency, along with the film thickness, depends, among other factors, on the dielectric constant, the work function, and the energy of the electron affinity.

Very thin but nonporous dielectric films required for tunnel diode components are most frequently obtained by oxidizing the metal of the lower electrode. The upper electrode can be made of another or (in some cases) the same metal.

If the metal electrodes on different sides of the film are made of the same metal, the shunt of the volt-ampere characteristic will not, theoretically, depend on the sign of the applied voltage. For different metals, rectification may occur as a result of the tunnel effect.

The presence of a dielectric film between the metal electrodes leads to the occurrence of a potential barrier which can be overcome by the quantum mechanical tunneling effect. An electron can go from one metal to the other, if there is a free energy level in the latter with the corresponding energy.

Without an external electric field, the number of electrons going from one metal to the other and back is the same. On applying an electric field, the energy levels in the first and second metals are shifted with respect to each other; some of the populated levels in the first metal turn out to be at the same height as the empty levels of the other metal, and the transfer of electrons from one electrode to the other begins. The coefficient of transparency of the barrier depends on the electron energy: the lower the electron energy, the lower the transparency of the barrier. In addition, the density of the permitted states increases with an increase in energy.

For structures with different metals, the asymmetry of the volt-ampere characteristic is connected with the magnitude and shape of the potential barrier at the interface of the electrode and the dielectric film. Asymmetrical barriers are, however, also observed in systems with the same metals as, for example, $Al\text{-}Al_2O_3\text{-}Al$. The reason for this is the occurrence of boundary layers with other properties, particularly layers with increased conductivity.

It is of interest to compare the situations where the metals are separated by a vacuum or dielectric. A tunnel current is possible across a vacuum gap if the width of the gap is several angstroms. On the other hand, a noticeable tunnel current flows across dielectric films even with a thickness of several tens of angstroms. This is explained through a lowering of the height of the potential barrier in comparison with the vacuum gap.

The current density is determined by the film resistance which, in turn, depends on the voltage, the dielectric constant of the film, the height and the shape of the potential barrier, and the film thickness. The resistance depends especially on the thickness of the dielectric film. For example, variation in the thickness from 40 to 50 Å leads to an increase in the film resistance by 10^8 times; variation in the thickness by 5 Å (from 35 to 40 Å) at 0.5 volts of applied voltage, causes a change in the resistance by 100 times.

Nonuniformity of thickness of a superthin dielectric film leads to the situation where the current density through the diode is different under different surface conditions. In practice, all the current will pass only through the thinnest parts of the film.

With a complex method of condensation in the vacuum, it is possible to obtain very thin dielectric films which are homogeneous with respect to thickness and composition. Therefore, oxide films built up at the expense of the metal of the lower electrode, are always used.

Inasmuch as the dielectric layer included between the metal electrodes is extraordinarily thin, it is necessary to have little diffusion of the metal into the dielectric film in order to prevent possible short-circuiting of the electrodes.

When manufacturing diodes based on the tunneling effect, aluminum electrodes or an aluminum electrode combined with an upper gold electrode, are often used (Figure 3-21). Aluminum films give dense, stable layers with good bonding to the substrate.

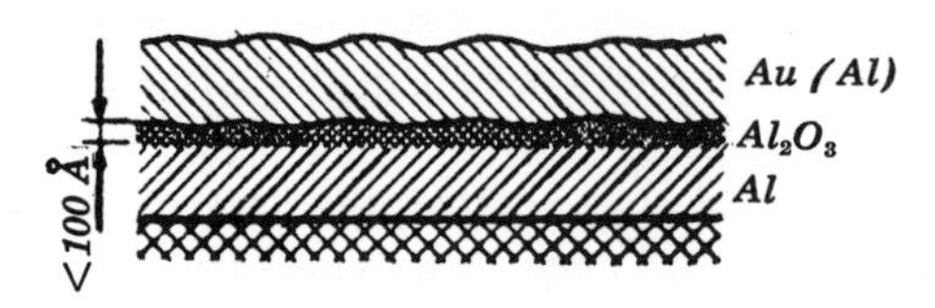

Fig. 3-21 Structure of a tunnel diode component.

Aluminum is usually evaporated with tungsten windings. Inasmuch as special purity of the metal of the lower electrode is important in connection with the oxidation, it is possible to evaporate aluminum from crucibles made of beryllium oxide or aluminum oxide.

An essential disadvantage of the tunnel diode components is instability of the resistance with time. Most frequently, this instability is caused by diffusion of the metal into the dielectric layer — which leads to a continuous decrease in the film resistance.

When using an Al_2O_3 film, the cause for variation of the tunnel current with time is an increase in thickness of the oxide layer, as a result of the slowly occurring process of oxidation of the aluminum.

3.5.2 SURFACE-BARRIER DIODE COMPONENTS

In cases where the thickness of the dielectric or semiconductor film in a diode component is somewhat greater than the superthin film of a tunnel diode, the rectification mechanism is determined by the passage of electrons over the contact potential barrier occurring at the metal-dielectric (semiconductor) interface.

The form and characteristics of the contact barrier under practical conditions depend essentially on the work function of the metal, the energy of the electron affinity, the energy spectrum of the surface states, and the uniformity of the film.

The surface-barrier phenomena are complicated for semiconductor and dielectric films obtained by deposition in a vacuum, as a result of structural distortions and the presence of adsorbed layers. The magnitude of the potential barrier can be different for one and the same metal-semiconductor pair, depending on the process conditions of manufacture of the contact or the mutual arrangement of the electrode and dielectric (semiconductor) film.

3.6 THIN FILM ACTIVE ELEMENTS

A variety of thin film active elements have been proposed, constructed, and demonstrated as workable. A high degree of refinement has been achieved in such devices as thin film magnetic memory elements, which are used with driving and sensing amplifiers, and thin film superconducting cryotrons, which are used in liquid helium environments. These devices are, however, of such characteristics that they are generally not considered as devices integral with thin film microelectronics. For this reason, emphasis is placed on the thin film active elements of the majority-carrier, transverse-field-modulated type, as exemplified by the thin film triode.

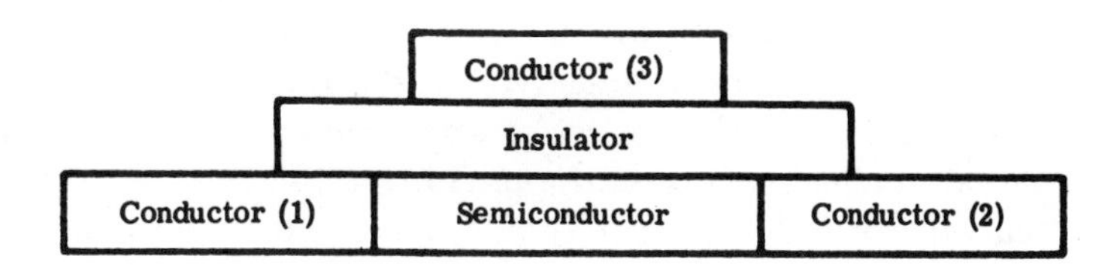

Fig. 3-22 Thin film triode design

Consider the configuration shown in Fig. 3-22 If a potential V is applied between the two electrodes (1) and (2), and if the contacts to the semiconductor are ohmic, a current I flows which depends on the carrier density in the semiconductor, N, and the carrier mobility, μ, as follows:

$$I = (A/L)\ V\mu Nq\,, \tag{3-11}$$

where:

A = cross sectional area of semiconductor = Wt
L = distance between conductors
W = width of semiconductor
t = thickness of semiconductor
q = charge on carrier.

N and μ refer here to the majority carriers, which constitute most of the carriers present.

The conductivity in the above configuration can be increased by increasing N or μ. Consider first the increasing of N. If the semiconductor is regarded as one plate of a capacitor, and the conductor (3) as the other plate, then, by making the conductor (3) positive with respect to the semiconductor through a voltage V_g, a net negative charge Q_s will be moved onto the surface of the semiconductor from conductor (3). This is given by

$$Q_s = -V_g K \frac{WL}{d} \qquad (3\text{-}12)$$

where d is the thickness of the insulator and K is its dielectric constant. As a result, if the semiconductor is n-type, the net negative charge available for conduction is increased, and therefore

$$N = N_o - \frac{Q_s}{WLt} = N_o + \frac{V_g K}{td} \qquad (3\text{-}13)$$

where N_o is the initial free charge density.

The following considerations are applicable to the analysis:

a) A time constant τ = RC, where R is the effective resistance due to the semiconductor film resistivity, is associated with the charging of the semiconductor surface. For high impedance devices (1 megohm), this time constant would be of the order of 10^{-5} seconds.
b) The interface between dielectric and semiconductor may have a large number of traps associated with it. Charges drawn to that surface would then find themselves immobilized with little or no contribution to the conductivity of the semiconductor.
c) When a potential, V_{12}, is applied between conductors (1) and (2), each point on the surface of the semiconductor will assume some related potential. If the conductivity of the semiconductor is independent of its position between conductors (1) and (2), the potential gradient will be uniform. However, in general, this condition will not prevail, for if an additional potential V_{13} between electrodes (3) and (1) is applied, the conductivity will increase more near electrode (1) than near electrode (2), thus changing the potential distribution.
d) The transconductance is highly temperature dependent.
e) The transconductance is highly photosensitive.
f) The transconductance is often so large as to lead one to infer values of mobility incompatible with independently measured values.
g) The transconductance first increases, then saturates, and finally decreases, as V_{13} increases.

A second mechanism for increasing N exists. It may be referred to as carrier generation in contrast to the carrier injection just discussed. Since a semiconductor has a limited number of free carriers available, an electric field can penetrate to substantial depth, thereby subjecting some of the bound electrons to field ionization. The electrons separated from their fixed sites are then indistinguishable from those electrons already in the conduction band. Some electrons can be excited more easily than most. They are loosely bound and are associated with impurities, lattice vacancies, dislocations, and the like. Assume three categories of bound carriers:

(a) lattice valence electrons, (b) impurity valence electrons, and (c) trapped electrons. Starting with an electrically neutral semiconductor (no net charge), the distinction between trapped electrons and impurity (donor) electrons is that for each trapped electron, a lattice valence electron is missing (a hole exists), whereas impurity electrons have no holes associated with them. When the impurity electrons are drawn into the conduction band, the impurity positive ion behaves like a trap. The lattice valence electrons are much more tightly bound and are not likely to be field excited to the conduction band.

Carrier generation, by transverse field excitation, is a very fast process, since it is quantum mechanical in nature and involves only the change in the energy state of an electron but not its physical displacement. There is no instantaneous change in the net charge of the semiconductor, and the excitation mechanism itself does not give rise to space charge effects, as does the carrier injection discussed earlier. Electric fields of the order of 10^5 V/cm will excite most electrons from impurity or trapping states which are of the order of 0.02 eV below the conduction band. For this mechanism to provide substantial conductivity modulation, many electrons must occupy states in this range. Traps must be filled before they can be emptied by the field. This last requirement may be satisfied through optical excitation. The modulation provided by the transverse field is enhanced by as much as a factor of five, by exposing the semiconductor to light. Experiments have shown that the photon energy primarily responsible, is just that required to excite valence band electrons to shallow traps.

Other possibilities for carrier generation exist. Among these are thermal excitation from traps; optical excitation of any bound electrons; and impact excitation, by hot electrons, of trapped and impurity electrons to the conduction band. None of these phenomena, however, qualify as transverse field effects.

Consider now the mobility, μ, of the carriers. Alteration of the local electric field by the application of a transverse field is unlikely to change the number of defect or impurity scattering centers. Two changes are likely to occur, however. The first is a concentration of mobile charge at the semiconductor-dielectric interface, where the mean free path is generally shorter than in the bulk. This will result in a decrease in the mobility. The second is the accumulation of a net negative charge near conductor 1; this "space charge" is tantamount to the existence of additional scattering centers, and thus reduces the mobility. Small increases in mobility can result from operation at reduced temperatures which reduce phonon scattering.

The configuration shown in Fig. 3-22 contains all the features of the thin film triode. Conductor (1) is called the "source" or "emitter," conductor (2) is called the "drain" or "collector," and conductor (3) is called the "gate" or "field plate." The conductors may be vacuum-deposited Ni, Au, Nichrome, Al, and others. Gold has been cited as being responsible for the early demise of some devices, when used as the gate, and aluminum introduces sometimes nonohmic contact problems. For the semiconductor, CdS, CdSe, CdTe, and Te may be used. The first three are n-type; whereas Te is p-type. The advent of the Te thin film triode has opened the way for production of complementary devices. The dielectric most commonly used is vacuum-deposited SiO. It is of the order of 2×10^{-5}

cm (2000Å) thick in some of the more stable devices. Gate voltages of about two volts will provide substantial modulation of the conductivity. Common values of the source-to-drain voltage are about 10 volts. It should be noted that the thin film triode is quite temperature sensitive. An example of a thin film amplifier is shown in Fig. 3-23. A more detailed description of the thin film triode is given in the following section.

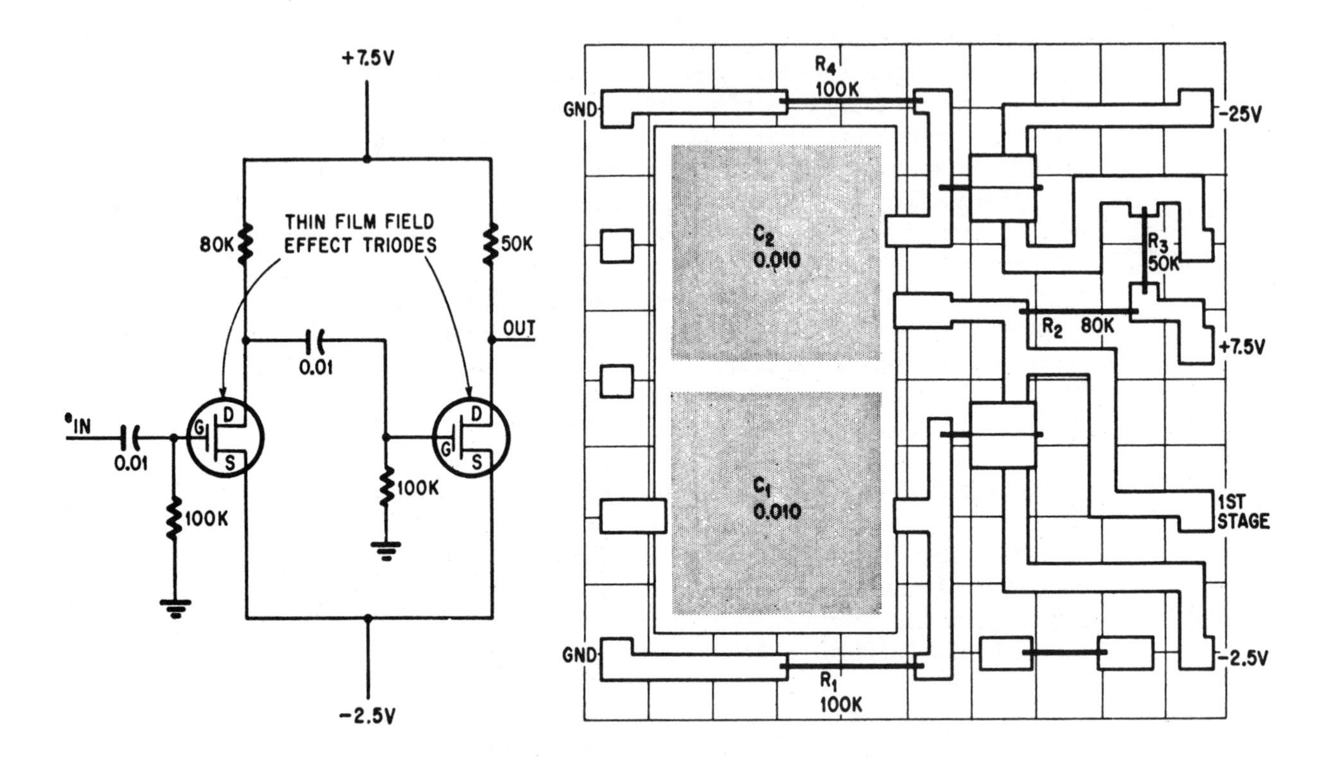

Fig. 3-23 Thin film amplifier with two silicon-monoxide aluminum capacitors, two thin film triodes, and several chromium and rhenium resistors

3.6.1 FILM FIELD EFFECT TRIODE

An amplifier using oscillation of the surface conductance of germanium films by a transverse electric field was first proposed by Shockley and Pearson in 1948, before point and planar transistors were invented. The semiconductor plate with two ohmic contacts was covered on the ends by a dielectric layer on which a third controlling metal electrode was applied. As a result, a capacitor was obtained in which one of the plates was a semiconductor and the other the controlling electrode. On applying a voltage to the controlling electrode, additional charge carriers were induced in the semiconductor plate, and this increased the current through the semiconductor between the two ohmic electrodes.

However, at the beginning of the investigation it was discovered that in many cases, the application of a transverse electric field has no effect on the conductance of the semiconductor, as a result of the surface energy states in the modulation mechanism. The development of transistor engineering at the beginning of the fifties lessened the interest in this idea, but it was returned to in 1959 in connection with progress in the passivation of the surface of a single crystal of silicon by thermal oxidation. This permitted insurance of low density of the surface states.

In operation, the film field effect triode has much in common with the metal-oxide-semiconductor (MOS) structures, but as a result of the film nature of this triode, there are essential differences. Figure 3-24a illustrates the bilateral structure of a thin-film triode as originally developed. Figure 3-24b shows the planar structure.

In the bilateral structure, the control and current electrodes are on opposite sides of the semiconductor film, and in the planar electrode, on one side. The planar and bilateral field triodes have identical operating characteristics, but preference is usually given to the planar structure; a semiconductor film requires high temperatures of the substrate during its application and subsequent annealing, and the metal films of the electrodes applied to the semiconductor film cannot withstand the heat treatment of the semiconductor and peel off. The planar structure is free of this disadvantage, since the application of the semiconductor film and treatment of it take place before applying the other layers — under the condition that the triode is applied first in a given microcircuit.

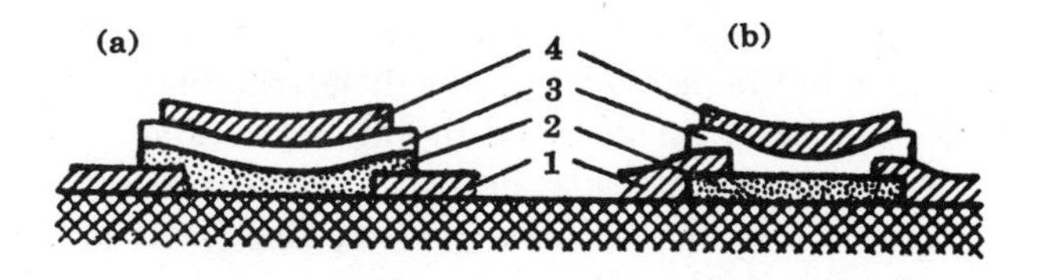

Fig. 3-24 Bilateral (a) and planner (b) structures of a thin-film field effect triode with an insulated gate: 1 — drain (source); 2 — semiconductor film; 3 — dielectric film; 4 — gate.

In a tantalum version of the bilateral structure (Figure 3-25), the gate is made of tantalum and is applied first. The surface of the tantalum film is then oxidized, thereby obtaining a dense and thin layer of insulating film on which the semiconductor film is applied. Tantalum and tantalum pentoxide tolerate the high-temperature treatment of the semiconductor film.

The thickness of the semiconductor film usually does not exceed 0.5 microns, the source-drain spacing = 40-50 microns, the thickness of the dielectric layer = 0.05-0.2 microns, the gate width somewhat overlaps the source-drain spacing, and the electrode width = 0.2 microns.

Considerable technical difficulties are encountered in manufacturing this triode: included are obtaining small current electrode spacings, alignment of the mask patterns with a tolerance of a few microns, and control of the gas effect on the surface of the layers in the structure.

Fig. 3-25 Tantalum modification of the bilateral structure of a thin-film field effect triode. Refer to Fig. 3-24 for legends.

The operation of a thin-film field effect triode is based on the mechanism of forming a conducting channel induced by the transverse electric field of the gate — the so-called "field effect" — at the semiconductor-dielectric interface. The gate field induces a surface charge in the semiconductor layer equal to the magnitude of the potential applied to the gate but opposite in sign. Due to the presence of the dielectric layer, there is no conductance current in the gate circuit (except the capacitive charge current), which permits applying a voltage of any polarity to the gate.

The occurrence of excess electrons induced by the gate field arises from two mechanisms: injection at the contact of the metal of the electrodes with the semiconductor, and excitation of charge carriers from the fine admixed levels by the transverse electric field of the gate. The predominant role of one of the mechanisms depends on the initial conductance of the semiconductor film. Theoretical and experimental research have demonstrated that both mechanisms are observed in film structures. A necessary condition for injection currents in the channel is a short source-drain electron flight time, in comparison with the dielectric relaxation time. On the other hand, the small thickness of the dielectric film (hundreds of angstroms) permits a high-intensity electrical field (up to 10^6 volts/cm) required for ionization of the carriers from the fine levels at acceptable values of the potential on the gate.

The injection mechanism can assume a predominant role in the case of applying high-resistance semiconductors when the specific impedance of the semiconductor is approximately 10^5 ohms-cm. In this case the source-drain spacing is approximately 10 microns, and the mobility of the carriers is units of cm^2/volt-second. This arrangement is typical for a triode created on the basis of polycrystalline semiconductor films.

The metal electrodes of the source and drain should be selected and manufactured in such a way that they form ohmic contact with the semiconductor film, since the concentration of the injected carriers depends exponentially on the magnitude of the potential barrier at the contact. Due to the charge carrier injection at the contact of the source (which acts as an infinite reservoir of electrons), the surface conducting channel is supplied with charge carriers. The injection level, that is, the amount of charge injected in the semiconductor layer is determined by the magnitude of the voltage on the gate. Injection becomes noticeable only when the density of the injected carriers begins to exceed the density of the basic carriers initially existing in the semiconductor, and thermally generated in the conduction zone from the admixture levels and from the valence zone.

The injected excess carriers in the form of a space charge move in the crystal lattice of the semiconductor, similarly to a space charge in a vacuum tube. The difference lies in the fact that in the solid state crystal lattice, the drift rate of the carriers in the field is limited by various scattering mechanisms, and it is characterized by the magnitude of the mobility. This injected space charge moves in the channel along the semiconductor-dielectric interface in the direction of the drain as a result of the longitudinal ''pulling'' electric field creating a drain current.

The above general concepts apply to the operation of a semiconductor field effect triode in which the semiconductor layer has an ideal crystalline structure. However, in a real semiconductor film, chemical impurities and lattice defects (grain boundaries, vacancies, dislocations, etc.) are unavoidable. On the diagram of the energy zones of a semiconductor, these correspond to local energy states in the forbidden zone. On dropping into these states the free electrons are bound; they lose mobility and do not contribute to conduction. The presence of this type of capture centers — traps — in a real film, leads to a sharp decrease in the current injected into the semiconductor layer.

For a gate field intensity of 10^5 volts/cm achieved in the triode, the depth of the conducting channel is on the order of hundreds of angstroms, and hence the physical processes corresponding to operation of the triode take place only in the thin surface layer of the semiconductor. Therefore, in analyzing the operation of the triode, it is meaningless to distinguish the concept of *traps* characterizing the volume of the semiconductor, and *surface states* inherent only in the surface and connected with breakdown of the crystal lattice and the presence of different types of defects and admixtures on the surface. It is possible, however, to discuss the total content of capture centers, implying the defects on the surface of the semiconductor film, on the surface of the grains forming it, and inside the grains. The traps can be deep or shallow, depending on the position in the forbidden zone of the energy levels relative to the stationary Fermi level, and they may be slow or fast depending on the lifetime of the electrons in these states.

It may be shown with Fermi statistics that shallow traps above the Fermi level, have the property that at any injection level, they trap some part of the free carriers and decrease the injected current.

Deep traps above the Fermi level capture all the electrons from the injected charge which are required to fill them. Thus, in the presence of deep traps the injection current becomes noticeable only after the injection level fills the deep traps. The corresdonding gate field intensity with high trap density in the films (above 10^{17} cm^{-3}), is too high for practical applications.

The fabrication of thin-film field effect triodes has become possible as a result of the short distance between the three electrodes, which allows high field intensities with acceptable potentials. At the same time, improvements in the film deposition process allowed lowering the trap density. This was achieved, principally, by: compensation for the deep traps through the introduction of fine donor admixtures, which, being almost entirely ionized at room temperature, function as shallow traps; the effect of the dielectric film vacuum deposited on a semiconductor film during the process of manufacturing the triodes; and the application of mono-crystalline semiconductor films.

Insulated-gate field effect triodes have been of two types: enhancement or depletion. In the absence of a gate bias, triodes of the enhancement type have low channel conductance which increases by several orders on supplying a positive gate bias. Triodes of the depletion type have a drain current which flows even in the absence of a gate bias, and to reduce this current to zero, it is necessary to supply a negative gate bias. Thus, for triodes of the enhancement type the gate bias required to cause a drain current (i.e., to trigger the triode), the so-called cutoff voltage is positive, and for triodes of the depletion type, it is negative. Triodes of the depletion type can operate for both polarities of the gate bias, but triodes of the enhancement type operate only with a positive bias. Triodes of the enhancement type can be used in amplifiers with a direct coupling between the cascades.

A triode of the depletion type differs from the enhancement type particularly by the nature of the dielectric film deposited on the semiconductor. Triodes with enhancement have an insulating film of CaF_2, and triodes of the depleted type, an insulated film of SiO.

The formation of a triode of one type or the other is connected with the creation of traps. If the trap density is high and the ratio of the lifetime of the basic carriers and the carriers in the traps is low, then almost all the basic carriers initially existing in the film are in the traps, and in the absence of a gate potential the conductance is low. It is necessary to apply a positive bias to the gate in order for the injected carriers to fill the traps still remaining empty, and for the conductance to increase. This case applies to a triode of the enrichment type. However, it has been discovered that in the presence of shallow trap levels (0.02-0.03 electron-volts), neutralization of their effect on the current can occur – not as a result of filling them, but as a result of the ionization mechanism in the large fields in a triode.

If the ratio of the lifetime of the basic carriers and the carriers in traps is high, the basic carriers are mobile, giving rise to drain current without a gate potential. In order to pump out the basic carriers and empty the channel, a negative gate potential is required. In this case, the triode is of the depletion type.

An insulated-gate field effect triode has a high input impedance and a volt-ampere characteristic with saturation similar to a vacuum pentode.

For a quantitative analysis of a field-effect transistor (FET), only majority carriers are assumed to exist in the semiconductor. The semiconductor layer is assumed to be homogeneous and thin compared with the insulator layer. The carrier mobility in the semiconductor layer is constant, and the source and drain contacts are ohmic. There is also an implicit assumption that the electric field is perpendicular to the surface of the semiconductor. This is the "gradual approximation" introduced by Shockley. The electric field can be perpendicular to the surface of the semiconductor only if the drain-to-field electric field is small. This condition is not satisfied under operating conditions particularly when the drain current has reached saturation.

The electrical characteristics of a typical FET are shown in Figure 3-26. The device is operating in the enhancement mode, where majority carriers are accumulated in the semiconductor by action of the gate voltage across the gate-semiconductor capacitor. In the simple theory, expressions are derived for the drain current and for mutual transconductance.

Assuming that the "gradual approximation" is valid, the voltage across the gate-semiconductor capacitor is given by

$$V_{cap} = V_G - V(x), \tag{3-14}$$

where V_G is the gate voltage referred to the source contact, and $V(x)$ is the voltage on the semiconductor also referred to the source potential. A drawing of the device structure is shown in Figure 3-27.

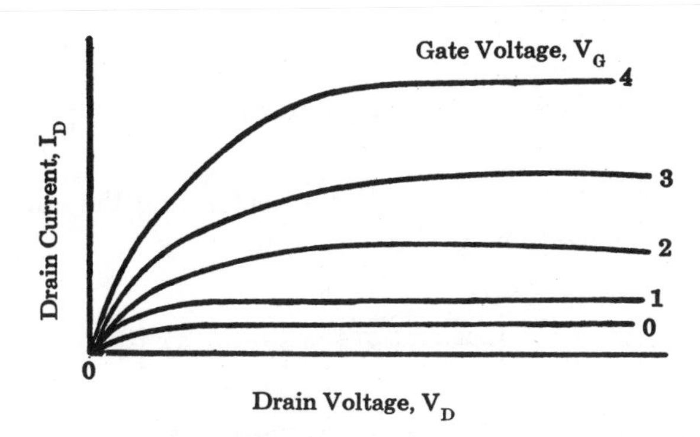

Fig. 3-26 Typical field-effect transistor characteristics (enhancement mode)

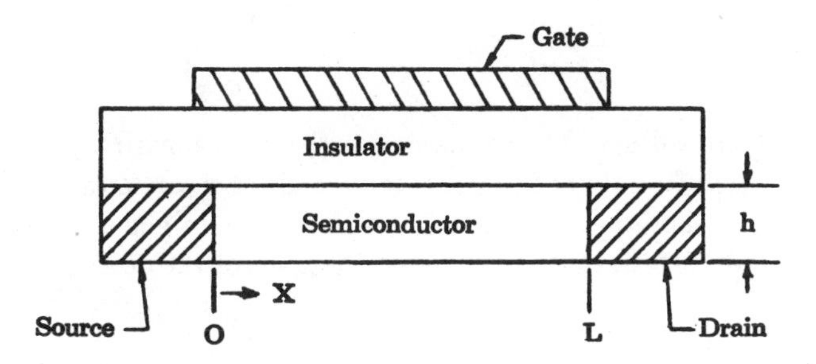

Fig. 3-27 FET device structure used for analysis.

The added charge accumulating on the plates of the capacitor per unit area will be

$$\Delta n(x) = \frac{C_G}{qWL}\left[V_G - V(x)\right], \tag{3-15}$$

where $\Delta n(x)$ is the charge per unit area, C_G is the capacitance of the gate-semiconductor capacitor, L is the source-to-drain spacing, W is the width of the device — i.e., the length of the source and drain contacts —, and q is the electronic charge.

The added charge carriers modulate the conductivity of the semiconductor, which becomes

$$\sigma(x) = q\mu \left[n_o + \frac{\Delta n(x)}{h} \right], \tag{3-16}$$

where $\sigma(x)$ is the local conductivity, μ is the charge carrier mobility, n_o is the equilibrium carrier concentration, and h is the thickness of the semiconductor. The drain current is given by the product of the local conductivity and local electric field

$$I_D = hW\sigma(x)\,E(x) = hW\sigma(x)\,\frac{dV(x)}{dx}, \tag{3-17}$$

where $E(x)$ is the local electric field. After performing the appropriate substitutions, the following integral is obtained.

$$\int_{I_D}^{L} dx = \frac{\mu C_G}{L} \int^{V_D} \left[\frac{qWLhn_o}{C_G} + V_G - V(x) \right] dV(x). \tag{3-18}$$

The quantity $qWLhn_o/C_G$ is a threshold voltage, the minimum voltage necessary on the gate to turn on the drain current. Therefore, it can be replaced by $-V_t$.

The simple integrations in Equation (3-18) yield the result

$$I_D = \frac{\mu C_G}{L^2} \left[(V_G - V_t)\,V_D - \frac{V_D^2}{2} \right], \tag{3-19}$$

where V_D is the drain voltage. This is the simple theoretical expression for the drain current as a function of the gate-semiconductor capacitance, the threshold voltage V_t, and the gate and drain voltages. It will be used to derive an expression for the mutual transconductance of the device.

When the drain current is saturated at constant gate voltage, $(\partial I_D/\partial V_D)_{VG} = 0$, or $V_G - V_t - V_D = 0$. Under this condition, a saturation drain voltage is defined as $V_D(sat) = V_G - V_t$. As a result, the saturation drain current is given by

$$I_D(sat) = \frac{\mu C_G V_D^2(sat)}{2L^2}, \text{ or } \frac{\mu C_G}{2L^2}(V_G - V_t)^2. \tag{3-20}$$

The mutual transconductance of the device at drain voltages above saturation is defined by the condition $g_m = (\partial I_D(sat)/\partial V_G)V_D$. The result is

$$g_m = \frac{\mu C_D}{L^2}(V_G - V_t). \tag{3-21}$$

It is evident from this expression that certain device design features should be considered. In order to obtain devices with high values of g_m, the carrier mobility and the gate-semiconductor capacitance should be large, and the source-to-drain spacing and the threshold voltage should be small. A small threshold voltage can be obtained with small semiconductor volume and low equilibrium carrier density. Thus, the source-to-drain spacing and the carrier density are found to be most critical for an FET of high gain.

From the materials standpoint, the most desirable semiconductor is one with low equilibrium carrier concentration and high carrier mobility. From a fabrication standpoint, the device needs to be small, and of particular importance is the source-drain spacing.

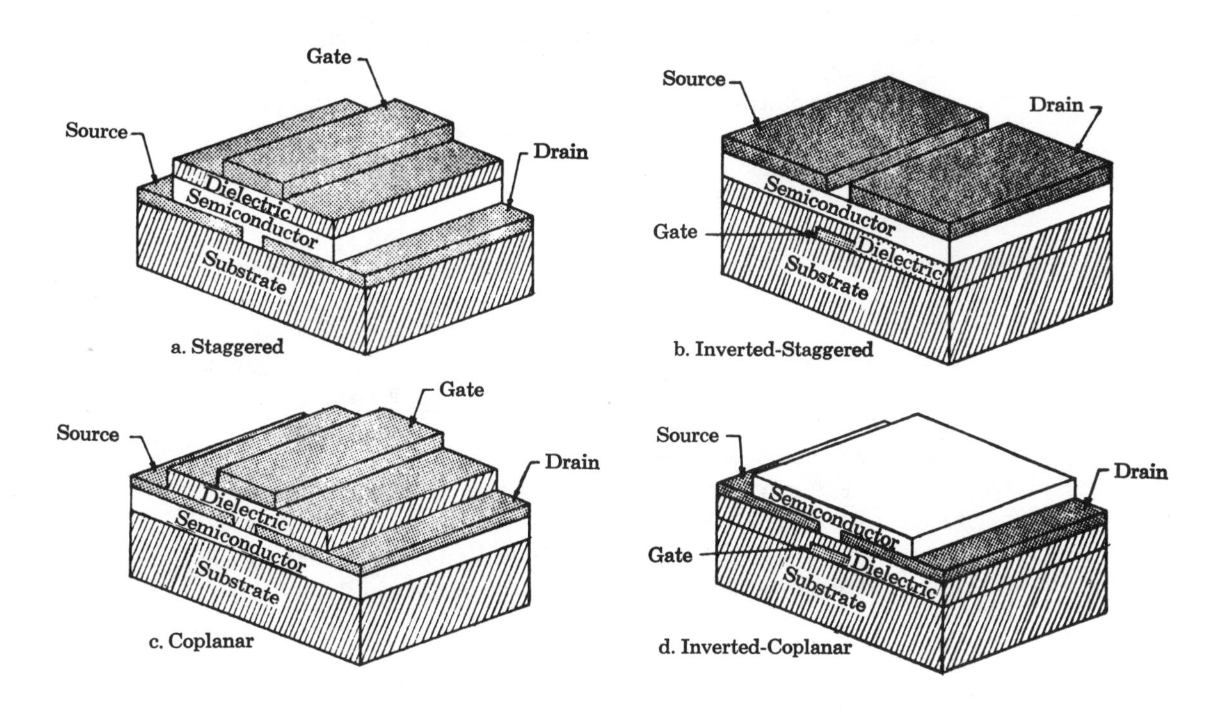

Fig. 3-28 Common thin-film transistor configurations.

Some modifications and additions have been made to the simple theory. The most important analysis is one in which the "gradual approximation" is not used. Among other things, it has been found that the saturation resistance of an FET, where the drain current does not saturate but increases slowly with drain voltage, is inversely proportional to the ratio h_{ins}/L. In this relation, h_{ins} is the thickness of the insulator layer.

The two basic types of field-effect transistors are the junction (or channel) FET and the insulated-gate FET. The junction FET uses single-crystalline bulk semiconductor material. One form of the insulated-gate FET (or IGFET) also uses single-crystalline bulk semiconductor material, while the other form of IGFET consists of a structure of thin film layers (usually polycrystalline).

The common configurations for the thin film IGFET or (TFT) are shown in Figure 3-28. They include the staggered, coplanar, inverted-staggered, and inverted-coplanar structures.

3.6.2 COMPLEMENTARY-SYMMETRY MOS TRANSISTOR CIRCUITS

The use of complementary-symmetry MOS (metal-oxide-semiconductor) transistor circuits offers a number of advantages for low-power, high-speed digital systems. These circuits, when fabricated in thin films of silicon on insulating substrates, such as sapphire or spinel, offer advantages of increases in speed and simpler circuit fabrication.

One of the advantages of complementary-symmetry MOS integrated circuits, namely, that of very low power dissipation in a standby condition (i.e., when the circuit is not being switched from a "0" to a "1" or vice versa), can be easily destroyed. Such factors as oxide charge, silicon-silicon dioxide interface states, silicon-sapphire interface states and poor quality silicon causing spike-like and/or erratic source and drain diffusions, can all cause increased leakage from source to drain at zero gate bias.

The problems of silicon-sapphire interface states and spike-like diffusions resulting from poor-quality silicon films, are so interconnected that it is extremely difficult to separate them. A diffusion spike across the channel from source to drain in an N-channel transistor behaves the same as does an induced N-channel at the silicon-sapphire interface. It can be shown that an interface state density of as low as 2×10^{10} donors/cm^2 will result in a leakage current of 10^{-7} A.

To determine the effect of the nucleation rate during film growth, several films were grown on sapphire at nucleation rates ranging from 0.4 μm/min to 15 μm/min. Growth rates were determined from the experimentally measured growth time of the first 1000 Å and thus are an average rate. All films were grown to a thickness of 1.0 μm. Subsequently, N-channel MOS transistors with channel lengths of 1.5 to 13 μm were fabricated in

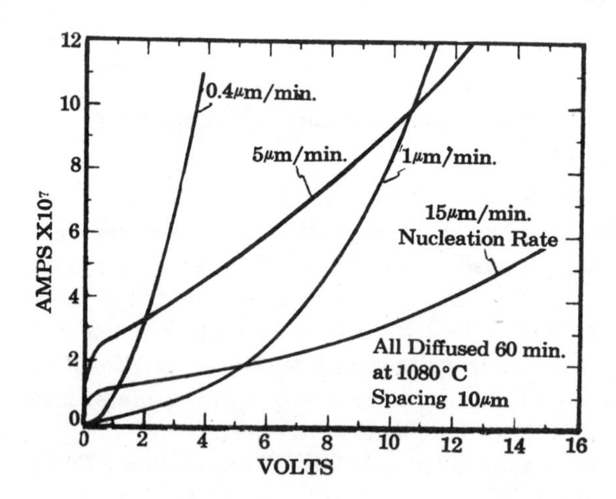

Fig. 3-29 Reverse-biased diode characteristics for units fabricated in films of silicon-on-sapphire nucleated at rates from 0.4 to 15 μm/minute.

the silicon. Each wafer was also divided into four pieces which were processed for different diffusion times to determine if the total diffusion depth affected the device characteristics. Figure 3-29 is a comparison of the reverse-biased diode characteristics for the four different nucleation rates. All samples shown in the diagram were diffused for 60 minutes at 1100°C, and all had a source-to-drain spacing of 10 μm.

Poorer-quality films grown at lower nucleation rates typically have soft breakdown characteristics and low transconductance, while films nucleated at the faster rates show good-quality devices.

One other factor which is important from the standpoint of device processing, is the surface quality of the films. Films grown at the lower nucleation rates are considerably rougher than those grown at high nucleation rates; in fact, it was extremely difficult to get a pinhole-free channel oxide grown on them.

In a low-temperature process for depositing silicon dioxide (at temperatures below 500°C), the deposited film may be doped with either phosphorus or boron in concentrations from 10^{15} to 10^{21}. With this technology it is possible to fabricate the entire complementary-symmetry MOS integrated circuit with only one heating step which diffuses the source and drain, at the same time that the channel oxide is grown. Wafers fabricated with this technology have yielded transistors with off currents ranging from 0.1 to 10×10^{-6}A.

3.7 THIN FILM HOT ELECTRON DEVICES

A hot electron is one with energy greater than that required to occupy a state within the conduction band. Figure 3-30 illustrates the basic construction of thin film hot electron devices. An applied voltage between electrodes (1) and (2) enables electrons of potential energy V_1 to penetrate the barrier, of peak height U_1 and thickness d_1, by tunneling, Schottky emission, or other mechanism. As a result, the electrons enter conductor (2) with sufficient kinetic energy to penetrate or surmount the barrier of height U_2 and reach conductor (3). There they must attain equilibrium with the other carriers, at potential V_3.

In order for a reasonable number of electrons to penetrate the first barrier, d_1 must be very small. When this requirement is satisfied, the interelectrode capacitance per unit area C_{12} is very large. Corresponding to a given operating frequency, the required densities may be so large as to result in the destruction of the device.

Hot electron devices may also be fabricated on single crystal semiconductors. These appear to have more practical features, but they require that they be made on single crystal films of considerable thickness.

3.8 DISTRIBUTED PARAMETER RESISTOR-CAPACITOR DEVICES

A good example of a new component made possible through microelectronics, is the thin film distributed parameter resistor-capacitor

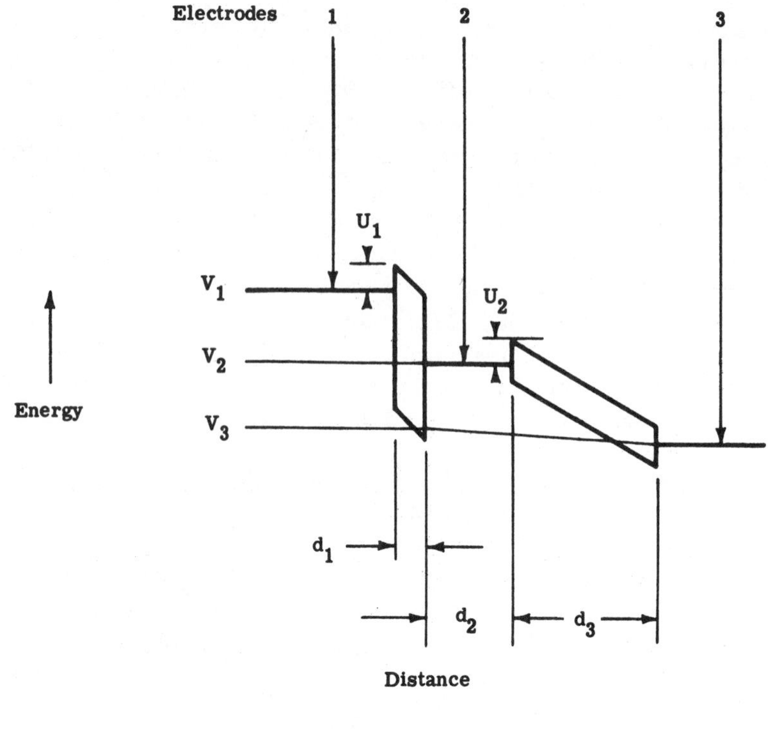

Fig. 3-30 Schematic representation of a hot electron device

(DPRC). This device is made by replacing the conductive plates of a capacitor by resistive sheets of nonzero sheet resistance. The device need not be rectangular — it can be tapered, or even circular. The number of terminals can vary from two to four. The result is, for all practical purposes, a versatile new device.

Although distributed parameter systems are not new to the electronics world, their use in conventional systems is generally avoided where possible, as their analytical complexity and physical size or complexity cause them to be economically undesirable. In the fabrication of thin film circuitry, however, the ease of fabrication of distributed RC networks has changed this trend. Thus, a substantial effort has been recently extended to effectively exploit the characteristics of distributed RC structures.

The initial effort directed toward use of RC distributed parameter networks, in thin film circuitry, was oriented toward linear RC distributed networks. These initial techniques sought an understanding of the network characteristics and thus fostered the use of the indefinite matrix in their analysis. Analyses were quickly extended to other structures, especially the exponentially tapered networks. This characterization led to applications, the bulk of which are oriented toward bandpass amplification. Having grasped the basic characteristics of distributed RC networks,

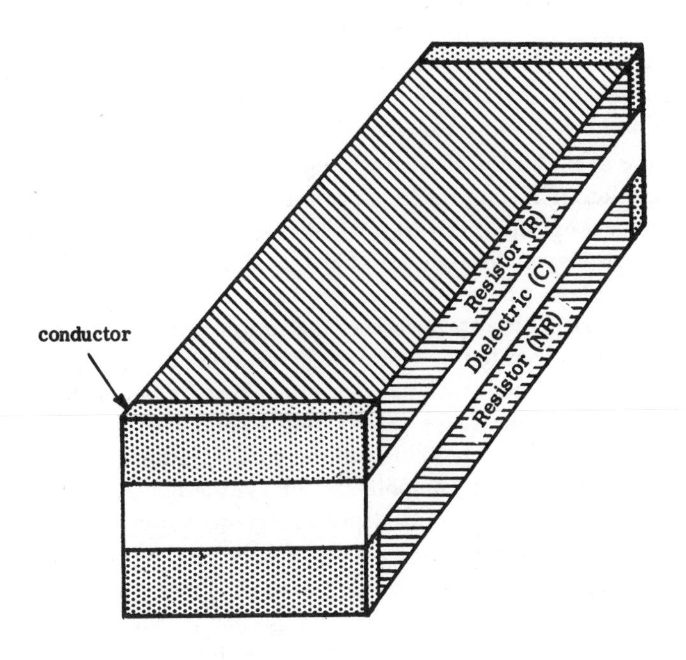

Fig. 3-31 Physical structure of R-C-NR distributed parameter network

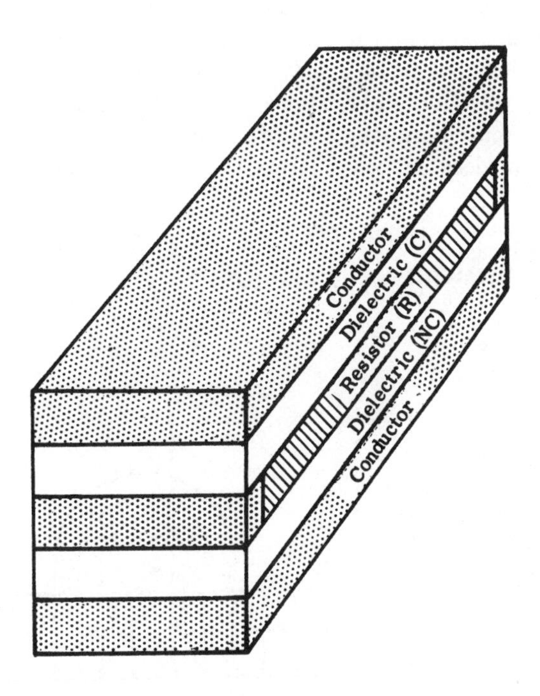

Fig. 3-32 Physical structure of C-R-NC distributed parameter network

synthesis of transfer functions utilizing distributed RC networks, as elements, followed. An investigation was performed on the possibility of finding distributed RC networks described by a rational polynomial, thereby allowing available synthesis techniques to be used in synthesizing a transfer function. This approach, however, forces constraints on the physical structure of the network, which make fabrication difficult. Another approach maps, through suitable transformations, the function into an auxiliary plane where available synthesis techniques can be used to realize the network. This approach uses the simple uniform RC network with its output open-circuited or shorted for components, as opposed to the R, L, and C used in conventional synthesis. But unlike conventional synthesis where R, L, and C are the available components, the simple R-C-NR network has over 20 possible unique configurations all of which can be considered as available components for synthesis.

Two four terminal RC distributed parameter networks which are of principal interest, are shown in Figs. 3-31 and 3-32. The R-C-NR network is constructed by depositing a dielectric of given thickness over a resistor, upon which another resistor is deposited. Though not shown in the figure, this resistor is generally overcoated with another dielectric for stability purposes. Including this overcoating, four depositions are required to fabricate the network. This is one more deposition than that required in the fabrication of a capacitor, as the third deposition of a capacitor is a conductor which does not require a protective overcoating.

The C-R-NC network of Fig. 3-32 is constructed by depositing a dielectric of predetermined thickness over a conductor; then a resistor, another dielectric of predetermined thickness and a conductor are deposited. As in the case of a typical capacitor, this structure need not be overcoated as the final deposition is a conductor. Therefore, this structure requires five depositions in its construction.

Of the many distributed resistance capacitance networks, the easiest to fabricate are those requiring essentially uniform film thicknesses (i.e., the film thickness does not vary across the width or along the length of the film).

3.9 THIN FILM MICROWAVE CIRCUITS

3.9.1 MICROSTRIP

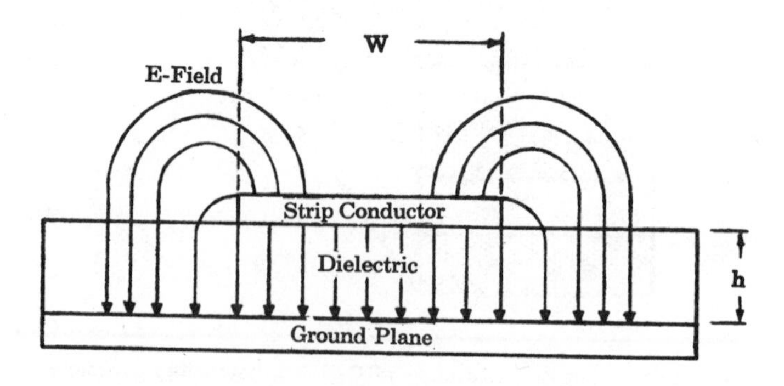

Fig. 3-33 Cross-section of the microstrip transmission line.

(a) Theory

The parameters of strip line (Figure 3-33) have been fairly well established. The characteristic impedance, Zo, wavelength, λg, and loss per wavelength can be determined as a function of frequency and other pertinent parameters for microstrip lines.

The characteristic impedance of the unshielded microstrip line is given by

$$Z_o = \frac{377h}{\sqrt{\epsilon r}\,W\ \left[1 + 1.735\, \epsilon r^{-0.0724} \frac{W}{h}^{-0.836}\right]} \qquad (3\text{-}22)$$

and the wavelength by $\frac{\lambda g}{\lambda TEM} \approx \left[\frac{\epsilon r}{1 + 0.6\,(\epsilon_{r-1})\ \frac{W}{h}\,0.0297}\right]^{1/2}$ (3-23)

when $\frac{W}{h} \geq 0.6$

where λg = guide wavelength

λo = free space wavelength

$\lambda TEM = \lambda_o/\sqrt{\epsilon r}$

ϵr = relative dielectric constant of insulating layer

W = width of conductor

h = thickness of dielectric substrate

The attenuation due to conductor loss is approximately

$$\alpha_c = \frac{\sqrt{\pi f \mu}}{2Z_0 W}\left[\frac{1}{\sqrt{\sigma_1}} + \frac{1}{\sqrt{\sigma_2}}\right] \text{ nepers/meter} \qquad (3\text{-}24)$$

where f = frequency in hertz

$\mu = 4\pi \times 10^{-7}$h/m

σ_1 = conductivity of strip in $(\text{ohm} - \text{m})^{-1}$

σ_2 = conductivity of groundplane in $(\text{ohm} - \text{m})^{-1}$

In the limiting case, as the impedance approaches a low value (when $W/h >> 1$),

$$\alpha_c = \frac{\sqrt{\pi f \epsilon_o \epsilon_r}}{2h}\left[\frac{1}{\sqrt{\sigma_1}} + \frac{1}{\sqrt{\sigma_2}}\right] \text{ nepers/meter} \qquad (3\text{-}25)$$

and from this equation it may be seen that large h (obtained by thick substrates) is required to minimize α_c.

The losses due to the insulating substrate can be neglected for hybrid circuits on ceramic. However, losses on silicon can not be neglected and may be approximated by

$$\alpha d_{\text{SEMICOND.}} \approx \frac{1/2\ Z_0 W}{\rho h}$$

where $\rho =$ Substrate resistivity in ohm-cm.

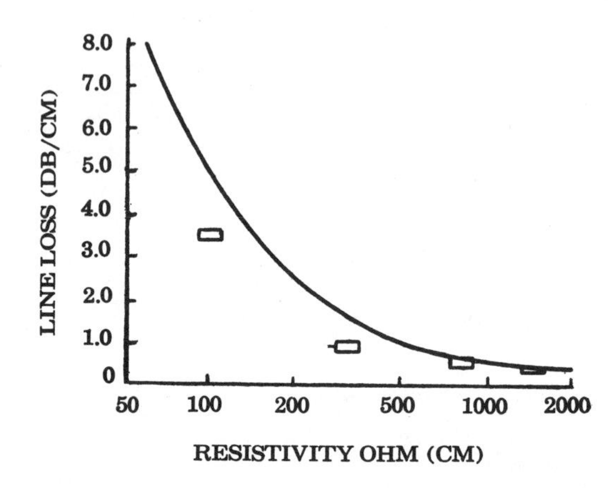

Fig. 3-34 Microstripline loss as a function of dielectric resistivity.

Figure 3-34 shows resistivity of silicon substrate versus line loss. However, as the temperature increases, charge carriers are thermally excited, lowering the resistivity of the material. Even at lower temperatures, the resistivity decreases due to the increase in mobility of the charge carriers with decreasing temperature. This curve is shown in Figure 3-35.

For circuits not requiring extremely high Q, microstrip on silicon can be very useful. If a higher Q is necessary it can be achieved, up to a limit, by using a higher dielectric substrate such as ceramic. A hybrid IC will then be necessary.

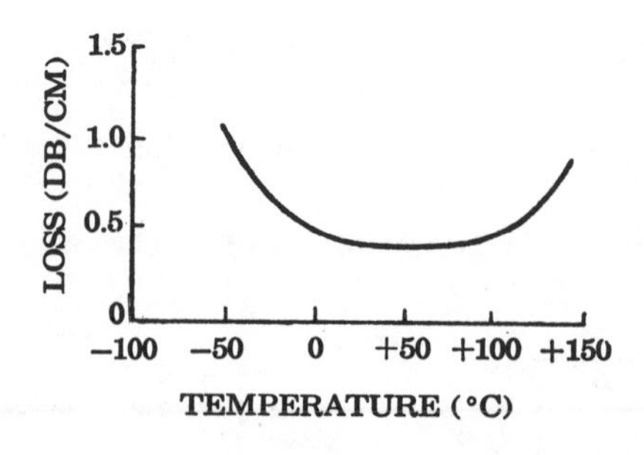

Fig. 3-35 Microstripline loss on 1500 ohm silicon substrate.

The Q (quality factor) of a microstrip resonator is an important design consideration. If the microstrip is an open or shorted section, the unloaded Qu of a quarter-wavelength resonator is $Qu = \frac{QdQc}{Qd+Qc}$ where Qc is the Q of the conductor, and Qd is the Q of the substrate. The Q of the substrate, (Q_d), is approximately the Q of the dielectric.

Calculated values of Q show that microstrip resonators have lower Q's than conventional cavity resonators, and that the penalty in performance is greatest in the monolithic IC type of microstrip.

(b) Selection of Substrates

Selection of a microwave IC substrate is critical in that the substrate is an integral part of the circuitry, and must be chosen for low dielectric loss. The chosen substrate must be homogeneous both within itself and from batch to batch. The thin-film must strongly adhere to the substrate, and even if the substrate is only 10 to 50 mils thick, neither the substrate nor the conductors should deform by temperature cycling. The substrate thermal conductivity should be high, and its surface must be free of pits for uniform transmission lines and short free capacitors to be deposited. Usually, substrate thickness is a tradeoff between conductor Q and the maximum allowable thermal resistance of the substrate.

(c) High Dielectric Substrates

The primary advantage of a substrate having a higher dielectric than ceramic, is size reduction. By increasing the dielectric constant of the substrate from 10 to 100, the guide wavelength of a 50-ohm microstrip line can be reduced by a factor of 0.35, and low loss and microstrip lines can be fabricated.

High-dielectric substrates, consisting of a temperature-compensated titanium dioxide homogeneous mixture, have been shown to have the properties required for reduced-size microwave integrated circuits. The variations of microstrip wavelength, characteristic impedance, and attenuation with geometry and dielectric constant are in good agreement with the theory. The low values of attenuation and guide wavelength make this material particularly attractive for low loss microwave circuitry.

Only smaller physical size may justify using high dielectric materials, because other factors tend to discourage its use. The rough surface of titanium dioxide makes it impossible to form thin film capacitors without shorting. The very small circuit dimensions, moreover, makes it difficult to make external connections. In order to get characteristic impedances at the higher levels, a thick substrate is required, and to add to the problem, the temperature conductivity is quite low: 0.05 for $Ti0_2$, 0.15 for ceramic and 1.0 for silicon.

(d) Circuits

Printed microstrip has made possible a variety of integrated components such as directional couplers. One type of directional coupler is shown in Figure 3-36a which uses the fringing effect to accomplish coupling. Tight coupling requires very close spacing between conductors (e. g. less than 0.5 mm) or very long sections (e. g. several wavelengths). It is possible that the circuit will resonate in the balanced twin strip mode. Another disadvantage of this type of coupling is the difficulty of controlling both the phase change coefficient and the coupling factor in a prescribed manner.

A second type of coupler, shown in Figure 3-36b, is known as the branched arm coupler. It has certain advantages over the parallel-line coupler, but if a high value of directivity over a large bandwidth is required, the number of branches becomes rather large (e. g. more than 3), and the impedance of the outer arms tends to become higher than can be realized in this or any similar form of construction (e. g. more than 150 ohms).

A third method (Figure 3-36c) employs two microstrip conductors back to back over a common ground plane. Coupling is accomplished by slotting the ground plane between the two strips. By the proper choice

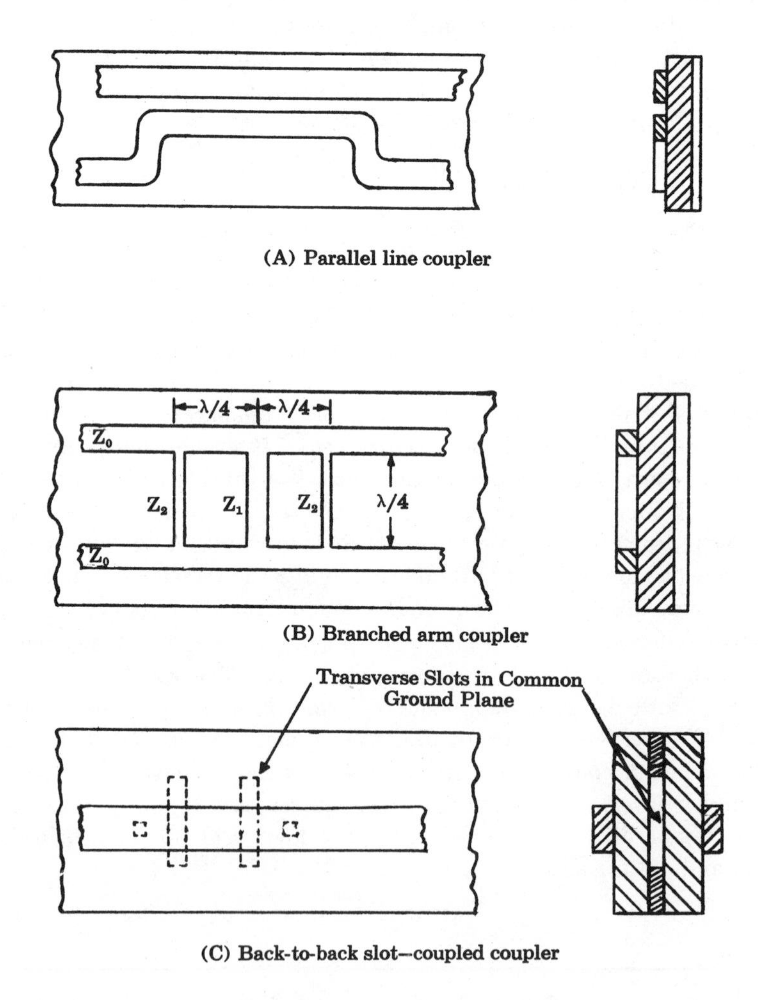

(A) Parallel line coupler

(B) Branched arm coupler

(C) Back-to-back slot—coupled coupler

Fig. 3-36 Differential forms of microstrip directional coupler.

of slots, a wide variation in coupling strength is possible, and either forward or reverse directivity can be obtained.

Microstrip filters may be one of three types. The parallel coupled resonator, shown in Figure 3-37, employs half wave sections which overlap adjacent section by one-quarter wave. (These sections need not be in a straight line. They can be in the form of a semicircle to conserve space.) The large amout of overlap results in both electrostatic and electromagnetic coupling. The coupling affects the image impedance of the lines, so that it is necessary to adjust the width of the sections to control the characteristic impedance.

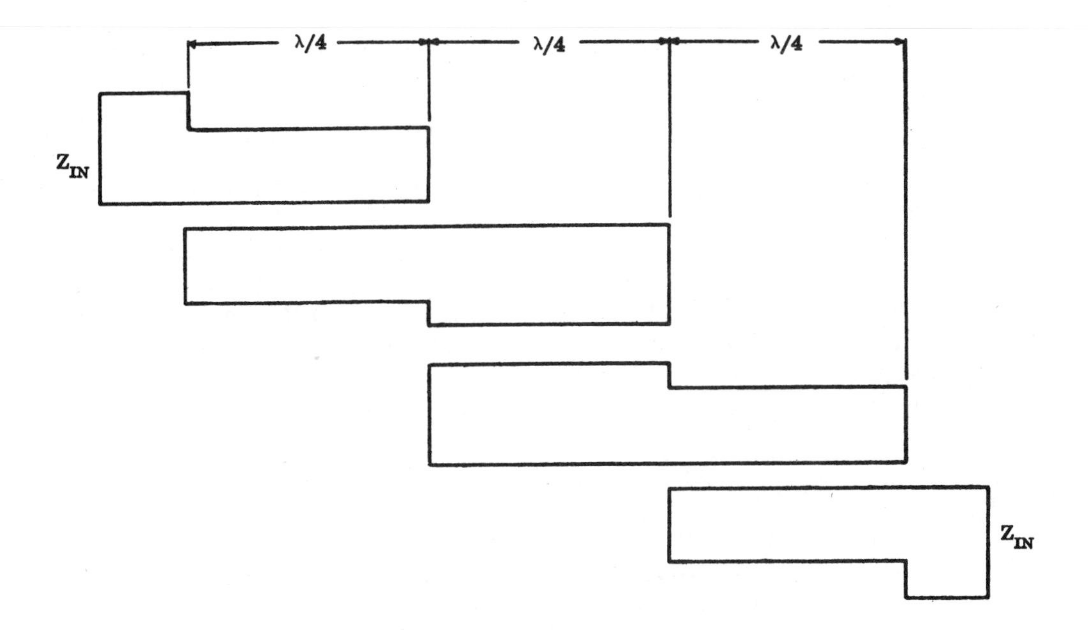

Fig. 3-37 Two-section parallel coupled resonator filter.

The direct-coupled filter, (Figure 3-38), has sections which are approximately one-half wave length long, and are coupled to each other at each end by gap capacitances. A computer program was written to determine gap widths and conductor lengths. The third type filter uses either open or shorted tuning stubs.

Couplers, filters, and transforming sections must be fabricated from transmission line sections having low loss and well defined characteristic impedance properties, using process and materials compatible with other circuitry components.

One problem is the loss characteristics. For short interconnection of 50-ohm circuits, a loss of 3-5 dB/cm can be tolerated. The design of filters and couplers requires that the loss be no greater than 0.5 to 1.0 dB/wavelength. Bandpass filters require a few hundredths of a dB per wavelength for steep skirts and low insertion loss.

An additional advantage can be gained by using microstrip on the same substrate as the active components. By using high resistivity silicon, microstrip transmission lines loss can, at room temperature, approach conductor losses only.

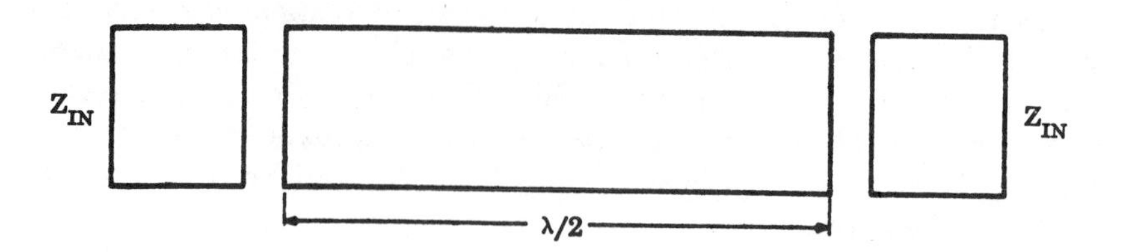

Fig. 3-38 One-section direct coupled resonator filter.

3.9.2 LUMPED ELEMENTS

If the dimensions of a circuit element are so much smaller than a wavelength that distributed effects play no part in its operation, it is called a lumped element. At microwave frequencies lumped elements are very small. It has been more practical to build large distributed circuits of waveguides and coaxial cables because the circuits have higher Q due to a higher storage volume. But in IC's these elements are not practical;

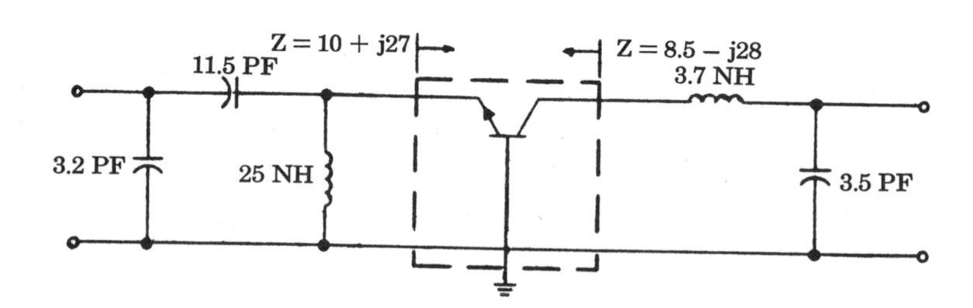

Fig. 3-39 RF circuit diagram for power amplifier.

consequently minute lumped elements are being fabricated with exacting precision. The phase shift over the element is zero because they are so short.

In the microwave region, inductors may be fabricated by thin-filming a spiral, loop, or a meandor line on a high dielectric substrate. At 2 GHz, a 5-turn spiral, 0.040 inch in diameter, on glass has an inductance of 10 nh and a Q of 45. A 0.060-inch coil has an inductance of 25 nh and a Q of 60.

Figure 3-39 is a thin-film lumped element circuit to match a transistor pellet to 50-ohm transmission lines. The entire amplifier occupies an area 130 mils by 110 mils. Figure 3-40 illustrates the relative sizes of the different elements of an integrated 2 GHz lumped element amplifier. A 2 GHz microstrip amplifier using the same transistor and designed for minimum size, is about 10 times as large.

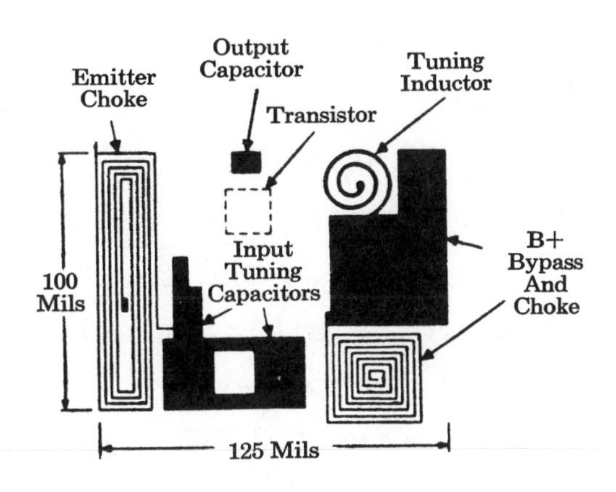

Fig. 3-40 Layout of 2-GHz integrated lumped circuit amplifier.

THIN-FILM ELECTROLUMINESCENCE

ELECTROLUMINESCENCE - DEFINITION AND PROPOSED MODELS

Electroluminescence is defined as the light generated by a crystalline solid as the result of an applied electric field. Depending upon the crystalline material and electrode geometry, light may be generated by an alternating (AC) or continuously applied (DC) field. One explanation for light emission may be related to the energy released upon recombination of minority and majority carriers in a solid at recombination centers. Alternately, the emission may be due to impact ionization of luminescent centers by charge carriers which have been accelerated by high electric fields within the solid.

DC Electroluminescence

Typical DC electroluminescence such as obtained from light-emitting diodes is achieved by injecting minority carriers into a crystalline material across a potential barrier. The minority carriers then recombine with the majority carriers, either directly or through luminescent centers, to produce radiation. Through use of the potential barrier, energy from the externally applied electric field can be selectively applied to injected minority carriers which, in turn can convert this energy to light upon recombination with majority carriers.

Typically, the potential barrier is achieved by diffusing appropriate deposits into a basic crystalline material to provide A p-n junction in that material. The forward-bias junctions require a several-volt potential for visible emission, and so result in low-electric-field devices, compared with those emitting light through impact ionization. Typically, efficient low-voltage junction emitters require a single crystal structure. Thus the active display area for a monolithic display is limited to the size of available single crystal substrates.

Larger area LED displays than 1-cm diameter, for instance, require a hybrid structure.

AC Electroluminescence

AC electroluminescence is produced by applying an alternating electic field across a luminescent crystalline material. Typically, the material is a high resistance II-IV compound (e.g., ZnS). Experimental observations of AC electroluminescence in ZnS powder layers indicate that the light output originates from discreet spots within each crystal. The spots coincide with each end of hairlike dicontinuities where hexagonal and cubic crystal structures meet. Alternate ends emit light on alternate half-cycles of exitation. The mixed crystal structure is a requirement for efficient luminescence in powder phosphors. In addition, the presence of Cu is a prerequisite. Usually, the final step in the preparation of efficient powder includes a wash in KCN in order to dissolve excess copper. Very possibly, a copper cyanide complex is left which contributes to electron emission from the copper sites. It is hypothesized that the hairline fissures are precipitated Cu_2S, which provides a source of carriers under high-field excitation.[2]

Several models have been proposed to explain the mechanism of AC electroluminescence in ZnS powders. The complexity of the emission system, coupled with the difficulty in direct observation and measurement of the phenomenon in the powder form, have precluded a definite model to date.

One model attributes the luminescence to impact ionization of the lattice by field-accelerated carriers originating from the tips of the conducting Cu_2S lines. Upon field reversal, the carriers return to the region of their origin and recombine under low field conditions to yield visible radiation.

Detailed observation of light emission vs. field polarity does not support the impact ionization model. Since the normal radiative combination from ZnS:Cu, Cl has been shown to be the result of a free electron and trapped hole, one would expect light emission to occur when the emitting spot is positive. Observations show that the emission occurs when the spot is negative.

Other observations also do not support the impact ionization model. Very weak electroluminescence has been detected for both DC and AC field excitation with voltages less than 2.2V across thin films of ZnS:Mn.[3] Another model proposed by Fisher[4] assumes that both holes and electrons tunnel from the Cu_2S inclusions. Tunnel emission can occur at less than bandgap energy (bandgap of hexagonal ZnS = 3.6V at room temperature). Selective trapping of the holes near their emission sites then allows radiative recombination to occur when the Cu_2S inclusion goes negative with an attendant emission of electrons near the trapped holes. This model then accommodates the low-voltage emission, the polarity of the emission site, and the time-delayed emission as a function of applied field.

The observations and theories described above pertain particularly to AC electroluminescence in bulk powder phosphors.

Thin-film AC emitters fabricated under this program have a number of distinctly different characteristics. In particular, copper is not a constituent in the film. There is no KCN wash used in the fabrication of the film. Observation of the emission under a microscope shows none of the hairline emission sources described for powder electroluminescence. The emission is distributed throughout the volume of the film, rather than emanating from a surface. Vlasenko and Popkov[5] report similar findings for thin-film electroluminescence in ZnS:Mn.

ELECTROLUMINESCENCE IN ZnS - GENERAL CHARACTERISTICS

Spectral Characteristics

The color of emission is determined primarily by the type of activator introduced into the ZnS lattice. For instance, for the ZnS:Cu, Cl, Mn system, excess copper shifts the emission toward blue, excess chlorine results in a green shift, and high concentration (1-2%) of manganese results in a characteristic Mn yellow-orange emission. Electroluminescent ZnS powder almost always contains copper. The concentration (0.1%) is a factor of 10 to 100 higher than that in photoluminescent phosphors. Apparently, the precipitated Cu_2S on the stacking faults between the hexagonal and cubic phases is a prerequisite for efficient electroluminescence. This is a significant difference between AC powders and the Sigmatron thin-film structures which contain no copper.

Operation of the ZnS:Cu, Cl powder phosphors at various frequencies results in a color shift with frequency. Increasing the frequency results in a blue shift, and lower frequencies provide green emission. A decrease in temperature has the same effect as an increase in frequency. Microscopic observations of powders indicate that radiative recombination takes place at different locations from the Cu_2S precipitate. This is interpreted as a shift in the range of the carriers as a function of frequency. Thus high frequency results in a shorter range of the carrier, with emission occurring nearer the copper precipitate. Since it is known that copper-rich doping results in a blue shift, it is inferred that the higher frequency, blue emission close to the precipitate is due to the higher copper concentration in the region of radiative recombinations.

Spectral shift with excitation frequency has been used to provide a multicolor display. However, for an x-y matrix scan, a frequency-stable spectral emission is desired to avoid color shift, with different repetition rates used in refreshing the display.

Manganese (Mn) activators provide a yellow-orange emission color that is virtually independent of frequency or temperature. The emission mechanism from the Mn^{+2} ion is different from that of copper or chlorine activators. The emission is the result of the relaxation of the Mn^{+2} from an excited energy state to a ground state. Excitation of the manganese ion can occur by resonance capture of electron-hole recombination energy or by direct-collision excitation (impact ionization). Vlasenko and Yaremko[6] investigated the AC-electroluminescence of ZnS:Mn films as a function of thickness between 0.04 and 2μ. Below 0.1μ the emission dropped very rapidly. They concluded that excitation mechanism was impact ionization by accelerated electrons.

As noted above, efficient luminescence with a Mn^{+2} activator requires a 10X higher concentration than with Cμ or Cl activators. The practical significance is that the Mn-activated phosphor system is less affected by impurity contamination and so is well adapted for producing large-area uniform luminescent films by vacuum deposition.

Brightness vs. Voltage Characteristics

Historically, the voltage-brightness characteristic was described by the empirical relation for powder phosphors[7]

$$\beta = \beta_o e^{\left(-\frac{c}{V^{\frac{1}{2}}}\right)}$$

Later work by Lehman [8], however, showed that the characteristic is a function of the particle size and approaches

$$\beta = \beta_o e^{-\left(\frac{V_o}{V}\right)}$$

as the particles become smaller. A "steep" brightness-voltage characteristic is desirable for an x-y display matrix in order to avoid visible crosstalk of non-addressed intersection due to capacitive cross-coupling.

A unique property of the thin-film phosphors is the steep brightness-voltage characteristic. This is attributed to volume emission from thin films, compared with surface emission for powders as observed through a microscope. For instance, Thornton reports no structure visible in the emission at resolutions less than one micron for thin films.[9]

Several investigators report a brightness characteristic similar to fine powder phosphor layers, i. e., $\beta \propto e^{-K/V}$ (Thorton, Nickerson and Goldberg). Vlasenko and Popkov [5] report a characteristic of $\beta \propto V^7$, while Goldman et al[10] report a characteristic of $\beta \propto V^{13}$. Films produced under this program exhibit a $\beta \propto V^n$ with an exponent for V between 9 and 15, depending on the fabrication parameters.

Time Dependence of Brightness Waveform

A typical characteristic of AC electroluminescence is a half-cycle delay in emission when the voltage waveform is first applied. No emission is observed if a single-voltage pulse is applied to a relaxed cell. The delay is due to untrapping processes and radiative recombination which occurs upon the second half-cycle of excitation. The buildup in brightness to steady-state excitation level requires 4 to 10 cycles as the trapping and untrapping processes reach an equilibrium. Depending upon the frequency of the AC excitation with respect to the rate and type of the trapping processes, the brightness waveform exhibits different peaks. At high frequency, a ripple of twice the excitation frequency results, superimposed on a constant brightness level. At lower frequencies, the brightness wave leads the voltage wave, and small secondary waves appear. For ZnS:Cu.Mn, only one peak per half-cycle is produced. Thus each phosphor system can exhibit different characteristics, depending upon its activators and excitation mechanisms. The brightness varies sublinearly with frequency for most powder phosphors.

Temperature Dependence

The major effect of temperature for the Mn-activated emitter is on the recombination kinetics, rather than the excitation mechanism by which carriers are initially generated. Below 200°K, there is trapping of all mobile charge carriers. A certain thermal activation is necessary to free electrons from trapping levels so that they are available for recombination upon the succeeding half-cycle of excitation. In the range of 200°-400°K, the effects of temperature on light output from powder ZnS:Mn are small (<3:1)[11]. Above 200°C, there is sufficient energy to free holes from trapping levels which result in nonradiative recombination and a decrease in light output. Temperature effects on the light-emitting film structure are detailed under "Device Characteristic." Essentially, similar characteristics to the above prevail. Minimum emission occurs near room temperature.

Crystal Structure Dependence

As previously described, an important prerequisite for most reported electroluminescence is the presence of a copper precipitate within the crystal. These occur at stacking faults between cubic and hexagonal phases in ZnS. Observation of electroluminescence from ZnS-activated powder particles imbedded in a matched index of refraction glass indicates that the emission originates from areas that coincide with crystal faults as revealed by viewing under polarized light[12]. This type of emission from local areas of the crystal appear in all electroluminescent powders, including manganese-doped and rare-earth-doped powders.

The stacking faults occur in planes perpendicular to the "C" axis of the hexagonal phase. Also, it is found that maximum electroluminescence, in the case of the single-crystal hexagonal ZnS, results when the applied field is perpendicular to the "C" axis[7]. Although it is possible to grow single crystals of cubic ZnS, apparently no electroluminescence has been reported from single crystals of this structure.

In the case of the light-emitting film structure, X-ray diffraction patterns indicate polycrystalline cubic structure with the "C" axis perpendicular to the substrate. Thus it appears that these light-emitting films are materially different from previously reported electroluminescent emitters.

Effect of Electrode Material on the Physics of Emission

For AC electroluminescence, the material of the contact is not critical. However, it is important to have an insulating or blocking contact between one electrode and the luminescent material, particularly in the case of thin-film emitters. In an investigation of electrical contact to thin films of ZnS:Cu, Cl, Harper found that Al and Mg were outstanding[13]. Both materials are capable of forming an insulating layer of oxide under anodic polarization. Most reports in the literature indicate that luminescence occurs when the blocking contact is positive. However, there are reports to the contrary in which strong DC electroluminescence was obtained for the Al contact negative and the counter SnO_2 contact positive[14]. Cusano[15] points out the importance of the Cu_2S rich side of a film structure being negative for the DC case.

Efficiency

Luminous efficiency becomes an important consideration for large-area arrays where the emitting area is a large percentage of the total display surface. For instance, a major problem in the use of light-emitting diodes (LED) for x-y arrays is the power

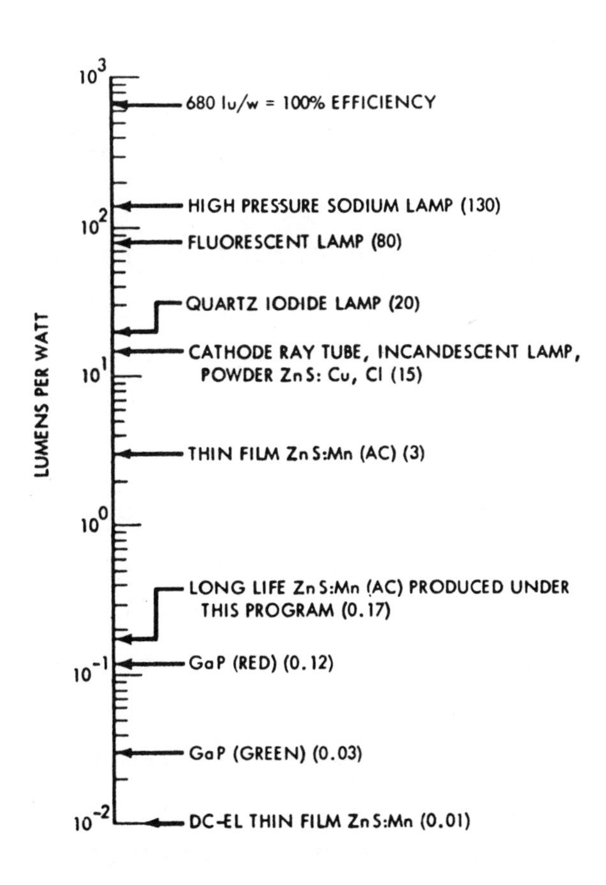

Fig.3-41 Luminous Efficiency of Various Light Emitters

dissipation required. Fig. 3-41 compares the luminous efficiency for various light sources. Note that luminous efficiency (lumens/watt) takes into account the relative color response of the human eye.

The highest efficiency for electroluminescence reported to date is 14 lumens per watt for a powder ZnS:Cu, Cl AC electroluminescent cell, as described by Lehmann[8]. This is equivalent to about 20 lu/w for the powder itself. Lehmann describes the effect of phosphor particle size on efficiency. A ten to one change in particle size (5μ - 50μ) results in a 3:1 change in efficiency, with the smaller particles producing the higher efficiency. Apparently, the 14 lu/w efficiency was obtained at 10 ft-L for 6μ-size powder. The frequency of operation was not stated.

The interpretation of the particle size effect assumes that the luminescence is a surface effect on each crystallite (as indicated by microscopic observation), while power dissipation is assumed to be a volume effect of the bulk properties of the powder, so that maximum surface area (at least to 6μ particle size) results in maximum light output for a given power input.

Estimates of the maximum electroluminescent efficiency possible, based on an impact ionization model, yield 38 lu/w[16]. Experimentally, it is found that maximum luminous efficiency in powder phosphors is obtained at about one-half maximum voltage, or less than one-tenth maximum brightness. In the case of thin-film phosphors with a much steeper voltage-brightness characteristic, maximum efficiency occurs near one-half maximum brightness. Efficiencies of 0.17 lu/w have been measured for light-emitting film structures under continuous 5 kHz sine wave excitation.

State of the Art

Fabrication

Thin-film electroluminescence has been investigated in at least six major independent laboratories during the past decade. Typically, the films have been fabricated in a two-step process. Initially, zinc sulphide phosphor containing activators is evaporated onto a tin oxide (SnO) coated glass substrate at room temperature. The tin oxide provides a transparent conductor and subsequently, a high-temperature heat treatment (500°-800°C for 10-30 minutes) promotes diffusion of the activators as well as improves the crystal structure throughout the deposited film. Variations on this general procedure have been reported by different investigators. For instance, Thornton maintains a proper concentration of the components in the system (Zn, S, activators) during heat treatment by burying the film in a bid of the activated powder[17]. Nickerson and Goldberg[18] recrystallize their films in the vacuum immediately after deposition. Also, extra activator material in the form of MnS is added to the ZnS:Cu, Cl source crucible.

Vlasenko and Popkov co-evaporated pure zinc sulphide powder and manganese metal from dual tantalum sources onto a room-temperature SnO-coated glass substrate. Subsequent heat treatment at 500°C yielded some of the more efficient thin-film emitters reported (3 lu/w)[5]. In addition, no other activators such as copper or chlorine were used. The absence of copper is a striking difference from most of the other reported work on luminescence in zinc sulphide. In fact, the presence of copper is so universal in most ZnS electroluminescent systems that the presence of a copper sulphide phase is included in the theory of emission as the source of charge carriers.

Koller and Coghill achieve recrystallization at atmospheric pressure through heat treatment in an atmosphere of H_2S and HCl[19].

Also, they described an alternate one-step evaporation technique which used a hot-wall evaporator and low pumping speeds so that high vapor pressures resulted. The hot-wall technique assured that any vapor tending to deposit on surfaces other than the substrate re-evaporated, and so favored deposition on the cooler substrate. In addition, a small flow of HCl was used during the evaporation in order to introduce a halogen activator.

Cusano describes a somewhat similar technique in which powders of the appropriate host and activator are slowly added to a hot-wall furnace (600°C) containing an H_2S atmosphere at a reduced pressure (0.5 Torr)[15]. In addition, Cusano reports a significant improvement in brightness through the use of a copper salt treatment of the exposed phosphor film prior to electroding. This treatment is especially important in the case of DC electroluminescence. One explanation is that the resultant copper sulphide layer is "p" type on the "n" type ZnS phosphor film, so that efficiency carrier injection results.

Spectral Characteristics

Depending on the activator incorporated in the zinc sulphide film, the color of emission can range from blue through red to the near infrared. Table 1 lists the various activators and emission peaks that have been reported in the literature. Except for the Cu, Cl activators, all the activators of Table 1 produce emission by transition between states of the metal activator ion similar to manganese described previously. It is within the past several years that these rare earth compounds have become readily available because of the emphasis on laser materials development.

The decay times associated with each activator are included as well. For visual readout, a decay time of less than 33 ms is adequate in order to avoid smear in a dynamic display. However, in the case where the emitter is to be used as an optoelectronic component (e.g., emitter in voltage isolated, optical coupler), the decay time is an important consideration. Table 1 indicates a considerable selection is available.

TABLE 1

Zinc Sulphide Thin-Film Electroluminescent Activators

ACTIVATOR	EMISSION WAVELENGTH (Å)	TIME CONSTANT (sec)
Cu,Cl	(Blue)	17
TmF_3	4750 Å	2
PrF_3	5000	24
ErF_3	5210	22
TbF_3	5430	220
Cu,Cl	(Green)	500
$DyF_3$6	5630	45
Mn	5800 (yellow)	1000
SmF_3	6500 (red)	230
NdF_3	6000/89000	<0.9/37

REFERENCES

1. P. Goldberg, Ed., Luminescense of Inorganic Solids, Academic Press, New York, 1966.

   M. Aven and J. S. Prener, Eds., Physics and Chemistry of II-VI Compounds, Amsterdam, The Netherlands: North-Holland

   P. R. Thorton, The Physics of Electroluminescent Devices, London, England: Span, 1967.

1a. E. J. Soxman, Electroluminescent Thin Film Research, JANAIR Report EL-1, 7 August 1965, AD 475 700L.

   E. J. Soxman, Electroluminescent Thin Film Research, JANAIR Report EL-2, 10 August 1966, AD 800 992L.

   E. J. Soxman and W. A. Smith, Jr., Electroluminescent Thin Film Research, JANAIR Report EL-3, 10 January 1967, AD 815 950 L.

   E. J. Soxman and W. A. Smith, Jr., Electroluminescent Thin Film Research, JANAIR Report El-4, 10 July 1967, AD 682 547.

   E. J. Soxman and Henry J. Hebert, High Contrast, Solid State Teletype Display, Technical Note ELTN-1, JANAIR Report 690309, 10 March 1969.

   E. J. Soxman, Electroluminescent Thin Film Research, EL-5, JANAIR Report 690410, 29 April 1969.

   E. J. Soxman, Electroluminescent Thin Film Research, EL-6 JANAIR Report 690513, 23 May 1969.

2. F. F. Morehead, "Electroluminescence," Physics and Chemistry of II-VI Compounds, p. 634 Ibid.

3. W. A. Thornton, Physical Review 116, 893, 1959; Physical Review 122, 58, 1961.

4. A. G. Fischer, J. Electrochemical Society 110, 733, 1963.

5. N. A. Vlasenko and Iu. A. Popkov, Optical Spectry (USSR), English Translation, 8, 39, 1960.

6. N. A. Vlasenko and A. M. Yaremko, Optical Spectry 18, USSR, 1965.

7. P. Zalm, Philips Research Report 11, 353, 417, 1956.

8. W. J. Lehman, J. Electrochemical Society 105, 585, 1958.

9. W. A. Thornton, J. Applied Physics 33, 3045, 1962.

10. A. G. Goldman, Soviet Physics - Doklady 9, 963, 1965

11. F. J. Morehead, J. Electrochemical Society 105, 461, 1958

12. A. G. Fischer, J. Electrochemical Society 109, 103, 1962.

13. W. J. Harper, J. Electrochemical Society 109, 103, 1962.

14. F. I. Vergunas, et al, Optical Spectry 16, 385, USSR, 1964.

15. D. A. Cusano, Luminescence of Organic and Inorganic Materials, Kallman and Spruch, Eds., John Wiley & Sons, New York.

16. F. J. Morehead, J. Electrochemical Society 107, 281, 1960

17. W. A. Thornton, U. S. Patent 3044902, 1959.

18. J. Nickerson and P. Goldberg, National Symposium on Vacuum Technology Transcript 10, 475, 1963.

19. L. Koller and H. Coghill, J. Electrochemical Society 107, 973, 1960.

CHAPTER 4

THICK FILM CIRCUITS

Thick film structures are prepared by screening and firing or by pyrolytic deposition. They generally contain only conductors, resistors, and capacitors. Other components must be added as discrete entities. The substrate is usually a ceramic wafer.

This technology is not new, but has been in existence for a number of years. A limitation of this technology is the inability to provide active components except by individual insertion with the attendant extra connections. There are also the usual limitations upon the value, tolerance, and stability of the resistors and capacitors. Attempts have been made to circumvent some of these problems by the development of new materials and new deposition techniques. In pyrolytic deposition, for example, chemical compounds containing the desired materials are thermally decomposed at the substrate over the entire surface. The pattern is then etched or machined into the surface layer.

A big advantage of the thick film technique is its high volume use and relatively low cost. It is also a flexible technique in which a variety of patterns can be achieved with little variation in the process.

The term "thick film" is defined as a conductive, resistive, or insulating film greater than 0.0001 inch thick that is produced by firing a thixotropic paste deposited on a substrate. (Thixotropic – a fluid which changes viscosity as a function of the rate at which it is sheared.) The paste is composed of powdered inorganic solids, such as metals and metal oxides, mixed with a powdered glass binder and suspended in an organic vehicle.

The pastes are deposited on a flat substrate – normally alumina – by means of the stencil screen process. In this deposition process, a fine mesh screen is employed to hold the pattern for the components to be deposited. The pattern is produced by photographic means, and the holes in the mesh are blocked by an emulsion wherever the inks are not to be deposited.

A different screening composition is used for each type of component —resistor, conductor, crossover. The ink contains the necessary polycrystalline solids to produce the desired electrical characteristics; that is, metals for conductors, high dielectric constant glasses for capacitors, and various resistivity value oxide and metals for resistors. A major ingredient is the glass frit which bonds the metals and oxides and provides adherence to the substrate when fired at temperatures of 750°C to 950°C.

The thick-film deposition process is an outgrowth of the common silk screen method of printing. The silk has been replaced by a woven mesh of fine stainless steel wires but the basic concept remains the same.

4.1 DESIGN AND LAYOUT CONSIDERATIONS

The central issue in converting a circuit schematic to a thick-film hybrid circuit is the matter of material selection and sequencing. Thus, conductor materials that fire around 900°C cannot be applied to a substrate after the application of resistors that fire at 750°C. Each successive firing cycle should have a peak temperature at least 50°C below the one preceding it. For most material combinations, a 100°C separation is needed. The more complex the design (i.e., designs containing multiple sheet resistivities, crossovers, overglazes, etc.) the more difficult is the sequencing process.

Once the schematic has been completed, and the resistor sizes and printed capacitors have been calculated, a layout is made on conventional grid paper. This layout is prepared 10 to 20 times normal size.

For each screening, a master pattern is required. Patterns are needed for: conductors, crossovers, capacitor dielectrics, resistors, and resistor protective glass encapsulants.

These masters are photographically reduced and a normal sized positive made for each. The nylon or stainless steel screens are then made in either of two ways.

In one method, the photo-sensitive emulsion is painted onto the bare screen and allowed to dry. The screen is then photographically exposed in contact with the master pattern. Next, the pattern is developed. The open area, through which the ink is to be squeezed, is "washed out" to remove any unexposed emulsion material.

The second technique for making the screened masks employs an indirect or transfer emulsion of light sensitive gelatinous film, backed by a polyester sheet. The photographic exposure, development, and washing out are performed on this gelatinous polyester film. When the developing processes are complete, the wet film is transferred to a clear raw screen to which gelatin adheres. After drying, the polyester backing is stripped off and the screen is ready for use. Finer lines and irregular patterns can be achieved with better definition using this technique. The lifetime of this type of screen is approximately 10% that of the direct emulsion type.

4.2 THICK-FILM DEPOSITION TECHNIQUES

In the deposition process, to reiterate, the inks are forced through a fine screen by using generally a polyurethane squeegee. A squeegee blade forces a viscous paste through the predetermined openings in a stencil-like mask to deposit a desired image. The mask serves as a metering device. Almost all production processes for thick-film hybrids employ a mask formed from a selectively exposed and removed photographic emulsion carried on a woven wire screen. An etched metal mask that eliminates both woven screen and emulsion has been investigated for production use. Each method utilizes a semi-flexible squeegee blade that forces the mask to conform to minor flatness deviations in the substrate as the squeegee passes over. Other things being equal, sharp squeegees produce higher sheet resistivities than do dull squeegees. Since squeegee blades wear significantly, a gradual increase in sheet resistivity can be expected over even modest sized production runs (on the order of 100 units). The shift is *not* negligible and can be as much as 30 to 50 percent.

Screen printing normally can be described as "off-contact" printing; that is, the screen is *not totally in contact* with the part being printed at any time. A conceptional layout of the off-contact printing technique is shown in Figure 4-1.

By contrast, contact printing requires that the screen be lifted abruptly and cleanly at the completion of the squeegee operation. This is difficult to do repeatedly and requires very precise and expensive equipment. However, contact printing can produce 2 mil line widths compared to 8 mil line widths for off-contact methods; this is a significant advantage, and an actual necessity where standard beam-lead device contact spacing of 6 mils is a design requirement.

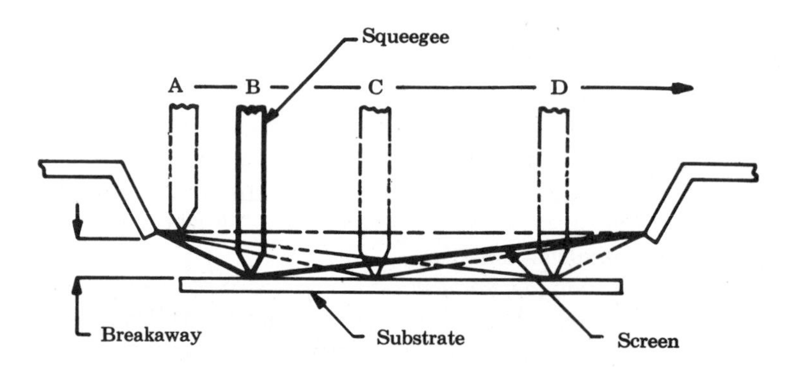

Fig. 4-1 Off-contact printing method.

To print, first a quantity of ink is deposited on the screen, usually with a spoon or spatula. In the "off-contact" process, the plane of the screen is fixed a distance above the substrate. When the squeegee is placed in contact with the screen and pressure is applied, the wire or mesh in the screen stretches or flexes sufficiently to permit the mesh directly under the squeegee to contact the surface of the substrate. As the sqeegee moves across the screen, the mesh snaps back into place behind the squeegee and is pushed down ahead.

In this "off-contact" printing, the snap-back behind the squeegee is responsible for the sharp, well-defined edge to the pattern. In the "contact" printing, some means of an abrupt separation between screen and substrate is also required for the same reason.

4.2.1 SCREEN MASKS

The screen mesh, wire diameter, and emulsion thickness establish the number and size of the metering chambers in a woven wire screen mask. The emulsion is generally applied to the surface of the screen after the desired pattern has been formed. It is pressed into the screen by a roller or some similar means, but most of the emulsion thickness remains on the surface of the screen. Some pattern distortion can result from this method, and the overall thickness of screen plus emulsion is not always very well controlled. This will affect the value of screened resistors. Forcing the emulsion completely into the screen might give a slight improvement of thickness control, but this would be countered by further pattern distortion due to the higher pressures required.

An alternate method of forming screen masks starts with an unexposed emulsion embedded in the screen. This gives good control over thickness. The pattern is then formed photographically, and does not undergo the distortion that would result if it were then pressed into the screen. Edge definition, however, turns out to be degraded with this method, probably because the curved, shiny surfaces of the screen wires reflect the exposing light to produce non-uniform exposure of the emulsion.

Other screen printing parameters include the material and shape of the squeegee blade, the squeegee pressure and angle of attack, and the spacing between substrate and screen before its deflection by the passing squeegee. Minimum line widths attainable with a woven screen cannot be less than the wire spacing, even in principle. In practice, the minimum line width must be several times this spacing. One might suppose that the use of finer meshes would allow the printing of finer lines, but this is so only up to a point. The finer mesh screens are less stiff, or taut, than are coarser meshes, and this eventually results in a degree of smearing that cancels any potential advantage.

4.2.2 ETCHED MASK

The etched mask does not carry an emulsion nor does it have the fundamental line width limitations of woven screens. It offers the potential of closer control over the deposition process, and the fine line widths needed for flip-chip and beam lead device bonding. Etched masks are more expensive than screen masks, however, and the woven wire screen mask has distinct economic advantages where lesser tolerances are allowable.

4.3 THICK FILM MATERIALS AND FABRICATION TECHNOLOGY

Thick-film pastes are readily available and contain such metals as silver, gold, platinum, palladium, rhodium, iridium, ruthenium, osmium, molybdenum, manganese, etc. They may contain different oxides, and a wide variety of glasses and solvents. There is literally no end to the number and variety of combinations available. A relatively few basic types of pastes, however, account for the great majority of thick-film devices. There is more than one supplier for most of these basic types, and each supplier's version has its own peculiarities.

4.3.1 CONDUCTORS

There is a multitude of criteria which must be considered in the selection of a compatible conductive composition. Although this imposing list of variables makes the material supplier's task difficult and sometimes impossible, these variables represent a major advantage of thick-fim technology to the user, since conductor materials can often be developed which have nearly ideal combinations of properties necessary to perform the functions required of them.

Thick-film conductors are fabricated from metal-glass compositions. These conductor formations consist of a finely divided suspension of such metal-oxides as palladium-silver, platinum-silver, or platinum-gold powder plus a glass frit in an organic vehicle which renders the paste suitable for screening.

Firing temperatures for conductors range from 500°C to 1000°C depending upon the ink used; the timing is set to optimize the conductivity, adhesion, and solderability of the conductor.

The conductive paste must offer high conductivity (0.01 ohms/square), high adhesion to the substrate, compatibility with the other printed materials, and be bondable. In practice, it is very difficult to achieve all these parameters at a satisfactory level in one paste, and it may be necessary to resort to printing more than one conductivity paste.

Substrate adhesion, for example, is generally in competition with line definition and sheet resistivity, and solderability usually competes with eutectic die bonding characteristics. Further, processing schedules (drying time, firing time, peak firing temperature) will usually affect these properties significantly.

The main parameters of these conductor inks that can be controlled are composition, particle size, and viscosity.

Silver: Silver is the least expensive of the common conductor materials. It exhibits good substrate adhesion, good solderability, and low sheet resistivity. Line definition is good also. The recommended firing temperature of most silver pastes is in the 500°C to 700°C range, which can be advantageous or not, depending on the application. The chief liability of silver pastes is the threat of silver ion migration. For this reason, silver is avoided in non-hermetic applications. A simple glass coating of the silver conductors drastically reduces the possibility of silver migration.

Gold: Gold pastes are second only to silver for low sheet resistivity. It is a superior material for eutectic and ultrasonic die bonding. Line definition is good and lead bonding characteristics are excellent. Gold conductors appear to be the most compatible conductor for interfacing with palladium-silver resistor pastes.

Substrate adhesion, however, is the poorest of the lot. Furthermore, the use of tin-lead solder on gold conductors inevitably introduces production and yield problems because of the severe scavenging action of molten solder on gold. Also the important properties of gold conductors are more sensitive to firing temperatures, than is the case with most other conductors.

Palladium-Gold: The addition of palladium to gold markedly improves substrate adhesion, but at the expense of poor line definition. Solderability is improved and lead bonding characteristics remain good.

Palladium-Silver: Palladium-silver has good substrate adhesion and lead bonding characteristics, is solderable, and can have firing schedules identical to that for the most commonly used resistor pastes. This permits co-firing of resistors and conductors, which is often desirable. The liabilities of the material are the impracticality of using eutectic die bonding processes, and the threat of silver ion migration. While palladium-silver can be soldered, the results are not as good as platinum-gold, and the solder should contain at least 2 percent silver to prevent leaching silver out of the conductor during tinning.

Platinum-Gold: Platinum-gold is the most widely used thick film conductor material even though it may be more expensive and more resistive than other types. Substrate adhesion and line definition are adequate for most purposes. The reasonable good balance of properties and resulting versatility is the reason for the popularity of platinum-gold. Close control of firing schedule is necessary to maintain this balance. In general, platinum-gold has very good soldering properties with the capability of being resoldered several times. Platinum-gold conductors are compatible with all resistive inks commonly used and can be co-fired with the platinum resistive inks. Direct eutectic bonding of silicon die is marginal. Gold-silicon preforms are used in order to secure a reliable silicon die bond to the platinum-gold.

Moly-Manganese (molybdenum-manganese) : Unlike the other conductor materials, moly-manganese is used primarily as an underlayer metallization in the fabrication of packages with ''hermetic'' seals. It is readily gold-plated so that it is used frequently as a metallization for ceramic hermetic packages rather than as a microelectronic conductor material.

Active Chip Bondings: Gold alloys which have received acceptance for active chip bonding are: 98Au/2si; 88Au/12Ge; and 80Au/20Sn.

4.3.2 ANOMALIES IN THICK FILM MATERIAL

The effect and importance of termination metals on high resistivity inks ($\rho_s > 100$ K/□) like the ruthenium oxide (Ru 0_2) – silver system is startling, especially in short resistors. Of those termination materials examined by the General Electric Company, the palladium silver produced the most uniform scaling, or freedom from resistivity variations as a function of resistor dimensions.

A uniform resistance distribution between the terminations will result in a straight line relation of voltage vs distance. However, a profile of a ruthenium oxide — silver system terminated with palladium gold or platinum gold exhibits three distinct resistance regions. Near each termination is found a small region having a resistivity higher than that of the bulk of the resistor. Therefore, the actual resistance is significantly higher than expected.

4.3.3 RESISTOR CHARACTERISTICS

The resistor is probably the most important thick-film circuit element. It is the most complex element chemically, and is subject to variations in processing by the circuit maker. The main parameters of a resistor are resistance, TCR (thermal coefficient of resistivity), stability, and noise. These parameters must be controlled to close numerical values both in the manufacture of the pastes, and in the conversion of pastes to resistive elements.

A wide range of resistance values can be obtained by either choosing a desired resistor paste or by varying the resistor geometry. The paste can be screened and fired to produce resistors of 0.1 ohm to 10 megohms although 1 ohm to 1 megohm is a more practical limit.

The original thick-film resistor was based on carbon. This was followed by compositions of silver and palladium suspended in a glass matrix. The final resistive properties of this metal thick film material, depend on a complex reaction occurring between the silver particles, palladium oxide particles, reduced palladium oxide and the glass matrix. This reaction is complicated by the fact that these materials are in very fine particle form, and hence have high surface free energy allowing reaction to occur that would not be expected from consideration of the bulk material.

The compositions of many thick-film resistor pastes are proprietary. Accordingly, they are not as readily categorized as are conductors. Two basic types can be distinguished by the sensitivity of sheet resistivity to the firing schedule. If the firing process produces a strong oxidation-reduction reaction, as it does with the most widely used composition (palladium-palladium oxide-silver), then firing sensitivity will be high. Other compositions, notably those based on a rare earth oxide, are relatively insensitive to firing because the oxidation-reduction reaction either does not occur, or is minimized. The dividing line is not so sharp as this view might suggest; most suppliers offer more than one series of their materials, over which the degree of firing sensitivity can vary substantially.

A resistor system which has been widely accepted is the palladium silver, palladium oxide cermet resistor. This resistor is formed as a metal-metal oxide glass cermet on firing at about 760°C in an oxidizing atmosphere. The reaction does not reach completion so that the final state of the resistor film parameters are dependent on the temperature profile and the time spent in the reaction zone.

The sheet resistivity is determined by the amount of metal present in the composition, and the amount of silver, determines the degree to which the reaction proceeds and controls the amount of palladium oxide in the final film.

4.3.3.1 Resistivity and Resistance

The resistance is expressed in ohms per square of a resistor film of a defined thickness. This is referred to as sheet resistivity. Whether a large or small square is printed, the resistance is the same for a given paste. Thus, one can for all practical purposes disregard the dimensions of the square. By printing two squares in a row, the resulting resistance doubles. The thickness of the thick film resistor, therefore, has an important bearing on the resulting sheet resistivity — the thinner the resistor film, the higher the resulting resistance value.

Final resistivity and the temperature coefficient of resistance will be determined by the final chemical composition of the resistor material. This final composition is determined by the firing profile.

4.3.3.2 Temperature Coefficient of Resistance

A thick film resistor's temperature coefficient of resistance (TCR) is the composite of two factors — the stress effect and composition effect. The stress (mechanical) component of TCR results from the making and breaking of electrical contacts plus certain electrostrictive effects. In contrast, the compositional effect is purely the result of chemical and physical composition on the temperature coefficients of the various electrical conducting phases that are present.

Conduction is complex and will proceed by metallic conduction in the alloy, semiconduction in the oxide, and quantum mechanical tunneling in the glass. The metallic conduction will exhibit a positive temperature coefficient while the other two conduction techniques will exhibit a negative temperature coefficient.

4.3.3.3 Voltage Coefficient of Resistance

Thick film resistors exhibit changes in resistance as a function of the applied voltage. Such changes have been shown to be principally a function of the resistor geometry and the resistivity of the thick film material. The following graph in Fig. 4-2 indicates *the voltage required to produce a 1% change in resistance* as a function of resistor length and film resistivity.

4.3.3.4 Resistor Materials

Resistor materials differ from the conductor materials primarily because the glass-to-metal ratio is considerably higher. In conductors, the metal content is high in order to achieve low resistivity. Conversely, in resistor materials, the metal content is low in order to achieve a controlled conduction to obtain a "semiconductor." Thus, the resistor pastes consist of conductor or semiconductor particles mixed with ground glass, organic binders, and solvents which, when combined, form a thixotropic paste.

Single phase carbon-based resins were originally used as resistor materials by the electronics industry. However, it was found that these early films were relatively unstable in ambients involving high temperature and humidity, thereby proving them to be unsatisfactory and unreliable for most of today's electronic requirements.

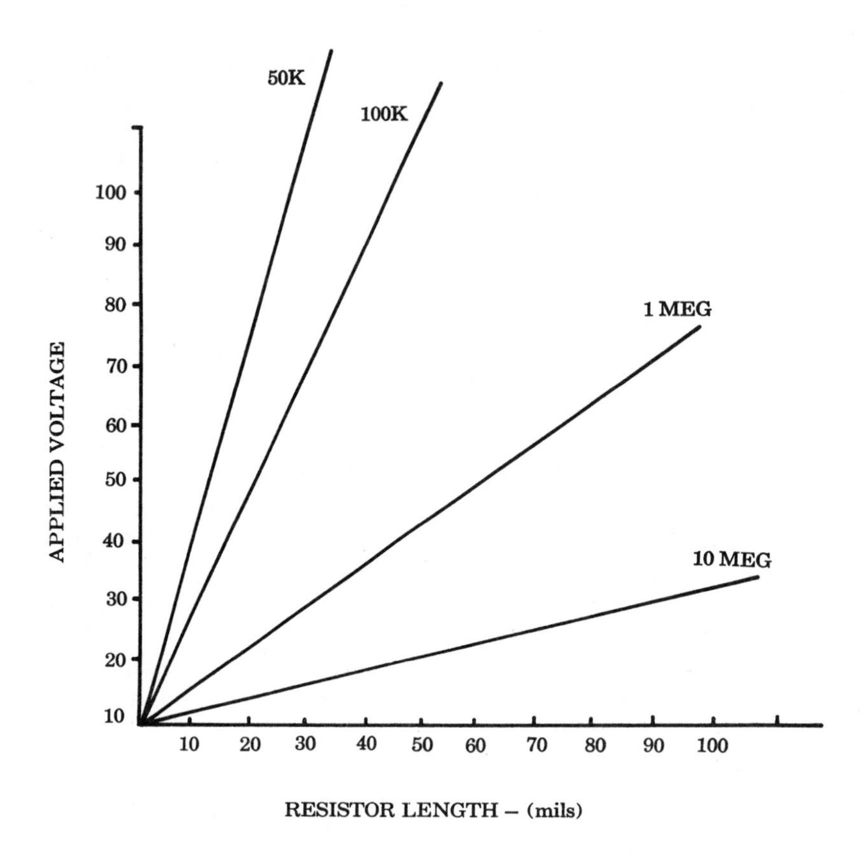

Fig. 4-2 Voltage coefficient of resistance effects.

Resistor materials used currently in thick films are based on multiphase mixtures generally falling into one of two classes:

a) Finely divided metal powders suspended with fritted milled glass in a solvent-vehicle system,

b) Metal resinates or modifications of metal resinates.

The most commonly available pastes (generally prepared by milling mixtures of two-thirds inorganic solids and one-third organic vehicle) fall into the first category and are manufactured by several companies. Characteristics vary among suppliers. The range of paste values typically includes values of from 1 ohm per square to 1 megohm per square and viscosities from 170 to 230 poises. Resistive glazes or cermet coatings (as they are also called), have been the subject of intensive investigation since the advent of microelectronics.

A number of different chemical compositional systems have been reported in the literature including the following:

1) Palladium-Palladium Oxide-Silver-Glass Cermet
2) Indium Oxide-Glass
3) Thallium Oxide-Glass
4) Tungsten-Tungstic Carbide-Glass
5) Precious Metal or Platinum Metal Group Combinations

To such compositions are added dopants, extenders, alloys, frit modifications, organic vehicles, various metal oxide particle sizes and shape distributions.

They are sometimes covered by overglazes for hermetic seal protection and underglazes for dielectric isolation.

Many parameters or characteristics, properties and conditions must be known, met, controlled and/or satisfied by a resistor material in order to be useful for microelectronics purposes:

1) Compatibility with substrates; 2) Firing conditions (atmosphere effects, temperature, belt speeds); 3) Load life; 4) Temperature Coefficient of Resistance (TCR); 5) Geometry of resistor; 6) Heat sinking; 7) Type of test; 8) TCR Tracking; 9) Stability (time at temperature); 10) Drift; 11) Type of ink blending; 12) Printing thickness; 13) Multiple printing; 14) Line width; 15) Noise; 16) Tolerances (as-processed vs as-trimmed); 17) Moisture effects; 18) Encapsulation behavior; 19) Reproducibility.

As already noted above, the cermet resistors are a mixture of precious metals and/or precious metal oxides and glass frit or glass-forming materials. These materials, in the form of fine powders, are mixed in an organic vehicle (providing proper flow) to make an ink suitable for screen stenciling deposition. The ink is screened on the ceramic substrate and fired, usually in air. Screen printing is a relatively inexpensive and rapid method of preparing resistors. Printed resistors are amenable to precise adjustments by mechanical means, such as abrasion.

The Pd-Ag-Pd0-glass cermet has been one of the most popular and is a mixture of Pd or Pd0, Ag, and glass, all of which are in form of fine powder (average particle size of 0.1 to 0.5 microns), flake, or frit. (Frit compositions are prepared by melting the batch components together, pouring into water, and ball-milling until the average particle diameter is about 5 microns). The precious metal normally comprises 30-60% by weight of the total. The glass is often a low-melting, heavy-metal flux similar to that in dispersion electrodes. The resistor is fired in air between 700-800°C, a temperature which generally corresponds to a region of maximum resistivity. The firing is often preceded by a bake to remove as much of the organic vehicle as possible. During the firing, peak temperature, profile and time are critical because they determine both resistivity and temperature coefficient of resistivity (TCR), as well as the spread of these parameters within the lot. The contacting electrodes are a dispersion type of Pt-Au or Pd-Ag. The thickness of the resistor is dependent on the flow properties of the paste, the mesh size of the screen, and the mechanical factors involved in applying the paste, such as squeegee pressure, angle and velocity.

A wide range of resistivities can be obtained by varying the metal content of the resistor compositions. This is shown in Fig. 4-3. The dependence of resistivity on gold or silver concentrations is very steep. Silver powder, for example, has a resistivity of less than one ohm per square per mil at 48% and over 100K ohms at 46%. Palladium, on the other hand, has useful resistivities over a range of 33 to 70%. When palladium and silver are used together, the concentration is even less precipitous. It is possible, therefore, to formulate mixtures of palladium and glass or palladium, silver, and glass over a range of resistivities, and obtain reproducible resistances in resistor manufacture.

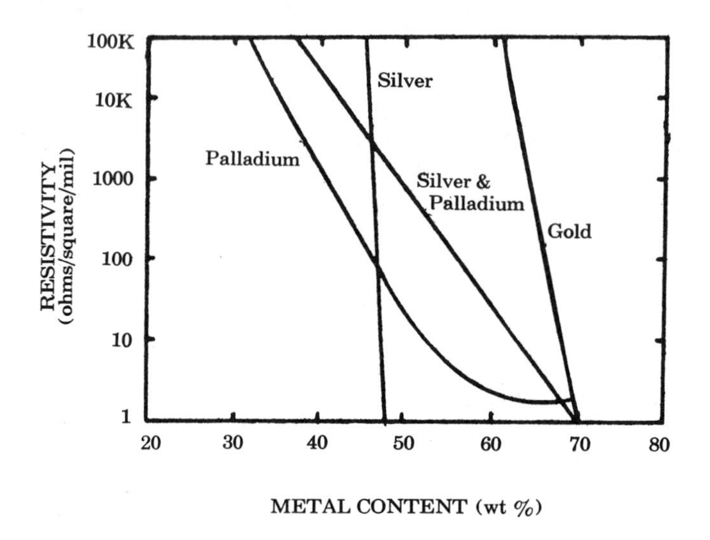

Fig. 4-3 Resistivity vs. concentration of metal powders in $PbO-B_2O_3-SiO_2$ frit

For some circuit applications, it may be necessary to blend resistor compositions to obtain intermediate resistance values. Moreover, this can be a source of varied problems and ink suppliers have various recommendations and some data to offer. When preparing blends of resistor compositions, the jar of each component should be stirred thoroughly before the required amount of composition is removed. When a small amount of one component is to be blended with a large amount of the other component, the smaller portion should first be blended with a comparable volume of the larger component, followed by good agitation. Subsequently, the larger component should be added to the smaller in increments and stirred to uniformity after each incremental addition, until the blending has been completed. Fig. 4-4 illustrates the results of blending two resistive ink compositions.

It is apparent from this figure that the resistivity is controlled by blending high-and low-resistivity inks. This must be done carefully, as blending can affect TCR. In addition, resistance is controlled by the number of squares designed into the resistor. As fired, resistors may vary 10% or 20% from design value. The final control of resistance is by adjustment. This may be accomplished by various techniques, including air abrasive

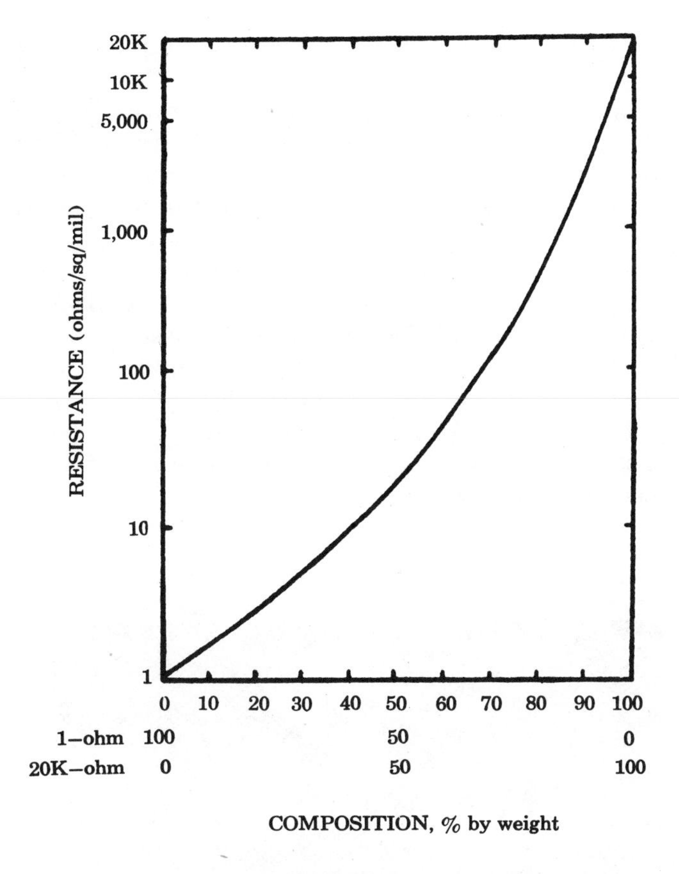

Fig. 4-4 Resistance of blends of 1-ohm and 20K-ohm compositions.

trimming, buffing, or electrical or thermal pulse (e.g. laser) adjustment. The effect of some adjustment techniques on resistor stability is not fully documented, but it is known to be adverse. In particular, buffing and electrical pulse adjustment of Pd-Ag-glass cermets can cause serious instabilities and much increased rate of drift on load life.

Palladium-Silver Cermet: Cermet resistors offer a practical range of resistivities from 50 ohms/square to 100k ohms/square.

Resistivity is determined by the percent of glass present. Some palladium-silver inks fire with sheet resistivities from 1 ohm/square to 1 megohm/square. The materials at the extreme ends are quite sensitive.

The cermet resistors are less sensitive to substrate surface roughness, so that special surface preparation such as underglazing is not required. Usually an overglaze is provided for maximum stability and isolation from the environment. The most commonly used cermet inks are palladium-silver-glass or palladium-oxide-glass compositions.

The palladium-silver was the first resistor system that was made commercially available to industry. This resistor ink consists of a mixture of silver, palladium and powdered glass, mixed with organic binders and solvents. After screening this ink on the substrate, it is dried at 90°C to 150°C for approximately one-half hour to remove the solvent. Then this paste is fired on the substrate in a continuous belt furnace at 690°C to approximately 760°C.

Upon firing, a chemical reaction starts somewhere above 500°C. The reaction rate appears to be logarithmic, beginning fast, and decreasing logarithmically toward completion. Since the firing time is never sufficient to achieve completion, the resulting resistivity is dependent on the time-temperature product during the firing cycle.

The temperature coefficient is to some extent substrate dependent, but in general lies between —100ppm/°C and +300ppm/°C. These are the extreme limits so that the manufacturer should be consulted for the TCR of a specific ink at a specific sheet resistivity, since this quality shifts with resistivity. The TCR of some inks have been observed to be mainly dependent on the firing time. Hence, it is possible to obtain a specific TCR by selecting the firing temperature, and then changing the firing time to obtain the desired resistivity.

Fig. 4-5 This rotary-type abrader automatically trims resistors to exact value while circuit is connected to resistance comparison bridge.

The palladium-silver system is the most difficult paste to process. This is due to the fact that the equilibrium between the palladium oxide and palladium alloy is unstable and depends on the exact firing cycle used.

Platinum Meal System: The platinum resistor system has overcome many of the problems encountered with the palladium-silver system. These pastes are insensitive to active hydrogen or reducing environments, and are expected to approach thin film resistor performance in precision applications.

Typical TCR's for resistor composition between 100 ohms/square and 100k ohms/square are less than ±100ppm/°C.

Noise and voltage coefficients for a given platinum system are a function of resistor size, geometry, and thickness. For 130 by 130 mil resistors, noise levels of —20 to —26db are obtained for 100 ohm/square, increasing to 0 to —5db for100k ohm/square.

Other Resistor Inks: Resistor systems free of palladium and silver have been introduced. These systems are primarily intended to provide stability in reactive, severe environments. The new systems offer a wide range of resistance from 30 ohm/square to 300k ohm/square, a TCR of less than ±100ppm/°C in the —55°C to 125°C temperature range, current noise (—30db at 100 ohm/square), and excellent load and no load stability (Δ R drift of less than 0.2% per 1000 hours at 150°C storage temperatures). Thallium Oxide requires a temperature profile of <600°, is relatively less expensive, and tends to absorb water.

4.3.3.5 Circuit Layout

Generally, the two major considerations in designing the thick film circuit are minimum size and fabrication simplicity. Minimum size is usually a goal in any design, while fabrication simplicity relates directly to low cost and reliability.

Thick film resistor values may be defined by

$$R=\frac{\rho l}{A} \tag{4-1}$$

where ρ is the resistivity in ohm-centimeters, l is the resistor length and A is the resistor cross-sectional area. The equation may be rewritten to define R in terms of sheet resistivity as

$$R = Rn \tag{4-2}$$

where n is defined as the number of squares, and R is the sheet resistivity per unit area or per square at a constant thickness (usually 1 mil for most thick film resistors).

If, for example, a thick film material having a sheet resistivity (R) of 10 kΩ/□ is employed and the resistor length-to-width ratio is 4, then the total resistance is

$$R = Rn$$

$$R = 10\ \frac{k\Omega}{\square} \times 4\ \square$$

$$R = 40\ k\Omega$$

By this artifice, the layout of resistor values may be based only on a proper relationship of the length-to-width ratio and the sheet resistivity value.

The case of a resistor design in which the aspect ratio is less than unity ($l < w$) can be handled in the same manner. For example, a material having a sheet resistivity (R) of 10 kΩ/□ and an aspect ratio of 0.25 has a total resistance of 2.5 kΩ. Figure 4-6 indicates these two examples of resistor design.

Since each resistor design attempts to achieve minimum area, it may be assumed that either the resistor length or width is chosen as a minimum value consistent with processing resolution. Power considerations may dictate more than this minimum area, but a minimum square prevails where the length and width are equal and as small as the process will tolerate. The minimum square concept is particularly convenient in calculating the total area required for the circuit resistors, but it leads to some confusion when the resistor aspect ratio is less than unity. In Fig.

4-6, for example, each resistor has four equal squares from an area consideration, but using the earlier derivation, resistor A has four squares while resistor B has 0.25 square. This contradiction is avoided if the earlier derivation is amended in the following manner:

$$\begin{aligned} R &= R n_s \\ n_s &= \frac{l}{w} (l > w) \end{aligned} \tag{4-3}$$

$$\begin{aligned} R &= R n_p \\ n_p &= \frac{w}{l} (l < w) \end{aligned} \tag{4-4}$$

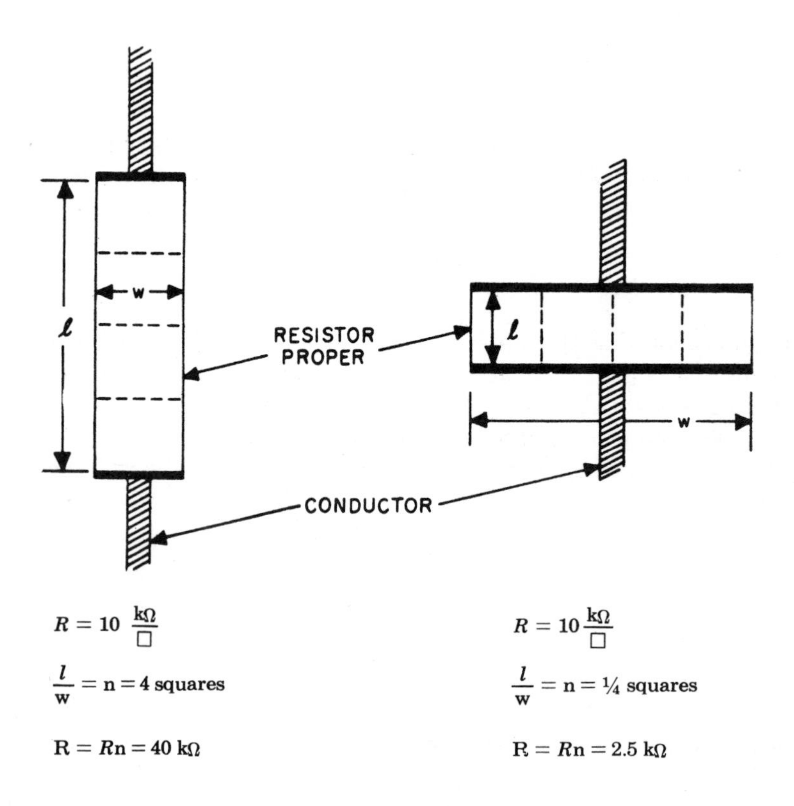

$R = 10 \frac{k\Omega}{\square}$

$\frac{l}{w} = n = 4$ squares

$R = Rn = 40$ kΩ

$R = 10 \frac{k\Omega}{\square}$

$\frac{l}{w} = n = 1/4$ squares

$R = Rn = 2.5$ kΩ

Fig. 4-6 Two thick film resistor configurations.

This allows employing the minimum-square concept and relating to electrical representation by treating the resistor with an aspect ratio greater than unity as having minimum squares *in series,* and the aspect ratio less than unity as a *parallel* arrangement of minimum squares.

From Fig. 4-6 it may be seen that minimum resistor area occurs when the paste resistivity is matched to the particular resistor value.

Not all pastes having different resistivity values can be cofired; hence, in any multiresistance circuit, minimum area is achieved by selecting a particular paste resistivity that best matches the various resistor values.

Table 4-1 illustrates a simple case of six resistors and the area required for two different paste sheet resistivities. If these were the only pastes available, minimum area would result for the 10-kΩ/□ paste.

TABLE 4-1

Example of Resistor Areas for Different Sheet Resistivity Values

Circuit Resistor Values		Paste Resistivity 1 kΩ/□	Paste Resistivity 10 kΩ/□
		No. of Minimum Squares	
1 each	1 kΩ	1	10
4 each	10 kΩ	40	4
1 each	50 kΩ	50	5
		91	19

Since it is unlikely that pastes of greatly different sheet resistivities can be conveniently cofired, an exercise or calculation as shown in Table 4-1 may be necessary for each circuit layout, in order to reduce the art work and number of screen printing operations. A computer program, may be used for providing the layout man with the optimum paste resistivity (from a list of available pastes) for a circuit with many resistors of different values.

The resistor area, of course, is also influenced by the power dissipation required of the resistor. Thick film material on a ceramic substrate will dissipate about 20 watts per square inch, or 20 microwatts per square mil. This power rating is based on a 10°C rise, and unless there is a large percentage of substrate occupied by the resistor, fringing effects provide a large safety factor.

Prior to the layout of any circuit, a second computer program may be used to analyze the circuit in detail, including the maximum power level of each resistor. This information may then be used to modify the results of the first program in the event that greater resistor area is required for power reasons.

For a minimum resistor width of 20 mils, the power rating for a minimum square is about 8 mw.

The ultimate capability of thick film resistor line widths is probably about 3 mils, while conductor and pad minimum line width might be reduced to about 1 mil. As better resolution is obtained, however, additional concern for power dissipation must be exercised. At some point, depending on the resistor value, the power dissipation consideration may play a more predominant role in establishing the minimum resistor area than the resolution.

By generalizing so that each circuit resistor is assumed to dissipate a power of V^2/R, where V is the maximum circuit voltage, a relationship between minimum area and resistor value, as shown in Fig. 4-7, is obtained

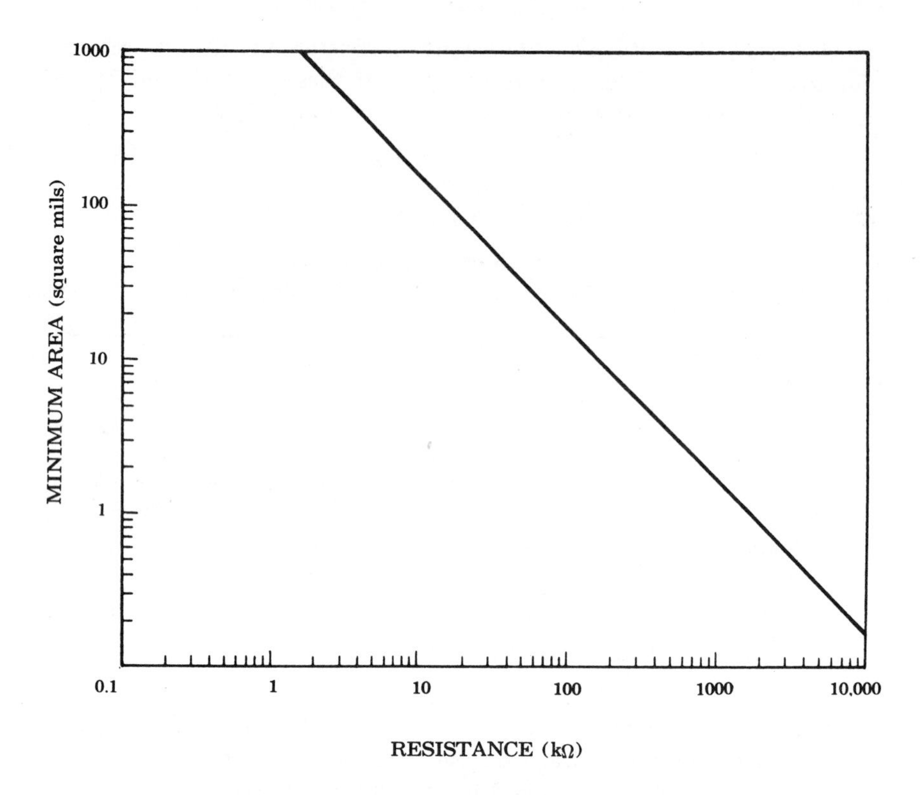

Fig. 4-7 Minimum resistor area required for adequate power dissipation in 6-volt system.

for a 6-volt system.

The effect of different resolution is shown by Fig. 4-8. Again, for a 6-volt system, this figure shows that, as better resolution is achieved, the minimum resistor value for which a single square can be used increases because of power requirements. It should be noted that as the resolution decreases, the dimensions of the minimum square also decrease.

At a resolution of 20 mils, Fig. 4-8 indicates that only for resistor values less than 3.5 kΩ, will an area greater than one square be required to satisfy the power dissipation.

As the resolution reaches 5 mils, all resistor values less than 72 kΩ may require more than one square. Since these figures are based on a particular supply voltage and assume maximum power dissipation, they are intended only as a guide and indicator of the relationship of resolution, power dissipation and area for resistor design.

Some conditioning of resistors by voltage or current pulsing may be desirable to obtain long-term stabilization, particularly against accidental over-voltage conditions that may occur during later use of the circuits.

Layout experience indicates that the substrate area allocated to resistors varies from 10 to 30 percent depending on the circuit type, and conductor and pad area vary from 20 to 30 percent. The unused substrate area is relatively constant at about 50 percent for a 20-mil resolution. Some area tradeoff between thick film conductors and wire-bond jumpers is possible, but overall reliability is increased if the number of wire bonds is minimized.

4.3.4 DIELECTRIC MATERIALS

Dielectric films for screened hybrid microcircuits consist of ceramic particles mechanically suspended in low dielectric glass, and have dielectric constants less than 100. Techniques and materials have been developed, however, to obtain high dielectric films which are deposited and recrystallized, in place, on alumina substrates for integration into a complete circuit. These films consist of barium titanate particles bound by smaller grained barium titanate which has been recrystallized out of glass. They have dielectric constants of 300 to 800.

In the past, screened hybrid microcircuits have used low dielectric thick films for capacitors. These consist of ceramic dielectric particles mechani-

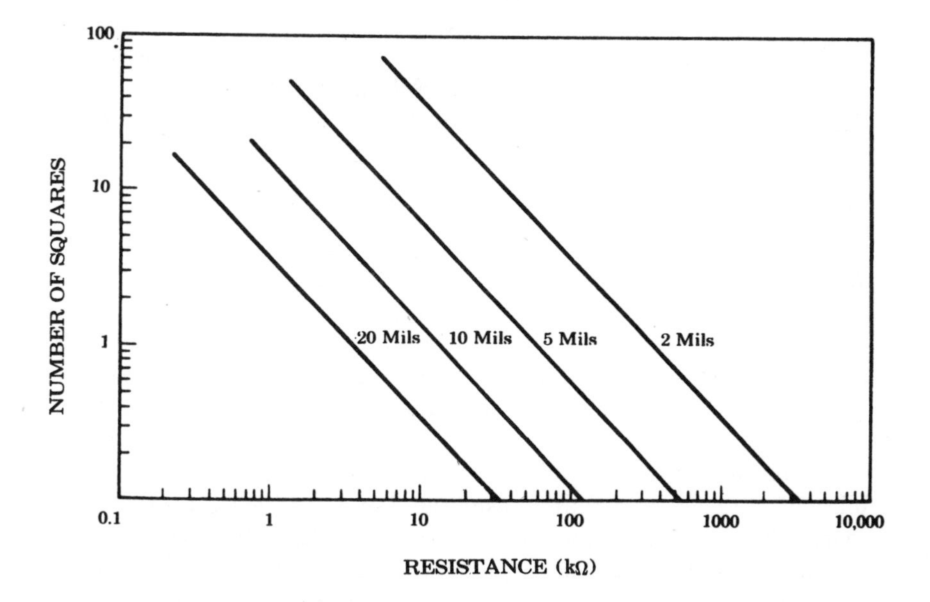

Fig. 4-8 Minimum number of squares for power dissipation at various resistor width resolutions.

cally suspended in a lead monosilicate glass matrix. Because of the small volume of high dielectric particles (35% by weight) which can be packed into the low dielectric glass, the dielectric constants of the thick films are less than 100. The mechanical suspension results in low-density structures and therefore high-dissipation factors.

Techniques and materials have been developed to obtain high dielectric films (K of 300 to 800). For these materials, high dielectric barium titanate particles are mixed in a barium titanate glass. The final dielectric material consists of barium titanate particles bound by smaller grained barium titanate which has recrystallized out of the glass. Therefore, the total barium titanate content is 90 to 95 percent. The films are deposited and recrystallized, in place, alumina substrates for integration into a complete circuit.

Crossover pastes have little or no metal content, and are used for underglazes, dielectric isolation, and environmental sealing. The most important properties are peak firing temperature and compatibility with resistor and conductor pastes. When used as an underglaze, the objective is usually to provide high adhesion of conductor areas, which enhances the bond strength of add-on parts and input/output leads. Since the underglaze is typically the first material applied to a substrate, high firing temperatures (800°C-1000°C) are desirable.

The most significant contribution of thick-film dielectrics has been to the field of crossover isolation. When used for dielectric isolation, the usual purpose is to provide circuit crossover capability. In this application, immediate firing temperatures (650°C-800°C) are needed because other materials may be printed and fired before and after the cross-over paste.

Thick-film resistors, because of their natural glass coating, do not necessarily require additional protection. It is common, though, to encapsulate the thick-film resistor with a protective glass coating. Special low-temperature glasses have been developed that are printed and then fired at 500°C to 650°C for a minute or two. These glasses increase the stability of the resistors ,provide additional mechanical protection, and protect the resistor surface against overspray during the air-abrasive-trimming process.

4.3.4.1 Thick-Film Capacitors

Thick-film capacitors are made by depositing layers of paste containing highly conductive metal powder alternating with insulating glass with ceramics such as barium titanate, followed by a platinum or gold top electrode.

Problems associated with these thick-film capacitors include the possibility of pin holes between the insulation layer and the electrodes causing shorts. The size of capacitance realized in this configuration is not adequate for many applications. The maximum size of screened capacitors are less than 1000pf with ±20% tolerances. In addition, the TCC's are poor in comparison with the TCC's of chip ceramic capacitors.

The term "monolithic ceramic capacitor" is sometimes used to refer to capacitors constructed of thin layers of ceramic dielectric sandwiched between even thinner layers of metallic electrodes which have been fused by virtue of true ceramic — metal and ceramic — ceramic seals, so that the entire structure, or "slug", appears as a single block. Figure 4-9 shows a cutaway view of a 6-layer monolithic ceramic capacitor slug.

Monolithic ceramic (multilayer) capacitors are made by screening metal electrode layers on a ceramic base. Between each electrode layer is a screened dielectric layer. The chemical composition of the ceramic dielectric plays the major role in determining a capacitor's electrical properties. For example, the dielectrics used generally contain about 80% to 95% Barium Titanate as the major constituent of the dielectric. This type of capacitor has a dielectric constant (k) between 1000 and 2000, dissipation factor (D.F.) between 0.005 and 0.025, insulation resistance in excess of 100,000 megohms, and temperature coefficient between ±7.5% and ±20% over the −55°C and +125°C. Alternate layers of electrodes project out at

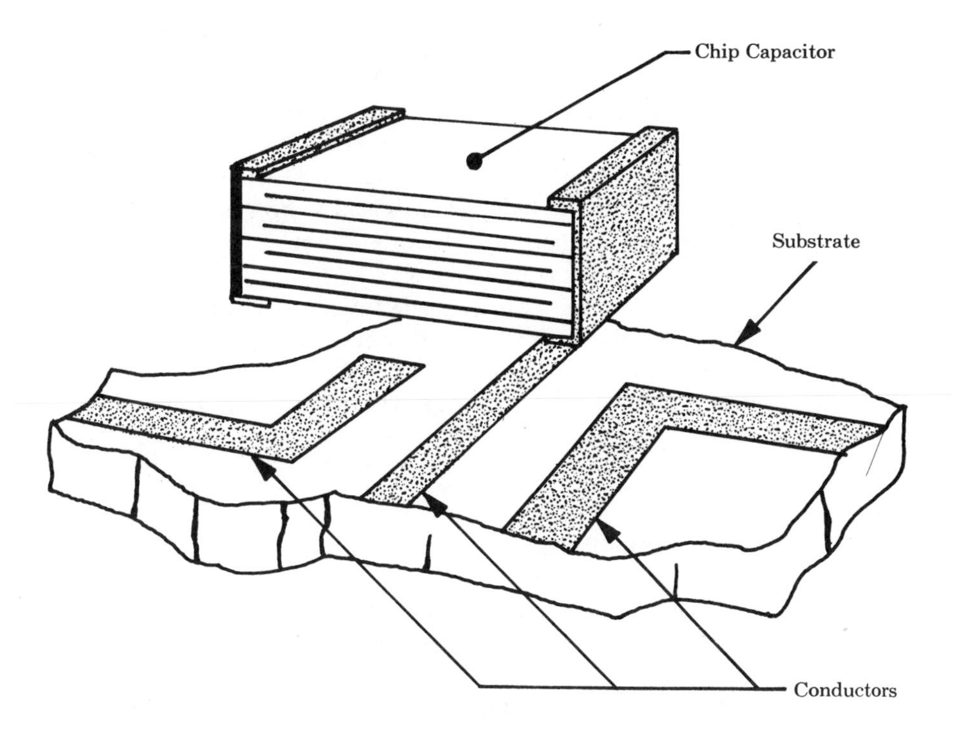

Fig. 4-9 Monolithic ceramic chip capacitor (multilayer)

opposite ends and are joined together by soldering. The chip is fastened to the hybrid circuit conductors by reflow soldering or ultrasonic bonding.

The units are finished off by metallizing the two ends where alternate electrodes stick out. Usually these metallized ends then are presoldered, so that the tiny chips can be attached to the hybrid circuit conductors simply by placing them down and reflow soldering. Welding and ultra-

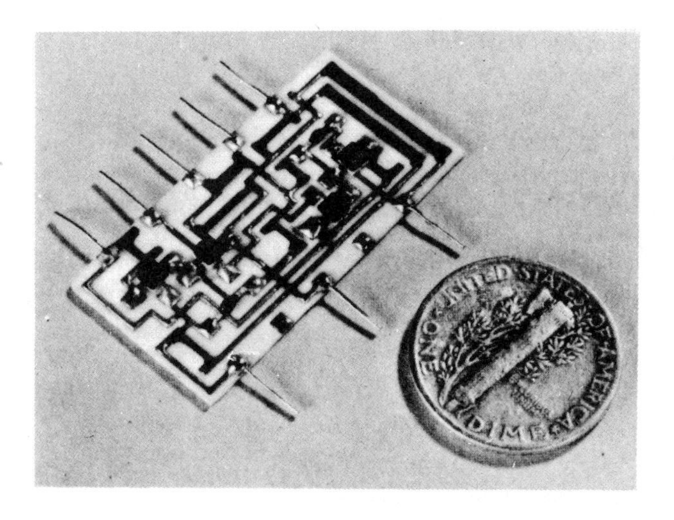

Fig. 4-10 A typical thick film circuit module.

sonic bonding also can be used.

For hybrids requiring capacitance in the 1000 pf to 1 μf range, the monolithic ceramics will remain the most attractive approach. There are numerous trade-offs involved here. To achieve the highest volumetric efficiency, a higher dielectric material (K value of 6000) must be used. Using higher dielectric material, however, will lead to greater capacitance instability with time and temperature. If a design calls for steady temperature characteristics, then the designer may be forced back down to a K of 30. Theoretically, dielectrics can be made thinner resulting in higher values of capacitance for the same volume. However, almost all film dielectrics tend to have pinholes that result in voltage breakdowns when the dielectric is thinner than about half a mil. The recommended minimum dielectric thickness is 1.5 mils.

4.4 PYROLYTIC TECHNIQUES

As noted previously, other methods for depositing thick films exist. The most important of these is the vapor or pyrolytic deposition process. Suitable chemicals are decomposed at the substrate surface and unwanted material is removed in a subsequent step by etching or sand-blasting. Vapor sources include metal-organic compounds and metal halides. One of the more common processes uses tin chloride in solution with a catalytic agent. This is sprayed onto a glass-covered alumina substrate and decomposed to give a thin oxide coating which is then patterned into film resistors. In this case, conductor patterns can be formed on top of the tin oxide by electroless plating of copper and dielectric patterns made by decomposition of ethylorthosilicate or by glass frit formulations. Components and circuits obtained by these and similar processes, are comparable to screen circuits.

4.5 BIAS SPUTTERING

When sputtering thick films, it has been found that a columnar structure is often obtained in place of a smooth dense material needed for good structural strength and low electrical resistivity. By applying a high negative bias (—500 V) to a substrate while sputtering, however, it has been possible to obtain a dense, fine-grained film approximately 6 microns thick.

The sputtering is performed in a vacuum chamber at a pressure of from 10 to 100 μ of argon.

The holder with the heated or cooled substrate is separated from the cathode, containing the metal to be deposited, by a gap measuring of 2 to 10 cm.

The substrate has applied to it, a biasing negative potential of up to 500 V. Application of the bias to the substrate causes argon ions to bombard the growing deposit. The ions dislodge metallic atoms from the deposit surface, and cause them to relocate and fill in the "valley" vacancies between the growing columns of materials.

This method has produced a reflective 6-μ thick film of tantalum on a carbon substrate heated to 300 C. The density was 16.2 g/cm^3, compared with 16.6 g/cm³ for bulk tantalum and 14 g/cm³ for Ta sputtered at zero target bias. The deposition rate is about as high as that experienced with unbiased sputtering – about 2 mg/cm²/hr.

Thin-film/thick-film comparison

	Thick film	Thin film
Size	small	small
Tolerance	good	best
T.C.R.	good	best
Temperature coefficient tracking	good	best
Power	high	medium
Design flexibility	good	fair
Resistance range	excellent	good
Resistivity range	widest	fair
Relative cost	low	moderate

Chip capacitors, resistors, and inductors behave basically like their standard size counterparts. However, they are limited in some performance characteristics — range of values, breakdown voltages, and power dissipation, for example.

The vast majority of chip capacitors are ceramic and of monolithic construction. They are manufactured by mixing ceramic powder in an organic binder (slurry) and casting it into thin layers typically ranging from about 3 mils thick down to about 1 mil or less. As shown in Fig. 4-12, metal electrodes are deposited onto ceramic layers, which are then stacked to form a laminated structure. The metal electrodes are arranged so that their terminations alternate from one edge of the capacitor to another. Sintering the part at high temperature changes it into a monolithic block that provides very high capacitance values in small mechanical volumes.

Ceramic capacitors fall into several broad classes based on dielectric properties. The dielectrics are classified into types according to temperature characteristics.

Tight-tolerance chip capacitors can be quite expensive compared to standard tolerance capacitors because, all other things equal, the tighter the tolerance, the lower the manufacturing yield. However, many chip capacitor manufacturers use sandblasting, a method of tightening a standard chip capacitor's tolerance that results in only a rather modest increase in its price. Sandblasting makes it easier for manufacturers to produce tight-tolerance capacitors without worrying about the excessive costs of low yields.

Sandblasting is performed by an automatic device that trims away part of the capacitor's monolithic layers of dielectric and electrodes (Fig. 4-13). A high value of capacitance is selected and the part is trimmed by sandblasting until the desired value and tolerance are reached. The sandblasted hole is then filled with a protective insulating material to seal and preserve the part's integrity. The amount of material that can be removed varies for each case size and value. Generally, this method is used for chips measuring 0.12″ x 0.10″ or larger.

Sandblasting doesn't affect the electrical and mechanical parameters of the finished parts.

Resistor chips are available as thick-film, thin-film, and bulk-metal devices. Designers are probably more familiar with the characteristics and performance of thick and thin-film resistors than they are with chip capacitors because of the rather extensive use of thick and thin-film resistor networks in today's electronic equipment. However, this may not be true for bulk-metal chip resistors, generally used only in applications requiring very special performance.

Film resistors are formed by depositing a film of conducting material on a non-conducting

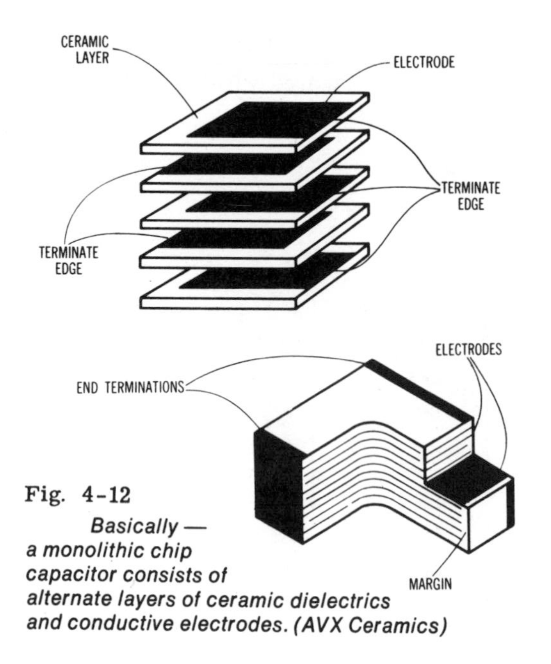

Fig. 4-12 *Basically — a monolithic chip capacitor consists of alternate layers of ceramic dielectrics and conductive electrodes. (AVX Ceramics)*

Fig. 4-13 *Standard tolerance ceramic chip capacitors can be sandblasted to obtain tighter tolerances. (Vitramon)*

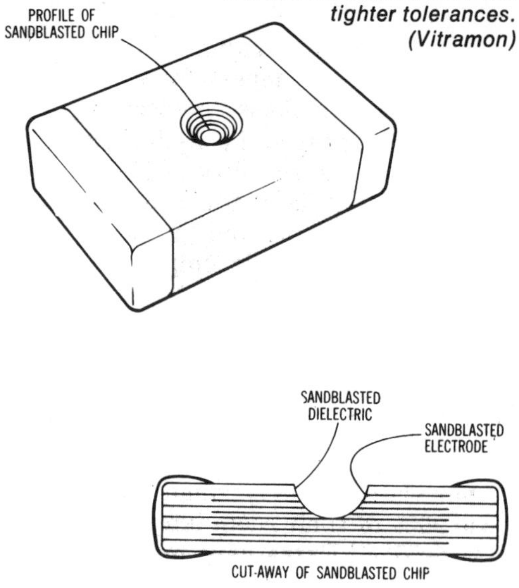

PROFILE OF SANDBLASTED CHIP

SANDBLASTED DIELECTRIC

SANDBLASTED ELECTRODE

CUT-AWAY OF SANDBLASTED CHIP

substrate.

Thick-film chip resistors are the most popular simply because they meet the performance requirements of most applications at the lowest cost. They are stock items available from most manufacturers in a range from 10Ω up to about 10 or 15 MΩ. However, some manufacturers offer them in ranges up to 100 MΩ. In fact, special high-megohm resistors are available up to as high as 1×10^{12} ohms.

A thin-film resistor is, by definition, one with conductive material less than one millionth of an inch thick. Thin-film chip resistors provide significantly better performance than the thick-film types but, obviously, at a higher cost. Their resistance range is very limited compared to that of thick-film chips, typically running from 10Ω to about 300 kΩ. Higher values are available usually only on request.

The trend towards miniaturization has resulted in dramatic size reductions in magnetic devices, just as it has with capacitors and resistors. This reduction has been more difficult with magnetic devices because of their considerable bulk as compared to the other devices. Now, however, miniature chip inductors, rf chokes, and even chip transformers are available for use in hybrid circuits and on pc boards. These magnetic devices are now small enough to be used in electronic watches and pacemakers.

Much more so than chip capacitors and chip resistors, chip inductors (and transformers) resemble the construction of their standard size brethren — wire wound on toroids, bobbins, pot cores, and so on, with the core made of iron, phenolic, and ferrite.

Chip inductor values typically range from as low as 0.01 to about 10,000 μH with standard tolerances of 5, 10, and 20%. Some chip inductors are capable of handling over 1A. Most, however, handle from about 500 or 600 mA, with power dissipation rated up to about 100 to 300 mW.

REFERENCES

1. Microelectronic Engineering Volume I Fabrication Technology, D. Abraham, Johns Hopkins University

2. Westervelt, D., Vondracek C., U. S. P., 3, 325, 590, 13/6/1967.

3. Microelectronic Device Data Handbook Vol. I, Arinc Research Corp., July 1968.

4. Borom, M., Lingwell, J., Pask, J., *J. Amer. Ceram. Soc.* 50 (2) 61, 1967.

5. Quantum Size Effect in Thin Bismuth Films, V. P. Duggal, R. Rup, P. Tripathi. *Applied physics Letters* v 9 n 8 Oct 15 1966 p 293-5.

6. Enhanced Tunneling through Dielectric Films Due to Ionic Defects. F. W. Schmidlin. *J Applied Physics* v 37 n 7 June 1966 p 2823-32.

7. Anisotropy Field H_k of Thin Ferromagnetic Films, Measured by Magnetoresistive Methods, G. Kneer. *IEEE—Trans on Magnetics* v Mag-2 n 4 Dec 1966 p 747-50.

8. Chemical Nickel-Iron Films, A. F. Schmeckenbecher. *Electrochem Soc*—J v 113 n 8 Aug 1966 p 778-82.

9. Distribution of Magnetization in Multilayer Films, W. Andrae. *IEEE—Trans on Magnetics* v MAG-2 n 3 Sept 1966 p 560-2.

10. Effect of Film Structure on Magnetoelastic Properties of Thin Permalloy Films, W. Metzdorf. *IEEE—Trans on Magnetics* v MAG-2 n 3 Sept 1966 p 575-9.

11. Nonlinear Behavior of Ferromagnetic Thin Films, A. J. Berteaud, H. Pascard. *J Applied Physics* v 37 n 5 Apr 1966 p 2035-43.

12. Hybrid Microcircuit Design and Procurement Guide J. W. Thornell, et al, Boeing Company, Renton, Washington, May 1970.

13. Kern, D., Young, D. U. S. P. 3, 303, 059, 7/2/1967.

14. Thick Film Conductor Function Inks and Pastes for Microelectronics Applications, John T. Milek, Hughes Aircraft Company, Culver City, California, Feb. 1, 1968.

15. Hunter, O., Brownell, W. *J. Amer. Ceram. Soc.* 50 (1) 19, 1967.

16. Flaschen, S., Gnaedinger, R. U. S. P. 3, 301, 706, 31/1/1967.

17. Hill, R. M., Bull. *Inst. Physics and Phys.* Soc. 18 (5) 147, 1967.

18. *''Physics of Thin Films,''* edited by G. Hass, Vol. I (1963). Vol. II edited by G. Hass and R.E.Thun (1965), Academic Press, New York.

19. T. P. Monaghan, ''The Electrical Properties of Deposited Glass Dielectric Capacitors for Compatible Integrated Circuits,'' Presented at Electrochemical Soc. Meeting, San Francisco, Calif., May 1965.

20. C. D. Phillips, ''A Compatible Thin Film Resistor — Capacitor Process for Silicon Integrated Circuits,'' Presented at Electrochemical Soc. Meeting, San Francisco, Calif., May 1965.

21. UASECOM Contract DA 28-043, AMC-01424(E), ''Multiple Integrated Circuit Techniques,'' Report No. 1, 15 June 1965 — *4 September 1965, Philco Corporation, Lansdale, Pennsylvania.

22. M. M. Attalla and R. W. Soshea, ''Hot-Carrier Triodes with Thin Film Metal Bases,'' *Solid State Electronics*, Vol. 6, p. 245 (1963).

23. Shaw, R., Heasley, J., *J. Amer. Ceram. Soc.* 50 (6) 297, 1967.

24. Bratten, R., Tien, T., *J. Amer. Ceram Soc.* 50 (2) 90, 1967.

25. Haerthing, G. H., *J. Amer. Ceram. Soc.* 50 (6) 329, 1967.

26. Allen, A. C., Cer. Ind., 88 (3) 56, 1967.

27. Eubank, W. R., et al, U. S. P. 3, 312, 922-923-924, 4/4/1967.

28. Perschy, J. A. *Electronics* p .74, 24/7/1967.

29. Simmons, J. G., 16th Meeting, Thin Film Group, Leatherhead, 1967.

30. Kennedy, T., Mackenzie, J., *Phys. Chem. Glass* 8 (5) 169, 1967.

31. F. W. Schenkel, ''Thin Film Capacitor Parameter Studies,'' *Proc. El. Comp. Conf.* 1964, p. 194, Washington, D. C.

32. Sputtered Thin Magnetic Films, W. N. Mayer. *IEEE—Trans on Magnetics* v MAG-2 n 3 Sept 1966 p 166-83.

33. Switching Properties of Multilayer Nickel-Iron Films with Ferrite Keeper, K. U. Stein, E. Feldtkeller, *IEEE—Trans on Magnetics* v MAG-2 n 3 Sept 1966 p 184-8.

34. Electrical Conduction between Metallic Microparticles, D. S. Herman, T. N. Rhodin. *J Applied Physics* v 37 n 4 Mar 15 1966 p 1594-1602.

35. Influence of Electric Field on Growth of Thin Metal Films, K. L. Chopra. *J Applied Physics* v 37 n 6 May 1966 p 2249-54.

36. Ionic Conductivity, Dielectric Constant, and Optical Properties of Anodic Oxide Films on Two Types of Sputtered Tantalum Films, D. Mills, L. Young, F. G. R. Zobel. *J Applied Physics* v 37 n 4 Mar 15 1966 p 1821-4.

37. Stripe Domains in Ni-Fe Films with Zero and Positive Magnetostriction, M. M. Hanson, D. I. Norman, D. S. Lo. *Applied Physics Letters* v 9 n 3 Aug 1 1966 p 99-100.

38. Structural Defects in Epitaxial Films — Role of Surfaces in Nucleation. S. Mendelson. *Matls Science & Eng* v 1 n 1 May 1966 p 42-64.

39. Coupling Between Flat Films, E. P. Valstyn. *IEEE—Trans on Magnetics* v MAG-2 n 3 Sept 1966 p 188-93.

40. Decrease of Barkhausen Effect and Increase of Aftereffect in Thin Iron Films at Low Temepratures, Observed with Kerr Effect, M. Lambeck. *IEEE—Trans on Magnetics* v MAG-2 n 3 Sept 1966 p 301-3.

41. Deposition of Permalloy Films on Organic-Polymer-Coated Substrates, J. W. Medernach, L. J. Bobb, N. A. Hunter. *J Applied Physics* v 37 n5 Apr 1966 p 2013-15.

42. High-Power Effects in Thin Films at Oblique Polarizing Fields, H. Pascard, A. J. Berteaud. *IEEE—Trans on Magnetics* v MAG-2 n 3 Sept 1966 p 283-6.

43. Magnetization Distribution in Flat and Cylindrical Films Subject to Nonuniform Hard Direction Felds, D. B. Dove, T. R. Long.*IEEE—Trans on Magnetics* v MAG-2 n 3 Sept 1966 p 194-7.

44. Waxman, A. S., Improved Stability in TfT Structures, Quarterly Report No. 2, RCA, N00019-C-0159, May, 1968.

45. Heitl, W., Puetz, W., *J. Amer. Cer. Soc.* 50 (7) 378, 1967.

46. D. W. Moore, ''*Introduction to Electron Beam Technology,*'' (R. Bakish, Ed., J. Wiley and Sons, 1962) Chap. 13.

47. Felley, E., Myers, M., *J. Amer. Ceram. Soc.* 50 (6) 335, 1967.

48. I. H. Pratt, W. L. Wade, W. Weintraub, ''Thickness and Composition of Vacuum Evaporated Nickel-Chromium Films,'' USAELRDL Tech. Rept. 2419, March 1964.

D. B. York, ''Properties of Evaporated Thin Films of SiO,'' *J. of Electrochemcial Society,* Vol. 110, No. 4, April 1963.

49. E. E. Smith and S. G. Ayling, ''Sputtered Dielectric Capacitors,'' *Proc. Elec-*

tronic Components Conference, Washington, May 1962, p. 82.

50. Hoogendoorna, H., Merrin, S. U. S. P. 3, 303, 399 7/2/1967.

51. Hamer, D. W., Reduced Titanate Chips for Thick Film Hybrid I. C.'s, Proc. 1968, Electronics Components Conf., McGregor & Werner, Washington, D.C.

52. Juhasz, C. and Anderson, J. C., Field-Effect Studies in InSb Films, The Radio and Electronic Engineer, 33, 223, April, 1967.

53. P. K. Weimer, ''T.F.T. – A New Thin Film Transistor,'' *Proc. I.R.E.*, June 1962.

54. R. P. Mandel, ''Thin Film Substrates,'' Aerojet-General Corp., Azusa, California, TM-529:63-1-731A (13 March 1963).

55. G. Riddle, ''Bandpass Amplifiers Using Distributed Parameter Networks,'' *Proc. East Coast Conf. on Aerospace and Navig. Electronics,* Baltimore, Md. (October 22, 1962).

56. D. A. McLean, N. Schwartz, and E. D. Tidd, ''Tantalum Film Technology,'' *Proc. I.E.E.E.*, p. 1450 (December 1964).

57. R. E. Hayes and A. R. V. Roberts, ''A Control System for the Evaporation of Silicon Monoxide Insulating Films,'' *J. Sci. Instrum. 39,* 428 (1962).

58. Zarzycki, J., Naudin, F. *Phys. Chem. Glass* 8 (1) 11, 1967.

59. ''Thin Film Silicon Microelectronics,'' North American Aviation/Autonetics Division, 3370 Miraloma Ave., Anaheim, Calif. (Nov. 1964).

60. Okamoto, H., and Aso, T., Formation of Thin Films of PdO and Their Electric Properties, *Japan J. Appl. Phys.*, 6, 779, 1967.

61. Peterson, D., ''Evaluation of Vapor-Plated Oxide Films for Capacitor Dielectric,'' *IEEE Trans. Comp. Pts.*, CP-10, No. 3, 119-122 (1963).

62. R. S. Clark and C. D. Orr, ''Reactively Sputtered Tantalum Resistors and Capacitors for Silicon Network,'' Proc. 1965 Elec. Comp. Conf., pp. 31-45, Washington, D. C., 1965.

63. E. H. Blevis, ''Hot Cathode and Radio Frequency Sputtering,'' Presented to Southwest Vacuum Group, 3 March 1965, Los Angeles, Calif.

64. J. W. Nickerson and R. Moseson, ''Application of Low Energy Sputtering for Thin Film Deposition,'' *Semiconductor Products and Solid State Technology,* December 1964.

65. Lockheed Microsystems Electronics, ''Thin Film Circuits,'' Lockheed Missiles and Space Co., Sunnyvale, Calif.

66. Laboratory Report AD 409422, ''Thin Films Formed by Electrochemical Reactions,'' General Telephone and Electronics, (January 31, 1963).

67. H. G. Dill, ''Designing Inductors for Thin Film Applications,'' *Electronic Design,* p. 52 (17 February 1964).

68. F. R. Gleason, ''Thin Film Microelectronic Inductors,'' *Proc. Nat'l. Elect. Conf.* (Chicago), Vol. 20 (October 1964).

69. Topfer, M. L., Danis, A. H., and Rapp, A. K., Thin-Film Polycrystalline Field-Effect Triode, Final Report, RCA, DA28-043-AMC-02432 (E), AD659788, Oct., 1967.

70. Forbes, D. W. A., *Glass Tech.* 8 (2) 32, 1967.

71. L. Holland, ''*The Properties of Glass Surfaces,*'' (J. Wiley and Sons, 1964) Chap. 5.

72. Waxman, A., Thin Film Transistors Don't Have to be Drifters, *Electronics,* 41, 88-93, March, 1968.

73. Wilson, H. L. and Gutierrez, W., Te TfT's Exceed 100 MHz and One-Watt Capabilities, *Proc. IEEE,* 55, 415, March, 1967.

74. O. C. Wells, "*Electron Beams in Microelectronics,*" Introduction to Electron Beam Technology, edited by R. Bakish, (John Wiley and Sons, New York, 1962).

75. Gadzhiev, N. D., and Talibi, M. A., Thin Film Selenium Transistors, *Radio Engineering & Electronic Physics (USSR),* 13, 144, 1968.

76. T. O. Poehler and W. Liben, "Induction Measurement of Semiconductor and Thin Film Resistivity," *Proc. I.E.E.E.*, Vol. 52, p. 731 (1964).

77. S. Dushman and J. M. Lafferty, "*Scientific Foundations of Vacuum Technique,*" (J. Wiley and Sons, 1962) 2nd ed.

78. M. Pirani and J. Yarwood, "*Principles of Vacuum Engineering,*" (Reinhold Publishing Co., 1961).

79. Technical Support Package for Tech Brief 68-10542 High Dielectric Thick Films for Screened Circuit Capacitors, Donald R. Ulrich, National Aeronautics and Space Administration, Washington, D. C.

80. Critical Fields of Superconducting Thin Films of Tin and Indium Alloys, R. Burton. *Cryogenics* v 6 n 5 Oct 1966 p 257-63.

81. Critical Currents and i/v Relations in Superconducting Alloy Films, R. Burton. *Cryogenics* v 6 n 3 June 1966 p 144-9.

82. Anomalous Optical Effects in Germanium Films, J. E. Davey, T. Pankey. *Optical Soc American—J* v 56 n 10 Oct 1966 p 1331-7.

84. Properties of Single-Crystal CoO Films on MgO, J. H. Greiner, A. E. Berkowitz, J. E. Weidenborner. *J Applied Physics* v 37 n 5 Apr 1966 p 2149-55.

85. Preparation and Hard Magnetic Properties of Co-Zn-P Films for High-Density Recording, R. D. Fisher. *IEEE—Trans on Magnetics* v MAG-2 n 4 Dec 1966 p 681-6.

86. Origin of Coupling in Multilayered Films, O. Massenet, F. Biragnet, H. Juret-Schke, R. Montmory, A. Yelon. *IEEE—Trans on Magnetics* v MAG-2 n 3 Sept 1966 p 553-6.

87. Optical Experiments with Magnetically Rotatable Diffraction Grating, E. U. Cohler, H. Rubinstein, C. Jones. *J Applied Physics* v 37 n 7 June 1966 p 2738-43.

88. Thick Film Resistors Functional Inks and Pastes For Microelectronics Applications, John T. Milek, Hughes Aircraft Company, Culver City, California, Feb. 1, 1968,

89. Thick Film Layout Considerations, Robert E. McMahon, et al, Massachusetts Institute of Technology, Lexington Massachusetts, 12 April 1968.

90. USAECOM Contract DA 28-043, AMC-01230(E), "High Performance Thin Films for Microcircuits," 1st Quarterly Report, 1 March 1965 – 31 May 1965, Radio Corp. of America, 1965.

91. USAECOM Contract DA 36-039, SC-90744, "Thin Films Formed by Electrochemical Reactions," Final Report, 1 June 1961 – 31 Jan. 1963, General Telephone and Electronics Laboratories, Inc., Bayside, N. Y., 1963.

92. USAECOM Contract DA 36-039, SC-90745, "Thin Films Formed by Electrochemical Reactions," Final Report, 1 June 1962 – 31 Jan. 1963, Texas Instruments, Inc., Dallas, Texas, 1963.

93. D. A. McLean, N. Schwartz and E. D. Tidd, "Tantalum-Film Technology," *Proc. of the I.E.E.E.*, Vol. 52, No. 12, pp. 1450-1462, December 1964.

94. Wilson, H. L., Fourth Quarterly and Final Summary Report on Thin-Film Monotronics, Melpar, Inc., N00019-67-C-0405, AD 833349L, April, 1968.

95. G. J. Tibol and W. M. Kaufman, "Plasma Anodized Thin Film Capacitors for Integrated Circuits," *Proc. of the I.E.E.E.*, Vol. 52, No. 12, pp. 1465-1468, December 1964.

96. P. J. Fopiano, ''The Gaseous Anodization of Aluminum,'' Proc. 1965 Elec. Comp. Conf., pp. 217-223, Washington, D. C., 1965.

97. M. C. Johnson, ''The Plasma Oxidation of Metals in Forming Electronic Circuit Components,'' Proc. 1964 Elec. Comp. Conf., pp. 1-4, Washington, D. C., 1964.

98. R. E. Whetmore and J. L. Vossen, ''Plasma Anodized Lanthanum Titanate Films,'' Proc. 1965 Elec. Comp. Conf., pp. 10-15, Washington, D C., 1965.

99. A. W. Fisher and J. A. Amick, ''Diffusion Characteristics of Doped Silicone Dioxide Layers Deposited from Premixed Nydrides,'' *RCA Review* 29 (Dec. 1968).

100. G. W. Cullen, et al., *Development of Thin-Film Active Devices on An Improved Insulating Substrate,* Interim Engineering Report Nos. 1 and 2, Contract No. N0039-68-C-2512.

101. Laznovsky, W., Thick Film Active Devices, Final Report, NAS 1-7340, June, 1968.

102. Sihvonen, Y. T., Parker, S. G., and Boyd, D. R., Printable Insulated-Gate Field-Effect Transistors, *J. Electrochem. Soc.*: *Solid State Science,* 114, 1, 96-102, Jan., 1967.

103. Survey of Glass Materials in Micro Electronics, S. M. Cox, Sept. 1968.

104. E. E. Eberhard, Latest Thin Film Circuit Techniques. *Electronics,* June 15th, 1962.

105. Witt, W., Huber, F., ann Laznovsky, W., Thick Film Field Effect Transistors Based on Silk-Screened CdS, Proc. 1967 Electronic Components Conf., McGregor & Werner, Inc., Washington, D. C., 1967.

106. A Survey of Semiconductor Materials and Processes for Thiek-Film Field-Effect Transistors Fabrication, F. W. Duncan, et al, Battelle Memorial Institute, Columbus, Ohio.

107. Patent Application Abstract — A Method For Fabricating Adherent Thick Layers of High-Conductance Metals On Oxide Surfaces, Kalus H. Behrndt, et al.

108. Wheeler, H. A., Transmission-line properties of parallel strips separated by a dielectric sheet. *IEEE Transactions on Microwave Theory and Techniques* MTTT-13 (1965): 2, pp. 172-185.

109. Mannersalo, K., Stubb, T., Certain aspects of thin-film microwave components. *Electricity in Finland* 42 (1969): 1, pp 23-27.

110. Investigation of Microwave Integrated Circuits Made by Thin Film Technique, *The State Institute for Technical Research, Finland,* 1969.

111. Peek, John R., ''Design of Thin Film Resistor and Capacitor Circuits,'' *Solid State Technology,* (1967), pp. 27-35.

112. Nester, H. H., T. E. Salzer. ''Thick Film Production Techniques,'' *Electronic Packaging and Production,* (1967).

113. Ing, S. W., Jr., R. . Morrison, and J. E. Sandor, ''Gas Permeation Study and Imperfection Detection in Thermally Grown and Deposited Thin Silicon Dioxide Films,'' *J. Electrochem. Soc.,* 109, 221-226 (1962).

114. Von Hippel, ''*Dielectric Materials and Applications,*'' pp. 1-30 (Wiley and Sons, Inc., New York, 1954).

115. O. J. Wied and M. Tierman, ''Planar Microminiaturized Nickel-Chromium Resistors,'' *Solid State Design,* p. 28 (November 1964).

116. R. P. Mandal, ''Some Physical Properties of Interest to Evaporated Thin Films Research,'' Aerojet-General Corporation, Azusa, California, TM 525: 64-0-790 (May 13, 1964).

117. W. J. Ostrander and C. W. Lewis, ''Electrical Properties of Metal-Dielectric Films,'' Volume 2, Trans. 8th *Nat'l. Vac. Symposium* (Macmillan Company, 1961), p. 881.

118. M. Beckerman and R. E. Thun, ''The Electrical and Structural Properties of Dielectric Metal and Mixtures,'' Vol. 2, Trans. 8th *Nat'l. Vac. Symp.* (Macmillan Company), p. 905.

119. W. Himes, B. F. Stout, and R. E. Thun, ''Production Equipment for Thin Film Circuitry Panels,'' Trans. 9th *Nat'l. Vac. Symp.* (Macmillan Company, 1962), p. 148.

120. E. H. Layer, ''Cermet Microelectronics,'' 1964 *Nat'l. Electronics Conference*, p. 191.

121. A. E. Lessor, L. I. Maissel, and R. E. Thun, ''Part I – Thin Film Networks,'' *I.E.E.E. Spectrum*, p. 73 (April 1964).

122. ''Thin Film Production Technique,'' Quarterly Progress Report Number 3 Contract Number N163-9142(x) for NAFI by IBM, Kingston, New York (April 15, 1962).

123. Leonard, W. F.; and Ramey, R. L.: Thin-Film Thickness from Theoretical Expressions for Conductivity and Isothermal Hall Effect, *J. Appl. Phys.* 35, October 1964, p. 2963.

124. Snow, E. H. and B. E. Deal, ''Polarization Effects on Silicone – A Review,'' *Transactions of the Metallurgical Society of AIME*, 242, 512-523 (1968).

125. High Performance Thin Films For Microcircuits, Morton L. Topfer, et al, Radio Corp of America, Aug. 1968.

126. Lathlaen, R. A. and D. A. Diehl, ''Stress in Thin Films of Silane Vapor Deposited Silicone Dioxide,'' 1968 Spring Meeting of the *Electrochem. Society*, Boston, Recent News Paper.

127. S. Dushman, ''*Scientific Foundations of Vacuum Technique,*'' pp. 757-764 (Wiley and Sons, Inc., New York, 1949).

128. G. A. Bassett, J. W. Menter, D. W. Pashley, ''*Structure and Properties of Thin Films,*'' edited by C. A. Neugebauer, J. B. Newkirk, D. A. Vermilyea, p. 11 (Wiley and Sons, Inc., 1958).

129. H. Mayer, ''*Structure and Properties of Thin Films,*'' pp. 225-247 (Wiley and Sons, Inc., New York, 1959).

130. P. K. Weimer, ''The TFT – A New Thin-Film Transistor,'' *Proc. I.R.E.*, Vol. 50, p. 1462 (June 1962).

131. W. Shockley and G. L. Pearson, ''Modulation of Conductance of Thin Films of Semiconductors by Surface Charges,'' *Phys. Rev.*, Vol. 74, pp. 232-233 (1948).

132. C. A. Hesselberth, ''Synthesis of Some Distributed RC Networks,'' Univ. of Illinois, Coordinated Science Laboratory, Report R-164, Urbana, Illinois.

133. R. W. Wyndrum, Jr., ''The Exact Synthesis of Distributed RC Networks,'' New York Univ. Laboratory for Electroscience Research, Tech. Report 400-76 (May 1963).

134. W. E. Flynt, Proc. Third Symp. on Electron Beam Technology, Alloyd Electronics Corp., Cambridge, Mass., p. 367 (March 1961).

135. A. R. Wolter, Symp. on Electron and Ion Beams, Royal Radar Establishment, Malvern, England (July 1964).

136. T. D. Schlabach and D. K. Rider, ''*Printed and Integrated Circuitry,*'' (McGraw-Hill Book Co., 1963), p. 334.

137. S. A. Bonis, Proc. Third Annual Microelectronics Symp., St. Louis Section I.E.E.E., p. IV-B-1 (April 1964).

138. C. F. Powell, I. E. Campbell, and B. W. Gonser, *''Vapor Plating,''* (J. Wiley and Sons, 1955).

139. R. M. Goldstein and F. W. Leanhard, ''Thin Film dielectric capacitors formed by reactive sputtering,'' *Proc. Elect. Comp Conf.* p. 312, 1967.

140. Varian Assoc., Palo Alto, Calif., Vacuum Prod. Div., Data Sheet on e-gun evaporation source.

141. B. Paul, ''Compilation of Evaporation Coefficients,'' *Am. Rocket Soc.*, V. 32, p. 1321 (1962).

142. R. E. Honig, ''Vapor Pressure Data for the Solid and Liquid Elements,'' *RCA Review*, V. 23, p. 567 (1962).

143. Hammond, M. L. and G. M. Bowers, ''Preparation and Properties of Vapor Deposited Silica,'' *Transactions of the Metallurgical Society of the AIME*, 242,546-550 (1968).

144. Nitsche, J. E., U. S. P. 3, 303, 115, 7/2/1967.

145. August, W. S., U. S. P. 3, 304, 362, 1967.

146. Topfer, M. L., Mitchell, J. H., and Schelhorn, R. L., High Performance Thin Films for Microcircuits, RCA, DA 28-043 AMC-01230 (E), March, 1968.

147. R. S. Clark and D. W. Brooks, ''Thin Film Monolithic Circuits — How and When to Use Them,'' *Electronic Design*, p. 64 (December 21, 1964).

148. Neugebauer, C. A., Temperature Dependance of the Field Effect Conductance in Thin Film Polycrystalline CdS Films, *J. of App. Phys.*, 39, 3177-86, June, 1968.

149. F. J. Hemmer, C. Feldman, and W. T. Layton, ''Thin Film Rhenium Resistors,'' 1964 *National Electronics Convention,* p. 201.

150. I.B.M., ''Cermet Resistor Films,'' Semi-Annual Report EWA-0083, IBM Kingston, New York (1 July to 31 December 1960).

151. F. Ruber, ''Thin Films of Titanium and Titanium Oxide for Microminiaturization,'' *I.E.E.E. Trans. on Comp. Parts,* 11 (2) 38, 1964.

152. A. E. Feuersanger, ''Applications of Insulating Titanium Dioxide Films Prepared by Chemical Vapor Reactions,'' *Proc. Nat. Elec. Conf.*, Vol. XX, pp. 188-189, 1964.

153. Hoffman, L. C., and Nakayama, T., Screen-Printed Capacitor Dielectrics, Microelectronics and Reliability, Pergamon Press, Great Britain, 131-35, 1968.

154. F. M. Smits, ''Measurement of Sheet Resistivities with the Four Point Probe,'' *Bell System Tech. J.* (Mar. 1958).

155. F. Vratny, N. Schwartz, ''Deposition of Tantalum Films by Substrate-electrode Bias Sputtering,'' *J. Vac. Science and Tech.*, Vol. 1, No. 2, Nov./Dec. 1964, p. 79.

156. R. W. Schmitt, ''Thin Films,'' *Intnl. Sci. and Technology*, p. 42 (Feb. 1962).

157. F. E. Cariou, V. A. Cajal and M. M. Gajary, ''Electron beam evaporation of metal oxide dielectric films,'' *Proc. Elect. Comp Conf.*, p. 60, 1967.

158. L. Holland, *''Vacuum Deposition of Thin Films,''* (John Wiley and Sons, 1958), p. 485.

159. Measuring Ferromagnetic Film Magnetostriction by Means of Magnetoresistance, T. C. Penn, F. G. West. *Rev Sci Instruments* v 37 n 9 Sept 1966 p 1137-9.

160. Measuring Ferromagnetic Film Magnetostriction by Means of Low Frequency rf Probes, T. C. Penn. *Rev Sci Instruments* v 37 n 9 Sept 1966 p 1134-6.

161. Magnetic Flux Reversal in Laminated Ni-Fe Films, F. B. Humphrey, R. Hasegawa, H. Clow. *IEEE—Tran on Magnetics* v MAG-2 n 3 Sept 1966 p 557-9.

162. Measurement of Tantalum Film Thickness by Anodization, C. B. Oliver. *Instn Elec Engrs*—Conference Publ 12 1965 paper 10.

163. Structure and Properties of Thin Films, C. A. Neugebauer. *Int Vacuum Congress,* 3rd—Trans v 1 Advances in Vacuum Science & Technology June 28-July 2 1965 p 29-40.

164. Some Electrical Properties of ''Natural'' Films Developed on Platinum Contacts, A. Fairweather, A. E. Parker. *Instn Elec Engrs*—Conference Publ 12 1965 paper 17.

165. Excitation of Spin Waves and Kerr Effect on Thin Ferromagnetic Films, A. Van Itterbeek, J. Witters, R. Vrambout. *Int Vacuum Congress,* 3rd Trans v 1 Advances in Vacuum Science & Technology June 28—July 2 1965 p 17-20.

166. Effect of NH_3 on Deposition from Alkaline Electroless Nickel and Cobalt Plating Baths, G. S. Alberts, R. H. Wright, C. C. Parker. *Electrochem Soc—J* v 113 n 7 July 1966 p 687-90.

167. Chemically Deposited Ferrite Films for Applications at Milimeter Wavelengths, W. Wade, R. A. Stern, T. Collins, W. J. Skudera, Jr. *Am Cer Soc*—Bul v 45 n 6 June 1966 p 571-3.

168. Research in Great Britain on Vacuum Deposited Conducting Resistive and Insulating Thin Films, J. R. Balmer. *Int Vacuum Congress,* 3rd Trans v 1 Advances iin Vacuum Science & Technology June 28-July 2 1965 p 21-7.

169. Deposition of Oxide Films by Reactive Evaporation, E. Ritter. *J Vacuum Science* N *Technology* v 3 n 4 July-Aug 1966 p 225-6.

170. Engineering Fundamentals of Thin-Film Microelectronics, N. K. Ivanov Yesipovich, Joint Publications Research Service, Washington, D. C., 9 December 1968.

171. Raue, E. J., U. S. P., 3, 305, 914, 8/2/1967.

PART II

INTEGRATED CIRCUITRY/ SEMICONDUCTOR DEVICES

CHAPTER 5

MATERIALS FOR SEMICONDUCTOR DEVICES OPTOELECTRONICS

The properties of solids are best described by the energy band theory of solids. Based on this theory and a simple understanding of the nature of crystalline solids, it is possible to explain the electrical, optical, thermal, and mechanical properties of the materials used in microelectronics. This chapter summarizes and unifies the widely diversified topics, in solid state physics, which are of most interest in the field of microelectronics. No attempt is made to provide a complete development of these basic topics, since they may be readily found in the available literature. This chapter seeks, instead, to provide the reader with an understanding of the relationships between the concepts explaining the properties of materials used in semiconductor device technology.

OVERVIEW

Any single atom possesses many permissible energy levels in which electrons can prevail. No electrons can exist at a level which is not one of the permissible ones. This situation also applies to atoms in combination, if the spaces between the atoms are sufficiently large—as in gases, for example.

In the case of solids, however, where atoms are brought into close proximity, the energy levels applicable to single isolated atoms break up into *bands* of energy levels. Within each band, discrete permissible levels prevail, rather than a continuous band. As a result, many more permissible energy levels exist, and these are grouped into separate bands. In considering electrical properties, however, only two bands are of interest. These are the *conduction* band and the *valence* band.

In the conduction band the level of energy of the electrons is high enough to enable the electrons to move readily under the influence of an external field. Through the motion of the electrons a current results, if the external field is applied in the form of a voltage. Solids which have many electrons in the conduction band are known, for this reason, as conductors.

In the valence band, the energy is of the same level as the energy of valence electrons. The electrons are here essentially attached to individual atoms and are not as free to move as are the conduction electrons. With the addition of energy, however, the electrons in the valence band may be raised to the conduction band. To accomplish this transfer, a gap known as the *forbidden energy band* must be bridged between the valence band and the conduction band. The energy difference across the forbidden region determines whether a solid acts as a conductor, semiconductor, or an insulator.

A conductor is a substance which contains many electrons in the conduction band at room temperature. No forbidden region exists between the valence band and the conduction band of the conductor, and the two bands may be looked upon as being essentially joined. A semiconductor, on the other hand, does contain a forbidden gap. Whereas a semiconductor may not, ordinarily, have electrons in the conduction band, the heat of room temperature provides sufficient energy to overcome the atomic force on a few valence electrons so that these appear in the conduction band. As a result, semiconductors will sustain some electric current at room temperature.

Insulators are materials in which the forbidden gap is too large to be bridged by applying energy to the electrons. Without accompanying extremely high temperatures, these materials will not conduct electricity.

In dealing with energy levels of materials, a useful concept is the *Fermi level.* The Fermi level is particularly applicable whenever two different materials are joined, or when a single material has two types of impurities. The Fermi level is a reference energy level from which all other energies may be conveniently measured. The Fermi level can be defined as that energy level at which the probability of finding an electron n energy units above the level is equal to the probability of finding an "absence" of an electron n energy units below the level. For a crystal of a pure semiconductor, this level falls midway in the forbidden region between the valence and the conduction bands, and follows from the condition that for every electron in the conduction band, there is an empty level in the valence band. It is not likely that an electron from a lower band will enter the valence band. Consequently, the state in which an electron in the conduction band corresponds to an empty level in the valence band is commensurate with the preceding definition of the Fermi level.

The main part of a transistor is a piece of single crystal germanium or silicon. It is essential that the semiconductor element be a part of a single crystal.

In a single crystal of a material, all the atoms are oriented in a definite and orderly manner relative to each other. Every atom in the crystal is related to its neighbors in precisely the same way. A crystal *lattice* denotes the geometrical configuration that the atoms assume in the crystal. In germanium and silicon, the lattice structure is called the *face-centered cubic* lattice. This type of lattice occurs when each atom has four neighbors which are all equally distant from it, and the neighbors are all equally spaced from each other. Such a crystal structure is also found in the diamond.

Solid material in the natural state contains generally many crystals. The material is thus composed of many units such as shown in the figure below. The boundary between such crystal units is referred to as *grain boundary.* The units are oriented relative to each other in various ways.

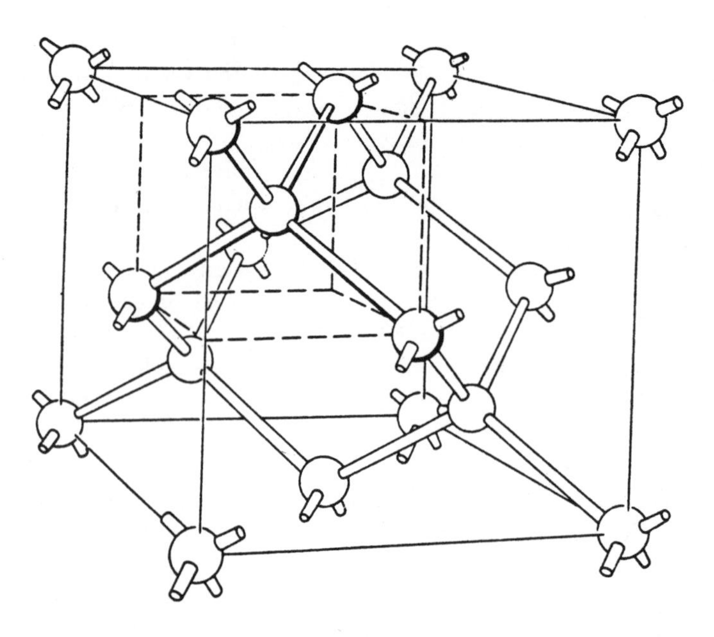

Crystal structure of germanium, silicon and diamond. These possess a tetrahedral bond arrangement.

A material which contains more than one crystal in this manner, is *polycrystalline.* Accordingly, solid materials are usually polycrystalline. A large single crystal piece of material may be produced through careful processing techniques, and this method is applied in the manufacture of transistors.

A crystal is held together by forces prevailing between atoms. In semiconductors such as germanium and silicon, these forces result from *electron-pair bonds.* The forces resulting from these bonds are referred to as *covalent forces.*

An electron-pair bond is produced when the motion of a valence electron of one atom is coordinated with the motion of a valence electron of another atom so as to give rise to an electrostatic force between the electrons. In this manner, an atom shares one of its valence electrons with a valence electron of another atom. The atoms of germanium and silicon have four valence electrons. In pure crystals of these materials, each atom shares one of these valence electrons with each of its four neighbors.

When any electrons transfer from the valence band to the conduction band, the crystal is considered to be ionized. This is due to the condition that when the transfer is made, the electron leaves the parent atom. This atom is then minus one unit of negative charge and is consequently ionized. Although heat may be applied to provide the added energy for such transfer of the electrons, the energy may also be derived from an electric field or incident light.

When an electron leaves the valence band for the conduction band, an empty energy level is produced in the valence band. Upon the application of an electric field, the affected conduction band electrons are induced to move, and such movement constitutes a current.

There is, however, still another method in which a current may be produced in a semiconductor. When a covalent bond is broken and the electron is removed, a vacancy is left in the parent atom since that atom has left only three valence electrons. This vacancy may be filled by an electron from a nearby covalent bond which becomes broken as a result. When this nearby electron arrives at the vacancy, it enters into a covalent bond. In changing position, this electron does not have a conduction band energy. Instead, its energy remained in the valence band. Once this nearby electron moves in this manner, however, a new vacancy results at the bond which the electron left. To fill this latter vacancy, another electron may, in turn, leave its bond. As a result, current is generated through electrons which are not in the conduction band, because their energies lie in the valence band. The interesting aspect to note here is that although an electron has moved from one bond to another, the resultant effect is that the vacancy has moved. Such movement of vacancies is referred to as *conduction by holes.*

The conduction by holes prevails in the valence band, since the electrons involved in this movement do not possess conduction band energies. Since the removal of an electron from a normal covalent bond results in localized positive charge, the hole may be thought of in terms of a positive charge. With this viewpoint in mind, the hole may be considered a particle which moves under the influence of an electric field – as the electron – but in opposite direction to that of the electrons. Whereas the hole may be regarded as possessing definite mass similar to the electron for analytical purposes, it must be remembered that the hole concept is only an analytical tool for describing the movement of the electrons.

In a crystal of pure semiconductor, therefore, current may be conducted by two processes. In the one process electrons are allowed to move essentially throughout the crystal. In the other process the movement of the electrons is relatively restricted, and the movement resembles the movement of holes. The holes may be thought of as positive charges with properties similar to the negative electrons.

When an electron is elevated to the conduction band, a hole is left at the region from which the electron moved. Both the electron and hole are made available for carrying current when a covalent bond is thus broken. The charge carriers resulting from the broken covalent bond, are called electron-hole pair, and these are the only charge carriers in a pure semiconductor. Such a semiconductor is said to be *intrinsic* to distinguish it from one containing impurities.

In the intrinsic semiconductor, the number of holes is equal to the number of conduction electrons, and therefore the Fermi level can be defined in terms of the two charge carriers. Accordingly, the Fermi level may be defined as the average energy of the charge carriers. This definition is equivalent to the one given earlier.

Current conduction in an intrinsic semiconductor depends on both temperature and the applied electric field. At absolute zero temperature, no conduction prevails through either method since no covalent bonds are broken. At raised temperatures, however, current will flow through electron and hole movement, depending on the applied electric field. By raising the temperature substantially, a pure semiconductor will behave as a conductor, since additional covalent bonds are broken for furnishing current carriers.

Under equilibrium conditions, thermal generation of electron-hole pairs must be accompanied by the reverse process, since the total number remains constant. This reverse process is called *recombination* and occurs whenever an electron fills a hole to eliminate an electron-hole pair. In the recombination process the electrons are temporarily held in *traps* during an intermediate step. These traps appear at crystal impurities and defects.

The net movement of current carriers from electrons and holes can occur by either *diffusion* or *drift*. Under diffusion there is temporarily an unequal distribution of charges. The diffusion current is always directed, thereby, so as to equalize the distribution. When an electric field is applied, drift occurs in which the available charges drift in the direction of the applied field.

Due to collisions between the electrons and the atoms, the movement of the individual charges is not a smooth process. The total effect of the many electrons involved, however, is that of smooth movement in the direction of the current flow.

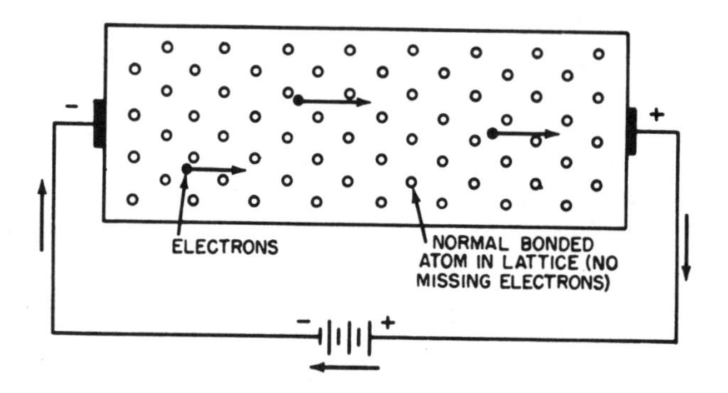

Current flow in N-type material.

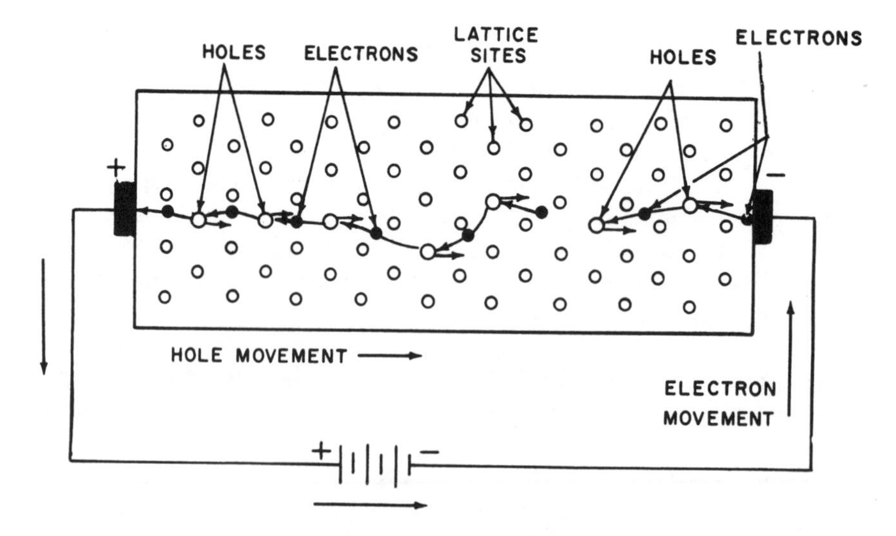

Current flow in P-type material.

5.1 ATOMIC PHENOMENA

To develop properly the concepts of the energy band theory, a review of certain principles from modern atomic physics may be of value. Consider an isolated hydrogen atom in free space. The classical Bohr theory of the atom indicates that the electron revolves about the nucleus in certain prescribed circular orbits. This restriction on the orbits of the electron, together with the nature of the electrostatic potential of the nucleus, results in the quantization of the total energy of the electron. Thus, it may possess only certain specific, discrete values. An electron in the lowest energy state, E_1, may be excited to a higher energy state, E_2, by radiation

Views of a diamond-type lattice from a <110> direction

of frequency f. The difference between the energies of the two states is $E = E_2 - E_1 = hf$, where h is Planck's constant. The excited electron will immediately return to the lowest energy state upon the emission of a photon of energy $E_2 - E_1$. These energy differences may be measured conveniently, since the transition frequencies are manifest in the optical spectrum of the atom.

Modern quantum theory explains this situation by asserting that each electron occupies a given quantum state which can be represented by an appropriate function—the "wave" function of the state, and which possesses a characteristic energy. The equation of motion for the electron, the Schrodinger wave equation, from which these unique relationships can be derived, is given by

$$-\frac{h^2}{8\pi^2 m}\nabla^2\psi + V\psi = E\psi \qquad (5\text{-}1)$$

where m is the electron mass, ∇^2 the Laplacian differential operator, E is the total energy, and V is the potential energy of the electron represented by the wave function ψ. For the isolated hydrogen atom, $V = -q^2/4\pi\epsilon_0 r$, where ϵ_0 is the permittivity of free space. By solving the differential equation, it is found that the energy may take on only the values $E = -mq^4/8n^2h^2\epsilon_0{}^2$; $n = 1, 2, 3, 4.\ldots$ Since n can have only integral values, the

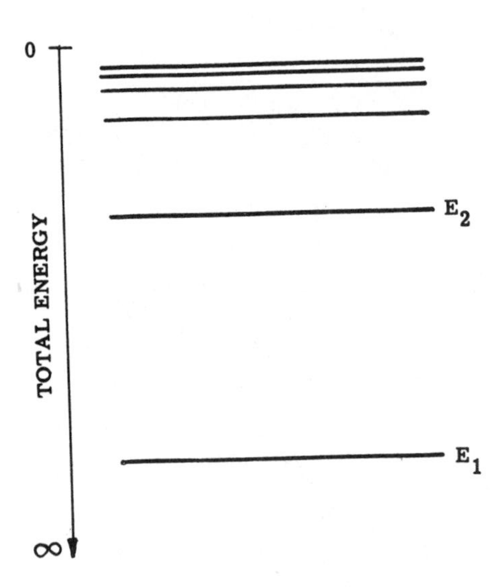

Fig. 5-1 Simple atomic energy level diagram

energy is quantized into discrete levels. This can be graphically represented by the energy level diagram shown in Fig. 5-1.

The complete description of the set of energy levels and states for an atom of many electrons requires the use of four quantum numbers:

n = total quantum number
a = angular momentum quantum number.
m = magnetic quantum number
m_s = spin quantum number.

The Pauli exclusion principle states that no two electrons may occupy the same quantum state. Thus, no two electrons in an atom can possess the same set of four quantum numbers. Only the quantum numbers, n and a, are associated with states of different energy, while m and m_s indicate how many states exist at each energy level. As long as two atoms are sufficiently isolated from each other, they may exist in the same quantum state, because their difference in position satisfies the exclusion principle.

5.2 ENERGY BAND CONCEPTS

When identical atoms are brought close to each other, as in a crystal, each of the discrete electronic energy levels of the individual atom splits into as many levels as there are atoms in the vicinity. Each set of levels lies within an energy range which is designated as a band, and each band is associated with one of the original atomic energy levels. The width of the energy band arising from the splitting of an energy level, is independent of the number of atoms in the crystal. There are no allowed states, with energies between the bands; hence, the energy between the bottom of one band and the top of the next is referred to as a "forbidden band" or "energy gap". The splitting of levels into bands as the distance between atoms is decreased, is shown in Fig. 5-2.

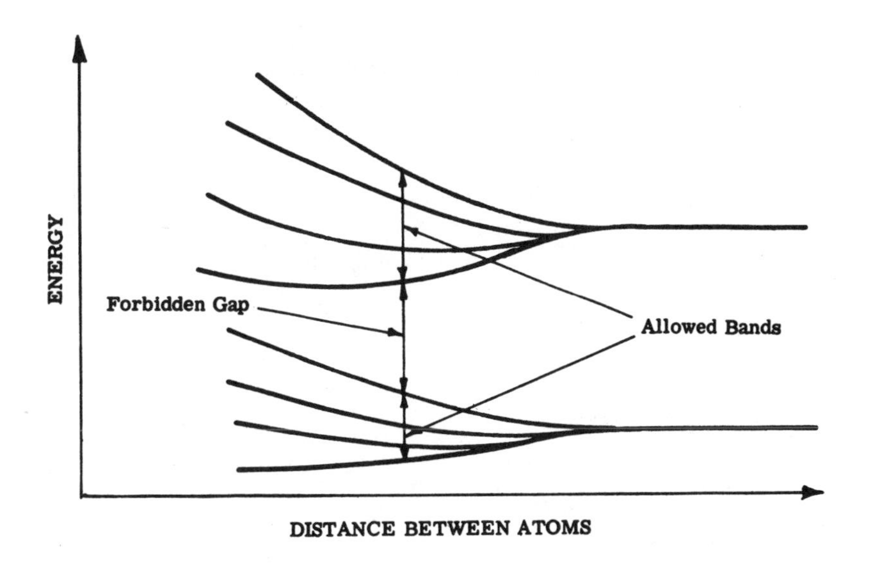

Fig. 5-2 Dependence of energy levels on interatomic distance

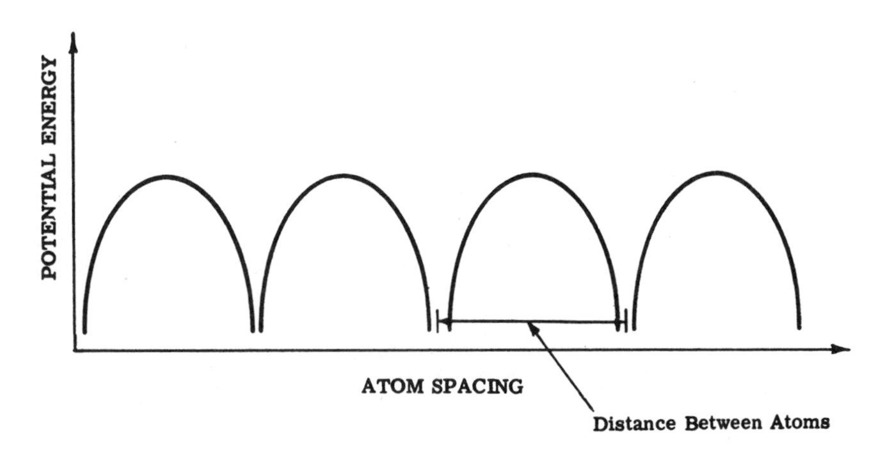

Fig. 5-3 Periodic potential associated with perfect crystals

If the collection of atoms, being described, is a crystalline solid, then they are in a regular periodic array. In its motion through the crystal, an electron is repelled by other electrons and attracted by nuclei. In the quantum mechanical description of the electron motion, this periodic array of atoms is represented by a periodic potential, as shown in Fig. 5-3. A functional representation of this potential V is then used in the wave equation, in place of the simple potential function, used in the case of the hydrogen atom. Solutions of Schrodinger's equation containing this periodic potential function do indicate the existence of allowed energy bands as opposed to the discrete levels in the case of the isolated hydrogen atom.

5.3 METALS, SEMICONDUCTORS, INSULATORS

The difference between many of the properties of metals, semiconductors, and insulators is directly related to differences in their band structure. In thermal equilibrium the lowest energy bands are completely filled in any solid material, just as the lowest discrete levels are filled in any isolated atom. Unless there is some form of excitation, all the highest allowed bands will be completely empty. However, for any material there must be an energy range in which transitions can occur to higher energy states. This situation can occur in two ways—somewhere on the energy scale there may be a partially filled band, or one band may be filled and the next higher completely empty.

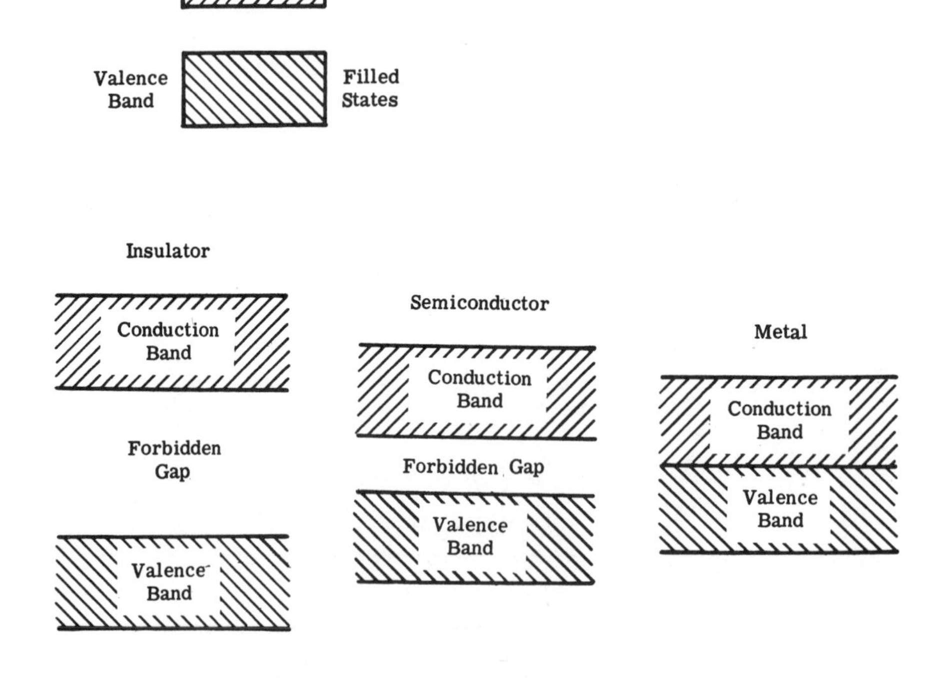

Fig. 5-4 Energy bands in solids

The case of a partially filled band, or a similar situation of several overlapping bands, is representative of common metals such as copper and silver. The case where one allowed band is filled, and the next higher band void, is characteristic of nonmetallic crystals including both semiconductors and insulators. A representation of energy bands in various types of solids is shown in Fig. 5-4.

Above the deep lying core energy levels, the energy bands can be divided into two classes; valence bands and the conduction bands. In all types of crystals a conduction band lies above a valence band, the two being separated by the forbidden band. In the case of the nonmetallic crystals, the only forbidden band of interest prevails between the highest valence band and the lowest conduction band. This is generally the range referred to when the term energy gap or forbidden band is used. The occupancy of these bands determines the electronic and optical properties of these materials.

The actual distribution of electrons among the various allowed states, under thermal equilibrium conditions, is governed by the Fermi-Dirac statistics. These statistics are based on the restriction on occupancy of states imposed by the Pauli principle. According to these statistics, derived from fundamental considerations in statistical mechanics, the probability that a quantum state with energy E is occupied by an electron is:

$$f(E,T) = 1/(\exp[(E - E_F)/kT] + 1) \tag{5-2}$$

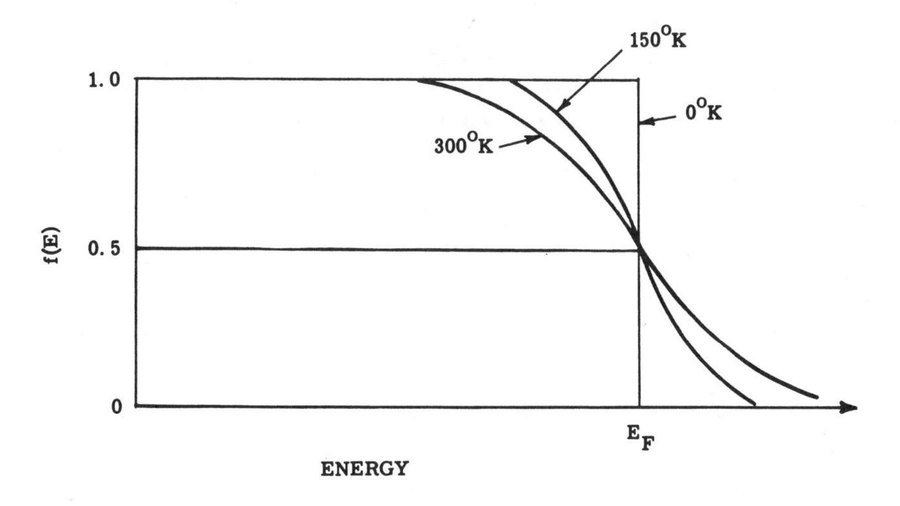

Fig. 5-5 Fermi-Dirac distribution functions for electrons

where E_F is the Fermi energy, i.e., that energy at which the probability of occupancy is 1/2, T is the absolute temperature, and K is Boltzmann's constant. This probability distribution is illustrated in Fig. 5-5 for various temperatures. At absolute zero the Fermi function (Eq. 5-2) has the following characteristics:

$$f(E,0) = f_o = \begin{cases} 0, & E > E_F \\ 1/2, & E = E_F \\ 1, & E < E_F \end{cases} \quad (5\text{-}3)$$

Thus, at absolute zero, all states below E_F are completely full, while all states with E greater than E_F, are empty. At any other temperature the cut-off condition is less distinct, although the total energy range in which the function is different from one or zero, is only about 2kT centered at E_F. Irrespective of the temperature, however, the Fermi probability function is equal to 1/2 at the Fermi energy.

In any nonmetallic solid, the Fermi energy lies in the forbidden band separating the highest valence band from the lowest conduction band. At absolute zero all the conduction states are empty and all valence states are fully occupied, so that in the valence band there are no higher energy states to which an electron can be excited. This implies that electrons in a fully occupied band cannot gain any energy from an applied electric field, and as a result they cannot contribute to the electrical conductivity. Hence, a nonmetallic crystal is a perfect insulator at absolute zero.

At any finite temperature, a number of electrons are thermally excited from the valence band into the conduction band, with the number depending on the width of the energy gap. In this situation electrons in the partially filled conduction band, can move up to higher energy states if a field is applied. Accordingly, the material will no longer be a perfect insulator. In metals, the Fermi level lies in an allowed band which is partially filled, and the material will be able to conduct electricity at any temperature. This difference in the temperature dependence of the electrical properties of metals and nonmetals, is an important distinguishing characteristic between these two classes of materials.

In the nonmetallic solid at absolute zero, all valence states are occupied and all conduction states are empty. If a single electron is removed from the valence band, the state left empty by removal of the electron, is known as a ''hole''. This vacant state is one into which other electrons in the same valence band can be excited. Conduction by available states of this type is known as hole conduction, and is in addition to the electron conduction by electrons in the conduction band.

Beside the possibility of thermally exciting electrons from the valence band to the conduction band in nonmetallic crystals, there is also the possibility that photons of sufficient energy will create carriers. The energy E of a photon of frequency f is given by $E = hf = hc/\lambda$, where h is Planck's constant, c is the velocity of light, and λ is the wavelength. If a nonmetallic crystal having an energy gap E_g, is illuminated with photons of energy E_g or greater, they will be absorbed, and will excite electrons from the valence band to the conduction band. Photons of energy less than E_g, hence, of wavelength longer than $\lambda_g = hc/E_g$, will not be absorbed, and the material will be transparent to such radiations. For example, silicon, with an energy gap of 1.08 ev, absorbs all radiation with wavelength shorter than 11,000 Å and is transparent to illumination of greater wavelengths. A typical transmission spectrum is shown in Fig. 5-6.

5.4 CRYSTAL STRUCTURE

Solid materials can also be classified according to the nature of their crystal structure and their crystalline perfection. A solid is considered to be a single crystal if all the atoms are in a regular array which is periodically repeated throughout the entire solid. A polycrystalline material is one composed of many small crystals generally randomly oriented with respect to each other. An amorphous solid is one in which there is complete disorder in the atomic arrangement.

A perfect single crystal is one in which there is no departure from the

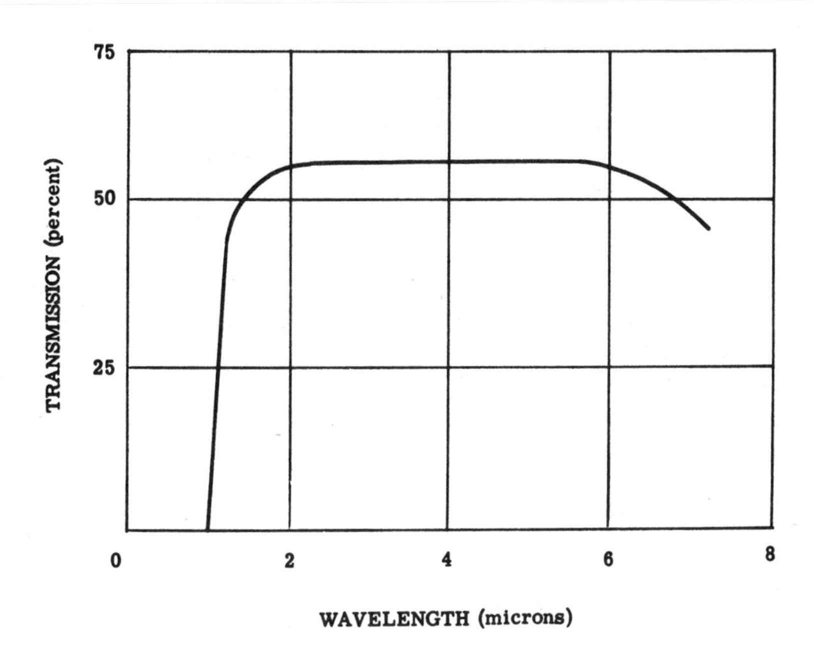

Fig. 5-6 Optical transmission spectrum of silicon

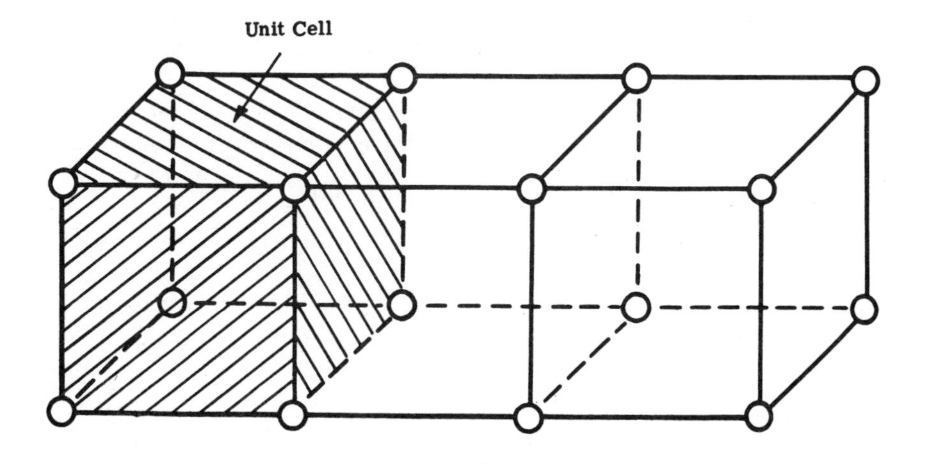

Fig. 5-7 Cubic crystal lattice

periodic arrangements of atoms. Crystals which are actually grown can approach this state, but always contain a small number of imperfections. It is these imperfections which determine many of the properties of materials of interest in microelectronics.

The perfect crystal can be described by a three-dimensional array of atoms, referred to as the "crystal lattice". Each crystal is composed of a number of identical cells designated unit cells. A primitive cell is another type of structural cell; it is the smallest parallelopiped which can be generated by linear translations along three different directions and it is chosen to contain the smallest number of atoms. The typical crystal lattice is shown in Fig. 5-7. This lattice is a cubic structure, which is one of the seven crystal systems. In order of decreasing symmetry, these systems include: 1) cubic, 2) tetragonal, 3) orthorhombic, 4) trigonal, 5) hexagonal, 6) monoclinic, and 7) triclinic.

The important imperfections in a solid include impurities, lattice defects, dislocations, and lattice vibrations. Chemical impurities may be added to a perfect crystal by either replacing an atom at a lattice site as a substitutional impurity, or inserting atoms between lattice points as interstitial impurities. Typical semiconductor dopants are deliberately added as substitutional impurities. Lattice defects are either vacancies caused by the removal of an atom from a lattice site, or interstitial atoms of the host element. The most severe type of lattice defect is a dislocation in which entire planes of atoms are displaced with respect to each other. A number of such dislocations can constitute a grain boundary, which is the boundary between two different crystals in a solid. A final form of imperfection is the lattice vibration due to absorption of heat by the crystal. A "phonon" is a quantized unit used to describe the energy of such vibrations. Since the vibrations are excited by thermal energy, the number of phonons increase with the temperature of the crystal.

5.5 ELECTRICAL CONDUCTIVITY

Ohm's law simply states that charge carriers in a solid, under the influence of a constant electric field, will attain a constant drift velocity which is proportional to the applied field. This drift velocity is due to the net retarding effect of the atoms in the solid, which prevent the carrier velocity from increasing indefinitely under the influence of the applied field. A consequence of this effect is that the current density J is proportional to the applied field E; hence,

$$J = \sigma E \tag{5-4}$$

where σ is the conductivity of the material. Furthermore, by definition

$$J = ne\bar{v} \tag{5-5}$$

where e is the electronic charge, n the number of electrons per cubic cm, and $\bar{v}$ the average carrier drift velocity. The force on a carrier of charge e, due to an electric field E, is given by,

$$F = eE = m^*a \tag{5-6}$$

where m^* is the effective mass of the charge carrier, and a is its acceleration. Assigning to the charge carrier an appropriate effective mass, enables one to circumvent the difficult problem of determining its detailed motion in the solid, where it is under the influence of a periodic potential. Thus, the charge carrier is treated, instead, as an isolated carrier. This effective mass is related to the curvature of the energy bands of a given material. In terms of the average velocity $\overline{v}$,

$$eE = m^*\overline{v}/t \tag{5-7}$$

where t is the mean time between collisions. Thus,

$$\overline{v} = (et/m^*)\,E = \mu E \tag{5-8}$$

where the constant of proportionality between the velocity and the field strength is defined as the mobility, μ, of the carrier. From the equations relating J, E, and $\overline{v}$, the conductivity σ is given by,

$$\sigma = ne^2t/m^*\,. \tag{5-9}$$

It may be seen from the expressions for conductivity and mobility, that an important factor in determining these parameters, is the mean free time between collisions. The nature of the collisions which determine this mean free time is highly dependent on the nature of the solid and environmental factors such as temperature. In the majority of cases scattering from the lattice, or phonon scattering, is the limiting factor on carrier mobility. If a large number of impurity atoms are present in a material, however, scattering from these may determine the mean free time. Scattering from structural imperfections may also limit the carrier mobility in some materials. Some imperfections tend to trap carriers which effectively reduce the available carriers or the average lifetime of a given free carrier. Finally, if some dimension of the specimen approaches the mean free path of the carriers, as is the case with thin films, scattering from the surface will reduce the mobility.

5.6 SEMICONDUCTOR PRINCIPLES

The properties of semiconductor materials and devices can be directly derived from the band structure of solids. In order to derive the electrical properties of these materials, the problem resides basically in determining the number of carriers in the material. The Fermi-Dirac distribution indicates that the probability for a quantum state of energy E to be occupied by an electron is,

$$f(E) = 1/[\exp(E - E_F)/kT + 1]\,. \tag{5-2}$$

If we multiply this by the density of states factor $D(E)dE$, which denotes the number of available states per unit volume with energies between E and E + dE, the result obtained is n(E)dE, the number of electrons per unit volume with energies between E and E + dE. To compute the electron density in the entire conduction band, i.e., in the energy range between E_c and E_b as shown in Fig. 5-8, a simple integration may be performed:

$$N = \int_{E_c}^{E_b} n(E)dE = \int_{E_c}^{E_b} D(E)f(E)dE . \qquad (5\text{-}10)$$

The density of states D(E) can be shown to be

$$D(E) = (4\pi/h^3)(2m_e)^{3/2}(E - E_c)^{1/2} , \qquad (5\text{-}11)$$

where m_e is the effective mass of the electron in the crystal. At room temperature, for levels E in the conduction band, $E - E_F >> kT$, so that

$$f(E) \approx \exp[-(E - E_F)/kT] , \qquad (5\text{-}12)$$

and hence the above integration yields,

$$N = (2/h^3)(2\pi m_e kT)^{3/2} \exp[-(E_c - E_F)/kT] . \qquad (5\text{-}13)$$

In the interest of brevity, this expression may be written as

$$N = N_c \exp[-(E_c - E_F)/kT] \qquad (5\text{-}14)$$

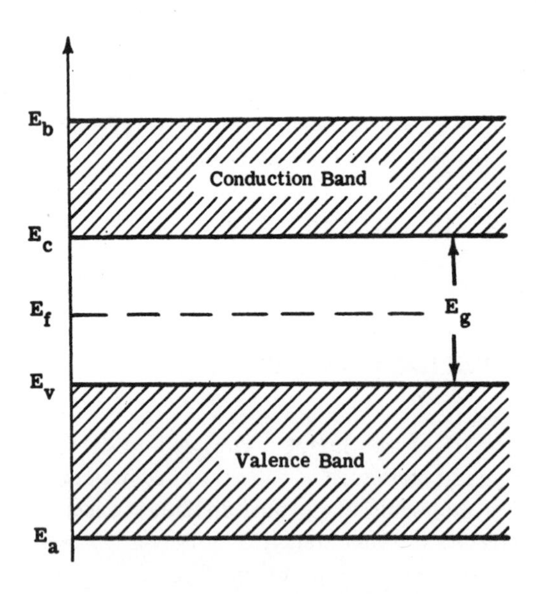

Fig. 5-8 Energy band diagram of semiconductors

where N_c is defined as the effective density of states at E_c, the energy of the lower edge of the conduction band. A similar calculation yields the hole density in the valence band:

$$P = (2/h^3)(2\pi m_h kT)^{3/2}\exp[-(E_F - E_v)/kT] = N_v \exp[-(E_F - E_v)/kT] \tag{5-15}$$

where m_h is the effective mass of a hole, and E_v is the energy of the upper edge of the valence band.

5.6.1 INTRINSIC SEMICONDUCTORS

In an intrinsic semiconductor – one containing only atoms of the same type, net charge neutrality requires that the number of mobile holes equal the number of mobile electrons. Let the subscript i indicate quantities in an intrinsic semiconductor. Then, Eqs. (5-14) and (5-15) become, for an intrinsic semiconductor:

$$N_i = N_c e^{-(E_c - E_{Fi})/kT} \tag{5-16}$$

$$P_i = N_v e^{-(E_{Fi} - E_v)/kT} \tag{5-17}$$

Setting the number of holes, P_i, equal to the number of electrons, N_i, the Fermi Energy in an intrinsic semiconductor is obtained as

$$E_{Fi} = (E_c + E_v)/2 + \frac{3kT}{4} ln \frac{m_h}{m_e} = (E_c + E_v)/2 + \frac{kT}{2} ln \frac{N_v}{N_c}. \tag{5-18}$$

Thus, at absolute zero, the Fermi level will be at the center of the energy gap, with some shifting at higher temperature caused by the difference in effective masses of electrons and holes.

The product of N_i and P_i is,

$$N_i P_i = 4(2\pi kT/h^2)^3 (m_e m_h)^{3/2} \exp[-(E_c - E_v)/kT] \tag{5-19}$$

or

$$N_i P_i = N_i^2 = 4(2\pi kT/h^2)^3 (m_e m_h)^{3/2} \exp[-E_g/kT] = P_i^2 \tag{5-20}$$

N_i is known as the "intrinsic carrier density." This indicates the important result that the product of electron and hole concentrations is a constant for a given material at a given temperature. In addition, the inverse exponential variation of N_i^2 with E_g, indicates that the electrical properties of an intrinsic semiconductor are highly dependent on the width of the energy gap E_g. For silicon this has a value of 1.08 ev at 300 °K, while for germanium the room temperature gap is 0.65 ev.

5.6.2 IMPURITY SEMICONDUCTORS

The addition of doping to semiconductors, will radically alter their electrical properties. In terms of the band theory of solids, the addition of dopants creates an allowed level in the formerly empty energy gap. This situation is illustrated in Fig. 5-9. If the dopant gives rise to an occupied level which can donate electrons to the conduction band, the level is designated a "donor" level, and the material is referred to as an "n-type" semiconductor, since the conductivity is primarily due to the negative electrons added to the conduction band. If, on the other hand, the dopant gives rise to an unoccupied level which can accept electrons from the valence band, thereby creating holes in that band, the level is called an acceptor level and the material is referred to as a "p-type" semiconductor, since the conductivity is primarily caused by the positive holes added to the valence band. Normally, the impurities introduced are chosen so that the donor levels will fall very close to the edge of the conduction band, and the acceptor levels close to the edge of the valence band.

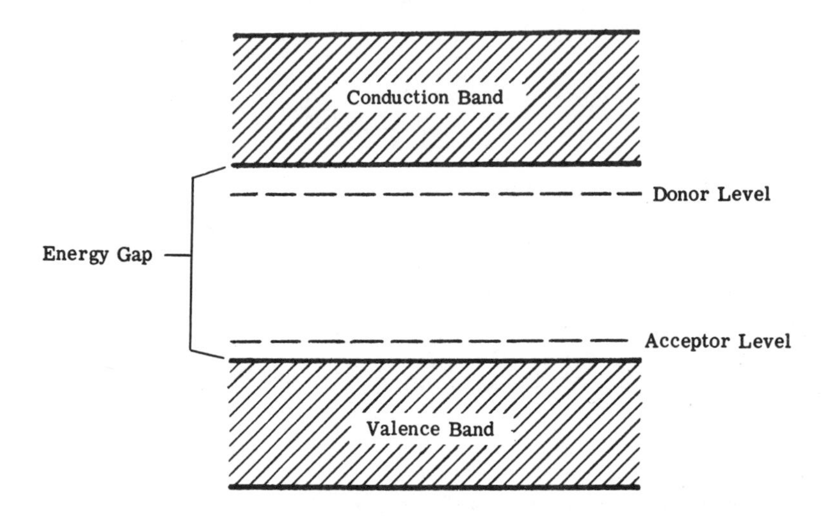

Fig. 5-9 Energy band diagram of doped semiconductors

In a manner similar to that for the intrinsic semiconductor, expressions for the Fermi levels of extrinsic semiconductors may be derived. Consider an n-type semiconductor and let N_d be the density of donor atoms. If the donor level is close to the edge of the conduction band, then at room temperature most of these donors will be thermally ionized, creating as many excess conduction electrons as donor centers. If the density of donor atoms N_d is much greater than the intrinsic carrier density N_i, as is typically the case, then the net number per unit volume of conduction electrons N_n is approximately equal to N_d, the number per unit volume added by the donors. Accordingly,

$$N_n \approx N_d = N_c \exp[-(E_c - E_{Fn})/kT] \qquad (5\text{-}21)$$

from which E_{Fn}, the Fermi level in an n-type semiconductor, is

$$E_{Fn} = E_c - kT \, ln \frac{N_c}{N_d}. \qquad (5\text{-}22)$$

An analogous expression can be derived for E_{Fp}, the Fermi level in a p-type semiconductor.

$$E_{Fp} = E_v + kT\,ln\frac{N_v}{N_a} \qquad (5\text{-}23)$$

where N_a is the density of acceptor atoms.

Interesting conclusions can be drawn about the position of the Fermi level in an extrinsic semiconductor compared to its intrinsic position, if the carrier density in a semiconductor is expressed in terms of its intrinsic carrier density. By combining Eq. (5-14) with Eq. (5-16), and Eq. (5-15) with Eq. (5-17),

$$N = N_i e^{(E_F - E_{Fi})/kT} \qquad (5\text{-}24)$$

$$P = P_i e^{(E_{Fi} - E_F)/kT}. \qquad (5\text{-}25)$$

For an n-type semiconductor $N_n > N_i$ because the donor has contributed extra mobile electrons. Hence, equations (5-14), and (5-24) reveal that $E_{Fn} > E_{Fi}$, i.e., that the Fermi level in an n-type semiconductor is shifted up from its intrinsic level toward the conduction band. For a p-type semiconductor, $P_p > P_i$ because the acceptor has provided extra mobile holes. Hence, $E_{Fp} < E_{Fi}$, implying that the Fermi level in a p-type semiconductor is shifted down from its intrinsic level toward the valence band.

5.7 PROPERTIES OF SEMICONDUCTOR MATERIALS

There are a number of physical properties of semiconductors which determine their usefulness in devices. Two types of semiconductors are of interest: the elemental semiconductors, and compound semiconductors. The first class includes germanium and silicon. The second class includes such materials as indium antimonide, gallium arsenide, gallium antimonide, and aluminum antimonide. The properties of greatest interest are energy band gap widths, impurity ionization energies, carrier mobilities, and dielectric constant. The band gap width determines the usable temperature range of the material, while high carrier mobility and low dielectric constant are essential in high frequency device structures. Impurity ionization energies also are of importance in determining the useful temperature range of the material.

Germanium and silicon are elemental semiconductors from Group IV of the Periodic Table of Elements. They are covalently bonded materials of a face-centered cubic crystal structure. Typical dopants in both Ge and Si include Phosphorous, Arsenic, Antimony, Boron, Aluminum, and Gallium. For Ge the typical impurity ionization energy is about 0.01 electron volts while for Si a typical figure is about 0.04 ev.

The III-V compound semiconductors, made by combining elements from Group III of the periodic table with those of Group V, are binary chemical compounds which form a crystal structure identical to the diamond structure of Ge and Si. While many other compounds exist, those listed above have been applied most. Impurity levels may be created in these materials by doping as in the elemental case, but a more important method resides in causing deviations from stoichiometry; i.e., creating a mixture of the materials that is not in proper proportion for the compound.

5.8 CURRENT FLOW IN A SEMICONDUCTOR

There are basically two mechanisms for current flow in a semiconductor: drift current, and diffusion current. The drift current is the usual current flow under the influence of an electric field and is expressed by,

$$J = \sigma E$$

where the conductivity $\sigma = q(\mu_n N + \mu_p P)$, and μ_n and μ_p are the electron and hole mobility, respectively. Thus, the flow of both electrons and holes are of importance in a semiconductor. The second mode of current flow is by the diffusion of carriers. Thus, if two regions of a semiconductor have different carrier concentrations, there will be a diffusion of charge from the region of high density to the region of lower density. The current density, J, associated with this mechanism, is proportional to the concentration gradient of the carrier density. This may be formally expressed as:

$$J = q(D_n \nabla N - D_p \nabla P) \tag{5-26}$$

where D_n and D_p are the diffusion constants of electrons and holes, respectively. An additional expression of importance in these considerations is the continuity equation,

$$\nabla \cdot J + \partial Q/\partial t = 0 \tag{5-27}$$

where Q is the total charge density. This equation is merely an expression of the fact that the total current flowing out of a given volume must be equal to the time rate of change of charge enclosed by that volume. With these equations it is possible to describe current flow in such semiconductor structures as the p-n junction.

5.9 THEORY OF THE P-N JUNCTION

The behavior of the p-n junction is determined by the nature of the energy band structure which results when two oppositely doped materials are brought together to form the junction. This is illustrated in Fig. 5-10. In the n-type material, where there is an excess of electrons, the latter are called majority carriers while holes are called minority carriers. In p-type material, holes are the majority carriers and electrons the minority carriers. The Fermi level is a chemical potential which must be constant throughout the material, and this requirement causes a distortion in the band system. The difference in energy or potential between the two sides of the junction brings about a charge flow across the transition region, although in thermal equilibrium, the net diffusion and drift current across the junction is zero.

An alternative explanation is that at the boundary between a p-type and an n-type material, the holes tend to diffuse into the p-side of the junction. Since the equilibrium current must be zero, a contact potential is built up which opposes majority carrier diffusion. The potential sets up a drift field which accelerates minority carriers across the junction. At

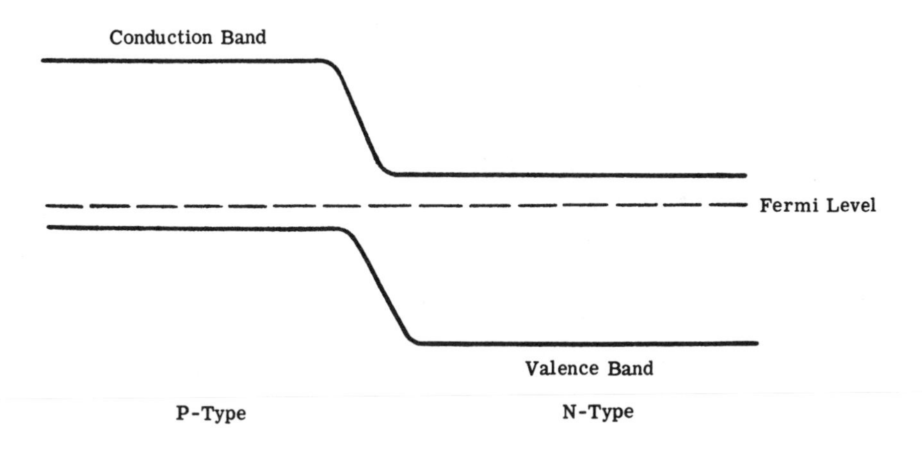

Fig. 5-10 Energy band diagram of P-N junctions

equilibrium, the current flow due to the diffusion current and drift current for each type of carrier is exactly balanced.

If an external potential, V, is applied across the junction, the potential that appears at the junction is the sum of the external potential and the built-in electrostatic potential. In the bulk of the n and p regions, the field and space charge can be regarded as negligible. Only at the p-n transition region is there a significant net charge and electric field. Essentially, most of the mobile carriers are swept out of the transition region by the field, and the space charge is the result of ionized dopant centers, which are immobile. Fig. 5-11, illustrates the space charge at an abrupt p-n junction, as well as the field and the resulting electrostatic potential. The width of the space charge region is related to the voltage and the impurity concentration, by electrostatic theory. This double layer of charge across the junction must be charged and discharged as a capacitor, the capacitance of which is referred to as the ''depletion layer capacitance''. The capacitance, C, is given by simple theory as,

$$C = \epsilon A/d \tag{5-28}$$

where ϵ is the dielectric constant, A is the junction area and d is the width of the depletion layer. For an abrupt junction this is given by,

$$d = [(2\epsilon/q)(V + V_j)(1/N_p + 1/N_n)]^{1/2}, \tag{5-29}$$

where V_j is the junction contact potential (0.8V for Si), and N_p and N_n are the net impurity densities for the p-type and n-type materials, respectively.

The voltage applied across the junction has little effect on the minority carrier flow into the junction, but there is a significant effect on the diffusion of majority carriers trying to overcome the potential barrier. The minority carrier densities on either side of the junction are related exponentially to the applied voltage. In terms of n_{po}, the number of mobile electrons present in the p-type material at equilibrium, the number n_p at a given bias voltage, V, can be written as

$$n_p = n_{po}\exp(eV/kT)\ . \tag{5-30}$$

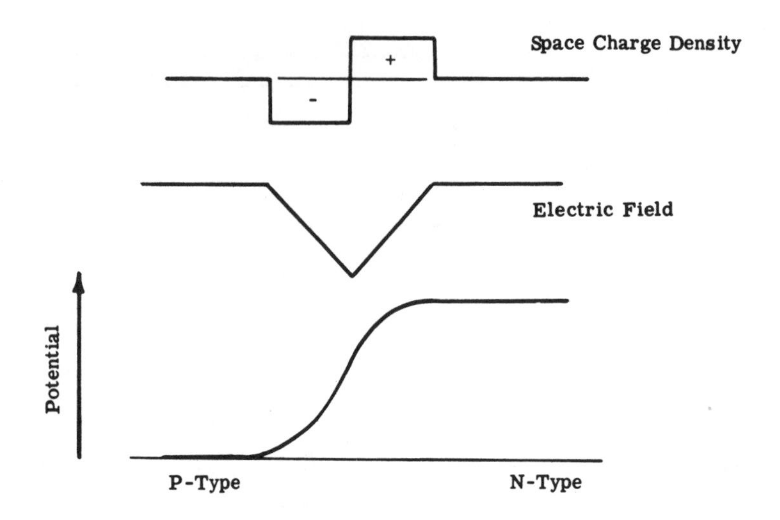

Fig. 5-11 Space charge, electric field and electrostatic potential at a P-N junction

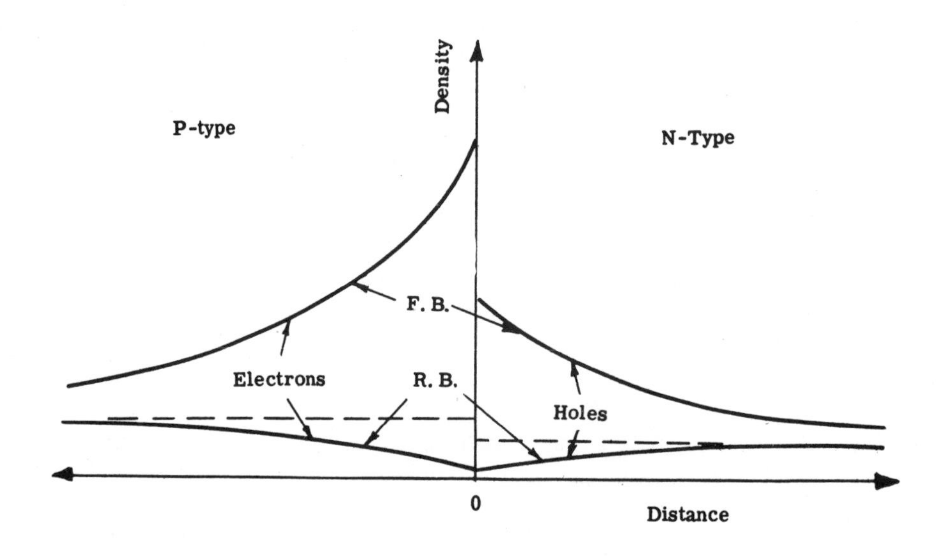

Fig. 5-12 Minority carrier densities at biased P-N junctions. (F. B. = forward bias; R. B. = reverse bias)

Solution of the continuity Eq. (5-27), for both electrons and holes leads to carrier densities as functions of distance in the material. This is depicted in Fig. 5-12. Assuming that outside of the junction region current flow is only by diffusion (the field is negligible outside transition region), it is possible to use the expressions discussed previously for diffusion current flow, in order to get the current flow. This is given by,

$$I = I_S[\exp(qV/kT) - 1] , \qquad (5\text{-}31)$$

where I_S is the saturation current for the junction. This is indicated in the typical diode characteristic illustrated in Fig. 5-13. At extremely high reverse voltages, the junction fields are high enough to cause collision ionization and an avalanche breakdown.

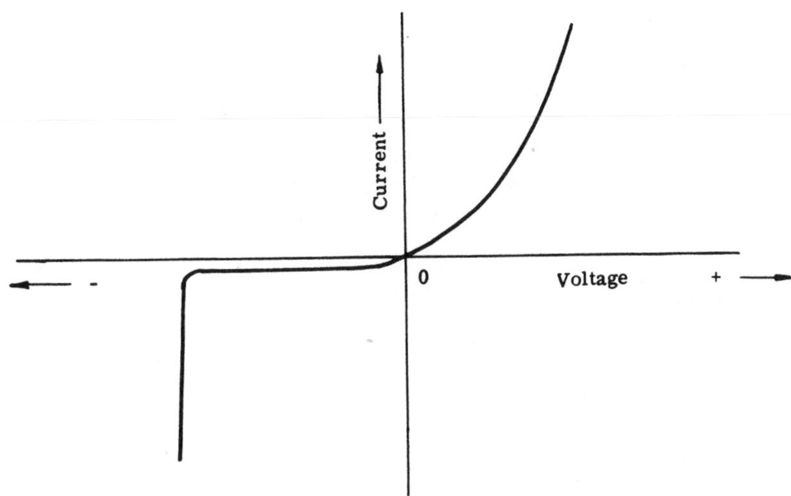

Fig. 5-13 Current - voltage characteristics of P-N junctions

5.10 JUNCTION TRANSISTOR OPERATION

A junction transistor is constructed with two junctions located in close proximity to each other as illustrated in Fig. 5-14. The emitter-base junction is biased in the forward direction, while the base-to-collector junction is biased in the reverse direction. Electrons in the emitter may easily flow into the base and then diffuse toward the collector junction. The flow over the potential barrier at the emitter can be varied by varying the emitter potential. The electrons that enter the base are minority carriers, and after they diffuse across the base, they are accelerated by the collector-to-base potential. If the base is thin, little recombination occurs in this region, and virtually all the carriers reach the collector. Since the carriers are injected through a low-impedance junction, and collected through a high-reverse impedance, high-voltage amplification will occur. The width of the base region is also an important factor in determining the high frequency characteristics of the device. Since the current flow across the base is by diffusion, the response of the device is restricted by the diffusion time. Transit time across the base is proportional to the square of the base width and inversely proportional to the diffusion constant. High-frequency devices have base widths in the range of a few microns or less.

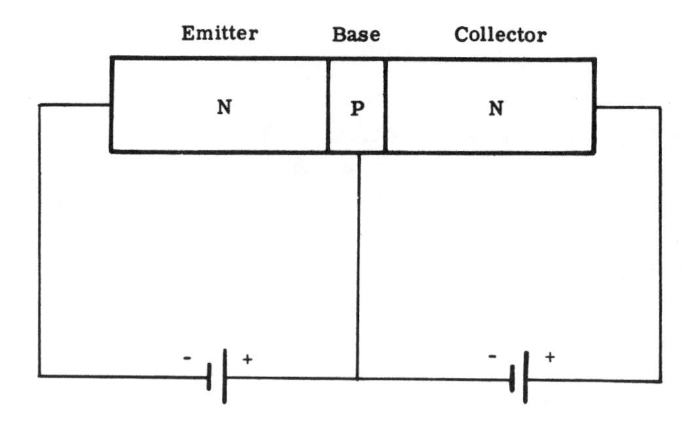

Fig. 5-14 NPN junction transistor structure

5.11 TUNNEL DIODE

A number of properties of semiconductors and semiconductor junctions have been used in the development of other types of devices. One of these is the tunnel or Esaki diode which is a negative resistance device capable of being used as an amplifying element. Its current-voltage characteristic is shown in Fig. 5-15. This device is an abrupt junction diode which is so heavily doped that the materials are degenerate. Thus, the Fermi level in these materials lies in the conduction band of the n-type region and in the valence band of the p-type material. As illustrated in Fig. 5-16, large numbers of electrons prevail in the conduction band of

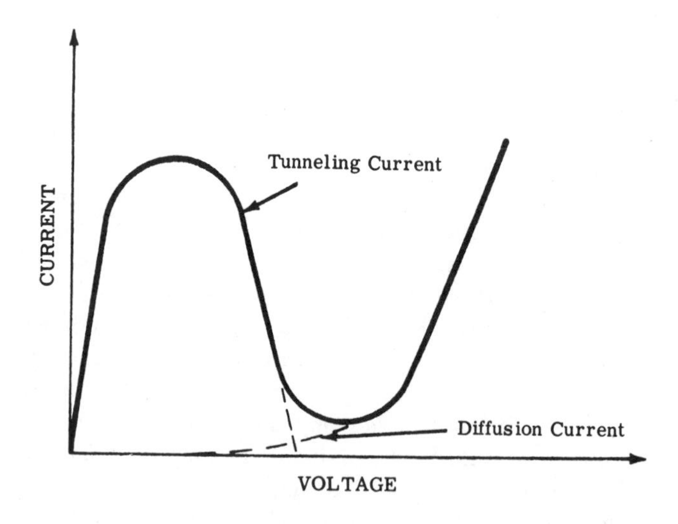

Fig. 5-15 Tunnel diode current - voltage characteristics

the n-type material, and many holes are available in the valence band of the p-type material. The width of the depletion layer is sufficiently small

(100Å) so that transitions across this region by tunneling are quite probable. This mechanism is thus available for conduction in addition to the usual diode diffusion current. The composite of these two currents are responsible for the resulting current-voltage characteristic of the device. In the region of interest for active elements, the conduction mechanism is tunneling, and therefore the device is not dependent on a diffusion time. The tunneling transitions are sufficiently fast to enable the device to be useful at microwave frequencies.

Josephson Tunnel-Junction

Josephson superconducting devices **of the junction-tunnel type are** based on the phenomena theoretically predicted by Brian Josephson in 1962. They possess two key properties: fast switching speeds and low power dissipation levels. The switching speeds of $\sim 10^{-11}$ second obtained for recent Josephson logic devices are faster than those of the most advanced semiconductor devices. In addition to having fast switching speeds, the devices must be small and closely packed, because the time required for electrical signals to propagate between devices can be the main factor limiting computer performance. (Electrical signals travel only $\sim$ 1.5 millimeters in the device switching time of $\sim 10^{-11}$ second; thus to achieve the highest performance most of the many thousands of logic and fast memory devices in a high-speed computer should be located within a few centimeters of each other.) The low power dissipation of Josephson devices ($< 10^{-6}$ watt per device) should allow them to be packed densely without incurring heat removal problems. For very fast semiconductor logic circuitry, sufficiently dense packaging is already difficult to achieve because the power dissipation of the transistors ($>$ 100 times higher than that of Josephson devices) exceeds the limits for which direct liquid cooling could be used to maintain safe operating temperatures. It is therefore necessary to use bulky heat sinks that result in lower packaging densities. For the potential of Josephson devices to be realized, novel materials and processes must be developed to enable them to be fabricated with the electrical characteristics desired for logic and memory devices and with the ability to withstand cooling to $\sim$ 4 K above absolute zero (by immersion in liquid helium), the temperature region at which these devices operate.

The type of Josephson device most suitable for high-speed computer applications is a SQUID (Superconducting QUantum Interference Device) composed of several superconducting tunneling junctions that share common electrodes. A schematic cross section through the center of such a device is shown in Fig. 5-15.1. The junction portions of the device consist of two electrodes separated by an ultrathin (5 nanometers) insulating layer. The electrodes are superconductors (metals that have an infinitely small electrical resistivity when cooled to below a characteristic temperature $T_c \approx 10$ K). The junction regions are defined by windows in a thicker insulating layer. The ultrathin insulating layer is typically an oxide grown on the base electrode. It is sufficiently thin that current flow can pass through it by electron tunneling; hence, it is referred to as a tunneling barrier. The current-voltage characteristic of such a junction has two branches, as shown in Fig. 5-15.2. A normal tunneling-current branch (b) for which a voltage occurs across the tunneling barrier, and a superconducting or Josephson tunneling-current branch (a) for which no voltage develops across the barrier. Because the Josephson current is very sensitive to magnetic fields, the junction can be switched from one branch to the

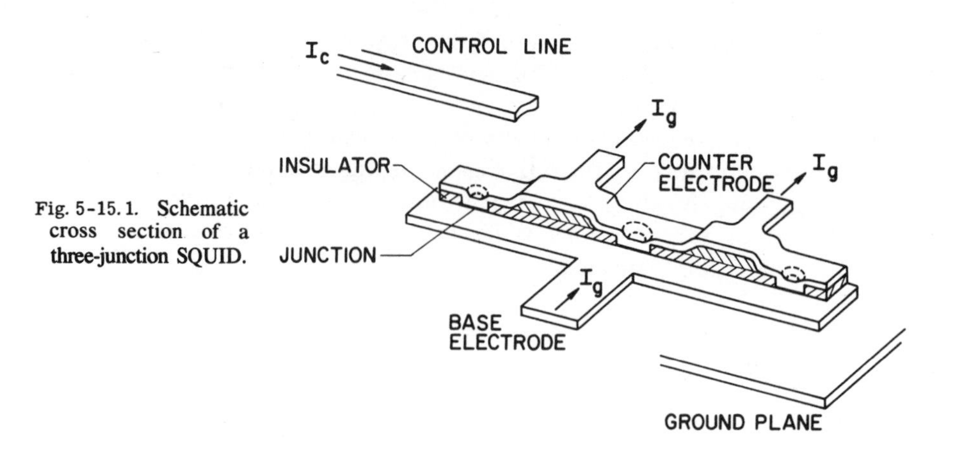

Fig. 5-15.1. Schematic cross section of a three-junction SQUID.

other by passing a small current through a thin film wire (the control line in Fig. 5-15.1) located in close proximity to the junction. The remaining element of the device is a superconducting ground plane which is used to confine electrical signals to the close proximity of the device, thereby permitting close packing of devices without incurring cross-coupling between them. The control line and ground plane are separated from the junction electrodes by insulating layers not shown in Fig. 5-15.1. In addition, a thin-film layer of a nonsuperconducting metal is used to provide load and damping resistors. Integrated logic and memory circuits can be obtained by using the electrode and control-line layers for device interconnections. Such circuits contain up to 14 thin-film layers that range in thickness from 0.02 to 2 micrometers, and are patterned in shapes with minimum dimensions of 2.5 μm by means of photolithography techniques similar to those developed for fabrication of semiconductor circuits. Most of the layers are prepared in vacuum by evaporating a source material and condensing the vapor on a silicon substrate containing any previously deposited patterned layers.

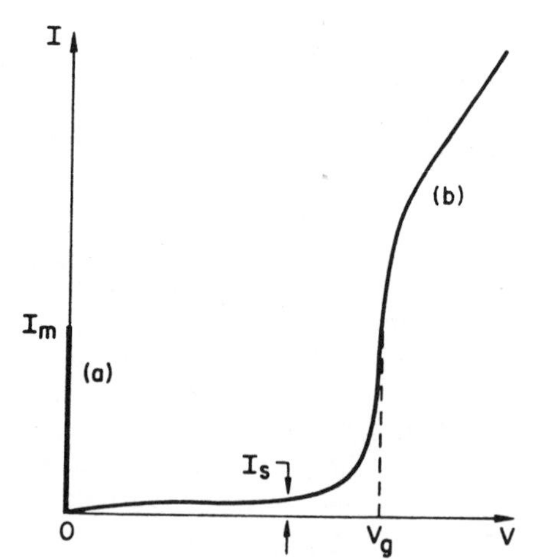

Fig. 5-15.2. Current-voltage characteristics of an interferometer similar to that shown in Fig. 5-15.1.

5.12 FIELD EFFECT TRANSISTOR

The field effect transistor is a device completely different in operation from the normal transistor. A typical field effect structure is shown in Fig. 5-17. The control electrode, or gate, modulates the flow of current from source to drain, by constricting the width of the conducting channel through the creation of depletion layers at the junctions which act as gates. In contrast with the usual transistor, this is a device with a high-input impedance and a low-output impedance.

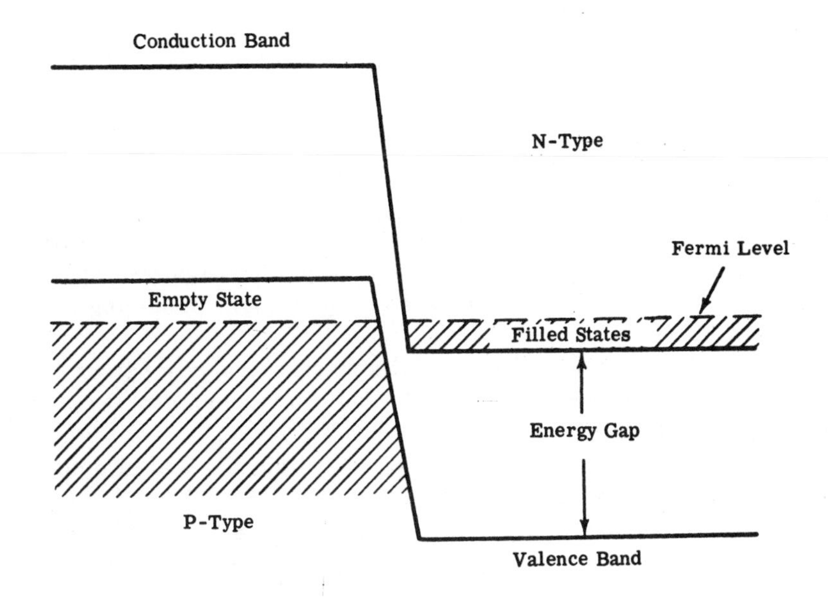

Fig. 5-16 Energy band diagram for tunnel diode at thermal equilibrium

Another use of the biased p-n junction is as a capacitor. Combining Eqs. (5-28) and (5-29), the capacitance per unit area of a p-n junction is found to be

$$C = [\epsilon qN/4V]^{1/2} \qquad (5\text{-}32)$$

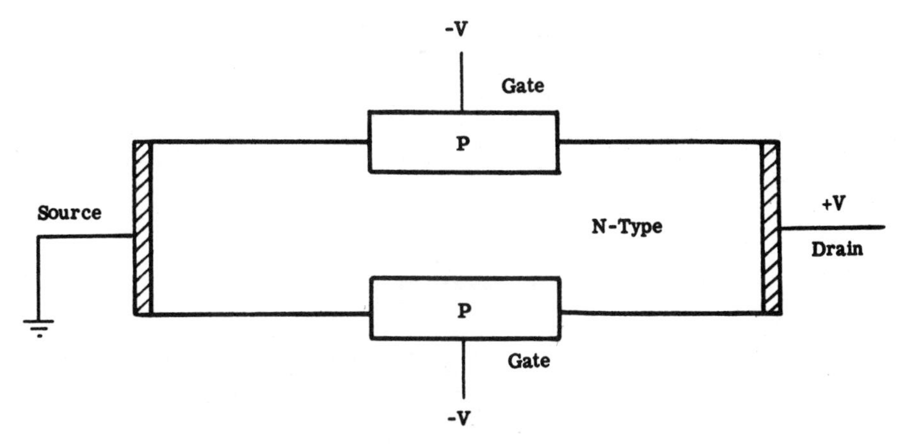

Fig. 5-17 Field effect transistor structure

where it is assumed that the p and n impurities are of equal density, N, and that $V_j << V$. This junction capacitance can be a useful capacitor in circuits, provided the junction is reverse biased. If the junction is forward biased, the capacitance will be shunted by the high conductance of the diode.

Finally, there is an entire class of photosensitive devices to be considered. All semiconductors are sensitive to illumination since photons will cause extra electrons to be excited into conduction states from either the valence band or other allowed levels. Included among these devices are photoconductors, photovoltaic cells, phototransistors, and solar batteries.

5.13 OPTOELECTRONICS

In the past, the optical effects of semiconductors have been undesirable in development. If a transistor case was somewhat transparent, spurious performance would result because the unwanted light impinging on the surface of the crystal, generated a charge.

However, these optical characteristics present in a very narrow band within the electromagnetic spectrum, have given rise to a new electronics industry: optoelectronics.

For some time, the development emphasis within the field of optoelectronics had been placed on sensors. The reason for this, is that as the silicon technology became highly developed, silicon devices have provided useful response in the infrared and visible range. With minor modifications in silicon processing, optical sensors such as photodiodes and phototransistors can be built. Furthermore, using IC technology, arrays of sensors can be built on a monolithic piece of silicon.

A delaying factor in the development of emitters, in the past, has been the fact that visible light emission does not take place in ordinary semiconductor materials, such as germanium and silicon. To get emission of the wavelength which is visible to the human eye, metallurgists and chemists had to turn their attention to compound materials, such as gallium phosphide or the more complex gallium arsenide phosphide, and various ceramics materials.

Light sensors are in use mainly in applications requiring no human intervention. For example, equipment with light sensors is used to inspect plywood in saw mills, and fabric rolls in knitting mills. Computer card readers are employing light sensing devices extensively, not only because of the efficiency and reliability of the optical devices, but also because they can scan information at a fast rate. Light sensors are also extensively used in cameras of all types.

5.13.1 TRANSMISSION AND ABSORPTION OF LIGHT

Transmission of radiant energy occurs through many media. Air will transmit energy with little attenuation at certain wavelengths. We are aware of the attenuation of visible light through fog and clouds. Not as well-known is the fact that the atmosphere transmits much more effectively at some wavelengths than at others. The atmosphere has windows for various spectral regions through which energy passes with little attenuation. Air, acting like an electronic bandpass filter, determines the effective operating wavelength for systems having long air paths. The short wavelength of light leads to the scattering and absorption of light in the atmosphere.

Absorption in materials is useful for the isolation and containment of radiant energy. Most systems are based, to one degree or another, on concepts of selectively absorbing energy in one place or transmitting it to another.

The concept of absorption is important in considering the way in which a semiconductor detects radiant energy. For detectors, light must enter the semiconductor material. The light energy is absorbed into the material, creating hole-electron pairs which are then collected or measured by their effect on the semiconductor device. Materials absorb light at different wavelengths in varying amounts.

Semiconductor materials can be classified by their absorption limits, and this aids in the selection of a material for designing a detector. For example, silicon readily absorbs radiant energy in wavelengths below 0.9 μm. Above 1 μm the silicon will absorb little light, becoming essentially transparent to the longer wavelengths. This absorption characteristic places a limit on the longest wavelength for a system using silicon devices.

The absorption effects in light emitters are vital in optimizing the output from a semiconductor light source. Semiconductor light emitters generate radiant energy within the material, and the material may be opaque or highly absorptive. The object, of course, is to get the generated light out of the device and to its intended target. Any absorption would only reduce the overall efficiency of the light emitter. Again careful selection of materials can minimize the effects of this absorption at various wavelengths.

The selection of optoelectronic materials involve a compromise of many considerations. Technological reasons may often predominate in the selection of a material. The great body of information and techniques available in silicon materials processing, causes Si detectors to remain a dominant factor in the optoelectronic market.

5.13.2 EFFECTS IN PHOTOSENSITIVE DEVICES

There are two primary effects in the operation of solid-state photosensitive devices: photoconduction and photo-voltaic effect. The phenomenon of photoconduction depends on the condition that whenever a conductor is exposed to radiation of the proper wavelength, the absorbed energy releases electrons at the surface, and electron-hole pairs are produced. In semiconductors, these electrons and holes remain separated long enough to provide current flow that increases the conductivity when voltage is applied.

The increase in conductivity depends upon the number of electrons available for conduction at any instant. As a result, the conductivity depends upon the intensity of the radiation as well as the area of the surface exposed to it.

The inherent spectral response of silicon is in the near infrared (Fig.

5-18), peaking slightly above 0.8 μm. As a result, these silicon radiation sensors are well-suited to operate with the generations of light-emitting devices, which are a product of the gallium arsenide, gallium phosphide and gallium arsenide phosphide technology. The output of tungsten lamps also falls in the region of maximum silicon-device sensitivity.

The second principal class of low-cost photoconductive devices are comprised of bulk compounds, such as cadmium sulfide, cadmium selenide and lead sulfide. These items are fabricated by pressing a powder onto a ceramic substrate and sintering it. Contacts are provided at each end of the sensitive area. These photoconductive cells are essentially a uniform

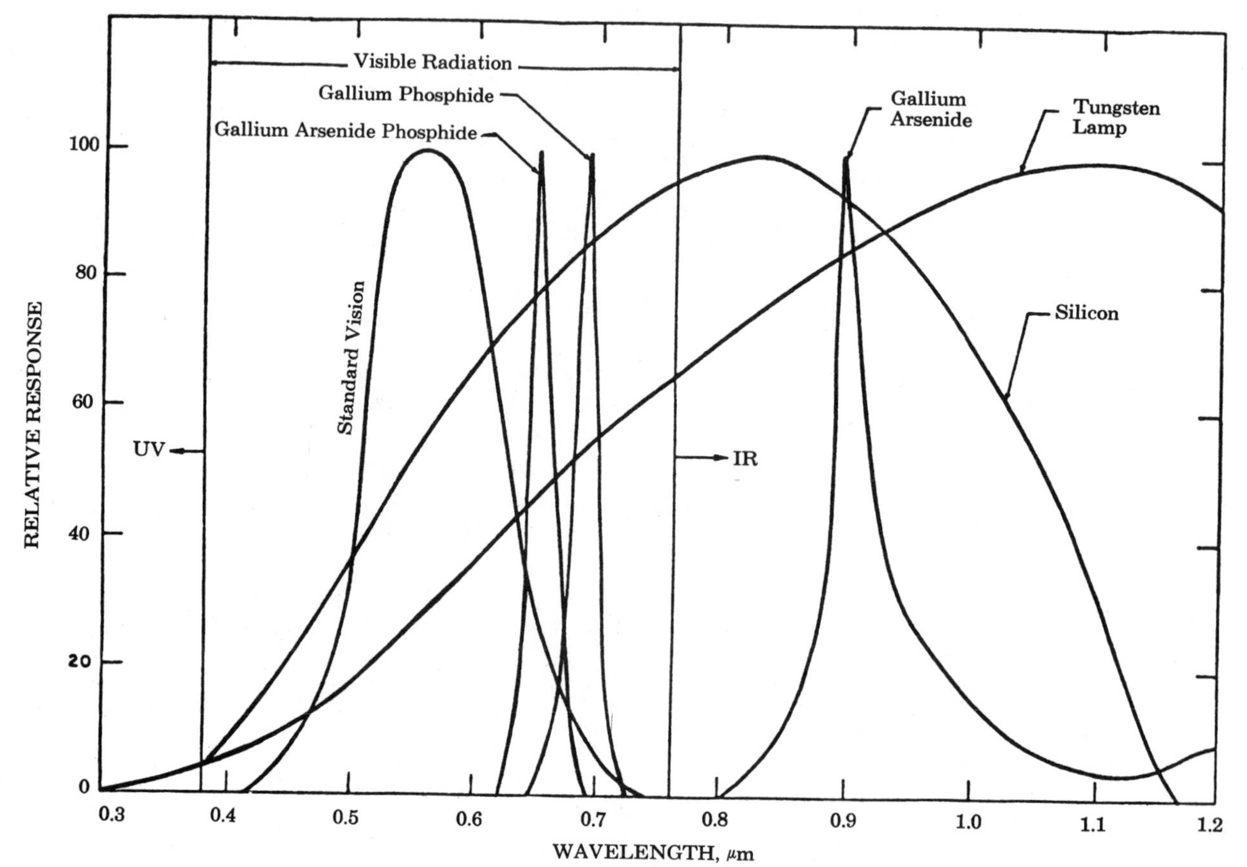

Fig. 5-18 The spectral response curves of the average human eye, silicon photosensitive devices, and solid-state and tungsten filaments normalized to a 100% value.

semiconducting mass with megohms of resistance in the dark – resistance that drops to a few thousand ohms or less in light.

The photoconductive cells have a relatively low frequency response – on the order of a few hundred or thousand cycles, while the response range of the silicon junction devices is in the megacycle region. Operating currents of the junction devices are on the order of microamperes or a few milliamperes, whereas those of the photoconductive cells can range up to 0.5 A. Unlike silicon, the response of the cadmium sulfide and selenide in the photoconductive cells lies in the visible region, so the cells are suitable for applications that use visible light, as in camera-exposure controls, for example.

Another basic effect that provides useful radiation-sensing devices is that of photogeneration. In this case, a photojunction device is operated in the mode in which radiation is applied to the junction, and a voltage is produced across it. These photovoltaic cells are usually produced as large-area silicon or selenium devices. For certain applications, as in tape readers and shaft encoders, they are made in a varitey of small silicon-cell arrays.

A photovoltaic cell differs from a photoconductor in that it generates a voltage when light is absorbed, rather than changing its conductivity. In the photovoltaic cell a pn junction is formed in the semiconductor. This creates a barrier for the separation of the hole-electron pairs. The field at the junction drives the electrons to the n-type side and the holes to the p-type side, so that a potential arises across the junction.

Photocells can be made from many different materials, such as germanium, cadmium telluride, indium phosphide, gallium arsenide and silicon, for example. The most common material is silicon, because of the extensively developed silicon processing techniques available. Generally, in the silicon photovoltaic cell, a p-type boron diffusion is made into an n-type silicon to form the pn junction. Solid metal ohmic contacts are fabricated to the n-type side and a grid pattern contact (permitting light absorption) is made to the p-type side. Usually an anti-reflective coating is deposited on the p-type surface to reduce the reflection losses from the surface. The manufacturing processes for these silicon pn photovoltaic devices are well developed. Photovoltaic components can convert solar energy to electricity, and they find use in card-type readers, cameras, and light sensitive switches.

Any semiconductor may be used as a photoconductor because its electrical conductivity is changed by the absorption of light. When light is absorbed by the photoconductive substance, hole-electron pairs are generated in proportion to the intensity of radiation. For this effect to occur, it is essential that the light be absorbed. Light which is reflected from the substance does not generate hole-electron pairs. Thus, the light to be absorbed must be of an energy equal to or greater than the band-gap of the material. However, once the holes/electrons are generated, they must have sufficient mobility and lifetime to cause increased conductivity. A field and electrodes must be provided to move the charge carriers within the material and create a current flow.

Widely used photoconductive materials are cadmium sulfide (CdS) and cadmium selenide (CdSe). There are devices made of thallous sulfide, lead telluride, and lead sulphide, as well as the special doping of CdS with copper, chlorine, and iodine. Some cells contain a rectifying pn junction or metal barrier contacts and can be considered simple photoresistors. The manufacturing processes and techniques used for photoconductors vary from sintering or firing operations to vapor, chemical, or vacuum deposition techniques. There are many more processing variations for photoconductors than for other semiconductor components.

Photoconductive components are used when applications require large sensitive areas, zero offset voltage, and large light-to-dark ratios, as well as relatively low cost. The speed of response for photoconductors is relatively slow. A memory or hysteresis effect may cause problems for some applications. The spectral response peaks sharply at 5500 Å for CdS and at about 7000 Å for CdSe.

Silicon and selenium are widely used low-cost photovoltaic cells for industrial and control applications. The spectral response of selenium peaks in the visible region, so that these cells, like the cadmium sulfide and selenide units, are used in photometric devices and camera-shutter controls, as well as other applications depending on visible light.

5.13.3 PERFORMANCE MEASUREMENTS

Measurement of detector performance requires an understanding of the light source. Tungsten bulbs give rise to a variety of problems from the emission characteristics of the radiating element and encapsulation system, to the mechanical stability of the filament. Semiconductor emitters offer narrow spectral output without filtering, but introduce new problems of size and optical coupling, with air and other transmission mediums such as glass fiber bundles.

Measurement of dc parameters common to semiconductor devices is not a serious problem. But with optoelectronic performance, the standard for light intensity presents a problem.

The concept of irradiance, or watts of radiant energy incident upon a surface, is important in the measurement of output from a wide variety of sources. The spectrum of a source is also important in determining how the radiant energy is related to various wavelengths.

The nature of a tungsten bulb causes it to shift its output power and spectral distribution as the temperature of the emitting wire is changed by current flow. Long-term calibration of light sources to $\pm 5\%$ can be achieved, but with difficulty. This is a major difference between optoelectronic device specifications and those of standard transistors and IC's, where limitations are imposed essentially by the accuracy of ac-dc instrumentation.

5.13.4 PHOTOJUNCTION RADIATION SENSORS

Photojunction devices in use include the photodiode, phototransistor and photoswitch (light-activated SCR).

The photodiode is usually a small-area device constructed with a glass or plastic lens that focuses radiation on the pn junction. It is operated in the back-biased mode, and consequently the operating currents are small – usually from tens of microamperes to a few hundred.

To obtain maximum sensitivity, some means of concentrating the available illumination on the device is frequently required, such as fiber-optic light pipes or an auxiliary optical system.

Photodiodes, like their standard junction counterparts, are subject to temperature and other environmental effects. The normal diode back-bias leakage current is called the dark current, since it is measured with no radiation on the junction.

Silicon diodes have dark currents that are three or four orders of magnitude less than germanium diodes. However, the dark current of silicon diodes increases exponentially with temperature, doubling for each 10° C. The forward current caused by radiation follows the same rule.

The reverse bias in a photodiode increases the field across the junction, so that the device operates as a current generator. Light absorbed by the cell creates hole-electron pairs which are swept across the junction barrier, giving rise to an excess of carriers. The bias extends the effective junction or depletion region, making the device efficient in converting photons. A photodiode has essentially a linear change of current with light intensity.

Photodiodes are made in a variety of configurations, taking advantage of the properties of the pn junction. As the resistivity of the base material is changed along with the bias voltages, it is possible to affect the junction characteristics and thereby, the light response of the device. As the depletion region of the junction is extended deeper into the base material, the holes and electrons have a shorter distance to travel to reach the separation barrier. There is, therefore, less chance of recombination. Also large depletion regions lead to lower junction capacitances, thereby increasing the speed of operation due to the shorter transit time or higher drift velocity.

Photodiodes are also made by taking advantage of the avalanche multiplication factor. As the bias voltage approaches the breakdown voltage, the holes and electrons created by the absorbed photons acquire enough energy to create other electron-hole pairs. As they collide with substrate atoms, these second order electron-hole pairs can then generate further pairs. This characteristic of avalanche photodiodes achieves high speed operation and multiplication of the light current.

Photodiodes can be made of any material in which a pn junction or barrier can be formed. The materials usually used, however, are silicon and germanium. Normally the junction is formed in a substrate by planar diffusion, and ohmic contacts are made to the n and p regions.

By selecting the initial material characteristics (dopant concentration, lifetime, and mobility), and the diffused impurity type and concentration, it is possible to optimize the photodiode for various characteristics such as speed of response, dark current, efficiency at specific wavelengths, and ac parameters .Photodiodes are also made by the metal barrier technique (Schottky barrier) where the metal-semiconductor interface forms an effective pn junction at the surface. Although many special types of photodiodes are made for specific applications, they are all basically diodes and follow diode theory. They are optimized for their photo response.

Phototransistors are more sensitive than photodiodes. If light is applied to the reverse-biased base-collector junction of a transistor, the absorbed energy creates electron-hole pairs that produce current flow. The current is proportional to the radiation intensity, and it is multiplied by the beta gain of the transistor. Thus the phototransistor is many times more sensitive than the simple diodes. However, the collector-to-base leakage current — the dark current — is amplified by the same beta factor and must be considered in phototransistor applications. Due to the low dark currents encountered with silicon, the base leads on many phototransistors, are not brought out — only the collector and emitter leads are. For these types, only light or other radiation can be used as an input.

Other silicon phototransistors have a base lead. But use of the base connection to adjust the device for optimum gain usually decreases the device's sensitivity. This occurs because the base-to-emitter resistor shunts some current around the base-emitter junction that is not amplified by the transistor beta factor. The open-base transistor consequently has greater optical gain.

The response of phototransistors to radiation inputs produces operating characteristics that are representative of transistor operation with an electrical input. Where phototransistors are designed for digital or switching applications, the response is given in terms of turn-on and turn-off times, typically in terms of a few microseconds.

In the phototransistor, the base current generator is simply a photodiode as described above. The transistor collector-base junction and the photodiode junction are identical, and both operate at the same reverse bias. The collector-base area is generally expanded to collect as much light as possible, while the emitter is located as far away as possible.

The bipolar phototransistor behaves much the same as a normal bipolar device; its h_{FE} decreases at low collector current. The increase in dark current versus temperature is much the same as in most transistors. The output current can be increased by increasing the active collector-base area, or increasing the transistor dc h_{FE} value (static forward-current transfer ratio). Enhancing the quantum efficiency of the collector-base diodes during material processing and assembly, also increases the output current.

The long lifetime values in silicon phototransistors lead to the absorption of radiant energy in the entire bulk of the silicon phototransistor. This causes the device to have a relatively long wavelength response, peaking near 0.9 μm – a close spectral match for tungsten light sources or infrared light emitters. A short lifetime material would reduce the amount of long wavelength energy absorbed, causing the device to have lower absolute response with a spectral peak at shorter wavelengths. Filters and optical coatings can also cause the peak relative response to occur at shorter wavelengths. Such spectral response shaping is widely used on phototransistors.

By controlling light input or irradiance, the transistor's light current can be controlled. Varying radiant energy will force the light sensor to provide a current compatible with following circuitry. Selection of load impedance and light intensity determines the operating point of a phototransistor. Operation of a phototransistor between the logic 0 and logic 1 state is a common technique for tape and card readers.

With simple optical lenses, phototransistor gain can easily be increased (Fig. 5-19). The lens, larger than the phototransistor, collects light and semifocuses it on the active area. (The light does not have to be in focus on the active area). Also, device operation is better if the focus is below the plane of the wafer. This way there is no image of the light source on the top surface, where contacts or minor surface blemishes could drastically change the apparent light sensitivity.

Silicon is the common material for phototransistors. The selection of silicon is due to the convenient spectral response as well as the large amount of knowledge gained through experience with transistors and ICs.

To improve the photosensitivity of phototransistors, device designers have developed Darlington phototransistors and other combination circuits. These devices have a chip that contains a silicon planar phototransistor plus a direct-coupled amplifier stage. The increase in sensitivities obtained

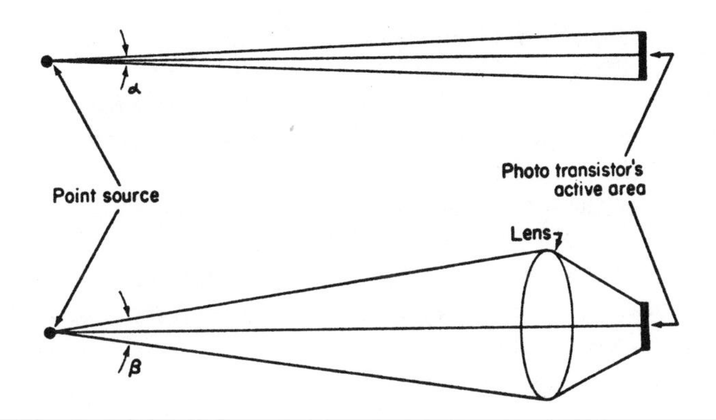

Fig. 5-19 Optical "gain" is possible in a phototransistor by using a lens. In case 1, the phototransistor intercepts all light in angle α, while in case 2, because of the lens, the same device has a higher collector current, angle β.

with these Darlingtons ranges from three to as high as 10 times that of conventional phototransistors. For many applications, the base lead of the Darlington is not used.

Phototransistors are packaged in a variety of configurations depending upon where they are to be used, and how much they're to be sold for. The clear plastic package is often the lowest cost device. Some transistors are packaged with a spherical lens in the top, while others have a flat lens. In general, the flat types have a wider angle of light-beam coverage, while the spherical-lens types are used to minimize light-beam crosstalk in arrays.

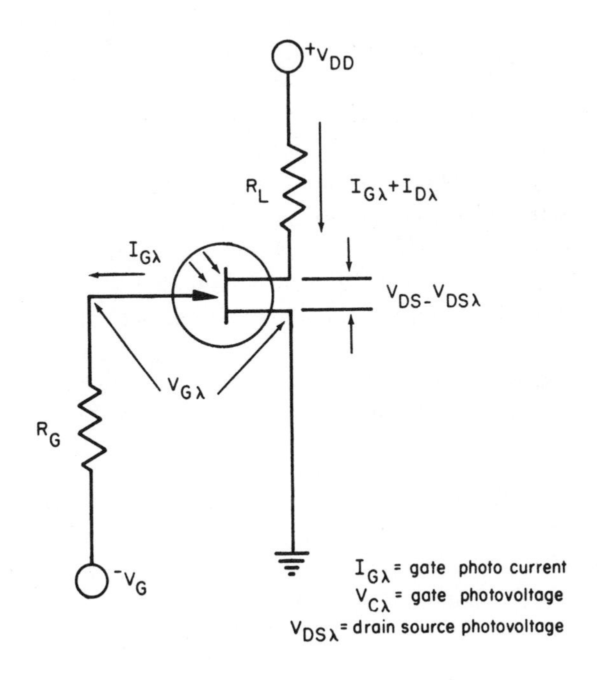

Fig. 5-20 Bias circuit of a photo-FET.

5.13.4.1 Field-Effect Transistors

Field-effect-transistors (FETS) are notably light sensitive. The gate junction acts as a photodiode, and the drain-to-gate junction is normally reversed biased through a gate resistor. (Fig. 5-20).

By a combination of bias and adjustment of the gate resistor, selection of a wide range of FET operation can be obtained.

The mechanical geometry of an FET is not optimum for light sensitivity. Most of the gate junction region is covered by metal source and drain contacts, so that care must be taken not to focus a small light source on the surface where it could be blocked by a contact.

5.13.4.2 PNPN Devices

Photo pnpn devices are similar to standard pnpn design with one collector junction expanded so that light triggering may be used. The isolation of the gate function by light is significant in that long strings of these devices may be placed in a series for very high voltage switching applications.

Device sensitivity and trigger characteristics tend to be in the general operating range of phototransistors. Their temperature sensitivity often makes precise trigger timing a problem.

5.13.4.3 SCR Photodevices

The highest current-carrying capacity of any of the junction photodevices is found in the light-activated SCR. Both an electrical signal and a light signal can trigger the device. As the usual SCR, once the device is triggered, it conducts until the voltage across it is removed or reversed.

When an impulse of light triggers the device ON, its internal resistance drops to the fraction of an ohm during turn-on time, and then it rises to a few ohms. The intensity of light or IR radiation needed to trigger the SCR ON decreases with an increase in temperature or device voltage. The addition of a current bias will also reduce the triggering level.

The required level of light necessary to fire the light-activated SCR is greatest when the device is first turned ON and decreases as the device warms up. This is because the lead current raises the junction temperature.

The principal advantages of the SCR devices are high current capabilities plus fast turn-on times. The device's turn-off time is fixed by its inherent recovery time.

Light-activated SCRs can be used for example, to replace relays, to drive higher capacity SCRs, and to provide switching logic functions.

5.13.5 PHOTOJUNCTION DEVICES AS VOLTAGE GENERATORS

Photojunction devices may be operated in either the back-biased photoconductive mode or in the photovoltaic mode, in which case no bias voltage is applied. However, if the junction is illuminated, a voltage appears across it.

For both selenium and silicon, open-circuit voltage of the photovoltaic cell is approximately the same; it is a logarithmic function of the illumination level. The relationship of voltage to light intensity is linear for low light levels. Except for very small cells, the open-circuit voltage is independent of cell area. However, it increases with illumination.

Photovoltaic cells (Fig. 5-21) produce an output current that is independent of the load under two conditions: low illumination and low load resistance.

In this respect, they act like constant-current generators. As both illumination and load resistance are increased, the variation of cell current with generated cell voltage becomes more linear.

As with any junction device, photovoltaic cells are affected by changes in cell temperature. The frequency response of photovoltaic cells is limited by the cell size, because of capacitance effects. However, as the diameter of the cell is reduced, the response time is faster, approaching the theoretical limit of 2 μs for silicon and 20 μs for selenium. The response can be increased, but it requires proper load matching and the neutralization of cell capacitance by peaking inductance.

While photovoltaic cells can be generally used in the same kinds of circuits as other photodiodes, it should be taken into account that the internal resistance of the cell is high with low illumination; it declines as the light level is increased.

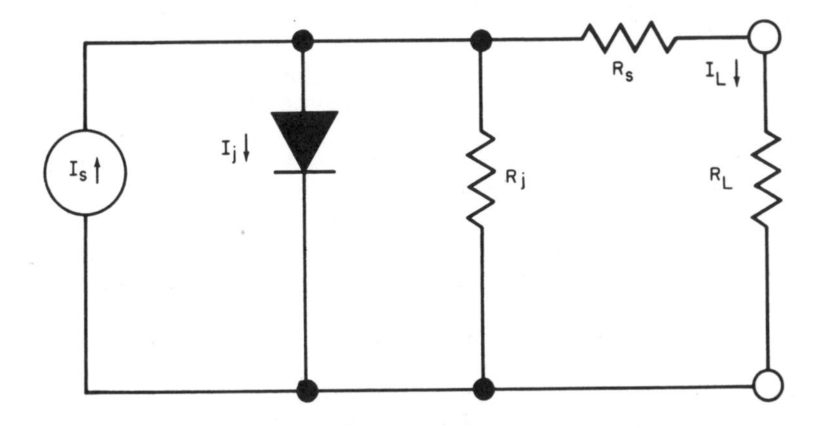

Fig. 5-21 **In this circuit equivalent of a photovoltaic cell, the pn junction, I_j, functions as a current source across the junction resistance, R_j, and also blocks current flow in the opposite direction.**

The key component of a photovoltaic cell is a semiconductor that absorbs light energy by exciting an electron from the valence band to the conduction band and leaving a positive hole behind. The electron and hole so generated eventually recombine, giving up the acquired energy to the lattice as heat or emitting light. In a photovoltaic cell, however, a region of high electric field is provided within the semiconductor, so that most of the photogenerated electrons and holes are separated on reaching this region and thus prevented from recombining. The flow of these charges through an external load produces useful work and in this way completes the process of direct conversion of light to electricity.

Intimate contact of two materials, at least one of which is a semiconductor, produces the high-field region if the chemical potentials of electrons in the

Fig.5-21.1
Schematic exploded view of a generalized solid-state photovoltaic cell.

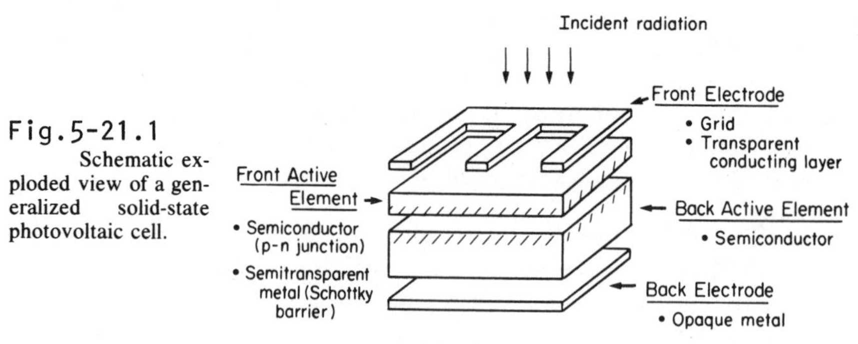

two materials are different. Such a structure—a semiconductor junction—can be obtained in various ways in solid-state devices. A *p-n* junction occurs between oppositely doped semiconductors, so that there are excess holes (*p* type) on one side of the junction and excess electrons (*n* type) on the other. The major components of the semiconductors on either side of the junction can be the same (homojunction) or different (heterojunction). Certain metal-semiconductor contacts (Schottky barriers) likewise produce high-field regions in the semiconductor side of the junction. Variations of these structures are achieved when a thin insulating layer is interposed between the two active regions, leading to semiconductor-insulator-semiconductor or metal-insulator-semiconductor configurations. All of these junctions have been used in promising photovoltaic devices, but the materials requirements of each depend on the predominant optical and electronic processes characteristic of that configuration.

The current output of solar cells is limited by the number of carriers (electrons and holes) generated by the incident light. Losses inevitably occur when light is prevented from reaching the active semiconductor, and these must be minimized by optimization of front electrode design and reduction of reflection losses. Absorption of light by the active semiconductor requires that the light energy be at least equal to the magnitude of the energy gap separating the valence band from the conduction band of the semiconductor (the band gap). The low-energy fraction of the available solar spectrum is therefore lost without being absorbed, since it falls below the band-gap energy of most semiconductors. A fraction of the light having the proper energy can also be lost if the semiconductor is not thick enough to allow complete light absorption—that is, thinner than the optical absorption length of the semiconductor.

The losses most sensitive to materials quality and device structure are those incurred when photogenerated electrons and holes recombine before they are separated in the high-field region. Fig. 5-21. 2 shows energy band diagrams of a *p-n* homojunction solar cell under three conditions. These diagrams present the electron energy as a function of depth into the cell. Solid lines indicate the energy at the edges of the conduction and valence bands, and the dashed line indicates the Fermi level—that is, the chemical potential of the electrons in the semiconductor. Fig. 5-21. 2A illustrates the condition when the terminals of an illuminated cell are directly connected to each other (short circuit) and shows the most important recombination paths. Bulk recombination of carriers (process 1) takes place away from surfaces and interfaces and generally occurs at impurity sites and structural defects in the semiconductor. Minimizing bulk recombination is important for efficient cell operation, since many carriers are generated outside the narrow high-field region and must diffuse to it in order to be separated. The diffusion length is the distance photogenerated carriers diffuse without recombina-

A: Short-circuit Condition

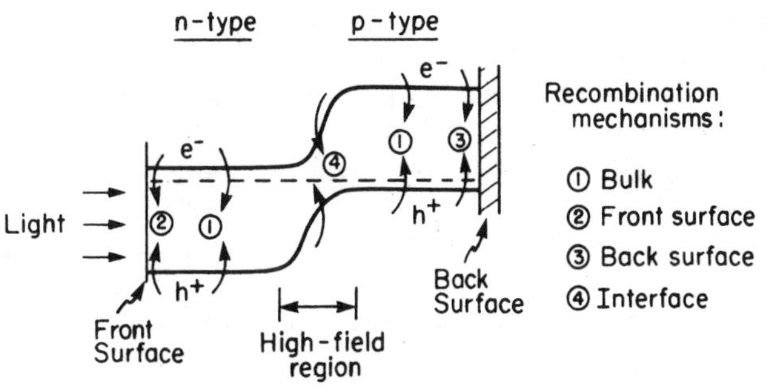

B: Operating Condition

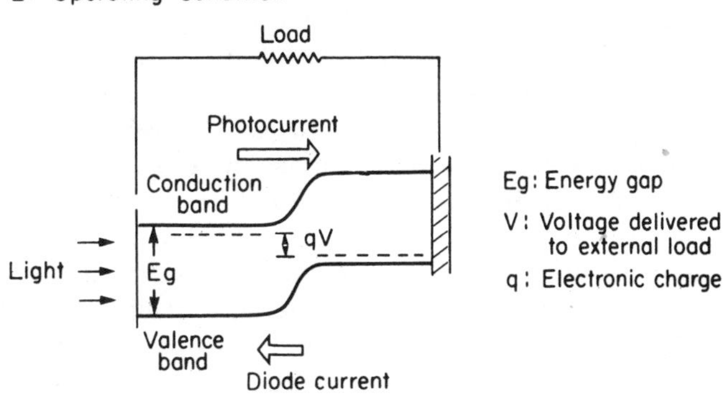

C: Maximum Open-circuit Voltage Condition

Fig.5-21.2 Electron energy diagram of a *p-n* homojunction photovoltaic cell: (A) short-circuit condition, (B) operating condition, and (C) maximum open-circuit voltage condition. Electrons and holes are denoted e^- and h^+, respectively.

tion and ideally is long enough to allow all such carriers to reach the high-field region. Weakly absorbing materials therefore require longer diffusion lengths for efficient carrier collection.

Unique electronic states arise in the vicinity of compositional and structural discontinuities in solids. These states often act as effective recombination centers; consequently, surfaces and interfaces are critical regions in photovoltaic devices. Recombination at the front (illuminated) surface (process 2) is a loss mechanism particularly important in *p-n* homojunction cells, where many carriers are generated in the front active element (Fig. 5-21.1). In these cells the effect of front-surface recombination can be reduced by decreasing the thickness of the front active element or by "passivating" the surface by chemical treatments or addition of a compatible, transparent semiconductor layer in front of the cell (heterostructure). In heterojunction cells, the front active element can be made of a semiconductor with a wide band gap, which will absorb only the small fraction of the incident radiation of energy larger than the band gap. In this way most carriers are generated in the back active element and thus are not susceptible to front-surface recombination.

Recombination at the back surface (process 3) becomes important when the thickness of the back active element is comparable to both the optical absorption length of the semiconductor and the carrier diffusion length. In this case a significant number of carriers are generated near the back surface, where the recombination rate is high. The effect of back-surface recombination can be reduced by using a back-surface-field structure, in which the region of the back active element adjacent to the electrode is heavily doped. This produces a field that effectively confines the photocarriers to the bulk of the back active element and keeps them away from the electrode.

Defects arising at the interface between two dissimilar materials provide another important recombination path.

Interface recombination (process 4) is an important loss mechanism in heterojunctions and Schottky barriers, and is reduced by matching certain key properties of the materials involved, such as lattice parameter, thermal expansion, energy of the conduction and valence band edges, and so on. A critical interface is also found at the grain boundaries of polycrystalline materials. Recombination at these centers is a crucial loss mechanism in many low-cost, thin-film devices.

5.13.6 VARIATIONS IN LIGHT SENSORS

Cadmium sulfide and cadmium selenide cells are usually low-cost photoconductive cells that change resistance from megohms to a few thousand ohms with the application of light. By design, this light-to-dark resistance ratio can be made to vary from 100 to 1 to over 10,000 to 1.

The change in cell resistance from dark to light conditions depends on the material used to fabricate the cell, the cell's doping, its geometry, the applied voltage, the illumination levels used, the cells's ''illumination history'' and its temperature. Spectral response is in the visible region.

In practice, variations in the cell illumination level, the cell voltage and the illumination history have marked effects on cell operation. Cell sensitivity for any voltage is greatest at the lower light levels, decreasing somewhat with an increase in light intensity. If the illumination level is held constant, the cell presents a constant resistance, and in this case, an increase in cell voltage increases cell current linearly.

The response of a photoconductive cell to a square pulse of light is similar to the response of an RC network to a square wave of voltage — both have delays in the rise time and also in the fall (decay) time. These delays increase in a photoconductive cell as the cell illumination level is lowered. It is characteristic of photoconductive cells that turn-on delay is substantially longer than the turn-off.

The rise time is also raised by an increase in the cell voltage. However, the decay time decreases, particularly for reduced levels of illumination. For best response, cell illumination should be high and cell voltage low.

The illumination history of a cell affects its response to light. If the cell is exposed to light over a long period of time, its sensitivity to change decreases and its resistance becomes higher. Even for shorter periods, a hysteresis, or memory effect that depends upon the cell's previous exposure to light is evident. This effect is larger for cadmium sulfide. The selenides tend to reach equilibrium more rapidly.

For continuous measurement of random light levels, the hysteresis effect limits the precision of the measurement. But for intermittent measurements, the effect of previous light history may be minimized by keeping the cell in a constant light environment between measurements.

As in semiconductors, photoconductive cells are affected by temperature variations. Cell dark current increases with temperature, and cell sensitivity decreases. Conversely, with a decrease in temperature, both cell dark current and sensitivity improve. The temperature-sensitivity variation is most marked at low light levels, but it is substantially decreased with high illumination.

Photoconductive cells generally require amplifiers for useful applications, but for the intermediate sizes, sensitive relays can be operated directly.

5.13.7 LIGHT SOURCES

When minority carriers recombine with majority carriers in a semiconductor, energy is released in the form of light, heat, or kinetic energy to other carriers. Although minority carriers can be introduced in several ways, the most effective way is carrier injection in a pn junction, the basis for semiconductor light sources.

Any pn junction, under forward bias, can emit some light, but before reasonable conversion efficiencies can be obtained, many factors must be optimized. For instance, light output can be severely reduced by competing nonradiative recombination processes, as well as by internal absorption and reflection losses. These losses can be minimized by choosing the right material in a state of high degree of purity and crystal perfection, which is properly doped, and by using geometrical structures and other methods to reduce absorption and reflection losses.

In an indirect gap semiconductor there is a difference in momentum between the initial and final states for a band-to-band transition, which must be taken up by a phonon cooperating in the electron-hole recombination. In effect, a three-body collision is required, hence the low probability of light emission from indirect gap materials. On the other hand, vertical band-to-band radiative transitions in direct gap material, such as GaAs, have a high probability

There are other possible radiative transitions in which the requirement to conserve momentum is relaxed. For example, the decay of bound excitons at the neutral center ZnO or N indirect gap semiconductor GaP leads to efficient red and green emission.

In selecting the semiconductor, the wavelength of light desired must be considered. To a good approximation, the wavelength of light emitted for band-to-band or shallow acceptor (or donor) level will be that corresponding to the band gap, or

$$E_g = h_v = \frac{hc}{\lambda}$$

By growing single crystal mixed compounds such as Ga(AsP) and GaAlAs, it is possible to obtain a material for a desired wavelength by varying the composition. In this way, for example, energy gaps from the $\sim$ 1.4 eV of GaAs to 2.26 eV corresponding to pure GaP can be obtained. Other possible light-emitting materials are the II-VIs (CdS, ZnTe, etc.). Silicon carbide, in which yellow, green, and blue emission has been observed, is also a possible material.

5.13.7.1 Types of Sources

In a simple planar junction in a flat device, much of the light generated is lost by internal reflection. The critical angle for total internal reflection at an interface between two media with index of refraction n_1 and n_2 is

$$\sin \theta_c = \frac{n_1}{n_2}$$

For the air-GaAs interface: $n_1 = 1$, $n_2 = 3.6$, and $\Theta_C = 16°$.

The fraction of total light reaching the front surface that lies within the critical angle is

$$F_C = (1 - \cos \theta_C).$$

For GaAs-air, this is 1—0.96, or 0.04 of the light generated. Unless an anti-reflective coating is used, an additional amount, given by

$$\left[1 - \frac{4n}{(1 + n^2)}\right]$$

or 30% is also reflected within the critical angle.

A typical flat device is shown in Figure 5-22. In the compound GaAs, light generation takes place almost entirely in the p-region and is of a slightly longer wavelength than the absorption edge of n-type material. Light can therefore be brought out through the n-region without excessive absorption, if it is not too thick.

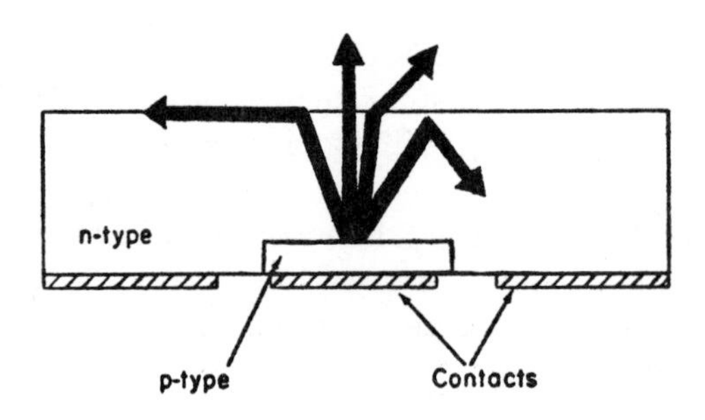

Fig. 5-22 Drawing of a simple flat, planar emitter shows internal reflection for angles of incidence greater than the critical angle, θ_c.

To eliminate the internal reflection, and thereby increase the light output by a factor of 25, a hemispherical dome can be placed over the junction. If the ratio of dome-to-junction diameter is at least as great as that of the index of refraction of GaAs to air, none of the light reaching the surface will exceed the critical angle (Fig. 5-23). In practice, the increased absorption loss in the thicker dome reduces the output so that a net gain of about a factor of 10 is achieved. Thus, a flat GaAs emitter will have about a 0.2% to 1.5% efficiency at 25°C and the addition of a GaAs dome raises the efficiency to about 2% to 8%. To achieve 25% efficiency in GaAs, the wavelength of emitted light output is shifted several hundred Å longer than the absorption edge, to reduce absorption in the dome. This is done by using Si as the dopant for both the n- and p-regions.

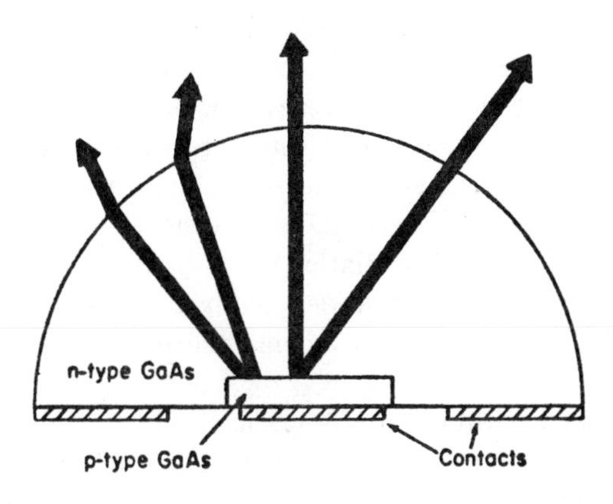

Fig. 5-23 **A high power dome emitter, showing how internal reflection is eliminated, and a net gain factor of about 10 is achieved.**

When Group II elements are added to GaAs, they replace Ga atoms to produce p-type material. Similarly, Group VI atoms replace As to form n-type material. The Group IV elements Si and Ge are amphoteric, going on both Ga and As sites, and pn junctions can be formed by changing the growth temperature to shift this ratio during epitaxial growth from Ga solutions.

The light emitting diode (LED) is a pn-junction diode that emits visible and invisible radiation when biased in the forward direction. The precise wavelength is determined by the doping in the three principal materials that are used to make the LED devices: gallium arsenide phosphide, gallium phosphide, and gallium arsenide. Gallium arsenide is also used to make laser diodes.

Both gallium arsenide and gallium arsenide phosphide diodes emit light from the junction only. Consequently, the light appears as a thin, flat beam, and the packages in which it appears are designed to produce a radiation pattern suitable for visual observation. For visible indicator diodes, the light is generally spread out, while for use in card readers, the package is small and the beam is generally concentrated in a spot.

Red transparent dye is frequently used in plastic or glass lenses to spread the visible radiation and produce a bigger light.

The invisible IR emitting diodes are used for industrial appilcations, such as intrusion detectors and production-line counters. Because the IR radiation of these LEDs matches the response curve of silicon transistors and diodes, tiny LEDs are assembled in linear arrays for punched card-reader applications. It should be noted that there is a spread in the output levels of diodes manufactured at the same time, as well as spread in the sensitivities of phototransistors or diodes assembled in arrays.

5.13.7.2 Coatings and Lenses

It is common practice to coat the surface of an LED with SiO_2 (silicon dioxide) ($n \approx 1.5$) or a higher index material, SiO or Si_3N_4 (silicon nitride). With the thickness of this coating equal to an odd number of quarter wavelengths, reflectance losses can be minimized. But, the critical angle remains the same as in the semiconductor-to-air interface due to the parallelism of the two interfaces of this thin film structure. Large increases of the critical angle within the LED chip (and therefore the total solid angle of externally obtainable radiation), are made by placing transparent masses of intermediate refractive index materials, such as epoxy or plastic, over the chip. This mass is usually a hemispherical dome surrounding the chip.

At the chip-plastic interface, the critical angle within the chip is increased due to the greater index of refraction (typically 1.4 to 1.5) of the dome medium. Most rays that enter into this surrounding medium are able to leave the hemisphere, since they strike the surface at near-normal incidence. This type of geometry gives a Lambertian-type of emission pattern. Thus, the total radiance appears to fall off as the cosine of the angle from the optical axis of the device.

Figure 5-24 shows a structure with a lens-type shape at the top that gives direction to the emitted radiation. Such a packaging scheme is useful when energy is transferred from the emitter to a small spatial area, such as that of a silicon detector. Lens-like structures improve bare-chip external efficiency by a factor of two or more.

Greater efficiency can be obtained from a structure in which the dome, rather than plastic, is made from semiconductor material itself. This gives a Lambertian-type emitter with no critical angle losses. Only surface reflec-

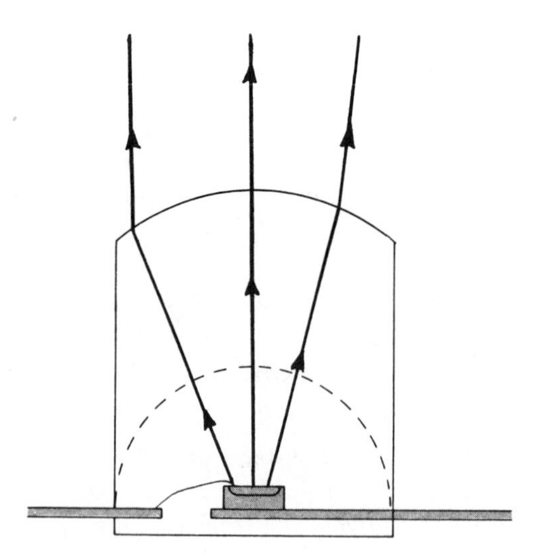

Fig. 5-24 A plastic-domed lens (broken line) gives emission over a large, solid angle, while a plastic shaped lens gives a more directional pattern.

tion losses occur. The domed semiconductor chip, however, is more costly than flat chips. If greater direction is desired from a device, it may be mounted in the focus of a parabolic reflector.

One way of improving the appearance of a visible LED is to increase the contrast. This is done on some red emitters by using a red (rather than clear) plastic covering. Transmission properties of the red plastic are such that it transmits the emitted wavelength and absorbs all other visible wavelengths. Thus the background observed by the viewer has less brightness than would be the case with clear plastic.

An important LED packaging consideration is that of heat sinking. The device's efficiency decreases as temperature increases. If high reliability is a consideration, then metal packages (for heat sinking) with glass lenses may be desirable.

Another device in which packaging is quite important is the coupled pair, consisting of an LED chip and a silicon photo-detector chip mounted in the same package. The main function is to transmit an input signal to an output terminal with an extremely high degree of isolation. The packaging goal is optimum light coupling. Chip proximity, index of refraction, and reflective techniques are employed to achieve this goal.

5.13.8 OPTICAL COUPLERS

There are many situations in which data must be transmitted between two circuits that must be electrically isolated from each other. While such isolation has, in the past, been provided by relays, isolation transformers and other arrangements, components called optically coupled isolators can also be used for this purpose (Fig. 5-25). These devices can be a combination of an IR light-emitting diode and a photosensitive transistor, diode or photoconductive cell. The input is applied to the LED terminals. The

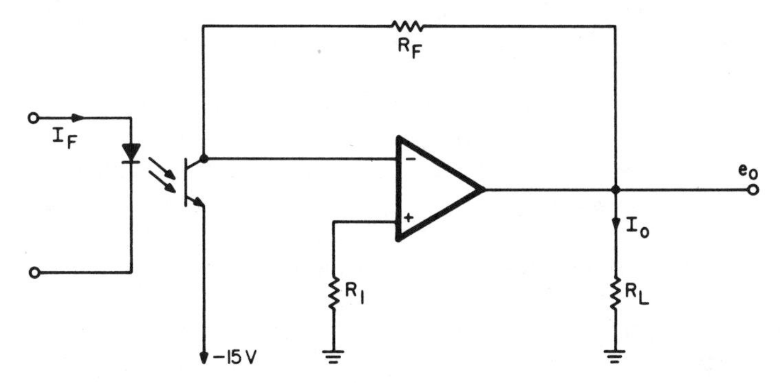

Fig. 5-25 Optically-coupled isolator. Pulse amplification as well a isolation can be achieved with this circuit using an isolator with an operational amplifier to amplify the pulse appearing at the anode of the emitter. Circuit gain is controlled by the feedback resistor, R_F.

LED output turns the photosensor ON, producing an isolated output. The coupling between the LED and the photosensor may be fiber optics or just a small glass lens.

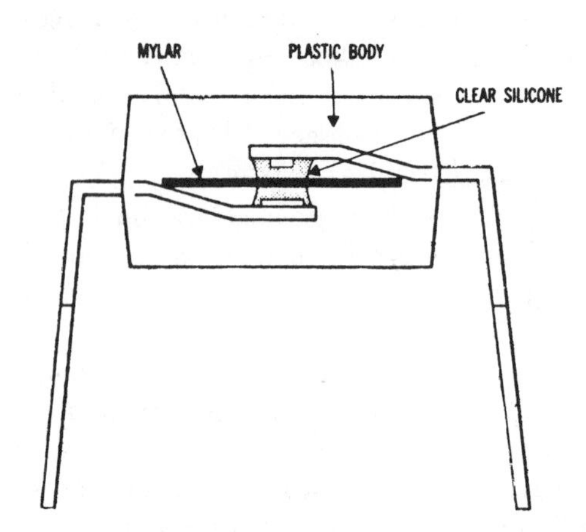

Cutaway drawing of a typical optocoupler DIP. The dielectric in this case consists of a Mylar plate embedded in clear silicone.

Basically, a coupler consists of an infrared emitting diode (IRED) positioned so its output falls on a photodetector. With a coupler, two circuits can be physically separated, with the coupler's light path providing the only link between them. Thus, a signal can be transmitted while electrical isolation is maintained. The photodetector component consists of a photodiode, phototransistor, photoDarlington, or photo SCR. Both components can be arranged in an integral package, with an insulating dielectric filling the space between the two. The transparent insulator can be a vacuum, air, epoxy, or glass.

Besides being offered in the integral version, with all components packaged in an enclosed unit, couplers are offered in air gap versions. Here, the components are supplied in a single package, but the user has access to the air gap between the emitter and the detector. A moving perforated tape, sheet of paper, or drops of fluid pass through the gap and intermittently break the optical path. This air gap type of device is used in flow sensing in industrial process control, and in controlling paper flow, formatting, and tone in business machines.

Generally, a gallium arsenide infrared emitting diode (IRED) is employed as the emitter, since its 900 nm wavelength output nearly coincides with the peak spectral response of the ordinary silicon photodetector. Couplers have some very unusual properties. Electrical isolation can reach 5 kv (or even 50 kv in some special applications), and isolation impedance (I/O) is commonly more than 10^{10} ohms in parallel with only 0.5 pf capacitance.

Choosing between the various types of detectors can be simplified if several basic considerations are observed. To make an effective selection, it is first necessary to consider the basic parameters of optocoupler operation.

Current transfer ratio (CTR) is one of these parameters. It is a figure, given in per cent, expressing the ratio of output current to input current. Higher input currents generally result in higher CTR's. But, in considering CTR values, it is important to remember that values can drop when the output device saturates. Nevertheless, values to 800% are not uncommon.

Isolation voltage is another parameter that must be considered. This is the voltage differential the device can tolerate without breakdown occurring across the gap, where a short circuit or arcing would occur. Values given in specification sheets can be misleading, because manufacturers do not always tell if the voltage given is ac or dc; and sometimes they do not say how long the device can tolerate such a voltage differential. But specifications in this area are in the process of being standardized, and improvements can be expected in the near future.

The onus is generally on the user to understand the implications of the specifications given. He should be sure what the guaranteed minimum value is for his application and how it is measured. Insufficient isolation can be a prime cause of breakdown.

Isolation voltage depends on the character of the dielectric used in the gap, as well as the variety of packaging employed. For the plastic DIP's com-

Typical optocoupler of the interrupter, or air gap, variety. The user has access to the gap between sensor and emitter through the slot in the top of the device.

monly used, the assured minimum isolation voltage ranges from 1 kv to 3.5 kv. For any plastic DIP, the isolation voltage is usually limited to about 5 kv maximum, because of the close pin spacing. To increase guaranteed isolation voltage, it is necessary to use other types of packaging that may include hermetic. Certain packages used in medical instrumentation applications can attain isolation voltages as high as 50 kv.

The third coupler operating parameter that deserves close attention is operating speed. Coupler switching speed depends primarily on load conditions. This is particularly important where TTL or other logic compatibility is desired. Although many manufacturers claim TTL compatibility for their couplers, this is not always justified. Not all such couplers meet TTL standards of load, supply, and temperature tolerances.

Operating speeds for the various types of IRED couplers range from 2.5 kHz to 5 MHz. Under optimum circuit conditions, the data rate can sometimes be extended to about 10 MHz. Within the next few years, the data rate will be raised to 25 MHz.

It is worth noting that optocouplers are not restricted to the use of IRED's as input components; visible LED's and neon and incandescent lamps have been used effectively as well.

In one common type of application, optocouplers are used to isolate a circuit's logic level from an analog or power level. Such a situation occurs in the interfacing of an A/D converter to a microprocessor. (See figures below.) Here, optocouplers with phototransistors as detectors can be used. Each output pin of the A/D is individually connected, through a separate optocoupler, to an input pin of the microprocessor. High potentials in the attached analog signal generator are effectively isolated from the 5 v operating voltage level of the microprocessor. The possibilities of data scrambling or of damage to the sensitive microprocessor circuitry are eliminated.

Optocouplers are finding an important application in traffic control signals. A traffic control generally has at its hub a logic system operating at 25 v. This logic system must be isolated from the higher voltage incandescent lights used in the actual traffic signals. In a typical system, various logic outputs are transmitted through an optocoupler having an SCR detector element. This provides the necessary isolation. The SCR is sufficiently powerful to trigger a triac power switch, which controls the incandescent bulb in the signal.

Switching of an ac motor into a 120 v line can be mediated by a ZVS (zero voltage switching) solid state relay, which contains an optocoupler of the phototransistor or photoSCR type. The function of the optocoupler is to isolate the low dc voltage control circuitry from the 120 vac line voltage. On the high voltage side of the circuit, a zero voltage switching circuit restricts motor turn-on to the region at or near the zero voltage crossing of the ac waveform. Under these circumstances, undesirable effects, such as RFI or hash on the line, are not produced.

Optocouplers with SCR detectors can also be used to replace reed relays in low level data acquisition, such as is carried out by semiconductor test equipment. Here, for example, V_C and V_B of a sample transistor can be separately tested using individual optocoupler subcircuits. On energizing the optocoupler, a signal passes from the emitter to the detector and activates the corresponding test circuit, permitting testing of each semiconductor parameter in turn. A measuring circuit and a readout are directly connected to the optocoupler outputs.

Optocouplers of the interrupter variety are commonly used in gasoline pumps. In one such application, a slotted disc is controlled by the gas pump motor. When the pump is operating, the disc rotates, interrupting the light beam of an optical coupler. Measuring the amount of gasoline pumped is then simply a matter of counting the total number of light pulses generated. This is done by appropriate electronic circuits.

Interrupter, or air gap, optocouplers are often used in automatic bowling alley scoring apparatus; to read out the perforated tape used in computer typesetting; and for formatting in various computer peripherals.

Similarly, interrupters are used in automatic labeling machinery to insure that the machine dispenses just one label at a time. A reflective spot associated with the emerging label triggers the optocoupler, which is designed to operate in the reflective mode.

Other optocouplers of the interrupter type are used in wattmeters and tachometers, and in scoring equipment for "fill-in-the-blank" questionnaires, such as are used in voting, OTB, and scholastic testing.

The range of optocoupler applications is quite wide, and includes the provision for biological protection in some instances. A shock hazard might exist in hand-held electrical appliances, where the low voltage control circuit interfaces with the high voltage power circuit. The use of an optocoupler at the interface can effectively eliminate the hazard.

Optocouplers provide one of the simplest means of providing essential electrical isolation. Innovations are continuing to widen the market and lower the price of these useful devices. Typical general purpose units are now selling for well under $1 each, and the price tag for high performance units is expected to drop below the $1 mark sometime within the next several years.

5.13.8.1 Circuit Applications

The application of optoelectronic components to selected circuits is illustrated in Figures 5-26 to 5-29.

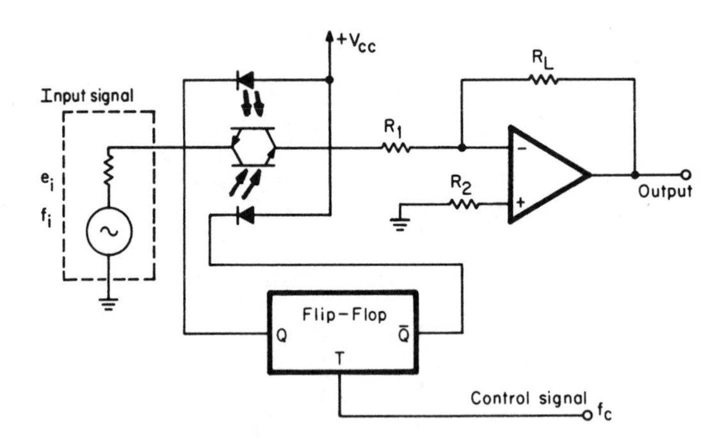

Fig. 5-26 Chopper circuits which use mechanical relays lack speed and have switching transients at the load. By using bipolar transistors or FETs as series and shunt switching elements, the speed can be improved, but capacitive coupling to the switching circuitry may still produce transient "spikes" on the output signal. By switching the input signal as shown, the switching circuitry can be isolated from the output, and thereby reduce output "spikes." The use of two couplers allows chopping of either positive or negative input signals with a frequency of one-half that of the input to the flip-flop.

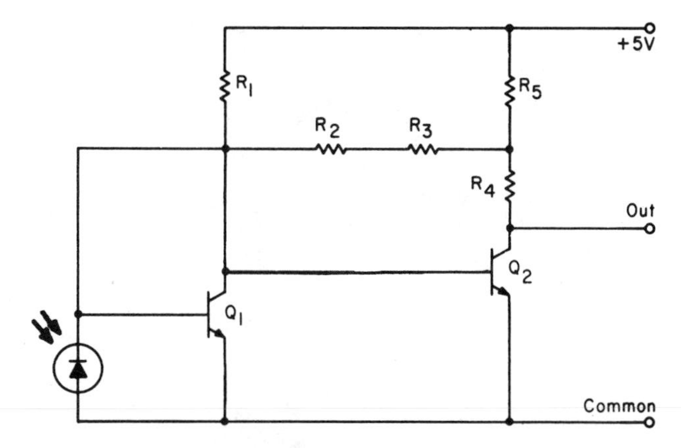

Fig. 5-27 **Schmitt trigger circuit** permits card reader output to be compatible with TTL or DTL logic functions. In the hybrid circuit, R_1, R_2, R_3, can be internally shunted to shift the switch point to correspond to the desired light level. Photovoltaic cell is operated in diode mode. Light incident on photocell causes Q_1 to switch OFF, in turn switching Q_2 ON. Q_2 ON lowers voltage at R_5, R_4 junction, changing bias on Q_1 through R_5, R_3, and R_2.

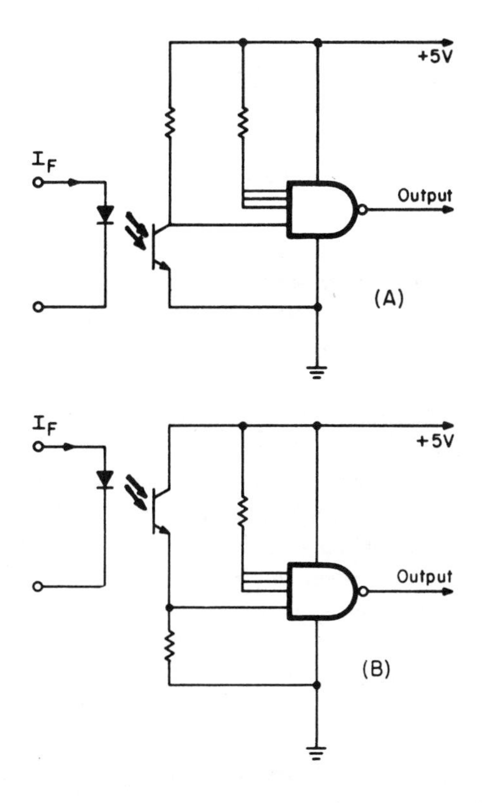

Fig. 5-28 **A Schmitt trigger** circuit using an optical couple to interface with TTL offering fast rise and fall times with an isolated signal input. Figure A provides a non-inverting function, while Fig. B provides an inverting function.

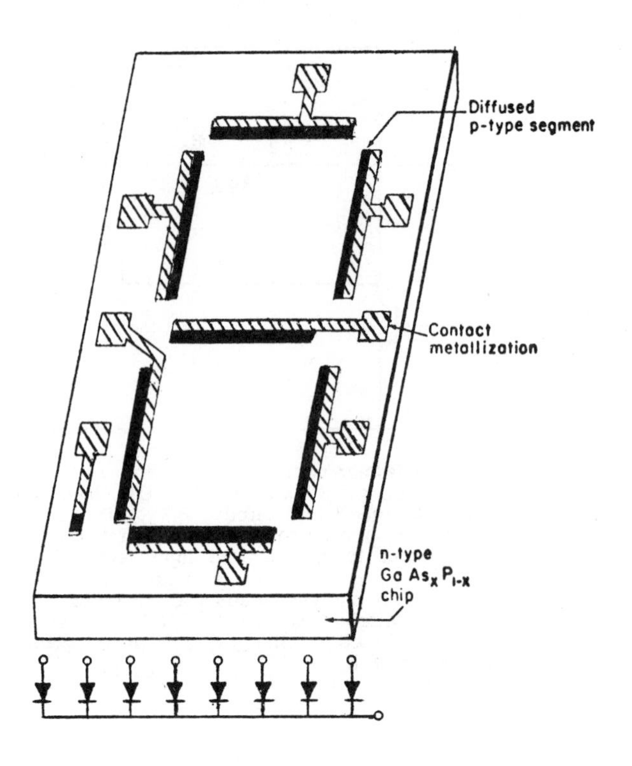

Fig. 5-29 Monolithic, 7-segment numeric displays with the eighth diode representing a decimal point on the display.

5.13.9 ELECTROOPTIC CERAMIC MATERIALS

In 1967, Land first reported that poled, small grain size, bismuth-doped lead zirconate titanate (PZT) ferroelectric ceramics possessed useful and controllable macroscopic uniaxial birefringence. Since then an increasing number of investigations of the electrical, optical and electrooptic properties of these materials have been carried out. Because of their low optical transparency, these bismuth-doped materials were only useful in thin polished plates of 0.1 mm thickness or less. Significantly higher optical transparency has been achieved in the lanthanum modified lead zicronate titanate (PLZT) solid solution system. As a result, much thicker (up to one centimeter) polished sections can be used in electrooptic devices, and many new applications utilizing the more transparent PLZT materials have become feasible.

The electrooptic properties of the PLZT ceramics are related to their ferroelectric properties. Rhombohedral PLZT material with a low coercive field exhibits a memory mode in which the birefringence varies as a function of the remanent polarization. Tetragonal material with a high coercive field exhibits a linear electrooptic effect when operated around one of its saturation remanence points. The electrooptic effects are achieved by varying the chemical composition of the ceramic material.

5.13.9.1 The PLZT Compositional System

The PLZT materials are basically lead zicronate titanate solid solution compositions, modified with substantial amounts of lanthanum oxide.

Materials in the PLZT system can exhibit a transverse electrooptic memory effect, a conventional transverse quadratic (Kerr) electrooptic effect, or a transverse linear (Pockels) electrooptic effect, depending on composition (lanthanum content and zirconium to titanium ratio) and hot-pressing conditions (temperature, time and pressure).

The memory compositions may be 7/65/35 (La/Zr/Ti) and 8/65/35, the linear compositions may be 8/40/60 and 12/40/60 and the quadratic compositions may be 9/65/35, 10/65/35 and 12/65/35.

The electrooptic memory effects, which are unique to ferroelectric ceramics, are related to variations of effective birefringence achieved by changing the remanent polarization. Memory occurs because both the effective birefringence and the remanent polarization depend on stable ferroelectric domain patterns in the absence of an applied electric field, and the memory state can be varied because both the birefringence and remanent polarization depend on the degree of domain alignment. Linear and quadratic electrooptic effects result from variations of effective birefringence with applied electric field. These effects are shown to arise from intrinsic electrooptic properties of the crystallites (linear electrooptic effect) or from field-enforced ferroelectic distortion of paraelectric materials (quadratic effect). Both the character and magnitude of all these effects strongly depend on material composition and grain size, as well as on temperature and light wavelength.

When birefringence of a ceramic plate is measured, one actually is measuring an effective birefringence which is both proportional to the intrinsic birefringence of a domain and dependent on an average over-all domain orientations in the ceramic. Memory occurs because the domain pattern is stable in the absence of an applied field. The memory state can be changed by varying the domain orientation through application of an electric field. In general, the more the domains are aligned along the field direction, the larger will be the measured parameter.

The linear electrooptic effect observed in certain poled PLZT ceramics is an intrinsic property of the ceramic crystalline material. The effect results from the averaged linear electrooptic states of all the ferroelectric domains, and not from reorientation of those domains. Domain reorientation is important only for initial alignment of domains in the virgin material Thereafter the domain pattern is fixed. Strong birefringence effects from domain alignment would occur when the ceramic reverses polarization direction.

Evidence that intrinsic crystal effects rather than domain reorientation account for the observed result comes from the lack of hysteresis in the birefringence during voltage cycling. The hysteresis is absent not only at room temperature but also at −60°C and +60°C.

The quadratic increase of the effective birefringence with applied field in ferroelectric materials arises from an increase in the intrinsic polarization, P, of the material as it is forced further into the ferroelectric phase by an increasing field. The reason the increase is quadratic is because the intrinsic domain birefringence varies as P^2.

5.13.9.2 Material Processing

The PLZT's, as the conventional lead zirconate titanates, are generally made by the mixed oxide (MO) process, starting with the individual raw material oxides of lead, lanthanum, zicronium and titanium. Careful selection of the raw materials to insure high purity (>99.5%) and a small particle size (<3 microns) is a critical factor for achieving the desired final material quality.

The oxides are mixed in a polyethylene ball mill with alumina media and distilled water for approximately one hour. The slurry is subsequently dried, crushed, blended and then calcined (chemically reacted) at 900°C for about one hour. A second one-hour milling is performed after calcining in order to break up the lightly sintered particles of PLZT which develop during the calcining process. A final drying operation prepares the powder for cold pressing into slug form prior to final hot pressing.

After cold pressing, the cylindrical pre-form slug is hot pressed to final density. An oxygen-enriched atmosphere introduced during hot pressing is beneficial in increasing optical transparency by removing residual porosity. Light scattering due to porosity, has been found to be a primary source of optical losses in the material.

The oxygen hot pressing technique allows oxygen to replace nitrogen in the open pores of the powder compact during the early stages of the hot press cycle. During the final stage of densification (isolated pore stage) the oxygen in the pores diffuses through the lattice and along grain boundaries which leads to eventual elimination of the pores. In the previous conventional hot press process, residual pores remained in the material because nitrogen in the pores could not readily diffuse out of the material.

5.13.9.3 Material Properties

Physical:—

PLZT materials, due to their method of fabrication, are polycrystalline ceramics. They are composed of very tiny individual crystallites bonded together but randomly oriented with respect to each other. The average size of these crystallites, within a given specimen, may vary from about one micron diameter to about fifteen microns, with a more typical size being about four microns average diameter. A typical microstructure is shown in Figure 5-30. The high uniformity of the grain size as illustrated in the figure, is characteristic of these materials and is a desirable feature from the viewpoint of material performance.

In addition to their optical quality and electrooptic versatility, electrooptic ceramics, and particularly PLZT materials, have certain constructional, cost, and functional advantages over electrooptic single crystals.

Electrical:—

Relative dielectric constant as a function of temperature, for various selected compositions in the PLZT system, is shown in Figure 5-31. The peaks in the dielectric constants are reduced in height and in temperature, as the lanthanum content is increased from 7 to 12 atom percent at a given Zr/Ti ratio.

Fig. 5-30 Etched microstructure of PLZT 9/65/35 illustrating the dense, pore-free structure.

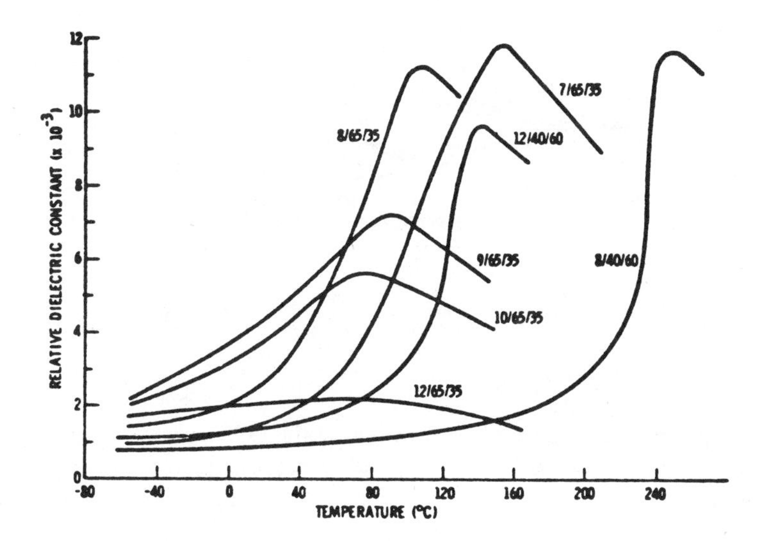

Fig. 5-31 Dielectric constant as a function of temperature for selected PLZT composition.

In general, temperature stability of the dielectric constant is achieved by a gradual decrease in maximum value. Composition 12/65/35 is the most temperature stable of the group of materials illustrated, possessing a value of 2000 ± 12% over the temperature range from —50°C to 150°C.

The temperature dependence of the dielectric dissipation factor on each of the selected compositions is given in Figure 5-32. Dielectric losses reach their maximum value at the Curie point and gradually decrease at higher and lower temperatures.

Remanent polarization as a function of temperature, for the selected compositions is shown in Figure 5-33. All of the compositions display a gradual decrease in remanent polarization for increasing temperature, with little noticeable change at the Curie point as defined by the maximum in the dielectric constant curve of Figure. 5-31. Compositions, 9/65/35, 10/65/35 and12/65/35 do not possess permanent remanent polarization at room temperature. However, under the influence of an electric field, they do exhibit substantial induced polarization arising from a paraelectric-to-ferroelectric phase change. The field induced polarization relaxes to essentially a zero value when the field is removed.

Optical:—

The variation of optical transmission as a function of wavelength is given in Figure 5-34 for 9/65/35. Optical absorption becomes extremely high in the violet end of the spectrum at about 0.37 μm. Transmission loss

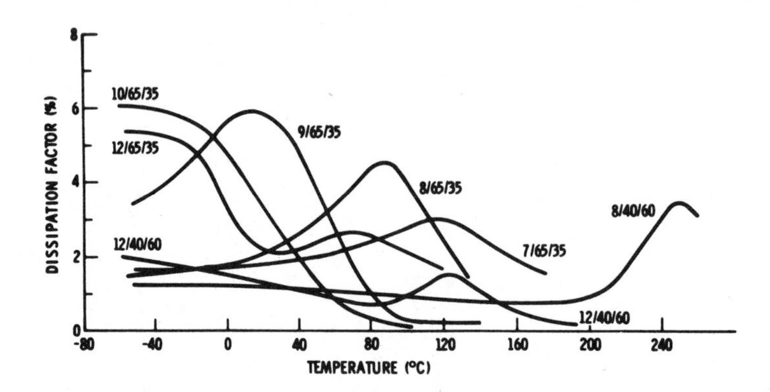

Fig. 5-32 Dissipation factor as a function of temperature for selected PLZT compositions.

due to single surface reflection is approximately 18% as calculated from a refractive index of n = 2.5. As the sample thickness becomes greater, the transmission is reduced at all wavelengths but it is most severely reduced

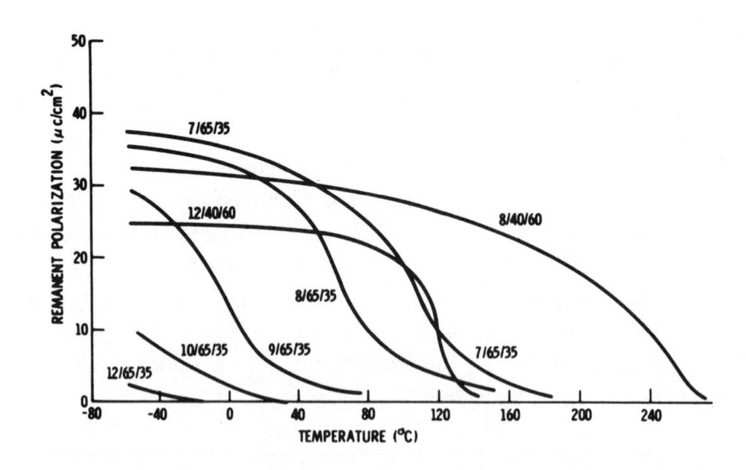

Fig. 5-33 Remanent polarization as a function of temperature for selected PLZT compositions.

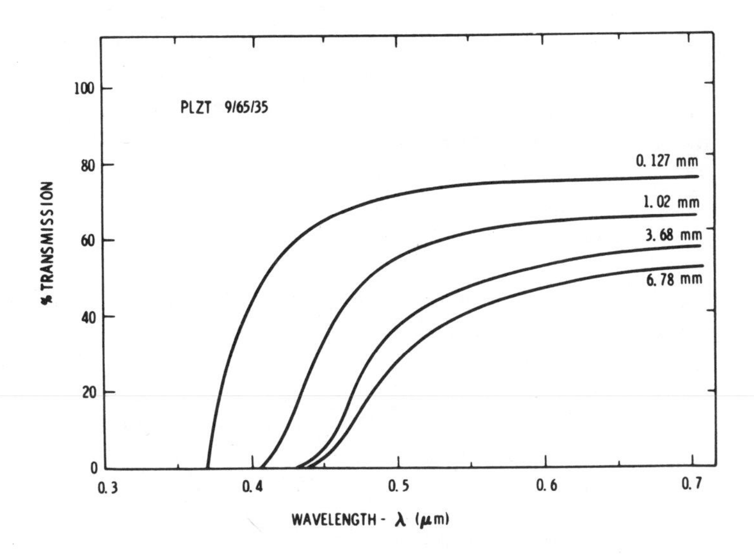

Fig. 5-34 Optical transmission as a function of wavelength and thickness of PLZT 9/65/35.

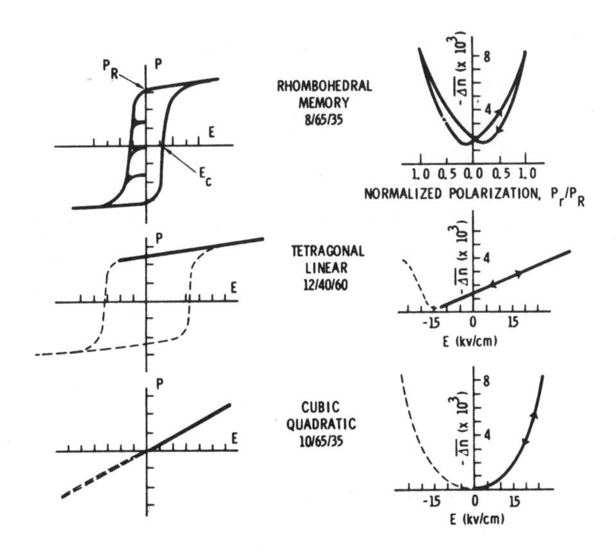

Fig. 5-35 Hysteresis loops and electrooptic birefringence characteristics for memory, linear and quadratic materials in the PLZT system, P scale = 10 μc/cm²/div, E scale = 5 kv/cm/div.

at the shorter wavelengths or violet end of the visible spectrum. This produces an apparent displacement in the high absorption region toward longer wavelengths as the thickness of the ceramic sample increases.

Electrooptic:—

The three different types of transverse electrooptic effects — memory, linear and quadratic — present in materials of the PLZT system, are shown in Figure 5-35. Hysteresis loops and corresponding electrooptic behavior are illustrated in this figure. The heavy, accented portions of the loops point out the usable portion of the loop for the indicated mode of operation. The resulting electrooptic effect is also accented.

5.13.9.4 Devices Based on Transverse Electrooptic Effects

Devices which use transverse electrooptic effects may be divided into two categories, according to the type of light source used:

(A) Devices which use a monochromatic light source vary the intensity of light transmitted by a polarizer-ceramic plate-analyzer network. These devices include light valves, shutters, modulators, memories, and displays.

(B) Devices which use white light vary the wavelength of light transmitted by a polarizer-ceramic plate-analyzer network. These devices are spectral filters which may be used as discrete single or multi-stage devices, or in linear or two-dimensional arrays.

A basic configuration of a discrete transverse electrooptic mode device applicable to both categories (A) and (B) above is shown in Figure 5-36. This device consists of a linear polarizer, a ceramic plate (with electrodes deposited on one or both surfaces), and a linear analyzer. Voltage is applied to the ceramic plate along the x_2 axis, the incident light is polarized at $\psi = 45°$, the analyzer is oriented at $\phi = 90°$, and light propagation is along the x_3 axis. In a highly transparent chemically prepared, oxygen hot-pressed PLZT, the light path length in the ceramic may be one cm or more. In such thick ceramic devices, the electrodes may be deposited on two opposing surfaces parallel to x_3 and normal to x_1 or x_2.

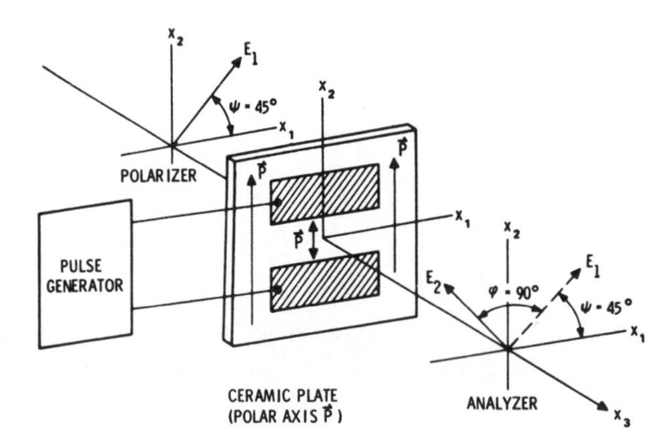

Fig. 5-36 Basic configuration for many electrooptic devices based on the transverse electrooptic effects in ferroelectric ceramics. The shaded areas represent electrodes deposited on the surface of a thin ceramic plate. Electric field and/or remanent polarization is along x_2, and is controlled by the pulse generator or other voltage source. Light propagation is along x_3.

Electrooptic Memory Devices:—

Discrete Light Valves or Shutters are natural applications of PLZT memory compositions. Multistate latching light valves are achieved by partial switching of the ceramic remanent polarization in increments.

Light Valve or Shutter Arrays have been designed for optical memory and display applications. Linear or two-dimensional arrays on a single ceramic plate are possible because of the unique localized switching properties of ferroelectric ceramics. Such arrays, however, involve problems in the design of means for electrically accessing individual valves or shutters. Although functional two-dimensional arrays have been designed and constructed, there are severe problems with half-select and disturb pulses which are inherent to ferroelectric materials because of their lack of a true switching threshold.

Linear Electrooptic Devices:—

Linear Modulators appear particularly attractive using PLZT compositions with relatively high transverse linear electrooptic coefficients.

Optical Voltage Sensors are a unique application of linear electrooptic ceramics. Used in circuits which cannot be accessed by wires for voltage measurements, these devices provide optical fiber light pipes for input and output of the voltage sensor.

Quadratic Electrooptic Devices:—

Momentary Open Light Valves constructed from quadratic electrooptic ceramics exhibit particularly high ON/OFF ratios. The principal reason for this, is the good extinction obtainable in the isotropic (OFF) state. Because of the high ON/OFF ratios (as high as 6000:1), these materials make good momentary on-normally off light valves and shutters. Shutter turn-on time is comparable with that of electrooptic memory materials.

Spectral Filters for visible wavelengths are another particularly appealing application of quadratic electrooptic ceramics. By cascading two or more stages, narrow bandwidth voltage-controlled spectral filters are feasible. Both memory and linear electrooptic ceramics may also be used for spectral filters, but, in general, the quadratic electrooptic materials are more transparent and, therefore, more applicable to multiple stage devices.

5.13.9.5 Strain-Biased Electrooptic Devices

Longitudinal Mode:—

The devices in this category are based on the observation that thin PLZT ceramic plates under strain have the retardation, Γ, vs. remanent polarization, P_r, characteristics shown in Figure 5-37. From the figure it may be seen that when P_r (the component of polarization through the plate thickness) is changed, corresponding changes in Γ can be observed, with the light propagating parallel to the remanent polarization direction (longitudinal mode).

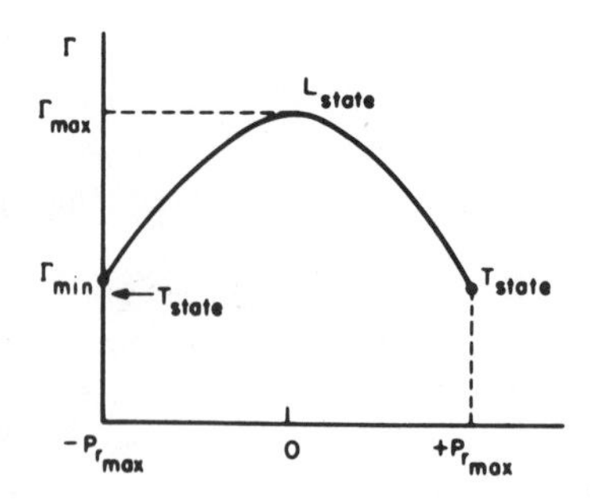

Fig. 5-37 Idealized characteristics of PLZT ceramics when operated in a longitudinal electrooptic mode under strain.

Large Area Polarization Switches:—

These devices consist of a relatively large area PLZT ceramic plate ($\sim$2 cm^2) in the structure shown in Figure 5-38. In this configuration half wavelength changes in retardation (for polarized light at 45° relative to the tension axis), can be obtained using a 125 μm thick plate of PLZT 6/65/35. Only about 125 volts applied to the plate are needed to get a condition of minimum $\Gamma(P_{r_{max}})$ and —60 volts are needed to get maximum Γ. In the controlled mode of operation mentioned above, the switching times are of the order of 100 μsec to 1 msec for latching operation. If larger voltages are used (>200 V), operation in times of the order of 10 microseconds is possible.

Transmission Mode Image Storage and Display Devices (Ferpics):—

The basic ferpic structure is shown in Figure 5-39 and differs from the one shown in Figure 5-38, by the addition of a photoconductive film between the ceramic and the top transparent electrode. The purpose of the photoconductive film is to selectively apply the electric field to the illuminated regions, and therefore store an image in the device as spatial changes in birefringence. The stored image can be made visible by placing the device between appropriately oriented crossed polarizers, and passing monochromatic light through the device (transmission mode). The device possesses gray scale capabilities due to the charge limited switching provided by the illuminated photoconductive film.

Reflection Mode Ferpics:—

Strain-biased devices operated in a reflection mode have been constructed. The device structure is shown in Figure 5-40. The writing and reading operations are shown in the figure.

The reflection mode device has three major advantages relative to the transmission mode devices: (1) Better resolution obtained by the use of thinner plates to get $\lambda/2$ change in retardation because the reading light passes twice through the device. (2) Isolation between the reading and writing light provided by the opaque film. (3) Because the projection light does not pass through the conductive film, this potential source of light loss is avoided.

Nonmemory Type:—

All longitudinal mode devices described above, can be operated in the nonmemory mode either by using PLZT ceramics in the rhombohedral phase with less than 8% La concentration, or using PLZT ceramics doped with more than 8% La.

The latter quadratic materials are more transparent than the memory type materials, but have the disadavntage of having a very large dielectric constant ($\epsilon > 3500$).

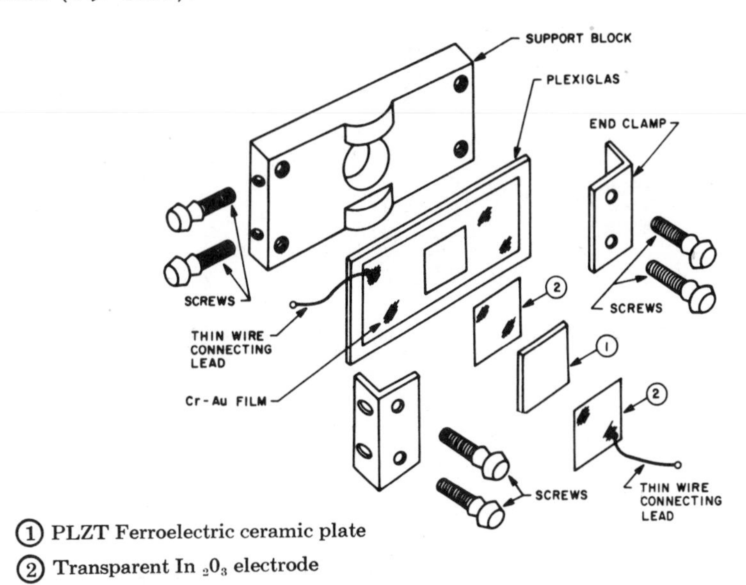

Fig. 5-38 Exploded view of a strain-biased polarization switch device.

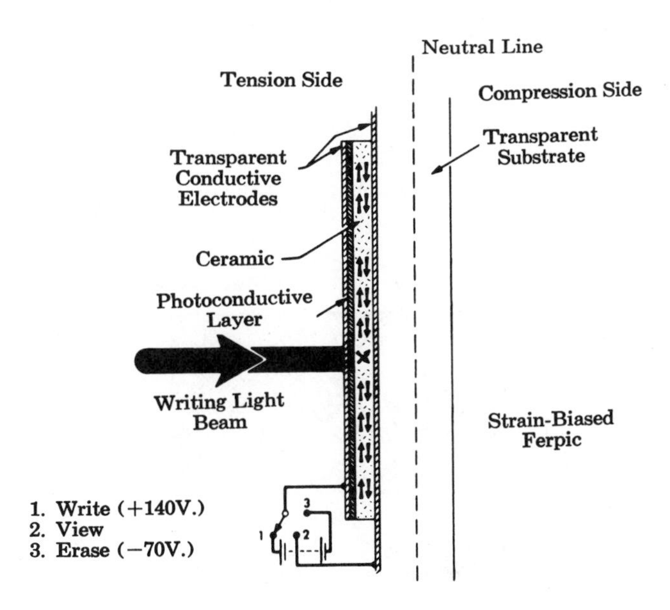

Fig. 5-39 Transmission mode strain-biased ferpic. A simplified view of the domain alignment that takes place during writing is shown.

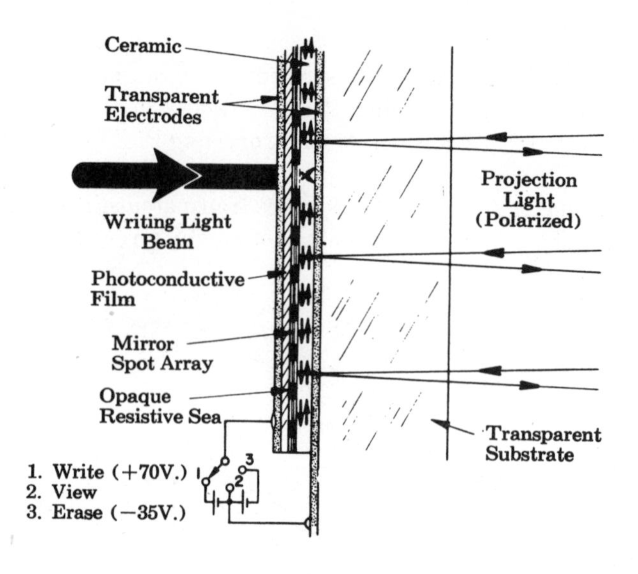

Fig. 5-40 Cross section of reflection mode strain-biased ferpic, with the writing and reading operations shown.

Transverse Mode:—

In these devices the applied electric field is in the plane of the plate and makes some angle between 0° and 90° with the direction of the tension axis, also in the plane of the plate.

*Memory Type, 45° Latching Light Gate:—*This device structure is shown in Figure 5-41 and consists of a conventional 4-electrode light gate (2 electrodes on top, 2 electrodes, immediately opposite, underneath), bonded to a Plexiglas substrate in such a way that the direction of the applied electric field is at 45° with respect to the tension axis in the plate.

This configuration provides the maximum change in retardation with the minimum change in remanent polarization, compared to a similar not bonded light gate.

*Nonmemory Type, Linear Electrooptic Modulators:—*The device structure is shown in Figure 5-42. In this case the electric field is applied parallel to the direction of the tension axis. If the plate is poled to maximum

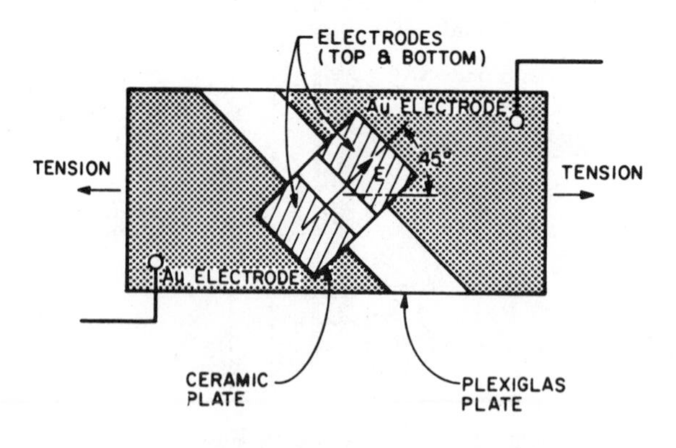

Fig. 5-41 Device structure for the 45° light gate.

P_r and an electric field of the same polarity as the poling field is applied to the electrodes, a linear electrooptic behavior is obtained.

Because the dielectric constant of the ceramic decreases with increasing strain, this configuration presents some advantages in terms of driving powers relative to the electrooptic modulators proposed using free ceramic plates.

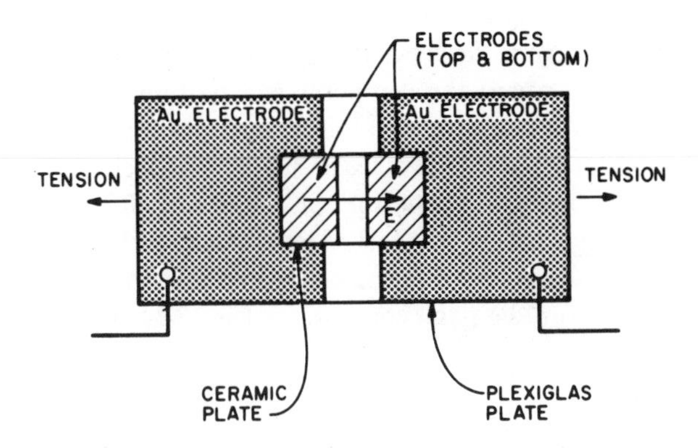

Fig. 5-42 Device structure for the transverse mode linear electrooptic modulator.

Performance Limitation In Strain-Biased Devices:–

When the strain-biased devices are illuminated with monochromatic light, the main limitation is the inability to obtain a large (ON-OFF) contrast ratio (>10 dB), when a large collecting aperture is used. This limitation is due to the depolarization of the reading light in the ceramic plate, and is a property of the ceramic material.

The contrast limitations of the strain-biased devices can be ameliorated by appropriate choice of the ceramic material and plate thickness, and by appropriate compensation for wide bandwidth operation (for example, by using two identical ceramic plates with equal retardation in the OFF condition and with their optic axes 90° with respect to each other).

Lifetime studies of PLZT ceramics in the polarization switch structure, have shown that the devices can operate for more than 10^8 cycles with no appreciable aging. More studies have to be done in a real-time animated display system to determine the ultimate lifetime capabilities of the devices. The lifetime capabilities of the ceramic material have been shown to be over 10^9 cycles.

5.13.10 OPTICAL MEMORIES

Holographic optical memories are attractive candidates for the next generation of read-write computer memories. Such memories may have capacities of 10^9 bits or more with transfers of individual blocks or pages (around 10^4 bits) of memory at μsec. or faster speeds. As shown in Figure 5-43, the four major components of an optical memory are (1) a digital deflector for the laser beam used for reading and writing; (2) a page composer for converting electrical signals into an optical pattern of the type needed to write into the storage medium; (3) an erasable, reusable optical storage medium; (4) a photodetector array for converting the optically readout signals into suitable electrical signals.

A page composer is usually an electrically addressed array of light valves. Ferroelectric and related materials, such as nematic liquid crystals, are usable for a page composer, since they exhibit large optical effects, and the electrical addressing signals required, are often within the capabilities of integrated circuit drivers. Because of the size and number of light valves required in a page composer, the composer can be fabricated on an integrated circuit chip with each light valve being directly addressed by an individual semiconductor driving circuit. Under such circumstances the light valve material need have neither a threshold field nor memory. Individual addressing is inherently fast. Also it permits the photodetecting function for the optical memory to be incorporated into the same integrated circuits that are used for the page composer. (Fig. 5-44).

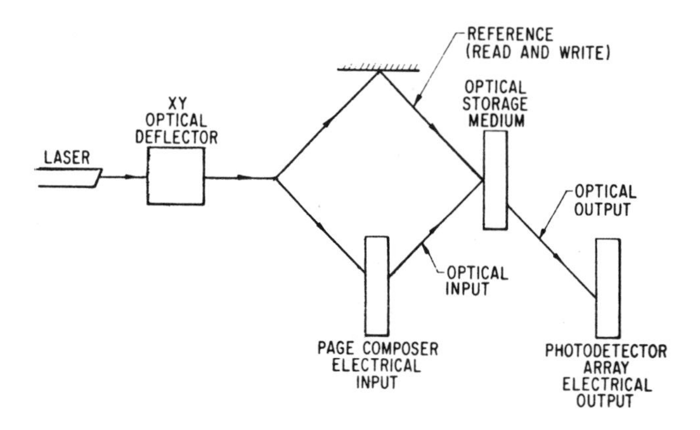

Fig. 5-43 Components of a holographic read-write optical memory.

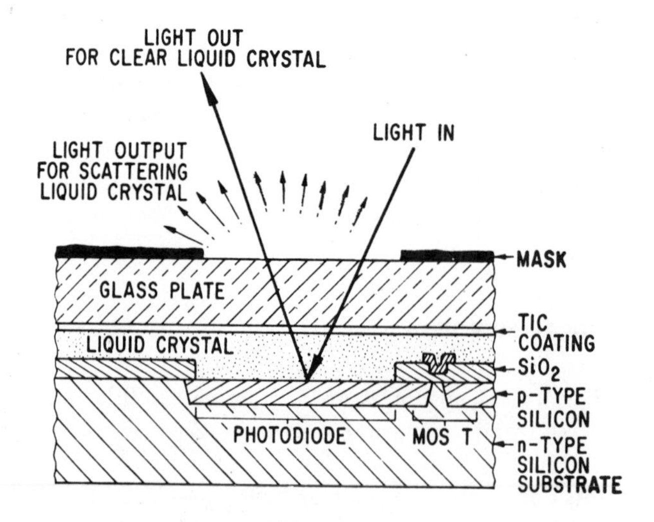

Fig. 5-44 Portion of an Individually addressed liquid crystal page composer.

5.13.10.1 Liquid Crystals

Nematic liquid crystals are related to solid ferroelectrics in that they exhibit a sharp order-disorder temperature phase transition, and they have large dielectric and optical effects. The application of a longitudinal electric field causes a normally transparent liquid crystal material to become a light scatterer. This can be used as the basis of a light valve for a page composer.

The fabrication of an individually addressed liquid crystal page composer takes advantage of the rheological properties of the liquid crystal material. One fabrication technique is to contain the liquid crystal between a transparent electrode mounted on a piece of glass and a silicon chip containing the integrated circuits, which act as the individual drivers. Figure 5-44 shows one such driver and its associated liquid crystal light valve. In this case, the back reflecting electrode also serves as part of the photodiode circuit. This permits the page composing and photodetection functions of the optical memory to be combined; an important advantage from a system's viewpoint.

The optical characteristics of the liquid crystal material are quite good. The on state is high transmission and a constant ratio of greater than 100:1 is achievable between the transmitting and light scattering states. Also, the contrast ratio remains high for widely diverging light. This is inherent because of the scattering effect.

A liquid-crystal display (LCD) is passive; it does not emit light regardless of the mode of operation. The two basic types of LCDs are dynamic-scattering and twisted-nematic (or field-effect). In turn, each can be obtained in either a reflective or transmissive mode. The major features of these four LCDs are shown in the accompanying table.

For all types, light falling upon the display is refracted or rotated by the thin film of the LC material. The display area is electrically controlled to produce a contrast over the unexcited area.

In dynamic-scattering (left-hand figure), many small domains of liquid crystal are formed with the application of the electric field. Multiple refractions at the boundaries of the domains cause the ballooning of a pencil beam of light, so an off-axis viewer sees a definite contrast. The unactivated area remains transparent, while the activated area appears frosty and opaque, and takes on the color of the light source.

The pencil beam that goes through the unactivated area can be seen only if the viewer looks down the beam itself. The lobe on the left side represents a polar diagram of light intensity, with a ±10° half-power angle typically. Note that light is only forward-scattered. To obtain the reflective mode, simply place a mirror

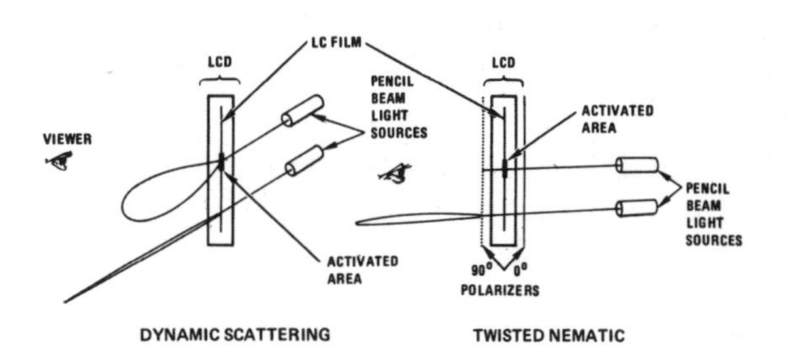

on the left side. The mirror permits the light source and viewer to be on the same side of the display. This can be detrimental if the viewer sees the image of the light source or cluttering images.

Light rotates in the twisted-nematic mode (right-hand figure) as opposed to being refracted in the dynamic-scattering mode. The LC material in the twisted-nematic cell aligns parallel to the inner surface. With the two surfaces rotated 90° during cell construction, polarized light rotates the same amount.

When unactivated, the pencil beam is first polarized, then rotated 90° by the liquid-crystal film; the light passes through the second polarizer. A viewer on the left who looks directly at the display sees the beam.

When the liquid-crystal film is activated, the rotating ability is destroyed by a breakup of the crystal order. The second polarizer now absorbs nearly all the light incident on it, and gives this area a black appearance. Inverse images can be achieved when both planes of polarization are made parallel.

The reflective mode is achieved by use of a reflector in place of the left polarizer. Again the viewer and the light source are on the same side. No image can be seen here (as with dynamic scattering) because of the polarizers.

A limited viewing angle, caused by excessive rotation, is an inherent property of the twisted nematic. The degree of rotation depends on the thickness of the liquid-crystal film. Therefore light passing obliquely through the film passes through more LC material. For a large enough angle, the light can be rotated beyond the optimum angle for complete transmittance. Reliable display interpretation should be expected at ±30° to the normal, although half of the light is absorbed at the first polarizer.

Dynamic scatter		Twisted nematic	
Reflective	Transmissive	Reflective	Transmissive
Efficient use of ambient light good down to moonlight Viewing angle sensitive to location of ambient light Background cluttered by image of viewer and surroundings (desirable on a woman's watch) Aesthetically interesting characteristics can be incorporated in colored dichroic mirror	Widest viewing angle Lowest cost Readable in total darkness with back light Changeable color by changing back-light color Back light (or front light) desirable in some form at all times	Most efficient use of ambient light No background image clutter as diffusive reflector can be used Narrower viewing angle Lowest power	Narrower viewing angle Readable in total darkness with back light No light spillage Changeable color by changing back-light color Back light in some form desirable at all times

If the efficiency of the LCD is high, a small percentage of the ambient light is internally trapped. But, of the portion trapped, only half the light can return out of the front side.

The liquid-crystal material does not back-scatter, and the display has immunity to high point light sources, like a spotlight, the sun or a window. But it is not so immune to continuous light sources, such as an open bright sky or a very large illuminated area. The loss of immunity is largely due to glare.

When glare, caused by first surface reflection, is directed toward the viewer, the display may wash out. This can be eliminated if the first surface is made to reflect a dark image. In one outdoor application, a user tilted an LCD-equipped traffic sign slightly forward, so that the first surface reflection was from the ground, not from the sky.

Liquid-crystal displays are high-impedance devices. High impedance, which lowers power requirements, also offers the user a choice of connector-assembly methods and styles—a feature not always offered with other displays.

Connector contact resistance of several hundred ohms does not affect the brightness of the LCD. One reason is that the transparent conductor on the LCD connector edge is usually either indium or tin oxide, both very hard and abrasive-resistant materials.

A pointed connector pin, which penetrates a connector edge that contains these materials, completes the physical interface. When 10 or more insertions are desired, or a high-vibration environment is expected, use gold-pressure contacts.

The connector edge should then be silk-screen-printed with gold to mate with the gold pin. Where screenable silver has been printed, wire can be bonded directly to the glass edge via indium-gold alloy solder. The user can also weld beam leads ultrasonically onto gold or silver pads printed on the glass.

At room temperature dc resistivity is typically

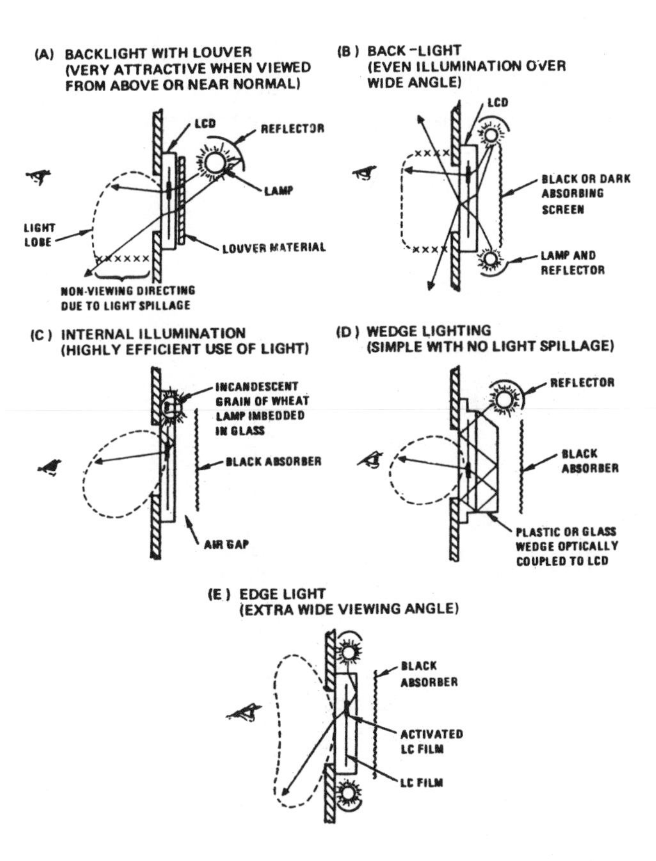

Five basic lighting techniques are used to back-light a dynamic-scattering, transmissive-mode LCD. One scheme (A) gives limited light spill when viewed from above or nearly head-on. Another (B) uses two lamp/reflectors to spread light over a wide angle. A third (C) uses light very efficiently. The fourth (D), called wedge-lighting, eliminates light spillage. And the fifth (E) offers the user an extra-wide viewing angle.

10^9 to 10^{10} Ω/cm, depending on whether a dynamic-scattering or twisted-nematic LCD is used. Cell requirements range from 50 to 5 μW, respectively, for a half-inch-high display.

The material resistivity of the LCD changes by almost three orders of magnitude over the operating temperature range. The dielectric constant, approximately 5, also varies by 50%.

Dynamic scattering displays usually specify 20 V ac. The display responds best to a square-wave drive, with a frequency in the range of from 20 to 200 Hz. The lower frequencies are used to excite the LCD when temperatures are low. However, the display won't respond to excitation frequencies above a so-called critical frequency—in this case, 400 Hz.

The LCD cell is black-boxed to simulate a leaky capacitor. This capacitor time constant is 1 ms at room temperature. Since an LCD presents a capacitive load to the drive electronics, power loss varies with drive frequency. Typical values of power loss are 1 μW for dynamic-scatter and twisted-nematic, referenced to a half-inch-high character.

Response time is also a function of temperature—about 200 ms at room temperature. This parameter changes by two orders of magnitude over the operating range. But display speed cannot be increased significantly by use of faster drive electronics.

Material viscosity represents a limiting factor of LCD response speed. The lower the ambient temperature that surrounds the display, the slower the display's speed.

Most dynamic-scattering displays use a mixture of materials called butylanilines, doped for conductivity with a proprietary ionic material. This combination gives the display an operating range of approximately −15 to 55 C.

Depending on ambient temperature, a complete reversal of LCD characteristics can occur. The effects are similar to the melting and freezing of water. The useful operating temperature range is limited to a band called the nematic range. Above this, the material becomes an isotropic liquid and ceases to exhibit anisotropic (liquid-crystal) properties.

Below the nematic range the material becomes a solid, and, once again, it shows isotropic properties. Ordinarily you don't have to worry about temperature-induced expansion. The material closely matches that of the surrounding glass envelope.

A malfunctioning display can be detected by one of the following characteristics:

A slippery feeling or peculiar odor. This means that the liquid-crystal material is leaking out of the glass sandwich. Bubbles develop in a few hours of operation, and sooner or later they void a segment.

Irregular dark and light areas when a display on standby is viewed through cross-polarizers. This indicates poor alignment of the liquid-crystal molecules relative to the glass walls. Hair-like lines, frosty surface or other irregularities show poor alignment and can lead to a visually disturbing appearance.

Excessively slow ON or OFF response. This

indicates an excessively thick liquid-crystal space, which normally remains stable unless caused by contamination.

Low resistance. This shows excessive doping or contamination of liquid-crystal material. The display should be rejected immediately, since this defect causes shortening of display life.

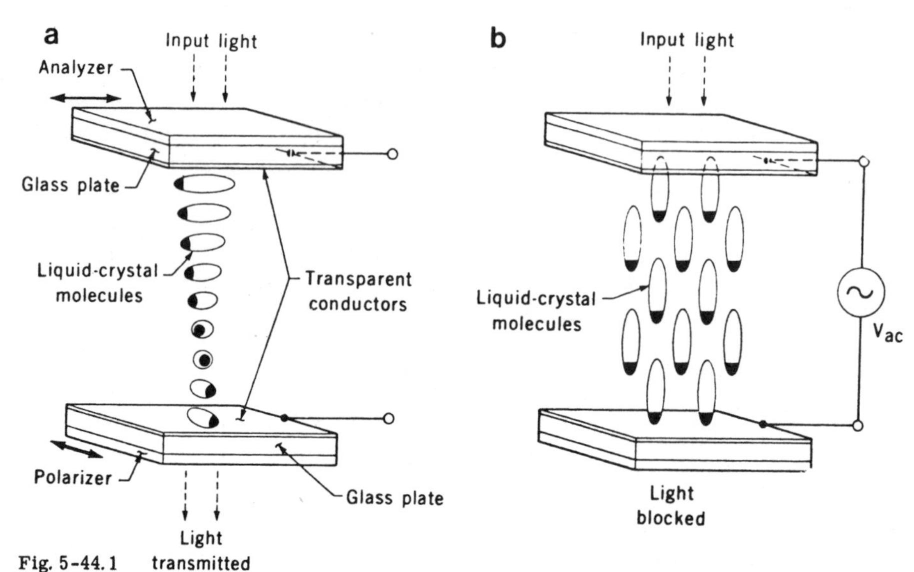

Fig. 5-44.1 (a) Twisted-nematic liquid-crystal cell with no voltage applied. (b) Orientation change of liquid-crystal molecules produced by applied a-c voltage.

The most common displays in present use are the "twisted nematic" type. As shown in Fig. 5-44.1a, a layer of nematic material (about 10 μm thick) is contained between the transparent conductive surfaces of two glass plates. This structure is then placed between two polarizers. Before assembly the glass plates are specially treated to make the molecules at the surface align in a particular direction almost parallel to the surface of the plate. One method for accomplishing this is to evaporate a thin layer (about 100 angstroms thick) of SiO onto the glass surface at an oblique angle. Since the alignment directions of the two plates are set at 90° with respect to each other this causes a gradual twist in direction of the liquid-crystal molecules between the plates as shown. If the input polarizer is oriented, for example, parallel to the alignment of the liquid-crystal molecules at the top plate, the plane of polarization of the light passing through the liquid crystal will be rotated by 90° because of the gradual twist of the liquid-crystal molecules. This light will then pass through the analyzer adjacent to the lower plate whose orientation is set perpendicular to the polarizer.

Since liquid-crystal molecules are used that have a strong dipole moment in the axial direction, if an a-c voltage (for example, 2 to 3 V rms) is applied across the cell, the molecules (except those held by surface forces at the glass plates) will to a large degree orient themselves perpendicular to the glass surfaces as shown in Fig. 5-44.1b. In this orientation the liquid-crystal layer can no longer produce a rotation in the polarization plane of the light, causing it to be blocked by the analyzer. In most display devices the ambient light itself is used as the source and, instead of viewing the changes in transmitted light, a reflector is placed at the output to return the light back through the cell, thus permitting viewing from the input side.

Many of the liquid-crystal materials

found useful in early device work have a molecular structure whose inner portion consists of two aromatic rings with an additional central group linking them together. One important class, the Schiff base materials (characterized by the −CH=N− central group) was studied extensively. The first compound of this class which could be used at room temperature (commonly known as MBBA) is in the nematic liquid-crystal state over the temperature range of 22° to 48°C. Another compound of this class (known as EBBA) is nematic over the range of 35° to 77°C. However, by mixing the two compounds in the correct proportion eutectic mixtures have been reported whose nematic range extends from 0° to 60°C. For use in twisted-nematic displays, Schiff base and other materials have been developed with an end group (such as −C≡N) attached to one of the aromatic rings which provides a strong axial dipole.

Materials of the above types suffer from decomposition in the presence of small amounts of water or when exposed to ultraviolet radiation, resulting in the breaking of the bonds between the two aromatic rings. In the past few years, however, new materials have been developed such as the biphenyls in which the aromatic rings are directly linked, making the molecules extremely stable in the presence of moisture, air, and light These also may have a CN group attached at one end to form a strong axial dipole and, depending on the particular alkyl or alkoxy group attached at the other end, allow different operating temperature ranges to be obtained. By using two or more compounds of this type, eutectic mixtures have been obtained that are nematic over a temperature range of −10° to 60°C. (Recently, new materials such as phenylcyclohexanes have been developed with comparable stability and temperature range.) It is of interest that because of the chemical stability and long life of present liquid-crystal materials (which may be considerably greater than 10^4 hours) the life of display devices is frequently limited by the deterioration of the plastic polarizing sheets usually employed which tend to degrade under conditions of high temperature and humidity.

To avoid electrolytic decomposition of liquid-crystal cells, it is usual to apply a-c voltage (for example, 100 Hz) rather than d-c. Because of the high resistivity of the materials (greater than 10^{10} ohm-cm) the power consumption is very small, being less than 1 microwatt per square centimeter (several orders of magnitude lower than for luminescent displays). Liquid-crystal materials are thus ideally suited for portable, battery-operated devices. An important limitation of liquid-crystal displays is their relatively slow response, the turn-on and turn-off times both being of the order of 0.1 second at room temperature Because of the increasing viscosity of the material at low temperature, the turn-off time at −10°C may be about 1 second. At a given temperature the turn-on time varies inversely as the square of the applied voltage. The turn-off time, which depends on the realignment forces between the molecules, increases with the square of the distance between electrodes.

To eliminate the polarizers and provide a wider viewing angle considerable present effort is being directed toward liquid-crystal materials in which a pleochroic dye is dissolved The elongated molecules of such a dye absorb light over a broad spectrum if they are oriented parallel to the electric vector of plane-polarized light but are relatively transparent if they are oriented perpendicular to this vector. Since the dye molecules tend to align in the same direction as the liquid-crystal molecules their orientation can be controlled by electric fields which act on the liquid-crystal material. In the preferred arrangement the dye molecules are incorporated in a cholesteric-type of liquid crystal whose molecules are made to align parallel to the cell walls. Unlike twisted-nematic cells, however, the molecules of the cholesteric material have an intrinsic twist angle which may be much greater (for example, several complete cycles instead of 90°) determined by the thickness of the layer rather than the orientation of the glass plates. In this state the dye molecules, oriented in all directions, absorb unpolarized light. If a voltage is applied, as in the case of the twisted-nematic cell, the liquid-crystal molecules (together with the dye molecules) orient themselves perpendicular to the glass surfaces, greatly reducing the light absorption. To obtain satisfactory operation, however, it is necessary to use dye molecules that are strongly oriented by the liquid crystal and are chemically stable. Very recently anthraquinone dyes have been reported which satisfy these conditions well.

Attempts to use *X-Y* addressing techniques for liquid-crystal displays have resulted in limited success. Aside from the threshold voltage not being very sharp, it varies with viewing angle and temperature. In addition, associated with their relatively sluggish response, liquid-crystal cells respond to the integrated effect of repetitive voltage pulses (in particular the rms voltage) rather than the peak value. In an *X-Y*-addressed array the response of unselected elements is thus almost as great as for selected elements since unselected elements are repetitively excited with "half-select" voltages (that is, either *X* or *Y* voltages alone), resulting in very low image contrast. With twisted-nematic cells *X-Y*-addressed arrays having good contrast over an acceptable viewing angle have not yet been developed with more than about ten rows. However, if collimated light can be employed, such as in projection systems, other optical effects in liquid crystals can be used whose threshold is much sharper, allowing images to be obtained from arrays with up to several hundred rows.

To overcome the *X-Y* addressing limitations of liquid crystals other approaches are being employed in which a highly nonlinear resistive element or a field-effect transistor is incorporated at each picture element to prevent voltage from appearing across unselected elements. In these arrangements a small capacitor is usually incorporated at each picture element as well. This is charged in accordance with the peak value of the *X-Y* signal voltages and then allowed to discharge through the relatively slow-responding liquid-crystal element over an extended time.

Characteristics of Good LCDs

Properly functioning displays meet the following criteria:

Wide operating temperature. A net operating band of at least 65 C should be expected from any LCD. The low end of the nematic band must approach 0 C. Check to make sure the response time is not too slow at the lower temperature and that current consumption is not excessive at the higher temperature.

A low critical frequency. This increases proportionately with temperature, and it is the upper frequency at which the display stops dynamic scattering. Excessive conductivity is one cause of a high critical frequency. For the dynamic-scattering mode, the critical frequency must be above the drive frequency at the lowest operating temperature. At room temperature, the critical frequency should be about 500 Hz and consistent (±50 Hz) from one display to another. Character-to-character consistency also should be ±50 Hz within each display.

Wide storage temperature. The liquid-crystal envelope must survive the storage temperature. Alignment of the display material can change, depending on storage history, at the lower temperature extremes. This is a particularly important parameter with twisted-nematic LCDs, since reliable hermetic seals have been developed only recently. The seal may leak at storage temperature extremes because of a mismatch in the thermal expansion of sealing materials. Storage temperatures of −50 to 100 C should be used.

Consistent upper and lower transition temperature. A variation of 1 C in upper transition temperature indicates a change in material purity, type or doping level. The transition temperature should be constant from batch to batch, and constant with time on any one batch. Failure is evident when the display temperature band narrows by 10% after an initial operation period of 100 hours.

Wide scattering lobe. An LCD's scattering lobe depends on the type of material, dopant and operating voltage used. If off-axis view is important, specify a wide-lobe LCD. Scattering lobe can be increased if the ac drive voltage is raised.

Speed of response. If the response speed rises above 200 ms, it becomes noticeable and may be objectionable. The response is proportional to the square of the display-film thickness. A high drive voltage causes the display to turn on quickly, but also to extinguish slowly. The opposite is true for a low-drive voltage.

Material color. The display should be transparent or have only a slight but bright yellow tinge. Reject displays with a dull yellow or brownish tinge, since this invariably indicates contamination.

Voltage. Display performance should be relatively insensitive to normal supply-voltage fluctuations. However, a 200% overvoltage may cause damage and shorten display life, if applied for more than a few minutes.

Electrochromics and Electrophoretics

Electrochromic displays make use of a material whose color can be changed reversibly by passing an electric current through it. Such coloration processes involve either a valence change of one of the constituent ions or in the formation of a color center associated with a lattice defect. One material that has been studied is deheptyl viologen dibromide, which is colorless in an aqueous solution. If voltage is applied across a cell containing this material an insoluble purple compound is formed on the cathode surface. Another material extensively studied is tungsten oxide. If a thin film of this material is coated on the transparent cathode of a cell containing an electrolyte that can supply H^+ ions (for example, H_2SO_4) the film will change from a transparent to a blue-colored state as a result of current flow through the cell. In this process it is believed that a tungsten bronze, H_xWO_3 ($x < 0.5$) is formed as a result of H^+ ions being injected into the film from the electrolyte together with electrons from the cathode. More recently, electrochromic action has been reported in iridium oxide films, the coloration here being attributed to the injection of hydroxyl (or other negative) ions from the electrolyte coupled with the extraction of electrons from the film by the anode. In the case of tungsten oxide and iridium oxide films, attempts have been made to replace the liquid electrolyte with a solid electrolyte or superionic conductor. Although some success has been achieved, with present materials limited life or reduced speed of response is obtained.

Electrochromic displays, like liquid-crystal displays, have the advantage of low-voltage operation (about 1 to 2 V). The switching time (dependent on the material and current density used) is of the order of 0.1 second. Although cells have been reported capable of 10^7 switching cycles, one of the problems encountered is deterioration due to unwanted electrochemical side effects, especially if the applied voltage is raised beyond a certain level. Typically an integrated charge transfer of several millicoulombs (or more) per square centimeter is required to produce a change in coloration, the power consumed in switching being several orders of magnitude greater than for liquid crystals. It should be noted that, because of the electrochemical processes occurring, a reverse voltage is built up during the coloration process. If the cell is short circuited the reverse current flow will cause decoloration (a situation which may be undesirable in *X-Y* addressing circuits). However, if the cell is maintained in the open-circuit condition after coloration it can remain in this state for hours or longer, thus providing a memory effect.

Electrophoretic displays make use of a thin layer of dyed fluid in which pigment particles of a strongly contrasting color or reflectivity are suspended. Depending on the materials used, as well as charge-control agents added, the particles may acquire either a positive or a negative charge with respect to the liquid. If a layer of such fluid is confined between two parallel electrodes (for example, 50 μm apart) and a d-c voltage (for example, 100 V) is applied, the particles will be drawn to one electrode, building up a coating on the surface. If this electrode is transparent, the color observed will be primarily due to the reflectivity of the particles (for example, yellow or white). After removal of the voltage the cell may remain in this state for hours. However, if a reverse voltage is applied the particles will be drawn to the opposite side of the cell and the color observed through the electrode will be that of the dyed fluid (for example, black or blue) which hides the particles.

Because of their high index of refraction and good light-scattering property titanium dioxide particles (which appear white) have frequently been used. To avoid sedimentation from occurring because of their high density compared to available fluids the particles have in some cases been coated with a resin to reduce their average density. Other particles that have been used are organic pigments, such as Hansa yellow and Diarylide yellow, whose density can be matched by a mixture of suitable fluids. The fluids used should have a high resistivity (for example, 10^{12} ohm-cm), should be chemically stable and, to enable high particle mobility, should have high dielectric constant and low viscosity. Some fluids that have been used are xylene, perchloroethylene, and trichlorotrifluoroethane.

To obtain satisfactory life times it is important to prevent flocculation or agglomeration of the particles, especially when they are compacted at the electrodes. Although to a limited degree this is prevented by the mutual repulsion of the charged particles, generally steric stabilizers are added to the solution. These provide long-chain molecular groups that attach to the particles and protrude outward, thus preventing the particles from approaching too closely.

To produce switching, relatively little integrated current flow is required (about 0.1 $\mu C/cm^2$), an advantage for low-power applications. The switching time is typically in the range of 10 to 20 msec, for a given material being proportional to the square of the electrode spacing and inversely proportional to the applied voltage. Because of their poor threshold characteristics, X-Y addressing techniques cannot be used effectively with electrophoretic cells unless some additional circuit component is added at each picture element. However, encouraging results have been obtained with cells incorporating an additional control-grid electrode which prevents particle migration unless both X and Y voltages are simultaneously applied.

5.13.10.2 Optical Storage Medium

A $Bi_4Ti_3O_{12}$ ferroelectric-photoconductor sandwich shown in Figure 5-45 may be used as an optical storage medium.

The write process for this medium consists of shining the optical pattern (derived from a page composer) onto the photoconductor and applying, in coincidence, a suitable voltage pulse to the transparent electrodes. In those regions where the light hits the photoconductor, the applied voltage hits the $Bi_4Ti_3O_{12}$ switching the c-axis component of polarization. In the dark regions, the applied voltage is developed across the photoconductor, and the $Bi_4Ti_3O_{12}$ remains in its intial c-axis state.

To erase the optical pattern stored in the sandwich, the photoconductor is flooded with light and a voltage pulse of the opposite polarity to that used for writing, is applied to the transparent electrodes. This switches all the $Bi_4Ti_3O_{12}$ back to its original c-axis polarization state.

The optical, and hence non-destructive read process, is achieved using the memory mode orientation. For this orientation, the light is shone through the sandwich at a small angle to the c-axis of the $Bi_4Ti_3O_{12}$ with no voltage applied to the electrodes. For birefringence readout the sandwich is tilted about the a-axis of the $Bi_4Ti_3O_{12}$. For phase readout the sandwich is tilted about the b-axis of the $Bi_4Ti_3O_{12}$. In the memory mode orientation, the $Bi_4Ti_3O_{12}$ is used in its natural platelet form.

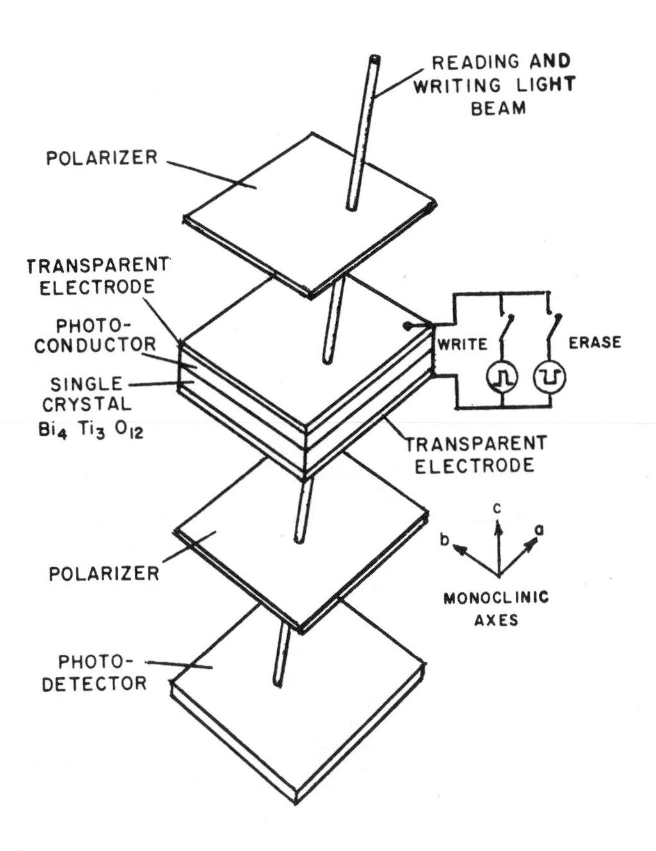

Fig. 5-45 Bismuth titanate-photoconductor optical storage medium.

5.13.11 FIBER OPTICS

In fiber optics, fibers act like waveguides for the optical radiation. The light is transmitted along the length of each fiber by means of multiple internal reflections.

The fibers have a core which is coated with a cladding of the same type of material as the core. The cladding, however, has an index of refraction that is less than that of the core. The difference in the indexes produces the internal reflections.

Plastic fibers cost considerably less than glass fibers. Also, the plastic fibers are softer and can be bent into substantially smaller radii without breaking the individual fibers.

But plastic fiber can't withstand the abrasion that glass can. And the plastic is not suitable at elevated temperatures.

For guiding light along a fiber bundle, the arrangement of the fibers with relation to one another is ordinarily ignored. However, to transmit images, the fibers at the output end must be arranged in exactly the same fashion as those at the input end.

One type of glass fiber, (Fig. 5-46 right) called selfoc has no discrete cladding. Instead, it has a refractive index that varies in a continuous fashion from a maximum at the center of the fiber to a minimum at the

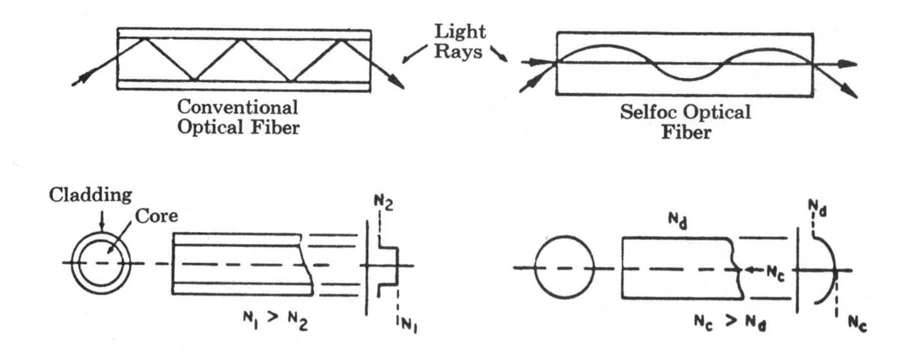

Fig. 5-46 The conventional optical fiber has a core surrounded by a cladding of lower refractive index. The Selfoc fiber has a refractive index of continuous variation, with a maximum at the center and minimum on the outside.

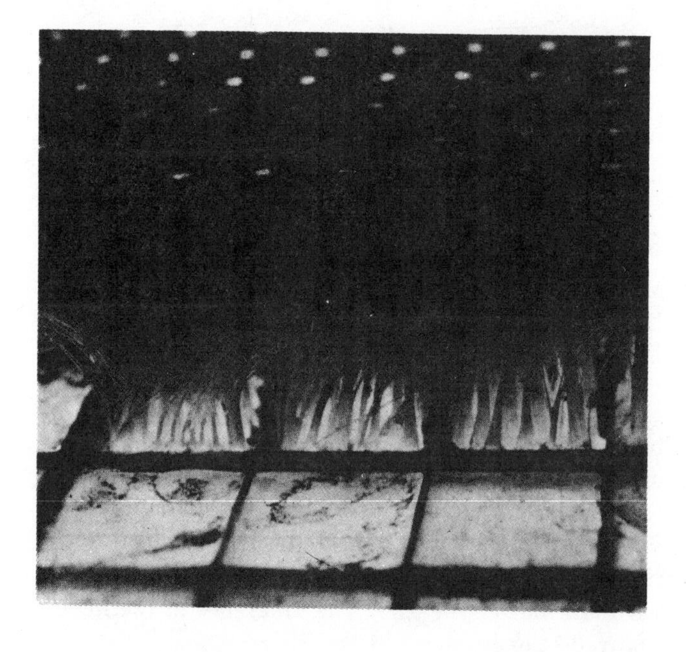

Fiber optic bundles in a memory carry light from gallium arsenide LED's to photosensors exposed to the bit pattern in a photographic mask. The fiber bundles are potted in position over photosensors.

circumference.

If an incident ray is applied in parallel to the optical axis, it passes along this axis. Off-axis rays advance down the fiber in a sinusoidal path.

The optical attenuation of plastic fibers is much greater than that for regular glass in the near-infrared, at 9 μm, where the radiation of gallium arsenide light-emitting diodes is centered.

Since the core glass has a higher index than the cladding, light trapped in the core bounces back and forth from wall to wall being totally reflected at grazing angles. The steeper the angle, the longer the zigzag path and the travel time down the guide; thus light signals propagating along different paths experience quite different delays and can cause substantial delay distortion in the message. The obvious corrective measure is to restrict the propagation to very shallow angles, for example, by curtailing the index difference between core and cladding. The wave nature of light helps in this effort insofar as it prohibits, through destructive interference, all but a few propagation angles associated with particular mode field patterns. The smaller the core, the fewer the modes thus permitted. If the core is only 3 to 4 wavelengths in diameter and the index difference a fraction of a percent, only one mode can propagate. Obviously, this condition completely avoids mode delay differences.

Fig. 5-48 illustrates the field pattern of the dominant mode in the cylindrical dielectric Waveguide designated HE_{11}. For a given index difference, here 1%, there is an optimum core diameter which best confines the field without permitting other modes to propogate. Unfortunately, this diameter is rather small, 3 wavelengths, and the tolerances associated with these minute dimensions complicate large-scale fabrication and field splicing. True, there is a possible trade-off between core diameter and index difference; reducing the index difference by a factor of 10, say, permits an increase of the core diameter by the square root of 10. On the other hand, the resulting 0.1% index difference is hardly capable of steering the wave around practical bends of a few centimeter radius.[5] So little can be gained in this direction.

ALTERNATIVE FIBER CONFIGURATIONS

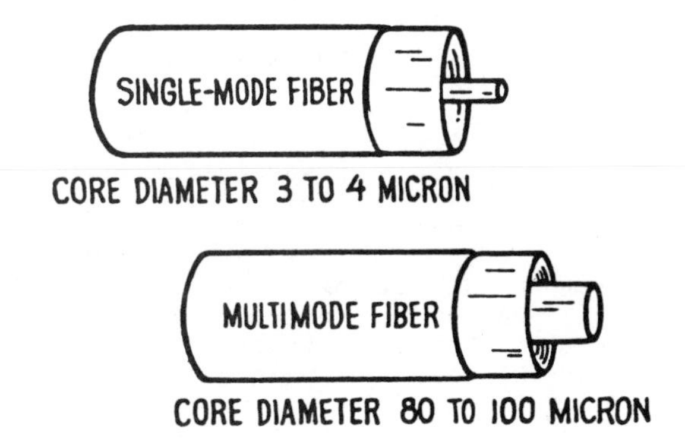

Fig.5-47

The two classical fiber structures.

MODE FIELDS AT DIFFERENT CORE DIAMETERS

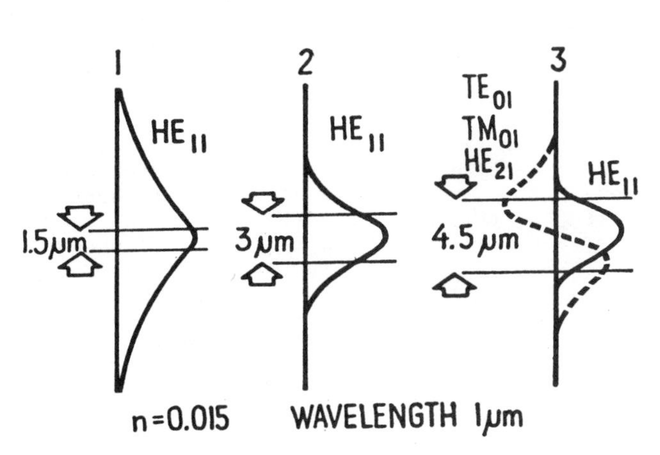

Fig.5-48 Field distribution in single mode fibers.

Coming back to to Fig.5-48: for cores small compared to the wavelength, the dominant mode propagates mostly in the cladding. In larger cores the mode is almost entirely concentrated in the core. Or, turning things around: if the wavelength decreases for a given core size, the mode field shifts from the cladding to the core. This transition is associated with a change in group velocity or delay as shown in Fig.5-49. It starts out with the plane-wave cladding delay and passing through a maximum finally approaches the plane-wave delay of the core. Because it is a function of the wavelength, this change in delay represents a true dispersion effect: an optical signal pulse consisting of a spectrum of optical components will broaden as a result of this. Take, for example, a typical presently available pulsed GaAs laser; its relative bandwidth is about 0.1% corresponding to a change of 0.003 in the diameter-to-wavelength ratio. This is shown in the blowup of Fig.5-49. If there were no change in index other than that resulting from the mode characteristic, the pulse spread would be pretty small, 10 ps/km or so. However, for a Silica fiber, for example, the plane-wave delay itself, either in the core or the cladding, is an even stronger function of the wavelength. The blowup in Fig.5-49 attempts to illustrate that. It shows the delay per unit length plotted versus the normalized frequency. The resulting total spread of a GaAs pulse[6] comes to about 100 ps/km. Of course, this is still fairly small, but let us remember this number for a later comparison

The present technology of glass purification is best developed in the case of synthetic silica. In spite of its high softening temperature, silica is therefore one of the most promising fiber materials.[1,7] One problem with silica, though, is the formation of an index step from core to cladding. Most dopants increase the refractive index of silica, so that they must be added in the core region, where as we have seen the optical wave intesity is high and losses associated with the doping process is most critical. A noteworthy exception is borosilicate, which when suitably quenched has a refractive index almost one hundredth below that of silica.[8] Hence, a cladding material of this kind permits the use of pure silica in the core.

SINGLE-MODE FIBER MADE FROM PURE SILICA

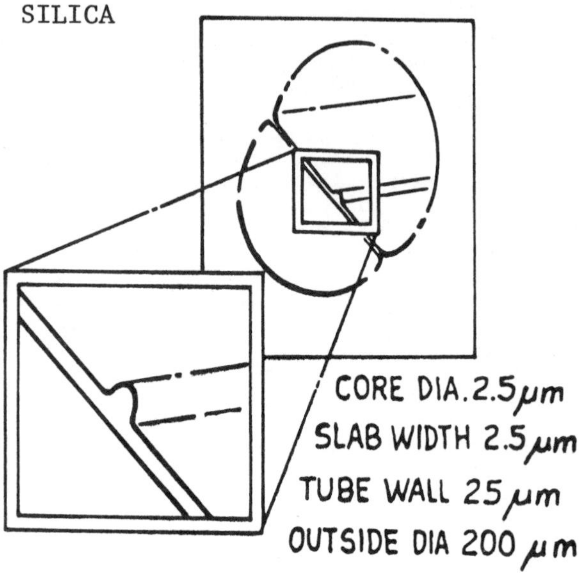

CORE DIA. 2.5 μm
SLAB WIDTH 2.5 μm
TUBE WALL 25 μm
OUTSIDE DIA 200 μm

Fig.5-50 Single material fiber.

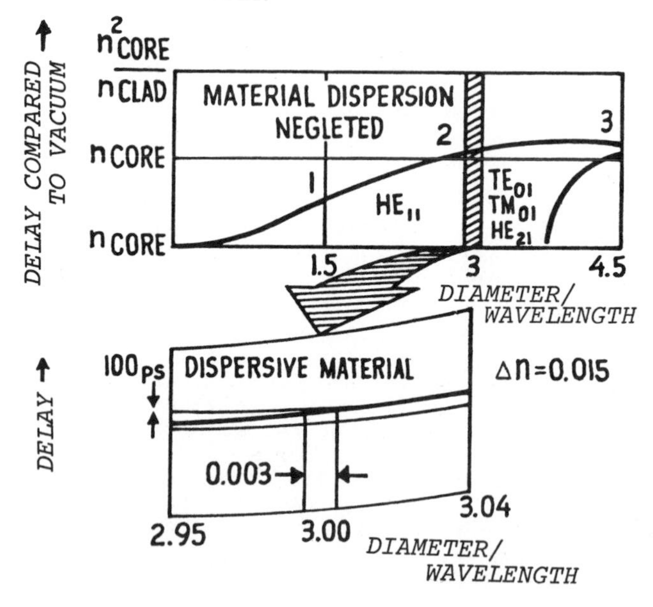

Fig.5-49. Group delay in single mode fiber.

Fig.5-50 illustates another particularly elegant solution.[9] By achieving guidance without an index step, it does away with any additives. The fiber is made from pure silica exclusively. It consists of a core, suspended on a slab in the center of a capillary tube. The tube has the sole purpose of providing strength and protection. The ratio between wavelength and slab thickness determines which core modes (or bouncing core rays)leak through the slab and which are trapped (or totally reflected) in the core. A 2.5-μm slab, for example, restricts propagating modes to angles below 8°. In this way, the slab assumes the function assigned to the core-cladding index difference in the case of the classical fiber structure: It curtails the propogation directions to the cone of angles desired. Note that this cone is independent of the index step between the core and the surrounding air. As far as single-mode operation is concerned, the same arguments hold as in the case of the classical fiber structure: If, in addition to the angle, the core diameter is kept below a certain limit, only one mode propagates. The fiber of Fig.5-50 was designed to meet precisely this condition, but obviously the general concept is equally suited for multimode operation.

Let us return now to a comparative discussion of the three basic modes of operation: single-mode, flat-index multimode and graded-index multimode. We have learned that single-mode operation eliminates the mode delay problem and reduces delay distortion to something like 0.1 ns/km for typical GaAs lasers, but that it does so at the expense of convenient fabrication and splicing tolerances. Because it is severely limited in core size and angle, it is inadequate for operation by an incoherent source like the luminescent diode. It is clear that if a similar or just adequate signal performance could be achieved by multimode operation the latter should be preferable. What we have in mind is a fiber that has a convenient core size and ample index difference, that accepts and transmits thousands of modes. A convenient core measures 50 to 100 μm in diameter. It turns out that, with these dimensions, one can almost forget about the discrete nature of the mode fields and their characteristic angles, since these angles fall so close together. Fig.5-51 illustrates a further significance of these angles using a two-dimensional model for simplicity. As a mode leaves the core refracted at the end face, it splits into two initially plane waves which separate to illuminate two distinct spots in the far-field plane. If n is the core index, the size of the outside angle is n times the inside angle. A relative core-cladding index difference Δ limits the outside angle to approximately $n(2\Delta)^{1/2}$. For Δ = 1%, for example, this value is 0.1. It is called the numerical aperture of the fiber. It represents a measure of the cone of light accepted from the luminescent diode. Naturally, one would like to choose Δ large to maximize this angle. On the other hand, a simple inspection of the propagation times of different rays or waves reveals that the (group) delay differences increase as ΔT, where T= 5 μs/km is the propagation time of a plane wave in the fiber core. For Δ = 1%, consequently ΔT = 50 ns/km.

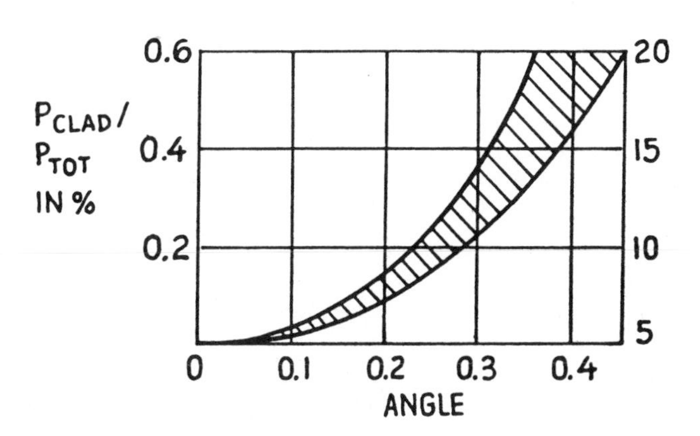

Fig.5-52 Attenuation versus mode angle.

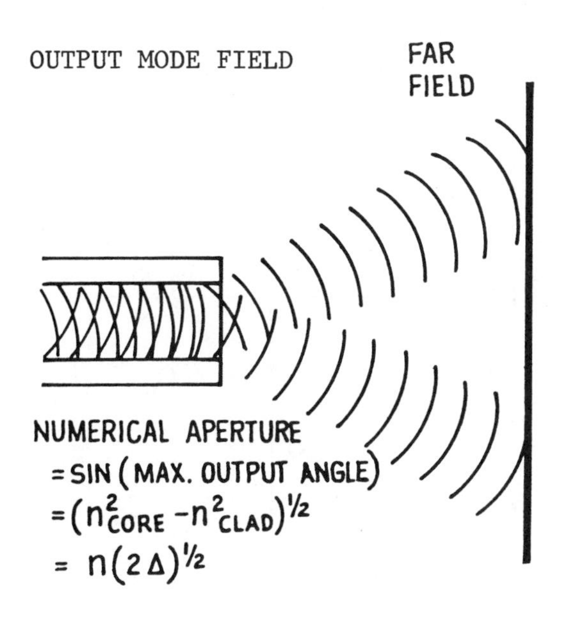

Fig.5-51 High order mode field.

Fortunately the delays measured are in general smaller than that. One reason is illustrated in Fig.5-52 by way of a typical example. This figure shows on the left side the power fraction of each mode that propagates in the cladding plotted versus the characteristic mode angle. This power increases roughly parabolically. Plotted on the right side is the loss of each mode as a result of a cladding loss of 2500 dB/km while the core is assumed to have a bulk loss of only 5 dB/km. In spite of the large cladding loss, small angles show a minimal increase over the core loss, but the loss for modes of larger angle is so high that they are lost along the transmission path and do not contribute to the delay distortion at the end. Of course, this effect also reduces the effective numerical aperture and therefore represents a dubious improvement of the fiber properties in general.

Imperfections and bends in the fiber tend to couple power among the modes so that there is a power flow from the lower to the higher modes, where some of it is lost (Fig.5-53). After a certain distance of propagation, this phenomenon produces a dynamic equilibrium between outflow and loss, associated with a steady-state power distribution among the modes. The distance within which the steady state is reached is called the coupling length. After this distance the overall loss approaches a fixed value.

Fig.5-54 illustrated the pulse distortion under such conditions. Since the light is switched back and forth between the modes, it takes some average propagation time to arrive at the end. The power distribution at the end is therefore centered around an average delay; it has an r.m.s. width which is proportional to the square root of the coupling length L and the fiber length Z. This square-root dependence can result in substantial signal improvements over the uncoupled case.[10]

STEADY STATE BUILD-UP

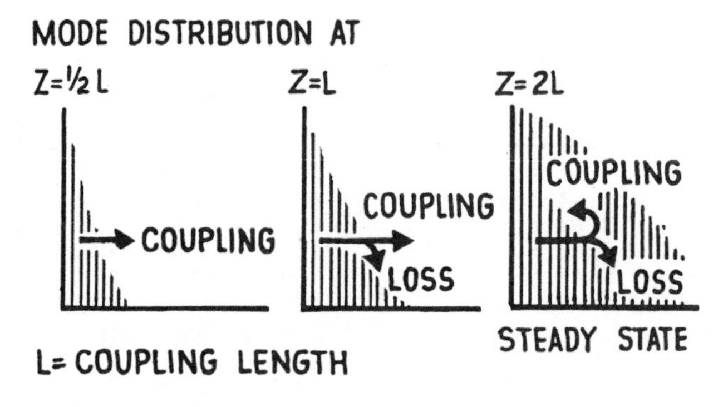

Fig.5-53.

Steady-state buildup for narrow-angle excitation.

Fig.5-55 shows pulse width measurements at three different wavelengths in 1 km of a fiber made by Corning Glass Works.[11] The broadening beyond 600 m clearly approaches a square-root dependence on the fiber length. For best results this concept of equalization

PULSE EVOLUTION AS A RESULT OF MODE COUPLING

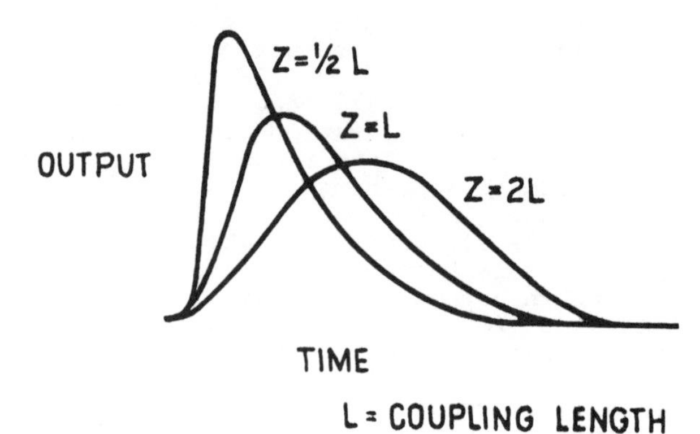

Fig.5-54. Pulse distortion in the presence of mode mixing.

by random coupling requires maximum interaction among low-loss modes and minimum interaction between these and lossy modes. This is a condition which may not be easy to optimize in practical multimode fibers.

If the core index is accessible to modification, either by doping silica or by using other glasses or glass mixtures, there is another possibility of equalizing signal transmission in a multimode fiber.[12] Fig.5-56 illustrates the concept. Instead of a step change, the refractive index of the fiber shown in Fig.5-56 exhibits a gentle grade until

INCREASE IN PULSE WIDTH VERSUS LENGTH

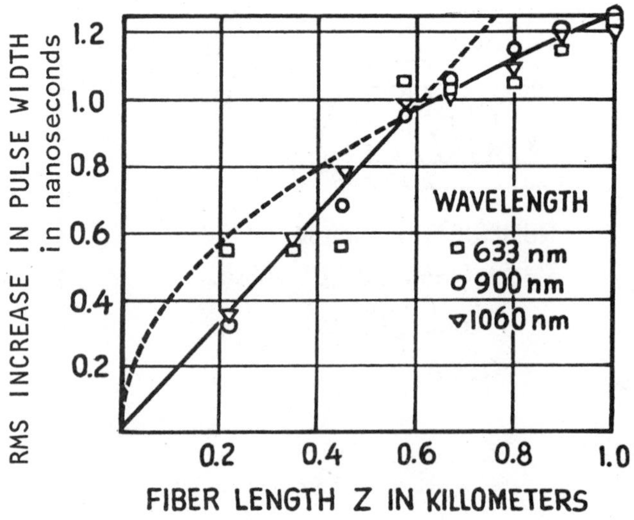

RMS INCREASE IN PULSE WIDTH in nanoseconds

FIBER LENGTH Z IN KILLOMETERS

WAVELENGTH
□ 633 nm
○ 900 nm
▽ 1060 nm

Fig.5-55 Pulse width versus fiber length (after Chinnock et al.[11]).

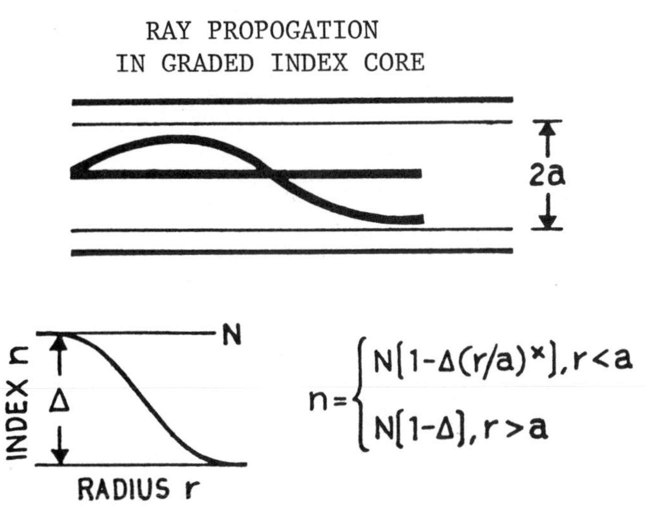

Fig.5-56

Ray propagation in graded-index fiber.

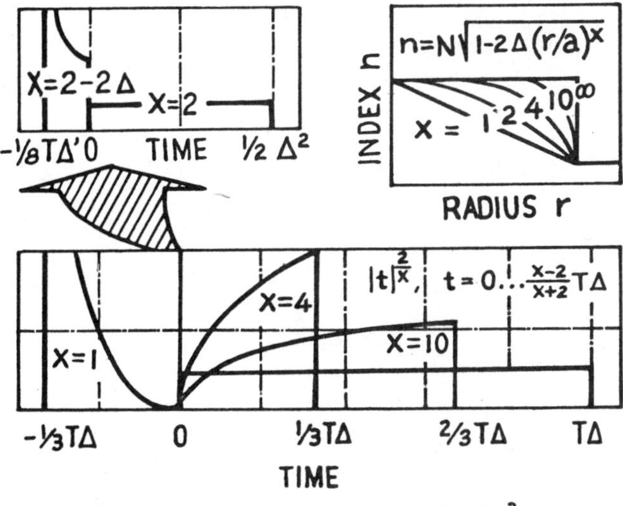

Fig.5-57.

Impulse response of graded-index fibers (after Gloge and Marcatili[13])

it reaches the cladding value. Rays deviating from the axial direction propagate into the regions of lower index, where their higher speed compensates for the larger distance they must travel.

To study the expected mode velocity equalization in such fibers,[13] let us make the following assumptions: The core index profile is circular symmetric. We inject narrow pulses of unit energy into all modes. The modes propagate without coupling and with equal loss. The number of modes is so large that individual mode pulses are not resolved at the output, but form a continuous response which we shall call the impulse response. We can describe the index profile in the core as a power x of the radius; typical profiles of this kind are shown in the upper right of Fig.5-57. The relative index difference Δ between center and cladding index is small, a few percent or so.

With these assumptions we find the respective impulse responses in the bottom graph of Fig.5-57. T is the propogation time of a light pulse in the fiber: about 5 μs/km. If the exponent x equals infinity, the profile is that of the uniform core with an abrupt index step. In this case, the impulse is rectangular and of width ΔT. For $\Delta = 2\%$, for example, $\Delta T = 100$ ns/km. A rounding of the profile shoulder results in a substantial narrowing of the impulse response as shown for the cases $x = 10$ and $x = 4$. For $x < 2$, the high-order modes tend to overtake the lower orders and arrive earlier. If x is in the vicinity of 2, all modes arrive at or near time "zero"; the impulse response resulting, for example, for $x = 2$ is illustrated in the blowup on the upper left. Its width is $\Delta^2 T/2$. If $\Delta = 2\%$, this width is 1 ns. For $x = 2 - 2\Delta$, the impulse response is four times narrower yet. As a matter of comparison, let us calculate for what index difference this width equals the bradening of a GaAs pulse in the single-mode fiber. We set $1/8\ \Delta^2 T$ equal to 0.1 ns/km and, solving for Δ, find $\Delta = 1.26\%$. We can make the core of the graded-index fiber as large as we want. Let us assume it is 100 μm in diameter. In this case, this fiber transmits close to 1000 modes. Of course, each of these modes is also subjected to the same dispersive broadening we found for the fundamental mode in the single-mode case. Hence, even the best graded-index fiber is always somewhat worse than the single-mode structure. Apart from this, it remains to be seen how accurately one can technically prepare a profile with $x = 2 - 2\Delta$. A deviation of the index of only 2 parts in 10,000 from the optimal distribution broadens the impulse response from 0.1 ns/km to 1 ns/km. On the other hand, pulse transmission measurements in graded-index fibers have shown encouraging results.

REFERENCES

1. R.D. Maurer, Glass Fibers for Optical Communications, Proc. IEEE, Vol. 61, pp. 452-462, April, 1973.

2. S.E. Miller, E.A.J. Marcatili and T. Li, Research Toward Optical Fiber Transmission Systems, Proc. IEEE Vol. 61, pp. 1703-1751, December, 1973.

3. D. Hondros and P. Debye, Electromagnetische Wellen and Dielektrischen Drahten, Ann. Phys., Vol. 32, pp. 465-476, 1910.

4. M.M. Ramsey, Fiber Optical Communications Within the United Kingdom, Opt-Electronics, Vol. 5, pp. 261-288, July, 1973

5. E. A. J. Marcatili and S. E. Miller, "Improved Relations Describing Directional Control in Electromagnetic Wave Guidance", Bell System Technical Journal, Vol. 48, pp. 2161-2187, September, 1969.

6. D. Gloge, Dispersion in Weakly Guiding Fibers, Appl. Opt., Vol. 10, pp. 2442-2445, November, 1971.

7. D. B. Keck, R. D. Maurer and P. C. Schultz, Appl. Phys. Lett., Vol. 22, pp. 307-309, April, 1973.

8. L. G. Van Uitert, D. A. Pinnow, D. C. Williams, T. C. Rich, R. E. Jaeger and W. H. Grodkiewicr, Borosilicate Glasses for Fiber Optical Waveguides, Mat. Res. Bull., Vol. 8, pp. 469-476, 1973.

9. P. Kaiser, E.A.J. Marcatili and S.E. Miller, A New Optical Fiber, Bell System Technical Journal, Vol. 52, pp. 265-269, February, 1973.

10. S. D. Personick, Time Dispersion in Dielectric Waveguides, Bell System Technical Journal, Vol. 50, pp. 843-859, March, 1971.

11. E. L. Chinnock, L. G. Cohen, W. S. Holden, R. D. Standley and D. B. Keck, The Length Dependence of Pulse Spreading in the CGW-Bell 10 Optical Fiber, Proc. IEEE, Vol. 61, pp. 1499-1500, October, 1973.

12. S. E. Miller, Light Propagation in Generalized Lens Like Media, Bell System Technical Journal, Vol. 44, pp. 2017-2064, November, 1965.

13. D. Gloge and E. A. J. Marcatili, Multimode Theory of Graded-Core Fibers, Bell System Technical Journal, Vol. 52, pp. 1563-1578, November, 1973.

FIBER OPTICS GLOSSARY

Absorption Loss: The loss of optical flux or energy caused by impurities in the transmission medium as well as intrinsic material absorption. Expressed in dB/km.

Acceptance Angle: The solid angle within which all incident light rays will enter the core of an optical fiber. Expressed in degrees.

Acceptance Cone: A cone with an included angle twice that of the acceptance angle.

Bundle: A collection of glass or plastic fibers that transmit data in the form of optical energy.

Cladding: The outer portion of an optical fiber having an index of refraction lower than that of the core.

Core: The center portion of an optical fiber. It is the light transmitting medium.

Coupling Loss: Attenuation of the optical signal due to coupling inefficiencies between the flux source and the optical fiber, or between fibers, or between the fiber and the detector in a receiver. Expressed in dB.

Dispersion: The undesirable effect of the broadening of optical pulses caused by lengthening of rise and fall times as the pulse travels along the fiber. Sometimes referred to as "pulse spreading", it results from either modal or material effects in the fiber that reduce bandwidth. Expressed in ns/km.

Fiber: A clear glass or plastic optical "cable", consisting of a core and cladding, designed to propagate optical energy. The diameter of a fiber can vary from about 10 to 1,000 μm, depending on type.

Flux: The rate of energy flow passing to, from, or through a surface or other geometric entity. Radiant flux is expressed in watts. Luminous flux is expressed in lumens. Flux is sometimes erroneously referred to as optical power.

Graded-Index Fiber: An optical fiber made with a refractive index that gets progressively lower as the diameter increases. See **step-index fiber.**

Index of Refraction: The physical property of a material that describes the behavior of optical energy passing through it. It is defined as the ratio of the velocity of light in a vacuum to the velocity of light in the material.

IRED: A diode that emits photons in the infrared spectrum when forward biased.

Modes: Each different path that a light ray can take to travel down an optical fiber.

Multimode Fiber: An optical fiber that propagates optical energy in more than one mode.

Numerical Aperture: A measure of the light-gathering capability of a fiber. Mathematically, it is expressed as the sine of the acceptance angle.

Optical Port: An opening through which optical energy can pass.

Refraction: The deflection from a straight path undergone by a light ray passing from one medium into another having a different index of refraction.

Responsivity: A measure of how much output current can be obtained from a photodetector for a given optical energy input. Expressed in A/W.

Single-Mode Fiber: An optical fiber that allows propagation of optical energy in only one mode.

Step-Index Fiber: An optical fiber made with an abrupt change in refractive index at the core-to-cladding interface. See **graded-index fiber.**

CHAPTER 6

OXIDATION

Oxidation is the cornerstone of the planar process which in turn is the foundation of silicon integrated device technology. Initially the role of the oxide was that of a mask during diffusion and a protective, passivating layer following fabrication. More recently oxide layers have been used as elements in active and passive devices on silicon functional blocks; this has led to an intensive re-examination of the oxidation process and the properties of the resulting oxide.

When silicon is exposed to an oxidizing gas, such as oxygen or water vapor, at an elevated temperature, an oxide forms on the surface. This oxide is a glassy amorphous form of quartz. Its composition is SiO_2, it possesses no crystalline structure, and its softening point is about 1500° C (which is about 70 degrees higher than the melting point of silicon). This oxide layer is unreactive to most acids, metals, and chlorine and bromine, but is attacked by hot alkaline solutions, cold hydrofluoric acid, hydrogen chloride and hydrogen fluoride at high temperature.

A film only a few atomic layers thick will passivate the surface of a silicon wafer, and render it inert to the effects of various chemical environments. After etching a silicon wafer in a mixture of nitric and hydrofluoric acids, a layer 8 angstroms thick remains on the surface. A treatment in boiling water or hot nitric acid leaves an oxide layer 23 angstroms thick. Such oxides are sufficiently thick to present difficulty to attempts to make electrical contacts to the underlying silicon.

Wet chemical etchants for silicon always contain an oxidizing agent (usually nitric acid) and an oxide etchant (hydrofluoric acid). Etching proceeds through a mechanism involving the oxidation of the silicon (in solution) followed by the dissolution of the newly formed oxide (in solution), a cycle which is repeated until the wafer is removed from solution. Upon drying, the wafer is always left with an oxide film.

Most oxides used as dopants for diffusion, form solutions with SiO_2. At elevated temperatures, boron oxide (B_2O) and phosphorus oxide (P_2O) will dissolve in SiO_2. If the SiO_2 layer is sufficiently thin, it is quickly saturated, and the dopant reaches the silicon surface rapidly. Diffusion into the silicon, then takes place. If a silicon wafer is heated to high temperature (900-1300° C) in the presence of a vaporized *elemental* dopant, a diffusion of the dopant into the semiconductor occurs. However, the silicon may become pitted as a result of chemical reaction with the dopant element, or it may alloy with the dopant, forming small pools on the wafer surface. Both of these effects tend to mitigate against uniformity of diffusion, and, as a result, a ragged junction of varying depth is obtained.

The preceding effects can be minimized if a thin oxide is first formed on the surface of the wafer. Due to the homogeneous nature of the resulting glass, which may be quite liquid at the diffusion temperature, localized alloy formation at the surface of the wafer does not occur. Thus, the diffusion front advances into the wafer at a uniform rate, giving a smooth junction. In addition, the concentration of the dopant in the wafer can be controlled by controlling the composition of the glass on the surface.

6.1 PROPERTIES OF SILICA GLASS

Silica glass has been described as one of twenty-two phases of silica. Thermodynamically, it is unstable below 1710° C, and should devitrify to a stable crystalline form. However, the rate of devitrification is negligible at temperatures beneath 1000° C, so that room temperature silica glass appears quite stable. Even at higher temperatures, transitions among the various crystal forms of silica (such as quartz, cristobalite, and tridymite) are so slow, that the term "sluggish" transformations has been used to describe the low rates. It may be necessary to maintain the temperature within the region of greatest stability of the desired phase for several days, before the conversion approaches completion. The melting of quartz or cristobalite is also very sluggish, and the retention of some crystalline regions in the manufacture of silica glass is difficult to avoid.

The properties of silica glass are sensitive to impurity content and structure. Nearly pure silica glass with density 2.2 gm/cm^3 is formed during the fabrication of silicon devices, but traces of impurities such as boron, phosphorus and aluminum are usually found in the oxide film following certain processes. The properties of each film are characteristic of the particular way in which it was formed and the processes to which it has been exposed. Accordingly, wide variations in the chemical and physical properties of the oxides on silicon, may be found.

6.1.2 STRUCTURE OF SILICA GLASS

Modern ideas of the structure of silica glass depict the amorphous silica structure as a random three-dimensional network, consisting of silicon-oxygen tetrahedra which are joined only at the corners and share no faces or edges. Each silicon atom in the network forms the center of the tetrahedron, the vertices of which are defined by the four associated oxygen atoms which themselves are shared between two silicon atoms. The average Si-O distance is 1.62 Å; the average 0-0 distance—a side of the tetrahedron – is 2.65 Å; the average Si-Si distance is 3.00 Å. The Si-O-Si angle is 143° ±17°. The tetrahedra of the silica glass network are similar to those of the quartz crystal structure, but form a random network instead of a regular lattice.

Even in silica glass, however, many of the tetrahedra cluster in the six-sided ring pattern characteristic of crystalline quartz. Contemporary thinking is that the vitreous state need not consist of a completely random network. Considerable short range order—even regions of crystallinity 10 to 100 Å in size—may exist. Thus, the terms "silica lattice," "vacancies" and "interstitial" have been used in describing the structure of glasses.

The loose, disordered networks of silica glass permit non-uniformities and holes to exist. Consequently the density of silica glass (2.2 gms/cm^3) is less than that of quartz (2.65 gms/cm^3). The constituent "molecules" of silica glass are clearly not structurally equivalent; the energy required to break the various "molecular" bonds differs so that, unlike quartz, silica glass has no sharp melting point. While all the oxygen atoms are shared by two silicon atoms in pure silica, and are called bridging oxygen atoms, some of the oxygen atoms in silica glass are associated with only one silicon atom and are called non-bridging. To preserve the SiO_2 ratio and electrical neutrality in the presence of non-bridging oxygen atoms, either an adjoining silicon atom is attached to the lattice by only two bridging oxygens, or an interstitial impurity cation is present, making the silica glass extrinsic.

It is of interest to note the relative sizes of the silicon and oxygen atoms. An approximate picture is that of grapefruit and basketballs. The grapefruit (relative size of silicon atoms) are surrounded by four basketballs (relative size of oxygen atoms). The silicon atoms cannot move without breaking four oxygen bonds; but bridging oxygen atoms are attached to only two silicon atoms by one electron each. Non-bridging oxygen atoms are attached to only one silicon atom. The oxygen atoms are thus freer to move around the lattice.

When elements other than silicon and oxygen are found in the network, the silica glass is said to be extrinsic. Electrically neutral elements may merely occupy holes in the network, but most impurity atoms are largely ionized and play an active role in determining the properties of silica glass. The more important atoms are the impurity cations (they are virtually completely ionized) which have been divided into two types—the network formers and the network modifiers. Some impurities, notably aluminum, can play both roles.

The network formers are capable of forming glasses by themselves and, in silica glass, can substitute for silicon in forming the network. Boron, phosphorus, and aluminum are the important network forming impurities and have small radii like silicon. The valence of these network formers is different from that of silicon. Boron, for example, has a valence of three and in B_2O_3 is surrounded by only three oxygen atoms. In silica, however, the boron cations will change to a coordination number (the number of nearest neighbor oxygen atoms) of four, creating a charge defect. Phosphorus will change from a coordination number of five to four. This creation of oxygen deficiencies or excesses in the silica glass network is similar to the role of donors and acceptors, introduced as substitutional impurities, into the semiconductor crystal lattice.

The network modifying cations are large cations such as sodium, potassium, lead, calcium and barium, and enter the silica glass network interstitially. Aluminum can also be in this group. A network modifier when introduced as an oxide, ionizes, giving up an oxygen atom to the network. The metal atom occupies an interstitial position in the network and the oxygen atom enters the network, producing two non-bridging oxygens where formerly there was one bridging oxygen. The effect of this action is to weaken the network. This is manifested by the lowered melting points of common extrinsic silica glasses such as sodium and lead glasses and by other variations in properties.

Particular extrinsic silica glasses which either occur in silicon device processing or are useful in silicon integrated devices include:

a) Borosilicate glass—formed during boron diffusion into silicon
b) Phosphorus glass—formed during phosphorus diffusion into silicon
c) Lead glass—formed in the low temperature process of "accelerated" oxidation
d) Aluminosilicate glass—studied as a dielectric for integrated circuit capacitors; may also result from heating an aluminum film deposited on silica glass
e) Wet silica glass—formed during steam oxidation, diffusion from an impurity oxide, or from heating silica glass in hydrogen.

6.1.3 VOLUME CONDUCTIVITY

Volume conductivity results from the mobility of electrons, holes, or ions. Ordinary metallic conductivity, which results from electron drift, is greater than 10^2 (ohm-cm)$^{-1}$ and decreases with increasing temperature. Semiconductors, in which both holes and electrons are mobile, have a conductivity from 10^{-5} to 10^3 (ohm-cm)$^{-1}$ which increases with temperature. Insulators generally are in the conductivity range below 10^{-16} (ohm-cm)$^{-1}$, have a positive temperature coefficient, and are primarily ionic conductors. Silica glasses are in the class of good insulators, conduction resulting from ion motion, although semiconducting and conducting silica glass compositions are known. Pure silica glass has a conductivity of less than 10^{-18} (ohm-cm)$^{-1}$ which can be either increased or decreased by the addition of impurities.

When a dc electric field is applied to silica glass, the current decreases with time for about an hour, depending on temperature and sample geometry. This transient is the polarization phenomenon which is characteristic of ionic conductors and can be used to distinguish them from electronic conductors. It results from the accumulation of positive sodium ions at the negative electrode which decreases the number of available current carriers.

Silicon glasses containing transition metal oxides have electronic conduction and a negative temperature coefficient of resistance. These semiconducting glasses have low carrier mobilities which results in conductivity of 10^{-6} (ohm-cm)$^{-1}$ at 350° C.

It has been observed that silica glass has a constant conductivity for low electric fields, but as the field increases the conductivity increases, probably due to impurity impact ionization.

6.2 METHODS OF OXIDE FORMATION

The term ''thermal oxides'' refers to those oxides formed from a thermally activated reaction of silicon with oxygen, water, or other oxygen bearing species. This definition includes not only the open tube thermal oxidation of silicon in steam or oxygen, but also the specialized techniques of high pressure oxidation and ''accelerated'' oxidation.

Anodic oxides are those oxides formed in a gaseous or liquid medium by the electric-field induced transport of mobile ions. Deposited oxides may be formed by the addition of appropriate layers to the surface, either by vacuum deposition or from a vapor.

The surface preparation of the silicon substrate is important in achieving amorphous oxide films. All oxide films described in this section, are assumed to have been grown or deposited on silicon surfaces prepared by chemical, mechanical, or other suitable surface polishing techniques.

6.2.1 STEAM OXIDATION OF SILICON

The reaction of high temperature water vapor with silicon is called steam oxidation, only when the quantity of vapor present does not limit the oxidation rate. The oxide formed by this reaction (as well as the oxides formed by all other methods to be described) is protective in that its presence inhibits further contact between the reactants. Several monolayers of oxide are formed by chemisorption of water. Further reaction requires the transport of one of the reactants through the already existing oxide. Steam oxidations are conventionally performed in either an open tube system or a high pressure bomb.

6.2.1.1 Open Tube Steam Oxidation

The apparatus for performing open tube steam oxidations is very similar to that used in open tube diffusions. For steam oxidation the original source is a flask of high purity water which is heated to a temperature that is sufficient to generate a flow of water vapor through a furnace tube as indicated in Fig. 6-1. The water bath temperature provides a measure of the partial pressure of the water vapor above the bath, which in turn is related to the flow of water vapor through the furnace. The lines leading from the flask to the furnace should be heated to avoid condensation.

Typical water bath temperatures are around 102°C. With high purity water the absence of condensation nuclei prevents the appearance of the large bubbles that characterize boiling water. While glass beads or other particles can also be used to prevent the formation of large vapor bubbles, they also introduce an unnecessary source of contamination.

The water temperature does not have an appreciable effect on the oxidation, as long as it is sufficient to maintain a partial pressure of approximately one atmosphere of water vapor, in the vicinity of the silicon. Higher temperatures increase the flow of water vapor past the silicon, but do not alter the water vapor concentration appreciably.

The relationship between oxide thickness, time and temperature, is shown in Fig. 6-2. The low temperature departure from parabolic growth can be clearly seen in the 900°C and lower curves. Departures from parabolic growth can also be seen at small values of t (less than 15 min) on the 900°C-1100°C curves, where the thickness of the oxide grown is comparable to that of the space charge region associated with the ion motion giving rise to its growth. Not until the film becomes thicker does the parabolic relationship apply. The curves of Fig. 6-2 apply to (111) oriented silicon surfaces with impurity content less than about 3×10^{19} cm^{-3}. Variations from these conditions introduce changes in the observed oxidation rates.

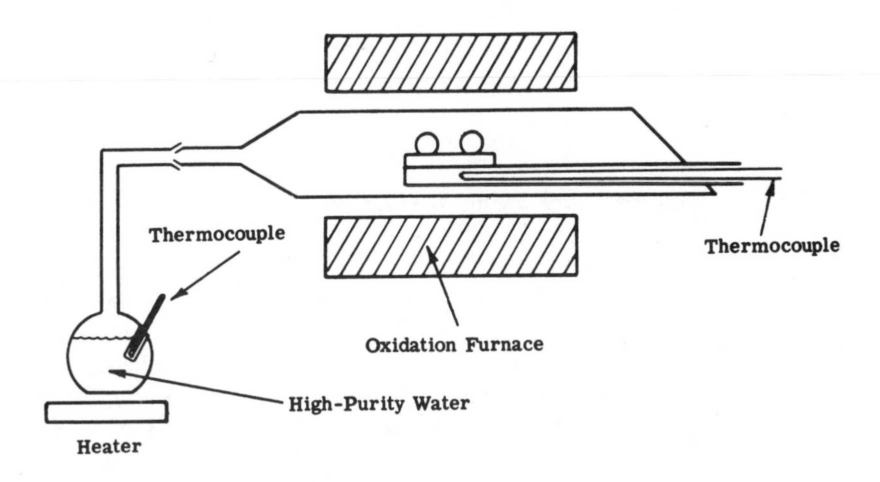

Fig. 6-1 Open tube steam oxidation apparatus

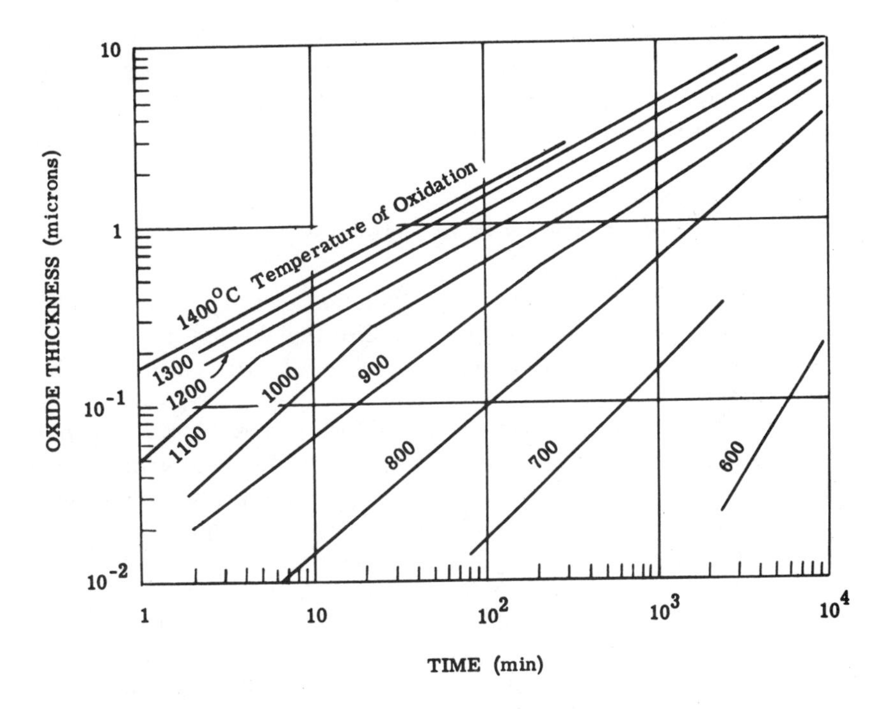

Fig. 6-2 Oxide growth in atmospheric steam

The role of high impurity concentration upon oxidation is pictured as influencing either, the interface reaction of oxidation, or the diffusing coefficient of the oxidizing species, through the oxide grown on the heavily doped silicon. Phosphorus influences primarily the interface reaction. It is rejected by the growing oxide in thermal oxidation so that the amount of phosphorus incorporated into the growing oxide is small (although sufficient for the etch rate of the surface oxide to exceed that of the substrate oxide by about 25%). For boron the reverse is true; the growing oxide incorporates large quantities of boron into its structure, permitting more rapid diffusion of the oxidizing species. Consequently, heavily boron doped silicon oxidizes faster in the high temperature parabolic region of growth, as well as at low temperatures.

6.2.1.2 High Pressure Steam Oxidation

The apparatus for the high pressure steam oxidation of silicon consists of a tight constant-volume enclosure, usually made of metal, into which a pre-determined quantity of high purity water is introduced along with the silicon. The inside of the ''bomb'' is often lined with gold or other inert material to prevent undesired reactions between the water and the walls of the chamber. The entire assembly is heated to the oxidation temperature, and from PVT data for water above the critical temperature, the pressure can be calculated as a function of the temperature. Any drop in steam pressure due to reaction with the silicon is assumed negligible.

The growth of oxide films on silicon in high pressure steam is linear in time and directly proportional to pressure over a certain range of time, temperature and pressure. The solubility of the oxide in steam at pressures of 100 atmospheres and above, causes departures from linearity for long oxidation times at this pressure. At still higher pressures, the silicon surface is etched rather than oxidized which results in a matte surface appearance.

6.2.2 OXIDATION OF SILICON IN DRY OXYGEN

The major difference between steam oxidation and dry oxygen oxidation is that water is no longer the species diffusing through the growing oxide film. The diffusion of some form of oxygen is presumed to be the rate limitation.

In dry oxygen oxidation, the species diffusing through the growing oxide layer is probably oxygen ions. The growth of such an oxide can be accelerated or retarded (and even stopped) by the application of an electric field. Conventional marker and radioactive tracer experiments have shown that the oxide grows at the oxide-silicon interface, rather than the oxide-gas interface.

The apparatus for dry oxygen is illustrated in Fig. 6-3. The drying unit can be a cold trap, chemical desiccant, or other apparatus capable of reducing the moisture content of the oxygen to a dew point of $-60°$ C or less. The filter should remove particles greater than 1/2 micron in size.

6.2.3 WET OXYGEN OXIDATION OF SILICON

Wet oxygen oxidation describes the systems in which the dry tank oxygen is passed through a water bath prior to being introduced into the oxidation furnace. The moisture content of the gas stream is determined by the water bath temperature and the flow rate. Since oxidation proceeds much more rapidly in water vapor than in dry oxygen, the water content of the carrier gas is the most important variable for determining oxide thickness for a given time and temperature. Over most of the range of typical water bath temperatures, the substitution of an inert carrier gas for oxygen does not greatly alter the oxide growth. Heating the lines between the water source and the furnace is again necessary and filtering of the carrier gas is again desirable, although bubbling the gas through the water bath is itself a type of filtering action which removes many of the particles that a submicron filter catches.

A major difference between the wet oxygen system and either the dry oxygen or the steam oxidation system, is that in the wet oxygen system, concentration of the oxidizing species is easily varied. In open tube steam oxidation, the oxidizing species is "water," and its partial pressure is approximately 1 atmosphere; in the dry oxygen system, the oxidizing species is oxygen at a pressure of 1 atmosphere. In the wet oxygen system, however, the oxidizing species is a mixture of oxygen and "water" whose ratio is determined by the mixing of water vapor with oxygen. In principle it is possible to vary the oxidation rate from that of 100% oxygen to that of nearly 100% water vapor. By using argon or nitrogen for the carrier gas and reducing the water bath temperature to about 16° C, the growth rate can actually be controlled to values less than that for dry oxygen. In practice, however, the water bath temperature is not generally cooled below room temperature, and the carrier gas flow is at least 200 cc/min.

6.2.4 ACCELERATED OXIDATION OF SILICON

If the rate limitation of the reaction between silicon and oxygen is determined by the diffusion of an oxidizing species through the previously formed oxide layer, one way to increase the oxidation rate is to create conditions that allow easier, more rapid diffusion of the oxidizing species in the oxide. In a perfect silica crystal, interstitial diffusion is hindered by the tight structure. Substitutional diffusion requires the breaking of two silicon-oxygen bonds which, in turn, requires considerable energy—higher than the activation energy observed for diffusion. Diffusion through voids in the structure is thought to be the dominant mechanism of oxygen diffusion in silica glass. A technique for increasing the number of voids is to introduce large impurity atoms such as lead into the oxide structure.

In addition to disrupting the order of the structure, the lead oxygen bonds are more easily broken so that substitutional diffusion is also enhanced. Regardless of the true mechanism, oxidation is found to proceed much more rapidly in the presence of lead; and hence the term "accelerated" oxidation.

The most successful system for the growth of lead accelerated oxides have been static systems in which the silicon is heated in air along with a source of lead oxide powder or paste. The growth rate depends on the concentration of lead in the growing oxide, as well as on the time and temperature of the silicon.

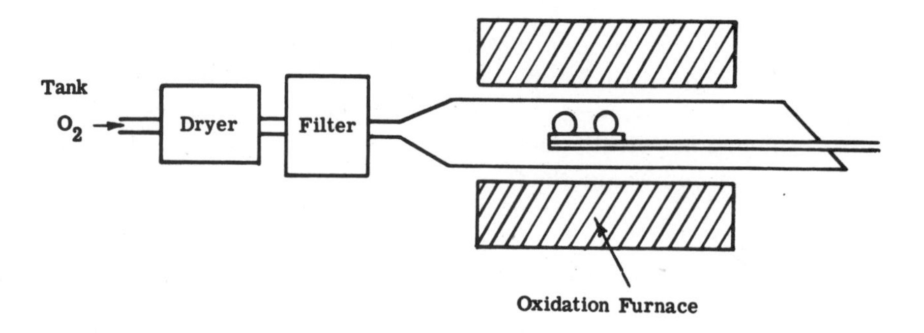

Fig. 6-3 Open tube oxidation in dry oxygen

6.2.5 ANODIC OXIDATION

Anodic oxidation includes both the oxidation of the surface of a silicon electrode in a liquid electrolyte, and the oxidation of silicon in a gaseous plasma. These two oxidation processes are termed "electrolytic anodization" and "gaseous anodization" respectively.

Silicon has a high affinity for electrons, and, in air at room temperature, is covered with a thin oxide film ($\approx$ 30 Å) which separates the reactants so that further growth of the oxide can occur only by transport (either diffusion or migration) of the reacting species through the film. To grow thick oxide films the transport mechanism must be enhanced.

The growth of anodic oxide films on silicon in a liquid electrolyte which does not dissolve the oxide, depends upon an electrostatic field in the oxide which enhances ion migration. The Si ions are the mobile species. Therefore, the growth of the anodic oxide can be described, to a first approximation, by the transport of Si ions across the oxide-silicon interface, through the oxide, and to the oxide-electrolyte interface where an oxidation reaction takes place. The ionic species in the electrolyte at the oxide-electrolyte interface markedly influences the growth rate, the limiting thickness, and the physical and chemical properties of the oxide.

Anodic polarization of n-type silicon below 100 V is affected by the potential barrier at the silicon-oxide interface. Most of the applied voltage is dropped across the depletion region in the silicon during the initial growth of the oxide, so that the field in the oxide which enhances ion transport, is lowered by the presence of the potential barrier. Shining light on the silicon surface increases the minority carrier concentration near the surface and reduces the potential barrier; hence the field in the oxide is higher. As the oxide thickness increases, the potential barrier in the silicon is lowered; and for oxides greater than approximately 400 Å, further growth depends upon the properties of the oxide alone.

Although the interfaces influence the kinetics of growth, the value of the ionic current (and its parameters) through the oxide mainly determine the oxide properties. Regulation of the total current is either at constant current or by constant voltage or a combination of the two.

Design of the electrolytic anodization cell depends upon the precision and reproducibility desired in the formation of the oxide. Also important in obtaining reproducible results, is the choice of electrolyte and its water content. The anodic cell should have uniform spacing between electrodes so that the current density is uniform and conducive to uniform oxide growth. For the current densities normally used, circulation of the solution is required to prevent overheating of the oxide. Such overheating can cause either variations in oxide thickness or breakdown. Temperature control of the system may be required if there is a large voltage drop in the electrolyte.

When anodizing n-type silicon, it is advantageous to control the amount of light which shines on the silicon anode. The use of focused light to define areas of oxide growth on n-type silicon has been shown to be capable of growing oxides with dimensional control on the order of 10 microns. With

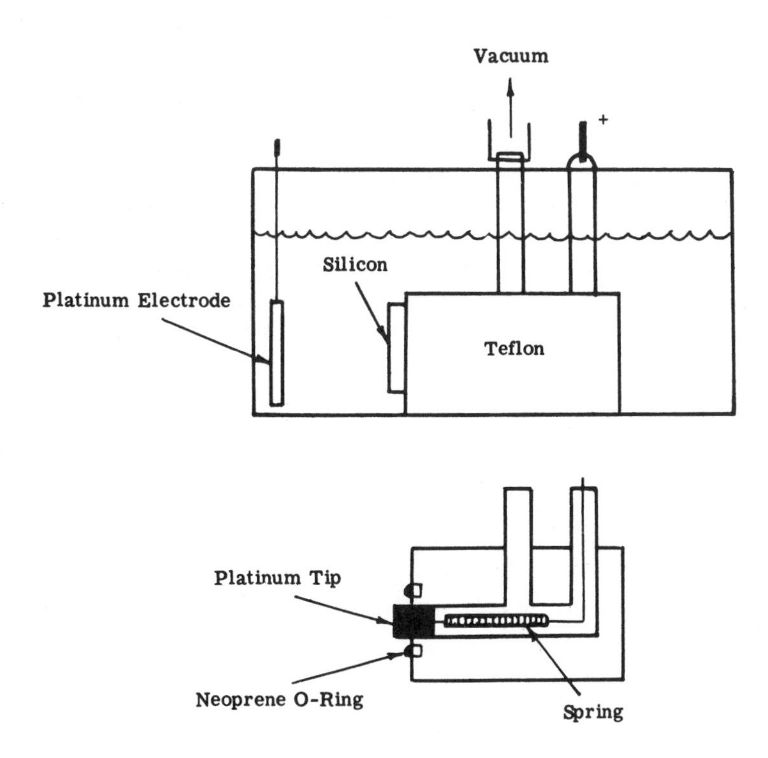

Fig. 6-4 Cell for the anodic oxidation of silicon

oxide thicknesses greater than 1000 Å, this resolution is reduced.

A cell design which can be provided with modifications to meet each of the aforementioned requirements is shown in Fig. 6-4. The main advantages of this cell are that the silicon anode needs no bonded or plated metallic contacts and that selective oxidation of only one side of the silicon wafer is possible. The body of the holder is teflon with machined teflon fittings for the vacuum line and for the insulation of the back contact to the wafer. A 1/4 inch diameter platinum rod 1/4 inch long, and attached to a spring, provides a pressure contact to the back surface of the wafer. A circular groove, machined outside the contact, provides an insert for an O-ring. Using a small vacuum pump, the silicon is held against the O-ring to provide a seal separating the electrolyte and the back contact. Wafers approximately 3/4 inch in diameter can be anodized to 500 volts with a parallel leakage current density (current flow around the wafer rather than through it) less than 140 μamps/cm^2 (determined by dividing the total leakage current by the area of the electrode). The most serious limitation of this cell design is the O-ring seal. The O-ring must not react with the electrolyte, and must not break down at high electric fields.

The single silicon anode cell design shown in Fig. 6-4 can be extended to a multiple anodization system. A flow type system in which the electrolyte is continuously replenished may be constructed, and oxides grown which are comparable to those grown in a static system.

6.2.5.1 Gaseous Anodization

Gaseous anodization of silicon is similar to electrolytic anodization except for the replacement of the liquid with a gas.. An oxygen plasma excited by 2450 M Hz supplies the negative oxygen ions at pressures of 0.4 to 1.5 torr. Negative oxygen ions are extracted from the plasma by a dc silicon anode. With this extracting potential, a parabolic growth rate occurs which depends upon pressure, temperature and the density of the plasma. The parabolic rate increases with an increase in the temperature of the silicon and with a decrease of the oxygen pressure.

6.2.6 DEPOSITED OXIDES

Deposited oxides include all those oxides formed by methods not requiring the silicon substrate to participate in the oxide forming reaction. The role of the silicon is only that of a substrate for the deposition. A subsidiary advantage of this method of preparation is that the substrate can be a material other than silicon so that the results may be applicable to germanium, gallium arsenide, etc.

Three general techniques are commonly employed to deposit oxide films: the first is a silicon compound decomposition called pyrolysis; the other two are reactive sputtering and vacuum evaporation.

6.2.6.1 Pyrolysis

Pyrolytically deposited oxides are prepared from the thermal decomposition of various silicon compounds. A typical reaction of the pyrolytic decomposition of an alkoxysilane at 700° C produces

$$SiO_2 + \text{gaseous organic radicals} + SiO \text{ and } C \text{ depending on the conditions}$$

The oxygen for this reaction must originate with the organic molecule itself rather than be introduced from external sources. Foreign oxygen or moisture in the system causes the deposited oxide to have a cloudy appearance and to etch erratically. The quantity of SiO present in the reaction products, depends upon the amount of oxygen available during the reaction. Less SiO appears with those alkoxysilane molecules containing three or four oxygen atoms. The amount of carbon depends upon furnace temperature and becomes objectionable above 750° C.

An alternative deposition reaction is that representing the oxidation of silane in oxygen:

$$SiH_4(g) + 2O_2(g) \rightleftarrows SiO_2(s) + 2H_2O(g)$$

$$g = \text{gas}, s = \text{solid}$$

This reaction is preferable to that described above in that no gaseous organic radicals are formed, and the temperature required for reaction is considerably lower. Although no external heat is required to catalyze the reaction, better quality films and superior uniformity are observed if the silicon substrate is heated to above 300° C. The apparatus for the pyrolytic decomposition of an alkoxysilane is shown in Fig. 6-5.

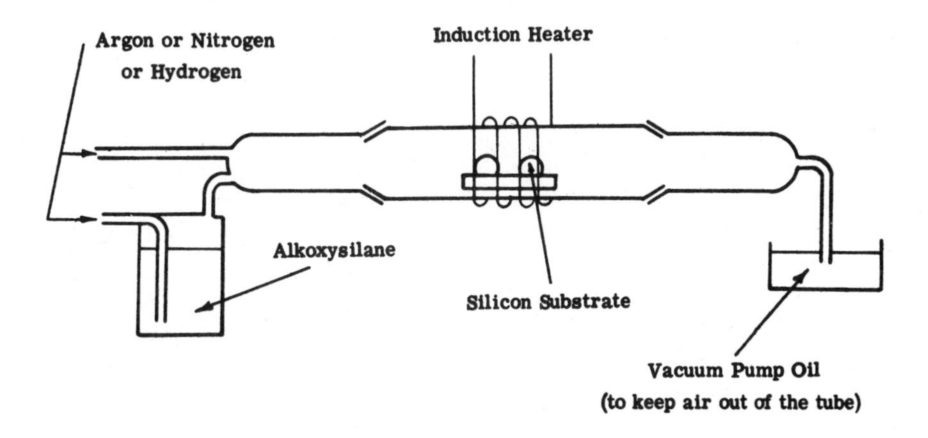

Fig. 6-5 Apparatus for the pyrolytic deposition of silicon oxide

6.2.6.2 Evaporation

There are a variety of methods used to vacuum deposit oxides. The two most common methods depend upon the evaporation of silicon monoxide. In the first method commercial powdered silicon monoxide is used. In the second, a silicon electrode is heated in an oxygen environment, forming a silicon monoxide surface which evaporates more readily due to its higher vapor pressure. The partial pressure of oxygen and the evaporation rate affect the composition of the deposited film. A third technique uses electron beam bombardment to achieve the high temperatures needed to evaporate silica glass directly.

6.2.6.3 Reactive Sputtering

Reactive sputtering is the phenomenon whereby an electrical discharge is passed between electrodes at low gas pressure, disintegrating the cathode under the bombardment of the ionized gas molecules. The cathode material leaves the electrode either as free atoms or as a composition resulting from a reaction of the cathode with residual gas molecules. Transport of the sputtered particles is due to their kinetic energy or to diffusion when the kinetic energy has been reduced through collisions with the residual gas.

The silicon to be sputtered and oxidized is made the cathode; the substrate to collect the deposited oxide, is made the anode. The electrodes are usually mounted horizontally in the vacuum chamber which is evacuated by a high speed oil diffusion pump. To oxidize the sputtered silicon, the discharge occurs either in oxygen or a mixture of oxygen and an inert gas such as argon or krypton, at pressures in the 0.02-0.2 torr range. Typically the cathode is at a negative potential of 1000 to 3000 volts with the anode grounded, and the current density is 1 ma/cm^2.

High purity oxygen and inert gas should be used for discharge; the cathode should be high purity silicon with surface contaminants removed. The partial pressure of reactive gas in inert gas will vary the stoichiometry of the deposit which will include Si, SiO and SiO_2. The formation of silicon oxides may occur through a reaction at the cathode, during transport, or during film deposition. The predominant reaction depends upon the pressure and ratio of oxygen to inert gas. For high ratios of oxygen to inert gas, reaction at the cathode predominates, while for low oxygen ratios, reaction during film deposition is the most likely process. The wide range of reactions possible during sputtering places a premium on the lowest ultimate pressure of residual gas obtainable in the vacuum.

Factors which can determine the physical properties of the thin oxide film are the substrate profile and stresses, in the film, which cause lift-off or peeling of the deposited layer. Preparation of the silicon substrate for sputtering carries the same importance as for thermal or anodic oxidation. Adherence of the deposited film does not depend upon heating the substrate as it does for evaporation of SiO films. One explanation for this assumes that the large amount of energy of the sputtered atom or molecule is sufficient to accomplish the same process which results from heating the substrate. The absorbed molecules are able to escape shallow traps which increases the probability of the absorbed molecule reaching a deeper trap on the substrate surface.

6.2.6.4 Carbon Dioxide Oxidation

Oxidation of silicon by CO_2 was investigated intially as a means of thermally oxidizing silicon in an epitaxial furnace so that layers of silicon and silicon oxide could be successively grown in the same apparatus. However, the formation rate of silicon oxide by the thermal reaction between the epitaxially deposited silicon layer and CO_2 is extremely slow compared to that of the oxide deposited from the reaction of $SiCl_4$, CO_2, and H_2. This latter process, not involving the substrate silicon, is most often used.

The apparatus for CO_2 oxidation is the same as that used for silicon epitaxy but with the addition of a CO_2 source. An epitaxially deposited layer of silicon is an ideal surface for oxide growth, being free of trace contaminants and/or oxide. If the process is used without first growing such an epitaxial layer, a gas etching of the surface is desirable prior to oxidation. Simple chemical cleaning of the surface is inadequate and often results in hazy appearing oxides as opposed to the clear oxides grown on the cleaner surfaces.

6.3 OXIDE PROPERTIES

The composition and structure of the oxide films, produced by the methods outlined in the preceding section, are generally described as amorphous glassy silicon oxide. The term "amorphous" here means only the absence of any long range structural order that would be detectable by electron diffraction patterns; there may be short range order.

The thin oxide films grown or deposited on polished silicon substrates have properties different from that of bulk silica glass. This is not surprising, since variations in the impurity content or structure of bulk silica glass cause changes in its measurable properties, and the different available methods of growth clearly permit variations in the impurity content of oxide films. Typical properties associated with the general methods of preparation are presented in this section.

6.3.1 DENSITY AND INDEX OF REFRACTION

The density of thin oxide films can be determined by weighing a silicon wafer before and after oxidation. The thickness of the film is also measured along with the sample area so that a direct calculation of density is possible.

The density of silica glass is listed as 2.20 gm/cm^3. The density of all the variously prepared oxide films, is close to this value with the exception of the wet electrolyte anodic oxides. Oxide density is probably not extremely sensitive to structure, although the presence of large concentrations of metal oxides or other impurities that cause imperfections can be expected to modify the structure sufficiently to influence the density. The room temperature anodic oxidation process has different growth kinetics than the other oxides, and consequently the structure may be grossly different.

Various methods have been reported for the measurement of index of refraction. Dry oxygen oxidation yields a film with density and index of refraction closest to that of silica glass. The presence of small quantities of water in silica glass apparently causes an increase in both density and index of refraction. Conversely, heating a steam grown film in a dry, inert atmosphere, causes a decrease in the index of refraction—toward the value of dry oxygen oxides.

No major distinction exists between the variously prepared films (or silica glass for that matter) with respect to density and index of refraction.

6.3.2 STRUCTURAL IMPERFECTIONS, POROSITY

Imperfections in bulk silica glass networks are likely to be either (1) oxygen vacancies (an oxygen ion is missing), (2) oxygen excesses or other foreign anions at interstitial positions in the glass or (3) metal ions at either substitutional or interstitial positions. Thin oxide films are subject to the same influences. The first two types of imperfections suggest that the oxygen available during growth is important in determining structural perfection; the second and third types of imperfections reflect a dependency upon impurity content.

Attempts to measure the effects of such imperfections have been made by observing the permeation of various gases through thin oxide films. In general much larger imperfections such as pores, crystalline boundaries, microchannels and microcracks have obscured the atomic scale imperfections. No well-defined differences appear between dry oxygen, steam, and pyrolytic oxides in the concentration of these grosser, structural imperfections.

6.3.3 RESISTIVITY, DIELECTRIC STRENGTH, DIELECTRIC CONSTANT

To measure the electrical properties of the various oxide films, the guard ring structure used in the direct deflection method of measuring the resistance of insulators, is quite conveniently employed. The test sample consists of an oxidized silicon wafer on top of which has been deposited a ring-dot pattern of aluminum or other metal for contact. The starting silicon acts as the bottom contact to the oxide film. The resistance of the silicon is in series with that of the oxide but is negligible.

Dielectric constant is computed from the capacitance and geometry measurements. Dielectric strength must be measured with the silicon positive with respect to the metal electrode. Otherwise a time dependent breakdown behavior is observed; i.e., the voltage at which breakdown occurs varies with $t^{-1/4}$. Typical dielectric strengths of oxide films are in the 10^6–10^7 volts/cm range. The dielectric constant varies from 3.78 for silica glass up to 10 for the lead accelerated oxide.

6.3.4 CHEMICAL PROPERTIES OF OXIDES

The etch rates of oxide films are sensitive to variations in the composition of the oxide. The most pronounced changes in etch rate occur when the composition of the oxide is altered to include other network formers such as P_2O_5. The oxide is then transformed to a new phase, and under such a transformation the physical properties change, as well as the etch rate. The difference in etch rate is known from photoengraving experience in which n^+ emitter regions, covered only with the oxide from the heavy phosphorus diffusion, etch much faster than adjacent, non-phosphorus diffused regions.

6.3.5 INVERSION LAYER PHENOMENA

An important property of an oxide film on silicon that has no counterpart in the published properties of silica glass, is the influence of the oxide film upon the electrical properties of the underlying silicon. The phenomena manifests itself primarily in the observation that a thermally oxidized silicon wafer generally tends to have a surface potential more n-type than that of the bulk silicon.

Reports of n-type surfaces being produced by thermal oxidation of silicon did not accompany the original public announcements of the planar process. Within the following two years, however, complaints of a high incidence of mysterious emitter-to-collector shorts on planar n-p-n transistors became audible. Remedial steps in the form of bake-out and annealing cycles have successfully eased the yield problem, but a complete understanding of the interaction is not agreed upon.

6.4 PROCESS TECHNIQUES

Oxidation of silicon is extremely simple. Indeed, avoiding oxidation requires much more technique. As specifications of oxide properties become more severe, however, the problems associated with obtaining the desired properties increase until the technology is unable to meet the demands made upon it. Thickness, uniformity and resistivity are not usually a source of oxide failure, given a certain amount of care in substrate preparation and operational cleanliness; interface electrical properties, however, are.

Cleanliness before, during, and even after oxidation is perhaps the most important requirement for satisfactory oxidation. Impurities present on the silicon surface prior to oxidation or during the oxide growth itself influence the homogeneity of the film and the interface electrical properties. Surface contamination following diffusion is objectionable.

6.4.1 PRE-OXIDATION CLEANING PROCEDURES

The need for a polished silicon surface has been previously noted. Without the smooth surface the structure of the oxide grown or deposited tends to develop crystalline regions of cristobalite which are undesirable for all the common uses of oxide films in integrated silicon device technology. Cristobalite is denser than silica glass and the boundaries between the amorphous regions and the denser crystalline regions are porous to both surface contamination and impurities during diffusions. Isolated regions of oxide mask failure during diffusion can sometimes be traced to the presence of localized regions of cristobalite in otherwise amorphous oxide.

Surface dislocations or irregularities may also serve as nucleation centers for the formation of cristobalite during oxidation. A surface etched with a preferential etch such as used to reveal dislocations, oxidizes nonuniformly and unsatisfactorily. To grow amorphous oxide films, a highly polished surface is a prerequisite.

Ultrasonic cleaning, vapor degreasing, and other techniques all have their advocates. A chemical oxidation step to form a hydrophyllic surface prior to oxidation seems to be generally preferred. Other important factors are the chemical purity of the acids and solvents employed in the cleaning procedure, the cleanliness of the containers used both in the cleaning operations and to store the chemicals, and the cleanliness of the tools used in transferring the wafers throughout the procedures. Any of these can contaminate the process and invalidate the entire cleaning operation.

6.4.2 CLEANLINESS DURING OXIDATION

The purity of the materials used in the oxidation process and that of the apparatus in which the oxidation is performed, clearly influence the impurity content of the oxide film. Silica glass is generally regarded as the most acceptable material for building thermal oxidation apparatus. It can be shaped to most any form desired. Its purity is demonstrated in that wafers can be oxidized in it without changing resistivity, unlike many other materials.

Purity of the source of oxygen is critical. In steam oxidation the water should be super pure ("intrinsic") for ideal conditions. In steam oxidation the conductivity of the water provides a convenient measure of source purity. In other sources for oxidation, the evaluation may not be so simple, but some account of it must be taken. Again solid particle contamination is particularly objectionable.

Atmospheric contamination by diffusion through the open tube apparatus previously described, can be avoided by the closed system illustrated in which a liquid seal separates the oxidation chamber from the outside atmosphere. Another technique becoming more common is that employing a "nitrogen curtain" to isolate the area of oxidation from the outside environment.

The presence of an electric field oriented perpendicular to and directed outward from the silicon surface during oxide growth, causes cationic impurities to be swept from the region adjacent to the oxide-silicon interface. This technique controls the ionic impurity distribution rather than the concentration. Following oxide growth in such an electric field, the cationic contaminants are bunched near the surface of the oxide where a light etch could presumably remove them. The direction of the field is that which has been previously found to accelerate oxide growth.

6.4.3 CLEANLINESS FOLLOWING OXIDATION

The influence of surface contamination upon the electrical properties of oxidized silicon devices has already been discussed. Surface ions originate from many sources and are hydrolyzed in the presence of water. Since the hydrophyllic oxide surface acquires water whenever exposed to room atmosphere, the presence of traces of water on the oxidized surface is almost unavoidable. To minimize the presence of water, oxidized wafers should be protected immediately upon completion of the oxidation process. Photoresist is often the most convenient coating to use for protection, since subsequent processing makes use of that type coating. Otherwise, sealing or storing in a controlled atmosphere seems desirable.

6.4.4 MEASURING THE THICKNESS OF OXIDE FILMS

Optical methods are most commonly used for measuring the thickness of oxide films on silicon. Weighing of samples before and after oxidation can be used but requires knowledge of the film density in order to convert weight into thickness. Electrical measurements such as breakdown voltage are useful for comparative purposes, but require a knowledge of dielectric strength in order to obtain thickness and depend upon film uniformity and perfection for accuracy and reproducibility. Optical measurements are generally free from these shortcomings, although some of them are, to an extent, destructive.

6.5 OXIDE MASKING

The opening of windows in the silicon dioxide layer used for device fabrication, forms a stencil-like structure through which diffusion and alloying can be carried out. This stencil-like structure or oxide mask is made by the following process (usually referred to as the planar process):

(1) The clean silicon wafer is placed in a furnace, where the oxide layer is grown.

(2) The oxide layer is coated with a photosensitive acid-resistant film.

(3) The film is stabilized with infrared light for about thirty minutes.

(4) A positive (the areas where diffusions are desired are black areas on a transparent film) of the configuration desired, is placed over the wafer with the film.

(5) Ultraviolet light is used to illuminate the film through the positive. Areas of the film exposed to the ultraviolet light are hardened, and are made impervious to acid etches; areas not exposed are removed by developing.

(6) The film is developed so that it is removed where not exposed. The result is an acid-resistant film stencil.

(7) The wafer is then placed in hydrofluoric acid, which removes the silicon dioxide where there is no film.

(8) A window is thus opened through the oxide, forming a mask for diffusion and alloying.

CHAPTER 7

DIFFUSION

ION IMPLANTATION

Diffusion is the method by which impurities are introduced into a crystal without destroying the crystal (as opposed to alloying, which destroys the crystal). Diffused devices may or may not contain epitaxial layers. The main problems associated with diffusion are concerned primarily with process control and with the achievement of specific depths and diffusion profiles. For example, when multiple diffusions are made on the same wafer, it is necessary to know the effect that additional temperature cycling has had on the previous diffusions. Although the mechanism of diffusion is important as a means of adding impurities to the semiconductor, the means of controlling the diffusion to produce a working device rests with the oxide. In this respect, the oxidation mechanism often commands greater attention than the actual diffusion, and for this reason a detailed description of oxidation is presented in the preceding chapter.

7.1 DIFFUSION APPARATUS

Diffusion may be carried out in an ''open tube'' furnace similar to that used in oxidation. It may consist of one or more furnaces aligned in tandem, each with its own temperature-controlling apparatus.

Wafers to be diffused are placed in the larger downstream furnace shown in Fig. 7-1. Diffusion is accomplished by passing diffusant gases through it. Bottled diffusant gases are fed directly into the diffusion chamber. Vaporized dopant from liquid sources is obtained by passing a carrier gas through a ''bubbler'' before it enters the furnace. Sources which are solid at room temperature are vaporzied upstream from the wafers in the small ''source'' furnace. Diffusion furnaces are constructed so that ''flat zones'' or zones of uniform temperature occur over some distance at the center of the furnace.

The ''open tube'' is adapted to large scale production since it can accommodate a larger number of wafers at one time and successive runs can be made closely upon one another. When diffusing boron, however, the ''closed box'' method is often used. Of the two methods, the mechanism involved in closed box diffusion is more clearly understood because the

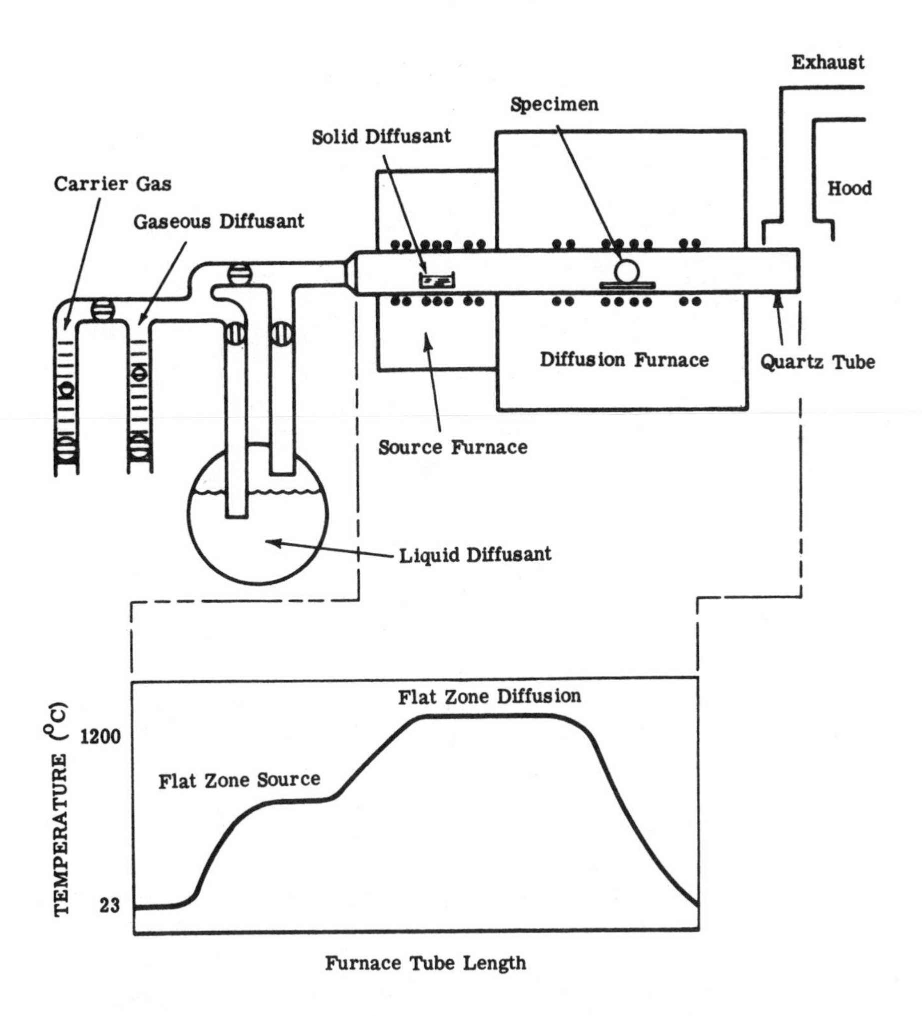

Fig. 7-1 Open tube, two zone diffusion furnace with temperature profile

simplicity of the experimental system lends itself more readily to experimental analysis. Fundamentally, however, the principles involved in both methods of diffusion are quite similar.

7.1.1 CLOSED BOX METHOD

An idealized closed box is shown in Fig. 7-2. The box is made of platinum and has a platinum lid. A platinum boat at one end of the box contains a glass consisting of, for example, 10 percent B_2O_3 in 90 percent SiO_2. At the other end are placed silicon wafers which have been oxidized for a very short time in a separate furnace. The lid is placed on the box and the box is put into the diffusion furnace held at the diffusion temperature.

The oxide initially present on the surface of the wafer contains no dopant (in this case boron) and the vapor pressure of the boron oxide in the source boat is small at room temperature. When the box is placed in a diffusion furnace held at a constant temperature, the vapor pressure exerted by the boron oxide in the source rises until it reaches a relatively high value.

If only the source were present in the closed box, an equilibrium value

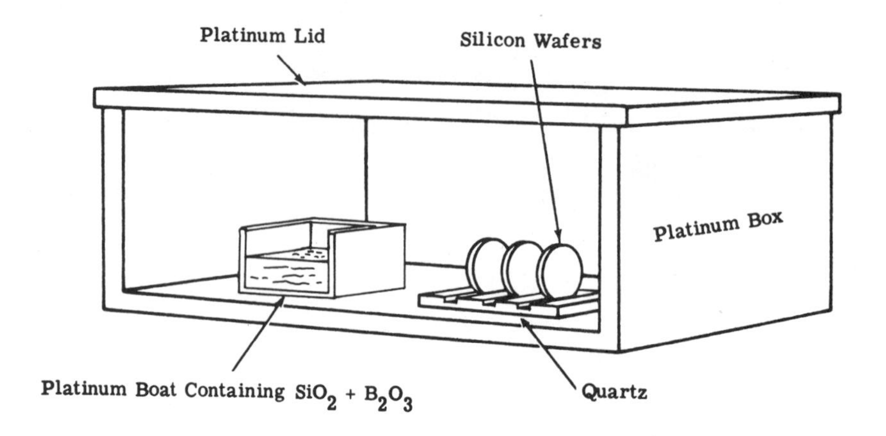

Fig. 7-2 An idealized closed box

of the vapor pressure due to boron oxide would soon be reached, and the amount of boron oxide in the vapor phase would reach a constant value. However, the additional silicon dioxide present in the box, in the form of a layer on the surface of the wafer, upsets the equilibrium. Assuming that none of the vaporized boron oxide is lost through leaks in the box, the equilibrium is soon re-established when the silicon dioxide layer on the surface of the wafer achieves the same composition as the source. In effect, boron oxide has been transferred from the source to the oxide on the surface of the wafer (the sink) by vapor transport. At equilibrium, both the source and the sink have the same composition of boron oxide and silicon dioxide, and the vapor pressure of the boron oxide is fixed.

The sink, which is now a borosilicate glass, loses some of its boron by diffusion into the silicon substrate at the interface between the oxide on the surface of the wafer and the silicon wafer. At the interface, an oxygen exchange between boron oxide and silicon (from the wafer) occurs. Boron oxide is thus reduced to elemental boron, and silicon is oxidized to silicon dioxide. This results in a slight unbalance of the equilibrium, which is quickly restored by the source, the latter being chosen to have a very large mass compared to that of the sink.

If no SiO_2 were present on the surface of the wafer before the diffusion, a layer would quickly have formed by virtue of the oxidizing nature of SiO_2 in the source. Hence, the SiO_2 layer on the surface of the wafer not only protects the surface from erosion, but also acts as source for diffusion into the wafer.

Closed box diffusions are normally carried out with sources that are solids at room temperature; examples are B_2O_3 and P_2O_5. The composition of the sink can be controlled by controlling the composition of the source glass. Once the box is "conditioned" by making a few experimental runs, it may be used repeatedly. A source containing 0.01 percent B_2O_3 will deplete by only 10 percent in 70 hours at a diffusion temperature of 1300° C. Since a typical diffusion run may be only of 15 minutes duration, it is evident that quite a few runs can be carried out before it is necessary to recharge the source.

7.1.2 OPEN TUBE DIFFUSIONS

Open tube diffusions may be performed with either solid, liquid, or gaseous sources. Examples are as follows:

Solids:

Boron Trioxide	B_2O_3
Phosphorus Pentoxide	P_2O_5
Phosphorus Nitride	P_3N_5
Red Phosphorus	P_4
Arsenic Trioxide	As_2O_3
Antimony Trioxide	Sb_2O_3
Gallium Trioxide	Ga_2O_3
Silicon doped with appropriate impurity	

Liquids:

Phosphorus Oxychloride	$POCl_3$
Phosphorus Trichloride	PCl_3
Boron Tribromide	BBr_3
Methyl Borate	$B(OCH_3)_3$

Gases:

Boron Trichloride	BCl_3
Diborane (Boroethane)	B_2H_6
Pentaborane	B_5H_9
Phosphine	PH_3
Arsine	AsH_3
Stibine	SbH_3

In all open tube diffusions, a mixture of "carrier gases," which may include reducing and oxidizing gases, is fed into the furnace tube from the front end. The carrier gases sweep the vaporized diffusant into the diffusion furnace, where the wafers are held at elevated temperature.

Solid sources are usually vaporized in a separate source furnace connected upstream from the diffusion furnace. Liquid sources are fed into the carrier gas stream by diverting part of the carrier gas flow through a bubbler containing the liquid. Gaseous sources are metered directly into the carrier gas stream.

Each diffusion source has certain difficulties associated with it. P_2O_5, for example, is extremely sensitive to the presence of water vapor in the atmosphere. Unless it is kept dry, it will pick up moisture which will affect its vapor pressure adversely. Red phosphorus is unreliable and often converts to another form at source temperatures needed to establish a reasonably high vapor pressure. Sources containing halogens frequently cause pitting. These include $POCl_3$, PCl_3, BCl_3, and BBr_3. In the latter two cases, it is necessary to add oxygen to the gas stream to prevent pits from forming. If oxygen is not present, a black deposit of unknown composition may occur on the surface of the wafer. The deposit is probably a boron compound which results from the transport of the reduced form of boron to the surface of the wafer. Pitting will also occur with Sb_2O_3. Boron diffusions from B_2O_3 usually result in the accumulation of boron deposits on the walls of the diffusion tube; these, in time, act as secondary sources of diffusion and cause difficulty in the control of the diffusion depth and concentration. It is necessary to add wet hydrogen when diffusing with Ga_2O_3. Wet hydrogen reacts with the source to give Ga_2O which then acts as the transport-diffusion medium. If wet hydrogen were not present, diffusion would not be evident.

The mechanism of diffusion in the open tube is similar to that in the closed box. In the open tube, however, one attempts to obtain a gas flow of homogeneous composition. The open tube has one principal advantage over closed box diffusions–the composition of the gas with respect to the dopant concentration can be varied at will. Furthermore, in an open tube it is possible to perform such operations as gas etching and oxidation before diffusion. In these respects, open tube diffusion techniques are far more versatile than closed box techniques.

7.1.3 OXIDE MASKING

Figure 7-3a shows a silicon wafer on which several thicknesses of oxide have been grown by thermal oxidation. The wafer has been subsequently diffused in a closed box using a glass of B_2O_3 in SiO_2 at an elevated temperature. It can be seen that masking of the silicon wafer from diffusion has occurred in the areas where the original oxide was 500 Å thick or greater. The highest surface concentration and greatest depth of diffusion of boron has occurred in the silicon where no oxide was present. The 25 Å thick film has been completely diffused with boron, and the silicon beneath it has been diffused to some extent. The surface concentration and diffusion depth, however, is less than in the previous case. At 500 Å, no diffusion into the silicon has occured, although the oxide layer has been completely diffused with dopant. At 200 Å a partial diffusion through the oxide layer has occurred.

If a wafer is prepared as shown in Fig. 7-3b and diffused as discussed above, preferential masking can be achieved for the fabrication of a transistor. Figure 7-3b represents the diffused base of a transistor, into which an emitter of opposite conductivity type is to be diffused. It is first necessary to remove the thin ''skin'' of boron diffused SiO_2 enveloping the surface of the wafer as shown in this figure. The diffused base region, with the diffused oxide removed, is shown in Fig. 7-3c. To diffuse the emitter, it is necessary to reform the oxide layer over the diffused base region to provide for phosphorus masking as shown in Fig. 7-3d. This regrown oxide layer must be made thick enough to mask selected areas of the base against phosphorus diffusion. Finally, the transistor fabrication can be completed by etching a hole through the oxide layer over the base, and repeating the diffusion process with a source of phosphorus diffusant. The resulting structure is shown in Fig 7-3e.

In the diagram shown, care has been taken to show the junctions resulting form the diffusions as having a ''U'' shape, with the arms terminating under the previously grown oxide on the surface. This ''U'' shape is the result of lateral diffusion into the wafer region below the oxide. Only under very unusual circumstances is the oxide ever removed from the

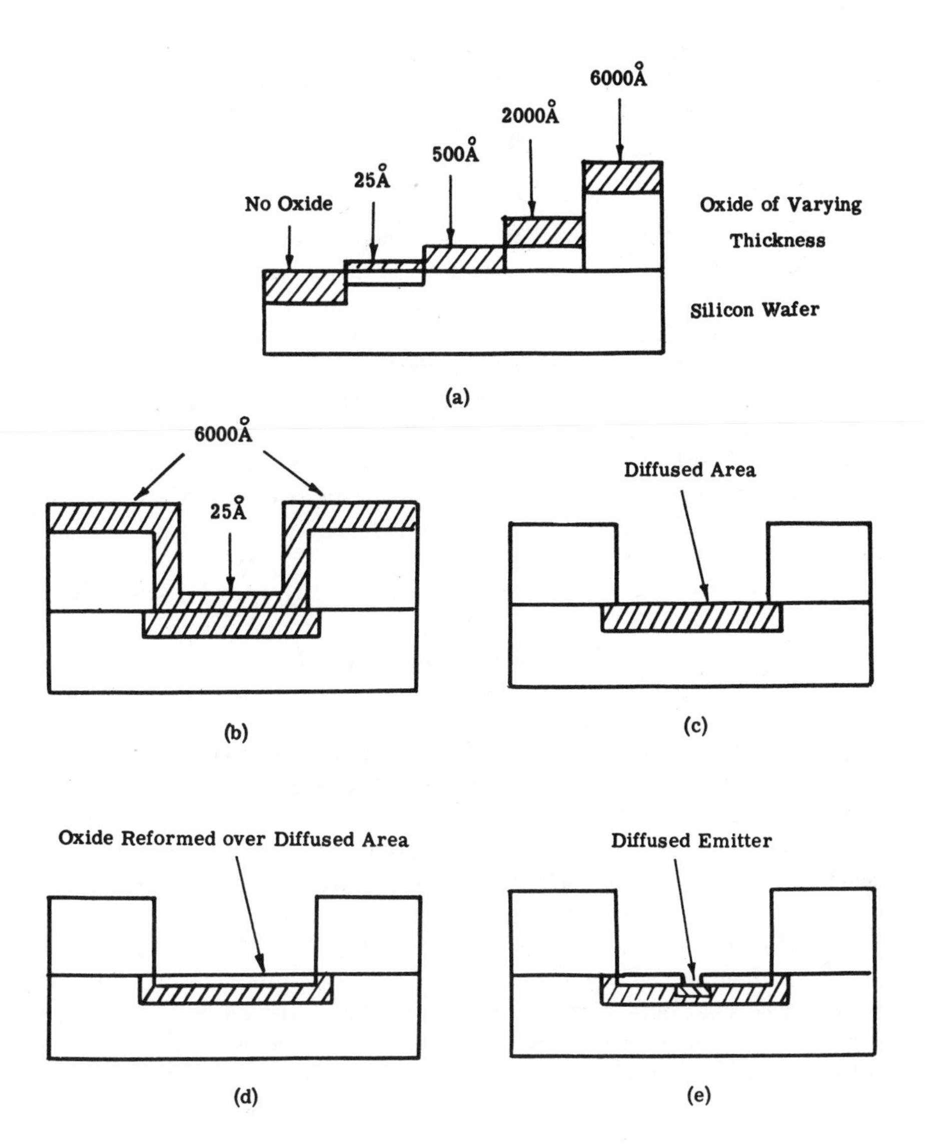

Fig. 7-3 Oxide masking (shaded areas indicate diffusion into the oxide and silicon)

region where the junction reaches the surface; its presence shields the junction from contamination during subsequent processing steps. If it were not present, electrically unstable devices would result.

7.1.4 SOLID SOLUBILITY LIMITS

Listed in Table 7-1 are Groups III, IV, and V of the elements, arranged in the form of the simplified Periodic Chart.

Table 7-1 : Periodic Groups of the Elements Used in Semiconductor Devices

GROUP III	GROUP IV	GROUP V
Boron	Carbon	Nitrogen

Aluminum	Silicon	Phosphorus
Gallium	Germanium	Arsenic
Indium	Tin	Antimony
Thallium	Lead	Bismuth

Silicon is a Group IV element. Like carbon, germanium, tin, and lead, it has a structure which is of the diamond cubic lattice type. It is, of course, a semiconductor, as are the diamond cubic forms of carbon, germanium, tin, and lead. A similar structure can be obtained with binary systems consisting of combinations of elements from Groups III and V. Binary systems possessing a diamond cubic lattice structure are said to be of the zinc blende (zinc sulfide) structure and are also semiconductors. Aluminum phosphide, gallium arsenide, gallium phosphide, and indium antimonide are examples of these. These materials remain semiconductors only as long as the diamond cubic or zinc blende lattice structure is not destroyed.

Dopants for silicon are chosen from among the Groups III and V elements. Group III supplies P-type dopants and Group V supplies N-type dopants. If the incorporation of a P-type dopant from Group III or an N-type dopant from Group V into the silicon lattice is accomplished by the diffusion, the dopant atoms occupy a site previously occupied by a silicon atom. The extent to which such substitutions can be made at any particular temperature is indicated by the "solid solubility limit," which is the maximum number of atoms that will diffuse into a solid solvent without causing a solid to melt. Semiconductor diffusion technology is concerned with the preparation of solid solutions of impurity elements in solid semiconductor blocks. The semiconductor is the solid solvent and the diffused dopant the solute.

Germanium and silicon are both Group IV elements with four outer valence electrons. Both form diamond cubic lattice structures and have atomic radii which are approximately equal. Because of these similarities, germanium and silicon have unlimited solubility in each other. Hence, diamond cubic structures can be obtained with any proportions of germanium in silicon and silicon in germanium.

As the differences in outer electronic shell structures and atomic radii increase, the solid solubility limit of one element in another decreases. In the design of semiconductor devices, it is necessary to know this limit. Figure 7-4 shows the solid solubility limits of some impurity elements in silicon as a function of temperature.

Boron has a maximum solubility of 6×10^{20} atoms per cc, while that of aluminum is only 2×10^{19}. In a PNP transistor, boron would, therefore, be chosen as the diffusant for the emitter because of its higher solubility, and only in special cases would aluminum be chosen. Similarly, arsenic, phosphorus, and antimony would be chosen for diffusing N-type emitters in an NPN structure. Thus, because of their high solid solubility limits, the elements arsenic, phosphorus, boron, and antimony are the primary elements used in semiconductor work.

A simple calculation further emphasizes the above points. Monocrystalline silicon contains 5×10^{22} atoms/cm^3. If the solid solubility limit of phosphorus is 1.5×10^{21} atoms/cm^3, then a silicon wafer will contain, at most, 3 percent phosphorus. On the other hand, aluminum,

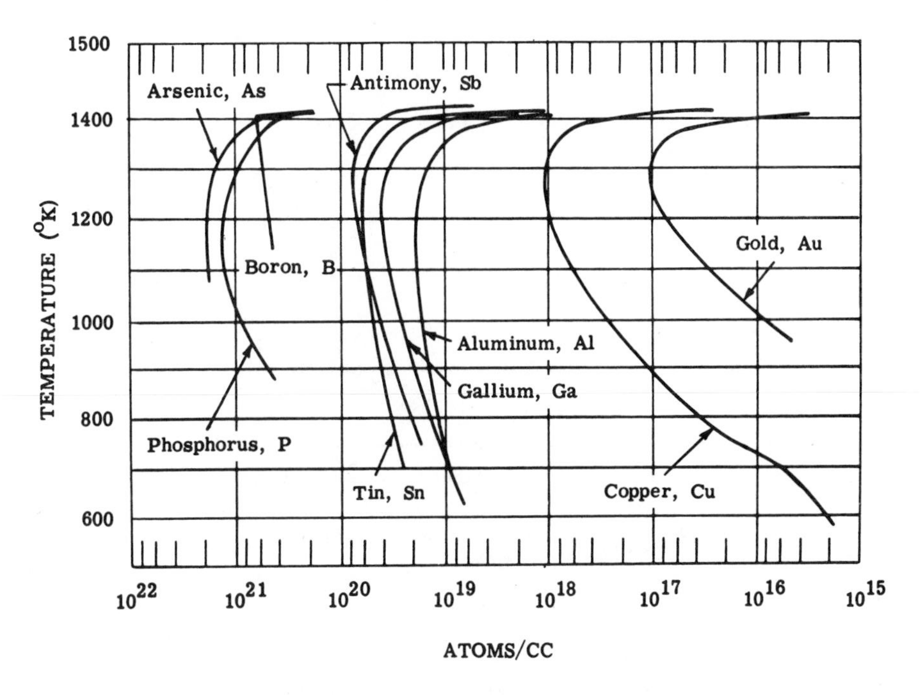

Fig. 7-4 Solid solubility limits of various elements in silicon

with a solid solubility limit of 2×10^{19} atoms/cm^3, can only comprise 0.04 percent of a silicon wafer.

7.1.5 FUNDAMENTAL CONCEPTS OF SOLID STATE DIFFUSION

Impurities can be introduced into silicon by several methods among which are alloying, epitaxy, and diffusion. Because of the ability to localize the diffiusant by masking techniques, diffusion has become one of the most

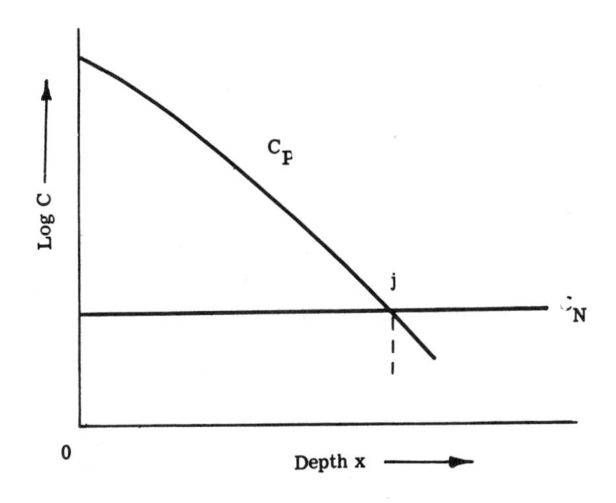

Fig. 7-5 Plot of total P-type diffusant concentration as a function of depth

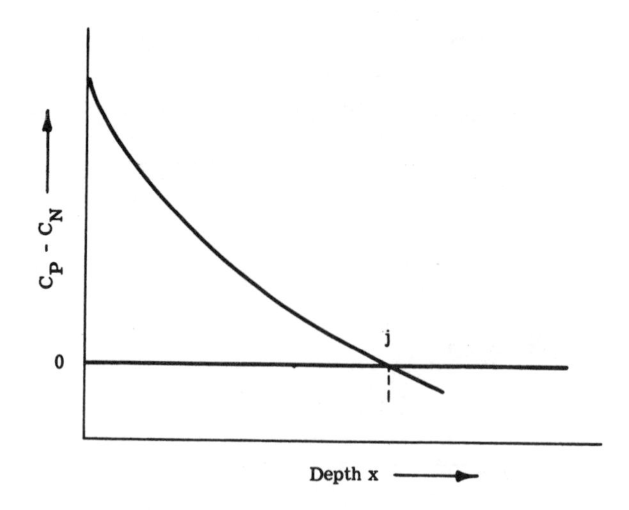

Fig. 7-6 Plot of excess P-type concentration as a function of depth

important methods. Much experimental and theoretical work has been done to gain a complete understanding of the diffusion mechanism, which can be exploited in the design of devices with special characteristics.

The primary aim of the diffusion technologist is the control of surface concentrations and diffusion profiles. Consider a wafer which is uniformly doped with an N-type element to a concentration of C_N. If a P-type element is diffused into it from one surface, the diffusion profile shown in Fig. 7-5 will be obtained. The N-type concentration C_N, is uniform throughout the wafer. It does not change with distance x from the surface and its profile is indicated by the horizontal line, C_N. The P-type concentration, however, decreases as the depth increases; its profile is indicated by the sloping line C_P in Fig. 7-5. At the point j, the N-type and P-type concentrations are equal, and a PN junction occurs. It is often convenient to plot, as in Fig. 7-6, the net concentration of excess P-type, $(C_P - C_N)$, or excess

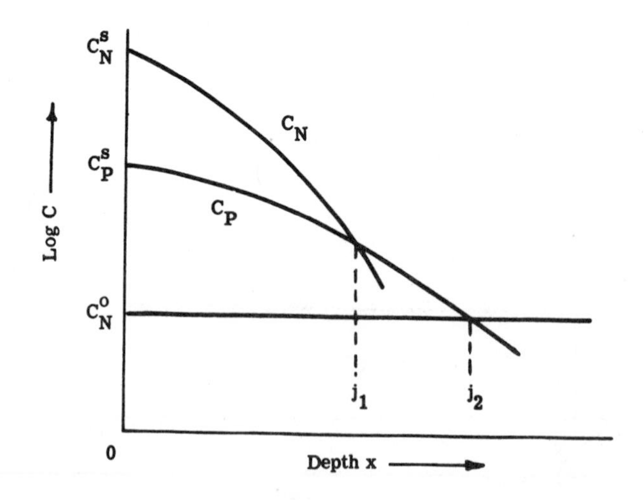

Fig. 7-7 Total N-type and P-type diffusant concentrations as a function of depth

N-type ($C_N - C_P$), impurity. The curve in Fig. 7-6 shows the excess P-type concentration as a function of depth.

If two diffusions are made into the same area of the wafer, the two profiles shown in Fig. 7-7 will be obtained. The total concentration of impurity at the surface of the wafer is equal to the total background concentration (C^o_N) plus the total P-type concentration plus the total N-type concentration, ($C^o_N + C^s_P + C^s_N$) at depth $x = 0$. Each time a diffusion is performed, the total impurity concentration at the surface increases. Two junctions occur, one at j_1 where $C^o_N + C_N = C_P$ and another at j_2 where $C_P = C^o_N$

If this information is plotted as excess N-type and P-type dopant, the curve in Fig. 7-8 results. Between the surface and the first junction, there is an excess of N-type diffusant; between j_1 and j_2 there is an excess of P-type diffusant; and between j_2 and the undiffused bulk of the wafer,

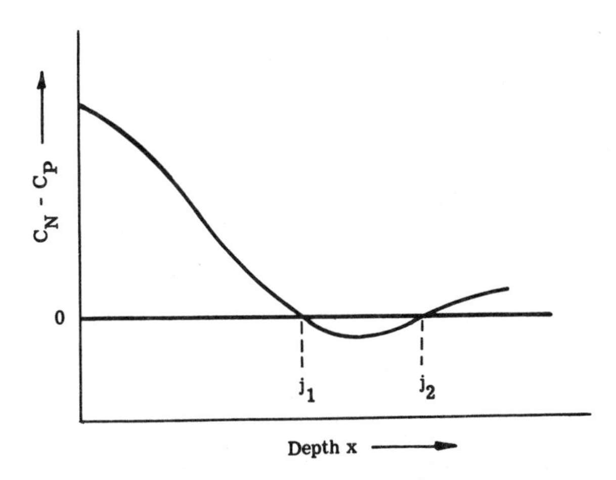

Fig. 7-8 Excess N-type and P-type concentrations as a function of depth

there is again an excess of N-type diffusant. Thus, this curve is a plot of an NPN structure and shows the concentration of excess donors and acceptors as a function of depth.

7.2 DISTRIBUTION PROFILES

Diffusion of a solute occurs whenever its concentration is not uniform; regions of high concentration degrade so that the entire system eventually will have a uniform concentration. This flow is governed by the usual diffusion law—that flow per unit area and unit time is proportional to the concentration gradient and is directed from regions of high concentration to regions of low concentration, or

$$J = -D \text{ grad } C \tag{7-1}$$

where J is the diffusion current in atoms per cm² per second, C the concentration, and D the diffusion constant. Assuming D to be independent of concentration and position, and applying this equation to an elementary volume,

$$\partial C/\partial t = D \nabla^2 C \tag{7-2}$$

or in the one-dimensional case, which is the one of interest here,

$$\partial C/\partial t = \frac{\partial^2 C}{\partial x^2} \quad . \tag{7-3}$$

These equations are known as Fick's Laws of Diffusion. It is these equations which control the diffusion of dopants in a crystal.

7.2.1 ERROR FUNCTION DISTRIBUTION

If the second law is solved for the boundary conditions (a) that none of the dopant is in the wafer initially, and (b) that the surface concentration does not change during the time of the diffusion, then the following equation results:

$$C_x = C_s \left[1 - \frac{2}{\sqrt{\pi}} \int_0^{x/2\sqrt{Dt}} e^{-\lambda^2} d\lambda \right] \tag{7-4}$$

where

C_x = concentration in a plane at any depth x in atoms/cm^3
C_s = concentration in the plane of the surface in atoms/cm^3
x = depth in cm
D = diffusion constant in cm^2/sec
t = time in seconds.

The integral $[2/\sqrt{\pi} \int_0^{x/2\sqrt{Dt}} e^{-\lambda^2} d\lambda]$ is commonly referred to as the "error function," erf $(x/2\sqrt{Dt})$, so that Eq. (7-4) may be more conveniently written as:

$$C_x = C_s[1 - \text{erf}(x/2\sqrt{Dt})] \tag{7-5}$$

The value of erf may be obtained to 15 decimal places from published tables of probability functions. The above equations describe the diffusion profile whether the diffusing medium and source are gaseous or liquid.

Various values of the surface concentration can be obtained by adjusting the concentration of the source. With any given source, however, the surface concentration does not change with time, but the diffusion profile and depth do. This fact is graphically illustrated in Fig. 7-9, in which it is seen that as the diffusion time increases, the curves tend to flatten out, even though they all intersect at $x = 0$. As the diffusion time t becomes very large, the distribution of impurity throughout the wafer tends to become uniform at the surface value C_s. It should be noted that

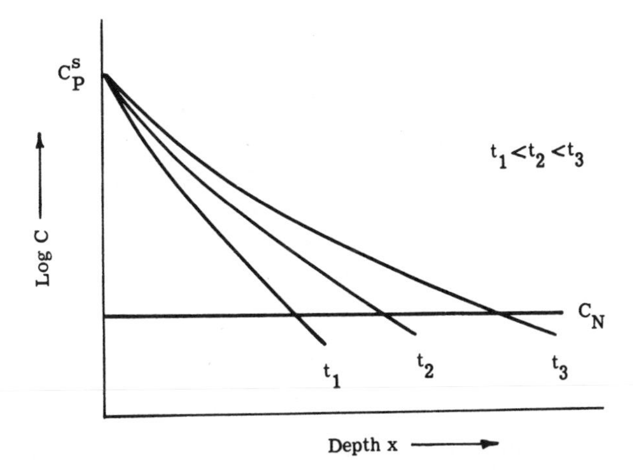

Fig. 7-9 An error function distribution showing the change in diffusion depth and concentration profile as a result of increased diffusion time, when the concentration of the external source is constant

if a dopant is diffused into a wafer that already contains some of the same dopant, then the resultant concentration at any depth is a simple summation of the individual concentrations.

7.2.2 GAUSSIAN DISTRIBUTION

Diffusions are often done in a two-step process. In the first step, called the "predeposition," the error function diffusion profile is obtained. In the second step, called the "drive-in," the source is removed, and the wafer is heated to the diffusion temperature in an inert atmosphere. A redistribution of the diffused impurity occurs. The new diffusion profile is called a "Gaussian distribution," and is described by Eq. (7-6). The boundary conditions imposed are (a) that no external source be present, and (b) the redistribution occurs from a finite number of atoms of diffusant deposited on the surface of the wafer.

$$C_x = Q/\sqrt{\pi Dt}\, \exp - x^2/4Dt \quad (7\text{-}6)$$

where Q = total number of atoms predeposited on the surface in atoms/cm^2. The quantity Q can be obtained by weighing or by radioactive tracer techniques. Typical Gaussian diffusion profiles at various diffusion times are shown in Fig. 7-10. The initial error-function profile is also shown for comparison. It is seen in this graph that, beginning at $t = 0$, the impurity atoms redistribute with increasing diffusion time, giving an increasingly lower surface concentration and an increasingly greater depth of penetration.

In practice, it is convenient to prepare a series of graphs, from which diffusion conditions can be determined, which will result in particular diffusion profiles. For each particular background concentration of interest, a series of predeposition runs are made in which the temperature is varied and all other conditions (time of diffusion, concentration of

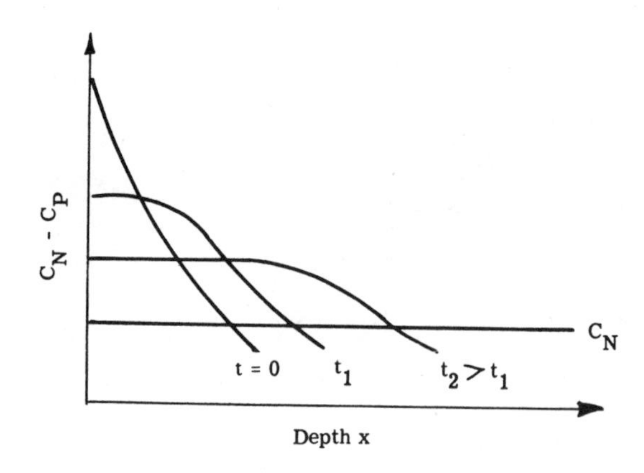

Fig. 7-10 A gaussian redistribution showing the charge in surface concentration, depth and diffusion profile, with increasing time of diffusion. External source not present

diffusant in vapor phase, wafer preparation, etc.) are held constant. The resulting profile for each run is an error function distribution, usually of high surface concentration and small depth (e.g., 10^{21} atoms/cm^3 at 0.1 microns). No attempt is made to determine either the surface concentration or depth after these diffusions. Instead, the sheet resistivity, ρ_s, is measured, and the graph in Fig. 7-11 is plotted.

The sheet resistivity of a thin diffused layer, as measured by the 4-point probe, is given by

$$\rho_s = \frac{V}{I} C(d/s) \qquad (7\text{-}7)$$

where:

ρ_s = sheet resistivity in ohms
V = voltage
I = current in amperes
$C(\frac{d}{s})$ = a correction factor which is dependent upon the diameter, (d), of the wafer and the distance, (s), between probes.

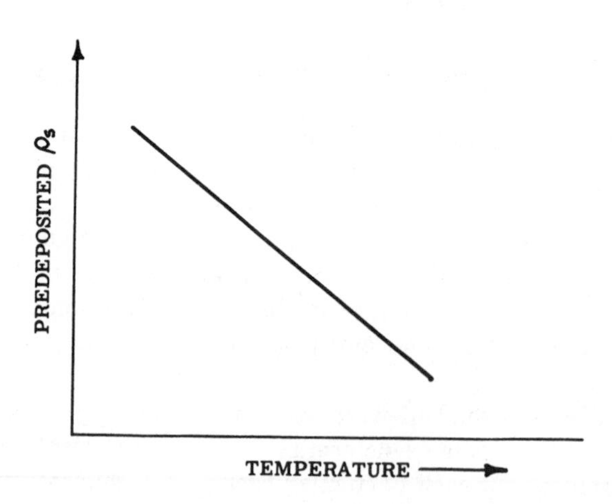

Fig. 7-11 Sheet resistivity after predeposition as a function of temperature

It is convenient to standardize wafer sizes so that the correction factor does not vary from wafer to wafer. It is important to note that the 4-point probe method of measuring ρ_s is accurate, provided the diffused layer is surrounded by nonconducting boundaries including the interface between the layer and the wafer.

After the predeposition runs are made and ρ_s measured, the thin diffused oxide layer, which is diffused with dopants and which was created during predeposition, is removed and the wafer is subjected to a second series of diffusion runs. In these, only the time of diffusion is varied; the temperature of diffusion is held constant. The resulting curves are shown in Fig. 7-12 and as shown, they are Gaussian.

Finally, the wafers are sectioned, the surface concentrations and diffusion depths measured, and a third graph is prepared relating final sheet resistivity to diffusion depth. A fourth graph relates diffusion depth to surface concentration.

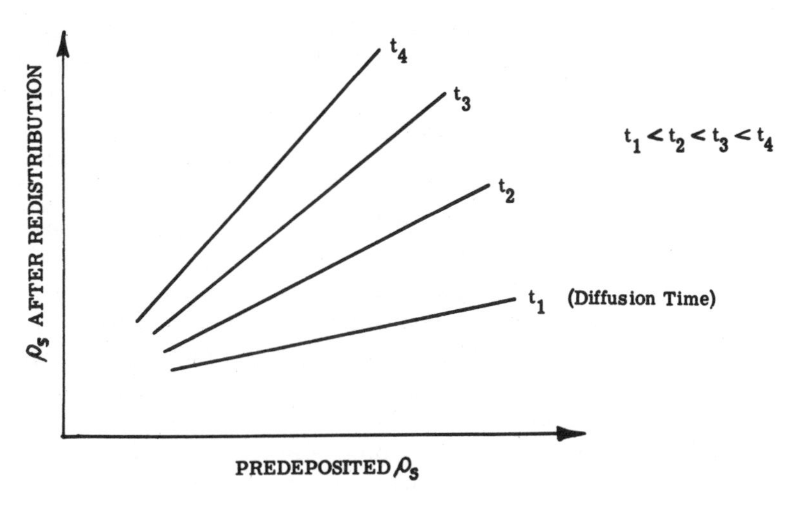

Fig. 7-12 Graph of predeposition sheet-resistivity for various drive-in diffusion times

Sheet resistivity measurements cannot be obtained on diffused devices in which small areas of diffusion are involved, as, for example, in the emitter or base regions of a transistor. An emitter region is usually of such small size that it is impossible to place an in-line 4-point probe array in the diffused area. In such a case, sheet resistivities are often measured on large diffused peripheral areas of the same wafer, on which the device is made, or on separate wafers processed along with wafers carrying diffused devices.

There are three principal reasons for using a two-step process in diffusion: a) the short predeposition times result in shallow penetrations of the oxide layer, which can easily be removed by a short dip in hydrofluoric acid; b) the diffusion depth during drive-in can be controlled more closely if a drive-in temperature is chosen so that the diffusion constant of the dopant is of low value; it is general practice to remove the wafers from the furnace after 90 percent of the drive-in time has expired, measure the depth on several of the wafers, and return the remainder to the furnace for a time sufficient to achieve the desired depth; and c) the Gaussian distribution profile is advantageous in base diffusions from device design considerations. If the Gaussian redistribution is always carried out under the same conditions of temperature, etc., the results of a test run can be used with good accuracy for this purpose. If the distance x_1 is achieved in time t_1, then the time necessary to achieve depth x_2 is given by $t_2 = x_2^2 t_1 / x_1^2$.

7.2.3 REOXIDATION DURING DRIVE-IN

After the thermal masking oxide is applied and the base diffusion is made through a hole in the oxide, the base diffused area is reoxidized, and a hole is etched over the base region to localize the emitter diffusion.

The reoxidation can be accomplished as a separate step after the base drive-in. However, it is not desirable to subject the device to elevated temperatures for longer than necessary, particularly when the wafer is multiply diffused, in order not to disturb the diffusion profiles of the previous diffusions. Therefore, the second thermal oxide is usually applied by providing the tube with an oxidizing atmosphere during drive-in. The redistribution of the predeposited thin skin of dopant atoms, thus occurs at the same time that the second thermal oxide is formed.

When boron diffused planar junctions are reoxidized at a fast rate, junction degradation may occur. This is exemplified by a reduction in breakdown voltage and a softening of the reverse characteristic. A wafer diffused with boron also contains a dopant of opposite conductivity type (usually phosphorus) from the bulk wafer, so that during reoxidation a redistribution of both dopants will occur. As the oxide front advances into the wafer, it is possible for the dopants at the surface to be either incorporated into the newly formed oxide or rejected by it. If they are incorporated into the oxide, as in the case of boron, a pile-up does not take place. If they are rejected, as occurs with phosphorus, they tend to pile-up at the silicon-SiO_2, interface. The effect is enhanced as the background doping of the bulk wafer is increased. Diffusant atoms have different affinities for SiO_2. Boron, for example, is as soluble in SiO_2 as it is in silicon. Phosphorus has a lower solubility in SiO_2 than in silicon and is rejected by the oxide. Aluminum, on the other hand, has higher solubility in SiO_2 than in silicon. Reoxidation of an aluminum diffused wafer would tend to deplete the surface of the silicon wafer with respect to aluminum.

This effect is often spoken of in terms of the K/D ratio where K is the rate constant of oxide growth defined by $x^2 = Kt$ (where x is the oxide thickness in cm and t is the diffusion time in seconds). The value of K varies with diffusion gas pressure and temperature. D is the diffusion coefficient of the dopant in silicon. The value of K can be determined for any particular drive-in cycle. When the K/D ratio is 1 or lower, neither pile-up or depletion of phosphorus will occur at the surface, and one can expect 75 percent of the junctions to be satisfactory. When the K/D ratio is 3, one can expect at least 90 percent of the junctions to be unsatisfactory from the point of view in question.

7.2.4 SURFACE CONCENTRATION

The surface concentration, C_s, and the diffusion constant, D, are basic parameters in the application of the above diffusion equations. Both are highly dependent on temperature. C_s, however, is a difficult quantity to measure directly. Charts have been compiled to facilitate this measurement. Using substrate background concentrations of 10^{14} to 10^{20}, these charts show the surface concentration as a function of the average conductivity of the diffused layer at various depths, ranging from the surface to close to the junction. The average conductivity is given by

$$\bar{\sigma} = 1/\rho_s(x_j - x) \qquad (7\text{-}8)$$

where

$\bar{\sigma}$ = average conductivity in ohm-cm^{-1}

ρ_s = sheet resistivity in ohms/square

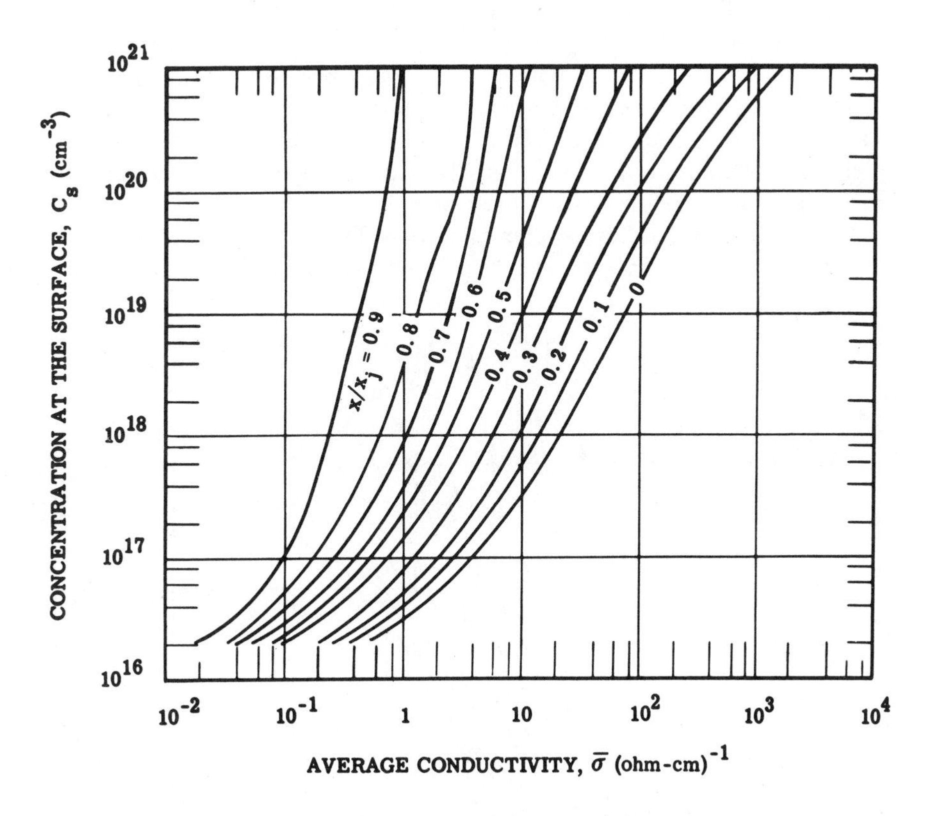

Fig. 7-13 Average conductivity of P-type complementary error function layers in N-type silicon

x_j = depth of junction in cm

x = depth at which C_s is measured in cm.

With the use of these charts, one need only measure ρ_s and x_j to obtain the value of the concentration, C_x, at any depth x from the surface, and thus a diffusion profile. Typical charts are shown in Fig. 7-13, for an error-function distribution, and in Fig. 7-14, for a Gaussian distribution. For both charts, the wafer background concentration, C_B, is N-type and has a value of 10^{16} cm^{-3}.

Another useful curve is shown in Fig. 7-15. Here, the background dopant concentration is plotted against resistivity in ohm-cm for P-type and N-type dopants. If the starting material is uniformly doped throughout its bulk, it is possible to measure background dopant concentration directly, using a 4-point probe by means of this curve.

7.2.5 DEPTH MEASUREMENTS

Because diffusion depths may be on the order of only a few microns, specialized techniques must be used in measuring the depths of the layers. Depth measurements are usually made by "angle lapping" or "cylindrical grooving" with the latter method providing the greater accuracy.

Angle lapping measurements are realized by lapping the surface of the specimen at a small angle (usually 2° to 5°), with the device shown in Fig. 7-16, to expose the junction cross section. The latter is then wet with a mixture of HF and HNO_3, which stains the P layer dark with respect to the N layer, leaving the junction clearly delineated. The specimen to be lapped is waxed onto a cylindrical lapping post which has one face beveled at the angle desired. The lapping post is placed in a loose fitting metal ring, the purpose of which is to hold the lapping post in a vertical position while lapping the specimen. The entire assembly is then lapped on a glass plate covered with slurry of fine particle abrasive, resulting in the exposure of the junction cross section. Finally, the depth is measured by counting the interference fringes formed, upon illumination with

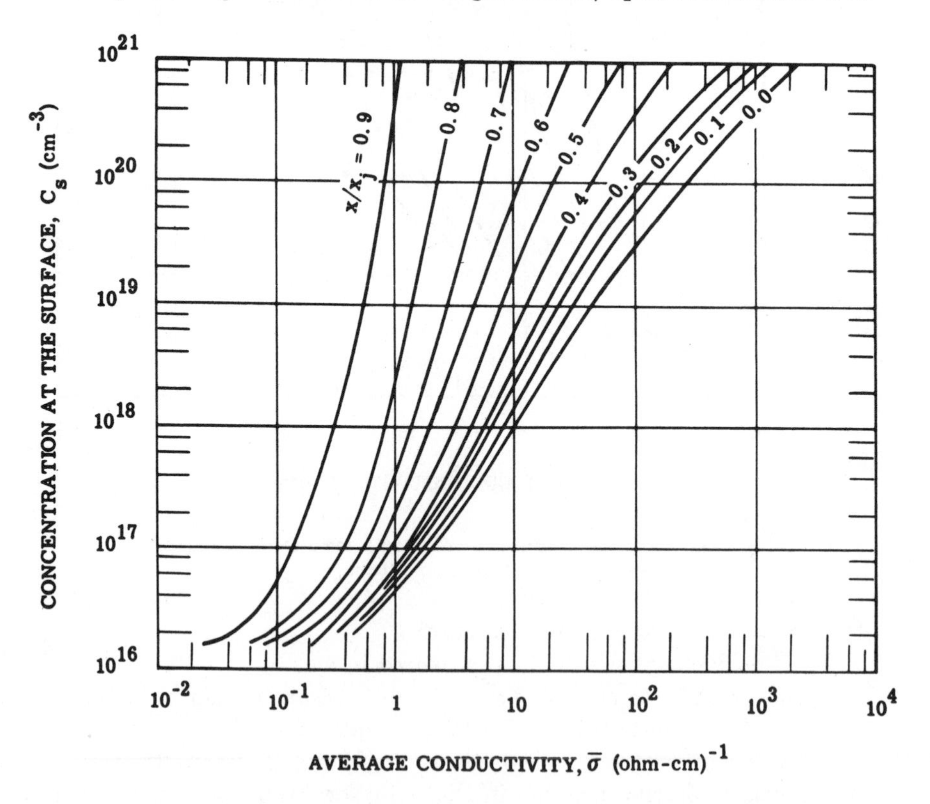

Fig. 7-14 Average conductivity of P-type gaussian layers in silicon

monochromatic light, in the air wedge between the lapped surface and a silver-coated glass plate with which it is placed in contact. The silver-coated plate and the monochromatic light source are usually incorporated into an interference microscope, through which the fringes may be observed. By a series of successive lappings, it is possible to locate a small diffused area, such as the emitter of a transistor, and to obtain a continuous series of depth measurements over its extent.

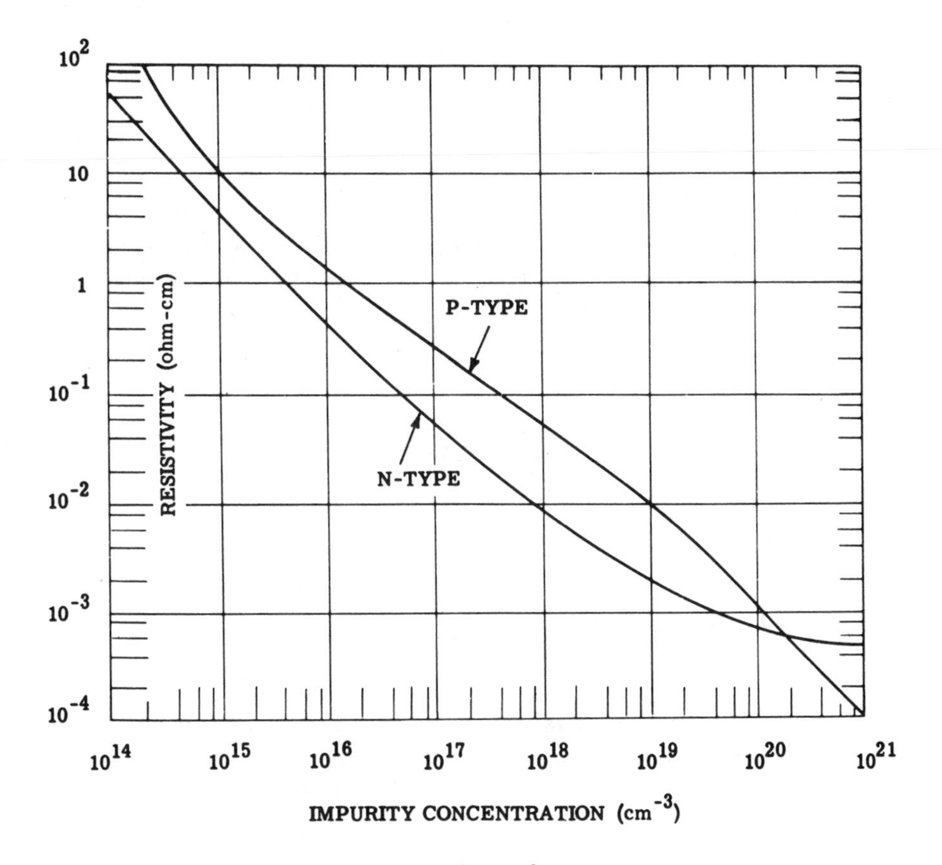

Fig. 7-15 Resistivity of silicon at 300°K as a function of acceptor and donor contribution

Cylindrical grooves are realized by rotating a cylinder, wet with an abrasive slurry, against the surface of the wafer. The resulting groove, shown in Fig. 7-17, is again stained as above and the depth measured with a calibrated microscope. The depth is given by $x_j = XY/2R$, where the symbols have the meanings shown in Fig. 7-17. Because of the large diameter of the cylinder (2.0 cm), it is not possible to locate and obtain diffusion depths on small specific areas. Consequently, the method cannot be used to obtain a continuous series of depth measurements over a small area.

7.2.6 THE DIFFUSION CONSTANT

The actual measurement of a diffusion profile may be obtained by several techniques, all of which are tedious and time consuming. In one technique, very thin layers are successively removed and the sheet resistivities measured until the junction is reached. From the sheet resistivity ρ_s, the conductivity is obtained by Eq. (7-8), and with the aid of curves of the type shown in Figs. 7-13 and 7-14, the surface concentration C_s may be realized from the conductivity. Finally, a plot of C_s versus depth gives the diffusion profile.

Similarly, the diffusion constant curves may be obtained, once the values of C_s, C_x, and x are known. Thus, from Eq. (7-5),

$$D = \tfrac{1}{4}t\left[\frac{x}{\text{Arg erf}\,[(1-(C_x/C_s)]}\right]^2 \tag{7-9}$$

Diffusion coefficient curves, plotted against reciprocal temperature are shown in Fig. 7-18. The diffusion constant is seen to vary exponentially with temperature. This temperature dependence is described by the empirically derived relation:

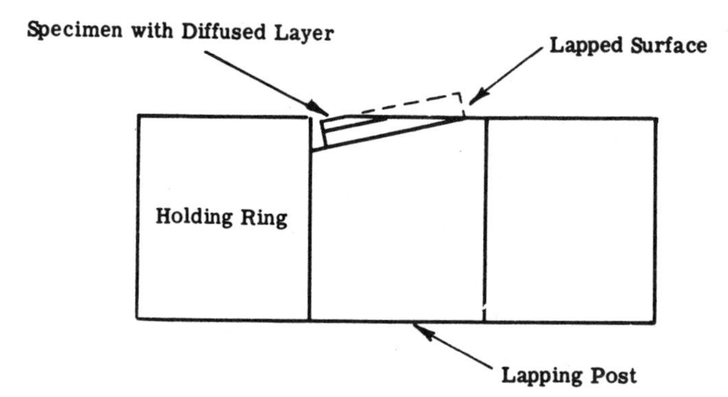

Fig. 7-16 Angle lapping device

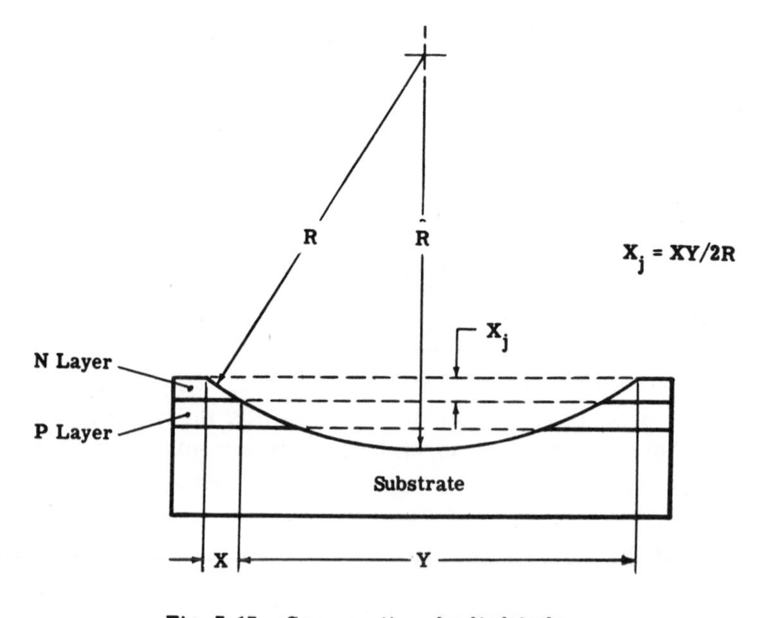

Fig. 7-17 Cross section of cylindrical groove

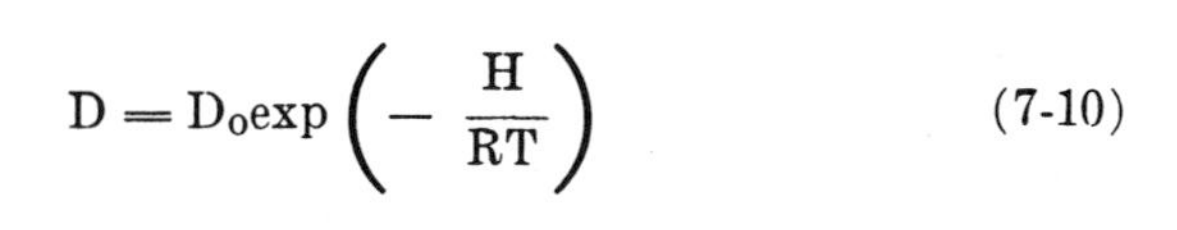

$$D = D_0 \exp\left(-\frac{H}{RT}\right) \quad (7\text{-}10)$$

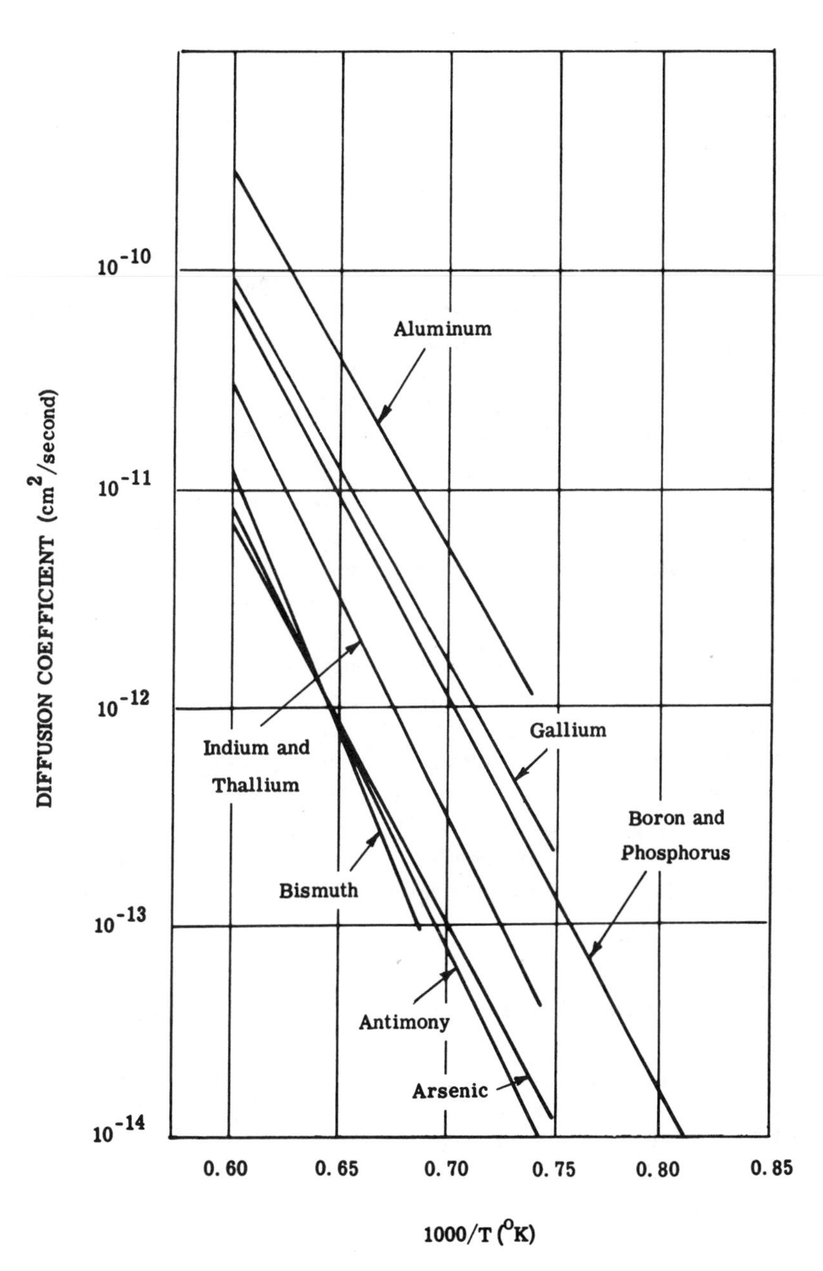

Fig. 7-18 Diffusion of donor and acceptor element into silicon by P-N junction method

where

D = diffusion constant in cm²/sec

D_0 = a constant which is temperature independent

H = activation energy for diffusion, in calories

R = ideal gas constant (1.99 cal/deg)
T = absolute temperature.

The above equation may be rewritten:

$$ln D = ln D_0 - \frac{H}{R}\frac{1}{T} \qquad (7\text{-}11)$$

If $ln D$ is plotted versus 1/T, the value of D_0 is obtained from the Y intercept of the line, and H/R, from which H can be computed, is obtained from the slope of the line. In Fig. 7-18 it should be noted that the slopes of the lines do not vary significantly, indicating the similarity of the activation energies H for the various dopants. At a given temperature, however, the values of the diffusion constants for the various elements show considerable variation. Displacement of the lines is caused by variations in the constant D_0. At 1106° C (0.725 on the X axis) the diffusion constant for boron is 4×10^{-13}. At 1155° C (0.70 on the X axis), the diffusion constant is 1.2×10^{-12}. A change of 50° C has caused the diffusion constant to vary three-fold. Thus, it is evident that unless close control is maintained over the temperature, little control can be exercised over the total impurity concentration, surface concentration, and the junction depth.

P and N diffused layers can be distinguished from each other by the use of a hot probe. If two probes are connected through a zero center galvanometer and touched to the surface of a sample and one of the probes is heated, a potential can be detected between them. The polarity of the cooler probe is positive for P-type and negative for N-type material. The effect is thermoelectric and can be produced by either heating or cooling one of the probes. It is convenient, in the laboratory, to use the tip of a small soldering iron as one of the probes. Commercial conductivity type measurement devices are available in which the hot probe is replaced by one which is thermoelectrically cooled.

7.3 CHOICE OF DIFFUSANTS

Boron (P) and Phosphorus (N). Boron and phosphorus have lower diffusion rates in SiO_2 than in silicon. Consequently, SiO_2 can be used to mask them. In silicon, their diffusion constants are quite similar, and they both have high solid solubility limits. Due to these properties, boron and phosphorus are widely used in NPN and PNP structures, and in resistor diffusions.

Arsenic (N). Arsenic is useful in certain applications where it is intended to minimize diffusion as a result of subsequent heat cycling. It has a high solid solubility in silicon, and a low diffusion constant in both silicon and SiO_2. It can be selectively masked. Arsenic is advantageously used as the background dopant or in epitaxial layers since it tends to remain relatively fixed in position. P-type wafers are often used as substrates on which N^+ and N epitaxial layers, doped with arsenic, are grown. An NPN structure is then diffused into the uppermost N layer.

Antimony (N). Antimony behaves very much like arsenic in SiO_2 and silicon. It is masked by the oxide and has a low diffusion constant. Antimony has a much lower solid solubility limit in silicon than arsenic. It is often difficult to obtain surface concentrations in excess of 10^{19} by diffusion of antimony.

Aluminum (P). Aluminum has the highest diffusion constant and lowest limit of solid solubility of any of the Groups III and V elements mentioned so far. Because of its high diffusion constant, it is seldom used in first or second diffusions where subsequent temperature cycling would redistribute it at a rapid rate. Aluminum has also a high diffusion constant in SiO_2. This tends to limit its usefulness as a dopant in diffused devices.

Due to these characteristics, aluminum is widely used in making contact and interconnections on diffused devices. Although aluminum is a p-type dopant, it can be used to make contact to both P and N areas. If it is intended to make contact to previously diffused P regions, aluminum can be deposited on the surface and alloyed and diffused into the silicon to make electrical contact with it. In the silicon, an enhancement of the P region occurs and the contact is ohmic. If, on the other hand, the region to which the contact is to be made is N, a rectifying contact will occur, if the net concentration of aluminum is greater than the concentration of the N dopant by virtue of the formation of a small junction.

In the formation of an electrical contact, a thin film of aluminum is evaporated and alloyed into the surface of the silicon. Upon cooling, the aluminum-silicon melt crystallizes and aluminum is rejected back to the surface of the wafer. A small amount of aluminum, about 5×10^{18} atoms/cc, is retained within the recrystallized silicon. If the background concentration of dopant is N-type and less than 5×10^{18}, a rectifying junction occurs and the contact is rectifying. If the background concentration is N-type and greater than 5×10^{18}, the aluminum is compensated by the background dopant, there is a net excess of N, and the contact is ohmic.

In an NPN structure where the collector is N-type and doped to about 10^{16}, or in a PNP structure where the base is N-type and doped to 10^{18}, it is customary to diffuse a small area of the collector or base, respectively, with additional N dopant during a subsequent diffusion. This raises the level of the N-type concentration so that an ohmic contact can be made to it by aluminum alloying. Diffusions of this type are known as "enhancement" diffusions and the regions are called N^+ regions.

Gallium (P). Gallium is seldom used as a dopant because it has a relatively low solid solubility limit in silicon, and a high diffusion rate in SiO_2. As a result, it cannot be effectively masked by SiO_2.

Gold (P). Unlike Group III and V elements, gold, a Group I element, occupies an interstitial position in the silicon lattice. It has a very low solid solubility limit in silicon, and diffuses at a rate much faster than any Group III or Group V element. Gold has the unique property of lowering the lifetime of minority carriers in silicon by the introduction of recombination centers, making it thereby useful in the fabrication of high-speed switching junctions.

Gold is one of the few elements diffused from an elemental source. Gold diffusions are usually accomplished by electroplating or vacuum evaporating a thin layer of gold on the back side of the wafer and subsequently heating the wafer to the diffusion temperature. After the diffusion, it is necessary to cool the wafer very quickly in order to prevent the precipitation of the gold from the lattice along crystal imperfections. Due to its high diffusion constant, the diffusion of gold is generally reserved as the last diffusion in a sequence.

7.4 SURFACE AND BULK REQUIREMENTS

From the previous discussions on oxidation and diffusion, it is evident that imperfections in the starting material can cause many problems during subsequent processing steps. It has been assumed that diffusions and oxidations were performed on flat, planar surfaces, and that the bulk of the wafer had a high order of crystallographic uniformity and perfection.

The method for realizing wafers of this quality, begins with crude silicon obtained from sand by reduction in an electric furnace with coke. The first step in the refining process involves the conversion of the crude silicon to a volatile compound (silicon tetrachloride, silicon tetrahydride, etc.) which is reduced or decomposed to yield higher purity silicon. The silicon is then melted and "zone-refined", in a horizontal furnace, to the highest purity, after which it is placed in a vertical furnace and doped with diffusant impurity. A monocrystalline ingot is then "pulled" from the melt by slowly rotating and withdrawing a seed of preferred orientation at the surface of the melt. The resulting product is a cylindrical bar or ingot, the entire body of which is one crystal, doped with an appropriate impurity. The ingot is then sliced to yield rough wafer blanks of small thickness. The wafers are lapped and polished, using successively finer abrasives, until a mirror finish is obtained. They are then ready for chemical processing.

Due to the abrasive action of the polishing grits, the crystalline structure of the surface is disturbed. The first step in a diffusion process always includes an etching operation, the function of which is to remove this disturbed layer and at the same time maintain the mirror surface finish obtained during polishing. The final polish with 0.01 micron grit leaves the surface with a microvariation of less than 25 Å. Wet chemical etching alone degrades the surface leaving a microvariation of about 200 Å. Gas etching techniques, using hydrogen chloride at elevated temperatures, have improved these results somewhat. Gas etching has the advantage that it can be done in the furnace immediately before oxidation, diffusion or epitaxy, thus, minimizing possible contamination through handling and transfer from one operation to the other.

It is often apparent that considerable discrepancies may exist between experiment and theory in diffusion work. It has been found, for example, that diffusion profiles and diffusion depths may not agree with predicted theoretical values if diffusions are made into layers which are highly dis-

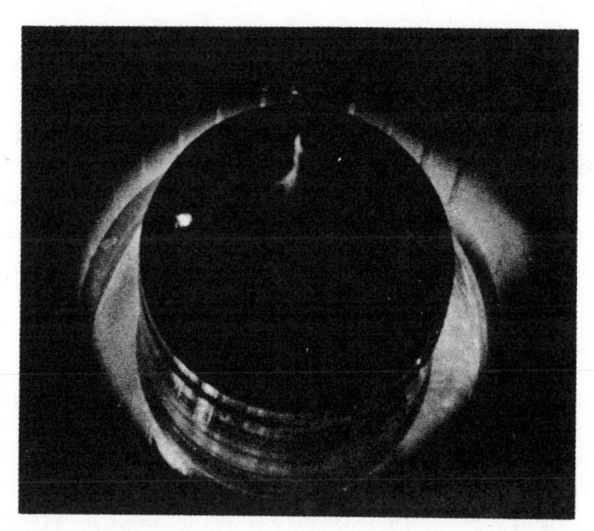

Fig. 7-19 Large diameter ($4\frac{1}{2}''$) silicon crystal in the process of being grown.

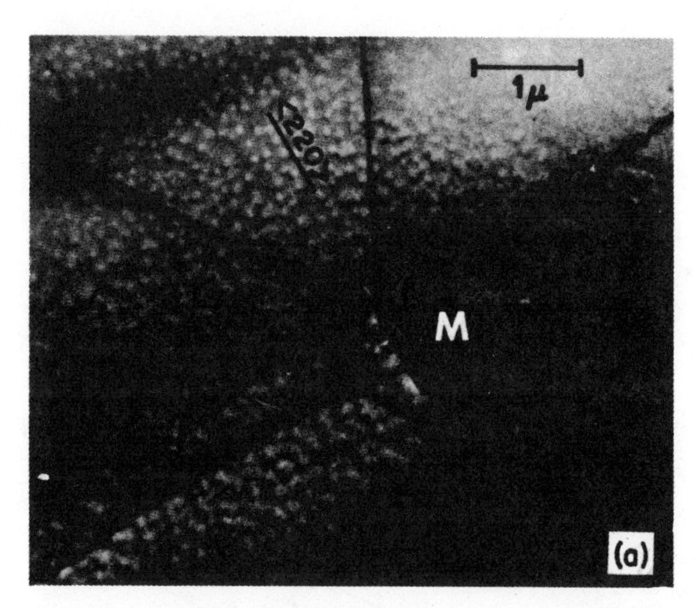

Fig. 7-20 Precipitates of phosphorus in silicon corresponding to shallow diffusion. A dislocation patch at the intersection of a precipitate platelet M is generated in (111) silicon.

turbed. It is possible in predeposition, to exceed a predicted inpurity concentration if diffusion is accompanied by precipitation along grain boundaries on other crystallographic defects. Resistivity measurements do not detect these precipitated, "electrically inactive," impurities. Such effects are minimized if highly oriented substrates are used.

Cleanliness is a second factor which must be taken into account in diffusion work. Foreign material such as dust, dirt, water vapor and residual resist, can all help to degrade the junction. A small amount of foreign matter present in the area to be diffused can increase the solid solubility limit to the point where base "punch-through" (the diffusion of an emitter spike through the base of the transistor), can occur. In addition, to the use of clean rooms and dust-free boxes, considerable care is usually exercised to protect wafers between processing steps and to clean the wafers immediately before an operation is performed. Whenever possible, cleaning is done in the same apparatus in which the step is performed, and without a lapse of time between cleaning and processing.

7.5 ELECTRICAL CONTACTS TO THE DIFFUSED LAYER

The extreme thinness of the diffused layer created obstacles, which required a search for new methods to make alloyed electrical contacts to the layer without puncturing or otherwise destroying it. Another problem required a solution just as urgently as that of the thin base layer. For high-frequency performance, these alloyed regions have to be as close together as possible – without electrically shorting. This problem was solved by a vacuum-evaporation technique. The evaporation technique allows the controlled deposition of thin layers of materials. It is also a method for

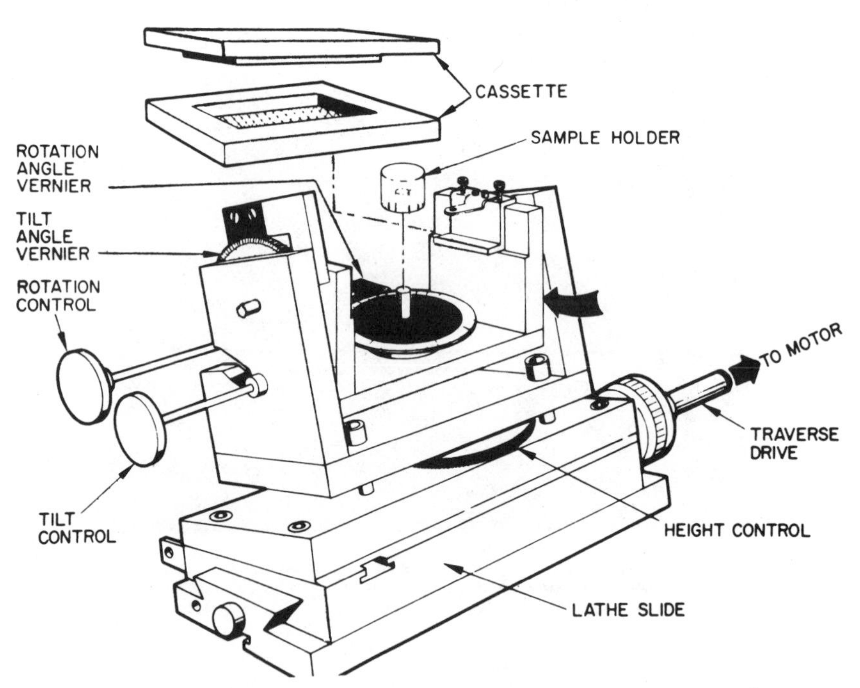

Fig. 7-21 Schematic diagram of Scanning Berg-Barrett camera used to obtain topographs of silicon wafers.

depositing highly reactive materials, such as aluminum, that are difficult to work with in air. Further, it allows greater cleanliness in technique and hence gives a more intimate contact between the layer and the surface upon which it has been cooled. In addition, the fact that the evaporation beams travel in straight lines permits the use of precisely dimensioned masks to create similarly precise evaporated patterns. An alloying operation follows the evaporation of the gold antimony or aluminum contact.

7.6 ION IMPLANTATION TECHNOLOGY

In this method, impurities are introduced into a crystal through the application of a high energy ion beam. In the early development stages of this method, radiation damage resulting from the use of high energy ion beams delayed the practical application of this technique. Subsequently, it was found that moderate thermal annealing, compatible with standard semiconductor processing, is sufficient to remove radiation which might otherwise lower device and circuit performance.

Another factor which lessened the immediate application of implantation was one of device geometry. In the past, relatively deep bipolar transistors were the standard throughout the integrated circuit industry. Since the penetration depth of an ion beam is directly related to its energy, and very energetic ion beams require large and costly acceleration equipment, deep geometries were generally incompatible with economical semiconductor fabrication using implantation. As a result, diffusion was the dominant integrated circuit process technology.

The greatest impact of ion-implantation has been in MOS circuits where there are often demands for stringent process controls. Significant improvements in frequency, extensive freedom to select gate threshold voltage independent of gate metal or dielectric and high – valve resistors are some of the advantages to be gained with ion-implanted MOS technology.

The success of ion implantation as an industrial processing tool has come about mainly through a growing number of applications. The main properities of ion implantation used in present device and circuit applications are: low temperature semiconductor doping, implantation masks from insulators, metals and photoresist, substantial accurate doping down to the lowest levels and effective control of the doping profile.

In ion implantation the dopant atoms are ionized (stripped of one or more of their electrons) and are accelerated to a high energy by passing them through a potential difference of tens of thousands of volts. At the end of their path they strike the silicon wafer and are embedded at various depths depending on their mass and their energy. The wafer can be selectively masked against the ions either by a patterned oxide layer, as in conventional diffusion, or by a photoresist pattern.

As the accelerated ions plow their way into the silicon crystal they cause considerable damage to the crystal lattice. It is possible to heal most of the damage, however, by annealing the crystal at a moderate temperature. Little diffusion takes place at the annealing temperature, so that the ion-implantation conditions can be chosen to obtain the desired distribution. For example, a very shallow, high concentration of dopant can be conveniently achieved by ion implantation. A more significant feature of the technique is the possibility of accurately controlling the concentration of the dopant. The ions bombarding the crystal each carry a charge, and by measuring the total charge that accumulates the number of impurities can be precisely determined. Hence ion implantation is used whenever the doping level must be very accurately controlled. Often ion implantation simply replaces the predeposit step of a diffusion process. Ion implantation is also used to introduce impurities that are difficult to predeposit from a high-temperature vapor. For example, the current exploration of the use of arsenic as a shallow n-type dopant in MOS devices coincides with the availability of suitable ion-implantation equipment.

A unique feature of ion implantation is its ability to introduce impurities through a thin oxide. This technique is particularly advantageous in adjusting the threshold voltage of MOS transistors. Either n-type or p-type dopants can be implanted through the gate oxide, resulting in either a decrease or an increase of the threshold voltage of the device. Thus by means of the ion-implantation technique it is possible to fabricate several different types of MOS transistors on the same wafer.

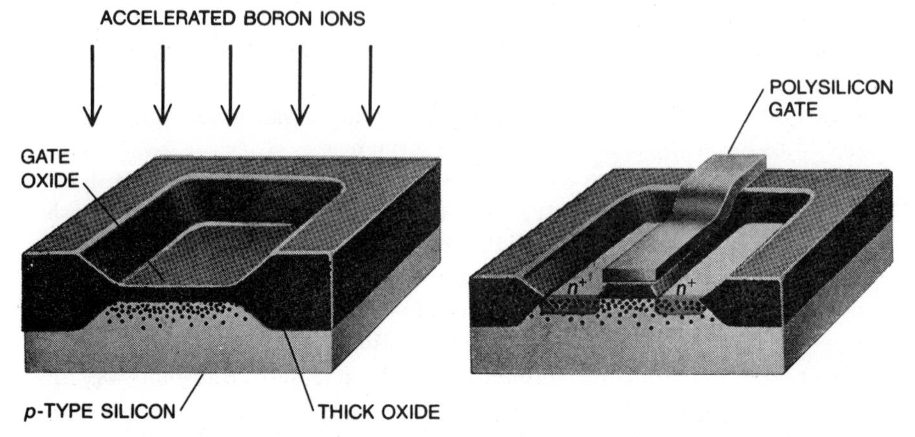

ION IMPLANTATION is employed to place a precisely controlled amount of dopant (in this case boron ions) below the gate oxide of an MOS transistor. By choosing a suitable acceleration voltage the ions can be made to just penetrate the gate oxide but not the thicker oxide (*left*). After the boron ions are implanted polycrystalline silicon is deposited and patterned to form the gate regions of the transistor. A thin layer of the oxide is then removed, and the source and drain regions of the transistor are formed by the diffusion of an n-type impurity (*right*).

7.6.1 ION-IMPLANTED RESISTORS

The development of ion implantation technology has provided an alternate fabrication technique for monolithic precision resistors which avoids some of the difficulties associated with thin-film technology. In most cases, the additional processing steps required for ion-implantation are non-critical and compatible with conventional planar processing. Compared to other monolithic or thin-film resistor technologies, ion-implanted resistors provide the following advantages:

(a) Availability of a wide range of sheet resistivities.

(b) Good control over absolute value and matching tolerances.

(c) Compatibility with conventional processing.

The availability of a wide range of sheet resistance values, particularly in the high resistivity region, is especially useful in the design of monolithic micropower circuits. Similarly, the good control of the absolute values and the close matching and tracking properties make implanted resistors suitable for precision attenuators or ladder networks.

Ion implantation offers an alternative to resistor fabrication, which avoids some of the shortcomings of the other techniques based on film or diffused components, (Fig. 7-22). In particular, boron implanted resistors having particularly low temperature sensitivity may be fabricated over a wide range of sheet resistivity. The use of boron for this application is particularly desirable since heavier ions (such as phosphorous and arsenic) would require implantation energies likely to be unavailable in production implantation equipment. The absolute value tolerance of ion implanted

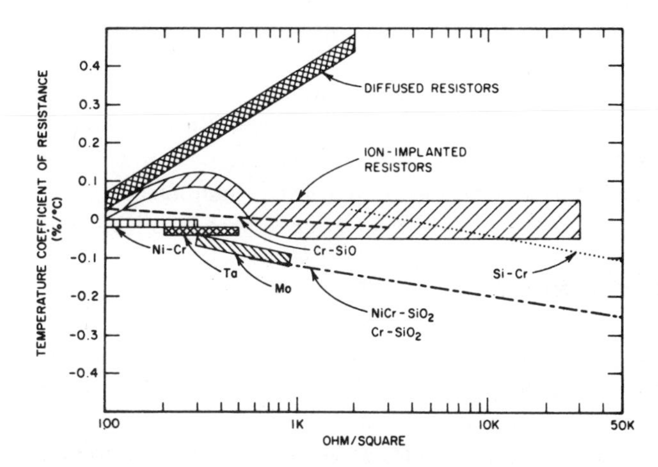

Fig. 7-22 Temperature coefficient vs. sheet resistivity for diffused, thin-film and ion-implanted resistors.

resistors is superior to that of both diffused and thin-film resistors because of the more precise doping control available through ion beam-monitoring. The matching tolerance obtained with ion implantation is comparable to that obtained with thin-film technology.

Boron implanted planar resistors are fabricated using the same preliminary processing as required for diffused resistors. A thermally grown oxide layer is selectively etched in order to define the resistor pattern.

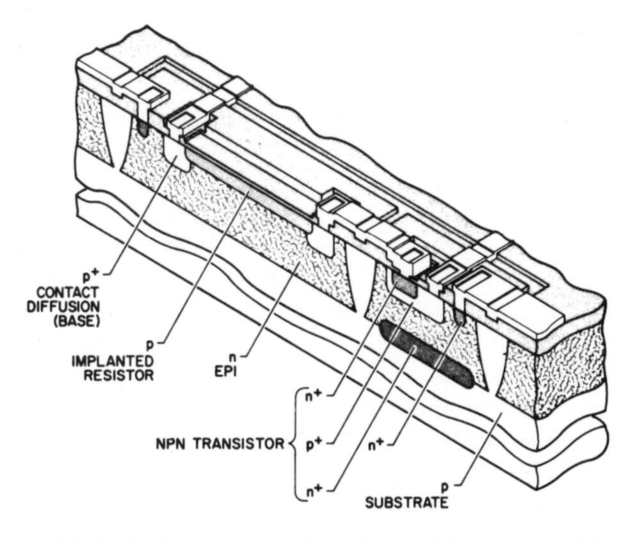

Fig. 7-23 Cross-section of ion-implanted resistor and bipolar transistor.

After that a thin protective oxide layer is grown on the exposed silicon. Impurity atoms are then introduced by implanting a controlled number of boron ions through the protective oxide layer. The distribution of impurity atoms is thus completely determined by the proper choice of the following process variables: (a) ion dose, (b) energy, (c) wafer orientation and (d) thickness of the protective oxide. Following implantation, the implanted layer is subjected to a heat treatment at a temperature well below that used in conventional diffusion technology. Therefore, this post-implantation heat treatment is compatible with any of the processing steps prior to implantation.

Figure 7-23 shows a cross-section of an ion implanted resistor together with a typical planar NPN transistor. The base diffusion step is used to simultaneously form the NPN base and the p-type contact beds for the resistors. After bipolar processing has been completed, the implanted resistors are fabricated with the additional processing outlined above.

7.6.2 ION-IMPLANTED MOS DEVICES

Ion implantation can be used in conjunction with conventional, MOS P-channel processing to produce both depletion-mode and enhancement-mode devices on the same monolithic IC. Used in place of conventional MOS load resistors, depletion-mode devices offer constant current operation with resulting performance improvements of speed, power, and operating power suppy voltage range. Circuits implemented with depletion-mode loads can be designed to operate from a single power supply. Additionally, linear MOS circuits can be integrated along with digital circuitry on the same chip.

7.6.3 DEPLETION DEVICE FUNDAMENTALS

A *depletion-mode* field-effect transistor is defined as one which exhibits substantial device current (I_{DSS}) with zero gate-to-source bias ($V_{GS}=0V$). An *enhancement-mode* field-effect transistor, on the other hand, is defined as one in which no device current flows (leakage only) when V_{GS} is zero volts. Conduction does not begin until V_{GS} reaches a voltage called the *threshold voltage,* V_T.

Junction FETs are all depletion-mode devices, whether N or P-channel. MOSFETs, on the other hand, tend to be enhancement-mode when P-channel processing is used, and depletion-mode with N-channel processing. Ion implantation can be used to controllably alter these natural tendencies. Figure 7-24 shows how both low-threshold, enhancement devices and depletion devices are produced on the same chip. The starting process is standard, <111> high-threshold P-channel. The first implantation step lowers the threshold of *all* MOS devices on the chip by two volts. A second implantation lowers the threshold of *selected* devices still further, through zero volts to +5V. This voltage is termed the *pinch-off voltage,* V_P, to be consistent with depletion device terminology. Selection of depletion devices is achieved by the addition of one extra masking step to the basic process.

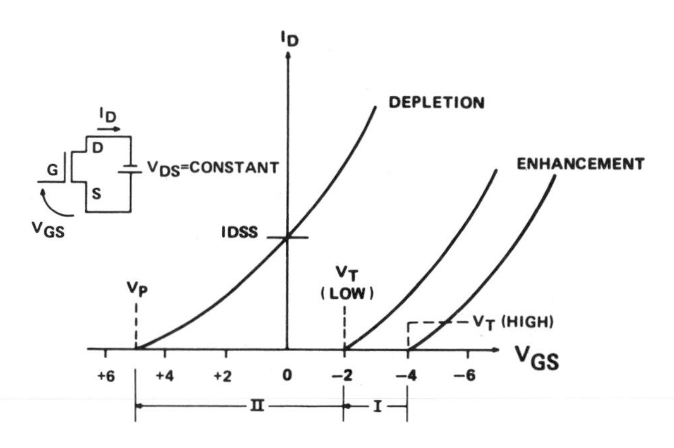

Fig. 7-24 Threshold voltage reduction by ion implantation.

7.6.4 TECHNIQUES FOR OBTAINING UNIFORM IMPLANTS

If ion implantation is to produce improved yields, then the implantation process must be extremely uniform throughout the surface of the wafer. Non-uniformities can arise from at least five different causes. These are: improper treatment of the neutral beam, non-linearities or other problems associated with the scanning system, unstable ion-source output, instabilities in accelerator or mass analyzer power supplies, and inadequate suppression of secondary electrons.

Figure 7-25 shows a typical ion implantation system including all the essentials: ion source, focusing electrodes, acceleration tube, mass analyzer, neutral trap, beam deflector, raster generator and wafer chamber.

The beam of ions travelling from the source to the target undergoes some neutralization due to collisions. The scanning system can produce no deflection on the neutral ions. Therefore, the target must not be allowed a direct path to the beam before the scanning system. Various methods are used to correct this problem – all producing generally the same effect. Typical is the 5° or so deflector between the mass analyzer and the raster generator. Of course, a high vacuum, preferably a few times 10^{-7} or less, will keep neutralization to a minimum. The ion beam is deflected but the neutral beam is trapped. Alternatively, the scanning system is operated with a bias so that the neutral beam simply lands off target.

The beam cannot be scanned in two planes at rates which produce a stationary pattern (or "Lissajou figure"). Also of importance, are the aberrations caused by the deflection system. A saw-tooth or triangular wave should be used which has a high degree of linearity. Stray capacitances invariably will cause rounding of the waveform at the extremes. Therefore, it is necessary to scan a width at least equal to the wafer diameter, plus two beam widths to avoid a greater dose at the edges of the wafer than at the center. It would appear best to scan at as high a rate as possible, at least in one plane, and to use a uniform (round) diffuse, "bell-shaped" beam.

If the output of the ion source varies, the wafer will carry a record of that variation implanted on it. "Averaging" by scanning the wafer over and over again will serve to obscure this "record." Some ion sources apparently cannot be made to produce exceptionally stable beams, and all

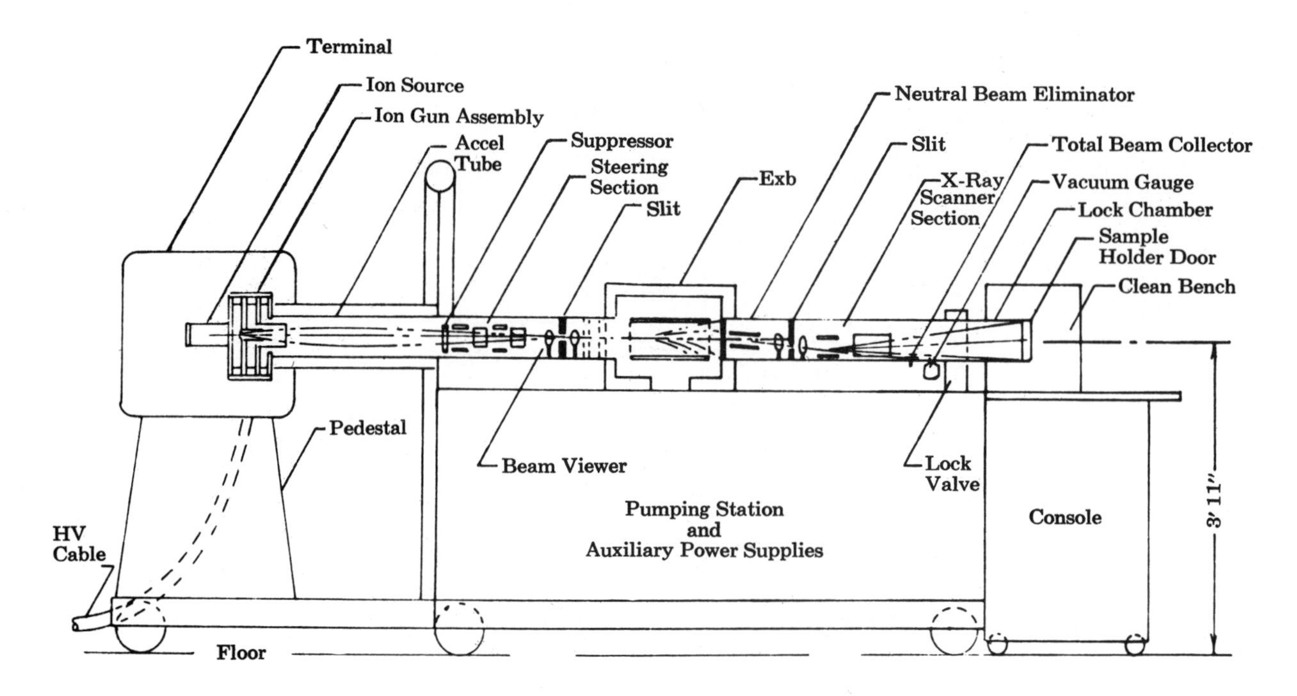

Fig. 7-25 Ion implanting system.

of them have instabilities when operated improperly. The rf, cold-cathode, and magnetron-type sources seem to be the most popular ones for this type of work. It should be realized, at the same time, that a great many different ion sources are available and new ones are being developed. The rf source, long used in high voltage accelerators for producing protons and deuterons, has found great usefulness in this field. It has been successfully used to produce high quality ion beams of a large number of elements all the way from hydrogen up to and including gold through the use of compounds containing the desired elements sometimes operating in combination with some "carrier" or sputtering gas. Some of the other sources utilize solid materials directly by evaporation and include an "oven" inside the source, which carries a tiny charge of the material and whose temperature can be controlled externally.

The acceleration voltage and mass analyzer parameters which must be carefully adjusted, must remain absolutely fixed. The solution to this problem is to use ultra-stable power supplies. Some mass analyzers have been built which utilize permanent magnets. One example of this is the 120 keV automated system at Hughes Aircraft MOS Division.

When an area of large dimensions must be covered — for example, more than an inch or so square—the adequate suppression of secondary electrons can become a problem. It becomes increasingly difficult to suppress secondaries near the center of the area as the raster dimensions are increased. Furthermore, the field near the edge of the suppressor electrode has a focusing effect on the beam, particularly at low energies, and tends to cause severe non-uniformities in that region.

One solution has been to put a suppressor "mask" in front of the wafer group which has a hole for each wafer. A grounded plate is placed in front of it, and the holes in both plates are registered with the wafers so as not to shadow any part of them from the beam. Chambers are made which scan one wafer at a time, but have holders which can accommodate large numbers of wafers.

Figure 7-26 shows an ion implantation chamber. It has a slice holder which accommodates 25 wafers 2″ dia. as standard equipment.

One solution to the problems of secondary electron suppression, scan uniformity and neutral beam elimination, has been the mechanical scanner. In this method, the wafers themselves move past a stationary ion beam at a constant rate. Since the beam is stationary, the suppressor electrode can be small. In this case the neutral beam would not have to be eliminated but could itself be used. (This would not be so in the case of cross-field analyzers). In the mechanical scanner the effective writing rate of the beam cannot be as fast as in electrical scanners, so "averaging" would take more time. Therefore, beam stability is even more important with the mechanical system. However, a great many wafers can be processed together, and better reproducibility can probably be achieved with this method.

Fig. 7-26 Ion implantation chamber.

CHAPTER 8

EPITAXY

Epitaxy is a uniquely defined form of crystal growing. When one crystal is grown upon another and the growth receives its crystallographic orientation from the crystal upon which it is grown, the growth is said to be epitaxial. We can distinguish between two forms of epitaxy. An *isoepitaxial* growth results when the seed and growth are of the same material, such as silicon grown on silicon. A *heteroepitaxial* growth is one where the seed and growth are of different materials, such as germanium on silicon or gallium arsenide on germanium. In a very restricted sense, a growth is epitaxial only if both the seed and growth are monocrystalline, the growth has been oriented crystallographically by the seed, and no lattice discontinuities exist at the seed-growth interface; in other words, if the new structure is itself monocrystalline.

In actual practice, the term epitaxy is used more loosely to describe any crystal growth process wherein both the seed and growth are monocrystalline, regardless of the crystallographic orientation of the growth with respect to the seed. When growths are isoepitaxial, the seed and growth usually have the same crystallographic orientation. When growths are heteroepitaxial, this is not always true.

8.1 PROCESS THEORY

A completely rigorous analysis of epitaxy is virtually impossible, since the mechanisms governing epitaxial growth are not completely understood. However, some simplified models for epitaxial growth have been advanced and are discussed below. Growth mechanisms can be divided into "direct" and "indirect" processes.

8.1.1 DIRECT PROCESSES

In the direct process, silicon is transferred from the source to the substrate with no intermediate reactions or events. Evaporation, sublimation and sputtering are examples of this technique. It is assumed that silicon atoms impinge on the substrate surface and are held there by interatomic forces. Under proper conditions the atoms are mobile and can add onto a step (discontinuity in the growing epitaxial layer) or can re-evaporate. The growth process, therefore, can be presumed to involve the lateral movement of steps across the substrate surface as depicted in Fig. 8-1. Steps can be formed either by the two dimensional nucleation of islands on the surface resulting from the initial deposition of silicon, or by a spiral dislocation existing on the substrate. Epitaxial growth occurs because of the influence of the substrate on the two dimensional nucleus, and because of good lattice matching as the steps proceed across the surface.

The rate of two dimensional nucleation can become large at high temperatures. This rate is dependent on the concentration of silicon in the vapor and on the free energy of formation of the nucleus, i.e.,

$$R_n \propto n_o \exp(-\Delta G/kT) \tag{8-1}$$

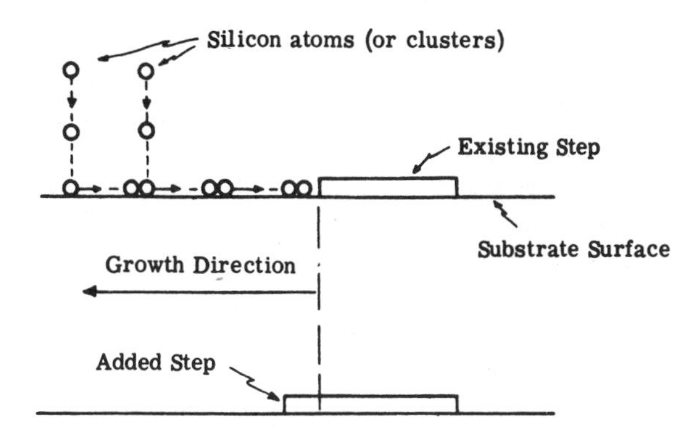

Fig. 8-1 Progress of a step during epitaxial growth

where R_n = two dimensional nucleation rate
n_o = concentration of silicon in the vapor
ΔG = free energy of formation of the nucleus,
T = temperature of deposition,
k = Boltzmann constant.

For large values of R_n, many steps are available for growth and good epitaxial growth would be expected to occur. Even for large values of n_o, R_n may still be small if the deposition temperature is reduced as seen from Eq. (8-1). Those steps formed (at low temperatures) quickly grow to a maximum extent and produce a ''singular'' or low index face.

Impurities tend to slow step motion, and can result in the formation of steps, many atoms high. Impurities may increase or decrease the nucleation rate, and they complicate the growth process.

8.1.2 INDIRECT PROCESSES

Indirect processes are those in which silicon atoms are obtained by the decomposition of a vapor of a silicon compound at the substrate surface. Examples of these are the hydrogen reduction of $SiCl_4$, $SiBr_4$, $SiHCl_3$ and the pyrolysis of SiH_4. There is some disagreement as to the exact decomposition reactions which produce silicon atoms during epitaxial growth. One hypothesis is that silicon is formed on the substrate by a surface controlled reaction. Another is that the silicon is separated from its compound in the gas phase, in a region a few microns from the substrate, resulting in individual atoms, groups of atoms, or overcooled droplets, with deposition occuring by diffusion.

For cases where the reaction is surface controlled, a classical heterogeneous reaction consisting of the following events is proposed:

a. Mass transfer of the reactants to the substrate surface,
b. Adsorption of the reactant onto the surface,
c. The reaction or series of reactions which occur on the surface,
d. Desorption of the by-product molecules,
e. Mass transfer of the by-product molecules to the main gas stream, and,
f. Addition of atoms to growth steps.

If one of the events in the heterogeneous process is much slower than the others, it will govern the growth rate. The mass transfer rate is influenced by the gas flow rate, but is relatively temperature independent. The other events, however, are strongly dependent on temperature, and practically independent of gas flow rate.

At low growth temperatures, the silicon compound adsorption rate seems to be the controlling factor and the reaction is controlled by the surface reaction rate. At high temperatures, the mass transfer rate of reactants is the controlling step.

In the alternate hypothesis of epitaxial deposition, silicon atoms are freed from their compounds adjacent to the substrate in a layer of gas which is in a state of laminar flow; transport to and from the substrate is governed by diffusion through this layer. After reaching the substrate, the silicon atoms are mobile and can align preferentially as before. Under certain conditions, the concentration of silicon atoms in the gaseous region over the substrate is not uniform. This can lead to imperfect epitaxial layers, e.g., ledges or pyramids are formed. One method of obtaining this instability is the reduction of the growth temperature which reduces the nucleation rate causing a rough irregular surface.

8.1.3 DEFECTS

Crystal imperfections and defects play an important role in the mechanical and electrical properties of semiconductor materials and devices. The cleanliness and perfection of the substrate wafer are particularly important in the prevention of such defects in epitaxial layers. Regardless of the precautions taken in precleaning the substrates before epitaxial growth, some mechanical imperfections, SiO_2, residual dust, and other impurities remain on the wafer surface. These impurities may be expected to affect growth in a number of ways: they may increase or decrease the rate of formation or nucleation of steps; they may cause the formation of imperfect nuclei; they may impair the movement of steps on the surface of the layer and cause dislocations; and, they may also become part of the crystal lattice causing lattice strain and distortion. These contaminants appear as dislocations, line defects, etch pits, stacking faults, hillocks or tri-pyramids, cones, and polycrystalline regions.

The most common type of defect in single crystal silicon is the edge dislocation which is caused by an imperfect array of atoms in localized regions within a crystal. A schematic of an edge dislocation in the silicon lattice is shown in Fig. 8-2. Individual atoms are represented by line intersections. Atoms present in a common crystallographic plane are connected by straight lines. The distortion in the lattice at A is caused by the termination of the row of atoms AA′ at A. If other planes of atoms exactly like the one shown are stacked directly below the plane of the paper, a plane of atoms, of which AA′ is a part, will trace a line perpendicular to the paper. This line is called an edge dislocation. Dislocations may act as nuclei for the precipitation of foreign atoms or impurities that may be present in the silicon lattice. These impurities migrate to dislocations to relieve part of the lattice strain present at the dislocation sites. The heterogeneous regions created, can impart poor electrical characteristics to devices containing them.

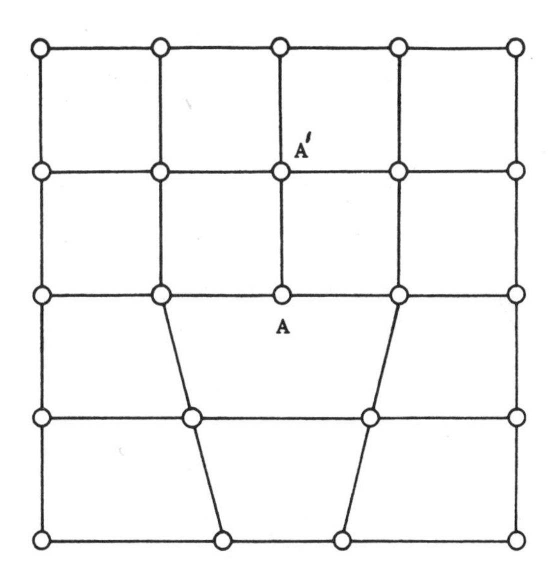

Fig. 8-2 Representation of an edge dislocation

The number of dislocations in the growing layer is substantially the same as the substrate, provided the wafers are thoroughly cleaned. By using low dislocation density substrate wafers that are properly cleaned, the degrading effects of dislocations can be minimized.

Another common type of defect in epitaxially grown silicon is the stacking fault which results when a plane of atoms is missing from the normal stacking sequence, or when an extra plane of atoms has been added. These defects propagate through the epitaxial layer and appear as triangles, semi-triangles, and lines, upon application of an appropriate etch.

Stacking faults are possibly caused either by the joining of two misoriented nuclei during initial film growth, or by localized surface defects in the substrate. If misoriented nuclei produce the stacking faults, other

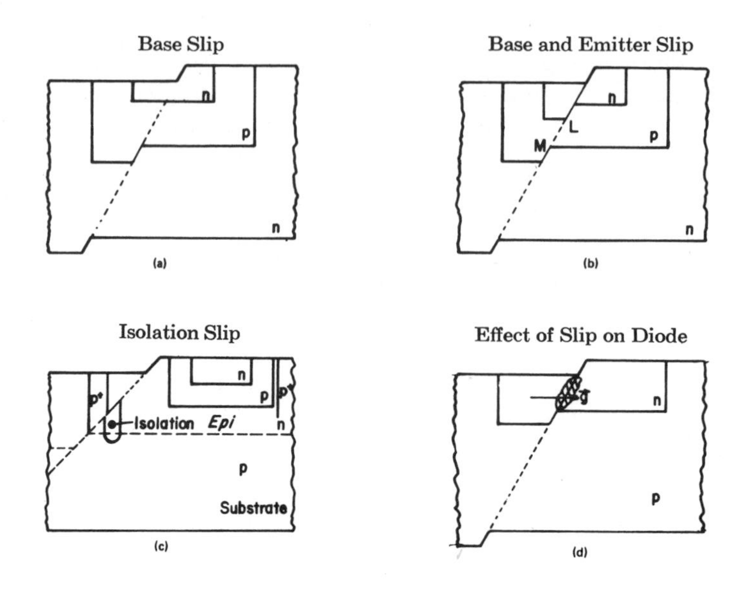

Effect of slip on vertical structures of transistors

defects may be produced, possibly as a result of the lattice strain introduced when the growing nuclei meet. This is the case when etch pits are present. If the layers grow out of the side of the pits, considerable distortion occurs when they meet and polycrystalline regions result. Stacking faults are ordinarily formed at the layer-substrate interface, but may be created in the growing layer.

The hillock or tri-pyramid appears with many variations. The defect may have a single high or low point at the center and may have two pyramids or only one. Stacking faults are often associated with these hillocks. The latter are thought to be formed in much the same manner as stacking faults, i.e., by an impurity (probably α-SiC or SiO_2) on the surface of the substrate. Upon repeated etching of the epitaxial layer, the hillock may terminate above the substrate surface or penetrate below it. The hillock, therefore, may be caused by an impurity within the substrate in addition to impurities on the surface. The formation of hillocks can be largely prevented by precleaning the substrate wafers in hydrogen at 1300° C for 40 to 90 minutes prior to epitaxial deposition.

High doping concentrations in the growing layer may cause pyramids or cones on the surface. The high doping concentrations probably reduce the rate of two dimensional nucleation. At maximum doping concentrations the epitaxial layer becomes completely covered with pyramids and cones and finally polycrystalline growth starts on top of these.

8.2 GROWTH TECHNIQUES

This section presents a detailed description of the technique for the epitaxial deposition of silicon. Primary attention is given to the deposition of silicon on silicon substrates by both vapor deposition and vacuum methods. The deposition of silicon on non-silicon insulating substrates is outlined briefly. Growth rates as a function of temperature, constituent pressure, and time are presented for the various deposition methods.

8.2.1 VAPOR DEPOSITION

In this method a variety of chemical reactions and deposition temperatures are used. Although each chemical reaction nominally yields the same results, certain processes produce higher quality epitaxial layers

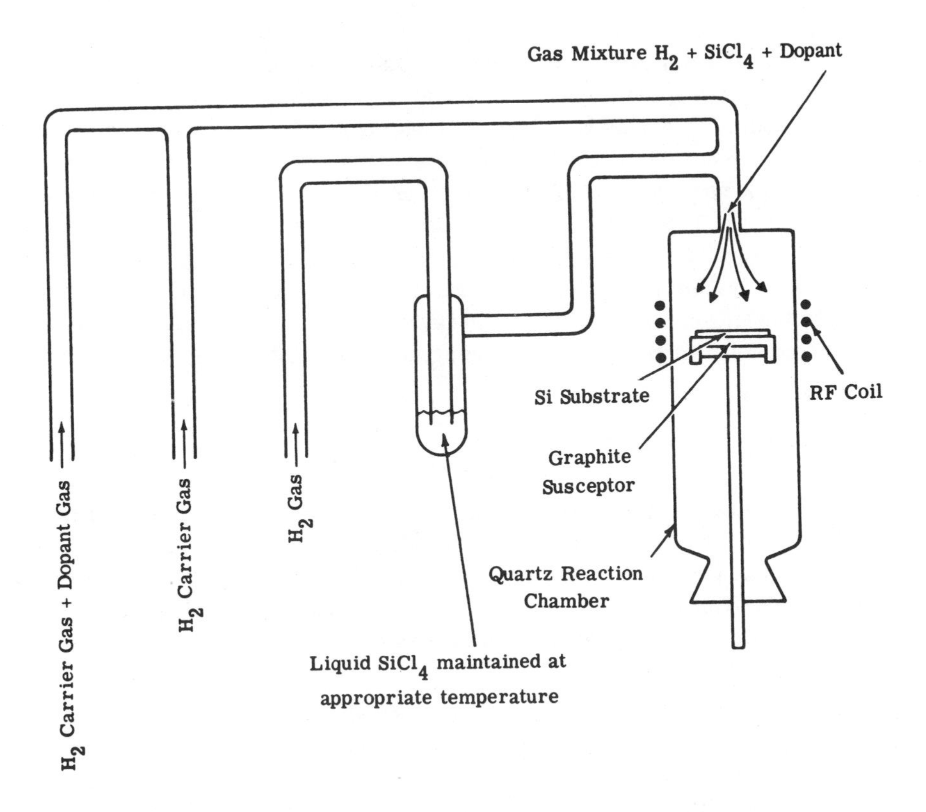

Fig. 8-3 Basic arrangement for growing doped epitaxial films of Si on Si

than others. The techniques include substrate preparation, etching and induction heating and susceptor requirements.

8.2.1.1 Substrate Preparation

A clean wafer, free of any deformation or imperfections, is necessary for good single-crystal epitaxial growth because even small irregularities are reproduced in the epitaxial layer. All growth methods require a series of mechanical lapping and polishing operations, with successive reduction in particle size of the abrasive. Each step removes enough material to insure freedom from the work damage, imposed by the preceding operation. A final slow chemical etch in an HF-HNO_3 solution removes oxides as well as the thin mechanically worked region. While this etch helps maintain planarity and produces a shiny, polished appearance, the resulting wafers are not without imperfections. Additional etching is performed in the epitaxial reactor. Water vapor or anhydrous HCl are most often used for this final etching.

After initial polishing the silicon substrate wafers are placed on a suitable susceptor in the reaction chamber where the epitaxial deposition is to take place. Vapor deposition techniques involve the passage of a vapor containing a silicon bearing compound over a heated seed crystal with conditions adjusted to favor epitaxial growth. The most frequently used source material is silicon tetrachloride which is a highly volatile liquid at room temperature. A typical deposition apparatus as shown in Fig. 8-3 is usually made of glass, and arranged so that a carrier gas passes through the vessel containing silicon tetrachloride to the heated crystal seed. The carrier gas acts as the transport medium for the vaporized silicon tetrachloride. When hydrogen is added to the gas stream, it reacts with the silicon tetrachloride on the surface of the Si wafer, reducing it to silicon which then deposits on the crystal seed epitaxially.

8.2.1.2 Doping Procedures

In epitaxy the impurity atoms are incorporated into the crystal lattice while the film is growing. The ratio of impurity atoms to silicon atoms, in the gas phase, is controlled so that the growing layer contains the desired impurity concentration. By changing either the impurity type or concentration, the electrical characteristics of the epitaxial layer can be varied over a wide range.

Two reproducible and stable doping techniques, solution doping and gas doping, involve the addition of appropriate Group III or V impurities to the gas stream during the epitaxial growth cycle.

Solution doping is used most successfully in doping n-type silicon to resistivities greater than 0.1 ohm-cm. For n-type solution doping of epitaxial layers grown by the hydrogen reduction of $SiCl_4$, volatile impurities, e.g., PCl_5 or $SbCl_5$, are added to the liquid $SiCl_4$ source. These impurities evaporate with the $SiCl_4$ and dope the epitaxial layer. Solution doping, however, requires a different solution for each doping level. This

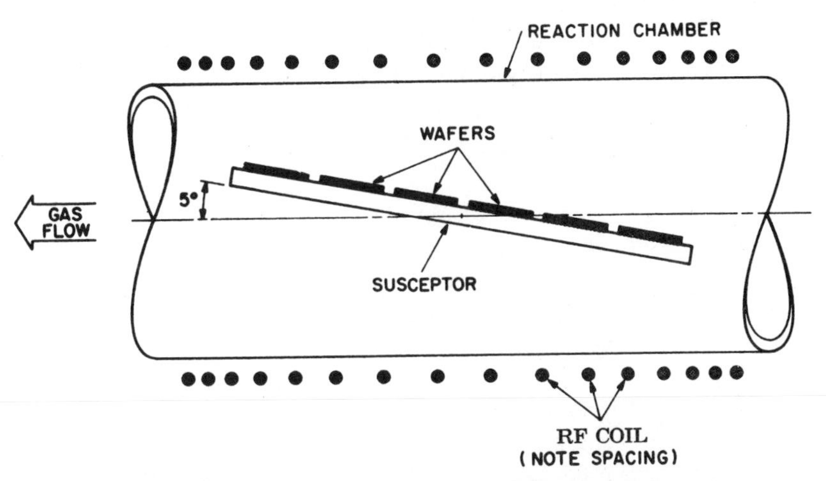

Fig. 8-3a The reaction chamber used in Fig. 8-3 consists of a long, cylindrical quartz tube. Single crystal silicon wafers are placed upon a graphite susceptor and placed into this reaction chamber. The controlled flow of gases grow the epitaxial layer.

lack of flexibility makes it less practical and economical for epitaxial depositions than gas doping.

The gas doping technique adds dopants to the system by separate gas streams. Many of these gases are quite poisonous and the proper safety precautions should be taken before using any of them. The common doping gases, B_2H_6, AsH_3, and PH_3 are diluted in H_2 and added to the system in desired concentrations. Phosphine can be regarded as being a stable compound for periods up to one year. However, B_2H_6 is stable for only a few months. For periods longer than this, the initial impurity analysis is no longer valid.

8.2.1.3 Susceptors

The susceptor is used to provide both support and heat for the substrate wafers during the deposition process. To prevent wafer contamination, the susceptor should be chemically inactive and have a minimum of outgassing.

The susceptor may be heated by induction or resistance heating. In induction heating the susceptor is placed inside a coil which carries r-f current. Eddy currents induced in the susceptor material cause the localized heating.

The r-f coil current is supplied by a conventional r-f oscillator through a voltage stepdown transformer. The susceptor should be about the same length as the work coil for most applications. Adjustment of the coil turns for a particular application allows the reflected impedance into the oscillator to be in the proper range. As an example, for a 450 kHz, 2 kw heater, the coil inductance is on the order of a microhenry.

The power generated by induction in a particular susceptor heats the substrate wafers by thermal conduction and radiation. The power lost by the susceptor is given by the Stefan-Boltzmann relation (assuming that the susceptor heat losses are by radiation only, and that the temperature is high enough so that the negligible heat is radiated to the susceptor by the surroundings):

$$P = \tau e \sigma T^4$$

where τ = surface area of the susceptor,
e = emissivity of the susceptor
σ = the Stefan-Boltzmann constant
[5.7×10^{-12} watts/cm^2°K^4],
T = susceptor temperature.

Graphite, coated with silicon or silica glass, pure silicon, and molybdenum-silicon glass compounds are materials normally used for susceptors. Silica glass has been used in some cases, but minute silica particles and deposited silicon tend to flake off the susceptor onto the substrate, causing imperfections in the deposited layer. Silica glass also increases the downstream doping of highly doped substrates. This is caused by the reaction of the silicon with the silica which forms the volatile monoxide and allows the substrate dopant to go into the gas stream. Susceptors of graphite are also objectionable because of volatile impurities in the porous graphite which contaminate the substrate wafers at the high reaction temperatures necessary for the epitaxial deposition of silicon. A pyrolytic graphite, however, overcomes this problem. It has negligible impurities, is free of voids, and is quite strong and durable.

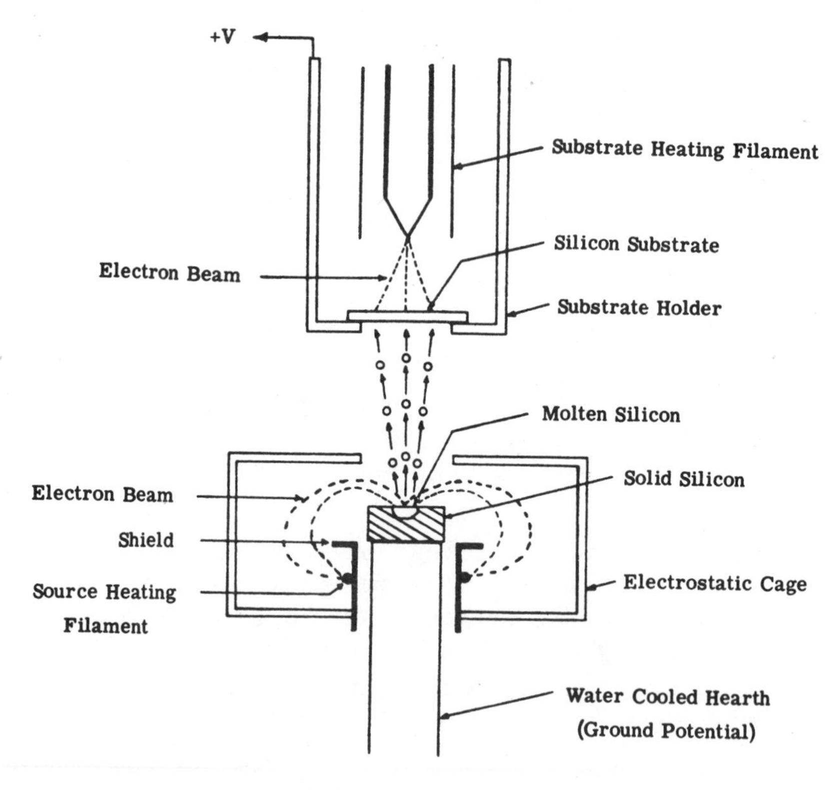

Fig. 8-4 Silicon evaporation and substrate heating by electron bombardment

8.2.2 VACUUM DEPOSITION

This process involves the growth of epitaxial layers by a direct transfer of silicon from source to substrate. Examples of this method are evaporation, sublimation and sputtering. The silicon atoms impinge on the substrate surface and stick. Under proper deposition conditions, the silicon atoms are mobile and align themselves according to the substrate crystal structure.

Although epitaxial films obtained by vacuum methods can be comparable in quality with those obtained by vapor deposition techniques, the vapor methods are better suited to mass production demands.

8.2.2.1 Evaporation

The most common method of evaporating silicon is by electron bombardment. The electron beam locally heats a small portion of a solid silicon block forming molten silicon. The molten silicon then acts as the vapor source; the solid silicon is supported on a water cooled hearth. A typical electron bombardment arrangement is shown in Fig. 8-4. The substrate is supplied with a positive potential and is heated by electron bombardment from a filament placed behind it. (Radiant heating may also be used but is less effective.)

The source of silicon is at ground potential and is heated by electron bombardment from another filament. The high negative potential surrounding the source, accelerates and directs the electrons to the desired area of the silicon. The focusing of the beam is a very sensitive function of the system geometry, but when correctly constructed, the heating is confined to an area less than 3 mm in diameter. Both the source and the substrate are shielded optically from the electron sources to limit contamination from filament outgassing and evaporation.

The molten silicon is supported by the solid silicon; this prevents contamination of the evaporating vapor since it is only in contact with silicon. This method also permits the temperature of the evaporating silicon to be maintained at several hundreds of °C above its melting point. The increased vapor pressure at these temperatures allows higher deposition rates. Deposition rates up to 4μ/min with minimum contamination may be obtained using systems similar to the type shown. In operation, the source silicon is evaporated for several minutes to allow the initial gas burst to be pumped away and to remove surface contamination. A thoroughly clean substrate is then heated rapidly to over 1200° C, to remove oxygen, and placed into position over the source. Deposition pressures of 10^{-7} to 10^{-10} torr are used.

8.2.2.2 Sublimation

Epitaxial growth of silicon by vacuum sublimation (evaporation directly from a solid) avoids the problem of retaining molten silicon, but lower deposition rates are realized. Suitable apparatus for growing silicon layers by this method, using induction heating, is shown in Fig. 8-5.

Source silicon vapor is transferred from the hot solid (about 1350° C), to the cooler substrate (about 1100° C), at a residual pressure of less than 10^{-8} torr. To prevent contamination, there are no metallic parts in the system. Dopant impurities are added to the films by using source silicon of known resistivity. The dopant transfer rate from the source reaches a steady-state value after several hours of heating.

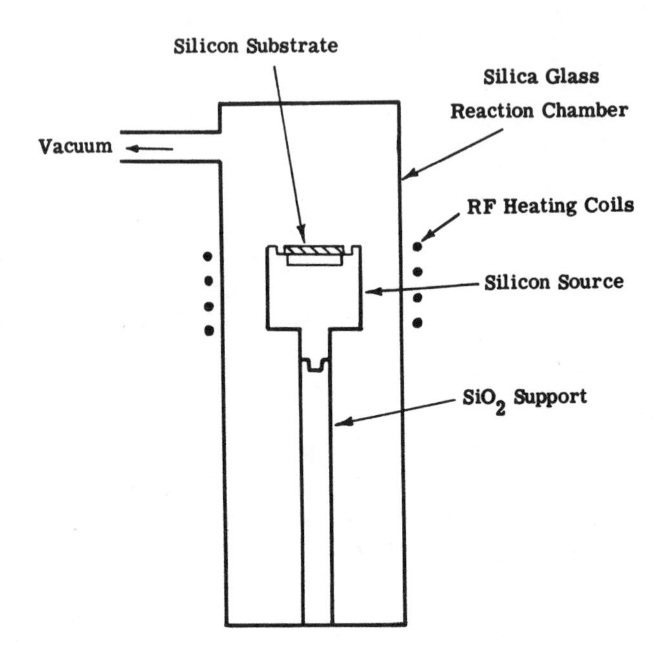

Fig. 8-5 Vacuum sublimation apparatus

Although reproducible results have been obtained with the sublimation technique, the physical arrangement requires a specially designed vacuum system, and is not easily adapted to the fabrication of multilayer devices.

8.2.3 EPITAXY ON NON-SILICON SUBSTRATES

The use of back biased p-n junctions for electrical isolation in monolithic integrated circuits can result in reduced operating speeds and more complex design. The epitaxial deposition of silicon on insulating non-silicon substrates is believed to be a possible means of improving isolation.

The feasibility for single crystal overgrowth on the metal-oxide insulators, is determined by the thermal coefficient of expansion of silicon and the substrate, and by the lattice fit. Substrates with thermal coefficients of expansion appreciably different from silicon, cause cracking in the layers upon cooling, and are unsuitable.

The deposition of silicon on non-silicon substrates is carried out by techniques described previously for ordinary epitaxy. However, films grown by the SiH_4 technique give more uniform results than those grown by the H_2-$SiCl_4$ process. Epitaxial single crystal silicon overgrowths may be obtained by chemical vapor phase techniques in large area single crystal form on sapphire (α-alumina), spinel ($Al_2O_3 \cdot MgO$) and beryllium oxide (BeO). Many new device capabilities seem possible with this technique of growing single crystal silicon on insulating substrates.

Another technique known as rheotaxy may be used as a possible improved isolation method. Silicon is deposited by vapor deposition techniques on a polycrystalline substrate whose surface is a thin fluid film of some oxide. Thus, initially, silicon atoms arriving from the gas phase see a surface of maximum atomic mobility. The silicon atoms are thereby able to align themselves into a single crystal

While good results have been obtained for devices fabricated by the rheotaxial process, it does not seem feasible for production situations because of the rigid standards that must be maintained for good single crystal growth.

Molecular Beam Epitaxy (MBE)

Molecular beam epitaxy is an ultrahigh vacuum technique for growing very thin eptitaxial layers of semiconductor crystals.

Reduced to its essentials, a system for MBE of GaAs consists of an ultrahigh vacuum system containing sources for atomic or molecular beams of Ga and As and a heated substrate wafer, as illustrated very schematically in Fig. 8-5. 1. The beam sources are usually containers for the liquid Ga or solid As. They have an orifice that faces the substrate wafer. When the container, or effusion oven as it is usually called, is heated, atoms of Ga or molecules of arsenic effuse from the orifice. The effusing species constitute a beam in which the mean free path is large compared to the distance between the oven orifice and the substrate wafer. If the orifice diameter is small compared to the mean free path of the gaseous components inside the effusion oven, the flux of Ga or As_4 at the target wafer may readily be shown to depend on the partial pressure of the species within the oven, the distance from orifice to substrate, the temperature, the species molecular weight, and the orifice area. Additional ovens, not shown in Fig. 8-5. 1, may be used to generate a beam of Al, for the growth of $Al_xGa_{1-x}As$, and to generate beams of impurity elements that can be used to make the epitaxial semiconductor *n* or *p* type. Most current MBE systems have about six effusion ovens. The beams may be shut off with shutters interposed between the substrate and the oven orifice, or the beam intensity may be varied by varying the oven temperature. Also illustrated in Fig. 8-5. 1 is an electron beam that impinges, at a glancing angle, on the growing surface of the crystal for in situ evaluation of surface morphology.

The successful use of MBE for epitaxy of GaAs, $Al_xGa_{1-x}As$, and other III-V compounds is a direct consequence of the behavior of group III atoms and group V molecules on striking the heated substrate surface.

For GaAs – and other III V compounds – there is a range of substrate temperatures over which virtually all of the group III element adsorbs on the surface. This holds for the entire usual temperature range of 450° to 650°C for the growth of GaAs. The surface lifetime of Ga on GaAs is greater than about 10 seconds, while the arsenic molecules desorb rapidly from a heated GaAs surface unless adsorbed Ga is present. In the latter case the surface lifetime of As increases as it bonds to the Ga. It decreases again when the excess Ga is consumed. The result of this is that one As atom remains on the surface for each Ga atom provided in the Ga beam. For the growth of GaAs the 1 : 1 ratio of Ga to As is maintained in the growing layer simply by having the As_4 flux be greater than the Ga flux. For epitaxy of $Al_xGa_{1-x}As$ the ratio of Al to Ga atoms in the solid is simply the ratio of the atom flux of each during growth, while the ratio of total group III (Al plus Ga) to As atoms in the solid is unity.

Achievement of high crystalline and semiconductor quality of the epitaxial layers also requires that clean ultrahigh vacuum conditions be maintained and that the substrate temperature be sufficiently high that the atoms adsorbing on the surface are mobile enough to migrate to the proper crystal sites. For GaAs, growth usually takes place with a substrate temperature above 450 °C, and for $Al_xGa_{1-x}As$ the temperature is usually above 550 ° or even 600 °C.

Molecular beam epitaxy is unique among crystal growth techniques in that it is possible to examine the crystal surface in some detail during the growth process. The electron beam shown in Fig. 8-5.1 is diffracted by the regular array of atoms that constitute the crystal structure near or at the surface of the crystal much as light is diffracted by a grating. The diffraction pattern of the electrons yields information about the arrangement of the atoms on the growing crystal surface. If the surface is microscopically rough, the diffraction pattern will be characteristic of the three-dimensional crystal since the beam must penetrate protuberances on the surface. If it is almost smooth on an atomic scale, the diffraction pattern will show the characteristic two-dimensional spacing of the atoms on that surface.

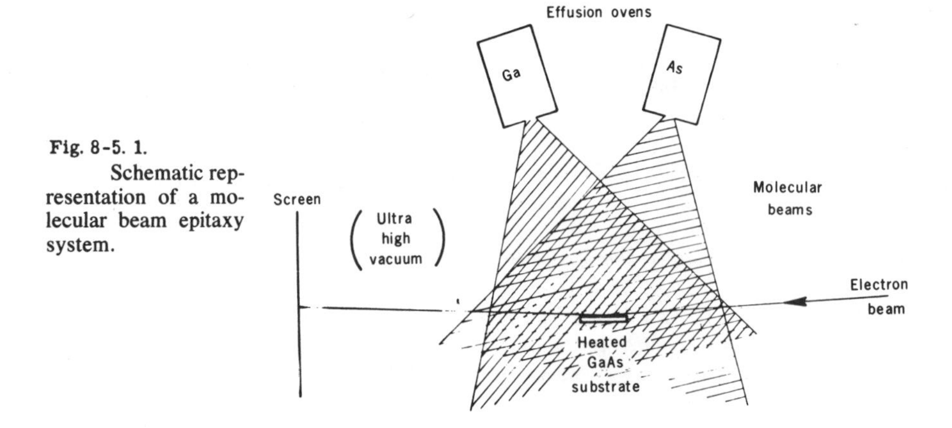

Fig. 8-5. 1. Schematic representation of a molecular beam epitaxy system.

8.3 EVALUATION TECHNIQUES AND RESULTS

This section describes methods for evaluating epitaxial layers. Techniques for determining impurity distributions, resistivity, and layer thickness are of prime importance. Stacking faults, imperfections occurring only in epitaxially grown layers, are also important considerations, including causes, control, and methods of measurement.

8.3.1 IMPURITY DISTRIBUTION

In the method referred to as double beveling, a sample from the epitaxial wafers is beveled at a slight angle to the surface and subjected to a diffusion of an impurity of opposite type to that in the epitaxial layer. A non-epitaxial silicon reference slice of known background impurity concentration, is diffused along with it. The diffused junction produced in the reference wafer, is delineated by a standard bevel-staining technique. The epitaxial sample is then bevel-stained with the axis of the first bevel 90° to the second.

The anodization technique presents a detailed profile of the impurity distribution in an epitaxial layer. Sheet resistivity measurements are made on the wafer before anodization. After anodization, the electrolytically formed oxide is removed by immersing the wafer in HF. The thickness of silicon removed, is calculated by weight loss measurements, and the sheet resistivity is remeasured. The mean bulk conductivity of the removed layer is

$$\sigma = \left(\frac{1}{\rho_{1s}} - \frac{1}{\rho_{2s}}\right)\frac{1}{\Delta t}, \tag{8-2}$$

where ρ_{1s} and ρ_{2s} = the sheet resistivities before and after anodization, respectively,

Δt = the thickness of the removed layer.

The impurity concentration profile across the wafer can then be determined from the conductivity profile.

In the capacitance-voltage technique, diodes are fabricated on the epitaxial wafer by shallow diffusing or alloying, and junction capacitance is measured as a function of the applied reverse voltage. For a junction between a heavily doped and a more lightly doped semiconductor, the slope of $(1/C^2)$ as a function of applied reverse voltage, is a simple function of the doping concentration, at the edge of the space charge region. The equations are

$$N = \frac{C^3}{\epsilon q}(-dC/dV)^{-1}, \tag{8-3}$$

$$w = \epsilon/c, \tag{8-4}$$

where N = dopant concentration at the edge of the space charge region,

w = space charge width below the junction,

C = specific junction capacitance (capacitance/junction area),

ϵ = absolute dielectric permittivity,

q = the electronic charge,

V = applied reverse voltage (dC/dV is always negative).

The procedure consists of measuring the capacitance at a given voltage, changing the capacitance bridge dial setting by a small increment (roughly 1%), and changing the applied reverse voltage to rebalance the bridge. This gives $\Delta C/\Delta V (\approx dC/dV)$ at the given values of C and V. The doping concentration N, space charge depth and the resistivity can then be determined. A doping profile is thus obtained from a series of capacitance-voltage measurements. This technique for measuring dopant profiles, does not subject the material to severe heating and cooling cycles. However, the epitaxial layer must be lightly doped so that the space charge or depletion region spreads primarily into the epitaxial layer. The range of measurable resistivities, furthermore, is limited by the breakdown voltage of the diode.

Neutron activation of silicon epitaxial films, is especially useful in determining the impurity profile of films lightly-doped with phosphorus or arsenic on heavily-doped substrates. The technique is not applicable for boron-doped films, however. The silicon film to be evaluated is placed in a reactor and bombarded with neutrons. After activation, the epitaxial layer is etched by a few drops of an HF-HNO_3 mixture. After etching, all of the etching solution is transferred to a stainless steel dish and filter paper, impregnated with an aqueous NaOH solution. The above procedure of etching is repeated several times to obtain samples at various depths. The thickness of the layers removed, may be calculated by weight-loss measurements. The sample dishes are heated to dryness by an infrared lamp, and the β decay activity is monitored by an end-window type counter. Decay curves of each fractional sample are compared with standards, and an impurity profile is then calculated.

Another method of determining impurity concentration in epitaxial layers, resides in the incorporation of a radioactive dopant into the reactant gases during the epitaxial deposition process. For example, radioactive phosphorus trichloride may be synthesized by the carbon reduction of phosphorus acid (containing β-emitting P^{32}) to yellow phosphorus in a nitrogen stream, followed by chlorination and reduction to phosphorus trichloride. This radioactive PCl_3 is then evaporated with the reactant gases, and epitaxial layers with β-emitting phosphorus are grown.

Grown p-n junctions, ranging from abrupt step junctions with high carrier concentrations on each side, to junctions with controlled grading and carrier concentrations of less than 10^{15} cm^{-3}, are necessary for the fabrication of epitaxial semiconductor devices. Any reasonable doping profile can be obtained on the silicon overgrowth layer if precise doping control is maintained in the epitaxial furnace. Figure 8-6 shows a typical doping profile for an abrupt junction and illustrates the exactness which can be obtained by precise doping control.

Diffusion of impurities and autodoping can make it difficult to obtain abrupt impurity profiles in epitaxial layers. Diffuson is inherent because of the high deposition temperatures required, and autodoping involves the chemical reversibility of the deposition reaction. The extent of these two effects depends, of course, on the method of deposition. The hydrogen reduction of $SiCl_4$, is a high temperature process ($\sim$ 1200° C) and diffusion can sometimes be a problem. Also, since the reduction reaction is chemically reversible, HCl or $SiCl_4$ could transport silicon and its dopants to various areas of the reduction system. This is particularly true when silicon is grown on a substrate of opposite conductivity or of high doping concentration. A small fraction of the dopant impurity in the substrate is transferred to the growing layer at the beginning of the growth process Impurities in the susceptor can also be transferred to the growing layer. The effects of diffusion and autodoping can be minimized by using lower deposition temperatures and a chemically irreversible process whose prod-

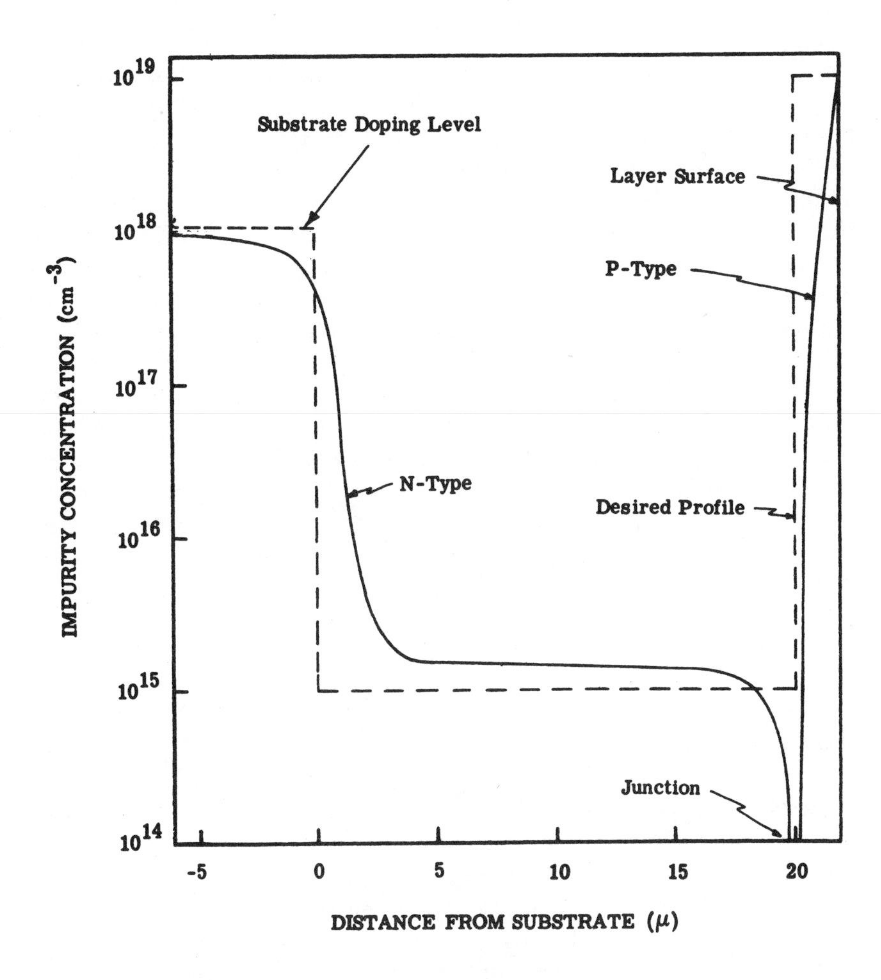

Fig. 8-6 Typical doping profile for an abrupt junction

ucts cannot transport the dopant impurities.

8.3.2 RESISTIVITY

The results obtained for impurity distributions are generally applicable to resistivity, since one is a direct measure of the other. Thus, the resistivity uniformity of epitaxial layers is determined by the impurity distribution within the layers. Several techniques are employed to determine the resistivity. The choice of a particular method is governed by the substrate and layer types, and the accuracy desired. While certain techniques are quite accurate, they are not suited for production situations where relative values are often all that are needed.

The most commonly used method for resistivity measurements of n-on-p or p-on-n epitaxial wafers, is the one based on the conventional four-point probe. For production situations where n-on-n^+ or p-on-p^+ wafers are to be evaluated, a high resistivity control wafer of opposite conductivity is processed with the group of substrates to be evaluated. The epitaxial layer grown on the control slice is effectively isolated by the p-n junction formed, and direct four-point probe measurements are possible.

The spreading resistance probe, or two-point probe, is a simplification of the four-point probe, and is applicable over a wide range of resistivities. The two probe points are placed on a wafer surface, a potential is applied between the two probe tips, and the resistance measured. The semiconductor-to-metal contact resistance of the probe tips is added to the measured resistance, but at very low currents this contact resistance is linearly dependent on carrier concentration. This effect is considered by a proportionality constant applicable over a wide range of doping levels. The applied probe voltage is kept at 30 mv to keep the maximum electric field below 1 kv/cm. This keeps the mobility of the majority carriers constant, prevents the injection of minority carriers from the probe tips, and keeps the temperature rise from Joule heating below 0.1° C.

After beveling and staining the sample of interest, the probes are moved down the bevel and a resistance profile as a function of distance from the surface is obtained. If a p-n junction is present, the corrected resistance given in Eq. 8-5 must be used:

$$R/R_o = 1 + \frac{2a}{\pi t}\left[ln\left(\frac{d}{2t}\right) - 0.116\right] \qquad (8\text{-}5)$$

where a = the contact area radius,
t = the distance from the p-n junction to probe position,
d = the distance between probe tips,
R = the measured resistance, and
R_o = the corrected resistance.

This correction factor must be used when the probe tips are above a p-n junction. After the junction is passed, no correction factor is needed, provided no other junctions are present. However, the correction factor is not applicable for a large abrupt change in impurity concentration of the same type, and another technique, e.g., the three-point probe, should be employed for this type of measurement.

The three-point probe is particularly adapted for measuring the resistivity of n-on-n^+ or p-on-p^+ epitaxial layers. The method consists of experimentally calibrating the reverse breakdown voltage of the point contact diode formed between melt-grown silicon of known resistivity and a metal probe. Measurements of reverse diode breakdown on epitaxial layers are then related to resistivity by means of this calibration curve. Readings depend on such variables as contact area, probe material and probe pressure. However, reproducible results are obtainable if measurement conditions are properly controlled.

The three-point probe technique does not require a knowledge of layer thickness, but the latter should always be greater than the depletion region at breakdown voltage. If the depletion region extends to the substrate, many different combinations of layer thickness and resistivity yield the same breakdown voltage.

The three-point probe is most often used to measure resistivities between 0.1 and 1.0 ohm-cm. Sampling and destructive testing are not necessary, and the procedures and equipment are simple. The technique is ideally suited to production situations, where large quantities of material must be evaluated quickly and inexpensively.

Another resistivity measurement involves the measurement of junction capacitance as a function of reverse-bias voltage. This has proven very useful for estimating layer resistivity values from direct measurements on the wafer. Shallow p-n junctions are formed on the epitaxial layers, and resistivity profiles may be determined from curves of resistivity as a function of impurity concentration.

The capacitance-voltage method may be employed for resistivity uniformity and control, but is time consuming and not readily applicable to production situations.

The infrared reflectivity of an epitaxial layer may be correlated with its dc resistivity as a possible means of measuring resistivity of epitaxial surface layers. The method appears promising, but requires further development. An accurate infrared method of resistivity measurement is quite convenient since both the thickness and resistivity of the epitaxial layer may be measured by the same instrument.

8.3.3 LAYER THICKNESS

The thickness of epitaxial silicon layers may be determined in several ways. Angle lapping and staining permits the measurement of multiple layers. Also, it can be used to delineate high-low junctions as well as p-n junctions. This technique involves mounting a chip of the wafer on a suitable fixture, lapping the chip at a small angle (1 to 5 degrees), staining the junction, usually under bright illumination, and measuring the thickness with a calibrated microscope. For junctions of opposite conductivity materials, concentrated HF solution containing 0.1% HNO_3 will stain the p-type material dark in the presence of light. Another stain consisting of 10% HF-water solution saturated with cupric sulphate will plate copper on the n-layer, but leave the p-region clear due to the difference in chemical potential of the layers. For n-on-n^+ or p-on-p^+ layers, a stronger acid is necessary for proper staining. Although this method provides accurate results, it is destructive.

Layer thickness can also be measured in some cases by an infrared interference technique. This method is not destructive, as is the previous technique, and is much faster in applicable situations. However it is limited to measurements of high resistivity layers (above 0.09 ohm-cm) on low resistivity substrates (less than 0.015 ohm-cm).

Infrared radiation will produce interference fringes which can be used to measure layer thickness, if reflection occurs both at the surface and at the layer-substrate interface. This condition is satisfied if the carrier concentration in the layer is low enough to permit transmission of the incident radiation through the layer, and the optical constants of the substrate differ sufficiently from those of the epi-layer so that reflection occurs at the layer-substrate interface.

Measurements are made by recording the intensity of reflected infrared radiation as a function of wavelength. Wavelengths in the 10- to 35-micron range are used to improve the infrared contrast on graded layers. The relationship between layer thickness and the wavelengths of interference fringe maxima can be determined from Fig. 8-7. The incident ray strikes the layer at A with an angle of incidence φ. Part of the incident radiation is reflected as ray 1 and the rest is refracted at an angle φ' until it strikes the interface B. Some of this radiation is absorbed by the substrate, but most of it is reflected to the layer surface, C, where it emerges as ray 2 parallel to ray 1. Varying the wavelength causes alternating bright and dark interference fringes in the reflected light whenever rays 1 and 2 differ in phase by an integral multiple of half wavelengths.

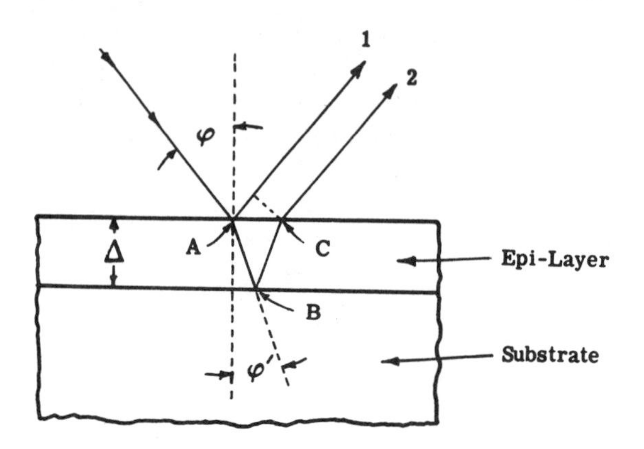

Fig. 8-7 Infrared reflection from an epitaxial layer

Non-uniform layer thickness or out-diffusion from the interface caused by high growth temperature and low growth rates may lead to poor interference curves. The best interference contrast is obtained from layers grown at the lowest temperature and the fastest rate.

The stacking fault method of measuring layer thickness is unique to epitaxially grown layers. Stacking faults in the growing epitaxial layer propagate throughout the layer; they create figures in the epitaxial layer upon application of an appropriate etch. A diagram of a common type of etch figure for an epi-layer grown on a (111) substrate is shown in Fig. 8-8. The figure is an equilateral triangle formed by the intersection of the three stacking fault planes.

The intersection of these three planes with the epi-layer forms a tetrahedron at the grown layer with one apex at the layer-substrate interface. This fact provides a means of measuring layer thickness. The thickness of the film is related geometrically to the length of one side of the equilateral triangle. If L is the length of one side and Δ is the thickness of the layer,

$$\Delta = L\sqrt{2/3} = 0.816L\,. \tag{8-6}$$

The stacking faults vary with substrate orientation, but each fault originating at the layer-substrate interface is geometrically related to the layer thickness regardless of orientation.

When using this technique, care should be taken to consider only those etch figures originating at the interface for measurements, since stacking faults can originate in the growing layer. Those originating at the interface will be the largest figures delineated by the etch. A detailed description of these crystalline defects, is given in the following section.

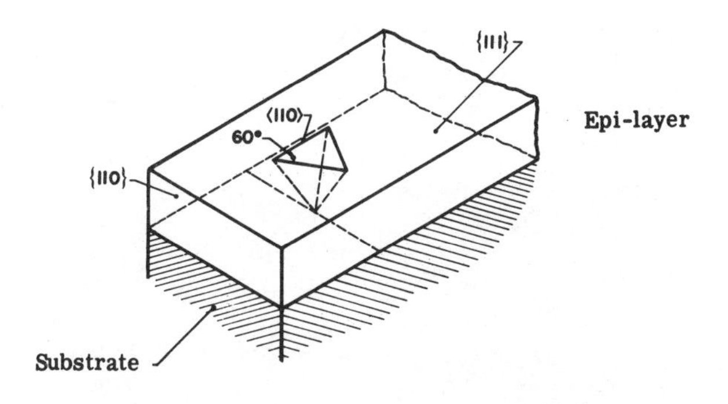

Fig. 8-8 Stacking fault formation on (111) substrate

8.3.4 STACKING FAULTS

These are caused in epitaxial silicon mainly by structural defects or work damage at the substrate surface, foreign impurities on the substrate surface, and localized stress areas throughout the crystal layer during the growth process. For mechanically lapped or polished substrates, stacking fault concentrations up to $10^7/cm^2$ are possible with the linear faults predominating. For substrate material free of mechanical damage, foreign impurities initially present on the substrate, and others introduced during the growth process, are mainly responsible for stacking fault formation. It is also suspected that regions of localized impurity segregation, slip lines, and microscratches on the substrate surface promote stacking fault formation.

A two-dimensional nucleation and growth model has been used to formulate a growth mechanism on the assumption that stacking faults form when the atomic layers are deposited in incorrect sequence. Whenever silicon atoms are deposited in incorrect sequence at one of the nucleating centers, a stacking fault forms on the (111) plane parallel to the substrate surface. This causes a small area to propagate which is crystallographically mismatched with respect to adjacent areas. As further atom layers are deposited, mismatch boundaries are propagated along inclined (111) planes, equivalent to stacking faults along these planes. The geometry of the stacking fault is determined by the shape of the mismatched boundary.

Another mechanism for fault formation is believed to be incomplete oxide removal from the substrate surface prior to epitaxial growth. Oxide patches can produce steps or inclusions at the substrate surface. Stacking faults are produced as the silicon crystal is forced to grow around these oxide steps to maintain coherence in the layer.

Large concentrations of stacking faults can cause poor p-n junction characteristics and polycrystalline epitaxial layers. The defects can be practically eliminated in epitaxial silicon films by precleaning the substrates in H_2 and etching with HCl prior to epitaxial growth. The substrate wafers are usually flushed in H_2 at 1200° C to remove any water vapor present, and after that HCl is added to the system. The HCl attacks the silicon surface and removes most of the work damage and foreign impurities. The use of substrates oriented slightly off the nominal direction also reduces stacking faults, probably because of the more stringent requirements for the nucleation of new growth.

Replication electron micrograph of a stacking fault.

Replication micrograph showing a linear stacking fault terminated by two dislocation-etch pits.

8.3.4.1 Detection Methods

Stacking faults have been shown by etch methods, by x-ray, optical transmission electron microscopy, and by electron diffraction.

The optical microscope is probably the simplest method for stacking fault detection in epitaxially grown silicon. Chemical etching of the grown film is usually necessary to make the stacking faults visible. The intersections of the higher energy stacking fault planes etch faster than the surrounding crystal, and they are revealed as grooves. A variety of etches may be used for this purpose.

In transmission electron microscopy epitaxial samples of 1 μ or less are required. A sample of silicon is mounted with wax on a teflon disk with the epitaxial layer downward. The disk is then placed in a plastic beaker filled with a 95% HNO_3–5% HF etching solution. The beaker is inclined at a 45° angle and rotated at 30 rpm to assure uniform etching. The substrate is dissolved entirely, and etching is continued until the proper thickness is obtained by examination of optical interference fringes. The thin foils obtained are removed from the teflon disk by dissolving the wax in an organic solvent.

Stacking faults in epitaxial silicon layers have also been studied by x-ray diffraction microscopy using the extinction contrast technique. This method is based on the change in diffraction contrast, produced by stacking faults, observed in topographs recorded by different Bragg reflections.

The interference contrast microscope is a nondestructive method of revealing stacking faults. Visible contrast is provided for height differences as small as 35 Å. While stacking fault markings and other surface features of the epitaxial layer are not revealed as clearly as in etching techniques, this method is quick and requires no further preparation or etching of the layer.

8.4 APPLICATION OF EPITAXY TO DEVICE STRUCTURES

8.4.1 MASKED AREA GROWTH

The epitaxial deposition of silicon in selected areas of the silicon wafer using appropriate masking techniques can be valuable for multilayer device fabrication. Separate mechanical masks of silica glass or silicon are generally unacceptable. The difficulty with silica glass is the unequal coefficients of thermal expansion of the mask and the substrate; silicon masks tend to stick to the substrate and to introduce unwanted impurities. Also, the precise geometries required for device fabrication are difficult to obtain with mechanical masks.

The most popular method of masking silicon substrates for multilayer growth is by growing or depositing a layer of SiO_2 on the silicon surface to be protected. Holes can be etched in these SiO_2 layers using standard photoresist techniques. The silicon wafer can thus be exposed to the epitaxial reaction in desired precise geometries for subsequent processing. The oxidation techniques used in epitaxial layer growth are discussed in detail in the chapter on oxidation.

The difference in thermal expansion coefficients between the oxide mask and Si can cause cracking of the oxide layer if the oxide film is much thicker than 1.5 μ, and especially if the wafer is subjected to a rapid, high temperature heat cycle. Furthermore, degradation layers around the edges of holes or window regions formed in these layers, can occur during a hydrogen heat treatment if they are poorly formed.

High growth rates (0.1 μ/min at 1150° C) tend to keep the oxide free of pinholes, cracks, crystalline inclusions, and other unwanted nucleation sites, thereby reducing polycrystalline silicon overgrowths on the oxide layers.

Another method which substantially reduces polycrystalline silicon overgrowth on the oxide is known as ribbon oxide masking. In the beginning, oxide masking techniques for selective epitaxy were copied from those used successfully in masked diffusion. The entire wafer was oxidized and window regions opened only in those small areas where epitaxial deposition was desired. Polycrystalline overgrowth on the oxide occurred during epitaxial deposition because of the large oxide areas. However, a much superior method for preferential epitaxy is to remove all the oxide except for encircling bands or ribbons of oxide around the windows. The ribbons are at least 1 mil wide. A large fraction of the wafer surface is now silicon and the conditions on the substrate are almost the same as in unmasked epitaxy.

8.4.2 ACTIVE DEVICE STRUCTURES

In diffused (as opposed to epi-layered) devices, impurity atoms are added to the silicon wafer by the diffusion of impurities from the surface of the wafer at elevated temperatures. Impurity compensation takes place when the devices require more than one junction. This compensation may limit the number of diffusions possible since the number of impurities allowable in silicon is governed by the particular solid-solubility limits of the impurity used. While three diffusions are possible, the bottom layer is then difficult to control, especially if it must be of high resistivity, i.e., low carrier concentration.

The impurity profiles for epi-layered devices, however, can be adjusted easily, since impurities are added to the growing films. Uniform impurity profiles, which are not practical in diffused layers, are readily obtainable by epitaxy. Also, epitaxial films containing the desired impurity concentration can be formed faster than diffused layers. The thickness of diffused junctions in multilayer structures is easier to control, however. The use of both epitaxial and diffusion techniques provides a basis for device fabrication that eliminates many of the compromises necessary in diffusion alone. Transistor structures with better frequency characteristics, lower base resistance and higher gain are made possible by epitaxy. Also, isolation techniques are greatly simplified. As a result, the reliability, cost and yield of integrated circuits have been greatly improved.

8.4.2.1 Planar Transistor

A conventional double-diffused planar transistor is shown in Fig. 8-9. The device needs a high collector-base breakdown voltage BV_{CBO} and a low collector-emitter saturation voltage $V_{CE(SAT)}$. The collector must be of high resistivity to obtain a high value of BV_{CBO} for the finished transistor. This requirement, however, produces a high series collector resistance which increases the $V_{CE(SAT)}$. The $V_{CE(SAT)}$ can be reduced to an acceptable level by decreasing the thickness of the high resistivity collector region, but at thicknesses below 120 microns the silicon wafer becomes too thin and brittle to handle, making it unacceptable for production. Also, in switching applications, when the device is in saturation, minority carriers are stored in the thick high resistivity collector region, limiting the device switching speed. The requirements for high BV_{CBO} (high resistivity) and low $V_{CE(SAT)}$ (low resistivity) are, therefore, not compatible in the simple double-diffusion process.

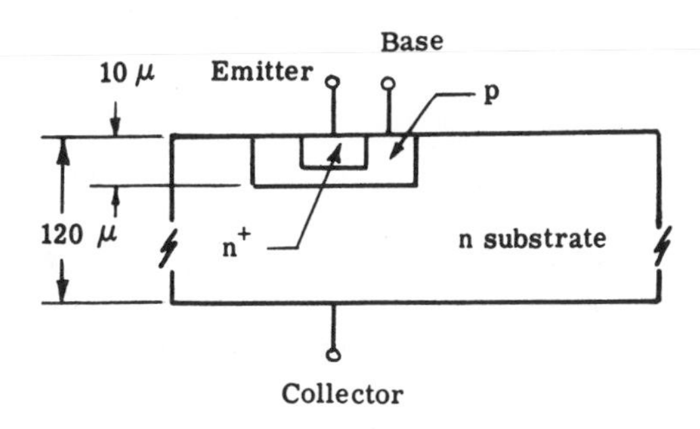

Fig. 8-9 Double - diffused planar transistor

The epitaxial process provides a simple solution to the preceding problem. In the fabrication of an epitaxial planar transistor as shown in Fig. 8-10, a high resistivity (typically 1 to 10 ohm-cm) layer is epitaxially deposited on a low resistivity (0.005 to 0.05 ohm-cm) substrate. The substrate is typically ten times thicker than the epi-layer. This epi-layer then forms the collector of the transistor. The p-type base and n-type emitter are added by a double diffusion. The base or the base and emitter may be deposited epitaxially, but because of the difficulty in controlling the base width during the emitter deposition or diffusion, this is seldom done. The collector region of the epitaxial planar transistor is now a high resistivity layer which provides a high value of BV_{CBO}. The low resistivity n^+ substrate is almost metallic and reduces the series collector resistance to a small value, thereby lowering $V_{CE(SAT)}$. The n^+ region is thick enough so that the wafer may be handled easily without breakage.

The advantages of this structure are a low saturation resistance which is almost independent of temperature, increased switching speeds, and improved linearity of characteristics. The saturation resistance is almost independent of temperature because the collector series component of the total saturation resistance is virtually eliminated. The increased switching speed is due to the proximity of the low resistivity layer to the collector-base junction. This low resistivity region limits the number and lifetime of minority carriers that are injected from the base, reducing storage time in switching circuits. For a specific circuit, the switching speed of the epitaxial structure is typically five times higher than the double diffused planar transistor. By diffusing gold into the base and collector regions of the transistor, the minority carrier lifetime can be further reduced.

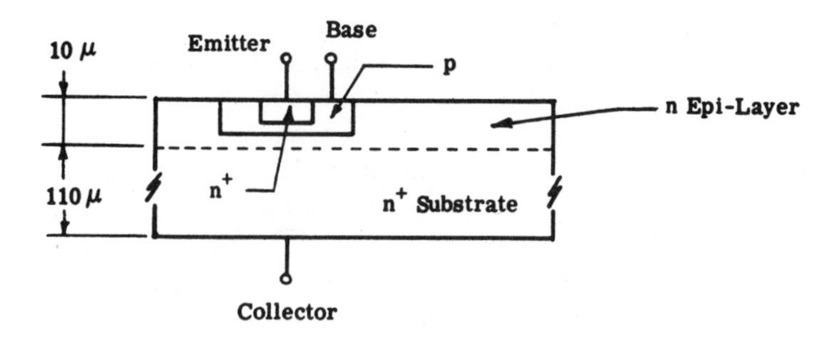

Fig. 8-10 Epitaxial - diffused planar transistor

8.4.2.2 Field Effect Transistor

The field effect transistor (FET) is an active device in which the current is carried mainly by majority carriers. Since one type of carrier is predominant, it is also known as a unipolar transistor. The majority carrier current is controlled by changing either the dimensions of a conductive channel, or the free carrier concentration in the channel. In the junction FET, the dimensions of the conductive channel are changed by reverse biased p-n junctions. In the surface FET, both the dimensions of the channel and the free carrier concentration are controlled by an applied electric field at the surface.

A double-diffused junction FET is shown in Fig. 8-11. The high resistivity requirement for the channel region is difficult to control by diffusion techniques alone. By epitaxially depositing a p-type layer, for example, on an n-type substrate, a constant resistivity profile is easily obtained for the channel regions. The top gate is diffused and defines the channel thick-

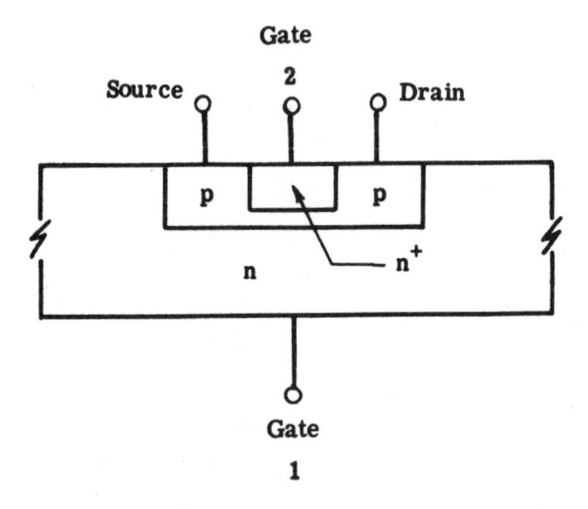

Fig. 8-11 Cross-section of simple double-diffused junction FET

ness. The p^+ diffusion for the source and drain contacts is then added. This diffusion provides a low resistance contact and keeps the depletion layer in the drain region from spreading to the surface at high drain voltages. The final structure appears in Fig. 8-12.

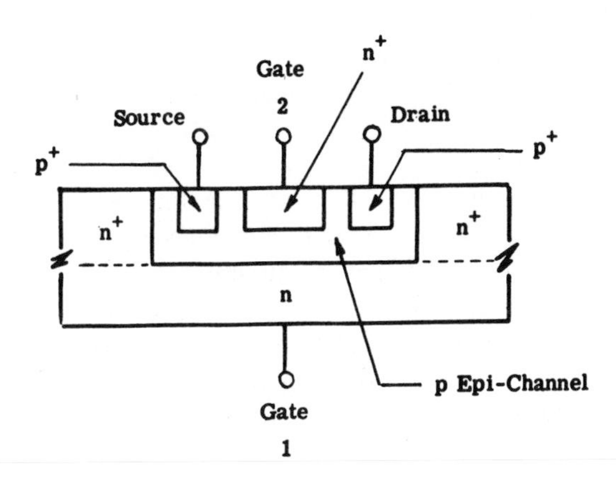

Fig. 8-12 Single epi-layer diffused junction FET

A second process for fabricating discrete junction FET's employs two epitaxial layers, as shown in Fig. 8-13. The first layer is the channel region and is sandwiched by the substrate and second epitaxial layer which are of opposite conductivity type. Precise doping control of the channel region is provided by epitaxy. The substrate and second epitaxial layer may be 0.5 ohm-cm n-type and the channel region 2 ohm-cm p-type. The lateral dimensions of the channel are defined by the source and drain diffusion. This method is somewhat faster than the first, since not as many diffusions are required.

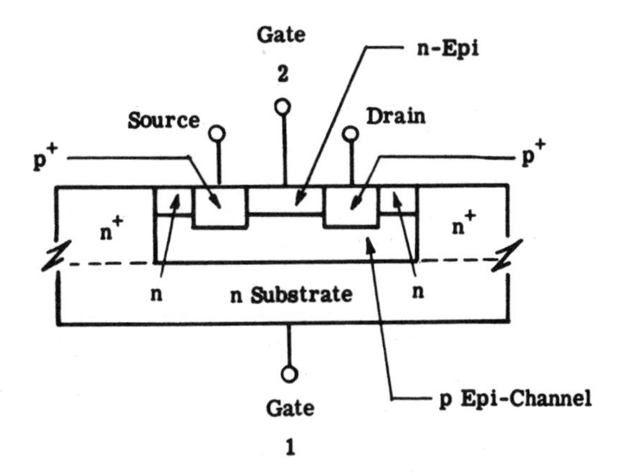

Fig. 8-13 Double epi-layer diffused junction FET

8.4.2.3 Variable Capacitance (Varactor) Diodes

The variation of capacitance with reverse voltage in p-n junction diodes is given by

$$C = kV^{-\alpha} \tag{8-7}$$

where C = specific capacitance (capacitance/junction area),
V = externally applied voltage plus contact potential,
k, α = constants

For abrupt step junctions, $\alpha = 0.50$, and for linearly graded junctions it is 0.33. (In a graded junction the impurity concentration changes linearly for some distance on either side of the junction.) If the impurity distribution is neither abrupt nor linear, α lies between 0.50 and 0.33. In the p-i-n diode—having a high resistivity or nearly intrinsic layer of silicon sandwiched between low resistivity p and n layers, the depletion layer punches through the intrinsic layer at a low voltage and gets only slightly thicker for higher voltages. When this occurs, the capacitance is almost independent of voltage and the device is similar to a fixed parallel-plate capacitor.

A wide range of doping profiles is easily obtainable through the use of epitaxy. A typical example of the variation in capacitance of a varactor diode is shown in Fig. 8-14.

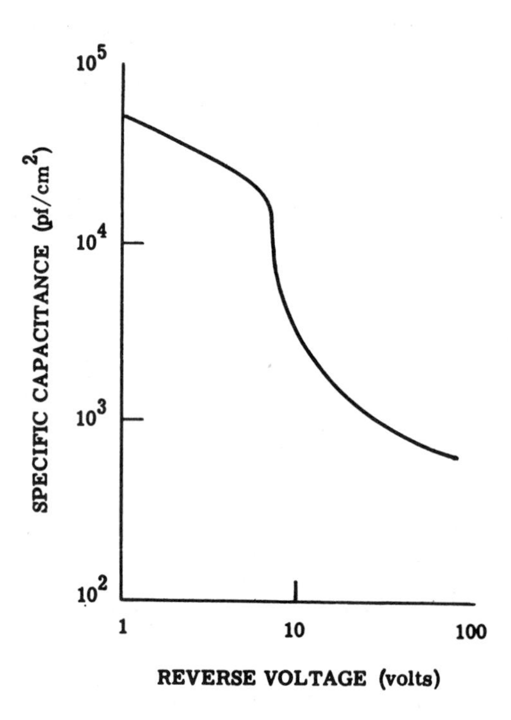

Fig. 8-14 Capacitance - voltage relationship for typical varactor diode

8.4.3 ISOLATION METHODS

The design and fabrication of high-yield, low-cost silicon integrated devices is based on the ability to obtain precisely shaped and sized zones of both conductivity types within single crystal substrates. These zones are used to fabricate various electrical components by a combination of diffusion and epitaxy. The various circuits on a particular wafer are isolated from each other by two methods; back-biased p-n junctions, and electrically inactive dielectric material. Both of these employ an epitaxial layer for improved performance.

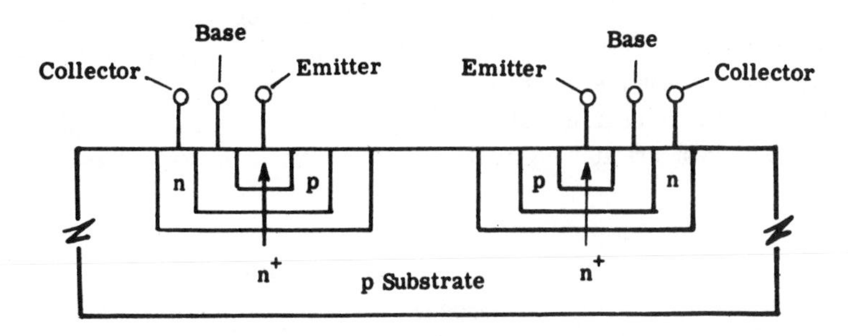

(a) Two Transistors on a Single Wafer

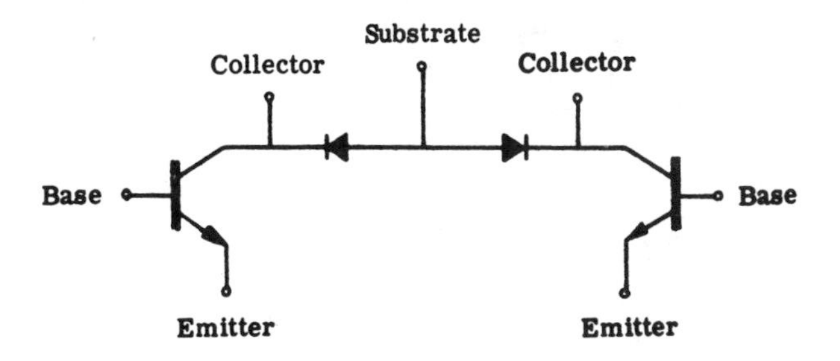

(b) Equivalent Circuit Showing Junction Isolation

Fig. 8-15 P-N junction isolation

8.4.3.1 P-N Junction Technique

Three p-n junctions and four-layers are usually required for integrated circuit fabrication. Isolation is provided by the p-n junction formed by the substrate and collector regions. This is shown in Fig. 8-15 for two transistors fabricated on the same wafer. Under reverse bias conditions, the isolation junction separates the transistor regions with a high dc resistance. The parasitic capacitance and leakage currents associated with the isolation junction can degrade the performance of circuits fabricated by this technique, especially at high frequencies and stringent operating conditions. However, the economy, ease of fabrication, and applicability to many situations make this technique a valuable tool for device fabrication.

8.4.3.1.1 DIFFUSED—COLLECTOR PROCESS

This method of p-n junction isolation is shown in Fig. 8-16. The starting wafer may be p-type with devices being fabricated by initial n-type diffusions into selected areas to form isolated collectors for the transistors, and isolated n-type islands for the subsequent diffusion of resistors, diodes, and capacitors. The initial n-diffusion is deep and difficult to control, as explained earlier. Also, this process requires certain compromises on device properties. The saturation voltage is determined by the depth and concentration of the n-region under the base, and cannot be optimized since the impurity concentration decreases as the collector-substrate junction is approached. Because of this profile, the collector-base capacitance is maximized and the BV_{CBO} is reduced, compared to that obtained by other techniques. Diffusion times of about 20 hours are required, and device properties at best are not equivalent to those of a discrete n-p-n device.

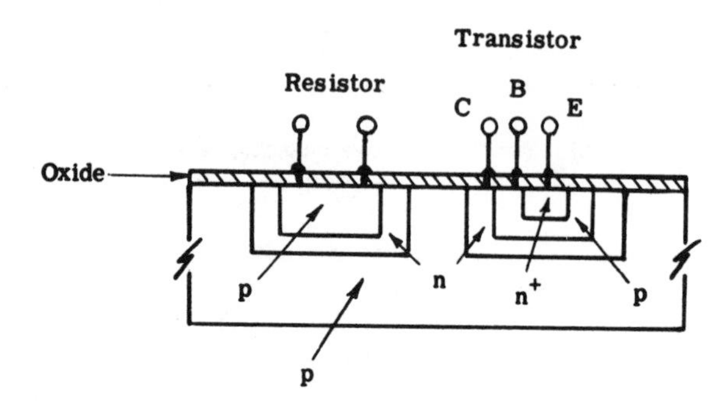

Fig. 8-16 Diffused collector isolation

8.4.3.1.2 TRIPLE—DIFFUSION PROCESS

This is a lengthy technique for obtaining p-n junction isolation. The configuration is shown in Fig. 8-17. The structure in the figure is obtained by masking selected areas of the top of a slice of n-type material, exposing the entire bottom of the slice, and diffusing p-type impurities from both sides to a depth of 1/2 the wafer thickness. Wafer thicknesses must be kept small to prevent diffusion times from being prohibitively long. However, for wafers of adequate thickness, minimum diffusion times are 40-60 hours. This technique has a uniformly doped collector region which allows the collector-base capacitance and the BV_{CBO} to be optimized for a desired value of $V_{CE(SAT)}$. However, since the collector-substrate capacitance is determined by the doping level in the collector, it cannot be minimized due to the minimum sheet resistance required for acceptable values of $V_{CE(SAT)}$.

8.4.3.1.3 EPITAXIAL—DIFFUSION PROCESS

The process has all the advantages of the two previous processes and none of the limitations. The method is shown in Fig. 8-18. The collector and substrate impurity profiles can be chosen independently. If a high-resistivity substrate is chosen with a moderate resistivity epitaxial collector layer, optimum values of collector-base junction capacitance, BV_{CBO} and $V_{CE(SAT)}$ can be obtained with a minimum collector-substrate junction capacitance. The results obtained by this process can be greatly improved by certain modifications described in the next section.

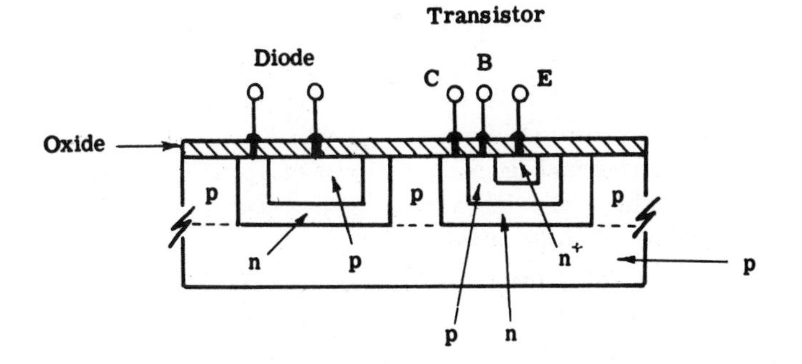

Fig. 8-17 Triple diffusion isolation

8.4.3.1.4 BURIED-LAYER, MOAT-DIFFUSED PROCESS

This technique consists of placing a low resistivity region at the interface between the substrate (typically high resistivity p-type) and the epi-

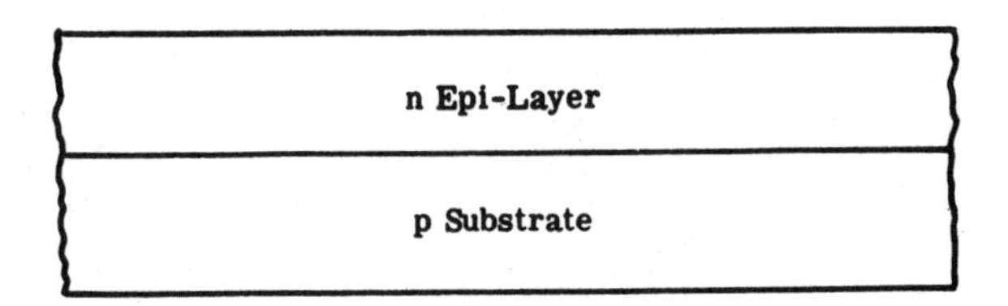

(a) Epitaxial Growth of Collector Region

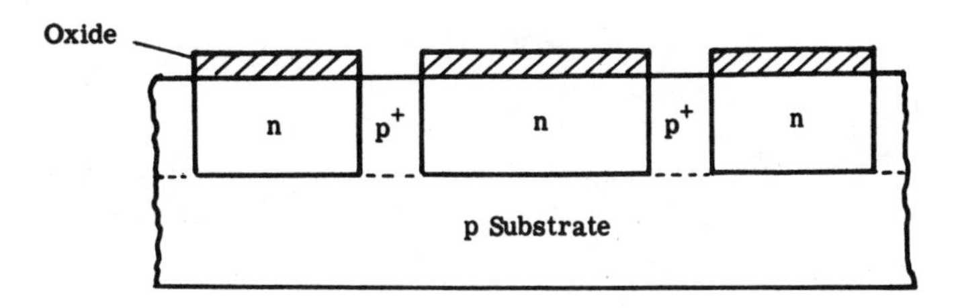

(b) Isolation Moat Diffusion

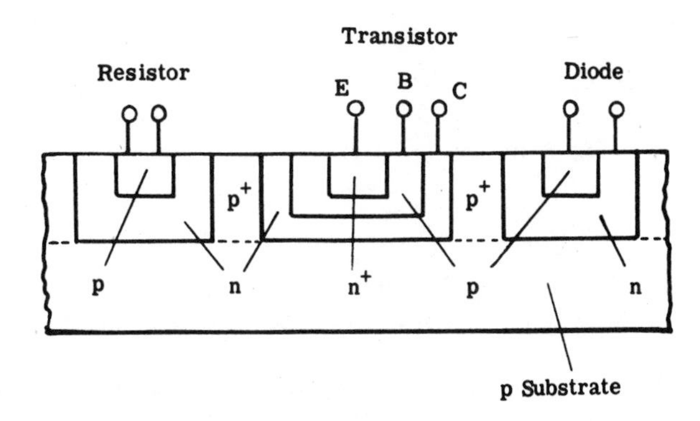

(c) Device fabrication by subsequent masking and diffusion

Fig. 8-18 Epitaxial - diffused isolation

taxial layer which might be n-type. The low resistivity region may be obtained either by prediffusion or by selective epitaxy. A typical buried-layer fabrication process is shown in Fig. 8-19 where the low resistivity region is obtained by prediffusion.

The starting wafer in Fig. 8-19 is p-type of approximately 10 ohm-cm. The n^+ layer is diffused with antimony because of its extremely small diffusion coefficient. This prevents the layer from spreading appreciably during subsequent fabrication procedures. If arsenic is used as the n^+ dopant, it is important that the surface concentration of the diffused layer be kept low (typically 10^{19} cm^{-3}), because of the tendency of arsenic to autodope or outdiffuse into the wafer during subsequent operations.

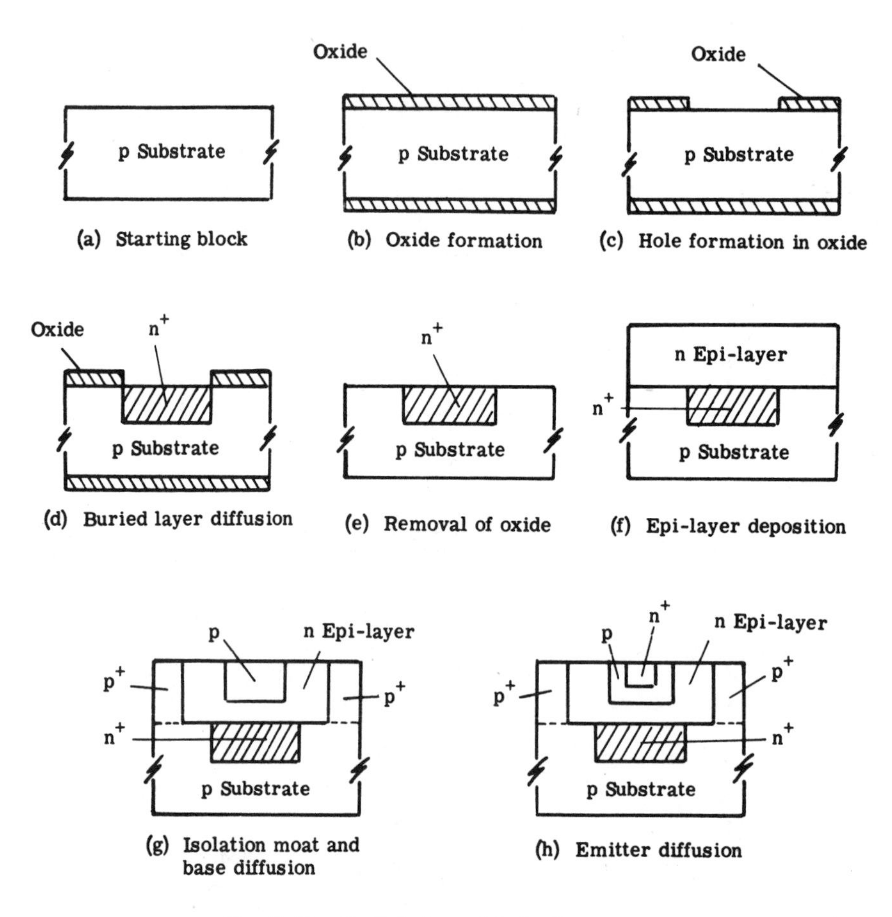

Fig. 8-19 Buried-layer fabrication process

Initially, an oxide is grown or deposited on the surface of the p-type substrate, and window regions are etched by a photoresist step. The oxide layer ($\sim 1\ \mu$) tends to remove work damage from the substrate, and, as a result, stacking fault concentrations in subsequent epitaxial layers are very small ($< 10/\text{cm}^2$). The n^+ dopant is diffused into the window regions to a depth of approximately two microns. After diffusion, the oxide layer is removed from the surface by a light HF etch. Also, the back of the wafer is etched to remove heavily doped silicon which reduces the probability of wafer contamination from out-diffusion or autodoping from the substrate.

An n-type epitaxial layer is then grown over the wafer as shown in Fig. 8-19f. The n-type layer may range from 0.5 ohm-cm to 5 ohm-cm depending on the BV_{CEO} required. The thickness of the layer is determined by transistor beta requirements. Typical thicknesses are between 15 and 25 μ.

The next step in the process is to isolate (electrically) the various wafer regions. This is done by a surface deposition and diffusion. Boron is chosen for the p^+ isolation moat. The n-epitaxial layer is now subdivided into isolated regions separated by back-to-back diode junction surfaces as shown in Fig. 8-19g. A column of n^+ material from the surface may also be required to reduce the transistor saturation resistance to small values and prevent the isolation diode surface from going into forward conduction and shunting the power supply through the device. The need for this n^+ channel depends on the resistivity of the n-type layer. In most practical situations, it is not necessary. Other devices, e.g. resistors and capacitors may be added during the base and emitter diffusions.

The buried-layer technique can be used to produce combinations of the following basic components: n-p-n or p-n-p transistors, junction field-effect transistors, Zener diodes, metal-oxide-semiconductor transistors, p-n junction capacitors, control rectifiers, dual emitter switches, and distributed RC networks. The method produces devices with high yields and is, therefore, economical. However, the technique has several disadvantages which are overcome by the method of dielectric isolation.

8.4.3.2 Dielectric Isolation Techniques

The isolation in the buried layer techniques is provided by the space charge layer of a p-n junction. The degree of isolation obtained is, therefore, limited by the capacitance and leakage current present in the junction. Also the isolation is polar and is affected by parasitic transistor action. The dielectric isolation technique overcomes these problems and presents other advantages not available with the buried layer techniques.

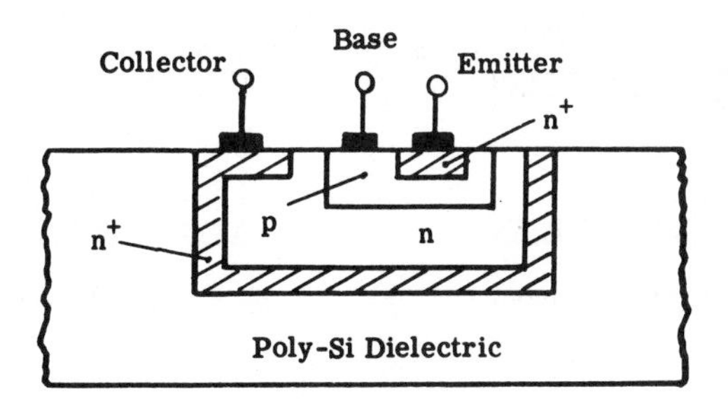

Fig. 8-20 Dielectric substrate

There are two methods for obtaining the dielectric isolation. The first method shown in Fig. 8-20 has a substrate formed entirely from dielectric material. Polycrystalline silicon grown to have dielectric material in the grain boundaries, acts as a nonconductor and is used as the substrate. This method is not as reproducible as the other method, but it has a lower parasitic capacitance. The other method of dielectric isolation is shown in Fig. 8-21. The substrate is deposited as a conductor, e.g., impurity doped polycrystalline silicon, and acts as a return for displacement current when connected to ground. Interaction from element to element is minimized and isolation is improved. The dielectric is deposited prior to the substrate. Thermally grown and deposited silicon oxides act as good dielectrics.

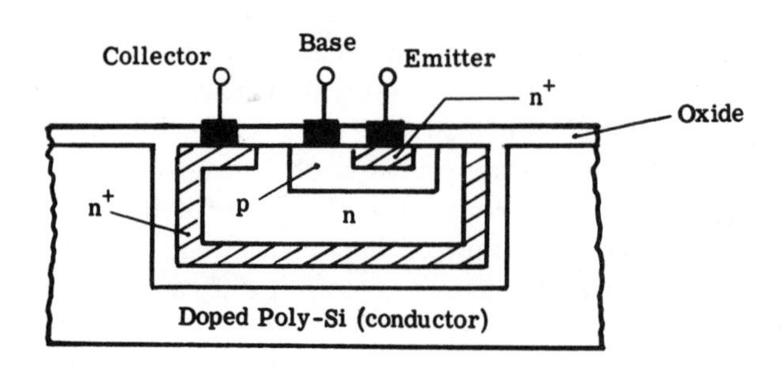

Fig. 8-21 Dielectric isolation and conducting substrate

A dielectric isolation technique using the latter method is outlined in Fig. 8-22. A thin ($\sim 4\ \mu$) n^+ layer is either epitaxially deposited on or diffused into an n-type substrate. The wafer is masked with oxide as shown in Fig. 8-22a and the open silicon areas are etched to a depth of perhaps 25 μ (Fig. 8-22b). About 1.5 μ of oxide for the dielectric is then grown or deposited on the etched regions, followed by about 175 μ of polycrystalline silicon (n-type < 0.01 ohm-cm). The original surface is lapped until the

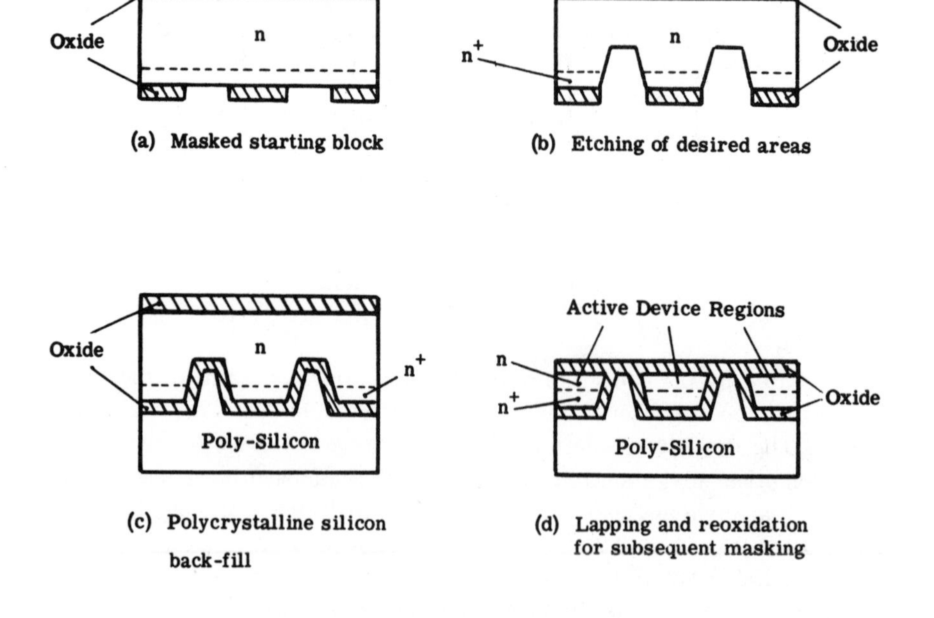

(a) Masked starting block

(b) Etching of desired areas

(c) Polycrystalline silicon back-fill

(d) Lapping and reoxidation for subsequent masking

Fig. 8-22 Standard oxide isolation process

oxide in the etched-out regions is reached. Regions of the original material are isolated from one another by the oxide dielectric. Semiconductor devices may then be fabricated in these isolated regions by the usual methods.

Another technique for achieving this result is shown in Fig. 8-23. The general procedure involves the growth of an oxide over the epitaxial layer and substrate, followed by the deposition of polycrystalline silicon on one side of the wafer. This polycrystalline silicon serves as a ''handle'' for the single crystal side of the wafer during lapping and etching. After the single crystal side of the wafer has been lapped and etched to a thickness

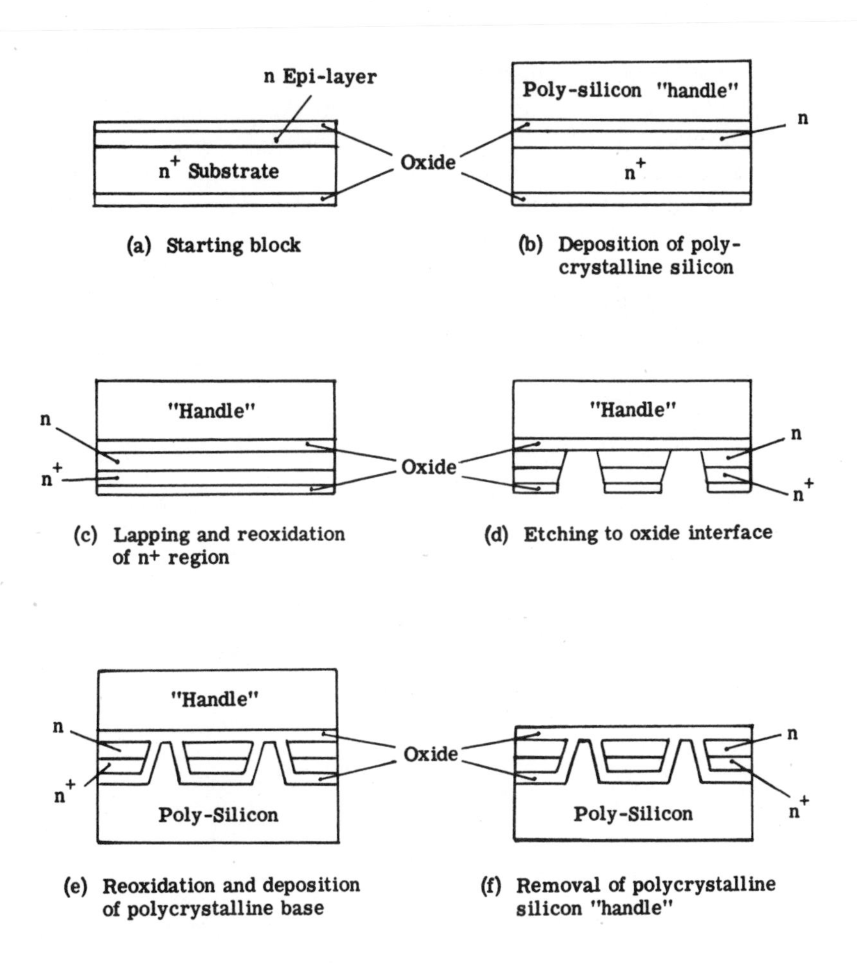

Fig. 8-23 Alternate oxide isolation process

of a few mils, it is selectively etched to the oxide interface to form isolated regions of single crystal silicon. An oxide layer is then grown over this, and the etched-out areas are filled with polycrystalline silicon. The top layer of polycrystalline silicon or ''handle'' is subsequently removed by etching to the oxide layer which is then used as a mask for device fabrication.

In this latter isolation technique, the top surface of the single crystal silicon is flat and has been protected by an oxide layer throughout all the isolation steps. There is little possibility of work damage in the areas to be used for device fabrication, in contrast to the preceding method where the active device areas are lapped. Thickness control of the active device regions is independent of the isolation process.

The etching of the single-crystal substrate silicon for the formation of the isolation regions can be accomplished with a variety of etches. The bottom of the etched-out regions should be as flat as possible and the lateral spread, or undercutting of the pattern wells, should be minimized.

Dielectric isolation techniques present several advantages over the buried-layer method. For transistors, collector substrate leakage is less than 10^{-11} amp, and the collector to substrate capacitance is reduced by a factor of 10 or more, depending upon design. The collector to substrate breakdown voltage is non-polar and is greater than 1000 volts. The collector saturation resistance and noise figure are equivalent to good conventional transistors. Parasitic transistor action is absent. Other devices are similarly improved. Dielectric isolation techniques should be used whenever requirements are stringent. However, ease of fabrication and economy of the buried layer technique make it desirable for most situations.

CHAPTER 9

MONOLITHIC CIRCUIT FABRICATION

This chapter describes the processing sequence during which the preceding techniques are applied to yield specific semiconductor micro-circuits. The first major processing step prepares the wafer through lapping, polishing, cleaning, and silicon wafer etching. Photolithography is then applied in the form of mask preparation, wafer photosensitization, contact printing, and silicon dioxide etching. Next, the wafer is processed through oxidation, diffusion, and epitaxy. Finally, electrical contacts and circuit element interconnections are formed through a metallization step.

9.1 WAFER PREPARATION

The processing sequence begins with a monocrystalline silicon wafer of from one-half to 2 inches or larger in diameter having a highly polished surface on one side. Most of the operations involved in the planar process are performed on the polished side of the wafer so that, unless special processing is involved, it is frequently unnecessary to polish the back side, which is left in the lapped state.

The condition of the polished surface is highly important to the success of the photolithographic and diffusion steps which follow. Attempts are made to obtain a surface as free from crystallographic defects and organic and inorganic contamination as possible, and one which is flat or nearly so.

An ingot of chemically purified polycrystalline silicon of 5 to 200 ohms-cm p-type is the usual starting material. This crystal is zone refined, i.e., a molten zone is passed slowly through the ingot from one end to the other. The preference for impurities to be in the liquid rather than in the solid zone, results in moving the impurities to the end (tail) of the crystal. This tail is removed and the resistivity of the material is

Growing a single-crystal ingot of doped silicon by withdrawing a small seed crystal of silicon from a quartz crucible containing pure molten silicon to which dopant has been added.

checked for intrinsic resistivity. If this resistivity is less than 100 ohms-cm, the material is zone refined again. Special material has been produced up to 1000 ohms-cm. Samples for lifetime measurements are also taken to insure that compensated doping has not produced an ''intrinsic'' reading.

Single-crystal silicon of desired resistivity is prepared from a zone-refined intrinsic polycrystalline ingot, a measured amount of ''p''- or ''n''-type dopant, and a low imperfection seed crystal with face cut on the desired crystal plane; the <111> plane, which has the greatest density, is the easiest to grow with least chance of polycrystal or twin being formed. A twin is two single crystals growing simultaneously due to (a) imperfections in the boat or seed, (b) mechanical jolting of apparatus, or (c) chemical contamination. The <110> plane is the easiest to scribe and break since its cleavage planes are 90° to each other. The <100> plane gives rise to the lowest number of ''surface states'' and is used, for example, in MOS devices. The polycrystalline intrinsic material is melted and impurities (dopant) are added. The seed is melted back and a single-crystal silicon of desired orientation, resistivity, and type is pulled by the *Czochralski process* (where the seed is dipped into molten silicon and the crystal slowly pulled out), or the *float zone process* (where the molten zone traverses the seed and ingot in a manner similar to zone refining).

The float zone material should give more uniform resistivity due to a constant volume liquid zone, while the Czochralski process involves continual depletion of the liquid, and also requires a quartz crucible which results in more oxygen being contained in the crystal. An oriented flat is ground on single-crystal ingot to assist in alignment during subsequent processing, and to make scribing and breaking easier by breaking along crystal planes.

The monocrystalline silicon ingot has a diameter essentially equal to the diameter of the finished wafer. The ingot, which may measure more than 8 inches in length, is cut into slices approximately 10 mils (0.010 inch) thick by a saw equipped with a diamond-impregnated blade. The rough wafer blanks are then placed on a lapping machine. Both "saw cut" surfaces are lapped simultaneously with an abrasive (usually aluminum oxide, Al_2O_3) having a particle size of 5 to 12 microns, until all visual traces of the saw cuts are removed. The surface finishes resulting from this operation are known as "lapped" surfaces.

Only one side of the wafer is selected for final polishing. The wafer is cemented onto a polishing fixture with a suitable wax or low melting point cement, and polished with successively finer polishing abrasives until a mirror finish is obtained. The polishing fixture usually consists of a post, to which the wafer is cemented, and a ring surrounding the post to keep it parallel to the polishing plate. The plate, which is covered with a soft fabric, may be of metal or glass, and has itself been polished flat. Polishing is commonly performed in three steps using aluminum oxide abrasive. The lapped surfaces are polished first with 1.0 micron abrasives, followed by 0.3 micron abrasive and 0.05 micron abrasives, in that order. Abrasives are added to the polishing cloth as a slurry in varying proportions of glycerin in water. After the final polish, the surface has a mirror finish.

Mechanically polished surfaces always contain highly disordered surface layers resulting from the abrading action of polishing abrasives, and must be further subjected to a chemical etch to remove the disordered layers. Though other techniques are available, such as electropolishing or high-temperature gas etching, which polish without introducing this condition, the mechanical polishing method is the most universally used for economical reasons.

After polishing, the wafers are subjected to a chemical etching cycle to remove this damaged surface layer. The etching conditions are chosen to minimize the introduction of surface irregularities and large deviations from the flatness obtained after polishing. A typical etching cycle consists of the following: (a) the wafers are first swabbed with cotton moistened in trichloroethylene to remove polishing abrasives clinging to the surface, then (b) ultrasonically cleaned in special detergents or methanol to remove water soluble residues and rinsed in water, (c) boiled in trichloroethylene to remove organic residues, (d) etched in a silicon etch to remove the damaged layer, and (e) rinsed to remove traces of the acid. The wafers are stored, after etching, either in an oven at 230° C or above in a closed container, or in a methanol or trichloroethylene filled jar.

The etching fixture used in this operation is arranged so that either the wafer or the solution is agitated in some manner, while etching is taking place, to minimize the kind of preferential attack which would lead to surface irregularities such as "orange peel" (waviness of the surface) or surface pitting. Typically, however, the etch chosen for removing the damaged layer is of the type known as a "slow polishing etch." It is chosen to be slow so that the predetermined depths can be etched by timing the process. Polishing etches are nonpreferential etches. They are also known as "leveling" etches and are characterized by the fact that they attack the surface uniformly regardless of whether or not there is present on the surface a crystallographic defect. A typical etch formula may have the following composition:

Nitric acid 69.6 percent	600 ml
Acetic acid, glacial	200 ml
Hydrofluoric acid 49 percent	80 ml
Hydrofluorosilicic acid 30 percent	16 ml

This solution will etch a polished silicon wafer at the rate of approximately 0.136 mils/min. (3.45 microns/min.) at room temperature. Five minutes are usually sufficient to etch well below the damaged surface layer.

9.2 PHOTOLITHOGRAPHY

Photolithographic techniques are widely used in planar microcircuit technology. Through their use, images of a desired circuit configuration can be conveniently reduced in size, and transferred to the silicon wafer with a high degree of accuracy. Since photolithographic processes are used several times during the circuit fabrication, it is desirable to describe the basic techniques rather than the specific applications.

In Chapter VI it has been shown that an oxide can be formed on the surface of a silicon wafer which acts as an effective mask against diffusion. It follows that diffusions can be confined to specific areas of a wafer if the oxide were preferentially etched away in that area. With the use of photolithographic techniques, preferential oxide etching can be accomplished quite readily. This is achieved by coating the wafer with a photosensitive lacquer-like film called a "photoresist" or "resist", and "contact printing" the circuit pattern onto the wafer. When the pattern is developed, certain areas of the resist will be washed away, and leave the oxide beneath exposed for etching. If an etch is used which is specific for silicon dioxide and does not attack silicon, then the oxide can be preferentially etched, to expose the silicon wafer beneath for diffusion.

Similarly, high conductivity metallic interconnection patterns can be formed on the wafer by photolithographic techniques. If a wafer is coated with a high conductivity metal, such as aluminum, the use of photolithographic techniques permit exposing the excess aluminum, so that it can be etched away and leave the wafer with an aluminum interconnection pattern.

The use of the term "mask" in semiconductor microelectronic technology may prove to be confusing unless careful attention is paid to the context in which the term is used. Thus, three kinds of "masks;" a "photographic mask," a "photoresist mask," and an "oxide mask," have been mentioned. A photographic mask is either a positive or negative image of a circuit pattern formed on a photosensitized glass plate. It is used for forming the resist mask. The resist mask is formed on the wafer in the photoresist. It is used for selectively etching the oxide or aluminum it covers. The oxide mask is the etched pattern formed in the oxide when the resist is removed, and is used for selective diffusion.

An example of masking follows for the p-type integrated circuit substrate.

Mask 1. Buried Layer: — Used to define an n+ region for the reduction of collector series resistance, and applied prior to the epitaxial deposition.

Mask 2. Isolation Diffusion: — Defines separate n-type islands in the epitaxial layer. The islands are isolated by reverse-bias p-n junction.

Mask 3. Base Diffusion: — Defines a base region of a transistor and the area for the diodes, resistors, and capacitors.

Mask 4. Emitter Diffusion: — Defines an emitter region for transistors, crossovers, and n+ collector contact areas.

Mask 5. Contact: — Defines metallization areas for ohmic contact.

Mask 6. Metallization: — Defines interconnecting conductor pattern, bonding pads, and ohmic contacts to circuit elements.

9.2.1 PHOTOGRAPHIC MASK

Semiconductor microcircuits are processed in batches, with each wafer carrying several identical individual circuits. To process a particular circuit, a set of precision photographic masks are required—one mask for each step in the process. Each mask carries a set or "array" of perhaps 200 or more identical patterns. Each pattern is reproduced with exacting precision on the mask. Each mask in the set must be precisely registered with all of the others, so that when all the masks are superimposed, the registration between masks is kept within narrow tolerances. These requirements stem largely from the small sizes of the finished circuits, and the small dimensions used in their fabrication. For example, diffused resistor line widths normally vary from one-half mil to three and one-half mils, depending on the final value of resistance desired. If a resistor one mil wide and 10 mils long is deposited at a value of 200 ohms per square, it will have a resistance of 2000 ohms. If the resistor were to be deposited with a tolerance of ± 5 percent, then the line width cannot vary more than ± 5 percent or ± 0.05 mil.

Photographic masks are prepared by photoreduction techniques. The original artwork is drawn 50 to 500 times larger than the final size, and photographically reduced to give a negative or positive in the final image size. The original artwork is deliberately made oversize for two practical reasons. The oversized layout can be prepared on precision machinery with tolerances as small as 0.1 mil. When the artwork is then reduced photographically, the errors are also reduced. Thus, assume a one mil line can be drawn 500 mils wide with a tolerance of ± 0.5 mil. When later reduced to final size, without introducing significant errors in the reduction process, the line will have a tolerance ± 0.001 mil. There are, of course, practical considerations which place a limit on the degree of accuracy with which the final image can be prepared. For example, if the initial artwork is made too large, the large reductions necessary may tend to introduce inaccuracies in the form of rounded off outer corners and filled in inner corners at pattern intersections. Similarly, if the image is so large for the reduction lens being used, that the ratio of field diameter to focal length is too large, image distortion will result in the reduction.

Original artwork may be prepared in anyone of a variety of methods described in Chapter II. Whether the artwork is laid out as a "positive" or "negative" of the final image, depends upon the number of photographic steps used between the preparation of the layout and the final reduction, as well as the manner in which the mask is to be used. Assume it is necessary to etch a hole in an oxide mask to diffuse a resistor. During the etching process, all of the oxide must be covered with resist except where the hole is to be etched. If a negative photoresist is used, the photographic mask must be transparent except where the hole is to be etched. The mask is then a clear glass plate with an opaque stripe, and may be referred to as a "positive." If only one reduction step was used, then the original artwork must have been a "negative" or reverse image. It thus consisted of a black or opaque sheet on which the stripe occurred as a clear line or hole.

It is a general practice to reduce the original artwork to final size in two steps. The first step reduces the image to 10 or 20 times of the final size. During the second reduction, the image is not only reduced to final size, but it is also reproduced in an array or matrix to permit the circuits to be processed on a batch scale.

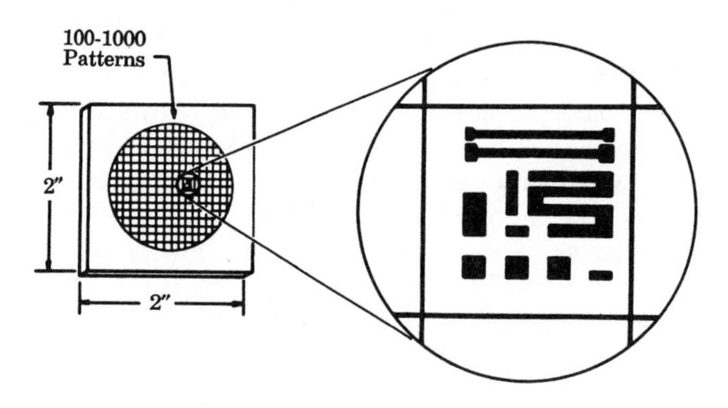

Fig. 9-1 Reproducing the desired pattern photographically on a glass plate.

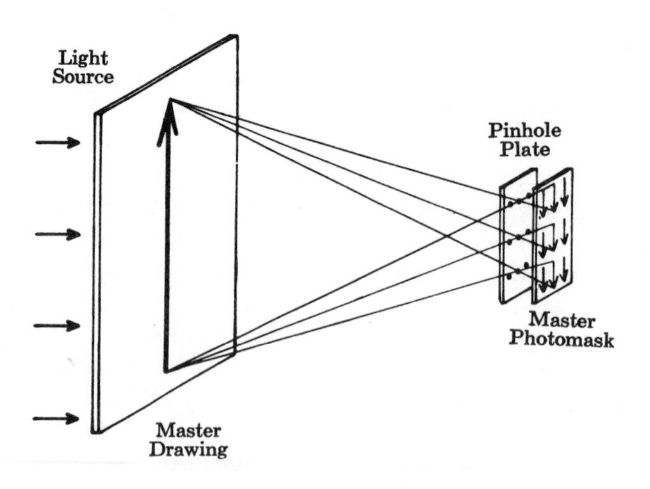

Fig. 9-2 The multiple pinhole method.

Three methods have been used to produce the matrix of circuit patterns from the single pattern. A simple method uses the pinhole camera principle, with one pinhole for each image of the array. Thus, a matrix of pinholes will produce a matrix of exposures from a single object. This method suffers from the usual defects of pinhole photography, e.g., small effective f-number and poor quality. The second method represents a substantial improvement over the first, in that lenses are inserted between the photographic plate and the pinholes. The lenses are molded into a plate of clear plastic, and the assembly results in the "Fly's Eye" camera shown in Fig. 9-3a. Although this system has undesirable off-axis distortion and degradation of resolution, it has the advantages of being fast, inexpensive, and capable of consistent mask reproduction. Good resolution between masks is obtained with both methods. Off-axis distortions, when they occur,

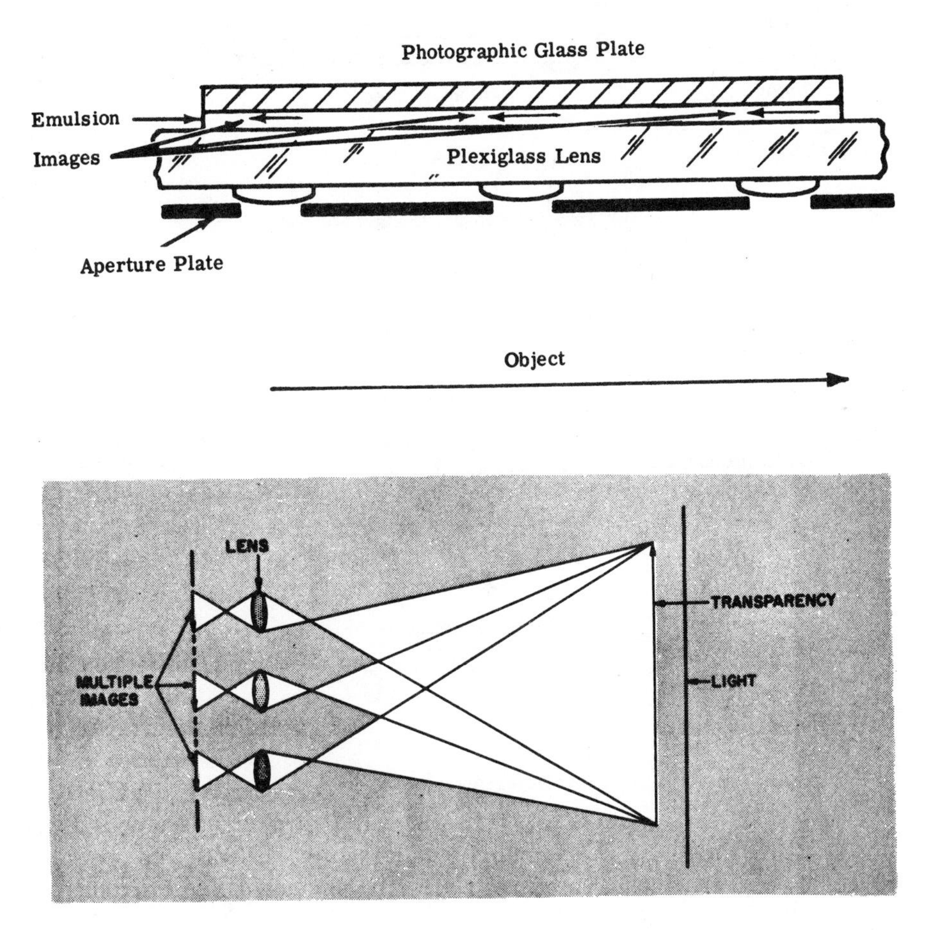

(a) Fly's Eye Principle

Fig. 9-3 Processes used to produce a matrix of patterns

are reproduced consistently and in some applications, are not objectionable. When simple planar diffused transistors are being manufactured, such distortions may not cause device degradation to the point where inconsistent device performance is obtained.

The step-and-repeat process, as its name indicates, produces a two-dimensional array of images by a multiplicity of exposures. Each exposure

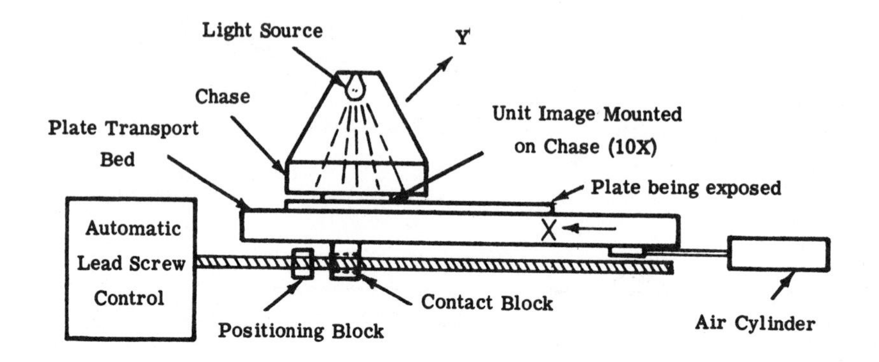

(b) Step and Repeat, Contact Method

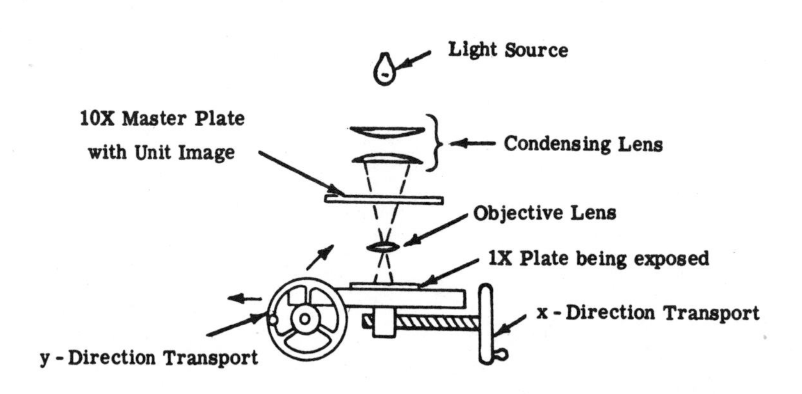

(c) Step and Repeat, Projection Method

Fig. 9-3 (continued) Processes used to produce a matrix of patterns

forms a single image. The exposure may be via the contact method, Fig. 9-3b or the projection method, Fig.9-3c. In either case, the original may be final size or larger. If a larger image is used, then it is usually photographically reduced during the projection step-and-repeat process. It is evident that the shifting of the photographic plate between exposures must be accomplished with great accuracy. This is usually performed with an X-Y coordinatograph possessing an accuracy of $\pm 1\mu$. The disadvantages of the step-and-repeat systems are: (a) high-quality coordinatographs are expensive and will eventually show wear; (b) it is time consuming to make a set of photo masks; and (c) successive masks of a set often have enough mismatch to cause difficulty. The optical quality of the result, however, is good.

As is usual in all photographic work, it is relatively easy to perform work of fair quality, but very good or very precise results require a very high degree of skill. Exposure times are critical. Optimum exposure depends on the line width, and therefore critical dimensions on the same plate should be of the same value, or a compromise exposure may be necessary.

Once the proper set of masks has been prepared, it is used in conjunction with photoresist to produce the various diffusion and interconnection patterns. For diffusion patterns, the silicon slices are first thermally oxidized (a 6000 Å thick oxide), and then stored at 300° to 400° C to prevent the condensation of moisture on the surface of the oxide. Coating with photoresist is accomplished on a ''spinner.'' Photoresist is applied through a filtering hypodermic syringe. The slices are spun slowly for about 10 seconds immediately after application of the resist and then spun at a high speed (6000 rpm or above) to level the resist and spin off the excess. The thickness of the coatings that are generated depend upon the viscosity of the resist and the rate of acceleration of the spinner. Generally, the layers are about one to two microns thick.

Control over the thickness and uniformity of the resist is very important in controlling the pattern dimensions during exposure. From this viewpoint it would appear that spinning is preferable to spray or dip coating, which tend to accentuate ''beading'' of the resist or the formation of layers of nonuniform thickness. ''Metal etch'' resists are generally preferred to ''photoresists'' since they can better withstand the chemical etches used to etch oxides. Recently introduced resists are less viscous and free of polymerized solids. In any case, before the resist is applied to the slices, it is additionally filtered in a Krueger filter to remove polymerized particles before it is placed in the syringe. Special care is used to avoid the formation of nonuniform coatings or coatings with voids or pinholes. Efforts are made to keep the slices clean and free from dust during the operation. Resist work is most often done in special dust-free ''clean rooms'' in which dust count, temperature, and humidity are controlled.

After applying the resist, the slices are air dried in a dust-free container for a least 30 minutes, and then further dried in an oven for 10 minutes at 100° C. They are then ready for exposure to the first of the photographic masks of the prepared set.

The exposure of the resist through the mask is accomplished by contact printing with a well-collimated and filtered ultraviolet light source. The exposure may take about 5 seconds. The mask and the resist-coated surface of the slice, are held parallel and in intimate contact by means of a vacuum clamping arrangement. A mask alignment instrument provides controlled translational and rotational movement of the mask, so that accurate mask alignment can be achieved. The same machine permits subsequent masks to be precisely aligned or registered over previous patterns used on the slice. Accurately situated registration marks at various locations on each pattern are useful in this connection. The marks on the mask are superimposed upon the identically located marks etched into the oxide during a previous step. The exposure also is performed in a dust-free environment to eliminate light scattering by dust particles. Instead of a collimated light source, an intense light from an uncollimated source may be used to undercut (with light) the areas masked by such scattering dust particles. The use of uncollimated light, however, for exposing, makes it more difficult to control exposed pattern dimensions because of this undercutting effect.

Wherever the mask is opaque, light is not transmitted to the resist immediately beneath it. During development of the resist, the unexposed areas of the resist are dissolved and washed away, while the exposed resist remains to serve as an etching mask. For any diffusion, the silicon must be exposed wherever the diffusion is to take place. Hence, the resist is left unexposed in these areas, and after development, the corresponding oxide region will be exposed to the etchant. The resist is developed by either spraying or bathing with resist developer, and then rinsing with a mixture of isopropyl alcohol and thinner. The slices are then baked in an oven at 120° C, for 30 minutes to one hour, to thoroughly cure the remaining resist and to improve its adhesion to the oxide. After this, the slices are etched in a slow buffered etch. Fast etches, such as hydrofluoric acid, may be used, but because of the faster etch rate, it is difficult to control the degree of etching. The masking resist is removed before diffusion by treating it with methylene chloride, hot sulfuric acid (H_2SO_4), or commercial "strippers." A typical ammonium bifluoride buffered oxide etch may have the composition:

Ammonium Bifluoride	15 gm
Hydrofluoric Acid (49 percent)	7 ml
Water	25 ml

At room temperature, this etch will remove oxide at the rate of about 800 Å/min.

Whichever etchant is used, changes in pattern dimensions are often observed after oxide etching. The degree of undercutting or poor edge definition at the edges of the protecting resist, during this stage, is dependent upon the quality of the resist, its degree of adhesion to the oxide, and the exposure time (assuming that the light is well collimated). If such effects are reproducible, they are often compensated for in the original artwork. The cleanliness of the developed pattern also determines the quality and accuracy of the etch, and this, in turn, affects the outcome of the diffusion. Resist particles left on the exposed silicon surface, as well as dust, tend to cause diffusion pipes which penetrate deeper than the overall diffusion layer. In the narrow active regions of a device, this can greatly degrade performance.

The last part of the wafer processing involves the creation of metal contacts at specific locations on the circuit pattern, and metal interconnections between devices in the circuit. An aluminum film is usually evaporated on the oxide layer of a slice, coated with resist, and then exposed to the photographic mask. The contact and interconnection patterns on the photographic mask are transparent and the background opaque. This leaves the aluminum protected by the resist, after developing, where the contacts and interconnections are situated. The unprotected areas of the aluminum are removed with an etch such as sodium hydroxide (10 percent NaOH at room temperature). Metal portions of metal-oxide-semiconductor capacitors may also be fabricated, at the same time, and in the same manner.

Photographic techniques are successful in achieving reproducible patterns on silicon. Other methods for producing finer images, such as ion beam implantation, are under investigation. Durable masks have been fabricated through the deposition of chromium films on glass slides using electron beam evaporation techniques from a fused chromium powder source. The films show good adhesion to glass slides, are free from pinholes, and are easily etched. Micropatterns may be prepared from these films by applying the conventional photoresist techniques followed by an etch and heat treatment. The masks have good line definitions, and can be used repeatedly (without being scratched or defaced) for obtaining contact exposures of photoresist film.

Plasma Processing

The use of cold discharge tube reactions in the semiconductor industry is by no means new. The first major use of plasma processing was in removing common organic photoresists In this process the slices are held upright in a carrier, much as in a furnace tube. The chamber is evacuated to a pressure of a few torrs and a radio-frequency generator is connected. A mixture of argon and oxygen is introduced into the chamber and generates a glow discharge of relatively low magnitude in the classical plasma discharge sense. This creates an oxidizing environment, which rapidly converts the organic photoresist to the volatile products carbon dioxide and water vapor. Here we have the underlying principle of all semiconductor dry etching processes. A solid material is converted to gaseous byproducts, which are removed by the vacuum pumping system.

The active species generated in a tube reactor must have a sufficient lifetime to reach the surfaces of interest. Some etches and depositions can be made only by generating the active species near the surface of the slice.
It has been shown that chemical vapor deposition can be promoted between two capacitor plates. The extension of this idea to a plasma reactor of the radial flow type was demonstrated and a sketch of such a reactor is shown in Fig. 9-3. 1

The scientific understanding of plasma processes was very sketchy in the early 1970's and certain areas are still not well understood. The world's first all-dry-processed IC was produced in 1975 by plasma development of photoresist; plasma etching of nitride, oxide, silicon, and aluminum; and ozone resist removal.

Nevertheless, virtually all the methods were worked out by determining the volatile products of the solids and choosing likely combinations of gas mixtures to produce them.

There is another plasma etching method. A high-power-density mi-

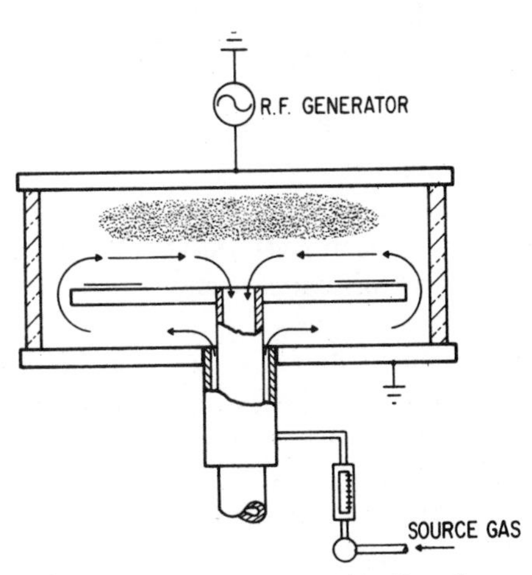

Fig. 9-3. 1 Radial flow reactor with slices flat on lower surface.

crowave discharge can produce metastable atoms with long lifetimes which etch material almost as if it were immersed in acid—that is, isotropically. There is some evidence that the etching species produced in this reaction is not the same as the species in the reactions described above.

A plethora of plasma etch gases and combinations can now be used to remove almost any material employed in the fabrication of silicon IC's, photomasks, and magnetic bubble structures. The emphasis has been on the small dimensions obtainable at low defect densities.

Solid materials deposited and removed with gases used for the plasma process.

Solid	Gases
Depositions	
Silicon nitride ($Si_xN_yH_z$)	Silane (SiH_4) and ammonia (NH_3)
Silicon oxide ($Si_xO_yH_z$)	Nitrous oxide (N_2O) and SiH_4
Amorphous silicon	SiH_4 and argon
Material removal	
Silicon oxide (SiO_2)	Silicon tetrafluoride (SiF_4)
	Carbon tetrafluoride (CF_4)
	C_3F_8, C_2F_6, C_5F_{12}, CHF_3
Silicon	CF_4 and O_2
	Carbon tetrachloride (CCl_4) and hydrogen chloride (HCl)
Silicon nitride (Si_3N_4)	CF_4
Vanadium, titanium, tantalum, molybdenum, tungsten	CF_4
Chrome and chrome oxide	CCl_4
Aluminum	CCl_4, boron trichloride (BCl_3)
Photoresist	Argon and O_2

9.3 WAFER PROCESSING

The fabrication of a planar diffused device requires a minimum of two diffusions. Frequently, as many as four or five are used. Equally as often, doped epitaxial layers are used in place of diffused layers. The choice of which processes are used, is determined largely by the geometry of the finished device. Epitaxial layers are often substituted for diffusions for very definite reasons. Diffused layers, by their very nature, have dopant impurity concentrations which vary from the surface, where they are highest, to the interior of the wafer, where they are lowest. Since diffusions are performed into the wafer from one surface, this concentration gradient or "diffusion profile" always prevails. Epitaxial layers, on the other hand, can be grown with uniform doping concentration levels so that such gradients do not occur. Assume, for example, it is desired to "bury" an N^+ region (of high concentration) beneath an N region (of low concentration), both of which are uniformly doped so that a concentration gradient is not present in either layer. This task is difficult to achieve by diffusion. Epitaxy would be preferred in such a case. If, in a similar device, the design dictates that only the N region be uniformly doped, then the N^+ region may be diffused into the parent wafer before the N region is grown epitaxially.

Buried N^+ regions are often used beneath transistors to reduce collector series resistances and are most often placed there by epitaxial means. N-type dopants, such as arsenic or antimony, which have low diffusion coefficients, are used to minimize redistribution during subsequent high-temperature processing.

The transistor base and emitter elements are almost always diffused. To minimize the total number of diffusions required to make a device, as many elements as possible are formed in any single run. When diffusing the base, for example, it is convenient to form the resistors at the same time. If a final sheet resistivity of 200 ohms per square is chosen for the base diffusion, then the resistors, which are diffused at the same time, will have a sheet resistivity of 200 ohms per square. The length and width of the resistor is chosen so that the necessary final value of resistance is obtained with this value of sheet resistance. To facilitate making ohmic contact to N diffused regions, N^+ enhancement diffusions are necessary. Since the emitter diffusion is a diffusion of high surface concentration, it is convenient to diffuse the emitters and the N^+ enhancement regions at the same time. Thus, only two diffusions are required where a total of four could have been used. It should be noted that whenever a new dopant layer is diffused into the wafer, all previous diffusions will tend to redistribute, causing a change to occur in their diffusion profiles. Thus, not only does the surface concentration and diffusion depth change, but also the lateral dimensions, of each of the elements of the device, change.

Lateral diffusion places a limit on the minimum spacing which can be achieved between components. When very close spacing is desired, the diffusions must be very shallow. A resistor diffused one mil wide and three microns deep will actually measure greater than one mil wide because of this tendency of lateral diffusion to occur. Lateral diffusions are disadvantageous, in this respect, because of the changes they cause in the geometry of the device.

Junctions which come to the surface and are exposed to various environments are susceptible to contamination. Such contaminations are, generally, undesirable and adversely affect the electrical properties of the junction. If care is taken to assure that the junction, which has migrated under the oxide at the surface of the wafer, is left protected by the oxide, then the protection afforded by the oxide to the junction is sufficient to prevent degradation of the operating characteristics of the device. Accordingly, the yield with which the devices can be made, is increased.

Two other diffusions, ''isolation diffusion'' and ''gold diffusion'', are frequently used in device fabrication. The purpose of the isolation diffusion is, as its name implies, to electrically isolate various ''islands'' into which components are diffused. Isolation is achieved by surrounding the island on all sides with opposite conductivity type material. When isolation is used in this manner, it is most often accomplished by diffusion, since epitaxial means are impractical in achieving these geometries. Gold diffusions are used to lower the recovery time of injected minority carriers after cut-off of a transistor switch. This is a special diffusion, which, again, cannot be accomplished by epitaxial means.

Gold diffusion temperatures of around 970° C are chosen in order to keep the gold solid solubility low. This prevents any appreciable increase in the collector saturation voltage (Fig. 9-4), and maintains lower magnitudes of the active device reverse currents, even though the active component areas are very small. The magnitude of the reverse current arising from carrier generation in the space-charge region, of a pn junction, may be expressed by the following:

$$I_r = qUWA \text{ amp.}, \tag{9-1}$$

where q is the electronic charge per carrier (1.6×10^{-19} coulombs); U is the carrier generation rate in carriers per cm^3-sec, and is directly related to the concentration of gold; W is the width of the space-charge region in cm; and A is the junction area in cm^2. If the solid solubility limit of the

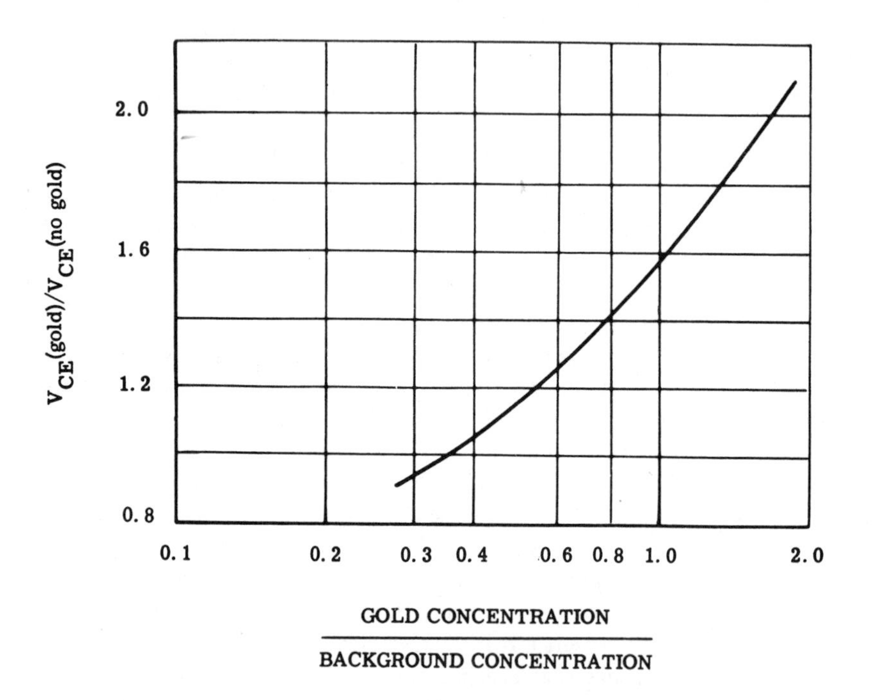

Fig. 9-4 Saturation voltage versus gold concentration

gold were to be approached in diffusion, U would be maximized at approximately 10^{17}. On the other hand, the recovery time (t_r) in the high resistivity side of a diffused junction is inversely proportional to the gold concentration, or to the number of recombination centers (N) per cm^3, and is expressed as:

$$t_r = \frac{2.53 \times 10^7}{N} \text{ sec.} \tag{9-2}$$

Figures 9-5 and 9-6 show the relationships of gold diffusion temperature to reverse current and recovery time. Newer techniques utilizing selective gold plating and diffusion within the emitter areas enable the fabricator to produce both high-frequency and low-frequency devices on the same silicon slice.

During the final drive-in, the exposed window through which the emitter diffusion occurred, is reoxidized leaving the surface of the wafer completely covered with oxide. In order to make ohmic contact to the various elements of the devices, windows are etched at appropriate points photolithographically. A thin film of aluminum is then evaporated over the entire surface of the wafer. This leaves both the unetched oxide and the etched windows coated with aluminum. If the wafer is heated to 576° C (the eutectic temperature of aluminum-silicon), aluminum will diffuse into the silicon, at the windows, and alloy with it. Some diffusion of aluminum will occur into the oxide, if the alloying were done at this stage. Since the latter diffusion is undesirable and can be easily avoided, it is common practice to remove the aluminum from all oxide surfaces before alloying is accomplished. This is again performed by photolithographic etching. After the first aluminization, resist is applied to the wafer and exposed to a pattern which leaves polymerized resist only in the areas of the windows. At the windows then, the aluminum beneath is protected during the subsequent etching. The undesired aluminum is then etched away in an alkaline solution (such as 10 percent NaOH). The resist is subsequently

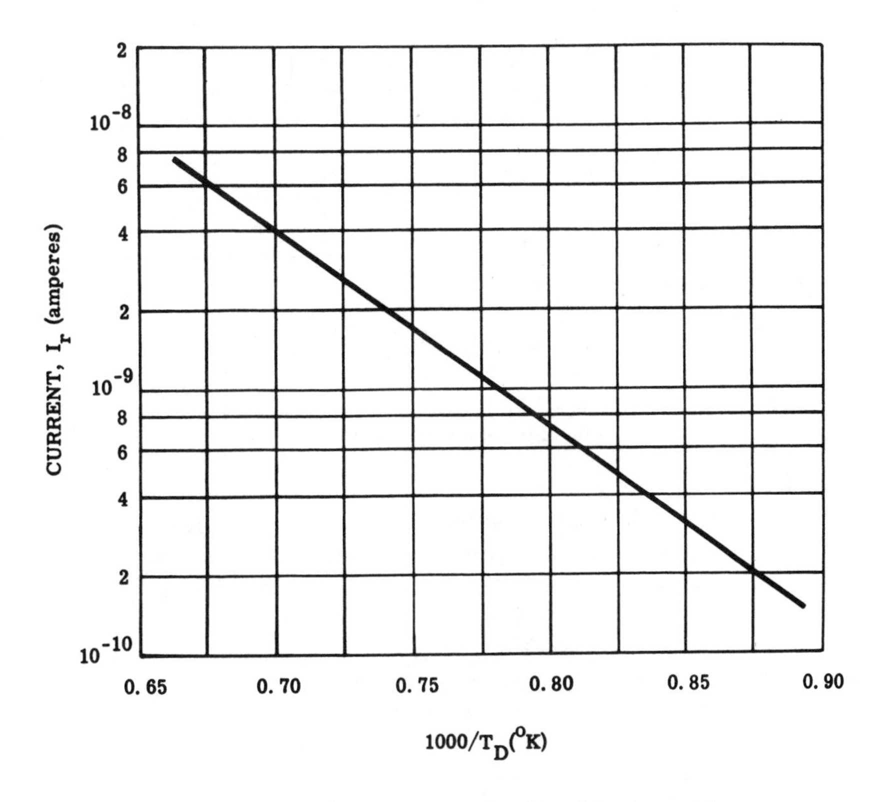

Fig. 9-5 Reverse current as a function of inverse gold diffusion temperature

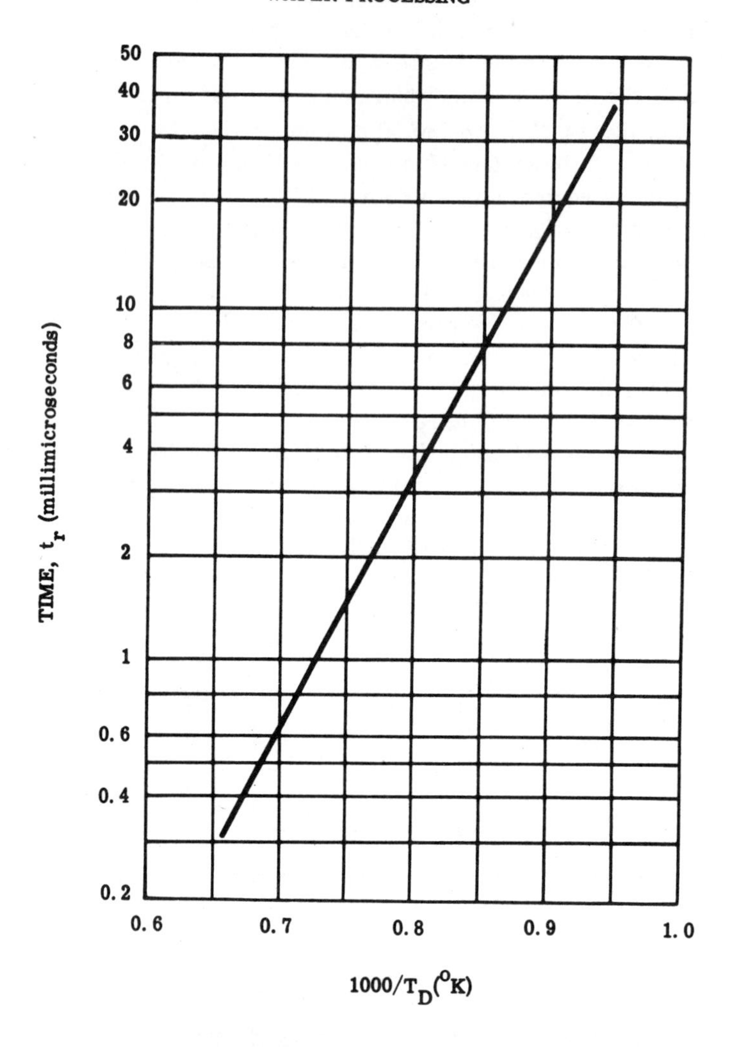

Fig. 9-6 **Reverse recovery time as a function of inverse gold diffusion temperature**

removed, and the aluminized windows are alloyed as described above.

The wafer processing ends with the formation of aluminum interconnection patterns and extension pads. The alloyed contact areas are interconnected on the surface of the wafer by a process identical to that just described. In performing this step, aluminum is again evaporated over the surface of the wafer, delineated with photoresist, etched, and cleaned to leave aluminum conductive paths in the desired form.

Frequently, capacitors are also formed during this step. If the oxide is used as the dielectric of the capacitor, one electrode can be formed out of this evaporated aluminum. This electrode is delineated by etching the aluminum. The semiconductor beneath, then becomes the second electrode.

It is quite possible to metallize the wafer in a one-step process. If a relatively thick evaporation of aluminum is deposited on the wafer, alloyed, and subsequently delineated by etching, to form both the ohmic contact pads and the interconnection pattern, only one step is required. This procedure has the disadvantage, however, of the possibility of forming high resistivity "knees" through thinning of the aluminum in the region where the interconnecting lead descends into the window, where the ohmic contact is formed. In the two-step process, the ohmic contact window is filled with aluminum first, thereby reducing the depth of the step. The tendency for the aluminum to thin out in this region during the last evaporation, is thus reduced.

9.3.1 PROCESS COORDINATION

OXIDATION

The planar process makes use of the fact that silicon dioxide exhibits the following properties:

(a) The oxide mask is an effective mask against boron and phosphorous.

(b) The oxide passivates the surface of silicon protecting the junction.

(c) The oxide is a good dielectric.

(d) The oxide may be etched with good resolution.

(e) The oxide forms a uniform, controlled thickness on the silicon with similar thermal expansion properties.

Oxidation is performed in an open-tube furnace at 900° to 1250°C by passing pure oxygen gas in a dry or wet state over the silicon surface. Layers from a few hundred to several thousand angstroms can be grown in a reasonable amount of time, although dry oxygen is considerably slower than wet oxygen. About 40 percent of the oxide layer is below the original silicon surface. The dry oxide is particularly useful for high-voltage and MOS devices due to its low surface state density. Dopants can diffuse laterally under $Si0_2$ (approximately 83 percent of the diffusion depth), resulting in the p-n junction being under a passivated surface.

DIFFUSION

Diffusion is a process of incorporating impurity atoms (dopant) into single-crystal silicon. By controlling conditions carefully, one can obtain various surface concentrations, depths, and profiles of p-n junctions. This profile depends upon the following:

(a) Level and distribution of dopant.

(b) Time and temperature of deposition and diffusion.

(c) Diffusion coefficient, solid solubility, and other characteristics of the dopants.

(d) Amount and form of dopant.

Each dopant material has individual characteristics, such as type (p or n) solid solubility and diffusion coefficient. Special characteristics are also to be noted as, for example, phosphorous is affected easily by moisture, while gold is a very fast diffuser and lowers the lifetime of silicon significantly.

Several entirely different diffusion systems can give identical electrical results; contrastingly, several seemingly similar systems can give different results. Although there are many diffusion systems, the one used most often is the *open-tube process.* Here, the source (dopant) as a gas, liquid, or solid at room temperature or elevated temperature, is flowed over the silicon wafer at 900° to 1250°C by a carrier gas, usually N_2 and/or O_2. Time, temperature, complementary error function or Gaussian conditions, and oxidizing or nonoxidizing ambient conditions determine the surface concentration, junction depth, and profile or gradient of the resulting diffusion. The effect of diffusing into a previously diffused wafer must be taken into consideration when selecting dopant, time, temperature, and other conditions, in order to obtain a proper relationship of p-n junctions to each other (base width, concentration gradient). This relationship determines resulting electrical parameters (e.g., Beta, storage time) of semiconductor devices, as much as initial device design (geometry of masks). Cleanliness is extremely important here, particularly in equipment, chemicals, wafers, and gas supply. Most of the gases used are toxic and must be handled with care. Several of the more important variables are:

(a) Time and temperature of diffusion (less than ±0.5°C temperature variation over the flat zone length is required).
(b) Control of amount, source, vapor pressure, and purity of dopants. (Dopants must be kept separate and prevented from intermixing with each other.)
(c) Uniform flow of dopant around wafer.
(d) Cleanliness. (Elimination of undesirable secondary sources, and effects.)

Evaluation of the diffusion requires electrical test equipment with a prober, a resistivity kit, bevel and stain facilities, and a microscope to determine surface concentrations, junction depths, and surface conditions.

EPITAXIAL DEPOSITION

The single-crystal silicon substrate establishes the crystal orientation of the epitaxial layer. The usual method is the hydrogen reduction of silicon tetrachloride at approximately 1200°C. The epitaxial layer is grown by introducing silicon tetrachloride diluted with H_2 and small amounts of H_2 containing phosphine or diborane. For n- and p-type layers, respectively, a good epitaxial system should be capable of 100-ohm-cm layers. Growth rates vary with time and temperature, and are of the order of 1 micron/minute. Cleanliness and reagent purity are most important. Starting materials must be free of imperfections, as this process worsens their condition. Most layers are in the resistivity range of 0.1 to 10 ohms-cm, and thickness of 3 to 30 microns. Good control of reagents, gas flow, etc., can result in variations of less than ±10 percent across a wafer. Caution must be taken due to the toxic and explosive nature of gases used.

ISOLATION PROCESSES

Some devices in integrated circuits must be electrically isolated from the rest of the circuit. This may be accomplished in several ways:

A. *Diffusion Isolation* (Fig. 9-7): Isolation is accomplished by diffusing p-type grid into a thin n epi-layer at high temperatures (1250°C) for a relatively long time (2 to 4 hours), until it reaches the p-type substrate.

B. *Dielectric Isolation* (Fig. 9-8):

(1) n on n+ (by buried or epitaxial layer) substrate is oxidized.

(2) The oxide is etched except where isolated islands of silicon are required.

(3) The silicon substrate is etched to form a moat around the islands.

(4) The silicon is reoxidized.

(5) Polycrystalline silicon is deposited to a thickness of several mils.

(6) The original n-layer is lapped down to the oxide-coated moat and reoxidized.

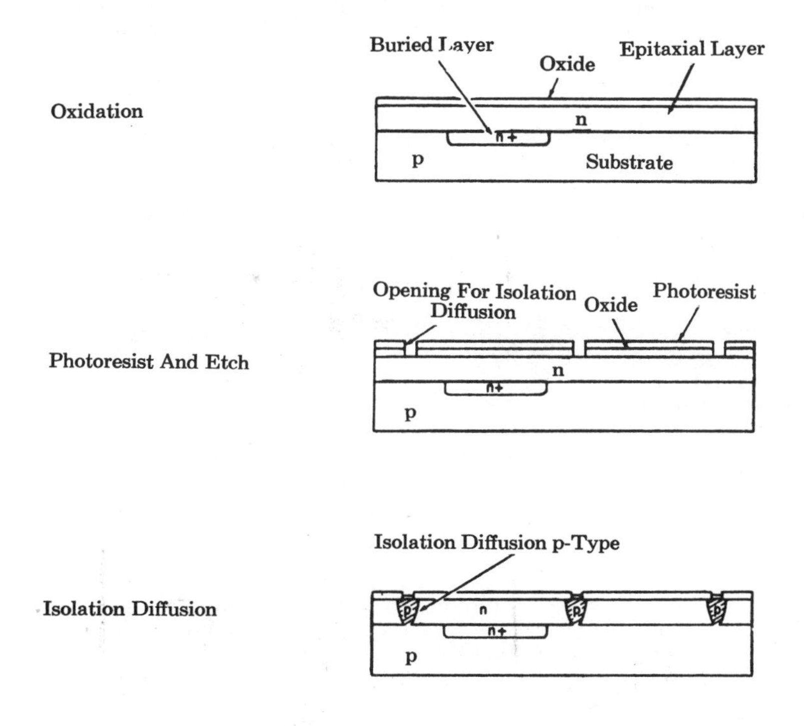

Fig. 9-7 Diffusion isolation.

(7) The wafer is inverted to yield a polycrystalline silicon slice with thin n on n+ islands isolated by SiO_2 tubs.

C. *Air Gap Isolation — Beam Leads* (Fig. 9-9):

(1) The wafer is prepared in the normal fashion up to the metallization step.

(2) Platinum is sputtered and alloyed into the contact windows. The platinum silicide that is formed gives good ohmic contact. The

unreacted platinum is removed by back sputtering.

(3) The oxide is etched, leaving an overhang to prevent shorts from the beams to the substrate.

(4) Titanium/platinum is sputtered, and gold evaporated for conductor paths. By photoresist masking, gold is plated up to 0.3 to 0.5 mil. The photoresist is removed and titanium/platinum are removed by back sputtering. Conductor paths remain where the

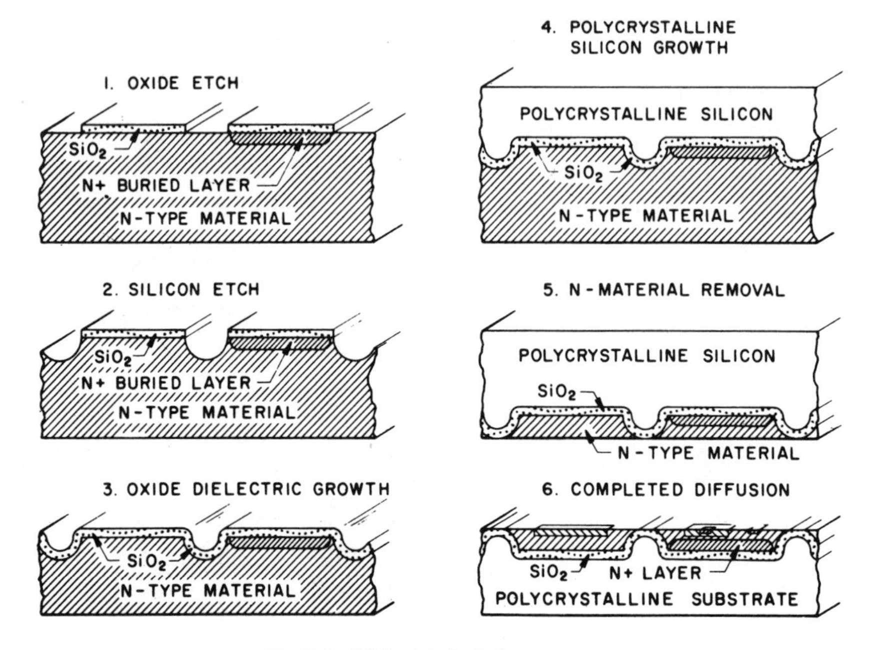

Fig. 9-8 Dielectric isolation.

gold was plated.

(5) Finally, the slice is thinned down to approximately 1 to 2 mils.

(6) Isolation paths are lined up for both the front and back of the wafer, with a backside or infrared aligner. The silicon is etched away leaving desired areas isolated by air, each isolated area held together with gold beams. Some beams are allowed to overhang the die for external connections.

METALLIZATION

Aluminum (Al) is the most commonly used metallization for integrated circuits because it forms an ohmic contact with p-type and heavily doped n-type silicon, adheres strongly to silicon and silicon dioxide, and is a good conductor. However, if gold (Au) wire is used for bonding, there is a possibility of a brittle Au-Al compound (purple plague) being formed at high temperatures. If Al wire is used, it must be bonded to Au-plated posts, where plague is less likely. Au metallization can be used to eliminate this problem. However, Au does not adhere to SiO_2. Thus, Cr, Mo, and Ti are used with Au. However, these metals do not make as good an ohmic contact as Al. In the beam lead process, the metallization system is more complex.

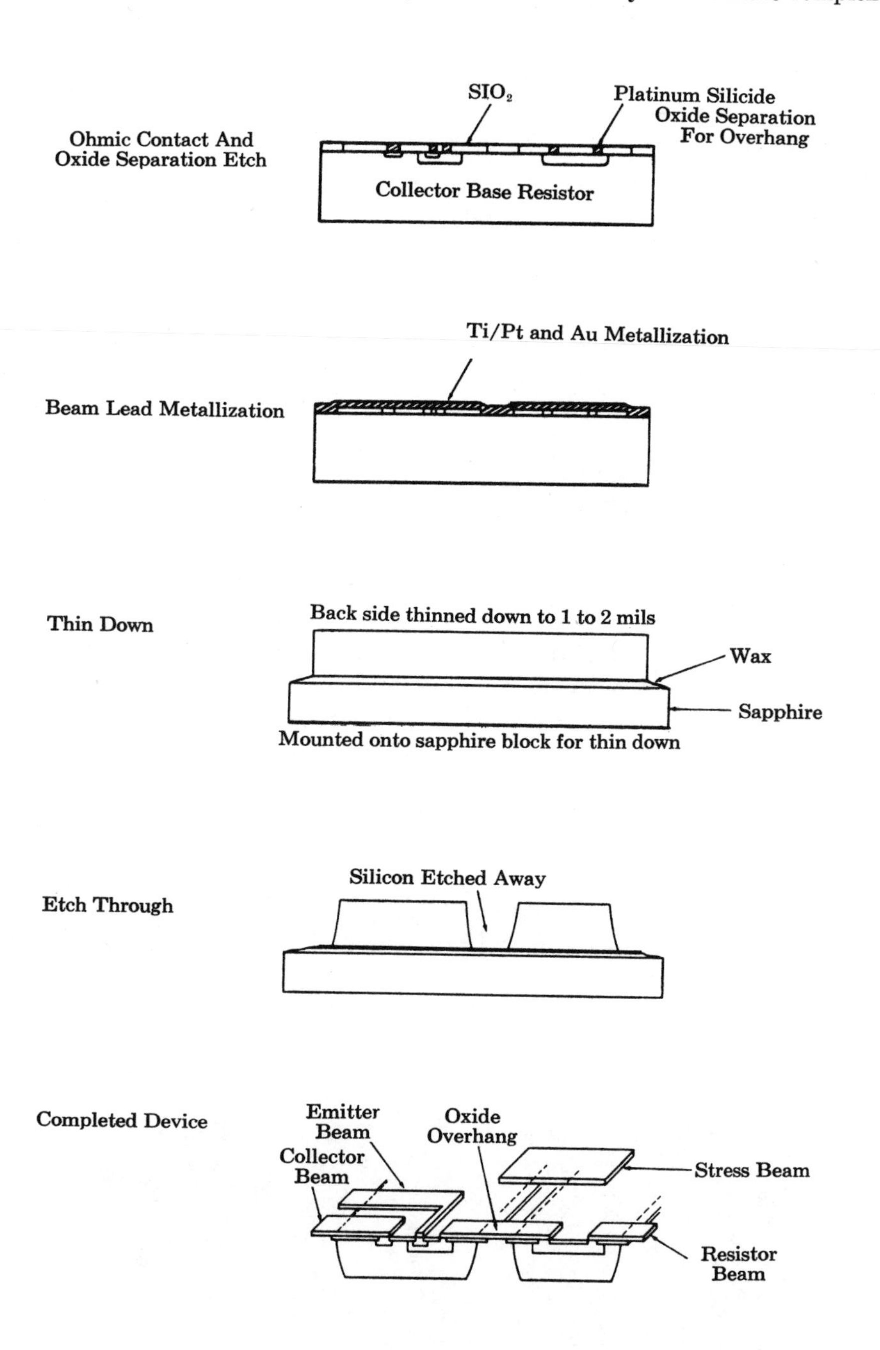

Fig. 9-9 Air gap isolation (beam lead)

Platinum is used to form platinum silicide (for ohmic contact) with silicon, titanium for forming bonds to SiO_2, and gold to conduct and serve as a bonding "wire", and to hold isolation regions together. The resulting Au-Au bonded system gives the best bond possible.

Ultra-pure metals are evaporated from tungsten coils in standard oil diffusion pump evaporators in a vacuum of at least 10^{-6} mmHg using a liquid nitrogen cold trap. If further cleanliness is necessary, such as for MOS devices, an oil-free vacuum system incorporating cryogenic absorption pumps and titanium chemical pumps can be used in conjunction with electron beam gun evaporation from a cooled crucible. In either case, equipment and material cleanliness are absolutely necessary here. The usual vacuum technology procedures apply. Some of the variables encountered are:

(a) Cleanliness of the metal substrate, the interior of the vacuum chamber, and the bell jar
(b) Substrate temperature control
(c) Amounts evaporated
(d) Time of evaporation
(e) Shutter variables
(f) Vacuum level employed
(g) Positioning accuracy.

These variables can affect the film's adhesion, thickness and uniformity (shadowing), electrical characteristics (resistance), and may subsequently affect bonding and reliability. The conductor pattern is then made using standard photoresist techniques and applicable etch methods. Sputtering and plating are two other techniques used less frequently for metallization.

To insure proper ohmic contact, the wafer is generally alloyed or sintered at or near the eutectic point for a few moments in an inert or reducing atmosphere. Up to this point, the wafer has been carried through the process at 5 to 8 mils thick to reduce breakage. Here, the wafer is thinned down to approximately 4 mils and a gold evaporation is performed as an aid to eutectic die mounting. All the devices on the wafer are electrically checked at this point, and bad ones are inked out. This is usually an automatic or semiautomatic operation.

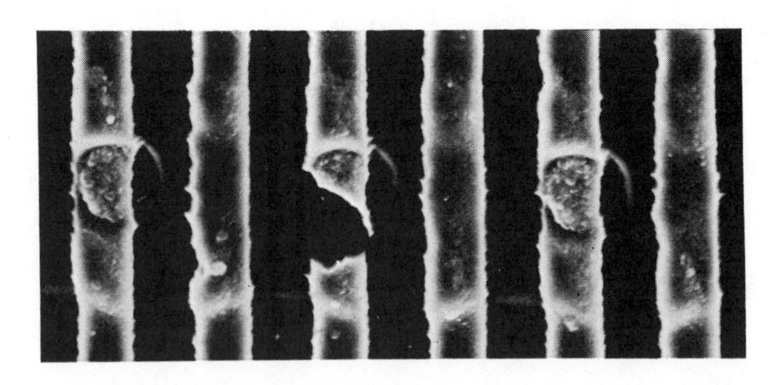

An entire microelectronic circuit is ruined by a single defect because a dust particle on the photomask used to define the aluminum layer has caused a break in an electrical connection. (Magnification is 1600 diameters)

SCRIBING AND BREAKING

Utmost care must be taken to scribe, especially in the case of dice <20 mils square. The ratio of edge to thickness is reduced drastically as the die size gets smaller, and the yield is reduced. Most slices are made with no oxide on the edge of the cell pattern, leaving a gridwork of bare silicon 2 to 4 mils wide. The diamond scribe point should ride in this area as it is difficult for the scribe point to penetrate the oxide area. Automatic machines are available for this step, but much is left to the operator in determining diamond tip quality, pressure, angle, and tracking. Machines exist for breaking, but, in many cases, the slice is rolled over small-diameter steel rods, and is usually mounted in wax or ice to improve breaking characteristics. At present, there is no uniform practice for breaking. Essentially, the larger and thinner the die, the easier it breaks. The devices which are not inked out in prior tests are selected, visually inspected, and sent on to final assembly.

DIE ASSEMBLY

The principal method used for die attachment is eutectic die bonding; this involves mounting a gold or n- or p-type doped gold-backed die with or without a corresponding preform onto a gold-plated substrate. The combined assembly is heated to slightly above the melting point of the eutectic mixture of Au/Si, and the die is scrubbed with tweezers or an ultrasonic fixture. The gold dissolves into the silicon until the composition is that of a gold silicon eutectic mixture, whereupon further dissolution causes freezing if the temperature is maintained constant and bonding is effected. The operation is usually performed on special die bonding equipment on a hot stage in an inert atmosphere. By using different composition preforms, varied bonding temperatures can be used.

For especially good ohmic contact, in the case of power transistors, for example, the back of the die has a nickel evaporation or plating, and a solder preform is used which melts and attaches to nickel but does not form a eutectic mixture. To freeze this, one must cool to below the freezing point of the solder.

Aluminum or gold wires are most frequently used for semiconductor work. Special equipment capable of handling wire as small as 0.5 mil in diameter is used. The equipment can locate the wire on an area as small as 1 square mil. The substrate containing the mounted die is heated in an inert atmosphere and the wire is bonded by one of the methods described in the chapter, "Interconnections".

ENVIRONMENTAL CONTROL

Cleanliness and ambient control are required at every step in the process. Distilled deionized water is used for all rinsing and cleaning with a minimum of 10 to 12 megohms-cm required, and 18 megohms-cm desired. Process and humidity control is maintained throughout the area with either local dust-free ambients maintained for critical operation, or the entire operation is performed in a clean room of acceptable class.

9.4 COMPLEMENTARY CIRCUITRY

The fabrication procedures described above are normally applicable to circuits containing NPN or PNP transistors. When complementary circuitry (i.e., circuits containing both NPN and PNP diffused devices) is designed for the same substrate, more diffusion steps are required, and the difficulty of fabrication increases. Thus, assume that a circuit containing an NPN transistor next to a PNP transistor on the same substrate, were to be fabricated. It is evident that they cannot be made in serial order because of the changes in diffusion profiles which the first made device will undergo while fabricating the second. If the devices are made entirely by diffusion, then it is evident that three N-type and three P-type diffusions will be required, each having a different surface concentration and distribution profile. If the starting substrate is chosen so that it will become the collector of one of the transistors, then the number of diffusions required can be reduced to five. But five diffusions is still a formidable number, and the trade-off required in device characteristics to make the technique practicable is sufficient to discourage this approach.

The concept of "oxide isolation" evolved out of attempts to solve this problem by achieving more complete electrical isolation between individual components on a single chip. A description of a typical application of this technique will make the method by which isolation is achieved quite evident.

The starting material is a slice of silicon, approximately 8 mils thick. The resistivity and conductivity type of the material is not of particular importance. For purposes of this discussion, assume it to be 10 ohm-cm N-type, arsenic or antimony doped. On one side of this wafer, two monocrystalline N-type epitaxial layers are grown as shown in Fig. 9-10a. The uppermost layer is about 4 to 5 microns thick and has an impurity concentration level (about 0.1 ohm-cm) equal to that required in the N^+ buried layer which is to be located beneath the collector of the transistor. Like the substrate, both the N and N^+ layers are arsenic or antimony doped in order to minimize impurity redistribution during subsequent processing.

Using photolithographic masking, depressions are etched into the surface of the wafer deep enough to penetrate through the N^+ and N epitaxial layers. In Fig. 9-10b, the holes are shown as being steep-sided for reasons of convenience. In practice, they have semicircular cross-sections and are etched to define mesas of epitaxial material. These mesas are called "bathtubs" or "tubs."

Next, as shown in Fig. 9-10c, the wafer is thermally oxidized so that the entire upper surface of the wafer is covered with oxide. This oxide layer will provide the electrical isolation between "tubs" in the finished wafer. Over the oxide, a polycrystalline silicon layer is grown as shown in Fig. 9-10d. The polycrystalline layer is grown in an epitaxial furnace and under conditions which would normally yield a monocrystalline layer. Because of the interposition of the oxide layer, however, the layer produced is polycrystalline (but fine grained). This polycrystalline layer becomes the new substrate when the original substrate is lapped and polished away. In Fig. 9-10e, the finished wafer is shown after this polishing operation. Most of the original substrate is now gone, and the wafer now consists of a poly-

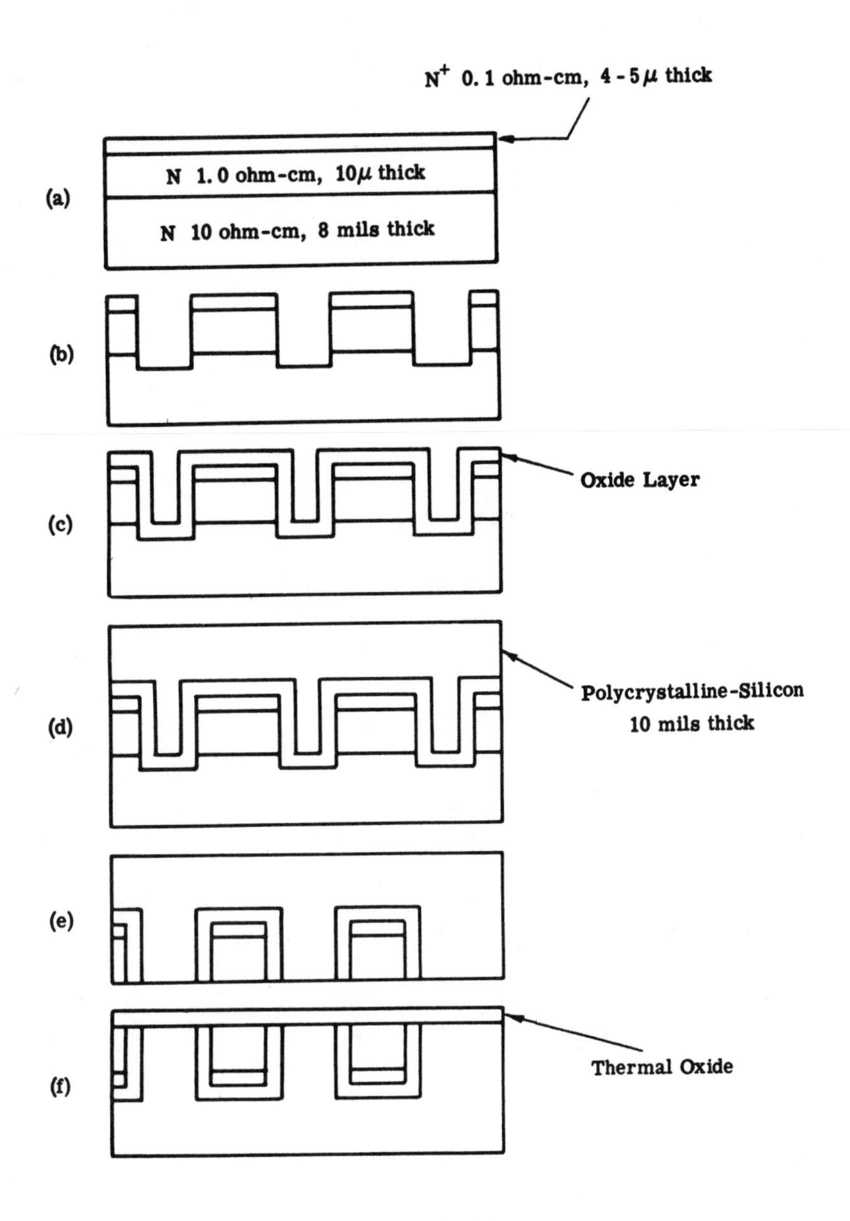

Fig. 9-10 A typical fabrication sequence used in obtaining "oxide isolated" monocrystalline islands for complementary circuitry

crystalline substrate carrying $N^+ - N$ islands of monocrystalline silicon which are isolated from each other and from the substrate by silicon dioxide. If the wafer is again thermally oxidized to provide for oxide masking for diffusion, each component of a circuit can be diffused into its own individual isolated monocrystalline "tub." Figure 9-10f shows the wafer inverted with respect to Fig. 9-10e. A thermal masking oxide has been grown over the surface of the wafer in preparation for selective diffusion.

The oxide isolation technique is an interesting innovation in planar diffusion technology from the designer's point of view. The method provides better isolation between components on the same chip than any other approach. Consequently, the design predictability is higher. Less time is consumed in fabricating the device because less consideration has to be given to the interaction between elements. In addition, each element of the device can be designed with optimum performance characteristics and fewer compromises.

It is possible, for example, using ''preferential'' epitaxial techniques, to grow epitaxial layers of opposite conductivity type, next to each other, on the same substrate before forming oxide isolated tubs. Preferential epitaxy is an epitaxial technique which limits growth of epitaxial layers of specific areas, of a substrate, through oxide masking. Thus, it is possible to grow $N^+ - N$ and $P^+ - P$ layers adjacent to each other on the same substrate. Subsequent applications of the oxide isolation and planar diffusion techniques can yield relatively complicated structures.

In Figs. 9-10b and 9-10c, before and after the isolating oxide is grown, and before the polycrystalline silicon is deposited, what will eventually become the bottoms of the tubs are exposed and applicable to a variety of design techniques. If, for example, a conductive film is deposited around the tub before or after the oxide is grown, such a film could be used to serve several functions. If deposited over the top of the isolating oxide, it could be used for interstage shielding. If deposited beneath the isolating oxide, it could be used to bring the collector contact to the surface, and aid in reducing the collector series resistance. If, after the isolating oxide is grown, holes are etched into adjacent tubs, interconnections, by means of deposited conductive films, can be made between tubs and brought to the surface where desired. When the polycrystalline substrate is grown, these interconnections are buried within the bulk of the device. Wherever they come to the surface, contact can be made to them by conventional enhancement diffusion and aluminization techniques.

In the application of the oxide isolation technique, the conductive films can be formed beneath isolating oxides by vacuum depositing metallic films on bared silicon, and reacting them to form conductive silicides. Molybdenum is useful in this respect. They can be formed over the top of isolating oxides by depositing silicon first, metal next, and reacting them in a similar manner. Similarly, thermal oxides can be replaced with vapor deposited or pyrolytic oxides, whenever the substrate is not available for oxidation.

The principal advantage of oxide isolation resides in the desirable feature of translating circuit design into hardware, with high predictability and a minimum of rework. When compared to simple planar diffusion technology, it has no disadvantages. However, the oxide isolation technique has one practical fabrication problem associated with the difficulty of polishing away the substrate satisfactorily in the final step. Polishing, and the etching which of necessity must follow, are difficult to achieve easily and quickly. Polishing must be accomplished on precision machinery and with great accuracy. The amount of material removed, so that the oxide isolated tubs are all of uniform thickness, is critical. Attempts have been made to apply means other than polishing, to accomplish the same result.

9.5 EXAMPLES OF SILICON PLANAR FABRICATION PROCESSES

To illustrate the essential steps involved in device and circuit fabrication, examples are given below for a diode, switching transistor, and basic integrated circuit. Many variations in the fabricating sequence of monolithic circuits are available, and particular functions may be accomplished through a variety of means. For purposes of illustration, however, a specific process and set of conditions will be assumed.

9.5.1 DIODE

The following sequence of steps will show the fabrication of a silicon planar diode:

1. *Material:* In this device, the material used is a single-crystal silicon slice. Orientation is <111> plane with resistivity of 0.3 to 0.5 ohm-cm p-type.

2. *Oxidation:* This step will grow 6000Å of oxide in 1 hour, which will be sufficient oxide to mask the following diffusion.

3. *Anode Mask:* Photoresist process to open the anode window for diffusion; approximately 5 mils diameter.

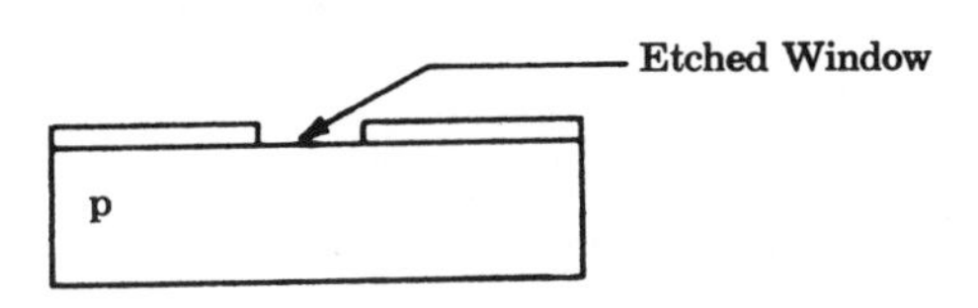

4. *n-Diffusion:* Complementary error function diffusion of phosphorus impurities into front and back side of slice. The concentration will be approximately 10^{21} atoms/cc and depth of 2 microns. This will produce high efficiency and good ohmic contact. Depth insures passivated junction and reasonable breakdown voltage. A low-temperature (900°C) oxidation is now made to help protect junction from contamination.

Second Layer of Oxide Diffusion

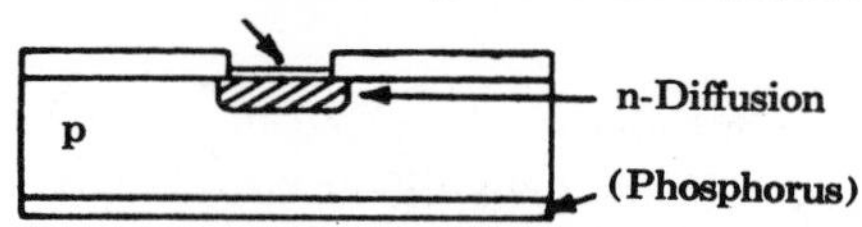

5. *Contact Mask:* Photoresist process to open window for metallization approximately 4 mils diameter.

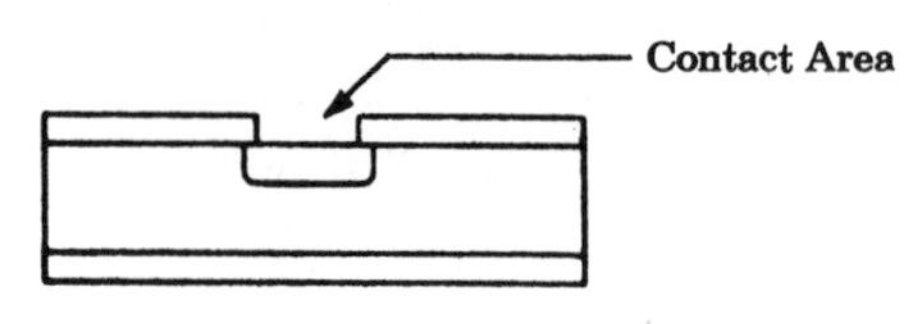

6. *Metallization:* Approximately 5000Å thick of metal (normally aluminum) is evaporated at 10^{-6} mmHg over entire surface of slice.

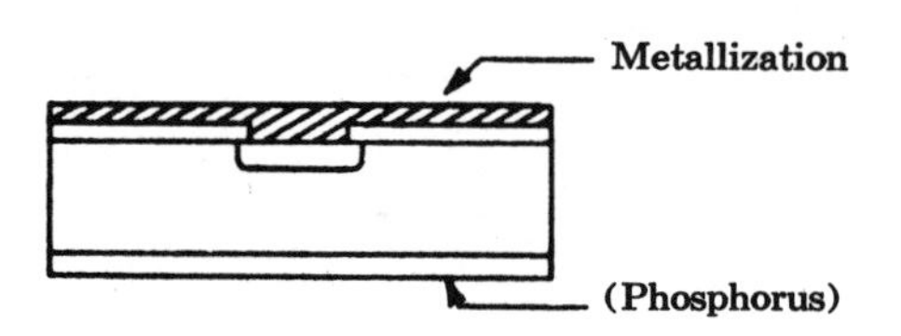

7. *Metallization Mask:* Photoresist process to etch away all unnecessary metal and form contact and bonding areas. Slice is then sintered at approximately 570°C. Back side phosphorus layer is etched off, and gold evaporated to insure good mounting.

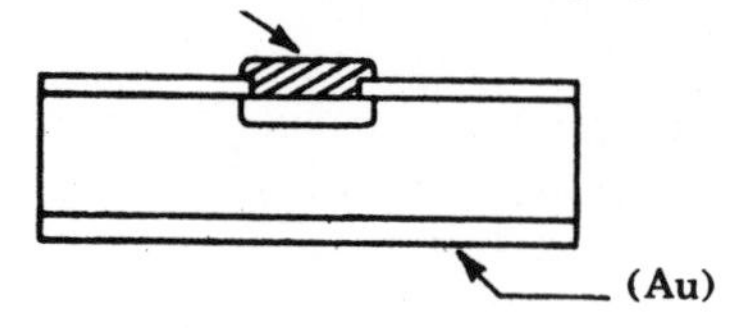

9.5.2 SWITCHING TRANSISTOR

The following steps show the process necessary to fabricate an n-p-n switching bipolar transistor. The general procedure is adaptable to all bipolar transistors.

1. *Material:* Starting material is a slice of 0.005 to 0.009 ohm-cm Sb-doped n-type <111> oriented single-crystal silicon. This low-resistivity material is used because it greatly decreases collector series resistance.

2. *Epi-Layer:* An n-type epitaxial layer is grown in an RF reactor at approximately 1200°C. Layer is 6 to 10 μ thick with resistivity of 0.1 to 0.3 ohm-cm. Resistivity and thickness of epi-layer must be optimized to provide maximum breakdown voltage and minimum series resistance.

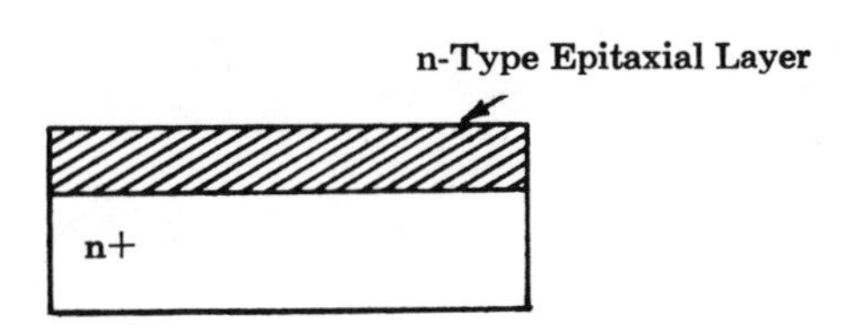

3. *Oxidation:* First oxidation is grown in 1100°C wet oxygen environment for 1 hour. This will produce 4000 to 6000Å which will be enough to mask the following diffusion.

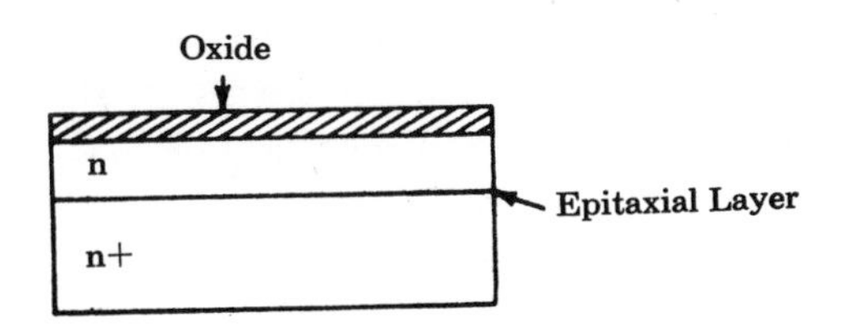

4. *Base Mask:* Photoresist will be applied to open the window for base diffusion. Window opening is 4 × 5 mils.

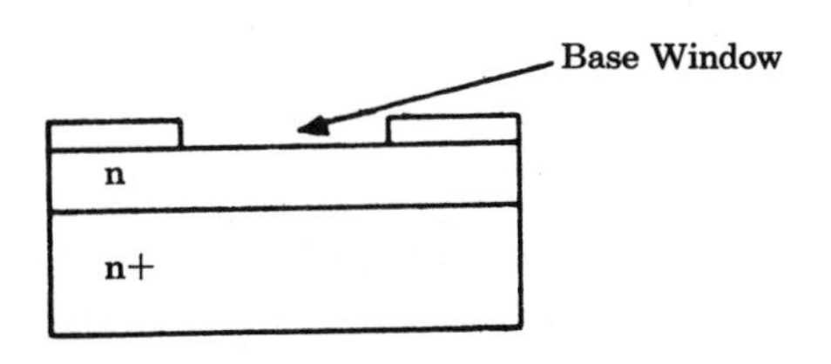

5. *Base Diffusion:* This p-type boron (BB_{r3}) diffusion will be a Gaussian distribution diffusion and will obtain surface concentration of 4×10^{19} atoms/cc and depth of 12,000Å. Deposition will be done at 1100°C for 10 minutes. Slice is then gold evaporated on back side, just prior to diffusion to reduce storage time (Ts). From here, slice is put into 1100°C oxidation for 45 minutes where diffusion to depth takes place. This oxide will also passivate junction and mask off emitter diffusion yet to come.

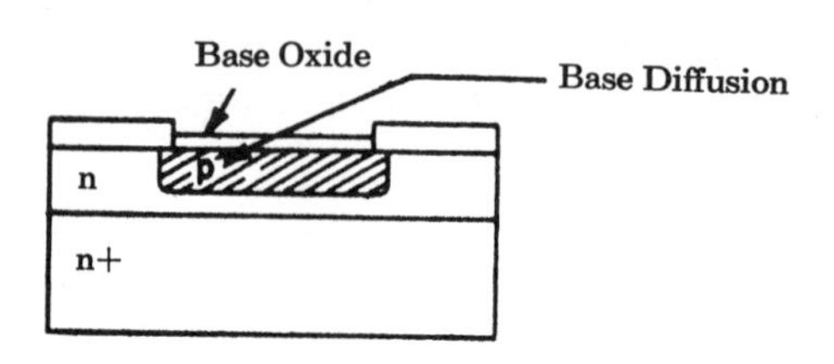

6. *Emitter Mask:* This photoresist is applied to open a window in base region for emitter diffusion (1×3 mils).

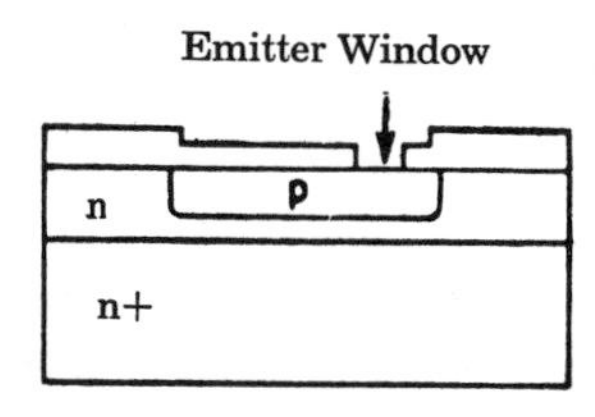

7. *Emitter Diffusion:* Here, an n-type ($POCl_3$) dopant is diffused into the emitter window. Diffusion is essentially an error function diffusion 9500Å deep with concentration of 10^{21} atoms/cc. Base width will be 2500Å which is necessary to obtain electrical parameters required for this device. (Beta, transit time, BVCEO.)

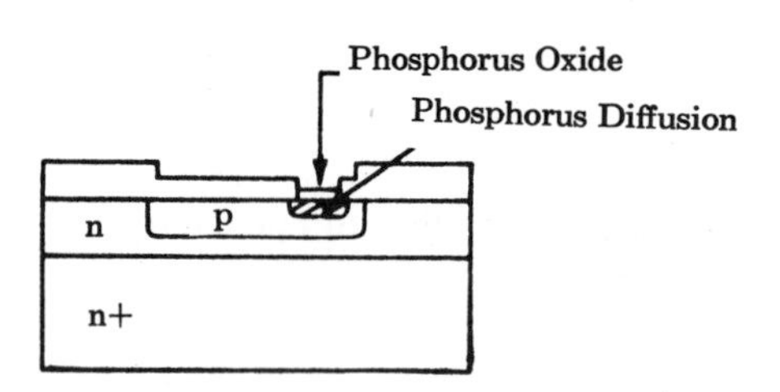

8. *Contact Mask:* Another mask is now applied which will open base and emitter contact areas (0.8×2.5 mils) for metallization. It is important here that all oxide be removed prior to metallization. Any oxide will hinder good ohmic contact.

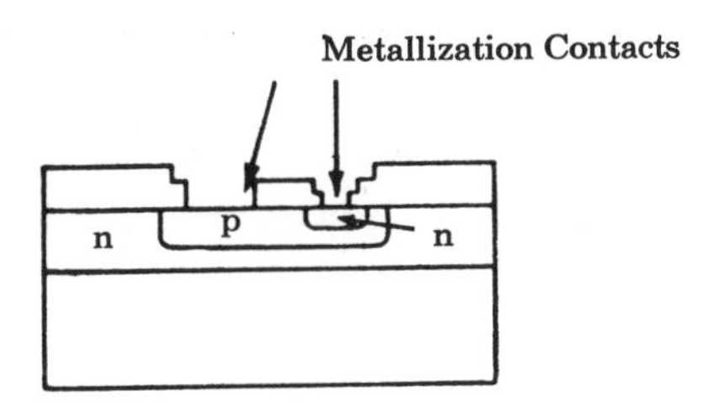

9. *Metallization:* Aluminum metal (5000Å) is now evaporated over entire surface of slice.

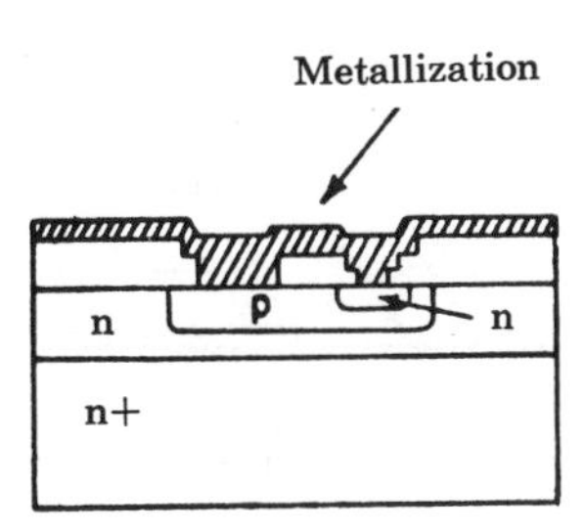

10. *Metallization Mask:* Final photoresist process is carried out on the metal which is then etched, leaving metal for contact and bonding areas. Aluminum is sintered at 570°C, back side is etched, and gold evaporated to insure good mounting.

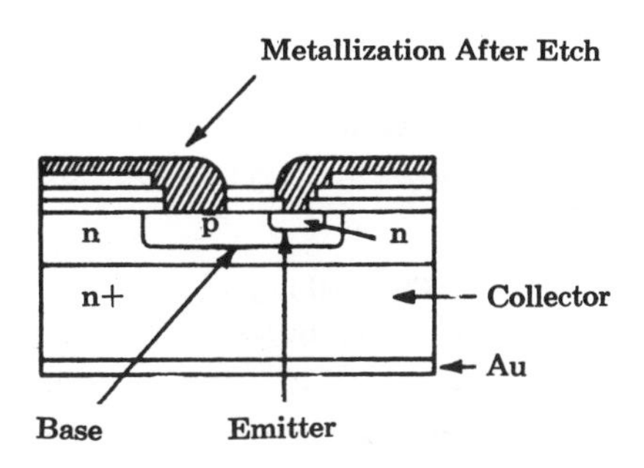

9.5.3 INTEGRATED CIRCUIT

The following steps will show the fabrication of a simplified circuit employing a transistor and resistor. This circuit can be expanded readily by adding capacitors, diodes, and resistors, if desired.

1. *Material:* A p-type (boron) slice 6 to 7 mils thick, 7 to 14 ohms-cm. High-resistivity material limits back diffusion into epitaxial layer.

Substrate

p

2. *Oxidation:* A 6000-Å oxide is grown to mask off buried layer diffusion.

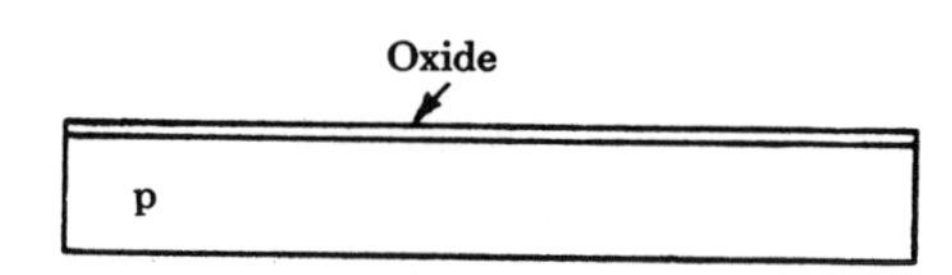

3. *Buried Layer Mask:* To open window in oxide to protect against diffusion.

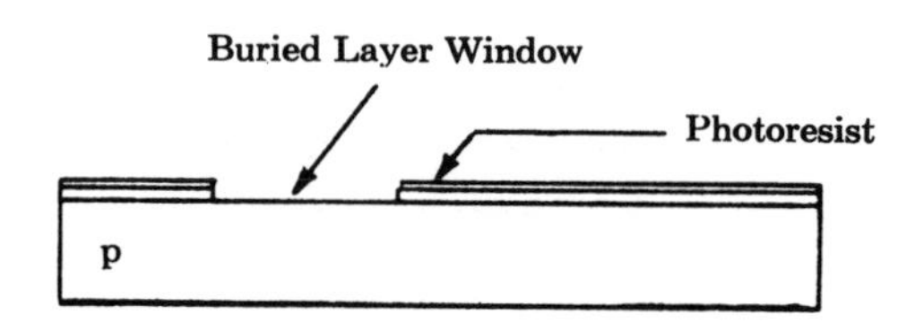

4. *Buried Layer:* High-concentration complementary error function, Co $= 10^{21}$ atoms/cc n-type (Sb) diffused in 3 microns. This diffusion is made to decrease collector series resistance.

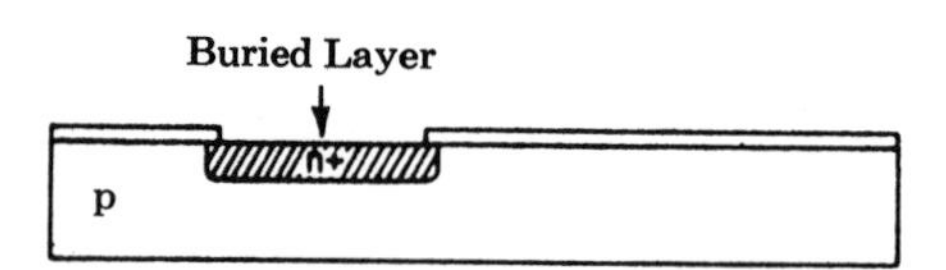

5. *Epitaxial Layer:* A thin 6- to 20-micron n-type (phosphorus) epitaxial layer is grown in a reactor. This layer is collector area for transistors.

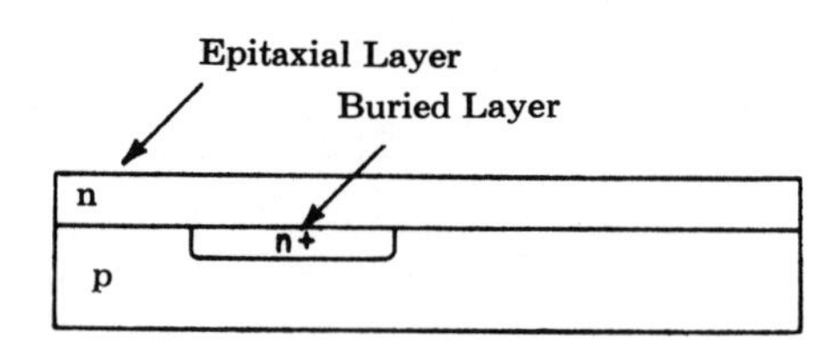

6. *Second Oxide:* 10,000-Å protecting epitaxial layer from deep isolation diffusion.

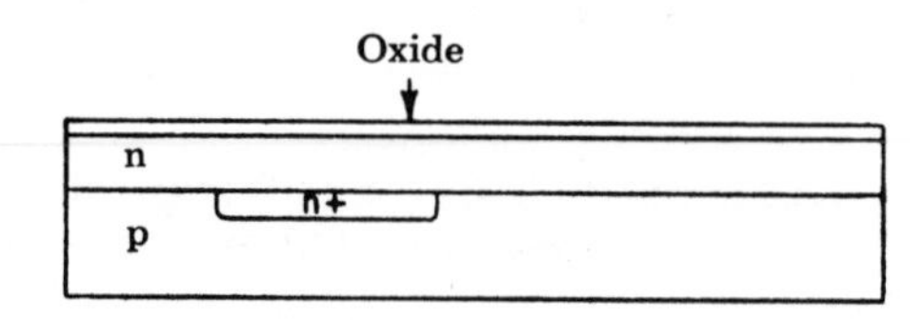

7. *Isolation Mask:* Opening of areas in oxide for isolation diffusion.

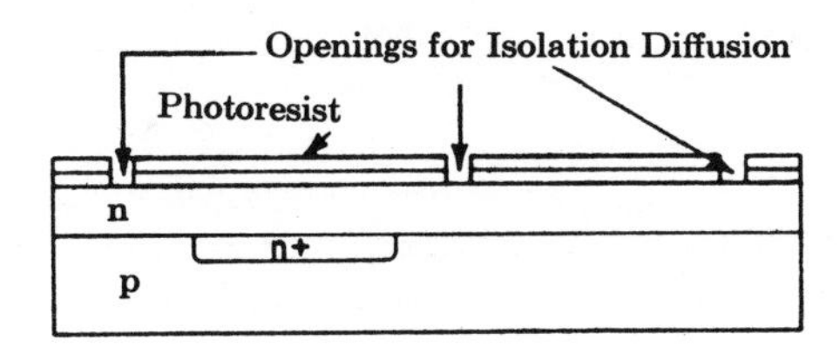

8. *Isolation Diffusion:* High-concentration pre-deposited p-type (boron) 10^{20} atoms/cc is diffused from surface through epitaxial layer to substrate, forming electrically isolated n regions..

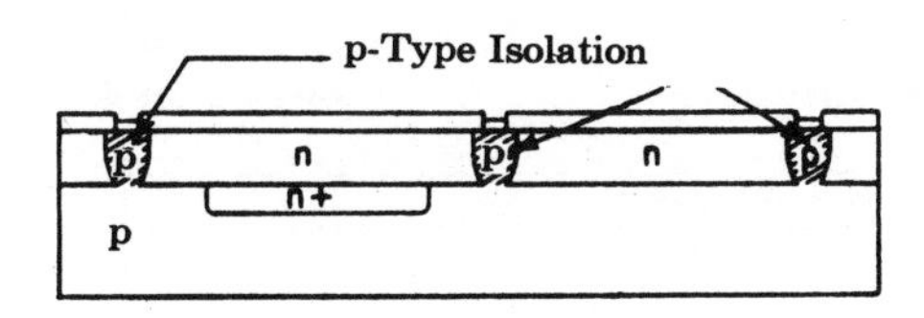

9. *Base Mask:* This step etches areas for base and resistor diffusion.

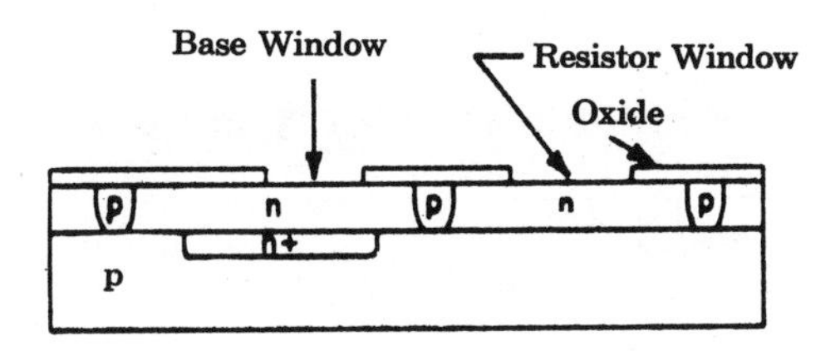

10. *Boron Diffusion:* This light-concentration diffusion is p-type BB_{r3} of 200 ohms/square sheet resistance. Sheet resistance is used instead of Co because of resistor value. Resistor and base mask are designed around 200 ohms/square diffusion. This is a difficult diffusion to control.

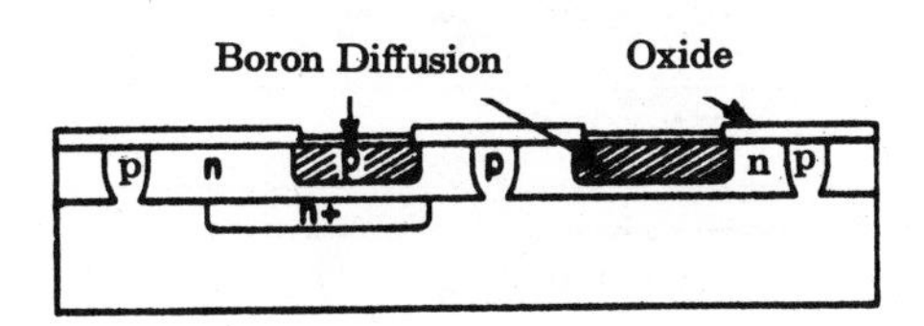

11. *Emitter Mask:* This mask will define areas for transistors, emitters, and collector diffusion.

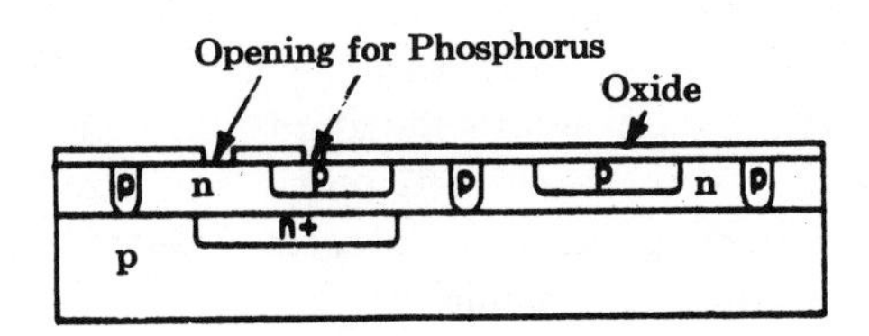

12. *Phosphorus Diffusion:* This is a high-concentration n-type diffusion. Its Co is on the order of 10^{21} and a micron in depth. High Co diffusion is essential in producing proper electrical parameters.

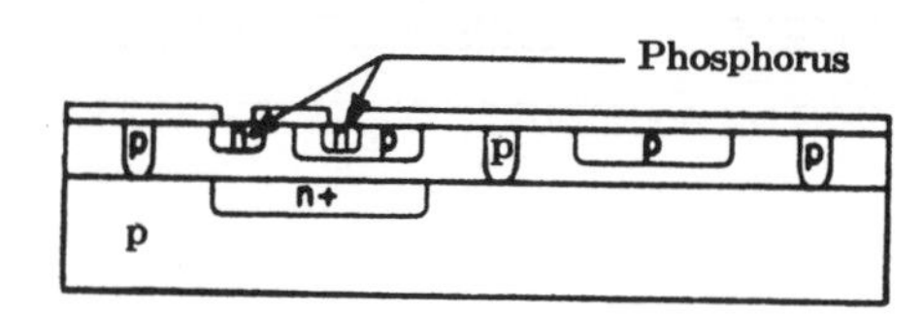

13. *Contact Mask:* Slice is photoresist processed for fifth time. Areas are etched away for ohmic contact.

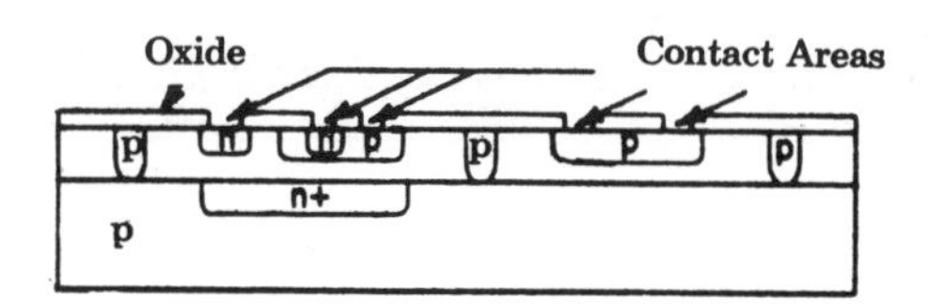

14. *Metallization:* Entire slice is covered with thin metal film.

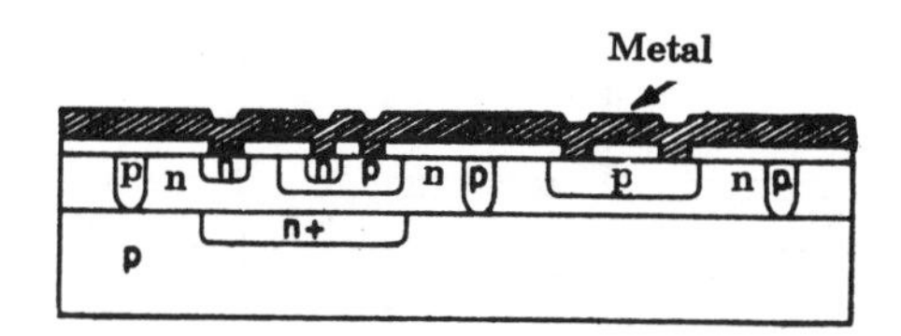

15. *Metallization Mask:* With one more series of photoresist process, the portion not needed for interconnections, contacts, and bonding areas is removed. Aluminum metallization is sintered at 570°C. Back side is etched and gold evaporated to insure good mounting.

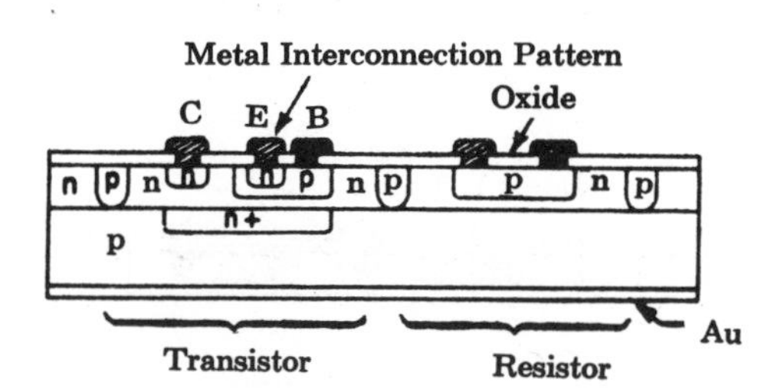

9.6 LINEAR INTEGRATED CIRCUITS

The most common linear integrated circuit is fabricated by monolithic silicon epitaxial construction, and will perform low level requirements from dc to 200 MHz. In addition to offering the system designer a savings in space and weight, an increased reliability is realized. In systems converted from discrete components to integrated devices, a reliability improvement of six to thirty is feasible.

The problems due to parasitic capacitances prevent the full frequency capability of integrated transistor from being fully realized. The restricted range of component values allowed by monolithic construction, in some instances, prevents the integrated circuit from performing as well as its discrete component counterpart.

9.6.1 BASIC CIRCUIT CONSIDERATIONS

Monolithic linear integrated circuits are fabricated by the same process which produces digital building blocks for logic systems. That is, the active and passive components are fabricated in a single structure, and all connections are made simultaneously by forming a metal interconnection pattern on the surface of a silicon bar. The process is illustrated in Figure 9-11.

In Figure 9-11, it can be noted that each component is completely surrounded by a dielectric material, silicon dioxide ($Si0_2$). One of the major problems stemming from monolithic construction is that individual components need to be isolated from each other. Several techniques such as using high resistivity substrate or reverse-biased PN junction have been used to achieve the required isolation. The latter method has been more predominant as it is easier to fabricate and provides more complete isolation. However, this method provides capacitive coupling to the substrate. Dielectric isolation, on the other hand, offers the possibility of making and using PNP and NPN devices on the same chip, and reduces capacitances as the oxide thickness also controls the capacitance per unit area.

The only disadvantage of using dielectric isolation over other methods is a slightly reduced power handling capability, as silicon dioxide has a lower thermal conductivity than silicon. However, this may not be too great a problem since the total thermal resistance can be controlled by the thickness of oxide used.

The greater range of circuit design requirements in systems utilizing linear circuits, essentially prohibits standardization which is common in logic systems that employ digital building blocks. In many instances, linear circuits are ''customized'' to meet a particular circuit application.

An approach to standardization is represented by the differential amplifier configuration in Figure 9-12. The configuration is basically that of a balanced differential amplifier in which the currents to the emitter-coupled differential pair or transistors are supplied from a controlled source. A wide use of linear integrated circuits for analog applications from dc to frequencies into the vhf region such as mixing, limiting, product detection, frequency multiplication, and amplitude modulation in addition to linear amplification, can be obtained from such a versatile configuration.

The differential amplifier may be used as a normal operational amplifier by ignoring one of the outputs. The single output and an inverting and noninverting input makes it possible for a positive or negative feedback to be applied. This feedback concept is well suited to integrated circuits because the complete circuit specification is dependent on tolerance of one or two components rather than components of integrated operational amplifier. Many different circuit functions can be performed with an operational amplifier by using different feedback networks around the amplifier. Linear integrated circuits with the operational amplifier configuration and utilizing such feedback are ideally suited for instrument and analog use or in control systems.

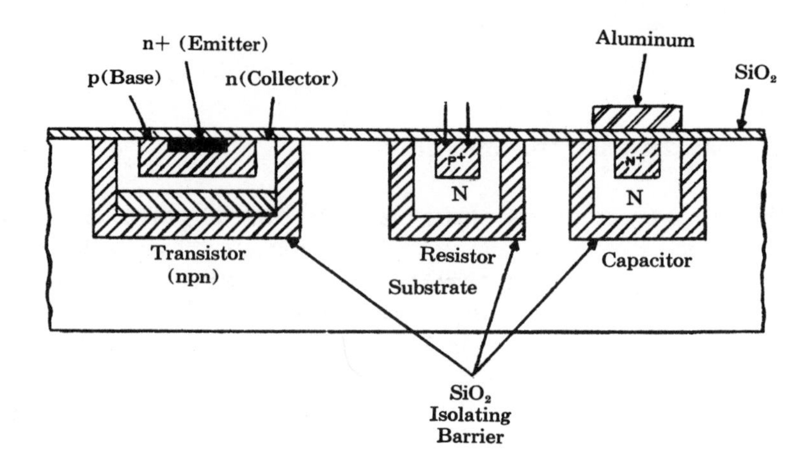

Fig. 9-11 Monolithic Linear Circuit.

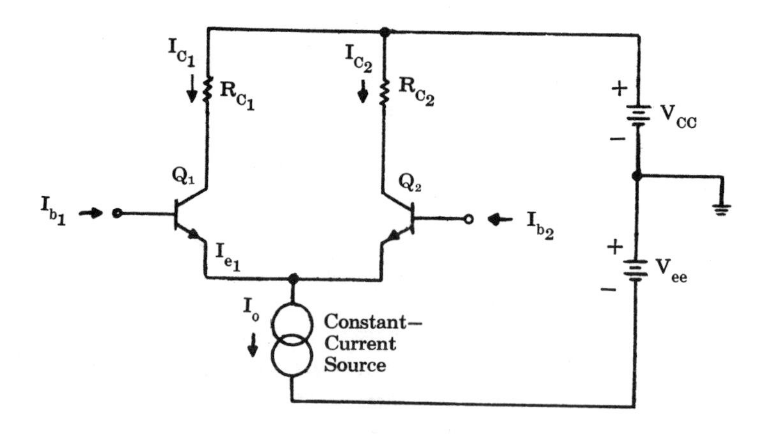

Fig. 9-12 Basic circuit configuration of a differential amplifier for standardization applications.

RF AND IF AMPLIFIERS:—The major advantage of integrated circuits is realized when many similar stages can be cascaded in a single package. In receiver applications such advantages can not be fully utilized when each stage of amplification is followed by a tuned coil. The lack of suitable inductive reactance compatible with integrated circuits, prevents full realization of cascading similar stages, and alternate frequency selective schemes must be considered. One approach has been the use of T-notch filters, composed of lumped RC networks. Another means of providing frequency discrimination is the use of a crystal filter followed by a high gain, wideband (video) amplifier. At frequencies above 30 MHz, distributed techniques may be used.

CHAPTER 10

CHARACTERISTICS OF SEMICONDUCTOR MICROELECTRONIC CIRCUITS

The preceding chapter related the methods used to fabricate semiconductor microelectronic circuits. These circuits are made up of a small number of components, and in this chapter the characteristics of these components are described.

There are basically four or five structures used in the fabrication of semiconductor microelectronic circuits. Of this group, emphasis has been placed on the 4-layer structure whereby an isolation region is diffused through an epitaxial layer, grown on the original substrate. This 4-layer structure is perhaps the most convenient to fabricate from the viewpoint of the manufacturer. However, its limitations lie in the fact that there always is an appreciable collector-substrate capacitance. This capacitance can be controlled, however, by the very small geometry which can be successfully generated by the present photolithographic technology. Nevertheless, as will be shown, this capacitance remains the largest parasitic capacitance in the circuit and, therefore, places a limitation on the highest useful frequency obtainable by the 4-layer structure.

10.1 TRANSISTORS

The transistor is the basic component in the semiconductor microelectronic circuit. Two of the three diffusions required to produce a circuit are specifically tailored to produce suitable transistors. Diodes and resistors are ''fallout'' from the transistor fabrication. The transistors associated with semiconductor microelectronic circuits are, by choice, either high frequency switching types which are gold diffused in order to increase their maximum frequency response, or of the non-gold diffused structure.

Typical impurity profiles are illustrated in Fig. 10-1. In the case of a non-gold diffused transistor, there is sufficient difference in resistivity between the base and collector of the transistor so that the junction may be considered an abrupt type of junction. In this case the reverse breakdown voltage obeys the following expression:

$$BV_{CBO} = \frac{k\epsilon_o}{2qN} E^2 \text{ volts,} \tag{10-1}$$

where BV_{CBO} = avalanche breakdown voltage in volts
k = dielectric constant of silicon (= 12)
ϵ_o = permittivity of free space (8.85×10^{-14} farad/cm)
N = the impurity concentration of the lightly doped or higher resistivity layer in atoms/cm³, and
E = electric field across the junction in volts per cm.

A less heavily doped collector-base transition region closely approximates a graded junction, and the collector-base breakdown, in this case, may be expressed by

$$BV_{CBO} = 1.71 \times 10^9 \, a^{-0.364} \text{ volts,} \tag{10-2}$$

where a = gradient constant in atoms/cm⁴, calculated roughly by dividing the difference of the impurity concentrations at the emitter-base junction and of the collector background, by the base width (W) of the transistor. Typical curves for step-junction breakdown voltage and graded junction breakdown voltage are shown in Fig. 10-2.

Since the collector region of the gold diffused transistor is more heavily doped, the collector-base breakdown voltage is usually lower for this type of transistor than that of the non-gold diffused transistor. This also applies to the case of emitter-collector breakdown voltage (BV_{CEO}), where BV_{CEO} is approximated by

$$BV_{CEO} \approx \frac{BV_{CBO}}{n\sqrt{\beta}} \text{ volts,} \tag{10-3}$$

where n = 4 for n-type silicon and 2 for p-type silicon,

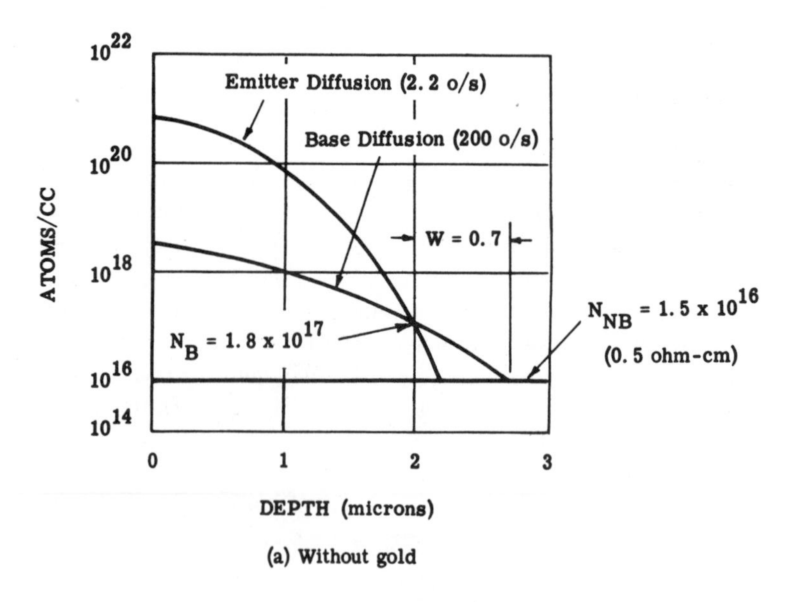

(a) Without gold

(BC = Background Concentration)

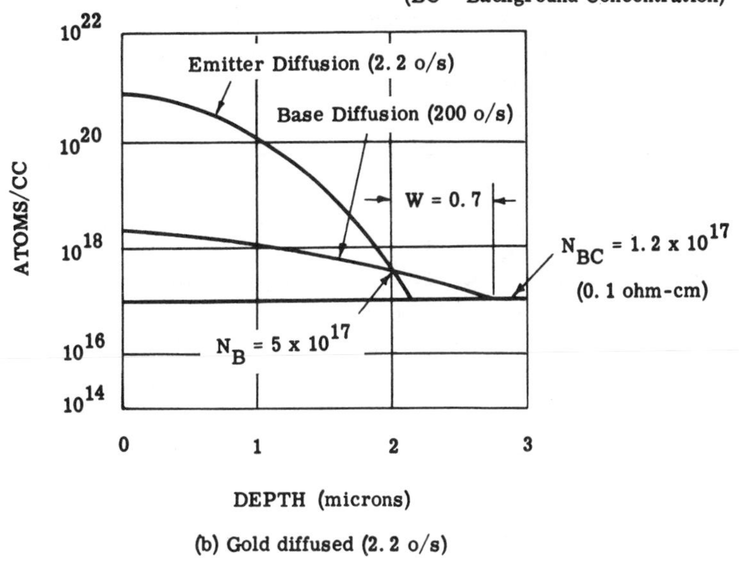

(b) Gold diffused (2.2 o/s)

Fig. 10-1 Typical impurity profiles for semiconductor microelectronic transistors

and β = the grounded-emitter current gain and is much lower for gold diffused transistors than for non-gold diffused.

As described earlier, leakage or reverse current of a junction is related to the generation of charge from recombination centers within the junction depletion layer, and also to the width of the depletion layer at a particular applied voltage. Since the gold diffusion introduces more recombination centers, it would ordinarily be expected that the leakage current would be higher for gold diffused transistors. However, the depletion layer width is smaller for a gold diffused junction and, therefore, the amount of leakage current for this type of junction does not increase appreciably. In both types, leakage currents are usually less than 10 nanoamps, at a reverse voltage of 10 volts, and vary slowly with voltage.

In considering transistor capacitance, the latter may be divided into that of the emitter-base which is forward biased, the collector-base which is reverse biased, and the collector-substrate which is also reverse biased. For forward bias conditions, the expression for capacitance is as follows:

$$C_{Te} \approx A_e \left(\frac{qK\epsilon_0 N'_B}{2V} \right)^{1/2} \text{farads} \qquad (10\text{-}4)$$

where C_{Te} = the emitter capacitance of a transistor under forward bias in farads

V = junction voltage

A_e = total area of the emitter in cm^2 which is composed of the area of the base-emitter junction including the sides of the emitter

q = the charge of an electron which is 1.6×10^{-19} coulombs

K $=$ the dielectric constant of silicon ($=$ 12).
ϵ_o $=$ permittivity of free space (8.85×10^{-14} farad/cm) and
$N'_B =$ the impurity concentration of the base side of the emitter junction in atoms/cm^3.

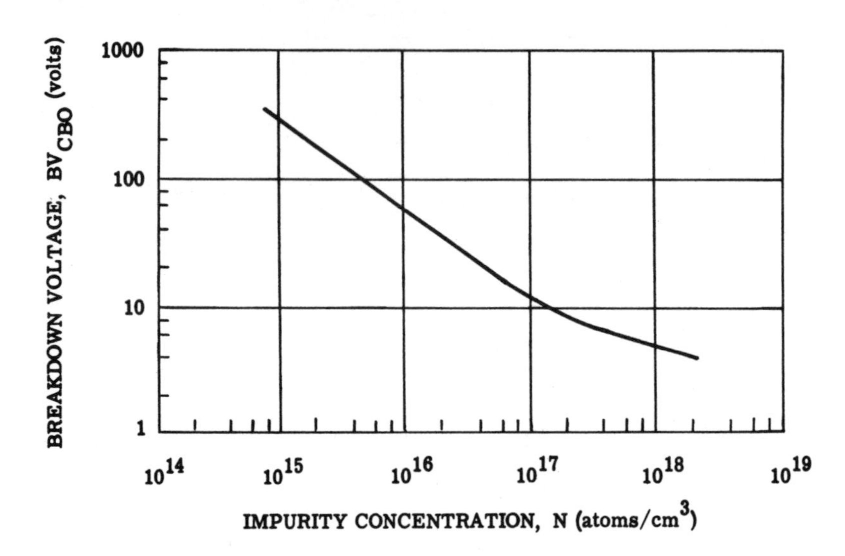

Fig. 10-2a Avalanche breakdown in step P-N junctions in silicon

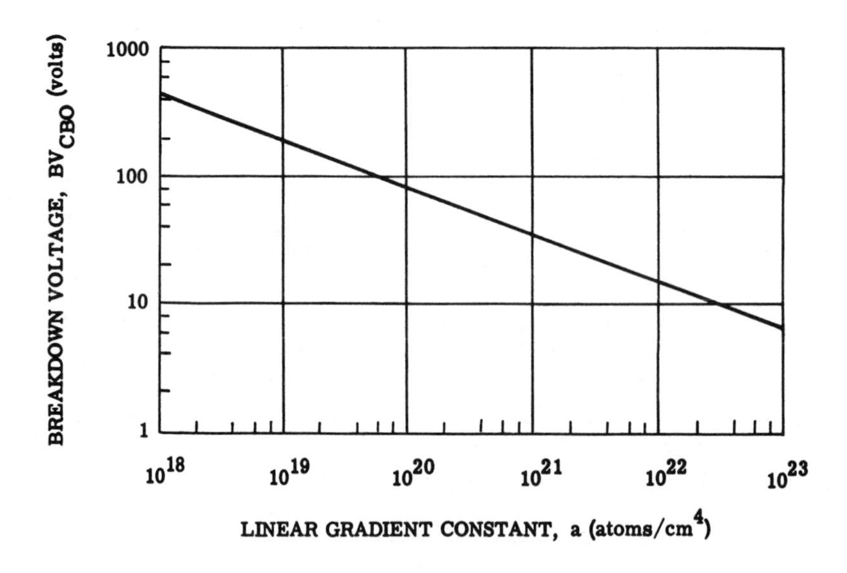

Fig. 10-2b Avalanche breakdown in graded P-N junctions in silicon

Thus, for the heavier doped gold diffused transistor, N'_B would increase and the emitter capacitance would be greater. Since the collector-base junction is a graded junction, its capacitance cannot be calculated from such a simple expression as used for the emitter. A set of curves especially developed for this purpose, however, can be used to compute junction capacitance in this case. Typical curves are shown in Fig. 10-3.

The capacitance for the collector, as was in the case of the emitter, would be higher for the gold diffused transistor, since the doping level is higher. For the substrate-to-collector capacitance the predominant factor is the area of the sides of the collector and the resistivity of the collector. For an epitaxial layer transistor such as the 4-layer type, the collector side area is at a minimum and collector capacitance would be lowest for this case.

The grounded-emitter current gain (β), of a transistor, is related to

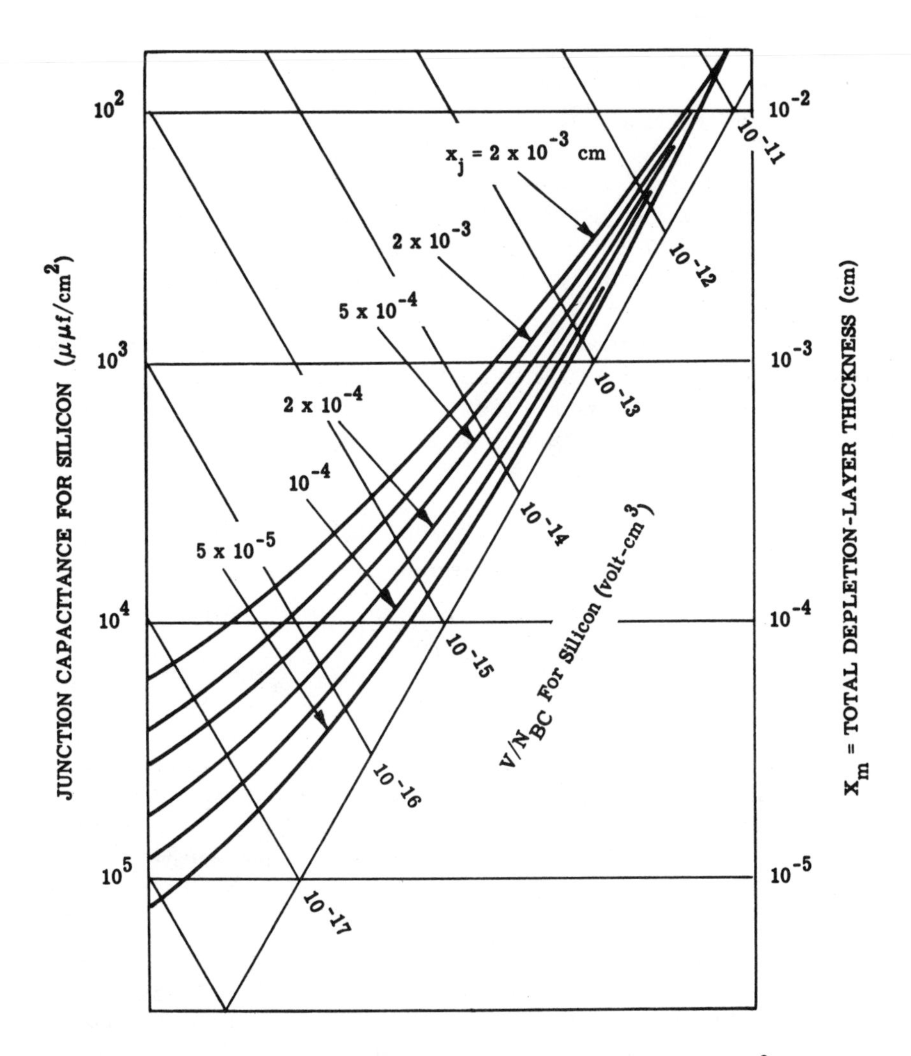

Fig. 10-3 Graded - junction curves for N_{BC}/N_o range of 3×10^{-3} to 3×10^{-2}. x_j = junction depth, N_o = surface concentration of dopant, N_{BC} = background concentration

the emitter efficiency, the base transport factor, the surface recombination effects, and the density of impurity centers at the emitter junction. Emitter efficiency is related to the ratio of the conductivities of the emitter and the base of the transistor. Accordingly, for a high emitter efficiency, it is desirable to have a heavily doped emitter and a lightly doped base under the emitter junction, in order to have a high ratio of conductivities. The base transport factor, B_F, is given by

$$B_F = \text{sech}\left(\frac{W}{L}\right) \qquad (10\text{-}5)$$

where W is the base width and L is the minority carrier diffusion length in the base. When $L >> W$, this factor approaches one. If the surface recombination velocities become significant, beta can drop off considerably, since charge injected by the emitter may recombine at the surface near the emitter. The last factor becomes important for gold diffused structures where the density of impurity centers, in the emitter junction, is of larger value, and beta is thereby affected by recombination at lower emitter currents. For both types of transistors, beta is roughly the same.

A factor of major importance in transistor design is the frequency response of the device. This response is determined by a series of factors including: the emitter delay-time constant; the base transit time; collector depletion layer transit time; and the collector time constant. Each of these factors can be related to the device fabrication and the resulting physical characteristics of the transistor.

The emitter delay time constant can be defined as the product of the emitter resistance and the emitter transition capacitance, or

$$\tau_e = r_e C_{Te}. \qquad (10\text{-}6)$$

The junction capacitance, C_{Te}, is given by Eq. (10-4). The resistance of the emitter, r_e can be expressed as

$$r_e = \frac{kT}{qI_e}. \qquad (10\text{-}7)$$

Since the emitter resistance is fixed by basic physical constants and the emitter current, I_e, any reduction in τ_e must be derived from reduction of transistor capacitance. This can be accomplished most easily by reducing the junction area, A_e. The only other parameter which can be altered is N'_B, the impurity concentration of the base side of the emitter junction. However, this value cannot be altered very significantly since the emitter and base impurity concentrations are fixed by the desired current gain (β) of the device. Thus, the only reduction in τ_e which is practical, is through reduction in emitter area. This reduction in emitter area has the additional effect of improving the emitter efficiency for low current operation, since the emitter efficiency is dependent on the emitter current density.

The base transit time is the second delay time which can be adjusted through device fabrication procedures. The base transit time, τ_B, is given by (for uniform base)

$$\tau_B = \frac{W^2}{2.43D} \qquad (10\text{-}8)$$

where W is the base width and D is the diffusion constant for the base. As indicated by this expression, the transit time is directly dependent on the width of the base region. By making a transistor with a graded inpurity distribution in the base, a built-in electric field is established in the base region which reduces the base transit time. The base transit time is then given by

$$\tau_B = \frac{W^2}{2.43D \, ln \, N'_B/N_{BC}} \tag{10-9}$$

where N_{BC} is the collector background impurity concentration and N'_B is the base concentration at the emitter junction. In the fabrication of a typical high frequency transistor with a 1μ base width, this additional factor may typically reduce the transit time to 1/2 to 1/10th of that for an equivalent device with a uniform impurity distribution in the base region.

The collector junction also introduces signal delay associated with the capacitance of the collector and the transit time of carriers across the junction depletion region. To reduce the latter effect, the resistivity of the higher resistance side of the collector junction must be reduced. The time delay associated with the collector capacitance is

$$\tau_e = (r_e + r_{es} + r_{cs}) \; C_c \tag{10-10}$$

where r_{es} is the series resistance of the emitter region, r_e is the resistance of the emitter junction, and r_{cs} is the collector series resistance. Since the emitter is heavily doped, the emitter series resistance is negligible. The emitter resistance, r_e, is also small unless the device is operated at very low currents. The only significant term, then, is the product of the collector series resistance and the collector capacitance. This term may be minimized by keeping the collector resistivity low, and reducing the area of the collector to lower its capacitance.

10.2 DIODES AND CAPACITORS

Aside from their common function, diodes also find wide use in microelectronic circuits both as non-linear elements and to fill the requirement for capacitors. Parameters important to diode operation, and which can be influenced by fabrication procedures include: capacitance, leakage current, breakdown voltage, recovery time, and forward voltage drop. Diodes may be obtained by directly fabricating a p-n junction for a circuit, or by using one junction of a transistor as a diode. The various possible diode connections when transistors are used are shown in Fig. 10-4.

In either method of producing a diode, the junction for the useful configurations will be an abrupt one, with breakdown voltage given by Eq. (10-1). The breakdown voltage is inversely proportional to the impurity concentration, and increases with increasing resistivity as shown in Fig. 10-2. Hence, to increase the breakdown voltage, the resistivity should be high. However, since the diodes are made by either using one junction of a transistor, or, if made separately, by fabrication during the same diffusion process as the transistors, the choice of resistivity is usually a compromise value. The highest breakdown voltage occurs at the collector-base junction.

The junction capacitance is an important parameter; it influences diode performance, and is also a source of capacitance in microelectronic circuits. The majority of such junctions can be regarded as step or abrupt junctions. The capacitance and depletion layer thickness of such a junction can be found from Poisson's equation

$$\nabla^2 V = \rho/\epsilon \tag{10-11}$$

where V is the voltage, ϵ the dielectric constant and ρ is the charge density. If the one dimensional form of this equation is integrated, solutions for both the n and p regions are obtained:

$$V_p = \frac{qN_A x^2}{2\epsilon} + A_1 x + B_1 \tag{10-12}$$

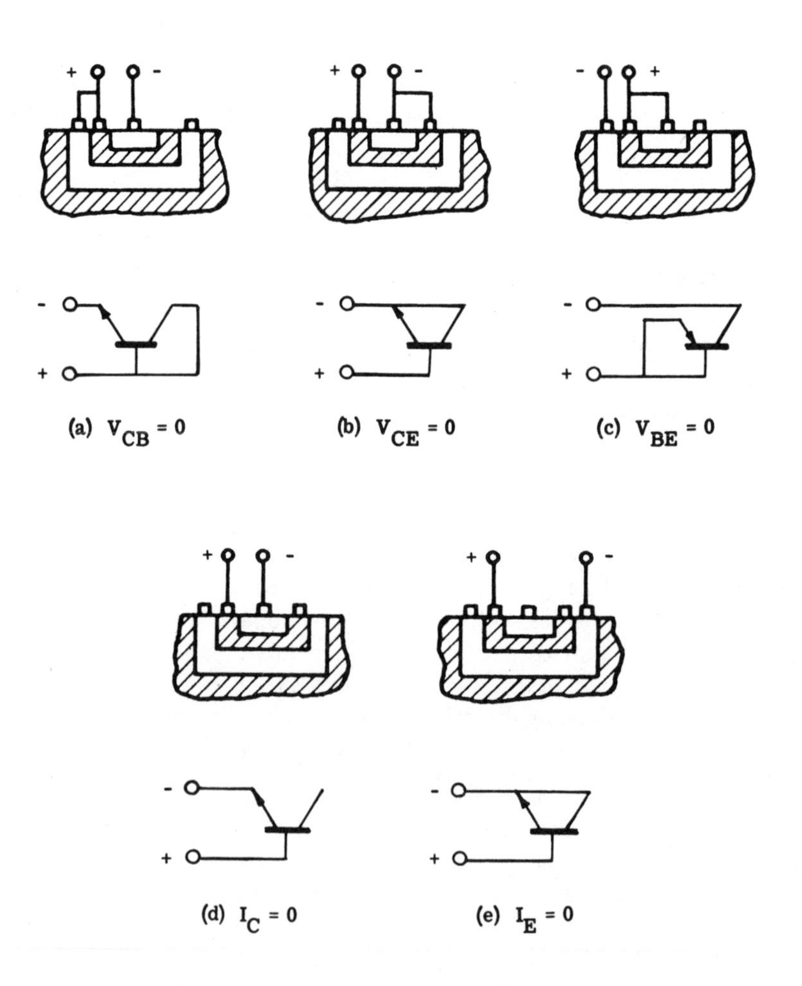

Fig. 10-4 Five possible diode connections using four-layer transistor structure

$$V_n = \frac{qN_D x^2}{2\epsilon} + A_2 x + B_2 \qquad (10\text{-}13)$$

where N_D is the density of donor impurities and N_A is the acceptor concentration. The constants A_1 and A_2 may be evaluated from the conditions that dV/dx is continuous at $x = 0$, and zero at $x = -x_1$ and x_2 (Fig. 10-5). Therefore

$$V_p = \frac{qN_A}{2\epsilon} x^2 + \frac{qN_A x_1}{\epsilon} x + B \qquad (10\text{-}14)$$

$$V_n = \frac{-qN_D}{2\epsilon} x^2 + \frac{qN_D x_2}{\epsilon} x + B \qquad (10\text{-}15)$$

The total potential difference across the junction is then

$$V = V_n (x_2) - V_p (-x_1) = \frac{q}{2\epsilon} (N_D x_2^2 + N_A x_1^2) \qquad (10\text{-}16)$$

From the boundary conditions, $N_A x_1 = N_D x_1$ and hence

$$V = \frac{qN_D}{2\epsilon} x_2 (x_1 + x_2) \qquad (10\text{-}17)$$

The total depletion layer width is given by

$$x_t = x_1 + x_2 \qquad (10\text{-}18)$$

Therefore, if one side of the junction is much more heavily doped, the

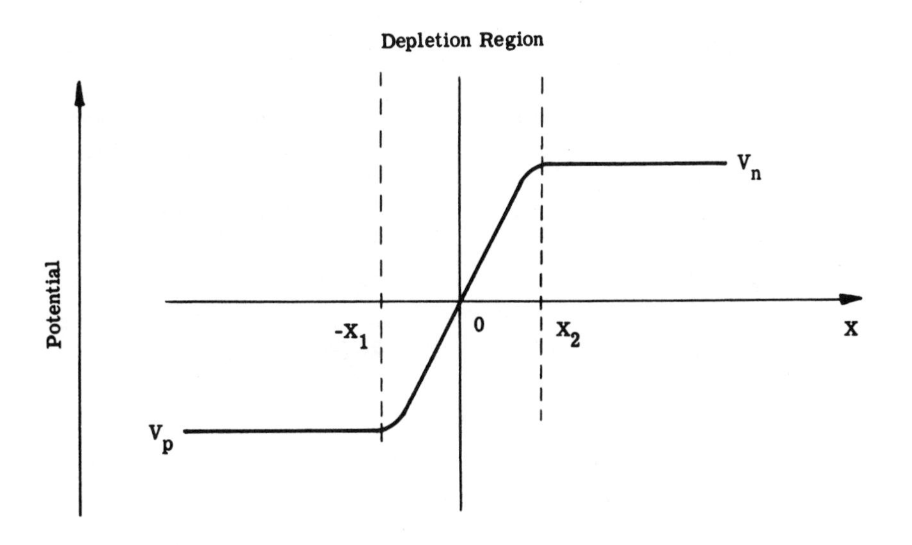

Fig. 10-5 Variation of potential across depletion layer of a P-N junction

depletion layer widths can be expressed as:

$$x_t = \left(\frac{2\epsilon V}{qN_D}\right)^{1/2} \qquad (\rho_n >> \rho_p) \qquad (10\text{-}19)$$

$$x_t = \left(\frac{2\epsilon V}{qN_A}\right)^{1/2} \qquad (\rho_p >> \rho_n) \qquad (10\text{-}20)$$

Considering the junction as a parallel plate capacitor, the capacitance is

$$C = A\left(\frac{q\,\epsilon\,N_A N_D}{2(N_A + N_D)V}\right)^{1/2} \qquad (10\text{-}21)$$

where A is the junction area. This formula is valid for an abrupt change in concentration. If the concentration change is gradual, then the exponent to be used in Eq. (10-21) should be decreased—a commonly used value is 1/3.

The potential, V, which appears in Eq. (10-21) is not only the applied voltage, but also the "built-in" difference in potential between the p and n regions. For silicon at 300° K this potential is approximately 0.65 volt. Therefore this value must be added to the applied external voltage for proper interpretation of Eq. (10-21). The results obtained by the space-charge approximation are not exact and some care must be taken in applying them. A more exact calculation indicates that the actual transition region is wider than that predicted by the preceding calculations, and increases at large forward bias with an exponential voltage dependent upon exp (qV/8KT). These more exact calculations also show that in regions of intermediate forward bias, the transition capacitance, due to free electrons and holes, becomes dominant over the fixed space-charge contribution.

If the junction capacitance of a diode is to be minimized, the impurity concentrations or the diode area must be reduced. In the case where the junction is to be used as a capacitor, these factors should be maximized. A problem associated with such a capacitor, is the dependence of capacitance on the junction voltage.

The fabrication procedures also influence the usefulness of a diode in high-speed switching applications. The recovery of a diode from a conducting state to a "cut-off" condition requires a certain finite amount of time. This time-lag in turning off a device is due to minority carrier storage; carriers will take a certain amount of time to recombine after a forward applied voltage is removed. The longer the lifetime in a region, the longer is the duration of the period before the diode returns to equilibrium. The recovery time, t_r is dependent on whether a step or graded junction, is involved, and is given by

$$t_r = 0.9\tau \text{ (step junction)} \qquad (10\text{-}22)$$

$$t_r = 0.5\tau \text{ (graded junction)}, \qquad (10\text{-}23)$$

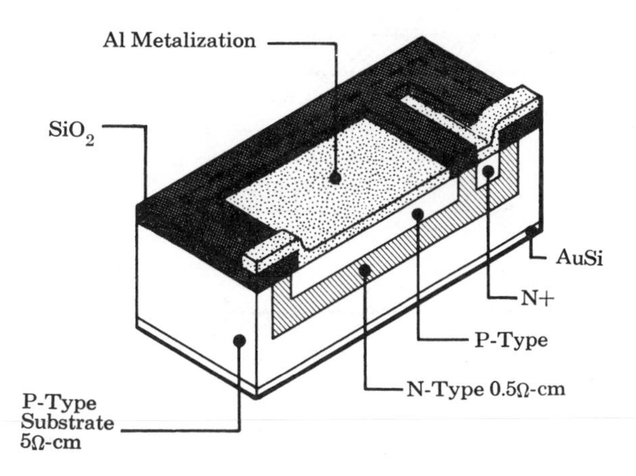

Fig. 10-6 Junction Capacitor

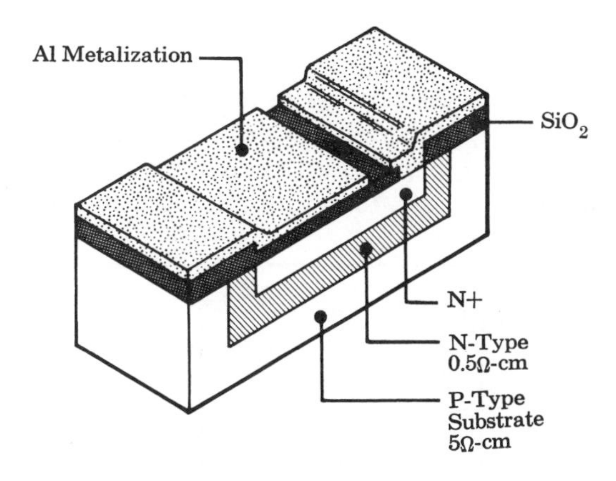

Fig. 10-7 Silicon Dioxide Capacitor.

where τ is the minority-carrier lifetime. In ordinary silicon the minority carrier lifetime is on the order of 10 μsec. In order to reduce the recovery time, large concentrations of gold impurity centers can be added to increase the speed of recombination.

In all silicon junction diodes, leakage currents prevail, and they determine the reverse bias characteristics. These currents are caused by both surface leakage and charge generation effects. Charge generation in the junction depletion layer is caused by generation-recombination centers in that region, and is dependent on the number of such centers which are present. Thus, if large numbers of such centers are introduced to shorten the recovery time, then the charge generation will be increased. Surface leakage can be reduced by proper treatment, but is still sizeable in silicon diodes.

Figure 10-8 shows a typical device structure and the capacities associated with the junctions, while Fig. 10-9 is the variation of capacity with voltage for three junctions found in all microelectronic circuits.

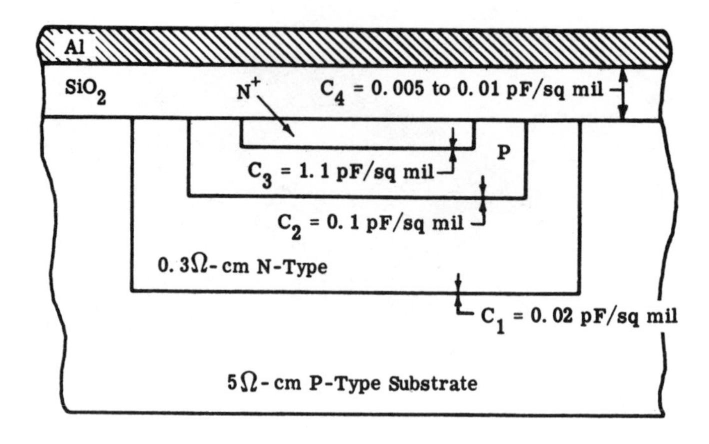

Fig. 10-8 Capacitance associated with each layer

10.3 RESISTORS

Integrated circuit resistors are usually fabricated during one of the transistor process steps; they are formed using one of the diffusions associated with transistor manufacture. In the preferred method, the transistor base diffusion is used for resistor formation — the sheet resistance

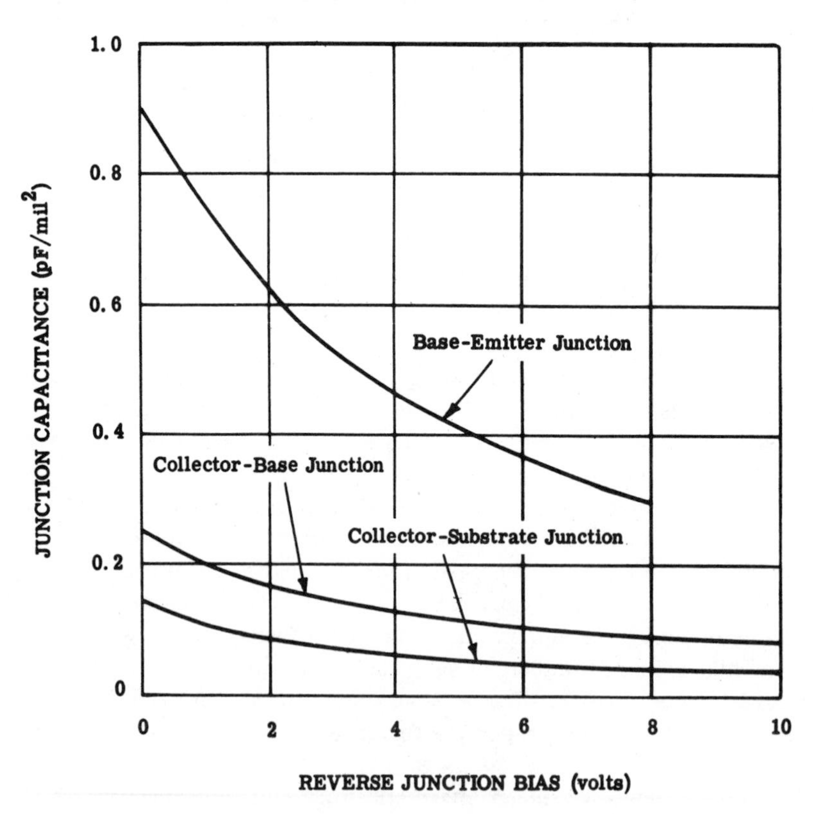

Fig. 10-9 Capacitance versus voltage for junctions found in microelectronic circuits

(ρ/t) for the base is usually about 200 ohms/square. The emitter diffusion can be used to form low value resistors. Using a p-type impurity in an n-type substrate, yields a thin diffused bar whose properties depend on the impurity concentration, diffusion depth, and the ratio of length to width. For a uniform layer, the resistance R is given by

$$R = \frac{\rho L}{tW} \tag{10-24}$$

where ρ is the resistivity, L the length, t the thickness, and W the width.

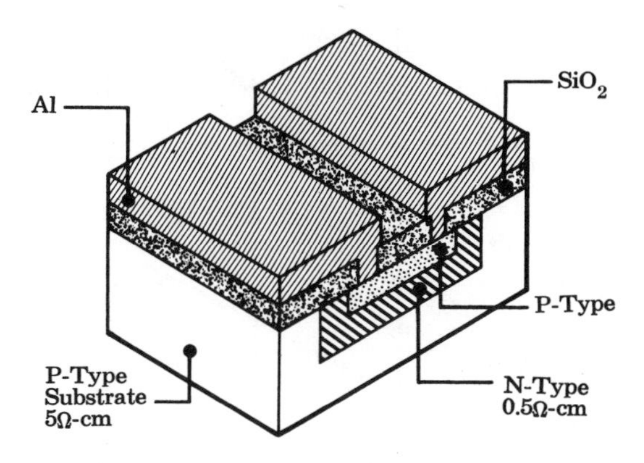

Fig. 10-10 Cross-section of a Diffused Resistor.

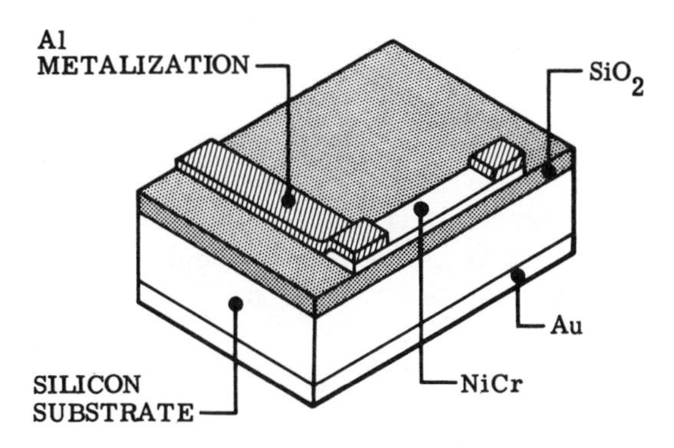

Fig. 10-11 Cross-section View of a Thin Film Resistor.

The doping concentration in the resistor is not uniform. It approximates the gaussian function if the resistor is diffused at the same time as the base, and approximates the erfc if it is generated at the same time as the emitter. An average resistivity ρ must therefore be used:

$$\bar{\rho} = \frac{1}{q\mu\bar{n}} \tag{10-25}$$

where $\bar{n}$ is the average dopant concentration, and is given by

$$\bar{n} = \frac{1}{x_j} \int_0^{x_j} n(x)\, dx\,. \tag{10-26}$$

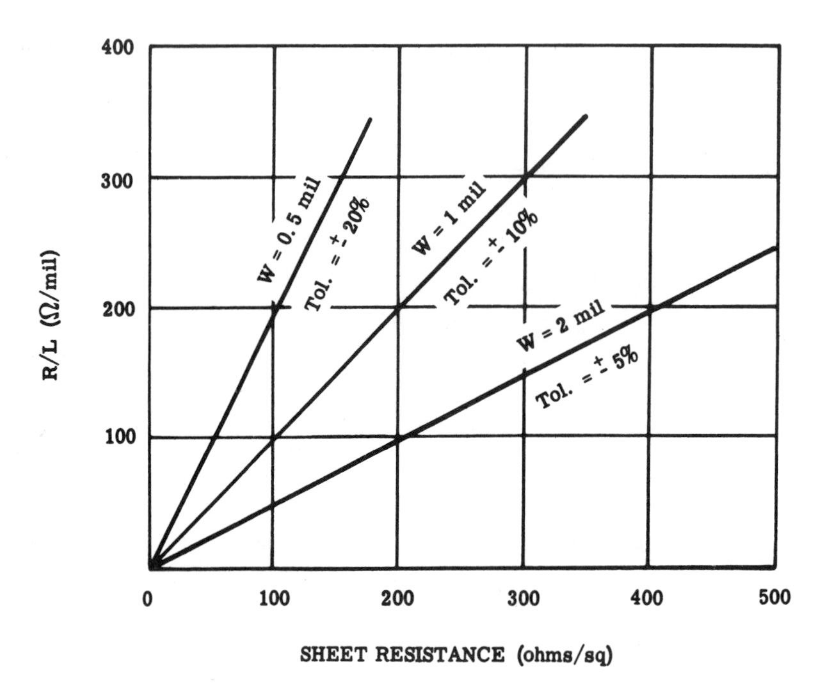

Fig. 10-12 **Resistance per mill of diffusion as a function of sheet resistance and line width, W**

Here x_j is the junction depth, dx is the thickness of a longitudinal slice of the resistor a distance x below the surface, and n(x) is the doping profile. The sheet resistance can be expressed as

$$R_s = \bar{\rho}/x_j\,. \tag{10-27}$$

If the resistor has been made during a base diffusion, the sheet resistance is about 200 ohms/square, and the junction depth is about 3μ. If it has been made during an emitter diffusion, the corresponding values are 2-3 ohm/square and $2\text{-}2.3\mu$, respectively.

Figure 10-12 illustrates typical resistance values as a function of sheet resistance and width (W), for the range of sheet resistance common in transistor base diffusions.

Since the resistor is a p-type layer on an n-type substrate and the junction is reverse biased, a certain amount of distributed capacitance is associated with this structure. This capacitance, which is of the order of 0.1 pf/mil^2, has the effect of limiting the frequency range within which such components are applicable. The frequency response of a typical resistor is shown in Fig. 10-13.

A typical semiconductor resistor made during a base diffusion is shown in Fig. 10-14. Not only is there a distributed capacitance at the resistor-collector junction, but the substrate-collector-resistor structure forms a

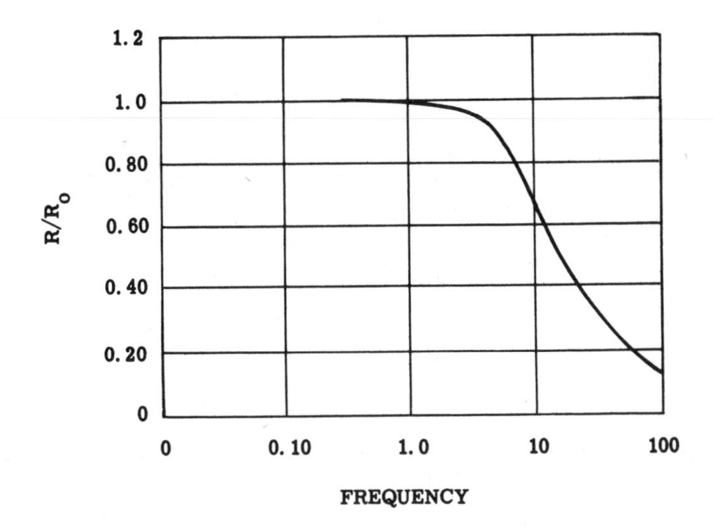

Fig. 10-13 Frequency response of a fully integrated diffused resistor

parasitic PNP transistor which, in some cases, will act as a low gain ($\beta = 0.5\text{-}5$) transistor. The isolation, which includes the p-type substrate, is biased to the most negative point in the circuit. If, for any reason, any part of the resistor becomes forward biased with respect to the N region, the correct bias situation prevails to produce transistor action. In this case the subtrate acts as the collector, and the forward biased part of the resistor acts as the emitter. To avoid this possibility, care must be taken to assure

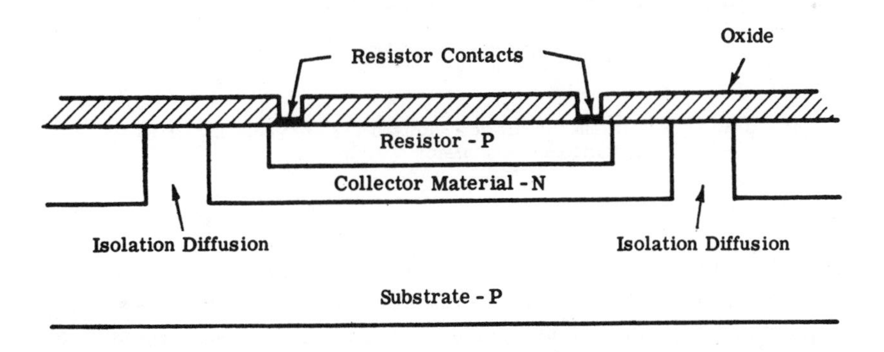

Fig. 10-14 Structure of a diffused resistor generated during base diffusion

that the resistor is, at all times, back-biased with respect to its surrounding N region. A simplified equivalent circuit for this parasitic transistor action is shown in Fig. 10-15. The distributed capacity C_1 is about 0.02 pf/mil^2 at 0 volts bias, while C_2 is 0.1 pf/mil^2 at 0 volts bias.

Another parameter which is determined by the fabrication is the temperature coefficient of resistance. The resistance was shown to be a function of the number of carriers, n, and their mobility, μ, and both of these factors are temperature dependent. Typically, the temperature coefficient of resistance for a 200 ohms/square base diffused resistor is about 2000 PPM/° C. These temperature coefficients are controlled by the doping of the diffused region.

If the base diffusion is used to make a resistor, the range of resistance values which is obtainable, with reasonable tolerance and area, is from approximately 25 ohms to 30,000 ohms. Both ends of this range can be extended if greater area is used. If the emitter diffusion is used, wherein the sheet resistance is about 2.5 ohms/square, the low end can be extended down to a few ohms.

10.4 PNPN TRANSISTORS

Silicon pnpn devices are of interest in semiconductor circuits because they exhibit current amplification factors greater than unity, and are also

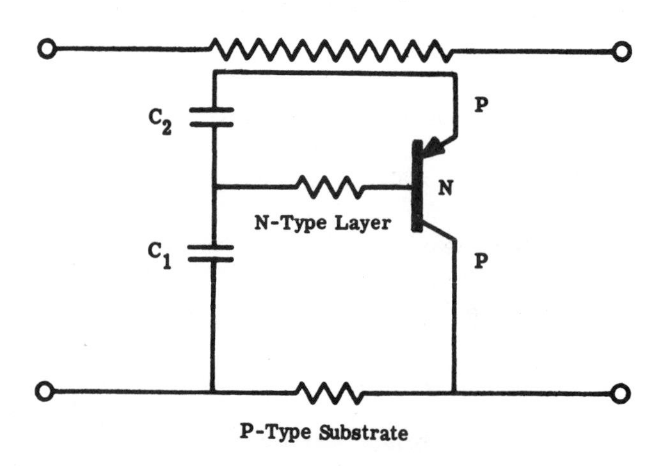

Fig. 10-15 Equivalent circuit of a diffused resistor showing parasitic capacitance and transistor

applicable to switching circuits. A schematic of the pnpn transistor is shown in Fig. 10-16. The devices can be operated as two-terminal elements (pnpn diodes) by allowing an open circuit to exist at the base contact. The operation then depends on the current gains (α_1 of the pnp section and α_2 of the npn section) and the current multiplication factors M_p and M_n at junction 2.

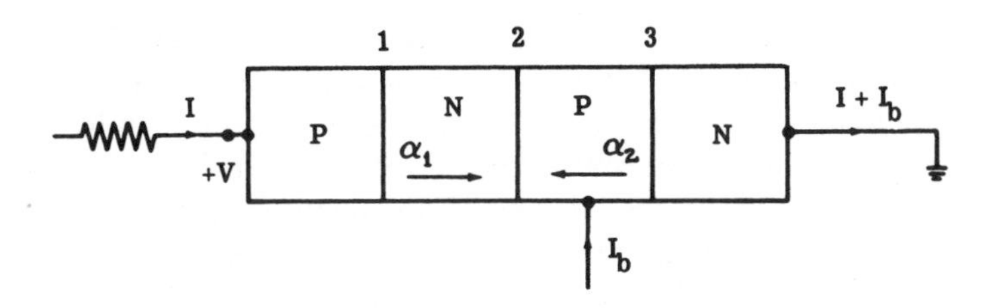

Fig. 10-16 Schematic of a PNPN transistor

The total current gain is given by $\alpha_T = \alpha_1 M_p + \alpha_2 M_n$ and the carrier multiplication factors are:

$$M_n = [1-(V/V_B)^{\beta_1}]^{-1} \tag{10-28}$$

$$M_p = [1-(V/V_B)^{\beta_2}]^{-1} \tag{10-29}$$

where V_B is the junction breakdown voltage and V is the applied voltage. The exponents β_1 and β_2 are approximately equal to 2 and 6, respectively, for graded silicon p-n junctions.

As a diode, the device operates with junctions 1 and 3 as emitters and junction 2 as a collector. At low currents, $\alpha_1 M_p + \alpha_2 M_n$ is less than one, and

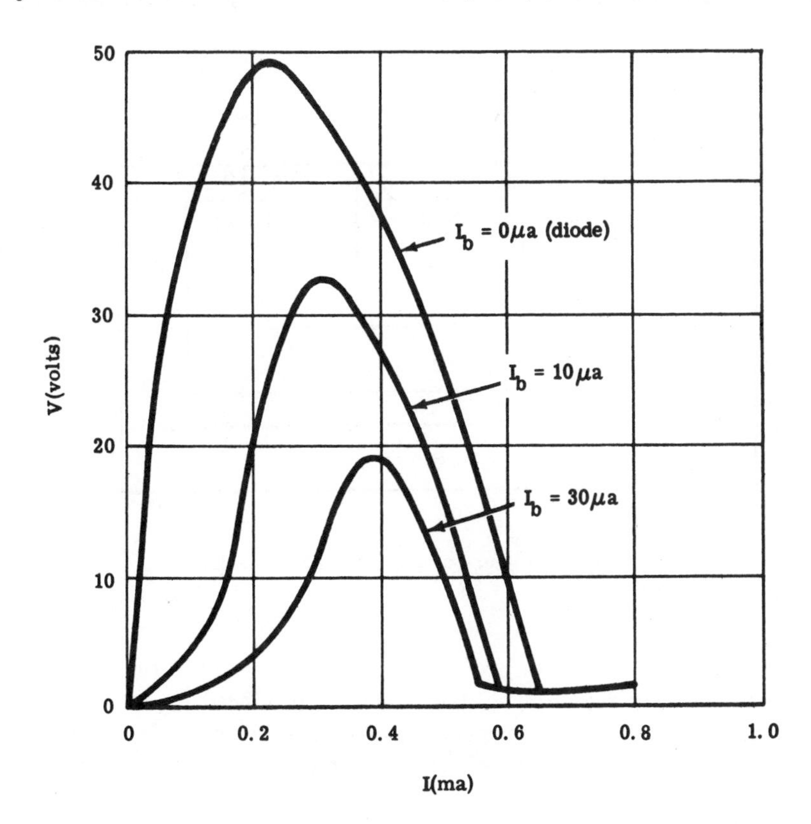

Fig. 10-17 Current - voltage characteristics of a PNPN device, for various base currents

junction 2 acts as a normal reverse biased junction. As the applied voltage is increased, carrier multiplication increases until $\alpha_1 M_p + \alpha_2 M_n$ reaches one and the junction 2 breaks down. This breakdown is followed by a negative resistance region which continues until a low impedance condition is attained. At these low voltages M_n and M_p are approximately equal to one so that $\alpha_1 M_p + \alpha_2 M_n$ is greater than one. In this region of operation, junction 2 is conducting or "turned on", implying that it operates in the forward bias condition.

A pnpn diode requires relatively high carrier multiplication in order to increase the total current gain to the extent that the device is turned on. An improved method for increasing the current gain, is to apply a base current via an additional base contact through emitter junction 3. The effect of this base current is to increase α_2 and hence result in more rapid increase in α_T to unity. The characteristics of both pnpn diodes and triodes with various base currents are shown in Fig. 10-17.

In order to minimize the required "turn-on" drive, it is necessary for the current gain of the transistor, with an external base drive, to be close to unity. High "turn-off" gain is attained if the total current gain just exceeds unity, and the transistor, with a drive-base, has a high current gain. Therefore, this design requires the section with an external base connection to have a high α, and the other section a low α. This can be accomplished by reducing the emitter injection efficiency or the base transport factor. Low injection efficiency can be obtained by making an emitter of high sheet resistance, in contrast with the usual design wherein

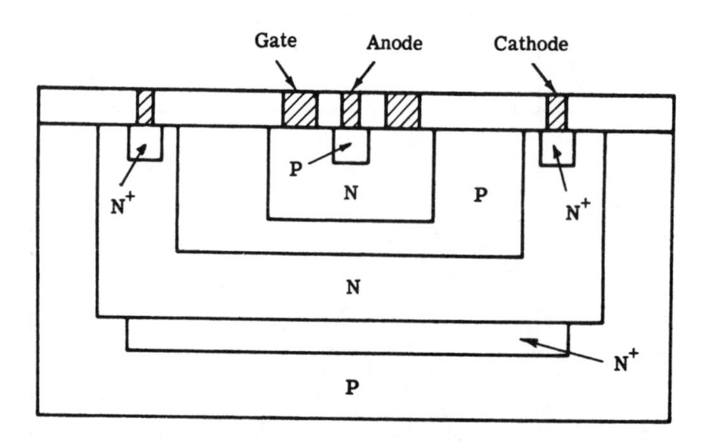

Fig. 10-18 Structure of a diffused PNPN device

the emitter resistance is made orders of magnitude less than the sheet resistance of the base region. The device can be produced by making the p-type emitter of a high sheet resistance layer. The rest of the structure is made, in this case, by conventional diffusion processes with its usual ratios between the doping densities.

An npn transistor and the P-isolation can form an npnp switch. This can be avoided by ensuring that the collector-isolation junction is reverse biased or by proper selection of the geometry and resistivity of the various layers. If an extra diffusion is used, pnpn switches can be made in a microelectronic circuit. The structure of such a device is shown in Fig. 10-18.

10.5 FIELD EFFECT TRANSISTORS

The junction field effect triode was the earliest successful field effect device. Earlier attempts to produce field effect structures with dielectric layers to insulate the semiconductor from the field plate had failed, due to the existence of large numbers of surface states at the semiconductor-dielectric interface. A schematic diagram of the diffused junction field effect triode is shown in Fig. 10-19. The device consists of a bar of n-type material with ohmic contacts at either end. These contacts are generally designated "source" and "drain," respectively. On two opposing faces, p-type regions are generated by the normal diffusion techniques used in forming bi-polar transistors. The current flowing in the n-type material

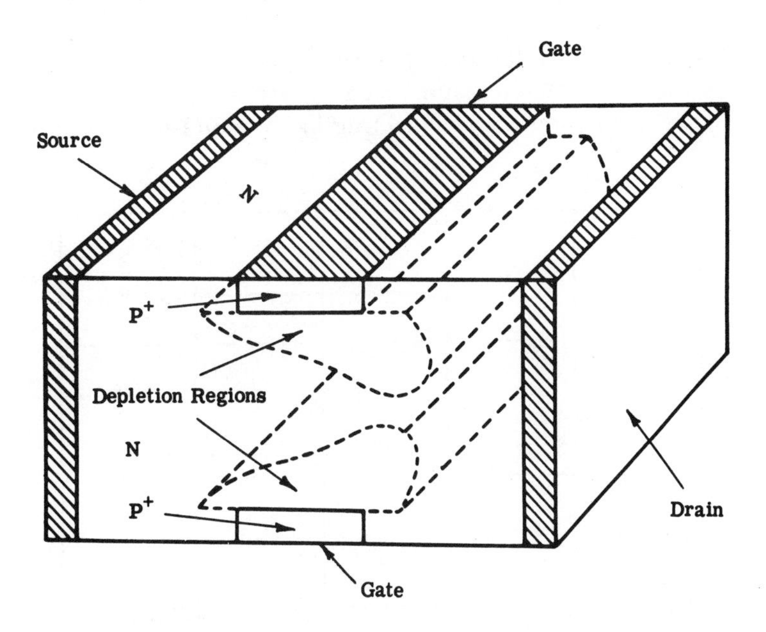

Fig. 10-19 Structure of a diffused junction field effect triode, showing penetration of depletion regions into current flow channels

from source to drain is that which would be delivered to an external load. The p-type regions serve as control electrodes to which the modulating signal is applied. The p-type layers are very heavily doped, and this condition is designated by the notation p^+.

A p-n junction exists between each of the p-type control "gates" and the n-type channel. Hence, a space charge region will exist at each of these junctions. As the penetration of the space charge region into the surrounding materials is inversely proportional to the impurity concentration of the materials, the space charge will principally extend into the n-type channel, rather than the heavily doped p-type gates. This situation is illustrated in Fig. 10-19. The penetration of the space charge region into the channel will increase as reverse bias is applied to the p-n junctions. The potential applied from source to drain will also cause a reverse bias condition. The latter bias will be non-uniform due to the drop along the channel, with a larger reverse bias existing near the drain. The space charge regions will extend further into the channel near the drain, and eventually constrain the current flow in the channel as the bias increases. When a high enough reverse bias exists, due to both gate bias and source-drain voltage, the space charge will "pinch-off" the channel, causing the current to saturate. Source-drain characteristics for this device are illustrated in Fig. 10-20, and are seen to closely resemble a pentode vacuum tube.

The transconductance represented by these curves is

$$g_m = \left.\frac{\partial I_D}{\partial V_G}\right|_{V_D = \text{CONST}} = 2\,\sigma_o \frac{a}{L} V_o^{\,1/2}\left[(V_D - V_G)^{1/2} - (V_S - V_G)^{1/2}\right] \qquad (10\text{-}30)$$

where V_D is the drain voltage, V_G the gate voltage, V_S source voltage, V_o is the "pinch-off" voltage, σ channel conductivity, a is distance from either gate to mid-channel, and L is channel length. The gain bandwidth product

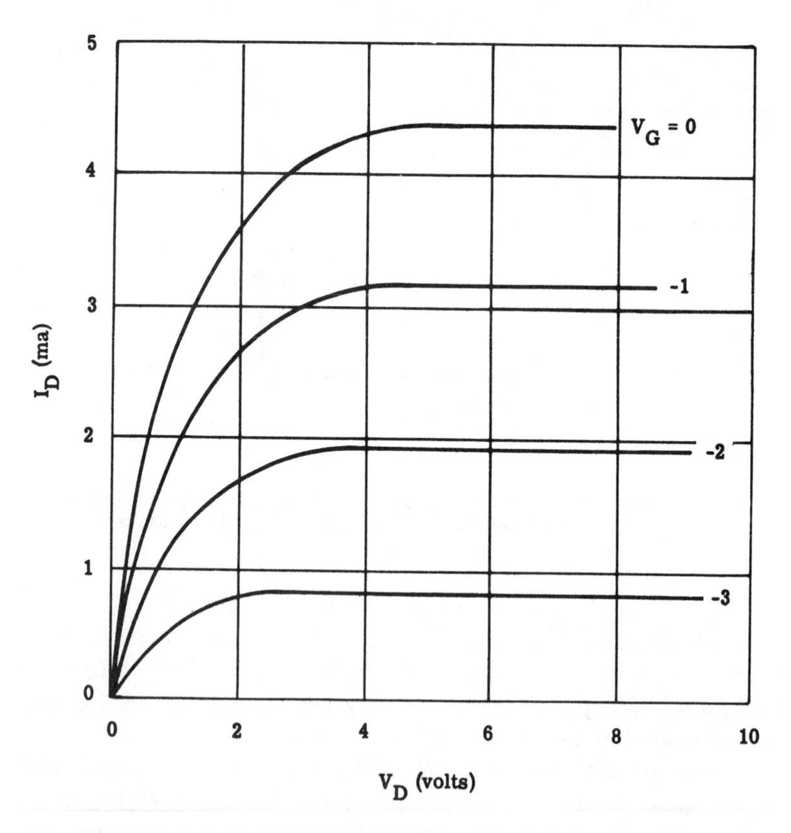

Fig. 10-20 Current - voltage characteristics for a diffused field effect transistor

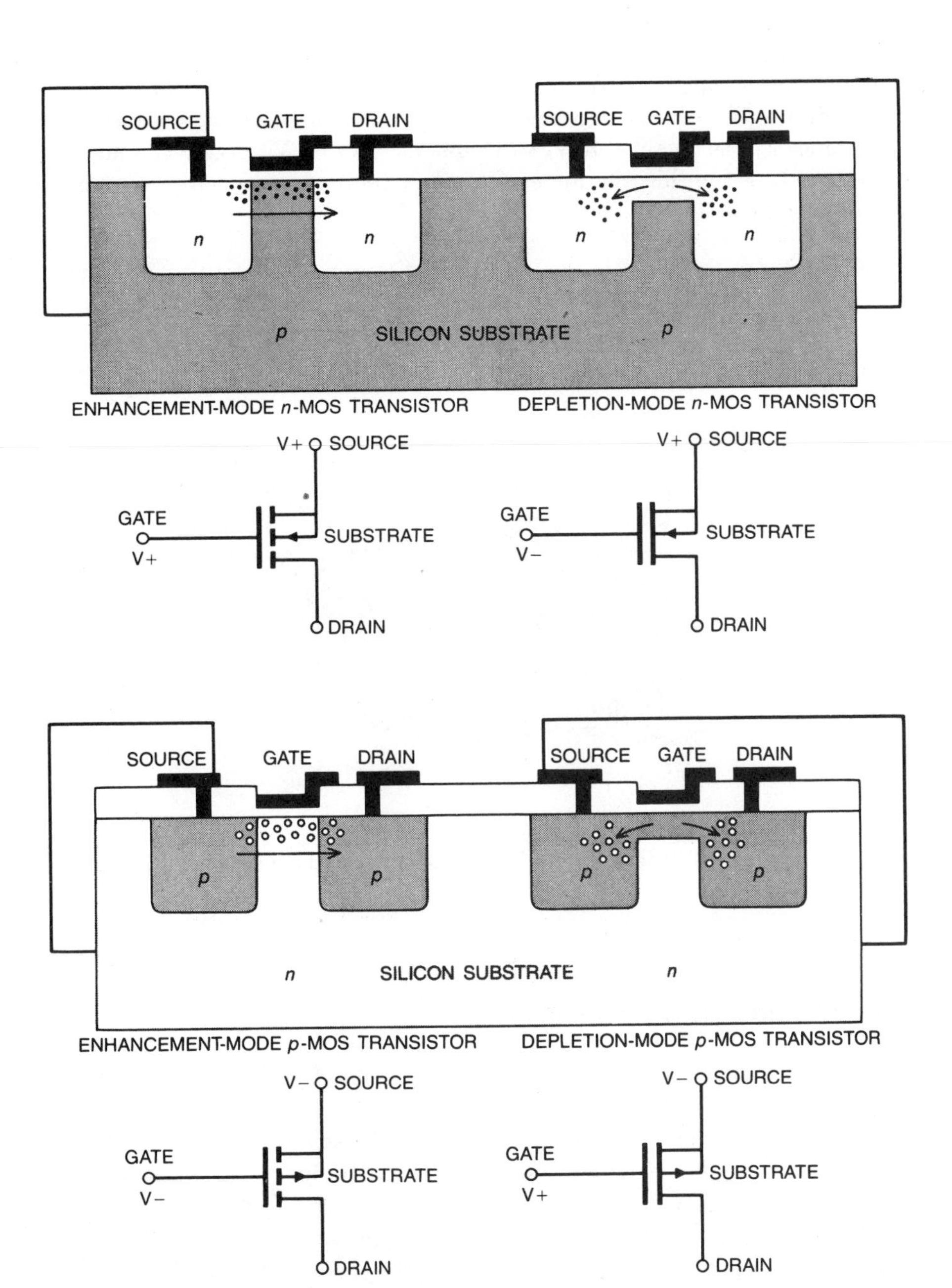

FIELD-EFFECT TRANSISTORS, made by the metal-oxide-semiconductor (MOS) technique, differ from bipolar ones in that only one kind of charge carrier is active in a single device. Those that employ electrons are called *n*-MOS transistors; those that employ holes are *p*-MOS transistors. In an *n*-MOS device (*top*) two islands of *n*-type silicon, called the source and the drain, are formed in a *p*-type substrate. Over the channel between the source and the drain is a metal electrode, the gate, that is prevented from making contact with the semiconductor by a thin layer of silicon dioxide. In an enhancement-mode transistor (*top left*) a positive potential at the drain exerts an attractive force on electrons available from the source, but the electrons cannot pass through the *p*-type channel, which is rich in holes. When a positive charge is applied to the gate, however, the resulting electric field attracts electrons to a thin layer at the surface of the channel and current flows from the source to the drain. In a depletion-mode *n*-MOS transistor (*top right*) there is a continuous channel of *n*-type silicon between the source and the drain, so that the transistor normally conducts; only when a negative voltage is applied to the gate is the current halted, since electrons are then expelled from the channel. By forming islands of *p*-type material in an *n*-type substrate the corresponding *p*-MOS devices (*bottom*) can be constructed; here the charge carriers are holes. Because MOS transistors require no isolation islands they can be packed more densely on a chip of silicon than bipolar transistors.

of the device can be estimated from the expression $g_m/2\pi C$ where the capacitance $C = 2\epsilon L/a$ represents the channel. The maximum transconductance will be obtained if $V_S = O$ and $V_D - V_G = V_o$, giving g_m (max) $= 2\sigma_o a/L$. From this, the upper frequency limit of operation, f, may be computed as

$$f = a^2 \sigma_o / 4\pi\epsilon L^2 \qquad (10\text{-}31)$$

where ϵ is the dielectric constant of the semiconductor.

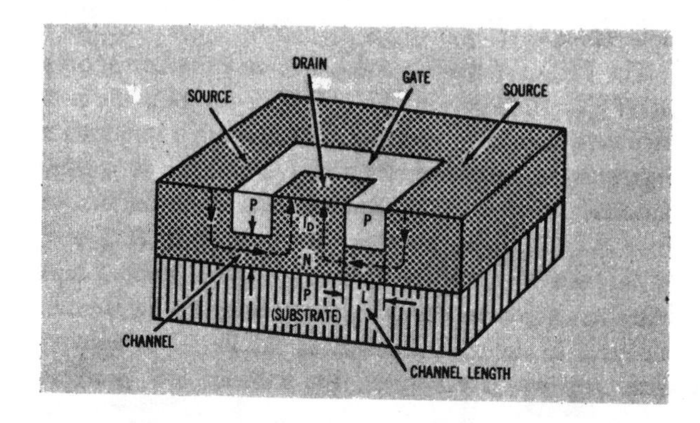

Fig. 10-21 Junction FET with Single-Ended Geometry.

10.6 MOS FIELD-EFFECT DEVICES

The metal-oxide-semiconductor (MOS) field-effect transistor (FET) uses an electric field applied through an insulated gate electrode to modulate the conductance of a channel layer in the semiconductor under the gate electrode. The channel consists of a lightly doped region contained between two highly doped regions called the "source" and "drain". The conductance of the channel is dependent upon the intensity of the field and hence upon the voltage applied to the gate electrode. Most MOS transistors now available are fabricated on a silicon substrate. The insulating oxide is usually silicon-dioxide. The MOS has high input resistance ($>10^{15}$ ohms), and at high frequencies the input impedance becomes highly capacitive. Unlike the junction transistor, which is a current-controlled device, the MOS is voltage- or charge-controlled.

There are two basic types of MOS devices: enhancement and depletion. Each of these types can be fabricated so that either electrons or holes are the majority carriers. Thus, four distinct types of MOS transistors exist.

10.6.1 N-CHANNEL ENHANCEMENT MOS

The majority carriers are electrons in the n-channel enhancement tran-

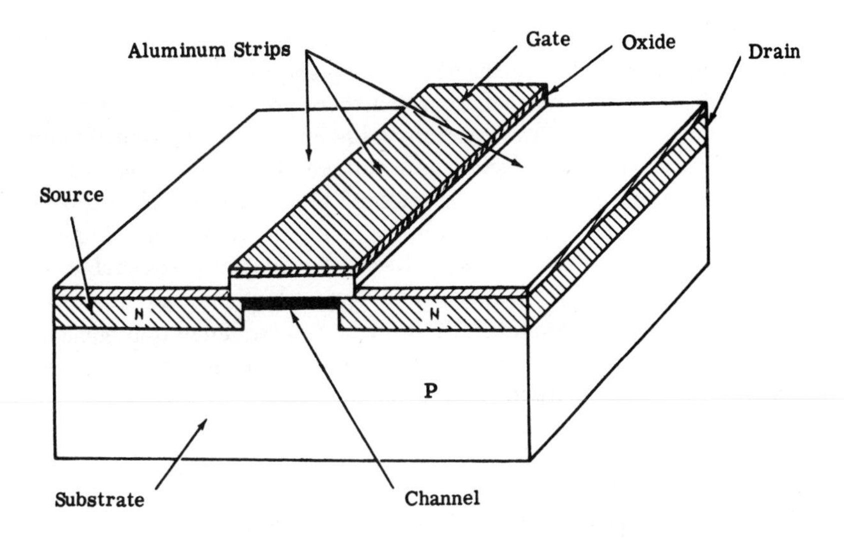

Fig. 10-22 Structure of an MOS field effect triode

sistor. A positive voltage (with respect to the source) applied at the gate induces a positive potential in the oxide at the oxide-semiconductor surface. This in turn attracts electrons to the semiconductor surface, and these carriers form the transistor's induced channel. An enhancement-type transistor structure is shown in Figure 10-22.

At zero gate voltage there is only a very small current flow ($\approx 10^{-9}$ amps) between source and drain. Since there is no gate voltage, there are no free carriers in the space between the two N+ regions. The current that does flow, is that of the reverse-biased junction formed by the P-type substrate and the N+ drain region.

When a positive voltage is applied to the gate electrode, minority carriers (electrons) are attracted from the P-region to the surface to form the channel. The surface has an excess of holes (empty valence states)

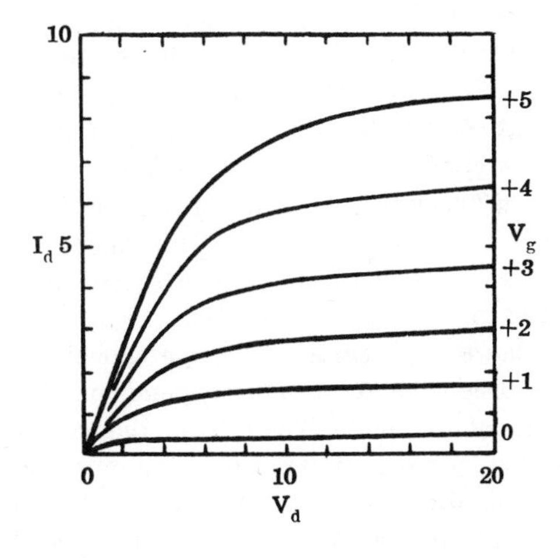

Fig. 10-23 Drain current vs drain voltage.
(Vertical: 1mA per division.
Horizontal: 2V per division).

since it is originally P-type, and as the electrons are drawn to the surface of the P-region underneath the oxide, these empty states are filled. Therefore, a gate voltage is reached at which there are just enough electrons to fill the empty valence states. At this value of gate voltage, the channel is intrinsic. For further increases in gate voltage, the channel becomes N-type. When this happens, the surface is referred to as "inverted." Figure 10-23 is an illustration of the drain current as a function of drain voltage for an enchancement field-effect transistor. Because of the nature of surface states, surfaces change easily from P-type to N-type, which makes it quite difficult to fabricate the enhancement FET; they have a tendency to become depletion-type devices. The surface resistivity must be kept low to prevent the channel from forming when there is no gate bias.

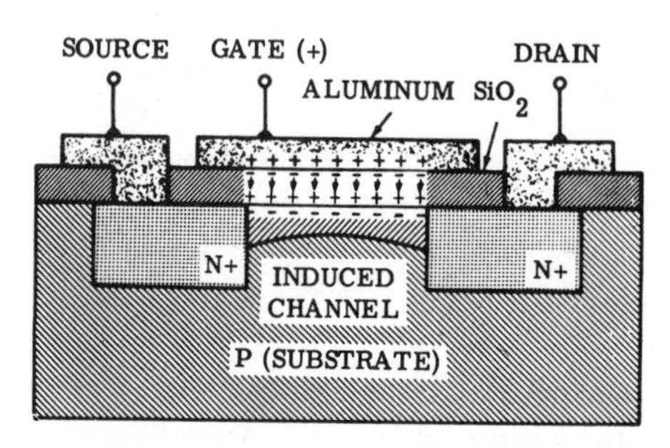

Fig. 10-24 Channel enhancement application of positive gate voltage causes redistribution of minority carriers in the substrate and results in the formation of a conductive channel between source and drain.

10.6.2 P-CHANNEL ENHANCEMENT MOS

The P-Channel enhancement MOS utilizes holes as the majority carriers. The substrate is N-type, and the induced channel is P-type. The drain and source regions are P+, and the drain-to-source voltage is negative. The channel is induced by application of a negative gate-to-source bias. The drain current as a function of drain and gate voltage is similar to that shown in Figure 10-23 except that all the voltages are reversed.

The metalized gate electrode shown in Figure 10-24 covers the entire channel region. It is characteristic of the enhancement-type transistor that the gate electrode overlaps both source and drain N+ regions. This is necessary to prevent a high resistance from appearing in series with the source and drain, since no carriers would be attracted into that part of the channel not included in the gate electric field.

This geometry, however, results in a large capacitance from gate to drain and gate to source. The exact value depends on the degree of overlap and the thickness of the oxide. The thicker the oxide, the smaller the capacitance value; yet, for sensitivity, the oxide must be thin enough to provide high gate fields for reasonably small gate voltages. To meet these opposite demands, the oxide is usually made thin over the channel and thick over the N+ regions (P+ regions for P-channel devices), as shown in Figure 10-25. Also illustrated is the effect of gate voltage on carriers for both P- and N-types.

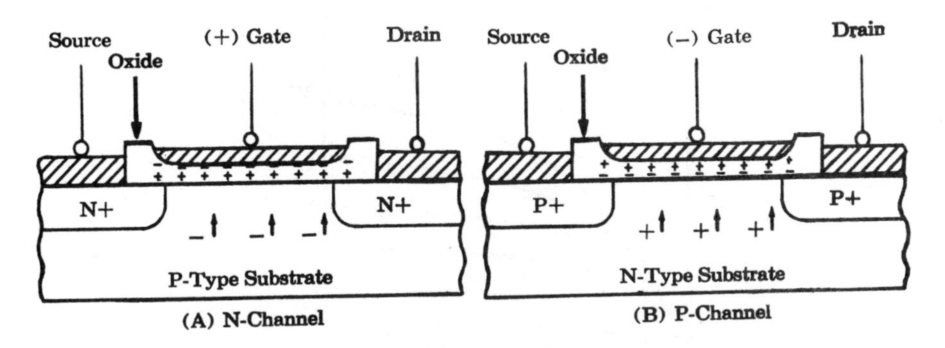

Fig. 10-25 Oxide thickness and charge on enhancement MOS transistor.

10.6.3 N-CHANNEL DEPLETION MOS

The depletion-type MOS is illustrated in Figure 10-26. The channel for this device is formed at the time when the oxide is fabricated on the surface. The degree of doping for the channel depends on the saturation current desired at zero gate voltage. The current that flows with zero gate voltage is higher in the depletion-type device, since the channel exists at all voltages. The drain current as a function of drain voltage is illustrated in Figure 10-27 for an N-channel depletion MOS transistor.

This device operates for negative as well as positive bias on the gate. It is unique in its applications and can be considered as operating either in the depletion mode or in the enhancement mode.

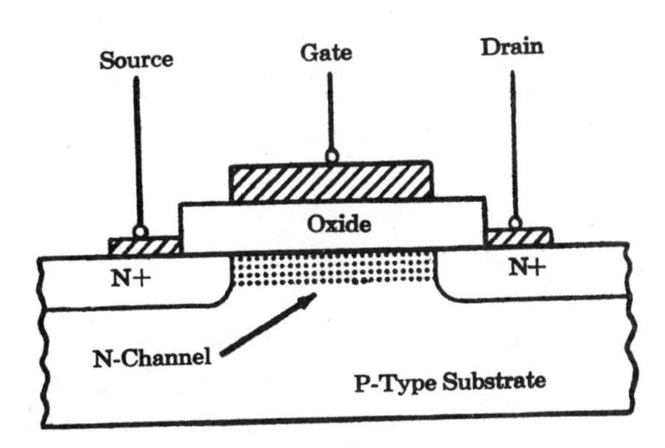

Fig. 10-26 Depletion-type MOS.

Operation in the depletion mode requires a negative gate-to-source bias. As the gate bias is made more negative, the channel is depleted of electrons because of the electric field induced in the oxide. The field at the oxide-semiconductor interface is positive and repels the electrons from the surface. The result is a decrease in current as a result of the decreasing conductance of the channel.

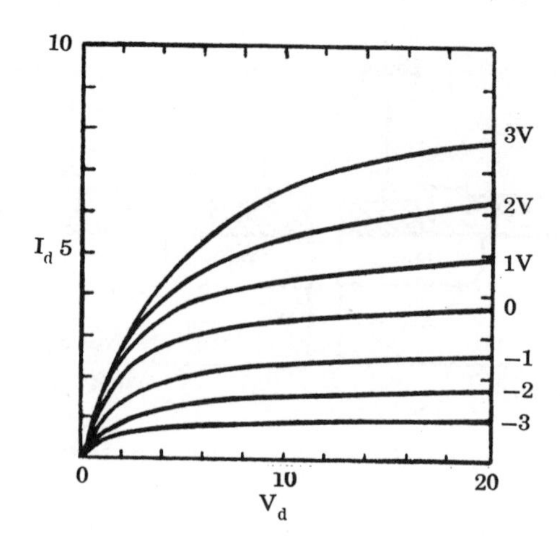

Fig. 10-27 Drain current vs drain voltage. (Vertical: 2mA per division. Horizontal: 5V per division).

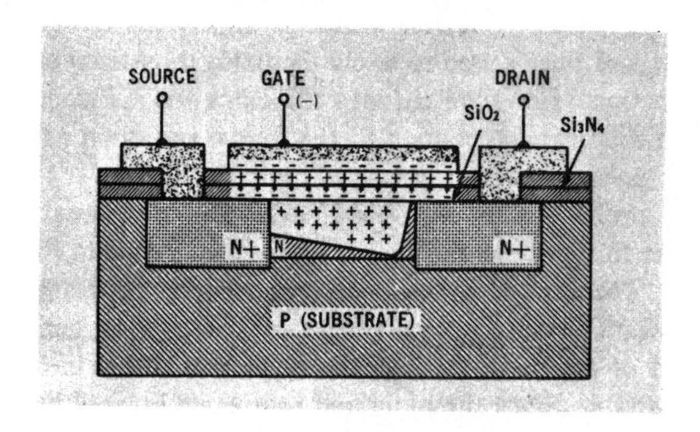

Fig. 10-28 Channel depletion phenomenon. Application of negative gate voltage causes redistribution of minority carriers in diffused channel and reduces effective channel thickness. This results in increased channel resistance.

10.6.4 P-CHANNEL DEPLETION MOS

The geometry for the P-channel is the same as that of the N-channel depletion MOS. The majority carriers, however, are holes. This device has all the characteristics of the N-channel except that all voltages are reversed.

The channel is not as highly doped as the drain and source regions. Because of the relatively high free-carrier concentration in the channel, it is not necessary that the gate electrode overlap the drain. This is referred to as "gate off-set". A small series resistance will be induced in the channel at the drain, when the transistor is operated in the enhancement mode. The only effect this has on device operation is to increase the drain voltage at which saturation occurs. The gate electrode does overlap the source. If it did not, the series resistance associated with the source would introduce degenerative feedback, which would have deleterious effects on device gain. The gate off-set significantly reduces the gate-to-drain feedback capacitance.

The two basic MOS structures are illustrated in Figure 10-29. These basic structures were first proposed by D. Hahng and M. M. Atalla.

The N+ regions of the structure shown in Figure 10-29 are obtained by high-temperature diffusion of phosphorous impurity into the P-type silicon substrate, by the use of oxide masking techniques. The N+ region used as the source can be internally connected to the P-type substrate,

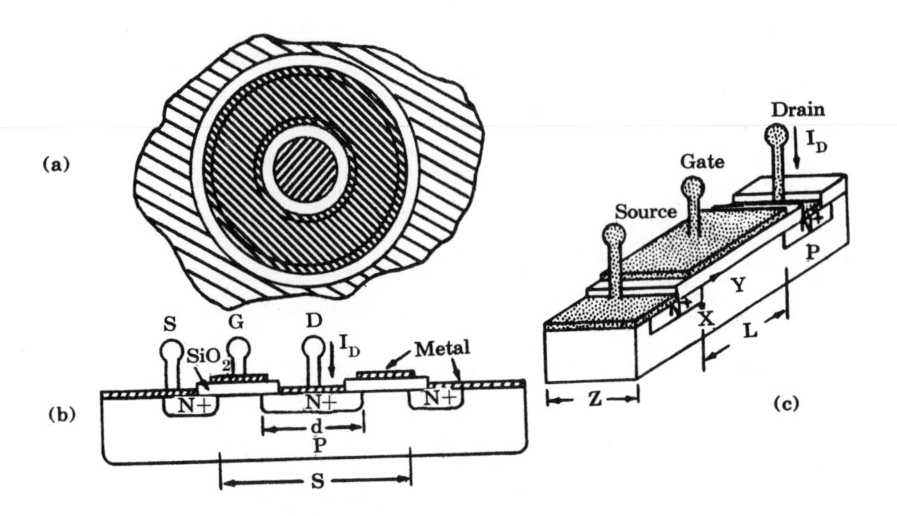

Fig. 10-29 Geometry of n-channel MOS transistors: (a) Top view. (b) Cross-sectional view. (c) The linear structure.

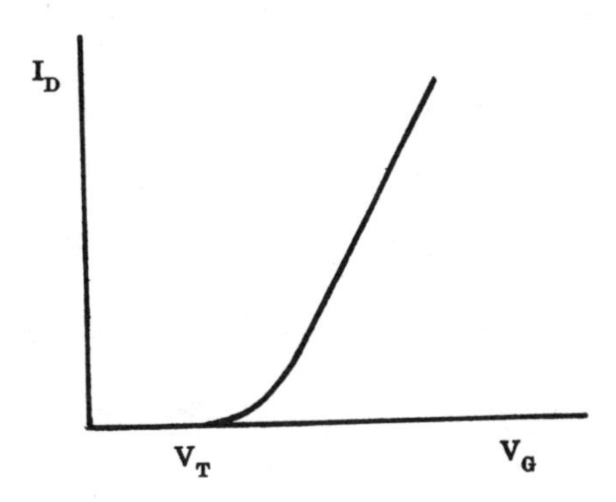

Fig. 10-30 Transfer characteristic of enhancement MOS transistor.

when the source electrode is fabricated. If it is not, the device is a four-terminal device, with the substrate acting as the fourth terminal. The other N+ region is the drain and is usually employed as the output terminal. The source acts as the common terminal for the input and output.

The input lead, called the gate, is evaporated over the oxide. In this particular structure, the oxide is silicon dioxide. Work has also been done with silicon nitride as the insulator. The SiO_2 layer is a determining factor of device characteristics and stability.

The oxide insulator can be grown thermally at high temperature or anodically at room temperature. Oxide thickness of the order of 1000 Å is

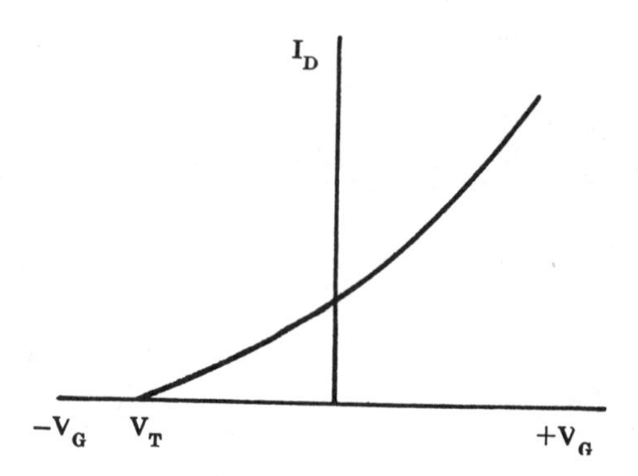

Fig. 10-31 Transfer characteristic for depletion MOS transistor.

generally used. Thick oxide reduces the transconductance, gain, and speed of the device for a given d-c operating point; and very thin oxide makes the reproducibility of the device difficult. The metal gate electrode and the metal electrodes that form the ohmic contacts to the source and drain regions are made by evaporation and photoresist techniques. Metals such as aluminum, silver, and gold are used.

The linear structure illustrated in Figure 10-29c is well suited for integrated-circuit work. It is easy to interconnect the devices, and there is a natural isolation between devices on the same substrate.

10.6.5 DRAIN CURRENT

The equation for drain current has been derived as

$$I_D = \frac{\bar{\mu}_n C_o}{L^2}\left[(V_G - V_T)\, V_D - \frac{V_D^2}{2}\right] \qquad (10\text{-}32)$$

where

$I_D =$ drain current
$V_G =$ gate-to-source voltage
$V_D =$ drain-to-source voltage
$V_T =$ the turn-on voltage if it is positive or turn-off voltage if it is negative
$C_o =$ the total capacitance of the oxide layer under the gate
$L =$ the channel length as indicated in Figure 10-29c
$\bar{\mu}_n =$ average surface mobility of electrons in the channel

Equation (10-32) is based on a simplified model of the channel and is valid only for gradual channels; it ceases to be a good approximation in the region of the channel where it is pinched off or nearly pinched off.

If V_T in equation (10-32) is positive and V_G is zero, the current will be negative. This cannot be the case in practice. What is indicated here, is that V_G must be larger than V_T in order to have current flow in the channel; thus, we are dealing with an enhancement transistor. This is easily seen in Figure 10-30, which illustrates the transfer characteristics of an enhancement MOS transistor.

From Figure 10-30 it can be seen that current does not flow until some value of gate voltage (V_T) is applied between gate and source. This is the voltage required to invert the channel. Current will not flow in the channel until the gate voltage exceeds the transistor turn-on voltage. The determining factor in the magnitude of V_T is the oxide thickness. For thin oxides, V_T is lower than for thicker oxides.

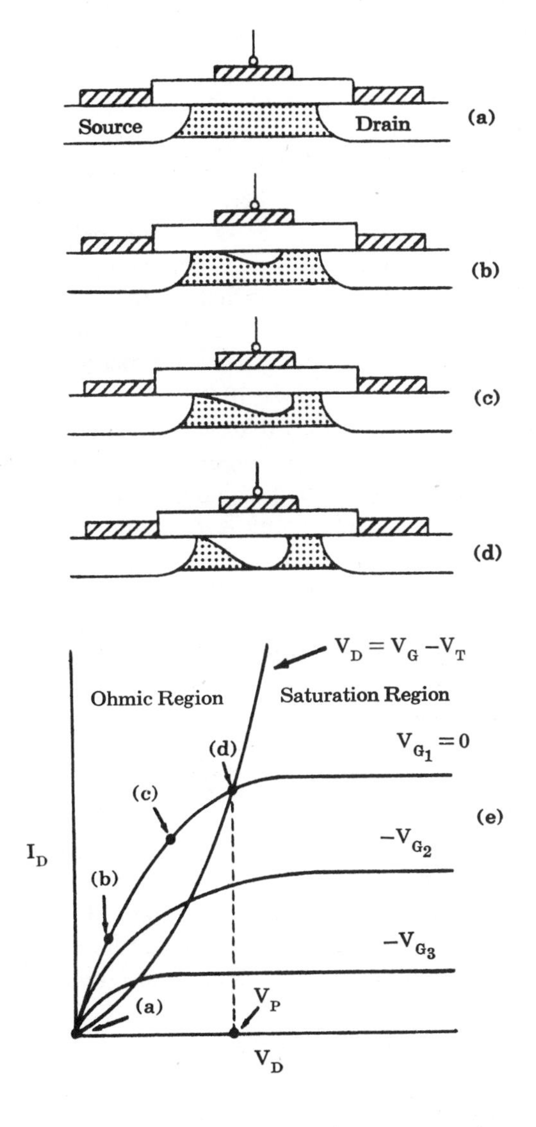

Fig. 10-32 Pinch-off for depetion MOS transistor.

If V_T is negative, the equation indicates a current at zero gate voltage. This is the case for the depletion-type transistors. The transfer characteristic for an N-channel depletion transistor is illustrated in Figure 10-31.

Note that for $V_G = 0$ in Figure 10-31, there is substantial drain current. In this case V_T is the voltage required to turn off the drain current.

10.6.5.1 Pinch-off

Equation (10-32) indicates a linear current-voltage relationship for the MOS. This is true only in the ohmic region of the V_D-I_D characteristic. As the drain voltage is increased beyond some value (assuming constant V_G), drain-current saturation takes place. At this point, the channel is pinched off. This is the pentode region of the V_D-I_D characteristic.

The gradual pinch-off of the channel of an N-type depletion MOS is illustrated in Figure 10-32. Assume that the gate electrode is shorted to the source ($V_G = 0$). For zero applied drain voltage, the channel is unconstricted, as shown in Figure 10-32a. As the drain voltage is increased from zero in a positive direction, current begins to flow in the channel. This is point (b) on the curve of Figure 10-32e.

The applied voltage is dropped across the channel, and the effect is to reverse-bias the gate electrode relative to the channel. Since the channel resistance is distributed, the voltage is distributed along the channel, and the most negative area is that closest to the drain. The result is an induced depletion region in the channel, with the greatest depletion occuring closest to the drain, as is indicated in Figure 10-32b.

As the drain voltage is increased to correspond to point (c) on the curve of Figure 10-32e, the depletion region is further enlarged. At point (d) the drain voltage equals the pinch-off voltage, and drain-current saturation takes place. At this point it is assumed that the depletion region extends across the channel. There is only a slight increase in drain current for further increases in drain voltage.

From Figure 10-32e it can be seen that the drain current can be divided into two segments: that which appears in the ohmic region and that which appears in the saturation region. These regions are divided by the locus of points determined by

$$V_D = V_G - V_T$$

For values of $V_G - V_T < V_D$, equation (10-32) is valid. However, in the saturation region, where $V_G - V_T \geq V_D$, this equation is no longer valid; the gradual-channel approximation fails. This condition corresponds approximately to the condition of maximum drain current from equation (10-32) and can be obtained from it as follows:

If we set

$$\left(\frac{\delta I_D}{\delta V_D}\right)_{V_G} = 0 \text{ and use equation (10-32),} \qquad (10\text{-}33)$$

the pinch-off condition is

$$V_D = V_{DS} = V_G - V_T \qquad (10\text{-}34)$$

where V_{DS} is the drain voltage at saturation.

The drain current at the pinch-off condition is given by

$$I_{DS} = \frac{\mu_n C_o}{2L^2}\left(V_G - V_T\right)^2 = \frac{\bar{\mu}_n C_o}{2L^2} V_{DS}^2 \qquad (10\text{-}35)$$

If the drain voltage is further increased beyond the pinch-off voltage given by equation (10-34) the pinch-off region lengthens into the channel. Most of the additional voltage applied to the drain beyond the pinch-off voltage appears across the length of the pinch-off region and causes little increase in the drain current.

In the saturation region the channel is said to be pinched off. Yet a constant current is being conducted from source to drain. In this region space-charge-dominated currents are generated in the drain area. This type of current flow differs from ohmic current flow since the channel drift field now controls the distribution of mobile charge as well as the charge velocity. It is analogous in this sense to space-charge current flow in conventional vacuum tubes, where the plate-to-cathode field gradient determines the space charge.

10.6.5.2 Determining V_T

A rapid experimental determination of the threshold voltage V_T can be obtained from two terminal characteristics. The gate electrode is connected directly to the drain for the enhancement-mode N-channel device, which has $V_T > 0$ and no built-in channel. For the depletion-mode N-channel device, a battery of $V_{GG} > |V_T|$ (positive side tied to the drain) is used that has a sufficiently high voltage to pinch off the built-in channel.

In these connections, the device is in the saturation region since $V_D = V_G + V_{GG} > V_{DS} = V_G - V_T$. Thus the two-terminal drain current can be obtained from equation (10-35) by the use of equation (10-34):

$$I_{DS} = (\bar{\mu}_n C_o/2L^2) \cdot (V_D - V_{GG} - V_T)^2, \qquad (10\text{-}36)$$

which shows that the onset of the drain current corresponds to a drain voltage of $(V_T + V_{GG})$. For devices of the enhancement type, the turn-off voltage V_T can be determined readily from a display of this characteristic without the use of a gate battery, since $V_T > 0$. For the depletion type, $V_T < 0$; and a gate battery of $V_{GG} > |V_T|$ must be used to determine the turn-on voltage V_T. This slight additional complication comes from the fact that an N-channel MOS transistor, which has a P-type substrate, cannot be biased with a large negative drain voltage, since then the drain junction would be forward-biased and the large forward-drain junction current would mask off the pinch-off point.

10.6.5.3 Determining the g_d and g_m

The low-frequency values of g_d, the drain conductance below saturation, can be determined from equation (10-37).

$$g_d = \left(\frac{\delta I_D}{\delta V_D}\right)_{V_G} = \frac{\bar{\mu}_n C_o}{L^2}\,(V_G - V_T - V_D), \qquad (10\text{-}37)$$

Thus g_d decreases linearly with drain voltage and becomes zero at saturation, where $V_D = V_{DS} = V_G - V_T$.

The transconductance (g_m) is

$$g_m = \left(\frac{I_D}{V_G}\right) V_D = \frac{\bar{\mu}_n C_o}{L^2}\,V_D \qquad (10\text{-}38)$$

and its maximum, which occurs at saturation, is given by

$$g_{ms} = \frac{\bar{\mu}_n C_o}{L^2}\,V_{DS} = \frac{\bar{\mu}_n C_o}{L^2}\left(V_G - V_T\right) \qquad (10\text{-}39)$$

Since $(V_G - V_T) = V_{DS}$ in the saturation region, one can use equation (10-35), substitute into equation (10-38), and rearrange to obtain

$$g_{ms} = \sqrt{\frac{2 I_{DS} \bar{\mu}_n C_o}{L^2}} \qquad (10\text{-}40)$$

The value of transconductance in equation (10-40) is also approximately the value of transconductance when the device is operated beyond the saturation voltage.

10.6.6 FREQUENCY LIMITATIONS OF AN MOS DEVICE

The upper frequency limit of an MOS device depends on τ, the transit time of a carrier from source to drain. The transit time is a function of device channel length (L), the surface mobility (μ), and the applied voltage from drain to source. The equation for transit time is

$$\tau = \frac{L^2}{\mu V_{DS}} \qquad (10\text{-}41)$$

The intrinsic gain-bandwidth product for the MOS is

$$GBW = \frac{g_m}{2\pi C_c} \qquad (10\text{-}42)$$

where g_m is the device transconductance and C_c is the active gate-to-channel capacitance (in saturation $C_c = 2/3\ C_o$). The gain-bandwidth product is directly related to the carrier transit time as

$$GBW = \frac{1}{2\pi\tau} \quad (10\text{-}43)$$

In actual practice, the upper frequency limit may be substantially lower than $1/2\pi\tau$. For a 10-micron channel, the upper frequency limit will be several hundred megacycles per second.

10.6.7 EQUIVALENT CIRCUIT OF THE MOS

An equivalent circuit of the MOS is illustrated in Figure 10-33. The equivalent circuit is composed of six resistors, five capacitors, a constant-current source, and two diodes.

Resistance r_{gs} represents the leakage from gate to source, and r_{gd} is the leakage from gate to drain. These values are very high, typically about 10^{15} ohms.

The series network formed by C_c and r_c is a lumped approximation of the distributed network of the active channel. The capacitance C_c is the sum of the small capacitors distributed over the active channel area; it is expressed as

$$C_c = \frac{\partial Q_c}{\partial V_g} \quad (10\text{-}44)$$

where Q_c is the total channel charge. C_c is usually a function of the applied gate and drain voltage. In saturation, this simplifies to

$$C_c = \frac{2}{3} A_c C_{ox} \quad (10\text{-}45)$$

where A_c is the gate area overlying the active channel and C_{ox} is the oxide capacitance per unit area. Typically, oxide thickness $T_{ox} = 1000$ angstroms, so that $C_{ox} = 10^{-8}$ Farads/Cm^2.

The capacitance C_c charges and discharges through the channel resistance r_c. The channel resistance, in turn, is composed of innumerable series and parallel resistors between the source, or drain, and the points in the channel where the individual channel-to-gate capacitances take effect. The high-frequency performance of the MOS is strongly dependent on the $r_c C_c$ time constant.

When the MOS is operated in the saturation region, it appears as a constant current source. For this reason the active portion of the circuit is depicted as a constant current source with a value of $g_m e_c$. The low frequency value of g_m can be obtained from the $V_D - I_D$ characteristic as

$$g_m = \left. \frac{\Delta I_D}{\Delta E_g} \right|_{V_D} \quad (10\text{-}46)$$

Resistance r_d in parallel with the current generator, $g_m e_c$, represents the dynamic output resistance of the transistor. It can be determined by the slope of the output characteristics as

$$r_d = \left.\frac{\Delta V_D}{\Delta I_D}\right|_{V_{sG}} \qquad (10\text{-}47)$$

In the pinch-off region, r_d is several orders of magnitude larger than any parasitic resistances. The parasitic resistances appear in series with the source and at the drain.

Resistances r_d' and r_s' represent those portions of the source-to-drain channel which are not controlled by the transistor's gate. These parasitic resistances are mostly caused by the metal-to-semiconductor contact of the source and drain regions. However, when the gate electrode is offset from the drain (depletion-mode transistor), that portion of the channel resistance not under the gate electrode is included in r_d'.

The only effect of r_d' is to shift the value of the external drain voltage required to obtain drain-current saturation. However, r_s' appears as a common element to the input and output circuit and therefore induces degenerative feedback. It lowers the external terminal transconductance, g_m, which is expressed as

$$g_m = \frac{g_{mo}}{1 + r_s' g_{mo}} \qquad (10\text{-}48)$$

where g_{mo} is the internal transconductance. To keep r_s' as small as possible, the gate electrode always overlaps the N+ source contact.

Capacitances C_{gd}, C_{ds}, and C_{gs} are the physical case and interlead capacitances between gate and drain, gate and source, and drain and source. Capacitances C_{gd} and C_{gs} also include any capacitances that are not dependent on voltage, such as that contributed by the physical overlap of the insulated gate over the source or drain. C_{gd} is reduced significantly when gate off-set is used for depletion-mode devices. Capacitance C_{ds} includes the capacitances of D_1 and D_2.

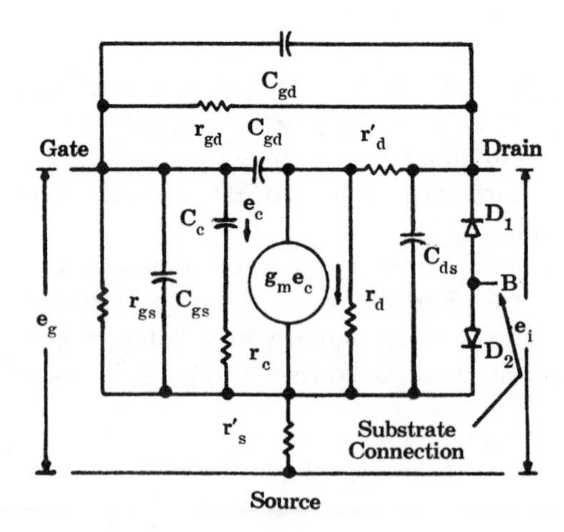

Fig. 10-33 Equivalent circuit for MOS FET operating in the pinch-off region.

Capacitance C'_{gd} represents the intrinsic gate-to-drain capacitance, which decreases as the channel voltage approaches the pinch-off region. This capacitor is quite significant since it determines the degree to which drain-current saturation is achieved.

Diode D_1 represents the junction formed between the N+ drain region and the semiconductor substrate; D_2 is the junction formed between the N+ source region and the substrate. D_1 and D_2 are back-to-back diodes in parallel with the channel. At high frequencies, the diodes contribute an equivalent series RC network that affects the output admittance.

MOSFET CIRCUITS USING SILICON ON INSULATING SUBSTRATES

The silicon on insulating substrate (SOIS) structure in conjunction with self-aligned gate processing can create either a high speed low power or a highly integrated MOSFET memory process.

Differentiation between silicon on an insulating substrate (SOIS) and silicon-on-sapphire (SOS) may be effected by reference to processing, final structure, and material properties. The SOIS structure is a thin silicon layer on a polycrystalline silicon backing with a chemically vapor deposited (C.V.D.) dielectric film in between; the structure is an unpatterned dielectrically isolated (D.I.) substrate in which the mono-crystalline layer is unconventionally thin. The SOS structure is a thin silicon epitaxial layer on a monocrystalline sapphire wafer. Because of chemical and mechanical interaction with the sapphire, the active silicon layer on sapphire must be thicker than 1.0 μ, and because of growth steps, pyramids, and other forms of surface texture an upper limit of approximately 3 μ is placed on the silicon thickness of SOS substrates. There is no material or process induced upper limit on SOIS active layer thickness; the lower limit is set by processing tolerances and is the subject of a continuing investigation. The most notable material difference between SOS and SOIS is crystal perfection. The SOIS structure contains less lattice distortion than the SOS structure. The processing of SOS substrates may be characterized with three unit operations:

(1) Sapphire crystal growth

(2) Wafer shaping (sawing, polishing, and cleaning)

(3) Heteroepitaxial growth of silicon

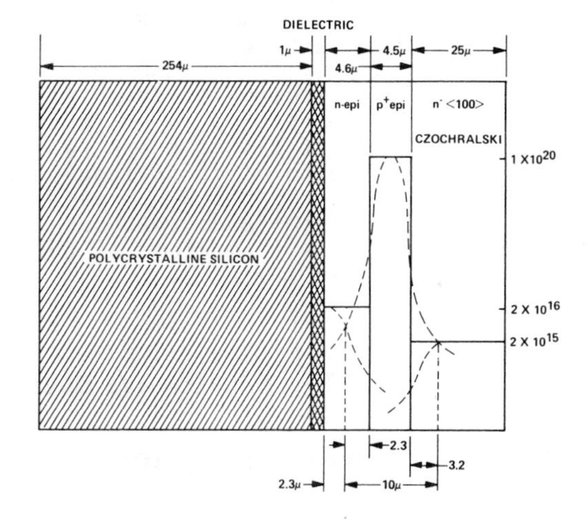

Fig.10-33.2 Sample cross-section showing undiffused and diffused schematic impurity profiles.

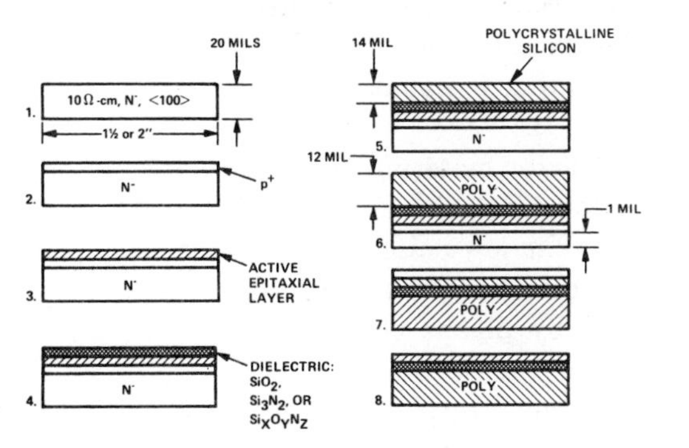

Fig.10-33.1.

Schematic of SOIS substrate fabrication

SOIS substrate fabrication is outlined in Fig.10-33.1. Discussion of the process is illustrated with the schematic shown in Fig. 10-33.2 The SOIS substrate cross-section exhibited in Fig.10-33.2 corresponds to step 6 of Fig.10-33.1. The polycrystalline silicon is grown at a temperature $1000 \leq T \leq 1150$ °C. Time at temperature varies from two hours at 1150 °C to four hours at 1000 °C. The time at temperature causes out-diffusion of the boron (B) from the p^+ layer, and the elevated temperature contributes to the room temperature warpage of the wafer. The thermal expansion coefficients of the various layers shown in Fig.10-33.2 are all different, and all the layers were deposited on the original n^- wafer at temperatures between 1000 and 1150 °C. The dielectric layer may be SiO_2, Si_3N_4, $Si_xO_yN_z$, or a composite of oxide and nitride. Consideration of thermal expansion and heat flow favors a Si_3N_4 layer, whereas control of interface state density requires a SiO_2 layer. The active layer of epitaxial silicon may be n, n^-, n^+, p, or p^-. The thickness of the active layer is corrected for out-diffusion of boron. The p^+ etch stop layer is fabricated by diffusion of B into the n^- wafer or by epitaxial growth of p^+ silicon on the n^- wafer. A one micron thick homogeneous p^+ layer doped to 7×10^{19} B/cc (without out-diffusion) would take 10 microns of thickness variation out of the substrate. The boron concentration dependence of the etch rate in KOH is shown in Fig.10-33.3. The <100> etch rate of silicon in KOH is approximately 1.0 μ/min for all dopants and resistivities other than B where $N_B > 10^{18}$ atoms/cc. Etch removal of the original wafer, with the etch terminating at the p^+ layer, is the key step in SOIS substrate processing. The starting material must be <100> oriented; n^- doping of the starting wafer is advantageous for epitaxial quality and p^+ layer evaluation.

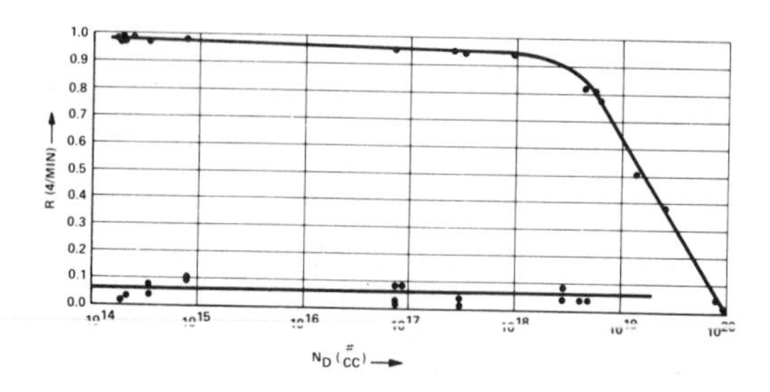

Fig.10-33.3 Etch rate of <100> and <111> silicon in KOH versus boron concentration with KOH composition and etch temperature held constant.

10.7 MICROWAVE DEVICES

10.7.1 TRANSISTORS

Germanium: — Advances in the alloy diffused planar process allows the same order of dimensions, tolerance control, and geometry in germanium that previously existed in the production of silicon transistors. Although germanium is the best material for achieving the highest maximum frequency of oscillation, f_{max}, in semiconductors, these transistors were limited by base and emitter stripes large enough to allow bonding to wires. Now, advances in the planarization technique allow bonding to expanded contacts, thus, freeing stripe size from the limitation.

Silicon: — Silicon remains the major semiconductor material used for power generation at high frequencies. In the past, low noise microwave transistors made of silicon could not compete with the low noise performance of germanium, but late developments in new geometries, have brought the noise figure of silicon devices to that offered by planar germanium devices.

Gallium Arsenide: — Gallium arsenide transistors are characterized by high noise figures. The development of these transistors as power devices has not been fully realized and the low noise requirement makes these transistors unsuited for receiver applications.

Field Effect Transistors: — The low noise and low cross-modulation distortion properties of insulated gate and junction FET's are very useful in many receiver applications. However, these devices are generally not applicable to microwave frequencies. The present useful frequency region extends to L-Band. In this frequency range, the FET performs well as an RF amplifier or mixer for UHF systems, and in high frequency IF stages of microwave receivers. A practical application of these devices is in stages requiring gain control, because these devices afford ease of control in various AGC modes of operation.

10.7.1.1 Frequency Limitations

An ultimate physical limitation exists which places an upper bound on the frequency and power capabilities of microwave transistors. The theoretical limit has been shown to be related to material parameters which are independent of device design. The relationship is given in Equation (10-49).

$$V_m f_T = K \tag{10-49}$$

where K equals 2×10^{11} volts/sec for silicon or 1×10^{11} volts/sec for germanium. V_m is the maximum allowable applied emitter-collector voltage. The parameter f_T is the charge-carrier transit-time cutoff frequency, more commonly called the gain-bandwidth frequency, and is defined

$$f_T = \frac{1}{2\pi\tau} \tag{10-50}$$

where τ is the transit time for a charge carrier to transverse the emitter-collector distance while moving at an average velocity. This relationship defines the upper cutoff frequency, since the minimum value of V_m must have some value, say one volt, that is sufficiently greater than thermal noise voltage to ensure normal transistor action of the base collector junction. Thus, for silicon devices, a theoretical upper limit of 200 GHz is determined, and for germanium, the highest obtainable frequency would be 100 GHz. In a practical sense, these frequencies would be less than that indicated because of such assumptions regarding uniform electric field stresses and carriers velocities which are made in deriving Equation (10-49).

For maximum power gain at high frequencies, the magnitude of $r_b{'}C_c$ becomes important. This is demonstrated in the familiar expression for the maximum available power of a common emitter stage given in Equation (10-51).

$$PG = \frac{f_T}{8\pi f^2 r_b{'} C_c} \tag{10-51}$$

where f = frequency of operation
f_T = gain bandwidth frequency in MHz
$r_b{'}$ = base spreading resistance in ohms
C_c = intrinsic collector capacitance in pfd.

The f^2 implies a 6-dB per-octave roll-off of power gain.

This equation may be solved for a particular frequency at which the stage power gain is equal to unity. The result is a frequency denoted as f_{max}. Thus,

$$f_{max} = \sqrt{\frac{f_T}{8\pi r_b{}' C_c}} \qquad (10\text{-}52)$$

Since the power gain is unity at this frequency, f_{max} represents the maximum frequency of oscillation of the transistor.

It is evident that extending f_{max} requires extending f_T, which is related to base width because the charge-carrier transit time is dependent upon the distance that the base vicinity carriers must travel. Thus, one of the most important frequency limitations is base width. Presently, this base width is controlled in the diffusion process where base widths as low as 0.25 micron have been realized. Thinning the base increases $r_b{}'$. In addition, the emitter junction area should be as small as possible to achieve the smallest emitter junction capacitance, consistent with the allowable current rating.

Geometries used in microwave power transistor create thermal problems, since the heat which must be dissipated is confined in a very small area. To overcome the problem of heat dissipation, power devices were evolved from scaling up the small area (7 mil^2) of the low power transistor. Scaling to a larger area must be done without sacrificing the high frequency parameters; i.e., f_T and the $r_b{}' C_c$ product must remain constant. However, the collector capacity will increase roughly in the ratio of device areas.

A successful approach is to make a larger number of interdigitated fingers in one structure. This has been accomplished, keeping the emitter and base stripe width and spacing to 0.2 mils. Electrically it is more efficient than using a number of small active areas spaced apart, but connected electrically by a metalization pattern over S_iO_2. Thermal resistance problems arising from the former technique can be minimized by choosing a suitable length-to-width ratio for the geometry.

10.7.1.2 High Frequency Field Effect Transistors

The metal-oxide semiconductor field effect transistor (MOS-FET) and junction FET's high frequency performance, has been extended to L-Band as a result of the technology advances in the fabrication of silicon microwave transistors. The FET is particularly well suited in many VHF and UHF receiver stages requiring high input impedance, low cross-modulation distortion, and low noise. The devices are especially well suited as IF and audio amplifiers, and can be controlled in either forward or reverse AGC modes. Present commercially available FET's have gain bandwidth products of 75 to 100 MHz, so that low noise, low distortion amplifiers and mixers in the VHF region are practical. Developmental FET's have been reported with larger gain bandwidth products that work satisfactorily at 1 GHz as amplifiers and oscillators.

For purposes of a brief review, a field effect transistor (FET) is essentially a semiconductor current path whose conductance is controlled by applying an electric field perpendicular to the current. A typical diffused field-effect transistor is shown in Figure 10-34a and its electrical equivalent diagram is shown in Figure 10-34b.

The devices consist of a single P or N type semiconductor bar with a contact at each end. N-type impurities are introduced into opposite sides of a P type bar as shown in Figure 10-34a. Contacts to the N type impurities provide the "gate" input. The contacts at each end of the bar are the drain and source connections. Comparing the FET to a vacuum tube, the "source" is analogous to the cathode, the "gate" corresponds to the grid, and the "drain" can be compared to the anode or plate. The field effect also has electrical properties similar to a vacuum tube, and as in vacuum tubes, the gain factor is expressed in transconductance, (gm). Amplification is achieved by the action of the majority of carriers within the device. For this reason, the FET is an unipolar device as compared to the bipolar junction transistor.

The design of a high frequency FET is similar to that of the interdigitated bipolar transistor, and it is constructed in much the same fashion, except in the location of the junction relative to the terminals. To produce an n-channel device, a p-type silicon wafer with a resistivity of 3 to 10 ohm/cm is used as a starting material. A number of heavily doped n-type stripes are diffused into the surface. The ends of alternate strips are connected together and form the drain and source contacts. An insulating layer of silicon dioxide or silicon nitride is placed over the areas between the diffused n stripes. The thickness of this insulating layer is typically one to two thousand angstroms. A metalized layer is then deposited on top of the dielectric insulating layer to form the gate electrode. A cross section of this structure is shown in Figure 10-35. Operation of a FET can be better understood by considering a semiconductor bar having the dimensions as shown in Figure 10-36. The conductances through the length of the bar depends on the dimensions of the bar, and the conductivity of the semiconductor. The conductance of the bar is proportional to the total number of current carriers present. The latter being a function of large concentration of impurity atoms contributed by doping. If this concentration is uniform in every part of the bar, the conductance of the bar is

$$G_{10} = \frac{q\mu pWT}{L} \tag{10-53}$$

where $q =$ electronic charge, 1.6019×10^{-19} coulombs
$\mu =$ carrier drift mobility $cm^2/volt\text{-}sec$
$p =$ impurity density, $atoms/cm^3$

As shown in Figure 10-34a, the diffused regions between the pn junction form the source and the drain terminals. The gate is a metallic layer evaporated on oxide surface which isolates the gate electrode from the semiconductor material, normally silicon. The region beneath the oxide layer forms a channel between the two diffused regions. Modulating this channel causes it to widen and extend into the high resistivity area formed

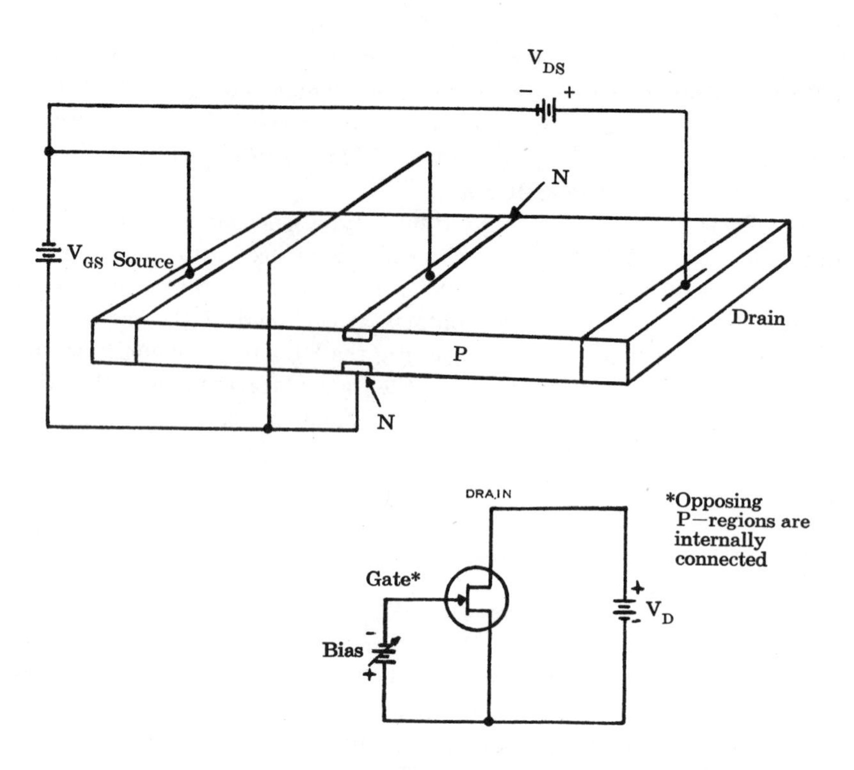

Fig. 10-34 Diffused FET transistor and equivalent electrical circuit for microwave applications

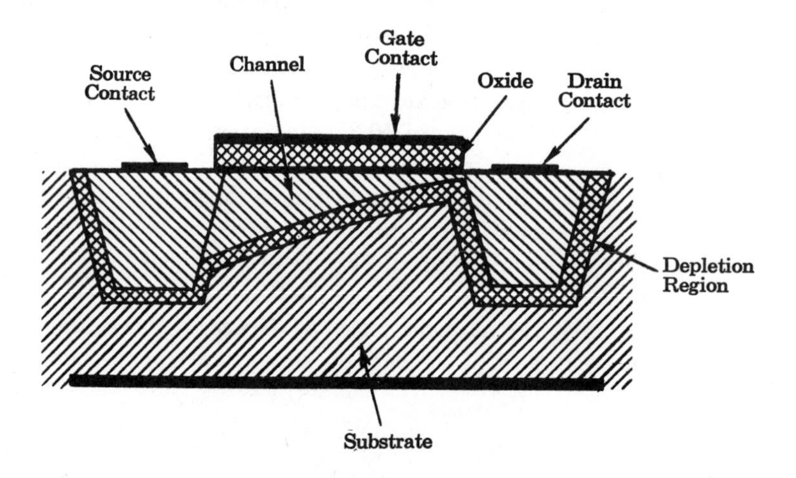

Fig. 10-35 Cross-section of junction FET device for microwave applications.

by the p-doped semiconductor. The modulation of this channel width is used to realize transconductance, in that the voltage applied to the gate terminal will vary the conductance (resistance) of the material in accordance with Equation (10-53). The FET described is known as the insulated gate or MOS (Metal Oxide Semiconductor) FET. It is a surface field-effect transistor, as compared to a unipolar or junction field-effect transistor.

In a typical VHF/UHF junction FET at high frequencies, performance is governed by series resistance within the chip. The equivalent circuit and exact equations for such an FET are given below with some practical approximations.

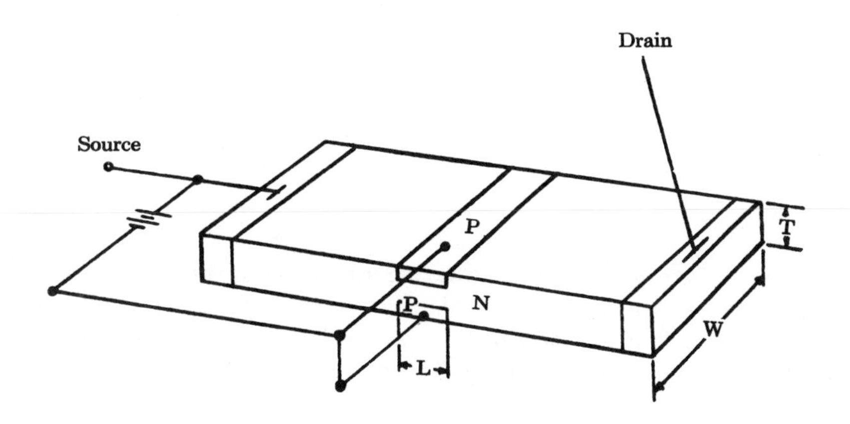

Fig. 10-36 Semiconductor bar.

Referring to Figure 10-37, equations for the common source VHF complex admittance parameters are approximated by:

$$
\begin{aligned}
y_{11} &= \omega^2 (C_{gs}+C_{gd})^2 R_g + j\omega (C_{gs}+C_{gd}) \\
y_{12} &= -\omega^2 C_{gd}^2 (R_g+R_d) - j\omega C_{gd} \\
y_{21} &= gf_s - j\omega C_{gd} \\
y_{22} &= gos + \omega^2 C_{gd}^2 + C_{gd}^2 (R_g+R_d) + j\omega C_{gd}
\end{aligned}
\tag{10-54}
$$

The input and output admittance transfer characteristics clearly show that the admittances increase as the square of the frequency. The power gain of an FET device may be defined as

$$P\,G_1 = 1/2[1 + (\frac{gm}{\omega C_3})^2] \tag{10-55}$$

where gm is the transconductance and C_3 is the combined external gate to drain capacitance and channel to gate capacitance. As seen in Equation (10-55), increasing the ratio of gm/C_3 will increase the power gain. The high input impedance is the result of the gate source path and involves a reverse biased junction. As this junction is large in area, it presents a large input capacitance, thus, limiting the high frequency response. The gm term is high frequency dependent and can be expressed as

$$gm\ (\omega) = \frac{gm_o}{1 + j\omega R C_G} \tag{10-56}$$

where gm_o = low frequency value of transconductance
R = series resistance of chip
C_G = gate to drain capacitance.

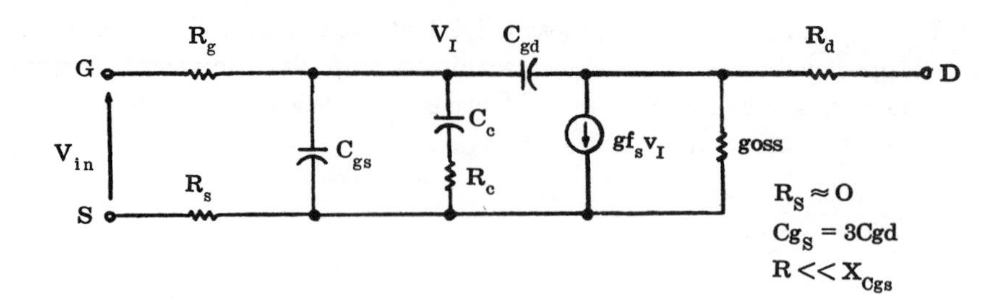

Fig. 10-37 Junction FET equivalent circuit.

Two factors to consider in achieving a high gm, is the requirement for large junction widening and a high gate area. The latter requirement necessitates a thin and narrow gate geometry. The gm cutoff frequency for the latest high frequency MOS-FET is in the order of one GHz and higher.

10.7.1.3 Noise Sources In Microwave Transistors

The amount of noise inherent in microwave semiconductor devices, is an important consideration if the device is intended for receiver applications. This is especially true for microwave transistors, as the noise contributed by the device adds significantly to the signal the transistor is amplifying. The noise figure, NF, expressed in decibels sets a figure of merit upon the noise performance of an active device. It is defined as follows:

$$NF = \frac{\text{total available noise power at output of device}}{\text{available noise power at output contributed by the generator source}} \qquad (10\text{-}57)$$

which is the ratio of a noisy network to that of a noise-free or ideal network. At high frequencies, two commonly occurring noise mechanisms contribute to making the microwave transistor noisy: thermal noise and shot noise. Within a junction transistor, three noise sources are formulated from these noise mechanisms; shot noise in the base emitter junction, thermal noise resulting from the base resistance, and shot noise in the collector-base junction.

Thermal noise phenomena, called ''white noise'' because of its similarity to the frequency spectrum of white light, is the result of spontaneous random fluctuations of free electrons in a conductor. These random fluctuations are present even in the absence of an electric field and are solely a function of temperature. The random behavior of electron agitation gives rise to an instantaneous value of current; however, over a long period of time, the average value of this current is zero. The mean-squared value of this current may be expressed in terms of the conductance associated with the conductor, and is given by

$$\overline{i^2}_{th} = 4KTG_{th} \Delta f \qquad (10\text{-}58)$$

where K is Boltzman's constant
T is temperature in degrees Kelvin
G is conductance in mhos
f is bandwidth in cycles following the conductor

For circuit applications, use can be made of Norton's constant current theorem and Equation (10-58) may be expressed as a mean-squared noise current generator paralleled by a noise-free conductance. This is shown in Figure 10-38. Since $i_{th} = V_{th}/R$, equation (10-58) may be expressed in terms of mean-squared voltage.

$$\overline{V^2}_{th} = 4KTR\Delta f \qquad (10\text{-}59)$$

This is the general expression for thermal noise voltage across the terminals of a device. Shot noise phenomena results from the random arrival of charges or carriers caused by diffusion when a current flows. The shot-noise energy associated with the stream of carriers is completely random with an uniform spectrum, and is proportional to the charge of an electron, the dc current and bandwidth. It may be expressed as

$$\overline{i^2}_{SH} = 2qIdc\Delta f \qquad (10\text{-}60)$$

where $q =$ charge (1.6×10^{-19} coulombs)
$Idc =$ dc current in amperes
$\Delta f =$ bandwidth in Hertz

Equation (10-60) is the same as the shot-noise due to the random emission of electrons from a heated surface such as the cathode of a vacuum tube.

The circuit representation of Equation (10-60) is the same as that for thermal noise, where the mean-squared value of shot-noise current can be represented by a constant current generator paralleled by a noise-free conductance, G_{SH}, as shown in Figure 10-39.

G_{SH} is the incremental conductance of a PN junction and is representative of the dynamic resistance of a diode, r_d. This relationship is demonstrated by

$$r_d = \frac{1}{G_{SH}} = \frac{KT}{qI_{dc}} = \frac{25}{I_{dc\ (ma)}} \text{ ohms} \qquad (10\text{-}61)$$

Since the shot-noise voltage, e_{SH}, is related to shot-noise current by the incremental conductance:

$$\overline{e_{SH}} = \frac{i_{SH}}{G_{SH}} \qquad (10\text{-}62)$$

the equation for shot-noise means-squared voltage may be derived from Equation (10-60) as

$$\overline{e_{SH}^2} = 2KTrd\Delta f \quad (10\text{-}63)$$

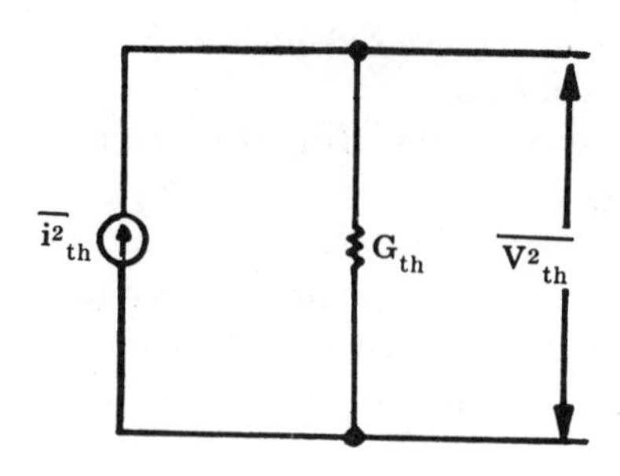

Fig. 10-38 Circuit presentation of thermal noise current.

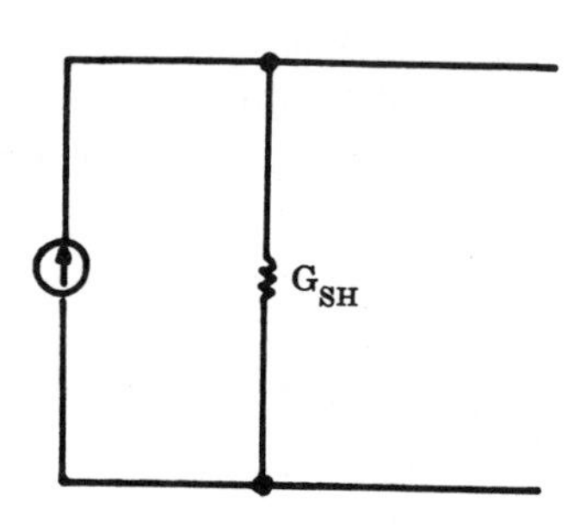

Fig. 10-39 Circuit presentation of shot noise current.

where the same constants which appear for thermal noise are present.

Equations (10-58) and (10-60) may be incorporated into a noise model for a junction transistor in a common-base configuration, such as shown in Figure 10-40. In this noise model,

$\overline{i_p^2}$ = collector shot-noise current generator (includes noise from I_{CO})
$\overline{i_e^2}$ = emitter shot-noise current generator
r_e = emitter resistance, $25/I_E$ ohms
r_b' = base spreading resistance
r_c = collector resistance
$\overline{e_b^2}$ = base thermal noise equivalent voltage generator
α = common-base current gain
R_g = source resistance
I_{CO} = dc collector cutoff current.

The model is valid above 1 kHz and has been shown to be valid for the common emitter and common base configurations. The noise figure, NF, of a high frequency transistor, derived from this noise model, is given by

$$NF = 1 + \frac{r_b'}{R_g} + \frac{r_e}{2R_g} + \frac{(r_b' + r_e + R_g)^2}{2\alpha_o R_g r_e} \left[\frac{1}{h_{FE}} + \left(\frac{f}{f\alpha}\right)^2 + \frac{I_{co}}{I_E} \right] \quad (10\text{-}64)$$

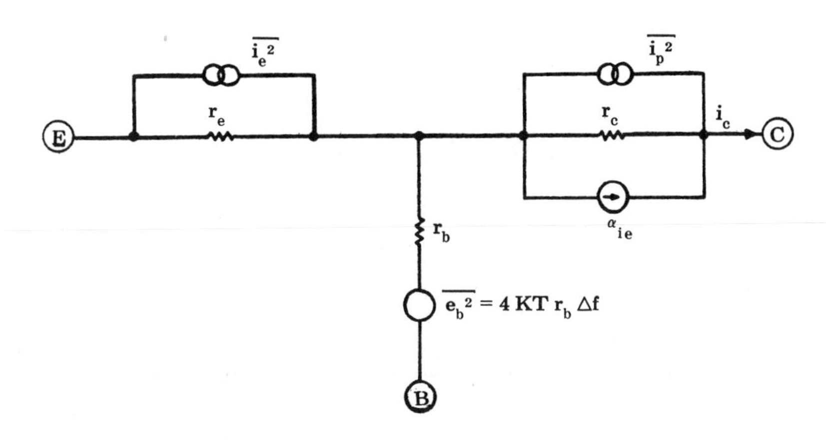

Fig. 10-40 Noise model for junction transistor.

The $f\alpha$ term expresses the alpha cutoff frequency, where α, the forward current gain, is reduced to 0.707 of its low frequency value, α_0. At this frequency, the $(\frac{f}{f\alpha})^2$ term becomes predominant and results in the noise figure asymptotically approaching a 6-dB per octave slope. This may be accounted for in the noise model by allowing α the forward current gain, to vary with frequency by replacing it with $\frac{\alpha_0}{1 + j\,f/f_c}$ where f_c is the break or corner frequency.

At higher frequencies where the transit time of carriers becomes appreciable, a new conductance is introduced in the expression for shot-noise. This conductance parallels the existing conductance corresponding to dynamic resistance of the diode, and is the result of carriers that cross the base-emitter junction towards the base, but return and recombine in the emitter region where they originated. The added conductance, G, is shown by

$$\overline{i_{SH}^2}\,(HF) = 2\,q\,I_{dc}\Delta f + 4\,KT\,(G - G_{SH})\,\Delta f \quad (10\text{-}65)$$

10.7.1.3.1 Field Effect Transistor

The field-effect transistor (FET) is subject to the same types of noise found in conventional transistors. The noise mechanisms in the FET are quite similar to those in the diffusion transistor, except for the part which capacitance plays. The noise model for the FET shown in Figure 10-41 accounts for the division of channel resistance.

In the noise model shown in Figure 10-41,

R_{CH} = channel resistance
r_d = dynamic drain resistance
r_s = incremental source resistance
C_{gd} = gate to drain resistance
C_{gs} = gate to source capacitance
e_{th} = equivalent noise voltage generator
i_p = gate to source noise current generator

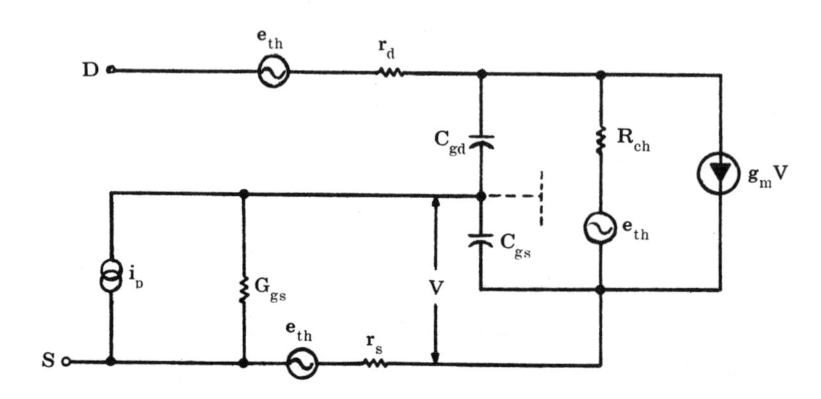

Fig. 10-41 Noise model for the FET.

The noise generating mechanisms operative in an FET are caused by the following physical phenomena:

(a) The gate leakage current, which generates full shot-noise

(b) A thermal noise voltage generated in the conducting channel which modulates the space-charge layer height

(c) Generation – recombination noise in the space-charge layer.

The generation-recombination noise in the space-charge layer modifies the thermal noise voltage in the conducting channel, and gives it a 1/f or flicker noise dependence. However, the 1/f noise does not become significant, unless the FET is operating at very low frequencies. For surface or insulated gate FET's the effect of shot noise created by the gate leakage current may be neglected as this current is small enough to be negligible.

For all frequencies above the region where the 1/f flicker noise is effective, the noise figure of the FET derived from the above noise model may be written as:

$$F = 1 + \frac{2\lambda\omega}{gm}(C_{gs} + C_{gd}) + \frac{\lambda}{gm}\left[\frac{G_S - \omega\,(C_{gs} + C_{gd})^2}{G_S}\right] \qquad (10\text{-}66)$$

where G_S = source conductance
$gm \cong \omega_C C_{gd}$
$\lambda \cong$ a constant that defines the division of channel resistance.

10.7.2 VARACTORS

In examining the varactor as a variable capacitance for use in circuit tuning, the total capacitance at the terminals of a varactor may be varied by adjusting the reverse bias from 0 Vdc to its zener breakdown voltage. The exact relationship is shown by

$$C = C_p + C_j = C_p + \frac{C_{jo}}{\left(1 + \frac{V_b}{\phi}\right)^n} \tag{10-67}$$

Where C_p = Capacitance of the package.
C_{jo} = Capacitance of the junction when V_b is zero.
V_b = Reverse bias in Vdc.
C_j = Capacitance of the junction at any value of V_b.
n = Exponent determined by junction characteristics. Equal to approximately 1/3 for diffused junctions and 1/2 for abrupt junction.
ϕ = Contact potential equal to 0.7 for silicon devices

If the junction capacitance is much larger than the package capacitance, Equation (10-67) becomes

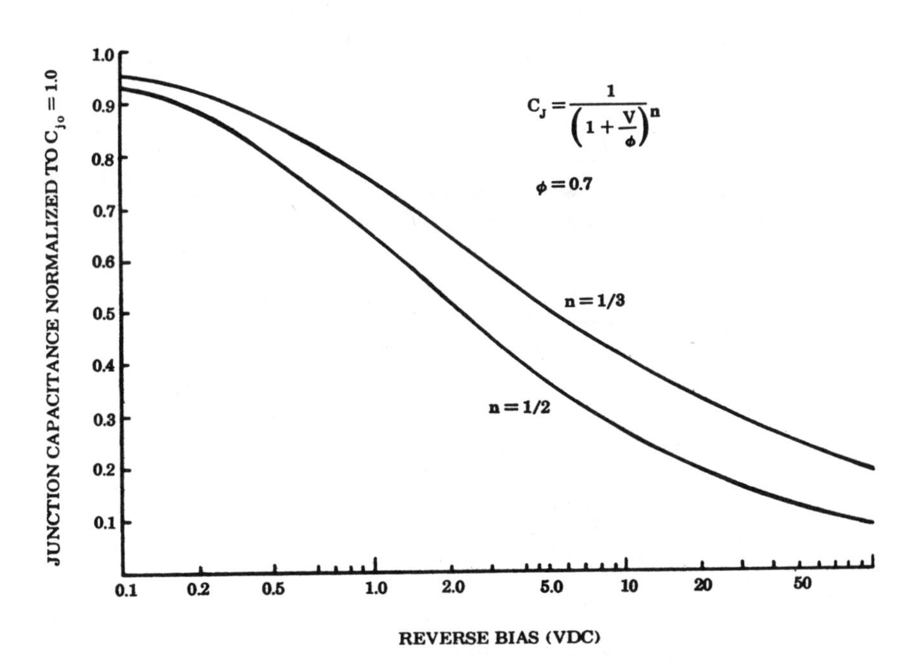

Fig. 10-42 Junction Capacitance vs Reverse Bias.

$$C \approx C_j = \frac{C_{jo}}{\left(1 + \frac{V_b}{\phi}\right)^n} \qquad (10\text{-}67a)$$

This equation is shown in Figure 10-42 for different values of n, representing a typical abrupt junction and a diffused junction varactor diode. This figure illustrates that for positive values of reverse bias, a voltage of 65 Vdc is required to cause the capacitance of the abrupt junction varactor to become one tenth of its value at zero bias. A voltage of 80 Vdc is required to cause the diffused junction varactor to become one fifth its value at zero bias. Any capacitance in parallel or series with the junction capacitance decreases the capacitance change ratio. Unfortunately, at microwave frequencies shunt capacitances may become appreciable. The package capacitance can be reduced to 25% or less by careful design. In cases where other parallel capacitances becomes too large, the varactor may be placed in series with an inductance large enough so that the junction capacitance change is regarded by the circuit as an inductive change.

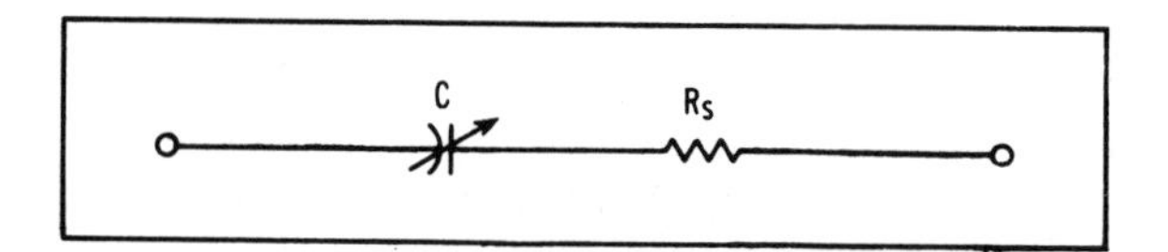

Fig. 10-43 Equivalent Circuit of a Varactor Diode.

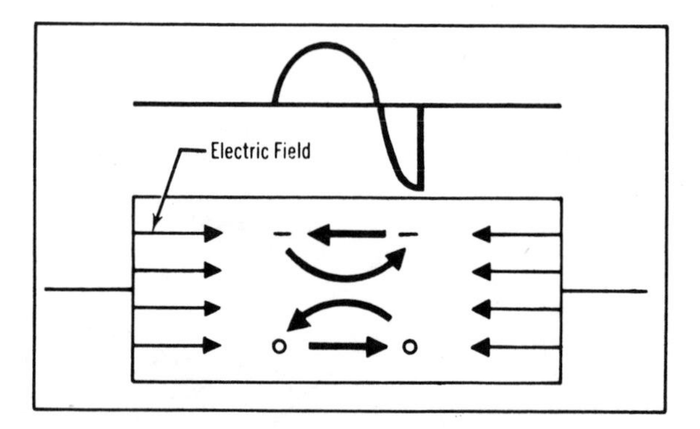

Fig. 10-44 When a forward voltage is applied, carriers are injected across the junction. However, before they can combine and result in a dc current flow, the applied voltage reverses and the carriers are returned to the point of origin in a bunch. This results in an abrupt cessation of reverse current and the waveform is rich in harmonics.

A varactor may be used for tuning, as long as the following requirements are met:

1. The effect of circuit capacitance is overcome either by the method outlined above, or by keeping it substantially lower than the junction capacitance C_j.

2. The Q of the circuit need not be extremely high. It should be noted that Q diminishes with increasing frequency, as shown by

$$Q = \frac{f_{co}}{f} \tag{10-68}$$

where

$$f = \text{frequency of operation}$$

$$f_{co} = \text{cutoff frequency} = \frac{1}{2\pi R_s C_j} \tag{10-69}$$

where

$$R_s = \text{series resistance}$$

3. The ac signal swing must be small compared to the reverse bias, otherwise it will cause the varactor capacitance to change enough to significantly detune the circuit. If the ac peak-to-peak voltage is more than 2 volts, the minimum reverse bias should be about 10 vdc. These factors are relative to the total circuit Q, and to whether the circuit being tuned is an oscillator, filter, or amplifier.

4. A dc voltage source capable of supplying adequate magnitudes of voltage must be available. A typical 28-volt supply for example, will give only about a 6.2- to -1 or 3.5- to -1 capacitance ratio, depending on the type of junction, and less when constant capacitances must be considered.

5. Total capacitance variation must be enough to accomplish the desired tuning. Assuming the resonating inductance to be constant, the frequency-capacitance relationship is

$$f_o = \frac{K}{\sqrt{C}} \tag{10-70}$$

where

$$K = \frac{1}{2\pi\sqrt{L}}$$

This results in a frequency versus varactor bias voltage of

$$f = m\left(\frac{V_b + \phi}{\phi}\right)^{n/2} \tag{10-71}$$

As can be seen from Figure 10-46, it takes a large reverse bias change to accomplish a relatively small frequency change even assuming that no external capacitance is affecting the varactor. Figure 10-47 shows the voltage bias versus frequency of a typical oscillator.

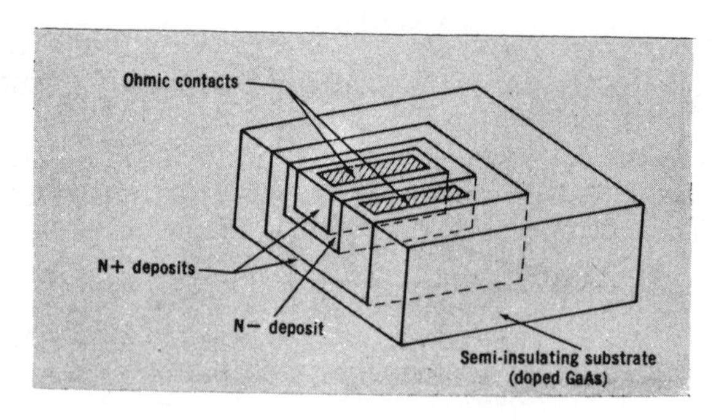

Fig. 10-45 Structure and Deposits of a Grown Planar Gunn Oscillator.

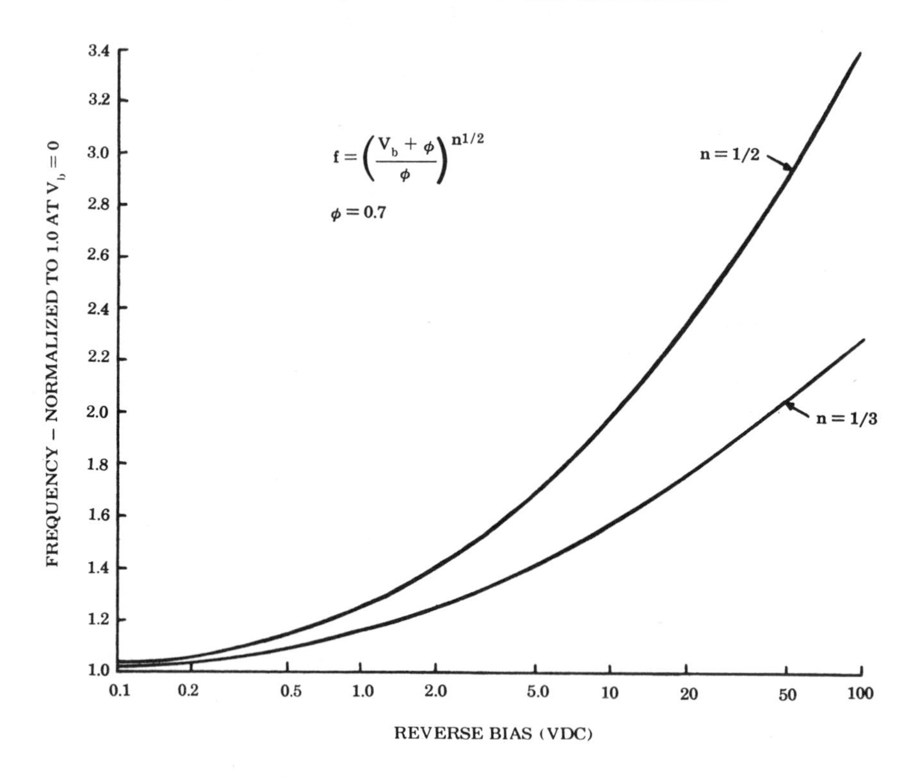

Fig. 10-46 Reverse Bias vs Frequency in a Varactor.

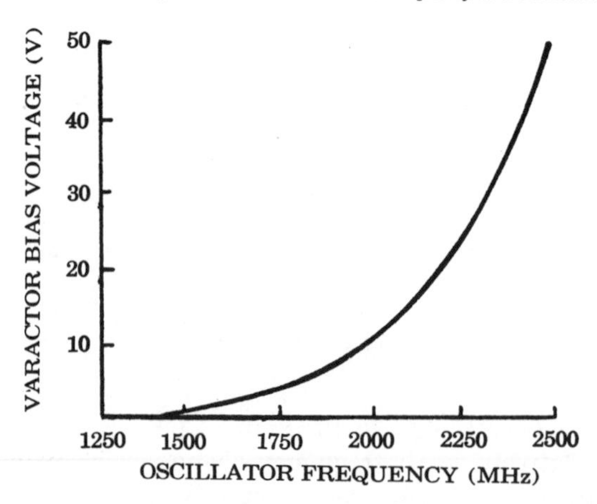

Fig. 10-47 Varactor Bias vs Frequency.

10.7.3 THE TUNNEL DIODE

The tunnel diode offers a unique physical mechanism for semiconductor operation, and at the same time offers a unique set of electrical characteristics. In comparison with transistors, the tunnel diode offers advantages of extremely high frequency operation, low noise, small size, low operating power levels, and high reliability.

The tunnel diode is a two-terminal device consisting of single PN junction. The essential difference between it and a converted diode is that the conductivity of the P and N material is more than 1000 times as high as the conductivity of the material used in conventional diodes.

Because of the very high conductivity of the P and N materials, the width of the junction (the depletion layer) is very small — of the order of 10^{-6} inches. Therefore, it is possible for electrons to tunnel through the junction, even though they do not have enough energy to surmount the potential barrier of the junction. Although tunneling is impossible in terms of classical physics, it can be explained in terms of quantum mechanics. For this reason the mechanism is commonly called quantum mechanical tunneling.

Under conditions of reverse bias of a conventional diode, there are no free holes or electrons to conduct charge across the junction. In the tunnel diode, however, a small reverse bias will cause the valence electrons of the semiconductor atoms near the junction to tunnel across the junction into the N region and, thus, the tunnel diode will conduct under reverse bias.

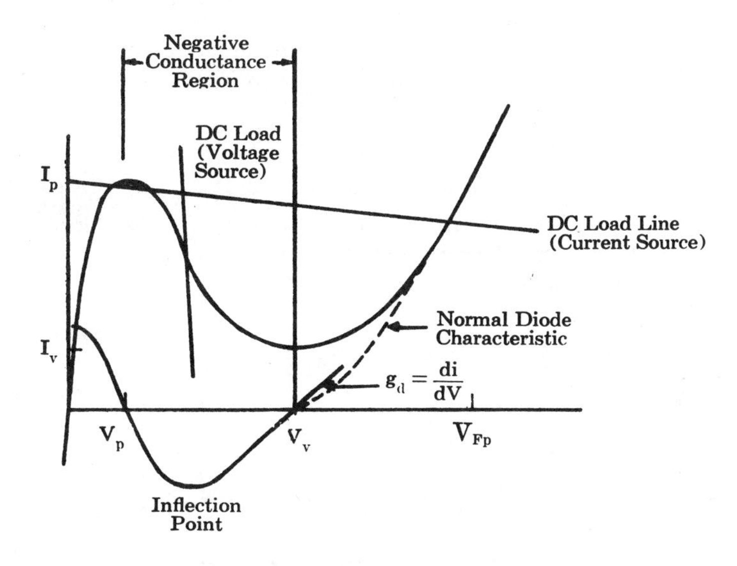

Fig. 10-48 Tunnel Diode V-I Characteristics

In the tunnel diode, a small forward bias will cause electrons in the N region to tunnel across the junction to the P region (appearing as valence electrons in the semiconductor atoms), and the tunnel diode will conduct, as shown in Figure 10-48. If the forward bias is increased, the energy of the free electrons in the N regions will become greater than the energy of the valence electrons in the P region, and consequently the tunneling current will decrease. This decrease in tunneling current with increasing forward bias, causes the negative conductance characteristic which is typical of the tunnel diode. As the forward bias is increased further, the free holes and electrons will have enough energy to flow over the potential barrier of the junction, in a manner identical to that of a conventional diode.

Quantum mechanical tunneling has a theoretical frequency limit of 10^7 MHz, and is inherently a much higher frequency mechanism than the drift and diffusion mechanisms involved in the operation of conventional diodes and transistors. In practice, the frequency limitation of the tunnel diode is determined by parasitic capacity, inductance, and resistance of the device.

In the characteristics presented graphically in Figure 10-48, the portion of the static curve from $V = 0$ to $V = V_p$ exhibits a low ac positive resistance. The portion from $V = V_p$ to $V = V_v$ is the negative resistance region as the current decreases with increasing voltage. The portion $V > V_v$ is characteristic of normal diode action. Also shown in Figure 10-48 is a plot of the conductance, $g_d = \frac{di}{dv}$. The voltage at which $\frac{di}{dv}$ becomes negative is called the inflection point. Other parameters are defined as follows:

I_p = the peak current before negative conductance begins

V_p = the voltage at which I_p occurs

I_v = the minimum current at the end of the negative conductance region

V_v = the voltage at which I_v occurs

V_{Fp} = the voltage in the conventional diode characteristic region at which the current is equal in value to I_p.

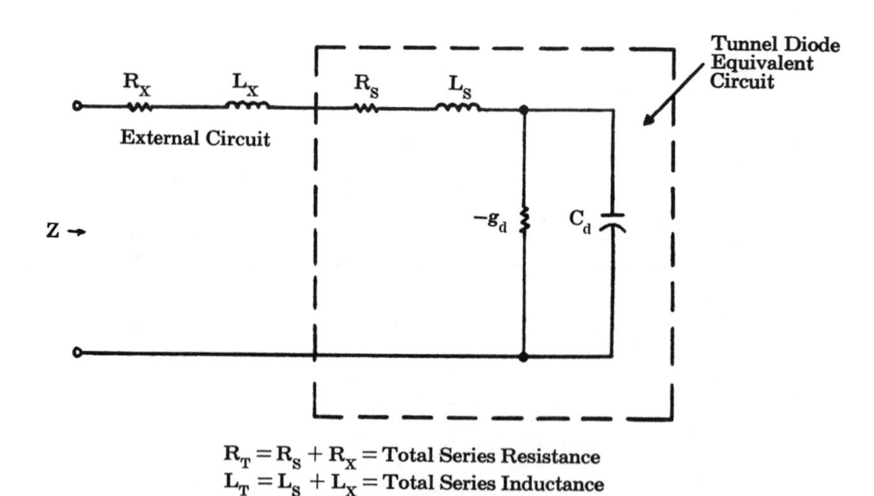

Fig. 10-49 The Tunnel Diode Equivalent Circuit and External Circuit.

Tunnel diodes are commonly constructed from two materials, germanium and gallium arsenide, and both exhibit the general characteristics shown in Figure 10-48.

The negative conductance region of the tunnel diode has many applications. It may be used for amplification, oscillation, frequency conversion, and switching. To be used as an active device, the tunnel diode must be biased into its negative conduction region with a constant voltage source. A constant current source bias is required in switching applications.

The operational stability of a tunnel diode circuit is determined by the total series resistance, $R_T = R_x + R_s$, as shown by Figure 10-49.

10.7.3.1 Tunnel Diode Amplifier

A positive conductance, by definition, dissipates energy. It follows, therefore, that a negative conductance will supply energy. This is the basis for negative resistance amplifiers.

Upon examination of the tunnel diode VI characteristics (Figure 10-48), it becomes evident that for amplifier operation, the "operating point" must be in the negative conductance region. Furthermore, to secure a stable operating point, the bias must be from a voltage source. The location of this operating point will depend on the magnitude of the anticipated signal swing, the required signal-to-noise ratio, and the operating temperature range. In most cases it is required to bias the tunnel diode at the point of maximum negative conductance.

The greatest bias problem is that the negative conductance region is non-linear; thus, it is necessary to match the diode conductance closely to the circuit conductance if high gain is to be achieved. Slight variations in the bias point with consequent variations in diode conductance, can cause large changes in circuit gain. Consequently, it is very important to ensure a very stable bias voltage. Some possible methods for obtaining stable voltage from low impedance sources are:

(a) the use of mercury cells; (b) the use of forward bias diodes as voltage regulators; and (c) the use of zener diodes as voltage regulators.

In oscillators where matching is not required, it is important that, at the lowest operating temperature, the device is driven from a voltage source. Oscillators have been operated successfully over a wide temperature range from —269°C to 300°C. But in amplifiers, matching must be maintained over the entire operating temperature range. Stable amplification may be accomplished by negative feedback, direct temperature compensation, or by deliberately making the bias network temperature sensitive.

The tunnel diode will not amplify above its resistive cutoff frequency (f_{ro}), and it should be noted that in a physical circuit, external components contribute to f_{ro}. In a transistor package, the frequency limit is about 1 GHz, due primarily to lead inductance, but with microstrip and microwave packaging this figure may be increased by an order of magnitude or more.

The tunnel diode may be placed in a parallel circuit to achieve current gain or it may be placed in a series circuit to obtain a voltage gain. It can be seen from Figure 10-50 that the voltage gain of the paralell circuit is unity. Current gain is accomplished when $g_g + g_l = -g_d$; the current in the $g_g + g_l$ branch increases with voltage at the same rate as the current in the $-g_d$ branch decreases with voltage. The net result is that a large current variation in the $g_g + g_l$ branch is achieved with a very small cur-

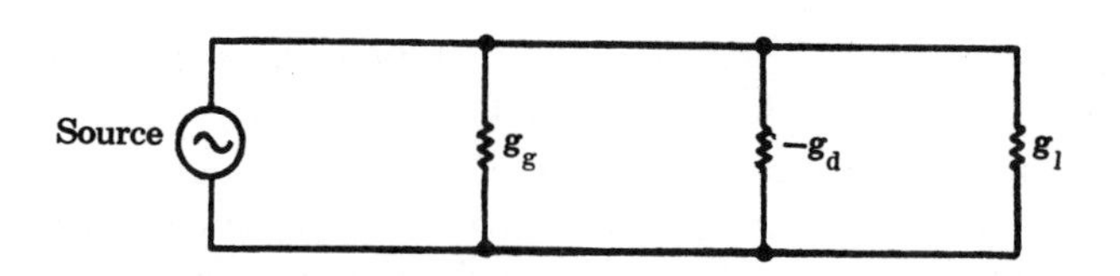

Fig. 10-50 Tunnel Diode Parallel Connection Amplifier Simplified.

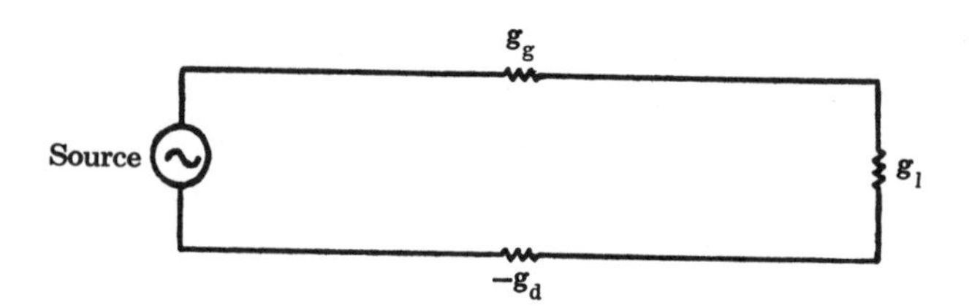

Fig. 10-51 Tunnel Diode Series Connection Amplifier.

rent variation in the source with a current gain of up to 30 dB. In the series connection shown in Figure 10-51, the current gain is unity.

Successful linear operation of the tunnel diode amplifier depends on the stability of the complete system, including in particular, the internal impedance of the bias supply and the signal source impedance. The basic amplifier circuit can be reduced to that shown in Figure 10-52 where $R_T = R_g + R_l + R_s$, $L_T = L_s + L_l$ is the total circuit inductance, and $-g_d$ is the negative conductance of the diode at the operating current and voltage.

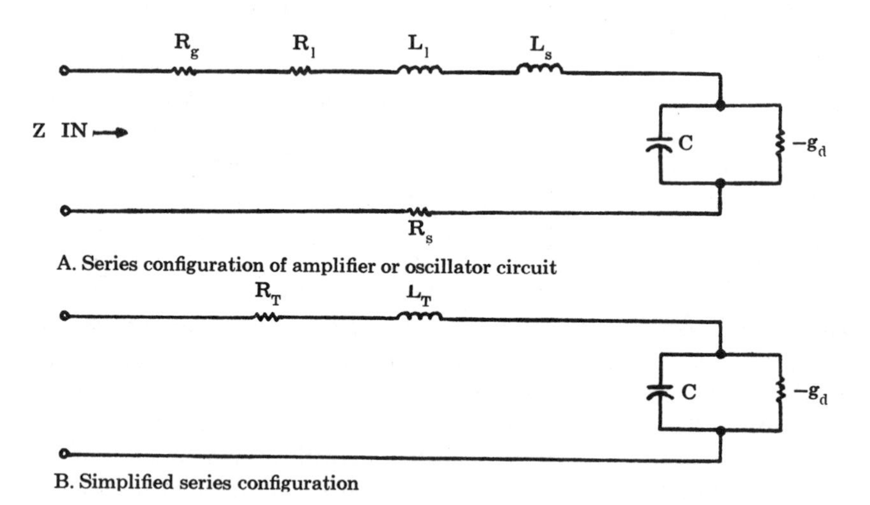

Fig. 10-52 Simplified Schematic for a Tunnel Diode Amplifier or Oscillator.

To determine the system stability, one can examine the distribution of the poles and zeros of the circuit determinant in the complex S-plane. If the zeros seen at the input fall in the right-half side of the S-plane, the system is unstable. Conversely, zeros in the left-half of the S-plane indicate a stable system. The criteria for circuit stability may be summarized as follows:

(a) The circuit inductance L_T must be less than $\frac{R_T C}{-g_d}$

(b) The sum of the positive conductances must be nearly equal to, but always greater than the negative conductance of the diode.

(c) The bias must be supplied by a stable voltage source (very low internal impedance).

(d) All of the above requirements must remain satisfied over the range of the supply voltages and temperature conditions.

10.7.3.2 The Tunnel Diode Oscillator

To operate as an oscillator, the total series resistance (R_T), of the tunnel diode circuit must be less than $\frac{L_T |-g_d|}{C}$, where L_T is the total series inductance, $-g_d$ is the negative conductance of the tunnel diode, and C is the capacitance of the tunnel diode. Oscillators are divided into two groups; the relaxation oscillator and the sinusoidal oscillator. The distinction between these two groups is that sinusoidal oscillators just barely satisfy the criterion for supplying the losses. Relaxation oscillators traverse large loops about the static characteristic (Figure 10-53).

The desirable features of tunnel diode oscillators include high frequency capability, low power consumption, good frequency stability when properly compensated, low noise, and circuit simplicity. The primary disadvantage is low output power and/or high harmonic content.

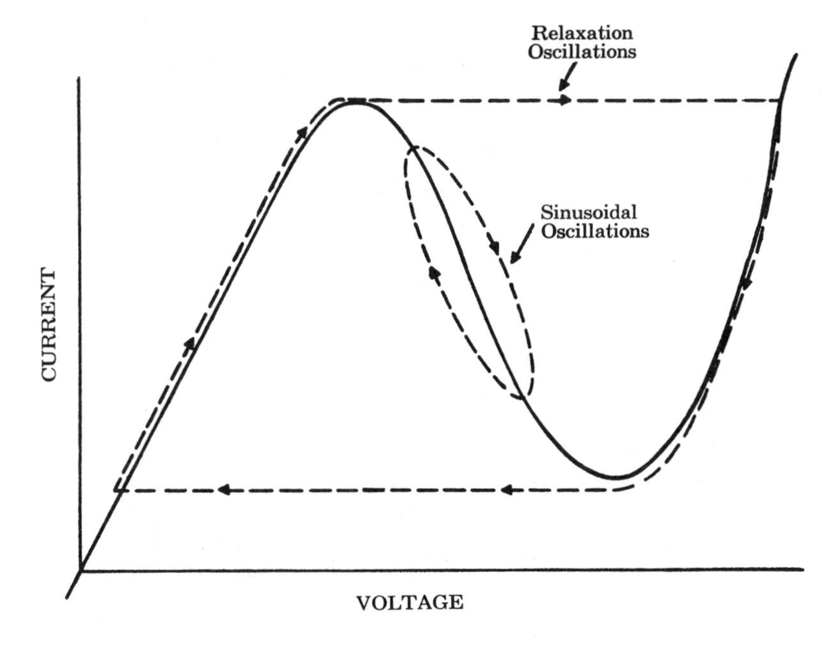

Fig. 10-53 Oscillator Limit Cycles.

10.7.3.2.1 Operating Parameters

The frequency limit of the circuit is determined by the resistive cutoff frequency, f_{ro}, of the device. At microwave frequencies two mode of oscillations are possible. They are above self-resonant frequency, f_{xo}, operation and below self-resonant frequency operation. If the resistive cutoff frequency is less than the self-resonant frequency, only one mode of operation is possible.

The equation for the self-resonant and resistive cutoff frequencies, f_{xo} and f_{ro}, respectively, are:

$$f_{xo} = \frac{1}{2\pi}\sqrt{\frac{1}{L_s\, C_d} - \frac{|-g_d|^2}{C_d^2}} \tag{10-72}$$

$$f_{ro} = \frac{|-g_d|}{2\pi C_d}\sqrt{\frac{1}{|-g_d|\, R_s} - 1} \tag{10-73}$$

Figure 10-54 shows the Nyquist impedance plots of two tunnel diodes, D1 and D2. Diode D2 reaches its resistive cutoff frequency before it reaches its self-resonant frequency; therefore, only one mode of oscillation is possible. Diode D2, on the other hand, may oscillate in the area between where the plot crosses the R axis ($f = f_{xo}$) and where it crosses the $j\omega$ axis ($f = f_{ro}$).

The frequency of operation of the series parallel circuit shown in Figure 10-55 is

$$fo = \frac{1}{2\pi}\sqrt{\frac{1}{L_T(C + C_1)} - \frac{|-g_d|^2}{C_1\,(C + C_1)}} \tag{10-74}$$

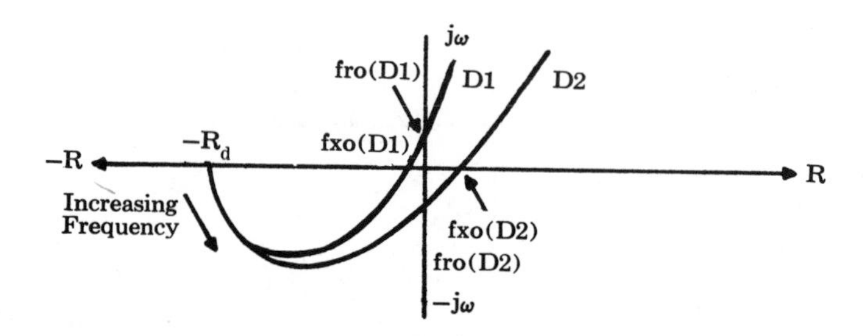

Fig. 10-54 Nyquist Plot of Tunnel Diode Impedance vs Frenquency for Two Tunnel Diodes D1 and D2.

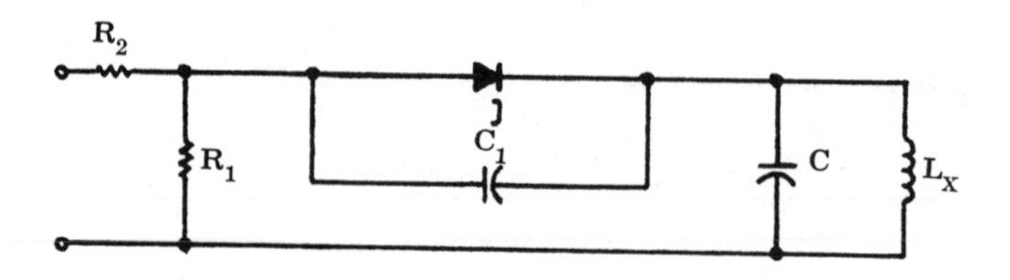

Fig. 10-55 Series–Parallel Circuit for a Tunnel Diode Oscillator.

It is dependent directly on the junction capacitance, C_1, which is in turn dependent directly on temperature and bias. For this reason, the bias and temperature considerations must be closely controlled.

Several approaches have been used to stabilize the frequency. One method is to make the frequency of oscillation relatively independent of junction capacitance. This can be achieved by weakly coupling the tunnel diode to a high Q resonant circuit, by tapping the diode into the circuit at a low voltage point on a capacitive voltage divider. If wideband tuning is desired, one capacitor of the voltage divider can be a varactor.

The frequency of operation of the tunnel diode oscillator changes in sharp steps as the bias voltage is varied, making it necessary to maintain a very constant bias.

10.7.3.3 The Tunnel Diode Converter

The following simultaneous functions must be performed by a single tunnel diode when it is used as a high gain self-oscillating converter:

(a) oscillation at the LO frequency

(b) amplification at the RF frequency

(c) mixing due to non-linearity

(d) amplification at the IF frequency.

On a mathematical basis:

The imaginary part of the external circuit admittance across the negative conductance of the diode should have zeros at the local oscillator and IF frequencies.

The real term of the external circuit admittance, Y, across $|-g_d|$ at the LO frequency must be smaller than the negative conductance of the diode.

The real part of the external admittance across $|-g_d|$ must be larger than the magnitude of $|-g_d|$ at the IF frequency.

The imaginary part of the external circuit admittance is shown plotted against frequency in Figure 10-56. This condition satisfies (a) above. A possible tunnel diode converter circuit is shown in Figure 10-57.

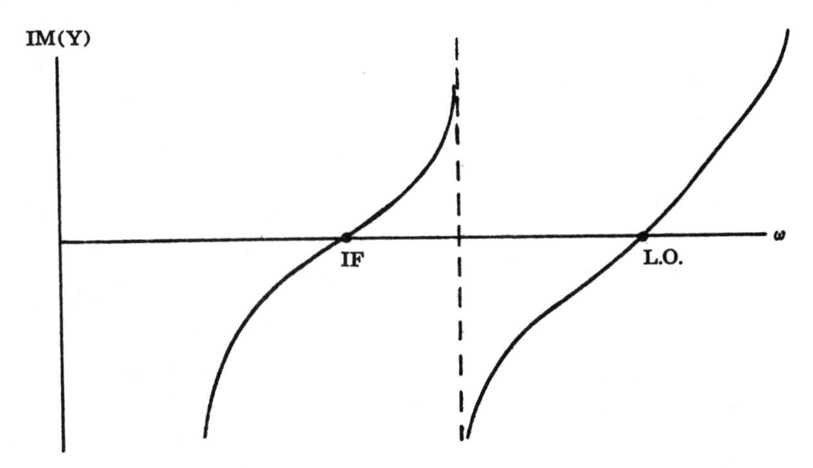

Fig. 10-56 Im(Y) as a Function of Frequency for Two Resonant Circuits.

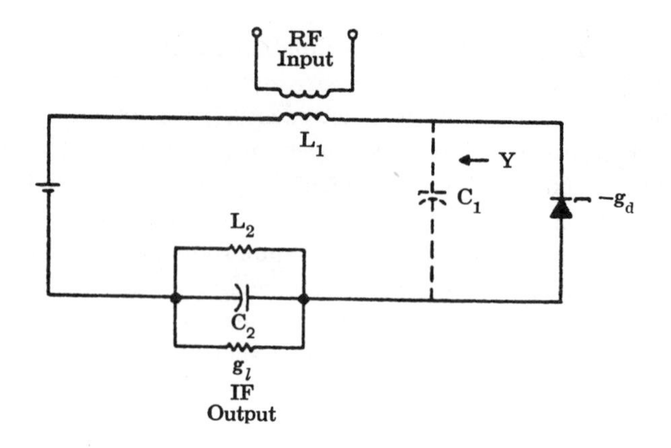

Fig. 10-57 Tunnel Diode Converter Circuit.

10.7.3.4 Tunnel Diode Noise

In the tunnel diode, the major contributor to noise is shot-noise. The noise figure equation is:

$$NF = 1 + \frac{20\, I_{DC}}{g_g} + \frac{T_1\, g_1}{T_g\, g_g} \qquad (10\text{-}75)$$

Where

I_{DC} is DC bias current through the tunnel diode
g_g is the conductance of the generator
g_1 is the conductance of the load
T_1 is the effective noise temperature of the load

It is evident that g_g should be large and g_1 should be small, but for highest gain, $g_g + g_1 = |-g_d|$.

Therefore, g_g should be nearly equal to $|-g_d|$.

10.7.4 AVALANCHE DIODES

In 1958, Read predicted microwave oscillations from avalanche diodes. At that time, special doping gradients were recommended in order to achieve a highly localized avalanche zone at one side of a high-field depletion region in a semiconductor diode. Since that time, a great amount of effort has been devoted to the development of solid-state microwave generators. In 1965, the generation of substantial power was obtained by avalanching silicon computer diodes. Later, high CW operation with good efficiency was obtained with GaAs varactor diodes and silicon devices.

A detailed discussion of how an avalanche diode generates power will not be given here; however, it is pointed out that avalanche effects, induced by reverse-bias, breakdown, and a transit time delay of carriers across the junction combine to yield negative resistance and gain.

Oscillations have been observed in p++-n+-n-n++ diodes over a wide frequency range extending to 8 GHz where transit time effects may have been present. However, the absence of a lower cutoff frequency showed that transit time was not essential for the oscillations. Instead, they could be attributed to a negative resistance which begins at a critical current density corresponding to carrier concentrations slightly smaller than the impurity concentration in the n+ zone. Therefore, space-charge effects must be considered. Inductance (related to the avalanche process), capacitance (related to the depletion layer), and negative conductance and a noise generator is shown in Figure 10-58. It has been established that a satisfactory small-signal equivalent circuit for the p-n junction avalanche diode is a parallel combination of the above elements. Probably, the most serious disadvantage of avalanche diodes is a high noise level.

The frequency of operation is dependent on the dc current at the avalanche condition. The characteristic of the avalanche diode, illustrated in Figure 10-59, shows that the current is controlled by limiting rather than by the dc bias voltage. Therefore, currents smaller than the limit current, and the resulting frequency shifts, are possible.

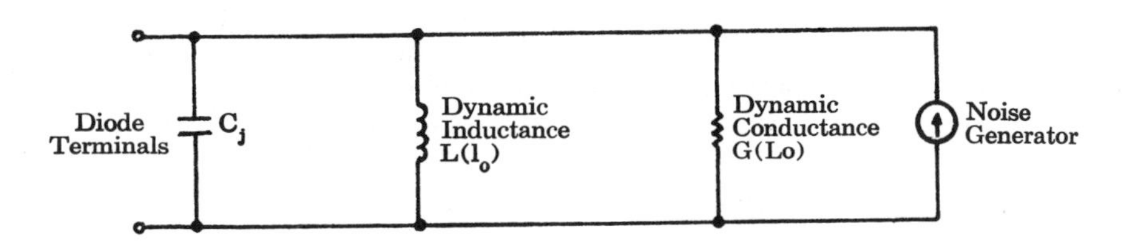

Fig. 10-58 Equivalent Circuit for the Avalanche Diode.

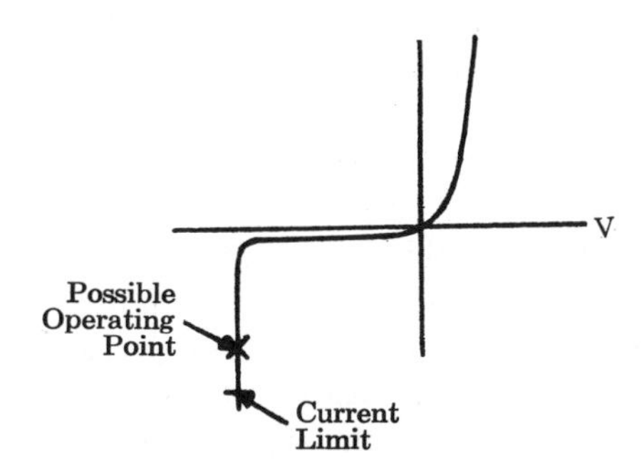

Fig. 10-59 V-I Characteristics of an Avalanche Diode.

The negative conductance of an avalanche diode is apparent over a wide frequency range; therefore, it can oscillate at several frequencies simultaneously over a wide range. The output at a specific frequency is dependent upon the tuning of the external circuit. Even the most sophisticated filtering techniques are generally unsatisfactory because several fundamentals are present, and harmonic and parametric oscillations may exist as well. To determine accurately the power content of any particular frequency component, it is essential to use signal substitution methods.

10.7.4.1 Noise Considerations

As indicated previously, noise is the major disadvantage of the avalanche process. Studies of avalanche noise in commercially available avalanche diodes reveal large noise variations from diode to diode. The spectral noise density passes through a sequence of maxima and minima with increasing avalanche current.

The maxima and minima are associated with microplasma especially at lower currents (100μA or less). A microplasma is a localized region of avalanche breakdown. At small magnitudes of current, e.g., of the order of 30μA, the avalanche discharge of the microplasma is unstable and switches on and off. As the current increases, the discharge becomes stable, the on-off switching ceases, and the spectral noise density decreases. Several microplasma may exist at different levels of current, so that as one becomes stable another may become unstable. When all the microplasma are stable (or inactive) the spectral noise density falls off sharply until the noise level of stable discharge is reached.

Microplasma-free uniform avalanche diodes have been prepared by the guard ring technique. The guard ring is a region of high breakdown voltage around the active area of the device. It must have a lower impurity gradient at the junction and a sufficiently large radius of curvature, so that the central region breaks down before the surface of the ring.

The spectral noise density has three contributors:

(1) $\frac{1}{\sqrt{I}}$ Noise, where I is the diode current

(2) Excess noise at low current densities

(3) Excess noise at high current densities

The $1/\sqrt{I}$ noise is independent of temperature from —196°C to 200°C. The series resistance (R_s) is, in the first approximation, independent of current; however, there is an increase in R_s at high currents. It is an inverse linear function of the breakdown diameter when plotted on log-log paper.

The excess noise at the higher current densities appears to be associated with thermal effects. Therefore, this excess noise can be reduced by improving heat dissipation methods. Experimental studies indicate that the spectral density is not white, but increases with decreasing frequency.

The excess noise at low current densities seems to be associated with non-uniformities of the avalanche breakdown. The non-uniformities causing the noise peaks consist of small variations of V_b which are likely caused by small local variations of the doping concentration in the starting material.

10.8 MOS COMPARED TO BIPOLAR CIRCUIT DESIGN

If circuit speed is a consideration, then bipolar design may be required, since MOS has an upper frequency limit which is approximately one quarter of the frequency capability of bipolar design. Consequently, when high speeds are required as in computers, for example, bipolar circuits may be the only possible choice.

MOS circuits have a significant advantage from the viewpoint of power dissipation. They operate in the nW/gate region. Bipolar structures have generally a lower limit of 1 m/gate. Therefore, if low power or low voltage operation is the main consideration, then MOS design is the proper choice.

The manufacture and use of MOS circuits has expanded extensively in recent years. Those advocating MOS design claim that lower system cost results from the use of smaller chips for an equivalent function. Smaller chips have accompanying lower material costs and improved yields by minimizing circuit loss to randomly distributed wafer defects. Lower cost of MOS circuits also results from the fewer process cycles required in MOS fabrication. The greater MOS chip density also appears attractive for LSI design.

With bipolar design, however, the required circuit complexity can generally be attained. Bipolar circuits are also fabricated on large production runs in established lines, and this feature tends to offset the advantage gained from the lower cost due to fewer process cycles in MOS fabrication. For LSI custom design, moreover, close calculations are required to determine if MOS chip costs are, in reality, lower in cost than the greater number of mass-produced, simpler bipolar functions. Also, with a large portion of the total circuitry on one MOS chip, maintenance costs are increased. The mixing of MOS and bipolar circuits to gain the advantages of both, may be feasible in specific designs.

REFERENCES

1. Control of Diffusion Induced Dislocations in Phosphorus Diffused Silicon, R. A. McDonald, G. G. Ehlenberger, T. F. Huffman. *Solid-State Electronics* v 9 n 8 Aug 1966 p 807-12.

2. Defect Structure on Silicon Surfaces after Thermal Oxidation, A. W. Fisher, J. A. Amick. *Electrochem Soc—J* v 113 n 10 Oct 1966 p 1054-60.

3. Effect of Dry Oxygen on Conductivity of Fluoridated Silicon Surfaces, M. Yamin, R. Lieberman. *Electrochem Soc—J* v 113 n 7 July 1966 p 720-3.

4. Infrared Interference Spectra Observed in Silicon Epitaxial Wafers, K. Sato, Y. Ishikawa, K. Sugawara. *Solid-State Electronics* v 9 n 8 Aug 1966 p 771-81.

5. Kinetics of Thermal Growth of Silicon Dioxide Films in Wafer Vapor-Oxygen-Argon Mixtures, T. Nakayama, F. C. Collins. *Electrochem Soc—J* v 113 n 7 July 1966 p 706-13.

6. Properties of Amorphous Silicon Nitride Films, S. M. Hu. *Electrochem Soc—J* v 113 n 7 July 1966 p 693-8.

7. Reactively Sputtered Silicon Dioxide Films, R. M. Valletta, J. A. Perri, J. Riseman. *Electrochem Technology* v 4 n 7-8 July Aug 1966 p 402-6.

8. Negative Photoconductivity Associated with Impurity Conduction in Germanium, E. A. Davis. *Solid-State Electronics* v 9 n 6 June 1966 p 605-8.

9. Microelectronic Engineering Volume I, Fabrication Technology, D. Abraham, Johns Hopkins University, 1966.

10. G. A. Lang and T. Stavish, "Chemical Polishing of Silicon with Anhydrous Hydrogen Chlorine," *RCA Review*, David Sarnoff Research Center, Princeton, N. J. (1963).

11. E. Tannenbaum, "Detailed Analysis of Thin Phosphorous Diffused Layers in P Type Silicon," *Solid State Electronics*, Vol. 2, pp. 123-132 (1961).

12. Integrated Silicon Device Technology, Vol. VII, Oxidation Research Triangle Institute, June 1965.

13. S. R. Hofstein, "Stabilization of MOS Devices", *Solid State Electronics* 10, 657 (1967).

14. G. O. Thomas, D. Kahng, and R. C. Manz, "Impurity Distribution in Epitaxial Silicon Films," *J. Electrochem. Soc.*, 109, p. 1055 (1962).

15. J. A. Brownson, "Three Point Probe Method for Electrical Characterization of Epitaxial Films," *J. Electrochem. Soc.*, 111, p. 919 (1962).

16. F. A. Trumbore, "Solid Solubilities of Impurity Elements in Germanium and Silicon," *Bell System Tech. J.*, pp. 205-233 (Jan. 1960).

17. Stern, A. P., ed.: Special Issue on Integrated Electronics, *Proc. I.E.E.E.* 52 December 1964, p. 1395.

18. "*Principles of Solid State Microelectronics,*" S. N. Levine, Holt, Rinehart, and Winston, Inc., New York, 1963.

19. "*Introduction to Integrated Semiconductor Circuits,*" Adi J. Khambata, John Wiley and Sons, 1963.

20. "Integrated Circuits — Technical Review and Business Analysis," Harvard Graduate School, Purnell Co., Boston, 1963.

21. M. P. Albert and J. F. Combs, ''Thickness Measurement of Epitaxial Films by the Infrared Interference Method,'' *J. Electrochem. Soc.*, 109, p. 713 (1962).

22. W. Shockley, ''A Unipolar Field-Effect Transistor,'' *Proc. I.R.E.*, Vol. 40, pp. 1365-1376 (November 1962).

23. C. S. Fuller, ''*Transistor Technology,*'' edited by F. J. Biondi, Vol. 3, Chapter 3 (D. Van Nostrand Co., 1958).

24. C. J. Forsch and L. Derick, ''Surface Protection and Selective Masking During Diffusion in Silicon,'' *J. Electrochem. Soc.*, pp. 547-552 (Sept. 1957).

25. Y. Tauri and S. Denda, Investigation of Resistor and Capacitor Elements in a Germanium Solid Circuit. *Solid State Electron.* 5, 49 (1962).

26. J. Tauc, ''*Photo and Thermoelectric Effects in Semiconductors,*'' pp. 59-144 (Pergamon Press, New York, 1962).

27. Galvanomagnetic Properties of Recrystallized Dendritic InSb Films, H. H. Wieder. *Solid-State Electronics* v 9 n 5 May 1966 p 373-82.

28. Preparation of High Mobility InSb Thin Films, J. A. Carroll, J. F. Spivak. *Solid-State Electronics* v 9 n 5 May 1966 p 383-7.

29. Effect of Junction Curvature on Breakdown Voltage in Semiconductors, S. M. Sze, G. Gibbons. *Solid-State Electronics* v 9 n 9 Sept 1966 p 831-45.

30. Microplasma Observations in Silicon Junctions Using Scanning Electron Beam, J. W. Gaylord. *Electrochem Soc—J* v 113 n 7 July 1966 p 753-4.

31. Surface Effects on p-n Junction – Characteristics of Surface Space-Charge Regions under Non-Equilibrium Conditions, A. S. Grove, D. J. Fitzgerald. *Solid-State Electronics* v 9 n 8 Aug 1966 p 783-806.

32. Effect of Pressure on Gunn Phenomena in Gallium Arsenide, M. P. Wasse, J. Lees, G. King. *Solid-State Electronics* v 9 n 6 June 1966 p 601-4.

33. Ga-As-Si Ternary Phase System, M. B. Panish. *Electrochem Soc—J* v 113 n 11 Nov 1966 p 1226-8.

34. D. Baker and J. R. Tillman, ''The Preparation of Very Flat Surfaces of Silicon by Electropolishing,'' *Solid-State Electronics,* 6 (Nov.-Dec. 1963).

35. K. E. Bean and P. S. Gleim, ''Vapor Etching Prior to Epitaxial Deposition of Silicon,'' Fall Meeting of Electrochem. Soc., New York, N. Y. (Sept.-Oct. 1963).

36. P. D. Payne, ''Technology of Transistor Mask Fabrication,'' *Semiconductor Products,* 5 (May 1962).

37. Microelectronic Device Data Handbook Vol. I, ARINC Research Corp., July 1968.

38. J. J. Murray and R. Maurer, ''Arrays of Microphotographs for Microelectronic Components,'' *Semiconductor Products,* Vol. 5, No. 2, 30 (1962).

39. I. M. MacKintosh, *Proc. I.R.E.*, ''The Electrical Characteristics of Silicon P-N-P-N- Triodes,'' Vol. 46, No. 6, pp. 1229-1235 (June 1958).

40. R. H. Norman and J. R. Nall, Micrologic Elements. *Electron Reliab. Microminiat.* 1, 251 (1962)

41. C. A. Lee, ''*Transistor Technology,*'' pp. 276-286 (Van Nostrand, Inc., New York, 1958).

42. M. V. Sullivan, D. L. Klein, R. M. Finne, L. A. Pompliano, and G. A. Klab, ''An Electropolishing Technique for Germanium and Silicon,'' *J. Electrochem. Soc.*, 110 (May 1963).

43. A. J. Dekker, "*Solid State Physics*," pp. 316-365 (Prentice-Hall, New York, 1962).

44. John C. Irvin, "Resistivity of Bulk Silicon and of Diffused Layers in Silicon," *The Bell System Tech. J.* (Mar. 1962).

45. F. Barson, W. J. Armstrong, and W. E. Mutter, "The Degradation of Planar Junctions During Reoxidation," oral presentation at the fall meeting of the Electrochem. Soc. (Oct. 1963).

46. M. M. Attala and E. Tannenbaum, "Impurity Redistribution and Junction Formation in Silicon," *Bell System Tech. J.*, Vol. 39, pp. 933-946 (1960).

47. H. K. Dicken, "Parasitics in Integrated Circuits," *Electronics* 36 (July 5, 1963).

48. C. Forge, "Driving the PIN Diode Switch", *Microwaves*, Vol. 6, No. 4, April 1967, pp. 30-36.

49. A. B. Phillips, "*Transistor Engineering*," McGraw-Hill Book Co. (1962).

50. L. B. Valdes, "*The Physical Theory of Transistors*," McGraw-Hill Book Company (1961).

51. W. W. Gaertner, "*Transistors – Principles, Design, and Application*," D. Van Nostrand Company (1960).

52. R. W. Warner and J. N. Fordemwalt, Editors, "*Integrated Circuits*," McGraw-Hill Book Company (1965).

53. Influence of Crystallographic Defects on Device Performance, J. M. Fairfield, G. H. Schwuttke. *Electrochem Soc—J* v 113 n 11 Nov 1966 p 1229-31.

54. Optically-Coupled Linear Circuit Techniques, E. L. Bonin. IEEE—Northeast Electronics Research & Eng Meeting—*NEREM Rec* v 7 1965 p 84-5.

55. Quantum Efficiency of GaAs Electroluminescent Diodes, A. H. Herzog. *Solid-State Electronics* v 9 n 7 July 1966.

56. Gallium Arsenide Double Injection Diode, I. J. Saunders. *Instn Elec Engrs—* Conference Publ 12 1965 paper 13.

57. Field and Thermionic-Field Emission in Schottky Barriers, F. A. Padovani, R. *Stratton. Solid-State Electronics* v 9 n 7 July 1966 p 695-707.

58. Computer Solution for Steady-State Behavior of PN-Junction Diode, D. E. Fulkerson, A. Nussbaum. *Solid-State Electronics* v 9 n 7 July 1966 p 709-19.

59. H. B. Bell and J. C. Day, "Selective Area Plating of Semiconductor Device Surfaces," Fall Meeting of the Electrochemical Society, New York, N. Y. 1963.

60. I. M. Mackintosh, P. F. Schmidt, and M. W. Larkin, "Integrated Complementary Devices Fabricated by Electrochemical Techniques," *Proc. I.E.E.E.*, Vol. 52, No. 12 (Dec. 1964).

61. D. McWilliams, C. Fa, G. Larchain, and O. Maxwell, Jr., "A New Dielectric Isolation Technique for High Quality Silicon Integrated Circuits," Spring Meeting of the Electrochem. Soc., Toronto, Canada.

62. D. S. King, G. R. Madlund, and W. J. Corrigan, "Methods of Isolation of Active Elements in Integrated Circuits," I.E.E.E. Electron Devices Meeting, Washington, D. C. (October 1962).

63. Floyd P. Hunter, "*Handbook of Semiconductor Electronics*," Ch. 20 (McGraw-Hill Book Co., 1962).

64. Study of Solid-State Integrated Microwave Circuits, Texas Instrument, Incorporated, June 1967.

65. H. C. Theurer, ''Epitaxial Silicon Films by the Hydrogen Reduction of $SiCl_4$,'' *J. Electrochem. Soc.*, 108, p. 649 (1961).

66. P. J. Holmes, ''*The Electrochemistry of Semiconductors,*'' Academic Press, pp. 370-371 (1962).

67. S. K. Tung, ''*Metallurgy of Semiconductors,*'' Interscience, p. 100 (1961).

68. W. E. Rudge, W. E. Harding, and W. E. Mutter, ''Fly's Eye Lens Technique for Generating Semiconductor Device Fabrication Masks,'' IBM J. Research Develop., 7 (April 1963).

69. C. H. Klute, G. B. Wetzel, R. J. Anstead, and M. L. Jones, ''Photo-Engraving Procedures for Silicon Device Fabrication,'' *Direct Current, London* 7 (May 1962).

70. L. E. Miller, ''Uniformity of Junctions in Diffused Silicon Devices,'' *Properties of Elemental and Compound Semiconductors,* 5, Metallurgical Society Conferences, pp. 303-321 (1959).

71. A. Goetzberger, ''Impurity-Induced Pipes Through Diffused Layers in Silicon,'' *Solid-State Electronics,* 5 (Mar.-Apr. 1962).

72. A. E. Bakanowski and J. H. Forester, ''Electrical Properties of Gold-Doped Diffused Silicon Computer Diodes,'' *Bell System Tech. J.,* 39 (Jan. 1960).

73. C. J. Uhl, ''Diffusion Process for Reducing Switching Times in Diffused Silicon Transistors,'' *The Western Electric Engineer,* 7 (Jan. 1963).

74. G. L. Schnable, W. J. Hillegas, Jr., and C. G. Thornton, ''Preferential Silicon Epitaxy with Oxide Masking,'' Meeting of Electrochem. Soc., New York (Sept.-Oct. 1963).

75. A. F. McKelvey, ''The Use of Preferential Etching and Preferential Epitaxy in Microelectronics,'' Proc. Third Annual Microelectronics Symposium of St. Louis Section I.E.E.E., St. Louis, Mo. (April 1964).

76. F. E. Everhart, O. C. Wells, and R. K. Matta, ''Evaluation of Passivated Integrated Circuits Using the Scanning Electron Microscope,'' Fall Meeting of the Electrochemical Society, New York, N. Y. (Sept.-Oct. 1963).

77. Breakdown Voltage of Planar Silicon Junctions, O. Leistiko, Jr., A. S. Grove. *Solid-State Electronics* v 9 n 9 Sept 1966 p 847-52.

78. Static Errors in Hall Effect Multiplying Devices, J. Jaworski, M. Nalecz, I. Zawicki. *Solid-State Electronics* v 9 n 5 May 1966 p 515-25.

79. Plastic Semiconductors and Their Impact on Consumer Industry, J. S. MacDougall. Wescon Tech Papers v 10 pt 3 *Electron Devices & Packaging* n 18.3 1966, 3p.

80. Physics of Gunn Effect and Its Relevance to Devices, D. E. McCumber. IEEE—Northeast Electronics Research & Eng Meeting—*NEREM Rec* v 7 1965 p 76-7.

81. New Hall Generator Applications, F. Kuhrt. *Solid-State Electronics* v 9 n 5 May 1966 p 567-70.

82. P. J. Etter and B. L. H. Wilson, Inductance from a Field Effect Tetrode. *Proc. I.R.E.*, Aug. 1962.

83. A. E. Brewster, The Tunnel Diode as a Solid State Circuit. *F. Brit. I.R.E.*, Dec. 1961.

84. W. Glendinning et al., Silicon Epitaxial Microcircuit. *Proc. I.R.E. 49,* No. 6, p. 1087 (1961).

85. John W. Beck, ''Boiling Water Oxidation Rates in Silicon,'' *J. Appl. Phys.*, Vol. 33, No. 7 (July 1962).

86. L. A. D'Asaro, *Solid State Electronics,* ''Diffusion and Oxide Masking in Silicon by the Box Method,'' Vol. 1, pp. 3-12 (March 1960).

87. J. R. Ligenza, ''Oxidation of Silicon by High Pressure Steam,'' *J. Electrochem. Soc.*, Vol. 109, No. 2 Feb. 1962).

88. Hofstein, S. R., ''Stabilization of MOS Devices'', Solid State Electronics, 10, 657-670 (1968).

89.Bruce E. Deal, ''The Oxidation of Silicon in Dry Oxygen, Wet Oxygen and Steam,'' *J. Electrochem. Soc.*, Vol. 110, No. 6, pp. 527-532 (June 1963).

90. H. Edagawa, Y. Morita, and S. Maekawa, ''Growth and Structure of Si Oxide Films on Si Surface,'' *Jap. J. Appl. Phys.*, Vol. 2, No. 12, pp. 765-775 (Dec. 1963).

91. B. H. Claussen and M. Flower, ''An Investigation of the Optical Properties and the Growth of Oxide Films on Silicon,'' *J. Electrochem. Soc.*, Vol. 110, No. 9, pp. 983-987 (Sept. 1963).

92. N. Karube, K. Yamamoto, M. Kamiyama, ''Thermal Oxidation of Silicon,'' *Jap. J. Appl. Phys.*, Vol. 2, No. 1, pp. 11-18 (Jan. 1963).

93. ''Tom Hyltin of TI and his Microwave IC's'', *EEE* (Jan. 1967), 74-79.

94. S. W. Ing, R. E. Morrison, and J. E. Sandor, ''Gas Permeation Study and Imperfection Detection of Thermally Grown and Deposited Thin Silicon Dioxide Films,'' *J. Electrochem. Soc.*, Vol. 109, No. 3, pp. 221-225 (Mar. 1962).

95. Edward Keonjiam, ''*Microelectronics,*'' (McGraw-Hill Book Co., 1963).

96. C. J. Frosch and L. Derick, ''Diffusion Control in Silicon by Carrier Gas Composition,'' *J. Electrochem. Soc.*, Vol. 105, No. 12 (Dec. 1958).

97. Ertel, Alfred. ''Monolithic IC Techniques Produce First All-Silicon X-Band Switch'', *Electronics*, (1967), 76-91.

98. Kile, C., Jr. ''Semiconductor Integrated Circuits'', *SCP and Solid State Technology*, (1967), pp. 30-38.

99. C. Kittel, ''*Introduction to Solid State Physics,*'' pp. 270-283 (Wiley and Sons, Inc., 1956)

100. S. L. Miller, ''Ionization Rates for Holes and Electrons in Si,'' *Phys. Rev.*, Vol. 105, pp. 1246-1249 (February 1957).

101. I. M. MacKintosh, ''Three-Terminal pnpn Transistor Switches,'' *I.R.E. Trans. on Electron Devices*, Vol. 5, pp. 10-12 (January 1958).

102. C. H. Taylor, ''Semiconductor Junction Capacitors,'' *Wireless World*, p. 193 (April 1962).

103. W. Adcock and J. S. Walker, ''Semiconductor Networks,'' *Electron. Reliab. Microminiat.* Vol. 1, p. 81 (1962).

104. ''*Integrated Circuits,*'' R. M. Warner and J. N. Fordemwalt, McGraw-Hill Book Co., New York, 1965).

105. W. D. Fuller, ''Production Techniques for Integrated Electronics,'' Lockeed Missiles and Space Co. 6-50-63-1.

106. ''*Transistor Engineering and Introduction to Integrated Semiconductor Circuits,*'' by A. B. Phillips, McGraw-Hill Book Company, 1962.

107. ''Integrated Silicon Device Technology,'' Vol. III — Photoengraving. Tech. Doc. Rep. No. ASD-TDR-63-316 January 1964. Prepared by Research Triangle Institute,

Durham, N. C. DDC No. AD-603-715.

108. W. Shockley, "*Electrons and Holes in Semiconductors,*" pp. 179-182, (D. Van Nostrand Co., New York, 1950).

109. USAECOM Contract, DA 36-039, SC-90806, "High Efficiency Transistor Structures," Report No. 8, July 1962 – March 1964, Texas Instruments, Inc., Dallas, Texas, 1964.

110. A. J. Decker, "*Solid State Physics,*" pp. 144-145 (Prentice-Hall, New York, 1962).

111. T. A. Ivesdal, "Integrated Circuit Components," *Electronic Progress* (Raytheon Co.), Vol. 8, p. 12 (1964).

112. H. L. Caswell, *Semiconductor Products and Solid State Technology,* p. 23 (December 1963).

113. GaSb Prepared from Nonstoichiometric Melts, F. J. Reid, R. D. Baxter, S. E. Miller. *Electrochem Soc—J* v 113 n 7 July 1966 p 713-16.

114. Anomalous Attenuation of Piezoelectrically Active Ultrasonic Waves in Photoexcited Cadmium Sulfide, E. Harnik. *J Applied Physics* v 37 n 7 June 1966 p 2563-7.

115. Tin and Zinc Diffusion into Gallium Arsenide from Doped Silicon Dioxide Layers, W. von Muench. *Solid-State Electronics* v 9 n 6 June 1966 p 619-24.

116. Theory of Magnetaplasma Resonances in n-Type Silicon and Germanium—2 P. R. Wallace. *Can J Physics* v 44 n 10 Oct 1966 p 2495-2507.

117. Surface Space-Charge Barriers on Semiconducting Barium Titanate, J. Holt. *Solid-State Electronics* v 9 n 8 Aug 1966 p 813-18.

118. Surface Photopotential of Single Crystals of Cadmium Sulphide in Alternating Electric Field, Y. L. Yousef, A. Mishriky, H. Mikhail. *Brit J Applied Physics* v 17 n 10 Oct 1966 p 1285-91.

119. Semiconducting Properties of Several III_B-V_BVI_B Ternary Materials and Their Metallurgical Aspects. K. Kurata, T. Hirai.*Solid-State Electronics* v 9 n 6 June 1966 p 633-40.

120. Metallurgy in Semiconductor Industry, R. Grubel. *J of Metals* v 19 n 1 Jan 1967 p 13-17.

121. Impurity Conduction in Diffused Germanium and Silicon Layers, J. S. Blakemore, R. O. Olson, T. H. Herder. *Solid-State Electronics* v 9 n 7 July 1966 p 673-80.

122. Hardness of AuSbz, M. B. McNeil. *Electrochem Technology* v 4 n 7-8 July-Aug 1966 p 446-7.

123. Hall Effect and Related Phenomena, A. C. Beer. *Solid-State Electronics* v 9 n 5 May 1966 p 339-51.

124. Effect of Cobalt Substitution on Electrical Conduction in Nickel Ferrite, R. Parker, B. A. Griffiths, D. Elwell. *Brit J Applied Physics* v 17 n 10 Oct 1966 p 1269-76.

125. Application of Relaxation Method to Calculation of Potential Distribution of Hall Plates, S. Gruetzmann. *Solid-State Electronics* v 9 n 5 May 1966 p 409-16.

126. Inhomogeneities in Electrical Properties of Gallium Arsenide Crystals, J. C. Bruce, R. E. Hunt, G. D. King, H. C. Wright. *Solid-State Electronics* v 9 n 9 Sept 1966 p 853-7.

127. Low-Temperature Alloy Contacts to Gallium Arsenide Using Metal Halide Fluxes, B. Schwartz, J. C. Sarace. *Solid-State Electronics* v 9 n 9 Sept 1966 p 859-62.

128. Stable Domain Propagation in Gunn Effect, P. N. Butcher, W. Fawcett. *Brit J Applied Physics* v 17 n 11 Nov 1966 p 1425-32.

129. Theory of Gunn Effect, H. Kroemer. IEEE—Northeast Electronics Research & Eng Meeting—*NEREM Rec* v 7 1965 p 208-9.

130. Controlled Doping of Germanium Layers Made by Evaporation-Condensation Method, W. Haidinger, J. C. Courvoisier, P. J. W. Jochems, L. J. Tummers. *Solid-State Electronics* v 9 n 7 July 1966 p 689-93.

131. Diffusion of Arsenic in Germanium from Germanium Arsenide Source—Prediffusion, G. F. Foxhall, L. E. Miller. *Electrochem Soc—J* v 113 n 7 July 1966 p 698-701.

132. Goetzberger, A., V. Heine, and E. N. Nicollian, ''Surface States in Silicone from Changes in the Oxide Coating'', *Appl. Phys. Letters,* 12, 95-97 (1968).

133. Semiconductor Processing Applied to Integrated Circuit Fabrication, R. A. Cohen, etal, Massachusetts Institute of Technology, Aug. 1968.

134. J. T. Wallmark, W. A. Bosenberg, E. C. Ross, D. Flatley, and H. Parker, ''MOS Field-Effect Transistor Technology'', Final Report, Contract NAS 1-5794, National Aeronautics and Space Administration, Washington, D. C., Aug. 1967.

135. D. Chakroborty and R. Coackley, ''Characterization of Varactor Diodes at Low Temperatures'', *The Radio and Electronic Engineer,* Vol. 33, No. 2, Feb. 1967, pp. 97-104.

136. S. Hamilton and R. Hall, ''Shunt Mode Harmonic Generation Using Step Recovery Diodes'', *The Microwave Journal,* Vol. 8, No. 4, April 1967, pp. 69-78.

137. Birk, R. P., ''Multilayer MOS Devices with Sputtered Silicone Dioxide'', *Transactions of the Metallurgical Society of AIME,* 242, 523-526 (1968).

138. Schlegel, E. S., ''A Bibliography of Metal-Insulator-Semiconductor Studies'', *IEEE Transactions on Electron Devices,* Ed-14, 728-749 (1967).

139. J. F. Allison, F. P. Heiman, and J. R. Burns, ''Silicone-on-Sapphire Complementary MOS Memory Cells'', *IEEE J. Solid State Circuits* SC-2, 208 (1967).

140. H. A. Waggener, R. C. Kragness, and A. L. Tyler, ''Improved Etching Technique Permits More Efficient Fabrication of Integrated Circuits'', *International Electron Device Meeting,* Washington, D. C., Oct. 1967.

141. R. M. Finne and D. L. Klein, ''A Water-Amine-Complexing Agent System for Etching Silicone'', *J. Electrochem. Soc.* 114, 965 (1967).

142. F. P. Heiman, ''Donor Surface States and Bulk Acceptor Traps in Silicone-on-Sapphire Films'', *Appl. Phys. Lett.* 11, 132 (1967).

143. P. Daly, S. Knight, M. Coulton, and R. Ekholdt, ''Lumped Elements in Microwave Integrated Circuits, *G-MTT Symposium Digest Cat. No. 17C66,* 1967, pp. 139-141.

144. G. Vendelin, ''High Dielectric Substrates for Microwave Hybrid Integrated Circuitry'', *G-MTT Symposium Digest Cat. No. 17C66,* 1967, pp. 125-128.

145. H. Soboa, ''Extending IC Technology to Microwave Equipment'', *Electronics,* Vol. 40, No. 6, 20 March 1967, pp. 112-124.

146. Texas Instruments Incorporated, *Molecular Electronics for Radar Applications,* by T. M. Hyltin and L. R. Pfeifer, Jr., Interim Engineering Report No. 4, Contract AF 33 (615)-1993, BPSN 4-6399-415906, Meȓa No. 3 and S-674159-415906, Jan. 1967.

147. Snow, E. H. and B. E. Deal, ''Polarization Effects in Insulating Films on Silicone – a Review'', *Trans. Met. Soc. AIME,* 242, 512-523 (1968).

148. Frohman-Bentchkowsky, D. and M. Lenzlinger, ''Charge Transport and Storage in MNOS Structures'', *International Electron Devices Meeting,* Washington, D. C., October 1968, Paper No. 24. 7.

149. Brown, D. M. and P. V. Gray, ''Si-SiO_2 Fast Interface State Measurements'', *Journal Electrochem. Soc.*, 115, 760-766 (1968).

150. Schlegel, E. S., G. L. Schnable, R. F. Schwarz, and J. P. Spratt, ''Behavior of Surface Ions on Semiconductor Devices'', Dec. 1968 issue of *IEE Trans Electron Devices.*

151. The Application of Ferroelectric Materials in Optical Memories, G. W. Taylor, *WESCON* 1971.

152. Electrooptic Ceramic Materials, G. H. Haertling, *WESCON* 1971.

153. A Course on Optoelectronics, *The Electronic Engineer*, 1971.

154. Electrooptic Devices Using Strain-Biased PLZT Ferroelectric Ceramics, J. R. Maldonado, *WESCON* 1971.

155. Ceramic Electrooptic Properties and Devices, P.D. Tracher and C. E. Land, *WESCON* 1971.

PART III

MICROELECTRONIC CIRCUIT CONFIGURATIONS

CHAPTER 11

CIRCUIT CONFIGURATIONS

11.1 DESIGN RULES AND PHILOSOPHY

To enable the circuit designer to synthesize circuits with the application of microelectronic techniques, it is essential that he have a thorough knowledge of the fabrication processes and problems which are involved in the production of microelectronic circuitry. It is, of course, also essential that he have full understanding of the operating characteristics of the circuit components. Without such knowledge, the circuit designer is handicapped in devising optimum circuitry to meet a specific set of requirements. It is for this reason, that the chapters on fabrication and component characteristics, precede this chapter on circuit design. In this particular aspect, microelectronic design differs from conventional circuit design, using discrete components, in which little attention has to be paid to the processes for fabricating the discrete components.

It is particularly important to be aware that microelectronics is not merely a matter of buying a series of black boxes in IC form, and then interconnecting them. Thus, successful microelectronics utilization cannot be achieved by subdividing an electronic equipment design into separate basic or standard IC circuits, obtaining these circuits in individual forms, and then interconnecting the individual circuits. This approach will not yield the advantages of minimum size, high reliability, and low cost available from microelectronics.

To benefit from microelectronics, it is necessary to treat the equipment or design as a whole or composite unit, and not as a combination of separate interconnected circuits, even if such circuits are individually in integrated form.

Integrated circuits, by their very nature, must be produced as complete units. Design errors, or even minor circuit changes will involve expensive re-tooling costs. The earliest integrated circuits were fabricated using successive design, whereby a circuit was built and tested to determine the corrections required in following circuits. In this "evolutionary" design procedure, a partial repetition of the total processing expense had to be incurred for each of the interim design steps. Using such an approach, the final circuit cost becomes extremely high, due to the extensive development costs.

The lack of flexibility in circuit configuration design, is a distinct disadvantage of integrated circuits. The high re-tooling costs resulting from last-minute circuit changes, arise mainly from the requirement to produce a new set of accurately registered, high resolution masks for the various steps in the fabrication processes. One answer to this problem is to apply thorough worst case design, using the rules of design pertinent to integrated circuits.

Successful microelectronic circuit design can be realized through any one or combination of the three principal techniques: thin film, semiconductor integrated circuit, and hybrid approach. Each technique has its own specific merits and disadvantages, and these are best summarized through the following design rules.

11.2 FILM CIRCUITS

11.2.1 DESIGN RULES FOR THIN FILM CIRCUITS

1. Circuits must be analyzed and tested very thoroughly to assure that they meet specifications before fabrication is begun. It is wise to breadboard the system with conventional components since the thin film process, in almost all cases, duplicates the breadboard. Changes in design can be very costly (even a change in the value of a resistance) once fabrication has begun. The costs of these changes vary according to how far along the fabrication process has progressed. If the electronic designer knows the fabrication process he will then be able to weigh the cost of a change and decide how and if it should be made.

2. Passive type components should be used whenever reasonable in preference to active components. An example might be resistive or capacitive coupling in digital circuitry rather than diodes. This is because active components are added on after the circuit is deposited.

3. The system should be designed to use a minimum number of different types of circuits. The fabrication costs depend more upon the number of different types of circuits, than the quantity of each circuit type. It is generally more economical, for example, to use a high speed flip flop in a low speed condition, rather than design a special low speed flip flop. Another technique is to design a general purpose type circuit which can be used for more than one function as, for example, a flip flop and shift register. Extra (spare) passive components can be deposited at practically no cost. Also, the types and placement of the active components can be changed without added cost.

4. Circuits should be designed to involve ratios of component values rather than the absolute component value whenever possible. Ratios of values can be held to much closer tolerances (by perhaps twice) than the tolerances specified for the absolute values of those components. This

is because the deposition thickness of the individual components fabricated on a circuit substrate are the same due to the fabrication process. The temperature tracking of these components is also better, by at least an order of magnitude, than the absolute temperature coefficient given for the component type.

5. Tunable components should be eliminated from all circuit design, if at all possible. It is better to increase complexity rather than employ a tunable component. Components may be trimmed to better than 1/2% to bring a value in line during fabrication, but once the circuit is assembled this is no longer possible..

6. Circuits and systems should be designed for minimum power dissipation. With a good thermal heat sink, it is possible to dissipate as much as one watt on a one inch square substrate packaged in a system. The volume dissipation should not exceed 5 watts/in^3.

7. Circuits should be designed to operate with low voltages. Although voltage ratings as high as 50 or 100 volts can be obtained, preferred levels are below approximately 25 volts.

8. The size of the circuit substrate should be made as large as practical so that a large amount of the circuitry can be put on one substrate. The basic reason for this is that as the circuit substrate becomes larger, the interconnections between substrates drops drastically. Interconnections between substrates represent volume and weight, and are the most unreliable part of a microelectronic assembly. The maximum practical sizes of a substrate is determined basically by the cost of fabrication, and cost of parts of the thin film module. This maximum size determination is very complex, and must be made with the aid of the fabricator. As many as 4 or 5 flip flop type circuits may be fabricated at a reasonable cost on a single substrate.

11.2.2 DESIGN RULES FOR THICK FILM CIRCUITS

a. Use straight-line patterns wherever possible to facilitate the screening operation.
b. Design all resistors running in the same direction to minimize screening variations.
c. Design for a 25 w/sq inch power dissipation with a maximum 50 w/sq inch after trimming when using alumina substrates, and 70 w/sq inch on Beryllia.
d. Interconnect wires should be kept to a minimum length, to avoid shorting against each other or to the circuit elements.
e. Conductive paths close to a substrate edge should be avoided to reduce problems with alignment in the screening operation.
f. Choose ink or paste resistivity to give best compromise between the required range of resistance values, power dissipation requirements, and available substrate area.

g. Keep the number of pastes or inks to a minimum (3 is a practical limit).
h. Thick film capacitors are not competitive with add-on capacitors.

11.2.3 THIN FILM VERSUS THICK FILM

a. When high frequency usage causes the conductors to affect circuit performance, use thin film circuitry.
b. Thin film offers greater circuit density.
c. Thin film resistors generally offer better temperature coefficient and tolerance, but less value range, and less power dissipation.
d. Thick film is generally less costly.

11.3 DESIGN RULES FOR SEMICONDUCTOR INTEGRATED CIRCUITS

1. Circuits must be analyzed and tested very thoroughly to be certain that they meet specifications before fabrication is begun. It is much more important that this criteria be met in semiconductor circuitry than in thin film circuitry, because of the higher initial costs of semiconductor circuit fabrication. Unfortunately, circuits breadboarded with conventional components do not fully represent the fabricated semiconductor circuit. This is primarily due to the interaction between components. Many semiconductor manufacturers fabricate both single semiconductor components and multiple semiconductor components on one chip to be used for breadboard purposes. These special components are believed to represent the true fabricated single semiconductor circuit more realistically than conventional components. Another design technique resides in analyzing the effect that the interaction between components has on the circuit operation. It is possible to breadboard circuitry in which component interaction is represented by conventional components. In any event, the electronic designer must understand the problems of circuit fabrication, to enable him to help the semiconductor fabricator with his design. The probability of obtaining the desired electrical characteristics in the first effort of circuit design to fabrication, for a simple circuit, is not better than 50%. Since prototype fabrication is expensive, a good understanding of all design processes is necessary to increase the probability of success for subsequent design procedures.

2. Circuits should be designed to employ active components rather than passive components whenever possible. For example, diode coupling in digital type circuitry is preferred to resistor coupling. Active components require very small areas, and the fabrication technology is more developed.

3. The system should be designed so as to use the minimum number of different circuit types. The reasons are similar to those mentioned for thin film circuit design. Since no add-on components are possible in semiconductor circuitry, the flexibility to change active component

types and active component placement (as in thin film circuitry) does not exist here. In order to overcome this inflexibility, extra components can be fabricated on a semiconductor wafer. By changing then only the evaporated lead pattern, a variety of similar type circuits can be fabricated from one type of semiconductor substrate.

4. Circuits should be designed to involve ratios of component values rather than the absolute component values whenever possible. The reasons for this are the same as those mentioned for thin film circuitry.

5. Tunable components should be eliminated from all circuit design. Passive components may be brought into line by trimming during fabrication, but this procedure is much more difficult than the trimming of thin film components.

6. Circuits should be designed to employ direct coupling rather than capacitive coupling. This is especially true with low frequency amplifiers because of the difficulties to fabricate large capacitors (above 500 $\mu\mu$f). Since similar type components are fabricated in parallel on the same semiconductor substrate, excellent temperature tracking of both active and passive components can be obtained. This feature permits the designing of balanced drift-free circuitry.

7. From the practical viewpoint, the number of components that can be fabricated on a single substrate is limited to approximately 15. This is due to the fact that circuit production yields drop radically as the number of circuit components increases. Since the substrate sizes are small, it is important that interconnection schemes and lead attachment problems be considered throughout the system design.

8. Circuits should be designed for minimum power dissipation. Due to the small size of the circuit, the circuit power dissipation should not exceed approximately 100 mw. A volume power dissipation of 5 watts/in^3 can be obtained with good thermal heat sinks. This consideration is not based on the physical limitations of parts, since diffused components will characteristically withstand high power and high temperature without being destroyed. The main consideration generally affecting maximum power dissipation, is the internal heating effect of the circuit, which causes it to drift or fall outside the expected performance temperature range. Such a situation may make it desirable for an effective thermal control element such as a heat sink to be used in direct contact with the circuit package, even in a very low power circuit.

9. Circuits should be designed to operate on low voltages. Levels above approximately 10 volts should be avoided. The power supply should be selected so that the voltage ratings will not be exceeded in any of the circuit parts. The LV_{CEO} rating of the transistor is, generally, the principal determining criteria. The isolation junction has characteristically the highest voltage breakdown, followed by collector-base junctions and emitter-to-base junctions, in that sequence. Cases may exist where advantage can be taken of these high voltage ratings if precau-

tions are taken to assure that no operating or transient conditions will allow any voltage rating to be exceeded.

Two additional topics are of importance when designing circuitry with semiconductor integrated devices. These deal with parasitic effects, and the choice of circuits.

11.3.1 PARASITIC EFFECTS AND ISOLATION CONSIDERATIONS

As pointed out earlier, parasitic capacitance effects prevail in semiconductor integrated circuit parts. The predominant parasitics are generally, associated with the isolation junction. Through efficient circuit design, the number of isolation areas may be reduced, resulting in a reduction of the undesired capacitance.

The circuit designer should exercise care to assure that, at critical circuit points, the effective shunt capacitance area involved, is minimized.

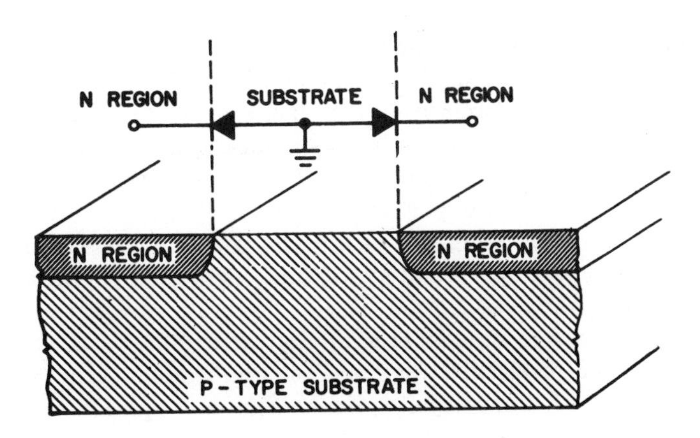

Fig. 11-1 The diode method of isolation.

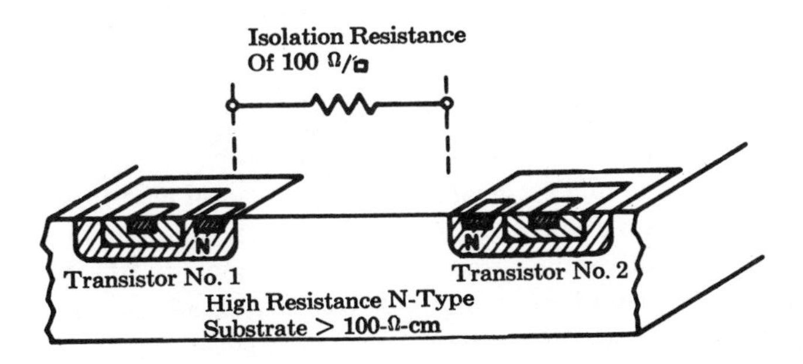

Fig. 11-2 Isolation is obtained through the resistance of the material.

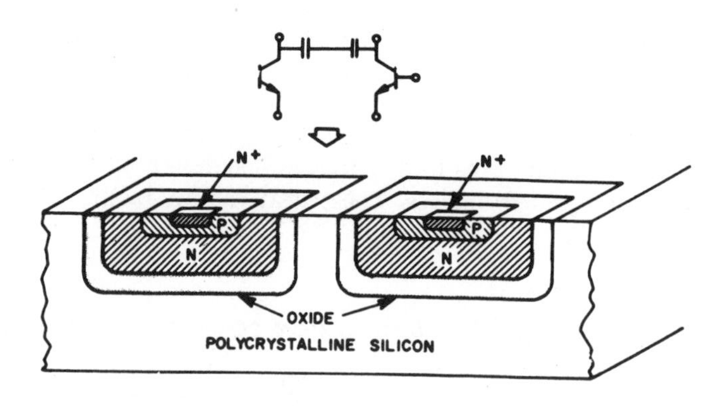

Fig. 11-3 Dielectric isolation (oxide).

This applies particularly to high frequency circuits. An example of such a situation resides in the diode NAND or low level logic circuit. The common anode point is at a relatively high impedance level, so that any capacitance existing at this location has a significant delaying effect. Accordingly, it is advisable to use a type of diode array and pull-up resistor which causes the lowest amount of parasitic capacitance. Thus, a multiple emitter diode-connected transistor used as a common anode diode array, results in highest switching speed and significantly less total capacitance, than could be realized with a collector base diode array. Also, the geometry of the resistor should be narrow to minimize capacitance.

The selection of circuits for amplifier applications, can be similarly performed, for the purpose of optimizing high frequency performance characteristics. Since capacitance exists at collector nodes due to the normal transistor junction plus the high isolation capacitance, it may often be advisable to obtain power gain through the use of common collector stages wherein the high collector capacitance has virtually no effect. Such an approach has, however, practical limitations, since voltage gain is eventually necessary in most cases. In some situations, though, it may be possible to defer voltage gain to a later portion of the system where hybrid, conventional circuits, or more suitable conditions exist for voltage gain.

Biasing of the isolation region is another consideration which must be observed. The voltage levels should be arranged so that the isolation junction is always nonconducting, and yet they should be low enough to avoid inducing fall-out due to voltage breakdown. A general rule to follow is to ascertain that the isolation area terminal is never allowed to rise to a potential more positive than any collector N-region in the integrated circuit. This will prevent isolation diode conduction and allow all of the resistors which are diffused into the circuit, to be placed in a single isolation area without unwanted currents—except for leakage-flowing between the elements.

11.3.2 CIRCUIT CHOICE

Semiconductor parts can be expected to match each other very closely, although absolute values have a relatively large spread. Accordingly, it is natural that balanced circuits, such as differential amplifiers, are very appropriate for integrated circuits.

Since large values of resistance are impractical with semiconductor techniques, it is much more reasonable to consider synthesizing large resistors through such means as transistor constant current sources. When applied to differential amplifiers, for example, such current sources may

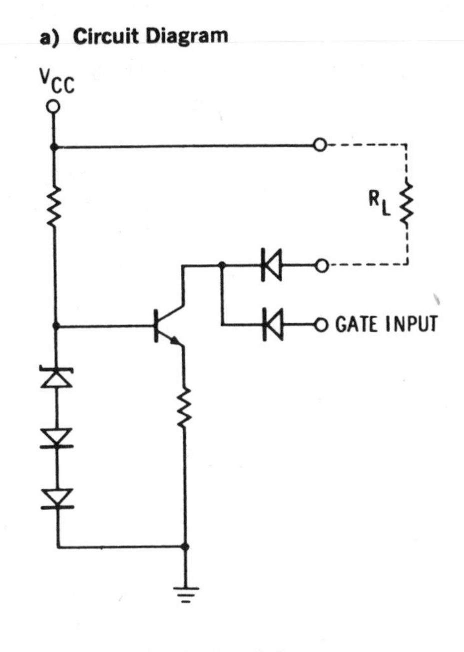

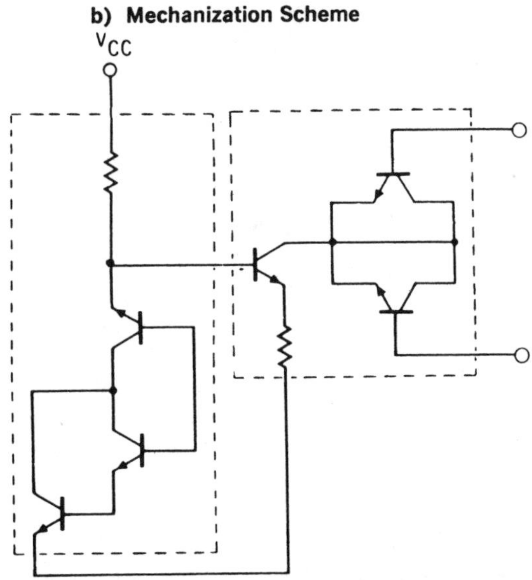

Fig. 11-4 Gated current source in two isolation areas

utilize the predictable positive temperature coefficient of a diffused resistor to compensate for the negative temperature coefficient of a junction, and thereby provide relatively constant current over a wide temperature range. A more conventional type of constant current source is also possible through the use of a temperature compensated zener reference formed from a base-emitter junction and two forward biased diodes, as shown in Fig. 11-4. The entire circuit can be constructed within two isolation areas including the means for switching the current at the output.

One way to minimize isolation areas, is to arrange a circuit so that a large number of components may be contained within a single isolation region. An examination of the equivalent circuits of the integrated parts would reveal that a resistor cannot be connected to a transistor collector within a single isolation area. This is because the bulk of the resistor would be bypassed to the collector node by the p-n junction. Collector resistors must always be in an isolation region separate from the transistor. It is possible, however, to have many resistors contained in a single isolation area along with other parts, provided this rule is not violated, and the isolation terminal is properly biased.

From considerations involving the construction of transistors, it may be seen that common collector transistors are desirable, whenever possible. In a similar manner, it can be shown that an array of common-cathode (collector-base) diodes can be connected to a transistor collector in a single

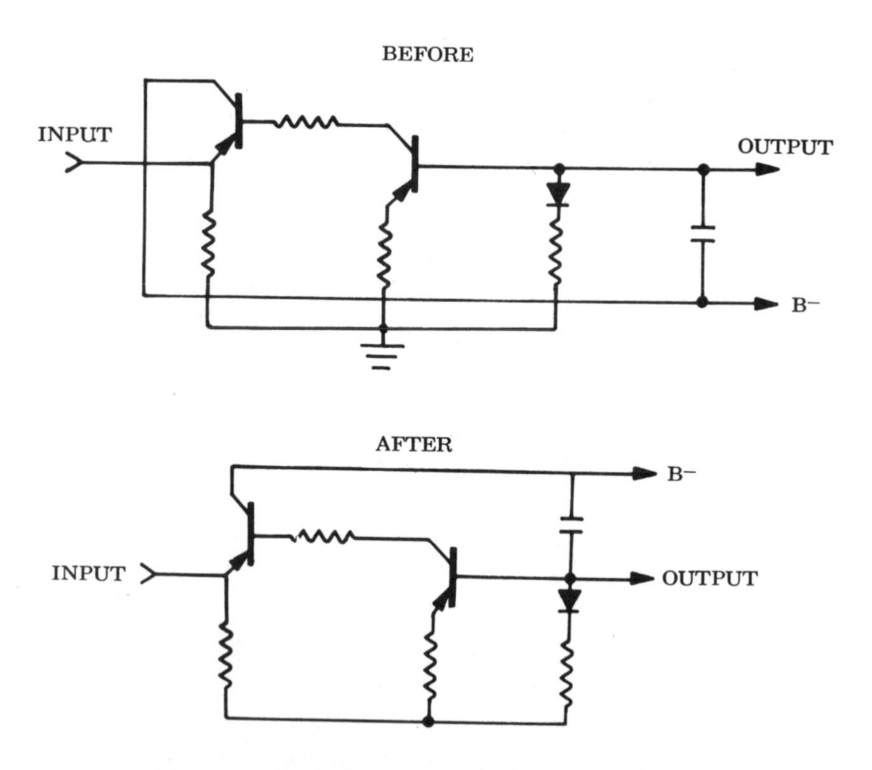

Fig. 11-5 Minimization of crossovers:—Before an attempt is made to reproduce a schematic in layout form, it is wise to eliminate or minimize crossovers. The resistors and conductors are fabricated in a single plane and cannot cross over one another without shorting. Careful rearrangement of a schematic will remove many unnecessary crossovers.

isolation region. Furthermore, a multiple emitter transistor, with its base and collector connected, may be used to form a common anode diode array in one isolation region. Circuit arrangements of these types are often usable for logic configurations. A variation of the common collector transistor configuration which is applicable to single isolation construction is the Darlington connection. Through this design it is possible to cascade a number of transistors having common collectors and incorporate emitter-resistors, where desired, without requiring more than a single isolation area.

11.4 HYBRID CIRCUITS

The term ''hybrid'' as used in microelectronics, denotes a number of different methods for synthesizing a circuit. Most of these are, in fact, packaging procedures for assembling miniature circuits. As used in this section, however, the name ''hybrid microelectronics'' denotes a marriage of the thin film and semiconductor microelectronics fabrication processes.

It is true that, according to this definition, the semiconductor microelectronic circuit is a hybrid since the conductors are thin films on the oxide layer. However, these conductors are such a small and noncritical part of the circuit, that it is reasonable to consider semiconductor microelectronics as purely the product of a semiconductor technology.

The hybrid circuit takes advantage of the best parts of the two technologies. The high quality transistors and diodes which can be made by diffusion in a silicon chip, are combined with the resistors and capacitors made by a thin film process, and deposited on the oxide layer. These components are finally interconnected by evaporated aluminum conductors. Since the semiconductor chip is made first, and the thin film devices are then deposited, the thin film processing must not affect the semiconductor devices. This imposes a number of limitations on the thin film processing procedures which must:

1. not alter the p-n junction components in the silicon substructure;
2. not deteriorate the passivation qualities of the SiO_2 layer on which they are formed;
3. be capable of the same pattern resolution as the diffused elements;
4. permit both resistors and capacitors to be formed in the same circuit;
5. allow resistor and capacitor trimming;
6. result in a materials system that can be hermetically sealed and not undergo metallurgical reactions at high temperatures (i.e., above 300° C and preferably to 500° C).

A serious problem associated with hybrids resides in the condition that the thin film components usually require more area than is available on the commonly used silicon chips — about 50 mils x 100 mils. It is desirable to use silicon circuit chips as small as possible. This allows the fabrication of a large number of circuits on a single semiconductor wafer, and improves the yield of circuits. For this reason, the area available for the thin film components, is rarely as large as one may desire. To make large valued resistors, either films of high sheet resistance values, or large aspect ratios are required. The line widths used are as small as one mil so that large values of aspect ratio are possible—though with poor tolerance unless trimming procedures are used.

Capacitors can be made by the evaporation or anodization process. The MOS (metal-oxide-semiconductor) capacitor can also be used in place of the all thin film capacitor. In this case an N^+ layer is formed by diffusion in the semiconductor, to act as one capacitor plate. The oxide layer is used as the dielectric, and a metal film is deposited on the oxide to form the second plate.

Since the hybrid process uses the best features of both the semiconductor and thin film technologies, the hybrid microelectronic circuit may be expected to have many advantages over either the semiconductor or thin film circuit. A number of these advantages are as follows:

1. Since it is not necessary to diffuse in the resistors and capacitors, transistors and diodes need no longer be compromises and their design can be optimized.
2. The design constraints imposed on the semiconductor circuit designer are considerably alleviated. Resistor and capacitor biasing problems are an example.
3. Parasitic capacitances to the substrate are about an order of magnitude smaller.
4. The temperature coefficient of resistance of diffused resistors is about 2000 PPM/°C, for thin film resistors about 50 PPM/°C.
5. Breakdown voltages are increased.
6. Sheet resistance values for thin film resistors can be as high as 5000 ohms/square for Cr/SiO cermet. It does not exceed 250 ohms/square for diffused resistors.
7. It is easy to trim thin film resistors to a tolerance of 1 percent, while it is very difficult to do this with diffused resistors.
8. The practical high frequency limit for thin film resistors is about 10,000 mcs, while it is 20 mcs for diffused resistors.
9. Diffused capacitors are nonlinear, thin film capacitors are linear.
10. The temperature coefficient of capacitance for thin film capacitors is about one order of magnitude smaller than for diffused capacitors.
11. The dissipation factor for diffused capacitors is substantially greater than for thin film capacitors.

11.4.1 STEP-BY-STEP DESIGN PROCEDURE

Below are the design steps involved in implementing a final circuit schematic, parts list, and preliminary physical layout to meet the function requirements:

1. Study and understand function requirements and concepts.
2. Draw up functional block diagram.
3. Identify electronic circuit sub-functions (gates, amplifiers, filters, etc.).
4. Design circuit sub-functions.
5. Assemble sub-functions into a preliminary schematic.
6. Establish the nominal values and tolerance required for each element.

a. *Resistors:* (1) Ohmic value and tolerance; (2) Required power dissipation; (3) Temperature coefficient.
b. *Capacitors:* (1) Capacitance value – with tolerance; (2) Working voltage; (3) Frequency; (4) Stability.
c. *Active discrete elements:* gain, breakdown voltage, switching characteristics, power dissipation.
d. *Active circuit elements:* functional requirements, input and output characteristics, frequency characteristics.
e. *Miscellaneous Elements – Inductors, etc.*

7. Choose basic technology: Consider total function requirements (including size, weight and projected production costs) and available production facilities, and make choice of thick film, thin film, a combination or other.
8. Select elements.
9. Prepare a parts list relating part type to its circuit symbol, fabrication techniques, value, tolerance and physical dimensions.
10. Analyze and redraw the preliminary schematic as required to minimize the number of crossovers.
11. Add up all of the areas for the elements given in the parts list. Multiply this area by the following factor (to allow for conductor patterns, etc.).

    a. 3.0 for a uniform mixture of large and small elements.
    b. 4.0 for a relatively large number of small elements.
    c. 2.0 for a relatively few number of large elements.

12. Add up the power dissipated in each circuit element. Calculate the minimum substrate size for this power dissipation and worst case ambient temperature.
13. Considering the physical and packaging requirements of the function, and the area estimate calculated from steps 2 and 3 (the larger estimate of the two should be used), choose an appropriate substrate size.
14. Before starting actual layout, determine which, if any, of the elements that may require unique placement. Examples of such requirements are high power dissipation, cross-talk or capacitive coupling restrictions, and requirements for trimming after assembly.
15. Using grid paper and colored pencil as required, make a 10 or 100-sized scaled layout.
    a. Locate any key elements.
    b. Draw out interconnect pattern and locate other circuit elements.
    c. Locate external pin connections and indicate interconnections required.

11.4.2 CIRCUIT DESIGN CHECK LIST

As a final step in the circuit design procedure, the final circuit configuration should be checked against the following criteria:

a. Are size and weight limitations defined?
b. Are cost guidelines defined?
c. Is power dissipation established for each element?

d. Are power supply requirements established?
e. Is the grounding system defined?
f. Are external test nodes indicated?
g. Do any elements generate interference?
h. Are any elements particularly sensitive to interference?
i. Do some elements require isolation (thermal or electrical) from one another?
j. Are all required element values specified? (Vendor data sheets are often inadequate).
k. Are all element tolerances specified?
l. Has worst case analysis been performed and confirmed?
m. Have elements been properly derated?
n. Are any matched elements required?
o. Are the preliminary thermal analysis results favorable?

11.5 EXAMPLES OF CIRCUIT DESIGN

A number of microelectronic circuits are presented in this section for the purpose of illustrating the application of the various design principles discussed in the preceding sections. The circuit designer has large opportunities to apply ingenuity in synthesizing microelectronic circuits that will meet a specific set of requirements.

11.5.1 BUFFER AMPLIFIER

A buffer amplifier of silicon monolithic design has been built with diffused resistors. The amplifier exhibits a high degree of temperature stability, and is built on a single chip with dimensions of approximately 60 by 90 mils. The unit is packaged in a ¼ by ⅛ inch flatpack with 14 leads.

The purpose of the buffer amplifier is to provide good isolation between the input and output. This is done by making the input resistance very high and the output resistance very low so that loading has a negligible effect on the source. A particular problem with dc amplifiers, of this type, is the drift in the dc characteristics of the amplifier components resulting from temperature variations. The silicon monolithic construction minimizes this problem because it results in close thermal coupling between the separate component regions. Input and output voltage equalization is also enhanced by the similarity of input and output junctions provided by the monolithic construction.

The basic circuit for the amplifier is shown in Figure 11-6a where Q_1 and Q_2 are n-p-n transistors with the same emitter resistance, R_1. The output from the collector of Q_1 is fed into the base of p-n-p phase inverting transistor, Q_3. The collector output of Q_3 is the output of the amplifier and is also fed back to the base of Q_2. The base-emitter voltage drops in Q_1 and Q_2 should be equal providing the base currents are equal. The input and output voltages are then identical.

Figure 11-6b shows a modification of the circuit which further improves the performance. Q_1 and Q_2 have been replaced by Darlington amplifiers to increase the input resistance of the amplifier. The diode, D_1, has been placed in series with the collector resistance, R_2, in order to offset the base-emitter voltage of Q_3 and also to provide for the same temperature dependence of this voltage. The double diode structure, D_2, in series with R_4 is used to provide both voltage compensation and tem-

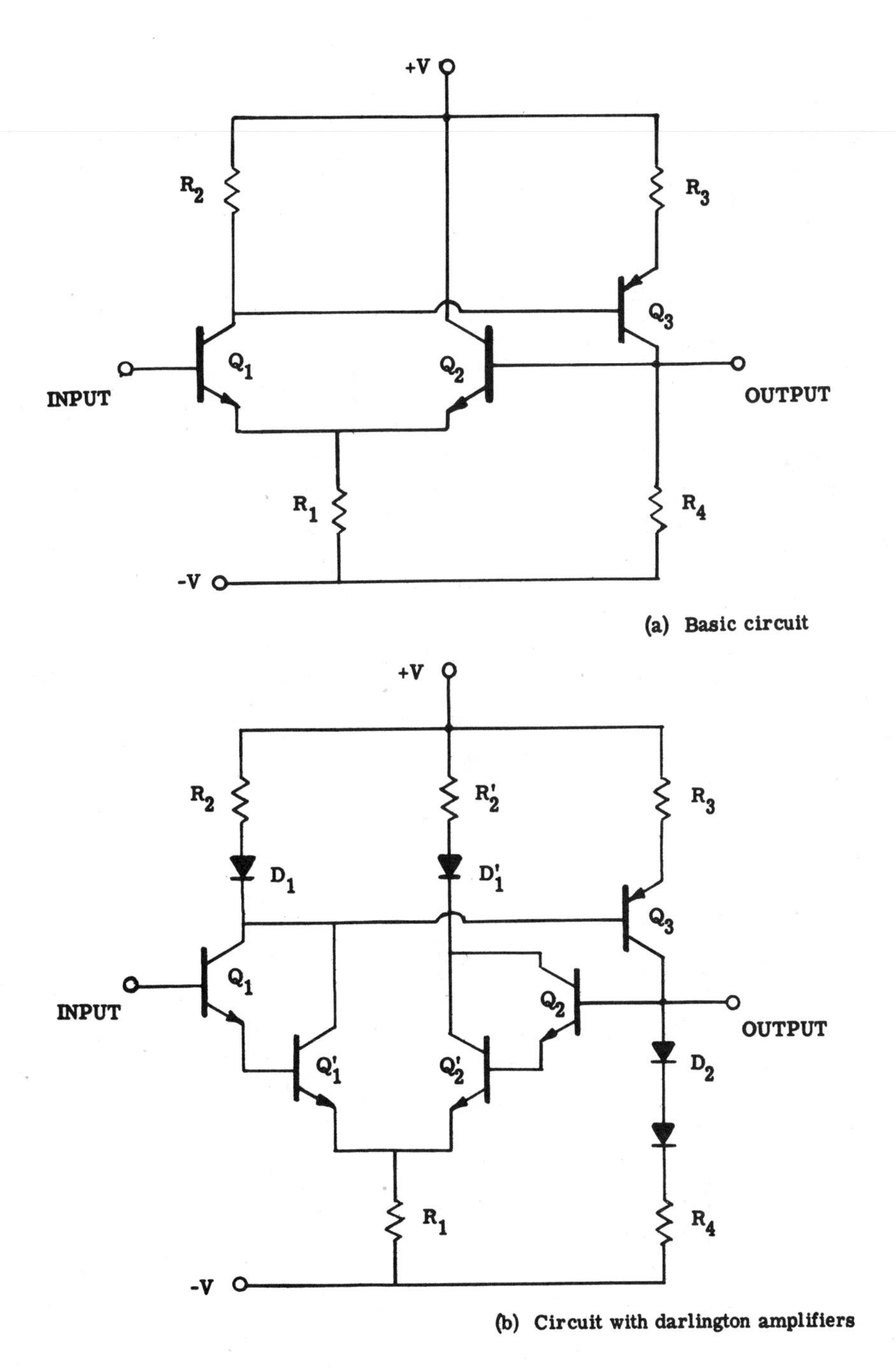

Fig. 11-6 Buffer amplifier

perature compensation for the equivalent emitter junctions of Q_2 and Q'_2. The dc currents and voltages in the amplifier circuit are by these means carefully balanced in order to achieve identical input and output voltage levels. The input impedance of the amplifier is essentially R_1 multiplied by twice the effective current gain of the input Darlington stage. The output impedance is lower than the value of R_4 because of the negative feedback action of R_1.

This buffer amplifier exhibits a remarkable degree of temperature stability for a dc amplifier. Such temperature stability is extremely difficult, if not impossible, to obtain using conventional circuit design techniques and components. This microelectronic device is a good example of how one can take advantage of certain unique aspects of the silicon monolithic design to realize enhanced performance over comparable conventional circuits.

11.5.2 INTEGRATED ANALOG SWITCH

This bi-directional switch consists of a single transistor and its drive circuit. The on-off impedance ratio is greater than 10^7 and it provides good isolation between the control and signal voltages which have a common ground. The unit has been fabricated into a single monolithic silicon block, and encapsulated in a 14 lead ¼" × ⅛" flatpack. The electrical circuit of the analog switch is shown in Fig. 11-7.

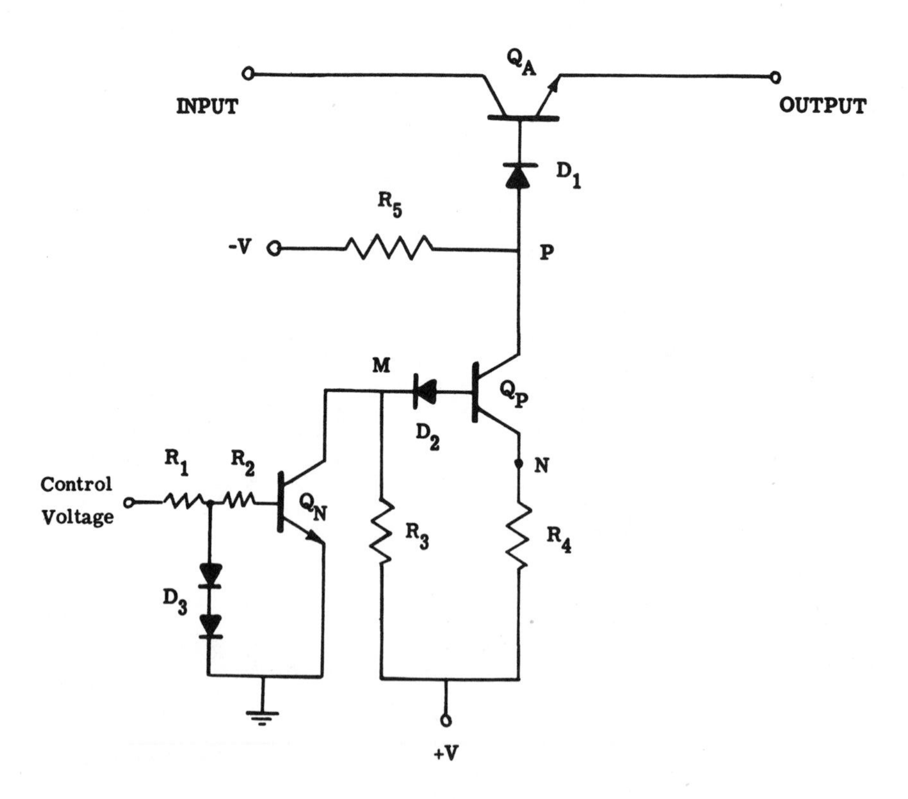

Fig. 11-7 Analog switch circuit

Transistor Q_A is the switch, and the remaining components make up the control circuitry that drives Q_A either "ON" or "OFF." Transistor Q_A is operated in the inverted mode for a lower offset voltage. When Q_N is driven into saturation by a positive control voltage, the potential at point M is about 0.2 volt and the potential at point N is fixed at about 1.4 volts by the forward biased D_2 and Q_P base-emitter junction. A constant current generator is thus formed by the potential difference between the supply voltage and the voltage at point N divided by resistor R_4. This generator supplies base current to Q_A causing it to turn "ON," and through R_5 swamping out the effect of the $-V$ potential.

In the absence of a control signal, the collector-emitter circuits of Q_N and, consequently, Q_P are open circuited. Therefore, the negative potential, $-V$, can reverse bias D_1 and the base-emitter junction of Q_A. Diode D_1 reduces the reverse bias applied to Q_A.

The design of this switch provides for controlling the base current in Q_A during the "ON" period such that the switch is just in saturation for a given controlled (collector) current, thus minimizing the opening (storage) time. It also provides for reverse biasing Q_A during the "OFF" period to minimize any current leakage through the open switch.

11.5.3 INTEGRATED LOW-LEVEL DC DIFFERENTIAL AMPLIFIER

Differential amplifiers, with a basic requirement for matching of transistors and resistors, are nearly ideal for fabrication as a monolithic functional block. The matching of characteristics in the block is insured by the uniformity of temperature in the structure, and by the simultaneous fabrication of the complementary component. The ease of matching in the monolith makes it possible for the amplifier to out-perform a discrete component circuit. An oven to maintain a constant temperature and a careful selection of components would be required if the discrete component circuit were to match the monolithic structure in performance. The amplifier shown in Fig. 11-8 has a voltage gain increase from 247 to 256 over a temperature range of 100° C.

11.5.4 LOW DRIFT DC AMPLIFIERS

Silicon diffused transistors exhibit a temperature coefficient of base-to-emitter voltage of about 2.3 mV/°C. In a single-transistor (unilateral) input stage, this coefficient is multiplied by the entire amplifier gain. To use such an amplifier for d-c amplification, therefore, temperature would have to be held constant, or it would be necessary to calibrate the amplifier with temperature. A better circuit configuration is obtained through the differential amplifier in Figure 11-9. Through canceling of the drift of one transistor with an equal and opposite drift of another, d-c drift can be reduced by several orders of magnitude.

A differential amplifier used as an input stage for a d-c amplifier helps reduce d-c drift. For a unilateral d-c amplifier, the performance of the discrete version and that of the integrated version are both subject to the d-c drift problem.

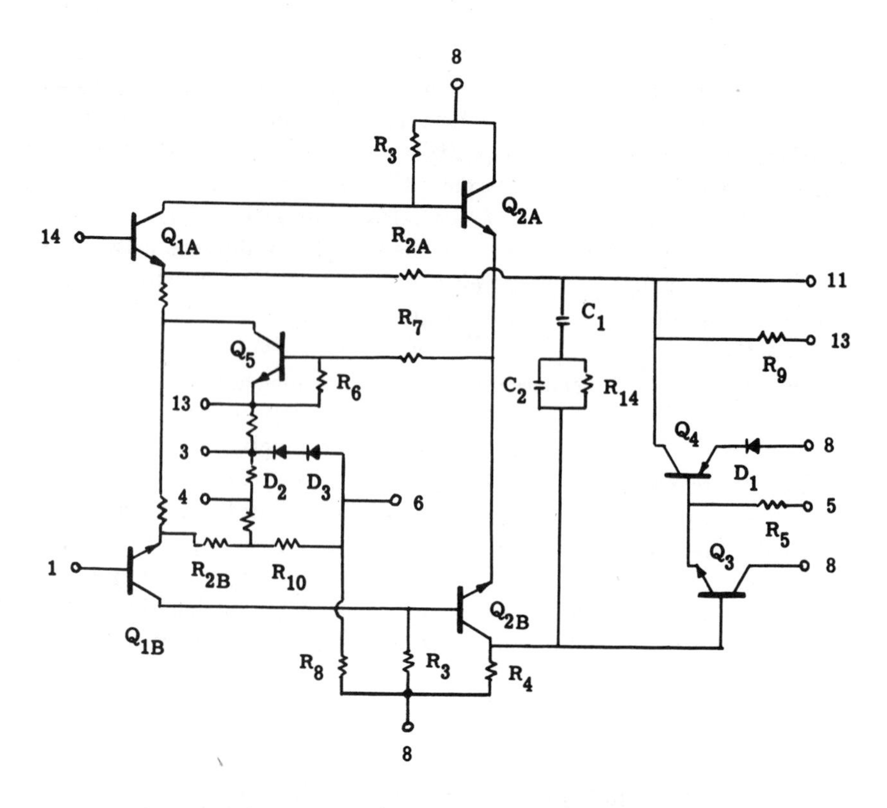

1. Input A (non-inverting)
3. No connection
4. No connection
5. Common
6. No connection
8. Positive supply voltage
13. Negative supply voltage
14. Input B (inverting)

Fig. 11-8 Low level DC differential amplifier

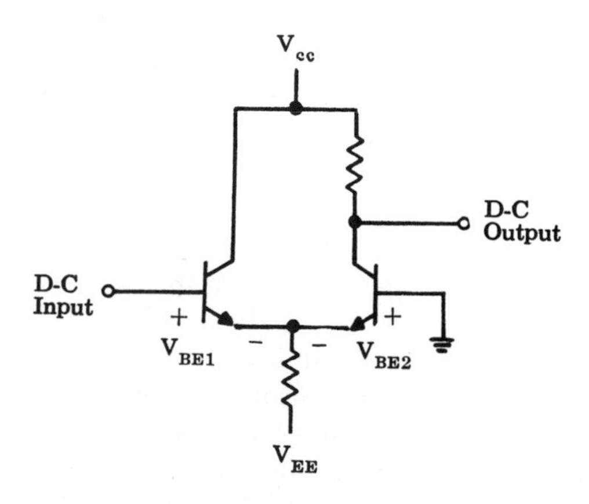

Fig. 11-9 Low-drift amplifier.

11.5.5 AC AMPLIFIERS

A well-known limitation of present integrated-circuit technology is its inability to produce inductors or large capacitors. Direct coupling should therefore, be used where possible. The following ground rules are recommended:

(1) Use a differential input stage.
(2) Design for as much d-c gain as can be tolerated with the specified amount of drift in the first stage.
(3) If gain is insufficient, add a second d-c amplifier stage, using an external coupling capacitor.
(4) Use overall feedback for gain stability.
(5) Obtain high impedance (FET's will provide this if it cannot be achieved otherwise) to permit the use of smaller coupling capacitors.

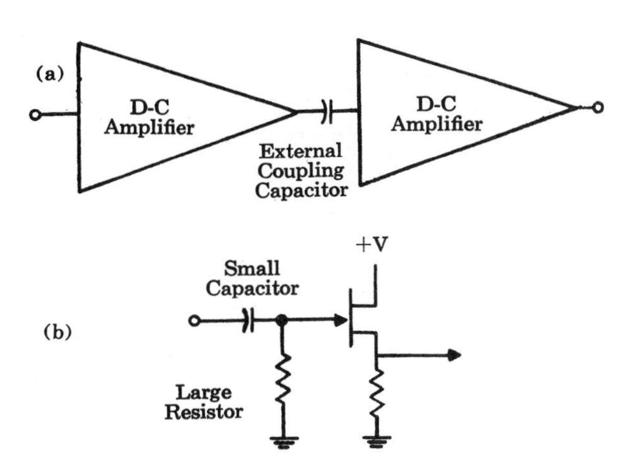

Fig. 11-10 A-C amplifier techniques.

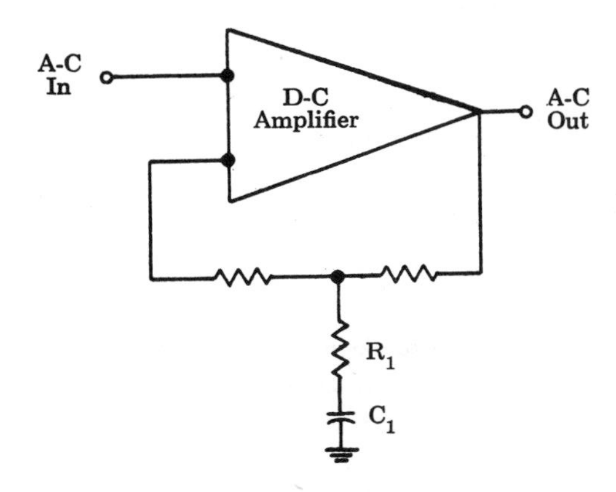

Fig. 11-11 High A-C gain, D-C stable circuit.

Figure 11-10 illustrates the application of these principles. The differential amplifier permits a gain of about 100 with sufficiently low drift to maintain a reasonable dynamic range. With the FET stage, a large external resistor is needed, but the small capacitor required can be integrated.

High a-c gain and good d-c stability are obtained in the circuit of Figure 11-11. Here a large amount of d-c feedback provides d-c stability, and a small amount of a-c feedback results in high a-c gain. The amount of a-c feedback is controlled by external capacitor C_1 and resistor R_1 (if R_1 is zero, a-c gain is maximized).

11.5.6 SIGNAL PROCESSING CIRCUITS

A major goal in the servo-control field is the elimination of transformers. The integrated choppers, demodulators, and quadrature rejection circuits illustrated in this section should be examined with that in mind.

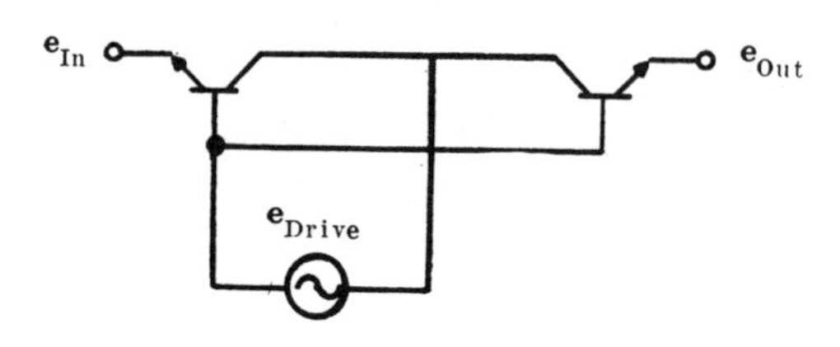

Fig. 11-12 Basic chopper circuit.

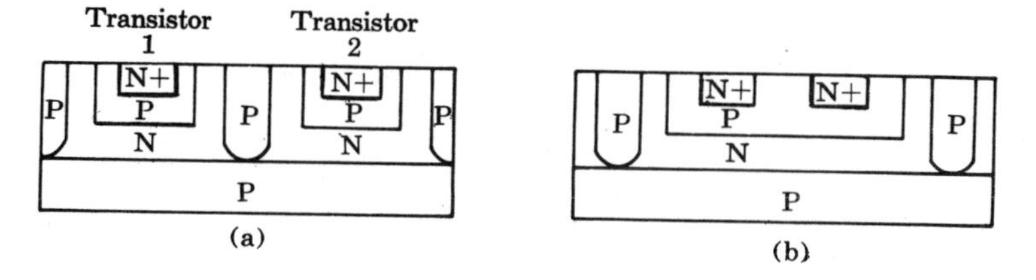

Fig. 11-13 Standard chopper configurations.

The two major factors in the design of choppers are offset current and voltage, and drift. (Offset causes an output in the absence of an input, and drift produces a change in output with time and temperature.) The inverter transistor connection reduces both offset and drift.

Figure 11-12 shows how even lower offsets can be achieved. Here the drift is determined by tracking of the transistor characteristics with temperature and time, and by the difference in junction temperature between the two units. Cancellation of offset is achieved by equal and opposite offset of another device. The integrated version has an advantage over the discrete version since the transistors will be matched and operating with small differences in temperature.

Offsets and drifts of standard integrated circuit pairs (Figure 11-13) in the inverted chopper connection average 50 μV offset and 1 μV/°C drift. Figure 11-13(a) shows two transistors diffused into a single die; and Figure 11-13(b) shows two emitters diffused into a single collector-base junction area, which provide even tighter control of drift and offset. Double-emitter devices have been reported to have offsets under 20 μV and drifts less than 0.2 μV/°C.

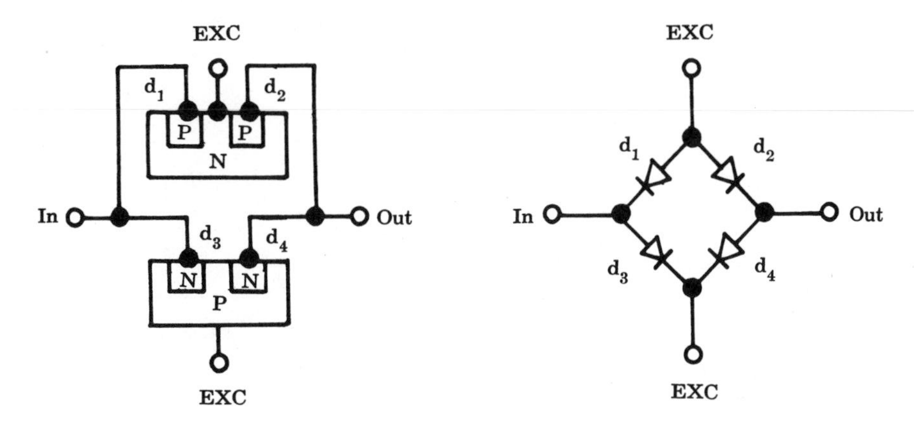

Fig. 11-14 Diode-quad switch construction.

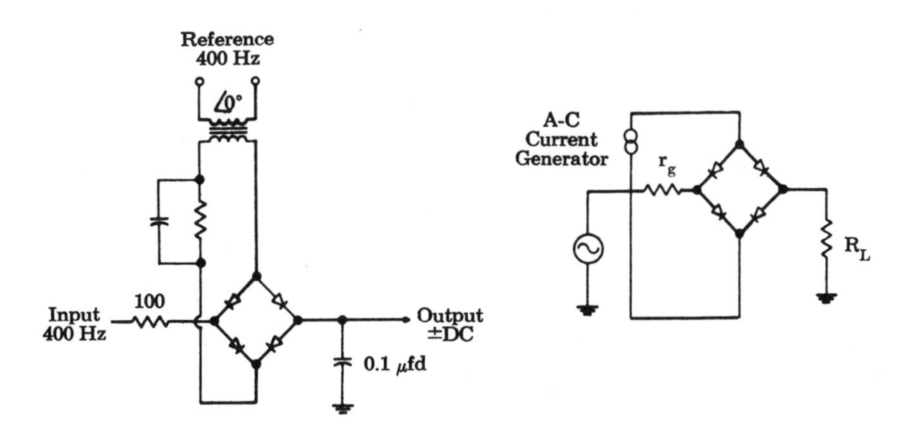

Fig. 11-15 Diode-quad demodulator and equivalent circuit.

11.5.7 DEMODULATORS

Although demodulators vary widely in configuration, only diode-quad types are discussed here.

In discrete diode-quad bilateral switches, the usual procedure is to select diodes in quads or pairs to achieve low offsets. In integrated circuits, two diodes can be diffused side by side into the same silicon substrate. An n-on-p diffused pair can be connected to a p-on-n pair to form a bridge illustrated in Figure 11-14. The p-on-n pair should match and track, as should the n-on-p pair. Figure 11-15 shows a diode quad in a demodulator. Quadrature rejection is accomplished by controlling the conduction angle. The d-c output is the average value of the in-phase sine wave. At that time the quadrature signal is passing through zero. The output will be zero for quadrature signals and nearly equal to the peak magnitude for in-phase (o°) signals. A paralleled R and C is used to achieve a peak-conducting circuit.

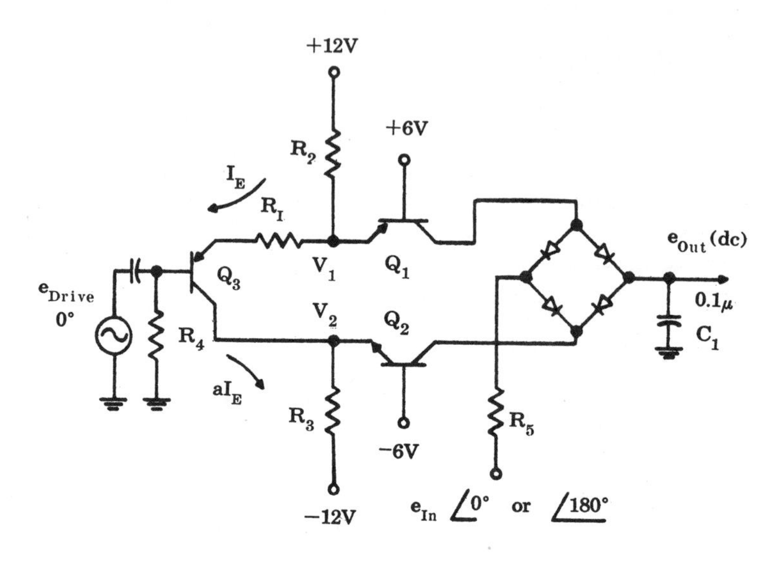

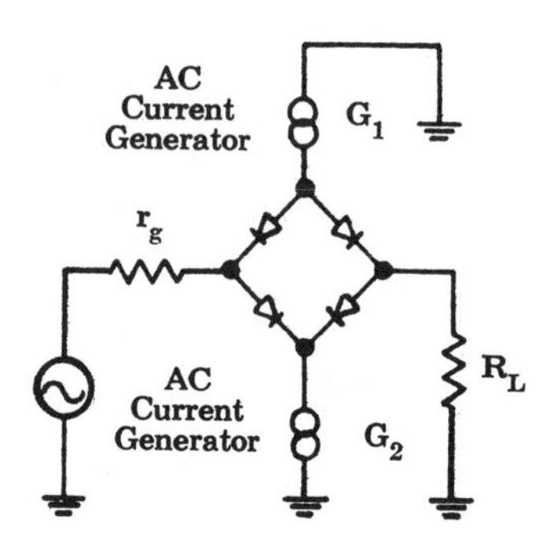

Fig. 11-18 Transformerless demodulator.

A transformerless version of the demodulator is shown in Figure 11-18. In this version, the isolation property of the transformer is approached by use of the high output-impedance of the common-base connection. When Q_1 and Q_2 are cut off, the impedance is essentially infinite. When Q_1 and Q_2 conduct, the drive nodes of the bridge are connected to the current generators with the output impedance of Q_1 and Q_2 across them.

The demodulator functions of the transformer and transformerless circuits are essentially the same. In the transformerless version, the bridge conducts for 30 to 50 degrees about the peaks of the cycle, causing C_1 to charge up to the average value of the peaks of the input signal through R_5. When the drive circuit cuts off, C_1 holds the charge until the next half cycle. The output is a bipolar d-c voltage, corresponding to the zero and 180-degree input signals.

11.5.8 COMMUNICATION CIRCUITS

R-f and i-f amplifiers, mixers, oscillators, and multipliers with good performances over a wide range of frequencies can be designed and fab-

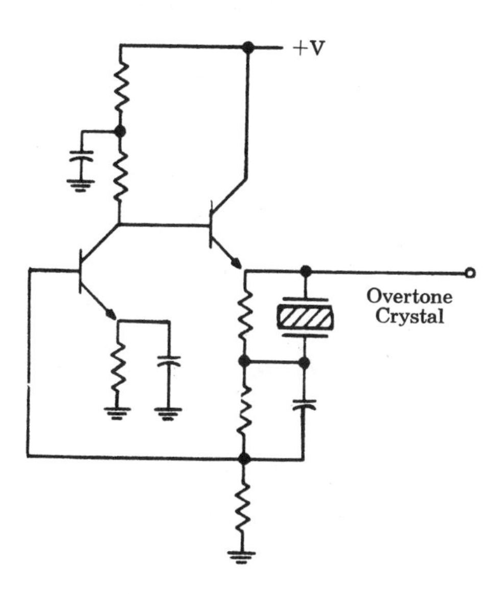

Fig. 11-19 Overtone crystal oscillator

ricated in integrated-circuit form. Monolithic integrated circuits can often match the performance of discrete versions up to about 10 MHz. Above this, parasitics begin to play a significant role.

For monolithic r-f circuits, impedances should be kept low or resistive isolation techniques should be used. In Vhf and uhf circuits, it is often desirable to separate the individual circuit elements on a high-frequency dielectric (e.g., ceramic), as in hybrid circuits.

Again, it is desirable to eliminate chokes and transformers. Figure 11-19 shows an overtone crystal oscillator without inductors. It can provide a frequency stability better than 0.005% from 0°C to 50°C. The crystal is not included in this monolithic integrated circuit.

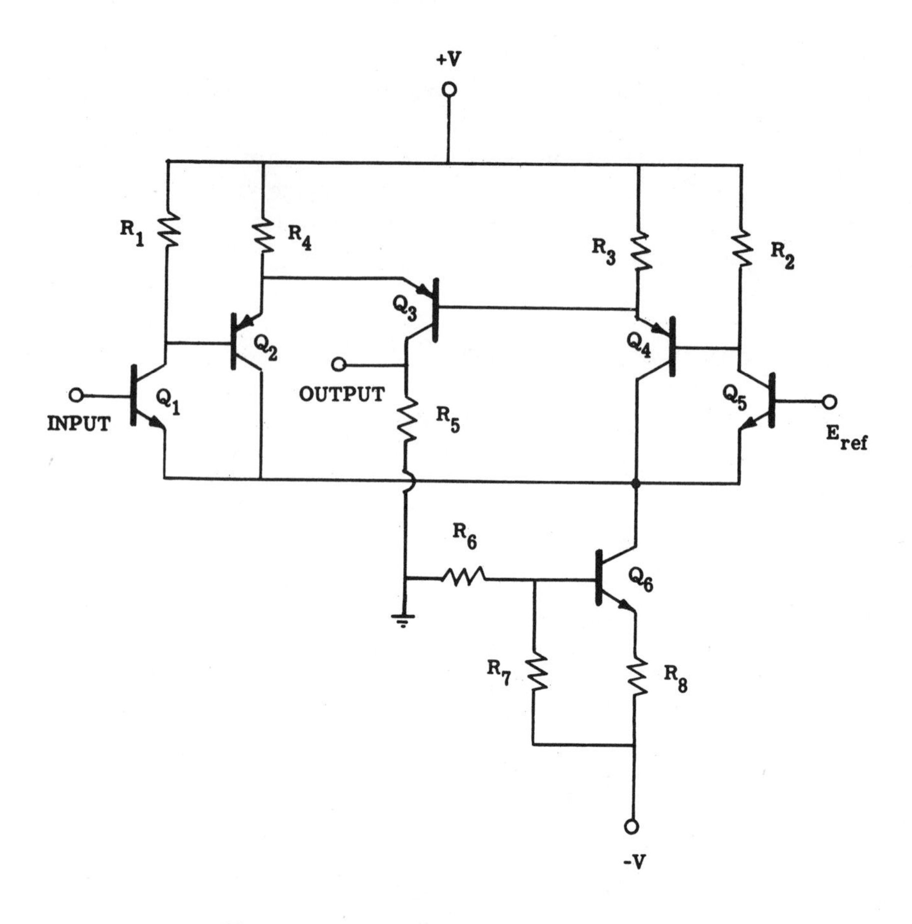

Fig. 11-20 DC comparator for A/D conversion

11.5.9 DC COMPARATOR FOR A/D CONVERSION

One form of analog to digital conversion for telemetry application utilizes a successive approximation method of conversion. With this method, the unknown voltage representing an analog quantity is compared successively to known voltages in a comparator circuit. If the known voltage exceeds that of the unknown, the circuit produces an output of sufficient amplitude to perform a logic function. Such a comparator circuit, fabricated as a multichip silicon integrated circuit, is shown in Fig. 11-20. The input voltage to the comparator varies between 0 and 5 V, and the known reference voltage for the comparator circuit is generated by a so-called ladder-adder network. The comparator is a differential amplifier which indicates when the ladder-adder voltage is greater than the unknown voltage input. Under these conditions, the comparator produces an output which removes the ladder-adder voltage from the comparator. Successively smaller ladder-adder voltages are compared to the input signal, and the sum of all those which are not rejected, indicates the magnitude of the input voltage. Using this method it is only necessary to try 6 voltage values in order to determine the input voltage level to within 2 percent of full scale.

11.5.10 MOS AMPLIFIER CIRCUITS

There are three basic single-stage circuit configurations in which the MOS is used. These are illustrated in Figure 11-21. Each has its advantages in a particular application.

11.5.10.1 COMMON SOURCE

The common-source configuration is illustrated in Figure 11-21a. The signal is applied between gate and source. This circuit has a high input impedance and a medium-to-high output impedance. The circuit provides a voltage gain greater than unity, which is given by

$$A = -\frac{g_m r_d R_L}{r_d + R_L} \qquad (11\text{-}1)$$

where g_m is the transconductance, r_d is the drain resistance, R_L is the effective load resistance, and the minus sign indicates the phase reversal from input to output. If an unbypassed resistance is introduced between the source and ground as shown in Figure 11-21a, degenerative feedback is induced in the circuit. The common-source voltage gain A' with an unbypassed source resistor is

$$A' = \frac{g_m\, r_d\, R_L}{r_d + (g_m\, r_d + 1)\, R_s + R_L} \qquad (11\text{-}2)$$

where R_s is the unbypassed source resistance. The output impedance, Z_o, is increased by the unbypassed resistor:

$$Z_o = r_d + (g_m\, r_d + 1)\, R_s \qquad (11\text{-}3)$$

11.5.10.2 COMMON-DRAIN CIRCUIT

The common drain circuit, illustrated in Figure 11-21b, is often referred to as a "source follower." The input impedance is very high, but the output impedance is low. There is no phase reversal between input and output. The distortion is low, and the voltage gain is less than unity.

This circuit finds application where an impedance transformation from high to low is required and where low input capacitance is desirable. The input is injected between gate and drain, and the output is taken between source and drain. This circuit provides 100% degenerative feedback. Its gain is given by

$$A' = \frac{R_s}{\left(\frac{\mu + 1}{\mu}\right) R_s + \frac{1}{gm}} \qquad (11\text{-}4)$$

Since the amplification factor, μ, is usually much greater than unity, this reduces to

$$A' \approx \frac{1}{1 + \frac{+1}{g_m R_s}} \quad (11\text{-}5)$$

The input resistance is R_G when R_G is connected to ground. If R_G is returned to the source contact of the MOS, the input resistance R_i is given by

$$R_i = \frac{R_G}{(1 - A')} \quad (11\text{-}6)$$

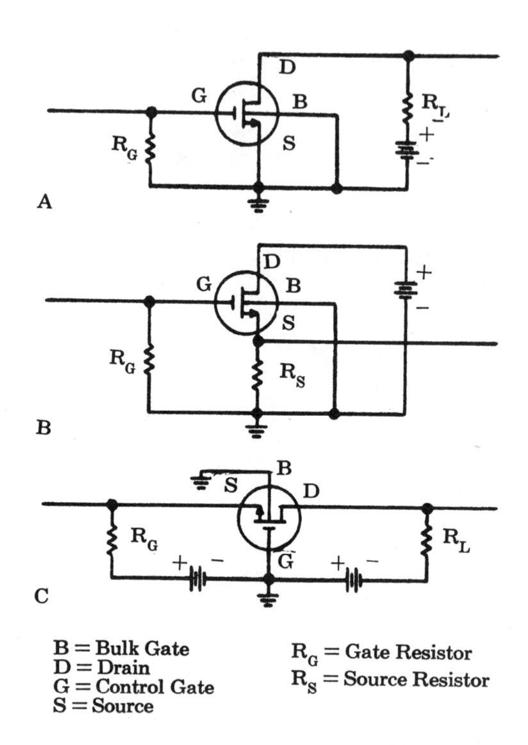

Fig. 11-21 The three basic amplifier configurations for FET:

(a) Common-source operation.
(b) Source-follower operation.
(c) Common-gate operation.

If the load is resistive, the input capacitance of the source follower is reduced by the negative feedback:

$$C_i = C_{gd} + (1 - A')C_{gs} \quad (11\text{-}7)$$

The output resistance is given by

$$R_o = \frac{r_d R_s}{(g_m r_d + 1) R_s + r_d} \quad (11\text{-}8)$$

Since $\mu = g_m r_d$ and is usually much greater than unity,

$$R_o \approx \frac{1}{g_m + G_s} \tag{11-9}$$

where

$$G_s = \frac{1}{R_s}$$

11.5.10.3 COMMON-GATE CIRCUIT

The common-gate circuit of Figure 11-21a is used to transform from a low to a high impedance. The input impedance of this configuration has approximately the same value as the output impedance of the source-follower circuit. The gain is given by:

$$A = \frac{(g_m r_d + 1) R_L}{(g_m r_d + 1) R_G + r_d + R_L} \tag{11-10}$$

11.5.10.4 MOS AS A VARIABLE RESISTOR

The MOS can be used as a variable resistor or voltage-controlled attenuator. A variation in gate voltage causes the drain-to-source resistance to vary. Within a certain range of gate voltage, this resistance change is linear. Figure 11-22 is an illustration of the variation of drain resistance with gate voltage. For high negative voltage, the characteristic is linear and the range is from 5kΩ to 1MΩ.

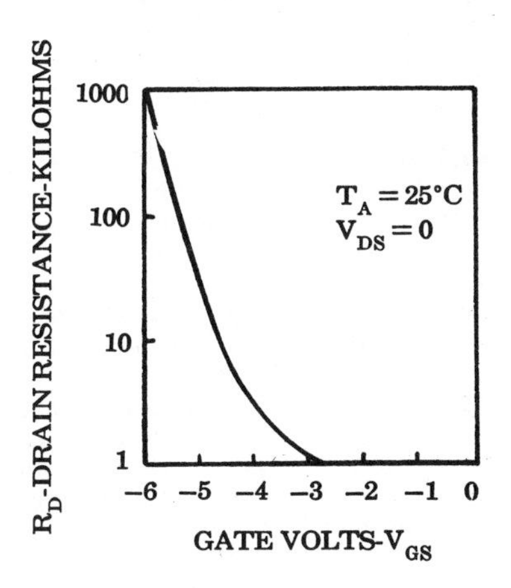

Fig. 11-22 Drain resistance characteristics.

11.5.10.5 MOS NOISE

In the MOS transistor, three mechanisms give rise to gate noise and thereby input noise in an amplifier. The main source of noise at high frequencies is thermal noise due to random fluctuations in the free-carrier concentration in the channel. Noise figures comparable to low-noise vacuum tubes, are obtained at frequencies above 50 MHz.

The low-frequency noise spectrum, which may extend up to tens of megacycles per second in some devices, is controlled by the fluctuations in the number of electrons occupying surface traps. It resembles an f^{-n} distribution. The value of n is generally between 1 and 2, and the spectrum extends down to very low frequencies. Shot noise is produced by fluctuations of the individual charges as they drift and diffuse toward the surface. Leakage noise is associated with the small flow of current through the oxide.

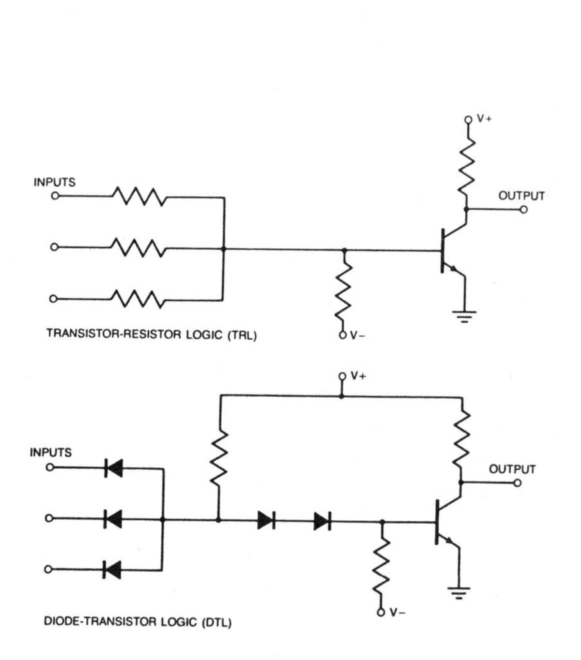

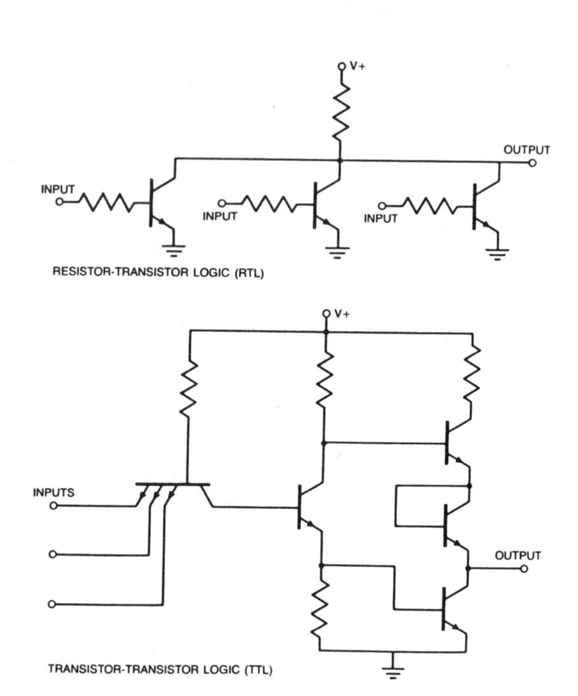

DIGITAL LOGIC CIRCUITS employing bipolar semiconductor devices have evolved toward a state in which transistors are adopted for almost all functions. Digital logic operates with signals that have only two recognizable levels (such as high voltage and low voltage); logic circuits accept such signals as inputs and transform them according to fixed rules to generate an output. The first families of bipolar logic circuits were constructed from discrete components. In transistor-resistor logic (TRL) the number of resistors was maximized, since they were the cheapest devices. In diode-transistor logic (DTL) performance was improved by substituting semiconductor diodes for many of the resistors. Resistor-transistor logic (RTL) was the first microelectronic technology; a transistor was supplied for each input and only a few resistors with small values were required. Transistor-transistor logic (TTL) is today the commonest form of bipolar microelectronic technology. Transistors are abundant and are coupled directly together. The circuit illustrated here includes a device, the multiple-emitter transistor, that has no equivalent among discrete components.

11.5.11 DIGITAL CIRCUITS

While design of digital circuits is similar to that of analog circuits, there are differences in the particulars of the design. Some of the more important digital characteristics are as follows:

(1) fan-in, (2) fan-out, (3) noise immunity (4) propagation delay, and (5) best logic type for a particular application. Design factors that must be considered before determining the desired diffusion profiles and

mask layout include the following: (1) reliability, (2) fan-out, (3) temperature range, (4) switching speed, (5) noise immunity, (6) power dissipation, and (7) production yield. Often a design that is best for one parameter acts to degrade another. For example, high switching speeds are not compatible with large fan-out and low power.

In preparing the specifications for the circuit it is necessary to determine the following:

(1) The type of transistor required, which includes setting the diffusion profiles and determining the geometrical layout
(2) The center values of the resistor
(3) The tolerance on transistor and resistor parameters

To determine these values it is necessary to state the basic black-box objectives of the digital circuit.

11.5.11.1 Logic Selection

Integrated circuits are available in a variety of logic types. Most digital processing can be performed by using IC's. The problem is usually not which IC to use, but rather which type of logic is best for a particular application. The user must be able to specify critical system-design parameters, and have a good understanding of the functions that each logic type can perform. With these data, he can choose the best logic type for a given application.

The most prevalent logic functions are the NOR function and the NAND function. These functions are performed with an OR gate followed by an inverter, and an AND gate followed by an inverter, respectively. The NOR function is performed by logic types such as DCTL and RTL for positive logic. With positive logic, the voltage level assigned to the "one" state is more positive than that assigned to the "zero" state. Negative logic is the opposite condition, i.e., the "one"-state voltage level is less positive than the "zero"-state voltage level.

Logic types such as DTL and TTL perform the NAND function for positive logic. Often the manufacturer will specify the logic function for both positive and negative logic on an IC data sheet. A gate that performs the NAND function for positive logic, will perform the NOR function for negative logic. The AND gate for positive logic performs the OR function for negative logic.

Some important factors that determine the selection of the logic type are the following:

- *Speed* — This factor is limited by the switching characteristics of the circuit and associated fabrication parasitics.
- *Noise Margin* — The magnitude of the smallest extraneous input signal that causes an error in the following chain of logic circuits. The value is usually referenced to the worst-case input-voltage level.
- *Noise Immunity* — The minimum noise margin divided by the maximum logic swing.

- *Fan-out* — The number of like loads the logic element is required to drive.
- *Power Dissipation* — The average d-c power required to operate the device.
- *Power-Speed Product* — The product of the device propagation delay (t_p) and its power dissipation.
- *Logic Capability* — This includes fan-in, inversion, and availability of complemented outputs.

Additional considerations are threshold levels and packaging. Also important is whether a particular circuit family includes triggered flip-flops, drivers, and other needed circuits, as well as logic gates.

The main function of any logic circuit is to provide a logic output, to one or a number of loads, in response to a logic input. An ideal logic-coupling network should provide high isolation between input and output, introduce as little signal delay as possible, keep the power dissipation as low as possible, and keep the number of components to a minimum.

11.5.11.2 Direct-Coupled-Transistor Logic (DCTL)

The simplicity of this logic type lends itself well to integrated-circuit fabrication. However, it has serious disadvantages resulting from high power dissipation, relatively small logic swing, and only fair noise immunity. Figure 11-23 illustrates a DCTL NAND/NOR gate.

DCTL has "current hogging" through which unequal distribution prevails of base current flowing into the bases in parallel from a common collector. Current hogging limits the fan-out and speed performance. In Fig. 11-23 the gate element should ideally supply equal base current to all loads. However, small variations in the base-to-emitter threshold characteristics, result in a wide variation in current supplied to the load.

This current waste limits the loading capability of DCTL gates, and limits their application, except where the base-to-emitter characteristics of the load transistors can be closely matched.

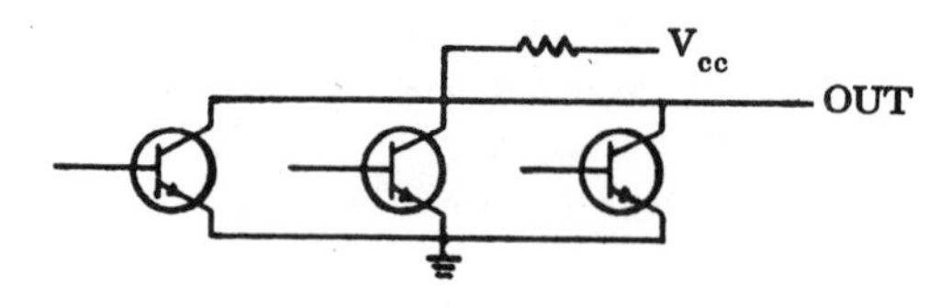

Fig. 11-23 DCTL gate.

11.5.11.3 Resistor-Transistor Logic (RTL)

To eliminate current hogging, DCTL can be modified by placing resistors or resistor-capacitor networks in the base leads. The resulting circuits are RTL and resistor-capacitor transistor logic (RCTL) circuits, respectively. Figure 11-24 is an RTL gate.

The addition of a base resistor increases the input resistance, and assures proper operation when more than one load is being driven. The addition of the base resistor to the circuit forces some of the input voltage to appear across it, as well as the base-emitter junction. As the value of this resistor is increased, the percentage of input voltage across the resistor increases, and the base currents are more evenly divided. This increases the fan-out capability, but reduces the speed somewhat since the addition of the base resistor to the circuit, causes longer storage and turn-off times.

The number of loads that an RTL output can drive is typically five, but special circuits are available that can drive considerably more. However, fan-out is limited because logic swing and noise immunity are adversely affected for high values of fan-out.

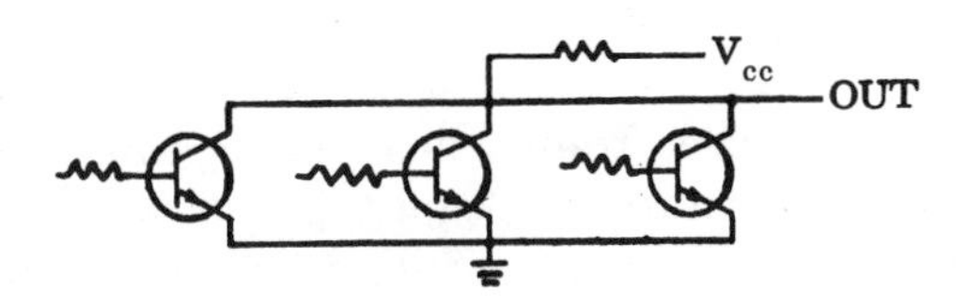

Fig. 11-24 RTL gate.

11.5.11.4 Diode-Transistor Logic (DTL)

DTL logic circuits are widely used. A typical DTL gate is illustrated in Figure 11-25. The diodes provide an additional input threshold, and thereby increase the impedance over that for RTL. In addition, DTL has a larger voltage swing and better noise margin than RTL. The circuit shown in Figure 11-25 requires two power supplies. However, some DTL circuits are operated with a single power supply. DTL has additional advantages such as lower power dissipation, larger drive capability, higher speed of operation, and better noise immunity, when compared with RTL.

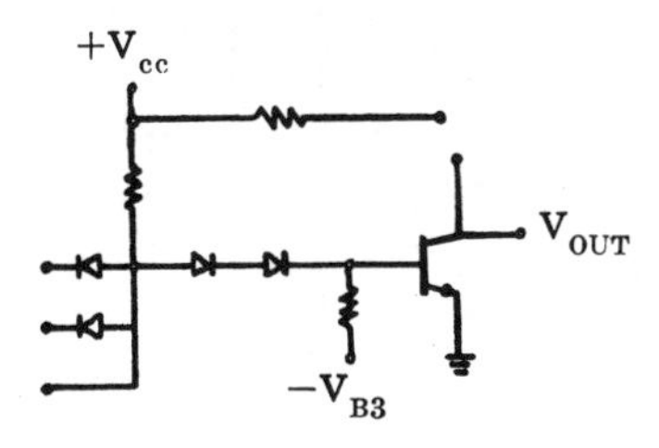

Fig. 11-25 Basic DTL gate circuit.

A modified unit of DTL is the Variable Threshold Logic (VTL). VTL operation is intended for use in circuits requiring very high noise immunity, such as relay and solenoid circuits. It is not intended for use in computer logic applications, because it has a poor power-speed product. It also requires two power supplies.

11.5.11.5 Current-Mode Logic (CML)

CML (also referred to as emitter-coupled logic, ECL) takes special advantage of monolithic fabrication techniques. CML uses the technique of current steering — switching well defined currents with small controlling voltages. Unlike other forms of logic discussed thus far, it is a non-saturating logic, and is therefore exceptionally fast. Propagation delays of 1 to 2 nsec are possible. It also features high fan-in and fan-out and high noise immunity.

A basic ECL gate is illustrated in Figure 11-26. Typically, the ECL circuit is designed with a differential-amplifier input and emitter-follower output, to restore d-c levels and provide a large output-drive capability. Two power supplies are required for this circuit. However, some CML circuits can be operated from a single power supply. CML circuits are not compatible with saturating logic types.

The ECL structure contains some interesting and important advantages. The circuit is reasonably independent of resistance tolerance, if resistor ratios are maintained. This is an important factor when one is considering diffused resistors. Another feature of this structure is the symmetry in its d-c input-output transfer characteristic, which results in approximately equal noise margins for both positive and negative noise. These margins are found to be substantially independent of temperature over a wide range.

11.5.11.6 Complementary Transistor Logic (CTL)

The CTL logic type is intended for high-speed digital applications similar to those of the CML. The CTL circuits use PNP emitter followers as the input AND diodes. Small currents at these inputs control the

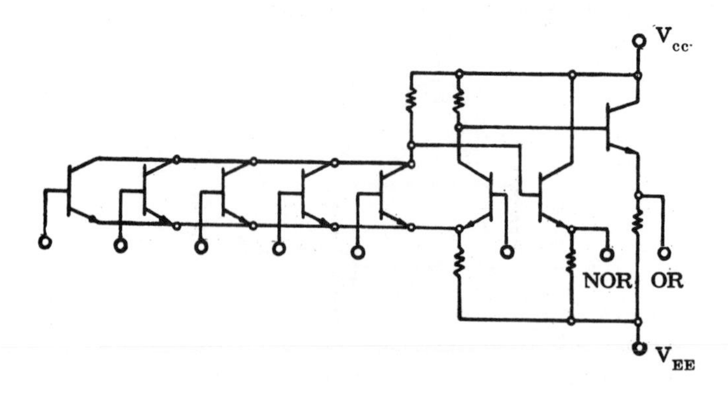

Fig. 11-26 ECL gate.

circuit. CTL exhibits high input impedance and low output impedance.

The emitter follower introduces a stability problem in both CTL and ECL. However, the problem is reported to be more severe in CTL, because there one emitter follower drives another emitter follower. In addition, the gain through the CTL AND/OR gate is less than unity. This limits the number of gates that may be cascaded. A CTL gate is illustrated in Figure 11-27.

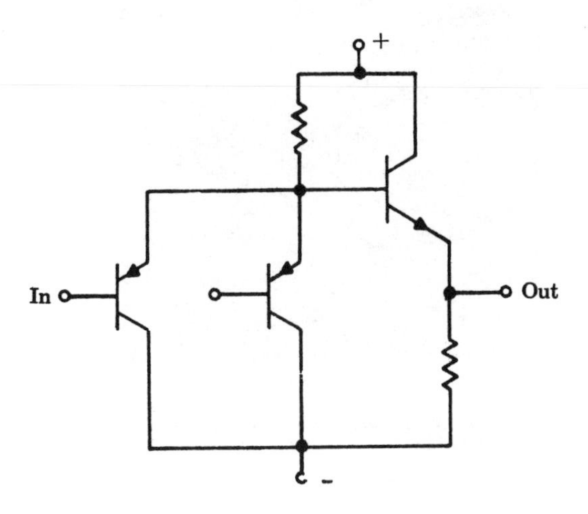

Fig. 11-27 CTL gate.

11.5.11.7 Transistor-Transistor Logic (TTL)

TTL is a modification of DTL, in which faster responding transistors are used in place of input diodes. A TTL gate is shown in Figure 11-28. This form of logic has a high speed – typically of the order of 3 nanoseconds. High speed can be maintained for high capacitive loads, because this logic form typically has both a pull-out and pull-down transistor in the output.

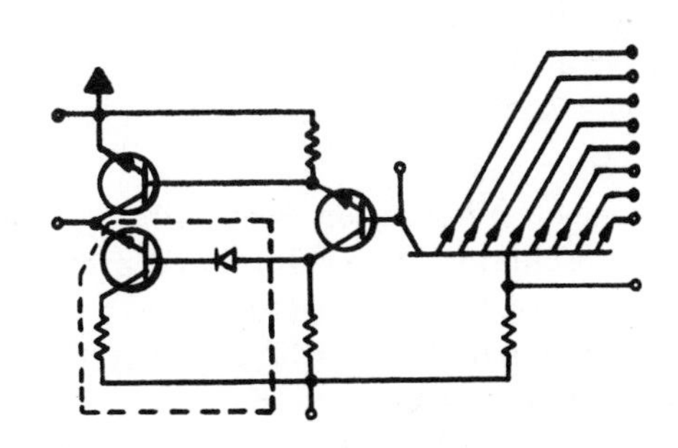

Fig. 11-28 TTL gate.

The threshold diode in Figure 11-28 can be in the base as shown, or in the emitter leg of the transistor enclosed by the dashed lines. When the diode is located in the base circuit, it reduces the speed of the circuit, because of substrate parasitic capacitance introduced at the collector node of the previous transistor. Locating the diode in the emitter removes this capacitance, and provides improved speed characteristic.

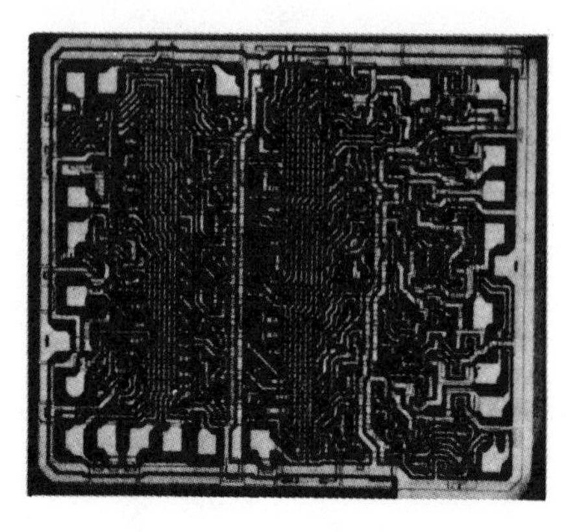

Fig. 11-29 Arithmetic logic for performing addition and subtraction.

11.5.11.8 Integrated Micropower Circuits

An extensive investigation has been carried out on low-power logic circuits for space research. Seven basic logic circuits were studied. They were Direct Coupled Transistor Logic (DCTL), Resistor-Transistor Logic (RTL), Resistor-Diode-Transistor Logic (RDTL), Emitter Coupled Transistor Logic (ECTL), Transistor-Transistor Logic (TTL), Diode Transistor Logic (DTL), and Complementary RDTL. The power drain capabilities of each circuit were considered as a function of fan-out and fan-in, switching speed, available spacecraft voltages, radiation damage, logic levels and operating temperature.

The comparative performance of each circuit regarding power drain and propagation time was made using worst-case design criteria. The most promising circuits were then tested thoroughly for temperature and load characteristics. The circuit configurations investigated were NAND and NOR gates, a universal flip-flop, and other applications utilizing flip-flops as shift accumulators. Circuits were tested over a temperature range from $-55°$ C to $+125°$ C.

Five gates for each configuration were built and connected in a closed loop to form a ring oscillator. Power drain and propagation time for each oscillator stage were used as the performance parameters. The results obtained, easily eliminated RTL, DCTL, and RDTL from further consideration. RTL showed excessive power drain and DCTL was subject to current "hogging." RDTL must be operated at low voltages to obtain acceptable results, making it sensitive to resistor and transistor beta variations. ECTL showed no improvement over DTL, and it is more difficult to design, in the gate and flip-flop configuration. Therefore, it was also eliminated from further consideration.

The three circuits investigated in more detail, Complementary RDTL, DTL, and TTL, are shown schematically in Fig 11-30.

Further investigation showed that the TTL circuits exhibited the most consistent performance characteristics over the temperature range. However, general system performance must be considered in addition to in-

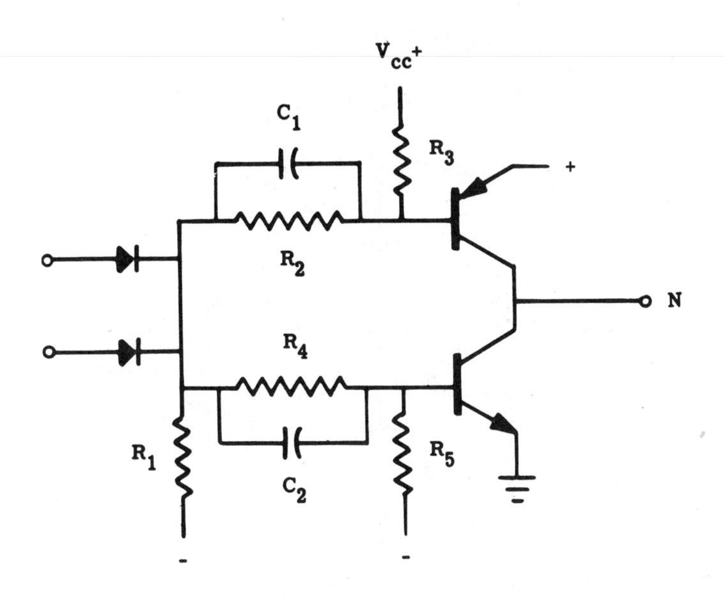

(a) CRDTL

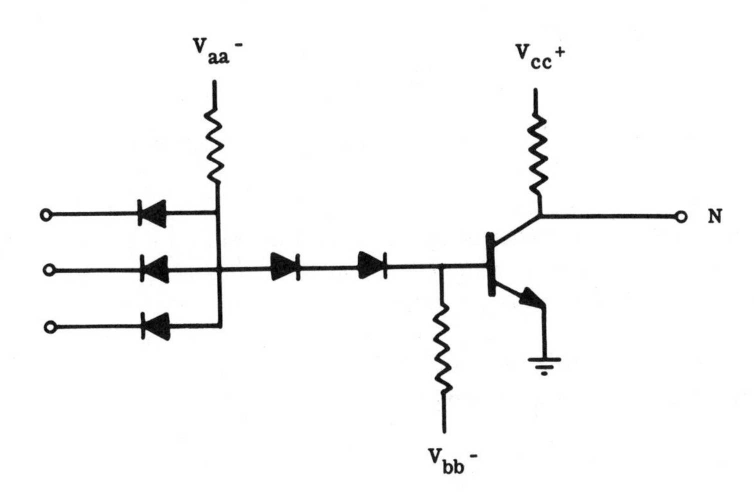

(b) DTL

Fig. 11-30 Micropower logic circuit types

dividual circuit performance and the TTL configuration has certain disadvantages. The complementing flip-flop is complex for the TTL circuit, requiring the equivalent of eight gates. Power drain is, consequently, high. This results from the small voltage swing of the gate which prohibits the use of ac steering. TTL is also more noise susceptible than other configurations because of the small turn-off voltage of the transistor base. Noise susceptibility also tends to be increased by the active gating elements because of the lack of a current limiting resistance element. The turn-off

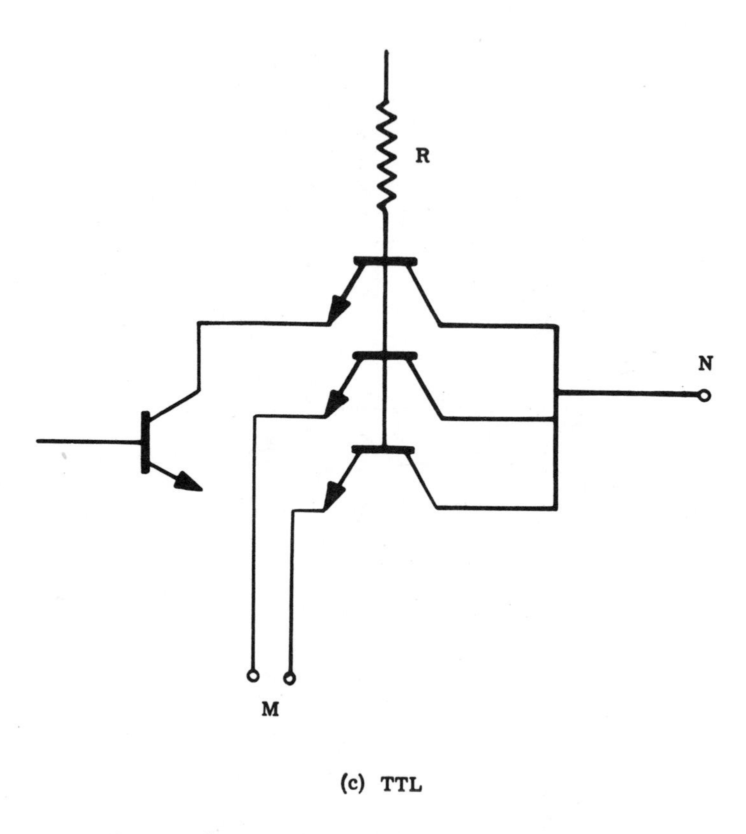

(c) TTL

Fig. 11-30 (continued) Micropower logic circuit types

voltage decreases with increasing temperature and the gate transistor requires a low offset voltage and a low inverse beta to minimize the turn-off voltage problem and prevent large reductions in base currents. These two requirements are difficult to obtain in the same device. Therefore, the advantage of low power drain for TTL may be outweighed by the inherent disadvantages in general system use.

Both DTL and Complementary RDTL permit ac coupled flip-flops because of larger output voltage swings. Satisfactory performance for DTL flip-flop operation over the temperature range is attained only by increasing the collector supply voltage, thereby increasing the power gain. NAND and NOR gates are not directly available in the basic DTL configuration. While flip-flops using ac steering are possible, their performance over a wide temperature range is poor.

The complementary RDTL configuration has the advantage of permitting NAND and NOR gates to be easily fabricated with similar design and performance characteristics. Additional NAND and NOR gates can be driven simultaneously, permitting greater design flexibility. Speedup of waveform fall time is obtained by transistors switching on for each logic level, thereby improving switching speeds. The worst case propagation time is degraded at high temperatures, although not as badly as DTL. Noise immunity is better than in the TTL, and turn off voltage is about half way between the values obtained with DTL and TTL. Complementary pair transistor beta requirements are slightly higher, but are not prohibitive.

The complementary RDTL flip-flop is not as complex as two gates and a considerable reduction in power drain is obtained over the other flip-flops considered. Also, ac set and reset features are possible. Considering circuit and system performance, complementary micropower RDTL seems to offer the best performance characteristics. At operating frequencies around 200 kc, DTL begins to gain in performance in relation to complementary RDTL, and at higher operating frequencies, DTL is probably the better of the two logic circuits.

Fig. 11-31 A 1 "x 2" module containing 320 TTL NAND gates. Interconnection of the 8 chips is by top metallization of the substrate only.

11.5.11.9 MOS Digital Circuits

The potential advantages of MOS devices in integrated digital arrays can be realized only if proper consideration is given to circuit choice. Although many circuit configurations are possible, only one or two take advantage of the unique properties of MOS devices to achieve a high figure of merit of functional complexity per unit at high yield.

Figure 11-33 shows an inverter that uses an MOS inverting transistor and a resistor. For small MOS structures, a high value of resistance is required (greater than 10,000 ohms). If the resistor is formed by the source and drain diffusion, it occupies too large an area. If a separate diffusion is used, or if the resistor is formed as a thin film over the diode, the processing is increased in complexity. In either case, the figure of merit for the circuit is reduced.

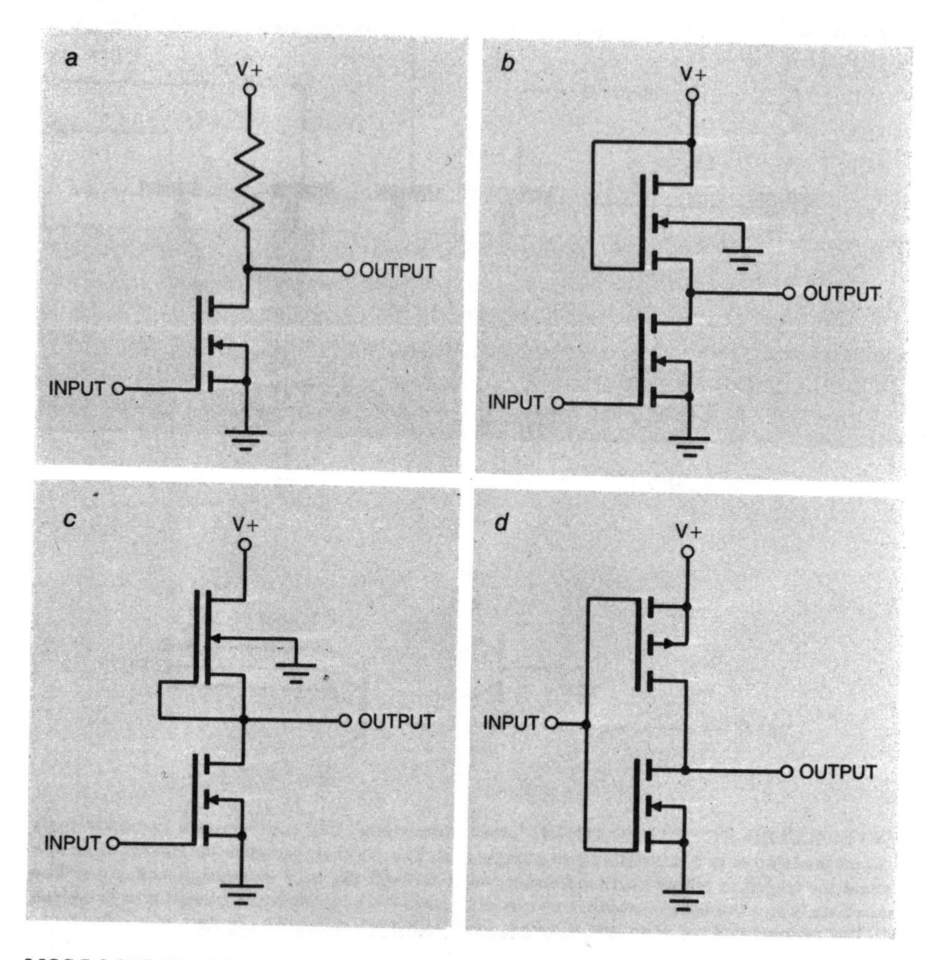

MOS LOGIC CIRCUITS were developed later than bipolar ones and for the most part have been built only in microelectronic form. Each of the circuits here performs the function of inverting a signal, so that if the input is high, the output is low, and vice versa. In each case the input is applied to the gate of an enhancement-mode *n*-MOS transistor; the circuits are distinguished by the choice of a load element needed to limit the current through this transistor. The simplest choice is a resistor (*a*), but it also gives the poorest performance. A second enhancement-mode *n*-MOS transistor (*b*) is the easiest to fabricate; a depletion-mode *n*-MOS transistor (*c*) gives the highest packing density. Finally, by adding an enhancement-mode *p*-channel device (*d*) the low power consumption of a complementary MOS circuit can be achieved.

Figure 11-34 shows an inverter that uses a depletion-type MOS transistor as a load for an enhancement-type MOS inverter. A depletion-type MOS transistor is one that conducts appreciable current with zero bias between gate and source.

This circuit has a number of advantages. First, its area is potentially very small because it consists only of MOS devices. Second, the shape of the load line that results is capable of improved speed as compared with an ohmic resistor. From a processing standpoint, however, it has the disadvantage that two types of MOS devices, with different threshold

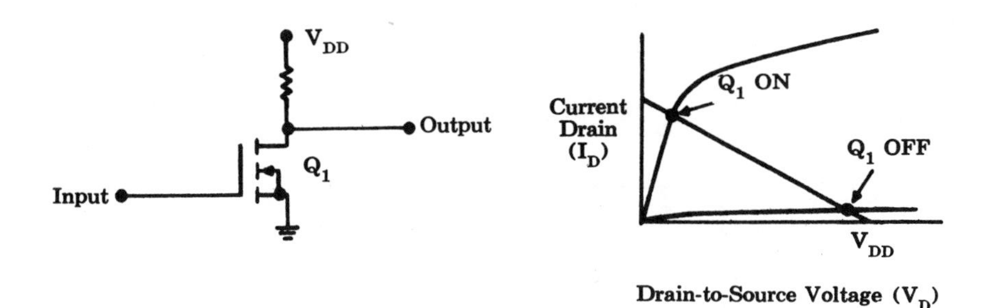

Fig. 11-33 Inverter using MOS transistor and a resistor with typical load line.

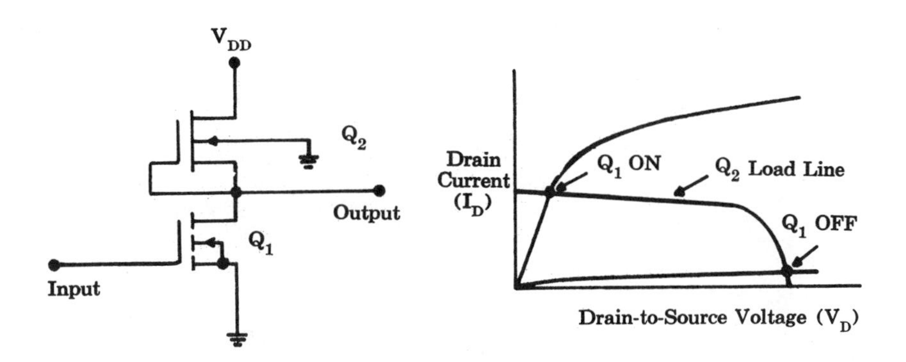

Fig. 11-34 Inverter using depletion-type MOS transistor as load.

voltages, are required. The resulting processing complexities tend to lower its figure of merit drastically.

Figure 11-35 shows an inverter that uses an enhancement-type MOS transistor connected as a source follower to serve as a load for the inverting MOS transistor. Although this circuit is a poor choice from a discrete-component point of view, it possesses a very high figure of merit considered in the light of digital integrated arrays. First, it uses the absolute minimum area because, to operate at all, the load MOS device must be smaller than the inverting MOS device. Second, the processing required to fabricate the circuit is also minimum because the load MOS device is made by the same process as that used for the inverter MOS device.The operation of the circuit depends·on the control of the ratio of the transconductance of the two devices. Although the absolute value of the transconductance depends on a number of factors, including the oxide thickness under the gate, the ratio is determined by the geometries of the two devices. When accurate masks are used, therefore, the ratio can be held constant even with variations in processing.

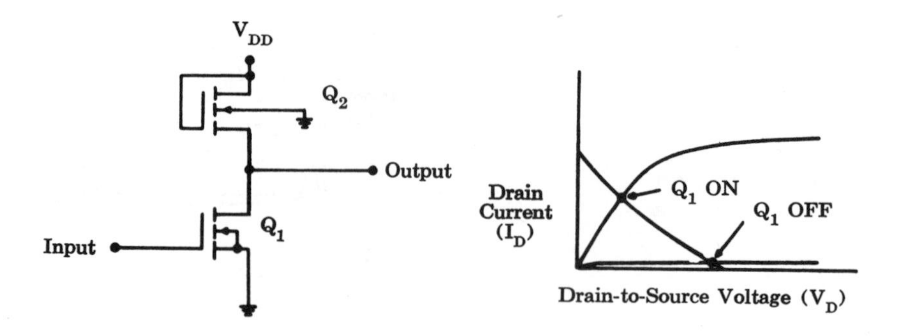

Fig. 11-35 Inverter using enhancement-type MOS transistor.

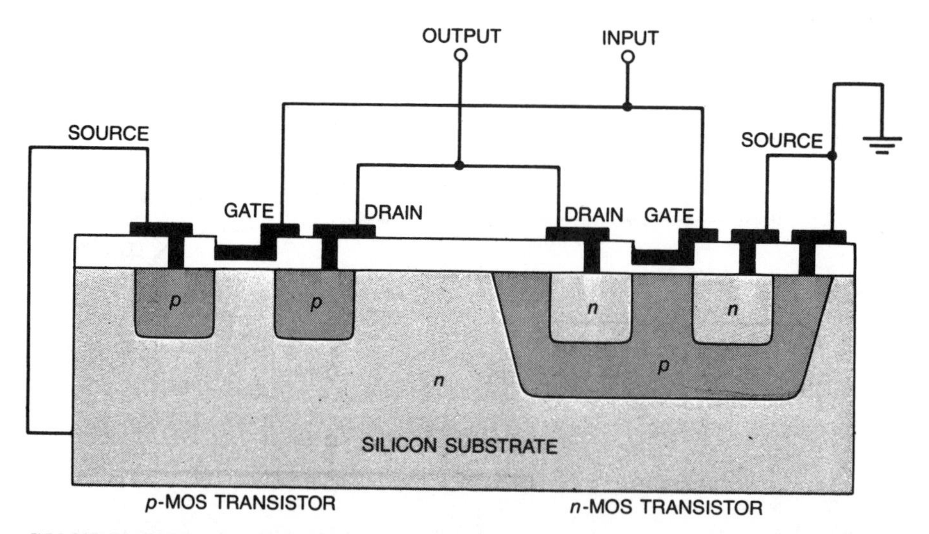

COMPLEMENTARY MOS DEVICE includes both *n*-MOS and *p*-MOS transistors on a single chip of silicon. If the circuit is fabricated in an *n*-type substrate, as it is here, then a *p*-channel transistor can be made in the normal way, but an *n*-channel device requires an island of *p*-type material. The need for such islands adds another step to the manufacturing process and also reduces the packing density, and thus the total number of transistors that can be fitted onto a single chip. Complementary MOS devices can be arranged to achieve low power consumption. In the circuit shown here the gates of both transistors are connected to a single input; since the two transistors require signals of opposite polarity for conduction they are never turned on at the same time and little current flows from the power supply ($V+$) to ground.

A final inverter circuit is shown in Figure 11-36. It consists of a pair of complementary MOS devices connected in series, with the gates connected to each other and driven by the input signal. When the input signal is zero, the P-channel device is on and the N-channel device is off. The reverse is true when the input signal is positive. Each device, when on, is required to supply a direct current equal to only the leakage current of the other device. During a transistion of the input signal, however, capacitive loads are charged through the low output impedance of one or the other of the two devices. Thus, although it requires very low

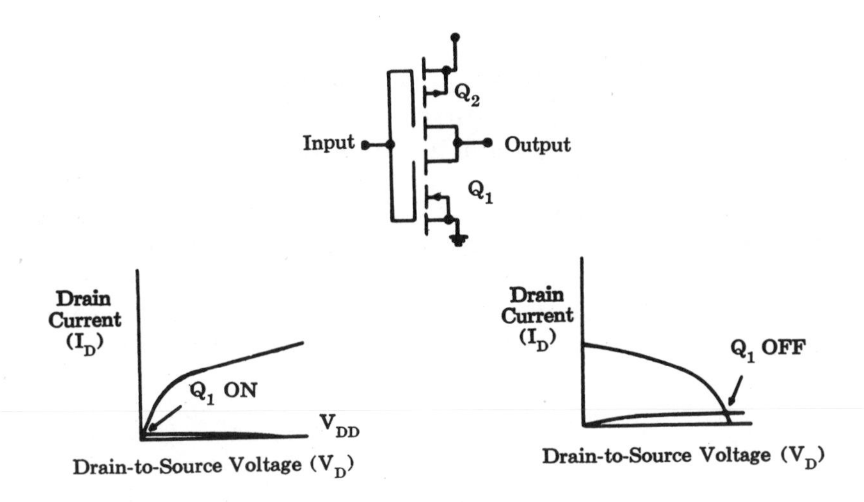

Fig. 11-36 Inverter and load lines with complementary MOS devices.

standby power, the circuit is inherently fast. The load lines in Figure 11-36 indicate the very low standby power. The figure of merit for the circuit is not as high as for the source-follower circuit shown in Figure 11-35, because slightly more area is required and the processing is considerably more involved. Despite these disadvantages, arrays of complementary MOS circuits are beginning to find applications where either very low standby power or high speed is important enough to compensate for the reduced figure of merit.

The total power dissipated, neglecting the standby power, in a complementary MOS circuit during switching is given by

$$P = C_L V_{DD}^2$$

where C_L is the total output capacitance and V_{DD} is the supply voltage.

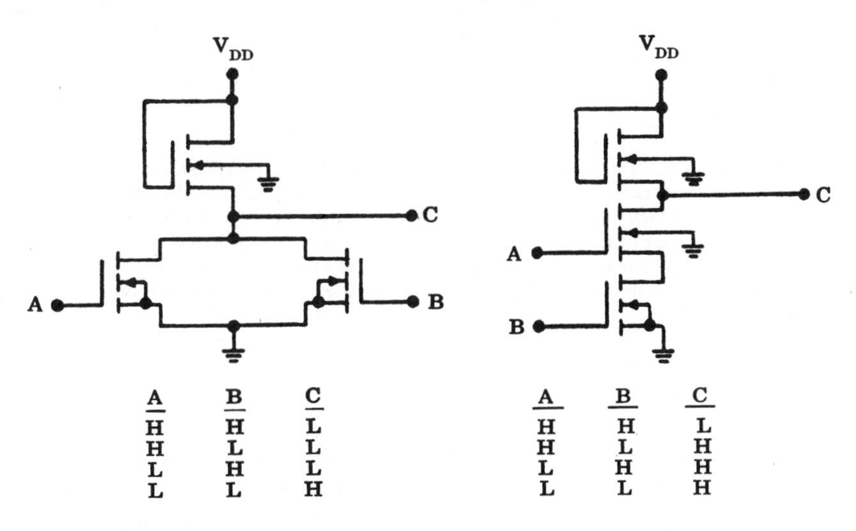

A	B	C
H	H	L
H	L	L
L	H	L
L	L	H

A	B	C
H	H	L
H	L	H
L	H	H
L	L	H

Fig. 11-37 NOR gate using parallel connection of MOS transistors.

Fig. 11-38 NAND gate using series connection of MOS transistors.

Single-Channel Arrays

The basic circuit shown in Figure 11-36 can be developed into a wide variety of digital circuit arrays. Parallel connection of the inverter transistor forms a NOR gate, as shown in Figure 11-37. Figure 11-38 shows how a NAND gate can be formed by series connection of the inverting transistors. When the NAND gate of Figure 11-38 is used in conjunction with the NOR gate of Figure 11-37, the resulting configuration offers a high degree of logic flexibility. However, the series connection has the disadvantage that the inverter transistors must be twice as large as their parallel counterparts, to maintain the same control of logic levels.

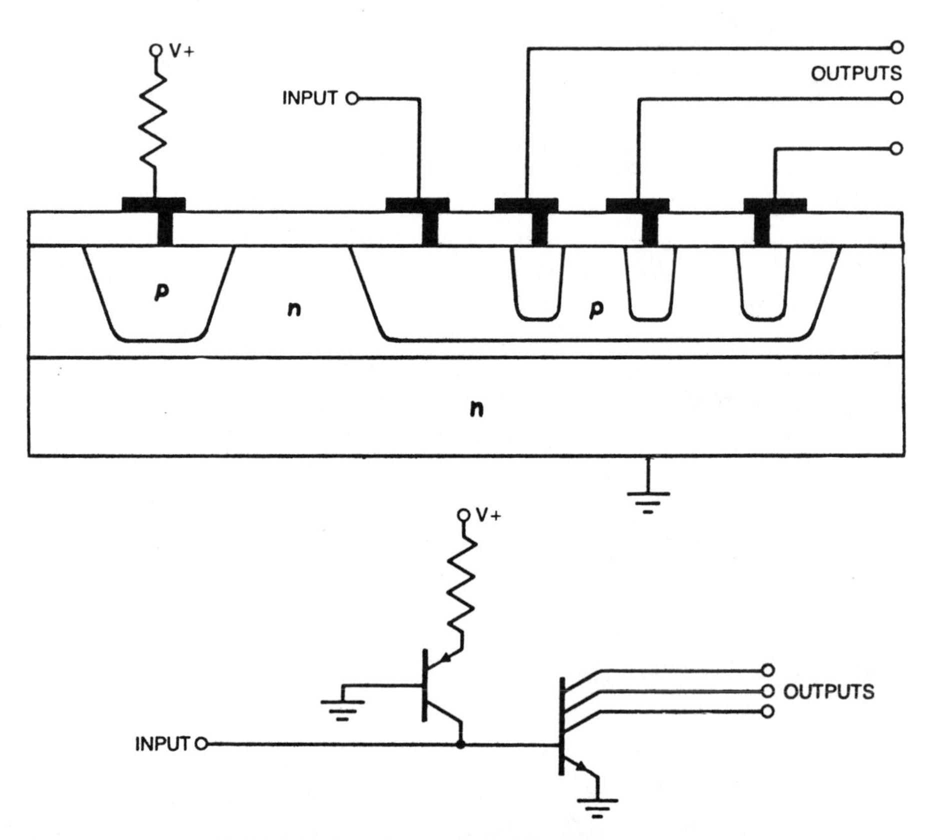

INTEGRATED-INJECTION LOGIC, often abbreviated I^2L, compresses a complete logic circuit made up of two transistors into a single unit. The substrate serves as the emitter of an *npn* transistor (*right*) in which current flows upward through the base to multiple collectors. The substrate is also the base of another transistor, a *pnp* device in which the current flow is lateral. By this arrangement of elements isolation islands that are normally required in bipolar technology are eliminated and packing densities similar to those of MOS circuits can be attained.

Dynamic Circuits

The NOR and NAND logic functions can be accomplished with dynamic circuit structures (Fig. 11-38. 2). The dynamic approach involves a precharging of load capacitor C_L during the clock (ϕ) pulse duration. Then C_L is either left charged or is discharged, depending on the input states, after the clock turns off and isolates C_L from the power supply. Either circuit in Fig. 11-38. 2 dissipates less power than corresponding static logic gates. Address decoders for random-access memories often use this technique.

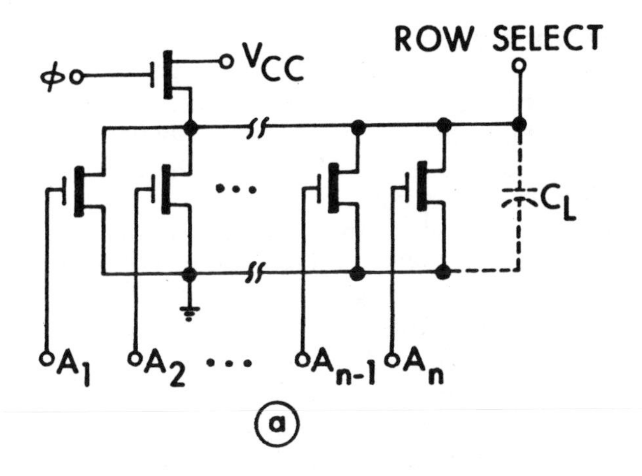

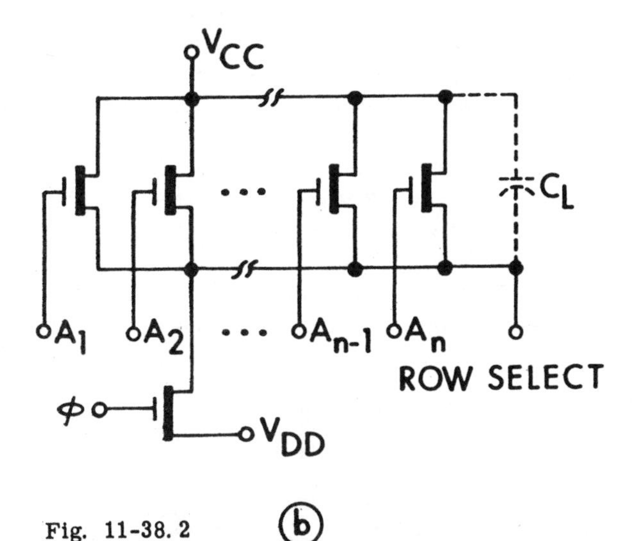

Fig. 11-38. 2

Dynamic circuit structures dissipate less power than static configurations. Shown here are a dynamic NOR gate using n-channel devices (a) and a dynamic NAND gate using p-channel transistors (b).

Memory

A simple way to obtain a memory function is to use the bistable multivibrator or flip-flop. Fig. 11-38. 3 shows the MOS version of a basic R-S flip-flop. The circuit uses p-channel MOS devices and consists of two cross-coupled inverters.

Transistors Q_1 and Q_4 provide access to the memory cell. The ON resistance ratio of Q_1 and either Q_2 or Q_3, depending on the cell state, determine the logic ONE, or most positive-voltage, level. The logic ZERO level is $-V + |V_T|$.

The circuit is often used in static MOS read/write random-access memories. A static memory holds data as long as power is applied. Hence data refresh can be avoided. The n-channel circuit is the same as the p-channel, except that the power-supply polarity is positive rather than negative.

To build a memory matrix, the device designer arranges memory cells in an array so that an addressed cell connects to a pair of "sense" lines. If a read cycle is required, information stored in the selected cell appears on the DATA and $\overline{\text{DATA}}$ lines. Output circuitry then senses the information. If a write cycle is required, the DATA and $\overline{\text{DATA}}$ lines are forced to the appropriate states, and the selected cell is set accordingly. This cell design requires six devices and 4-1/2 interconnections.

The high input impedance of MOS inverters permits the use of dynamic memory-cell designs. In a dynamic cell a parasitic capacitance stores charge that is periodically refreshed. The refresh compensates for the small leakage current at the storage nodes. By comparison, a static cell continuously refreshes the charge through load resistors.

Fig. 11-38. 4 shows the evolution of dynamic MOS RAM cells to higher densities. Cell sizes are distinguished by two factors: One is the number of transistors used to form the memory function; the other is the number of interconnect lines needed to connect the cells to the rest of the circuitry.

The dynamic cell in Fig. 11-38. 4a is identical to the static cell previously described, except that the dc load devices have been removed. As a result, the cell has two less transistors and one less interconnect and is correspondingly smaller. The substrate connection counts as a half interconnect. This dynamic cell maintains the differential data-sense feature offered by the static cell.

A simpler dynamic cell (Fig. 11-38. 4b) has one less transistor and one more interconnect. This cell is used in the 1103 type of 1024-bit dynamic RAM. One interconnect can be eliminated for increased density if we make the READ DATA and WRITE DATA lines common, as in Fig. 11-38. 4c. Each access of a cell inverts the polarity of the data on that cell's entire column.

For larger memories, the one-transistor cell of Fig. 11-38. 4d provides still higher densities. This cell uses a series transfer device to isolate the storage capacitor and a ROW SELECT line to control the transfer device. Only 2-1/2 interconnects are needed in the one-transistor cell.

MOS devices can also be used for serial-access memories, or shift-register functions. The basic dynamic shift-register cell uses two MOS inverters coupled together through transfer devices (Fig. 11-38. 5). These transfer devices isolate one cell half from the other during shifting so that data entering the cell do not interfere with data leaving.

The use of high-voltage clocks minimizes the size of the transfer and clocked-load devices for a given required ON resistance. The size reduction occurs because the ON resistance decreases as the gate-to-source voltage increases.

Capacitor C_1 or C_2 can be charged to a voltage of either ($V_{CLOCK} - 2V_T$) or V_{DD}, whichever is lower. Operation of the load and transfer devices in the nonsaturated mode means that $|V_{CLOCK}| > |V_{DD} + 2V_T|$. As a result, the voltage on C_1 or C_2 can be maximized without dissipation of excessive power.

When ϕ_{IN} is low, data stored on capacitor C_1 transfer to C_2 through Q_3. During the time ϕ_{OUT} is low, data transfer to the output structure through Q_6. Hence data shift from one cell to the next with each pair of input-output clock pulses.

Since load devices Q_2 and Q_5 are turned on and off by the clocks, dc current flows between V_{DD} and the substrate only when a clock is low. Consequently the power consumed varies with the duty cycle of the clocks.

The use of dynamic logic for serial memories necessitates refreshing to prevent loss of data, just as in dynamic random-access memories. Here shifting accomplishes the refresh, and a minimum clock-frequency constraint must be observed on the use of the device. Operation at less than the minimum clock rate can allow the charge stored on C_1 or C_2 to leak away, and the cell "forgets" the information it was storing.

For static shift registers, a more complex cell is used (Fig. 11-38. 6). The cell contains two cascaded inverters and feedback to accomplish a latching function. Serial transfer device Q_5 gates the feedback path and device Q_3 gates the feed-forward path. Devices Q_1 and Q_3 isolate each cell half from the other.

When the clock input is in the active state (either high or low, depending on the particular device) ϕ_1 is low—about -12 V. Data then transfer through Q_1 to C_1 from either the previous cell or the data input buffers. During this time ϕ_2 and ϕ_3 are high—about +5 V—and C_2 is isolated from the input of the cell.

When the clock input returns to the inactive state, ϕ_1 goes high, ϕ_2 goes low and ϕ_3 begins to fall. C_1 is now isolated from the cell input, and the data transfer to C_2.

At this point the cell has completed a full clock cycle and data have shifted from the cell input to the cell output. So far the cell operates in much the same way as a dynamic shift register cell. The static cell differs in the operation of the load devices, Q_L. These are not clocked, as in the dynamic cell, but are always conducting.

The static feedback path is enabled only at relatively low frequencies, since an interval of about 5 μs occurs before ϕ_3 goes low. Were ϕ_3 to go low as quickly as ϕ_2, the cell would not have time to "flip" and would be bypassed through Q_5.

Note that Q_3 is on only during the inactive clock state. As a result, the device cannot retain

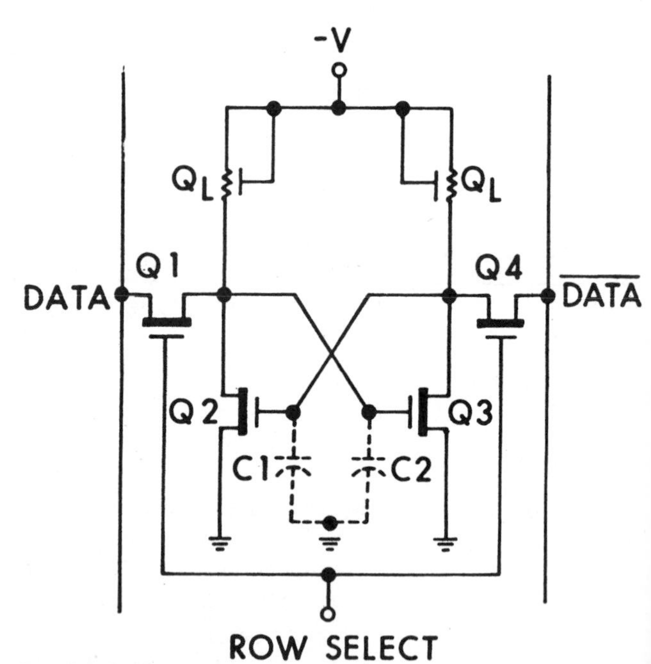

Fig. 11-38. 3

The MOS version of a static RAM cell consists of a simple R-S flip-flop. The RAM cell uses six devices and requires 4-1/2 interconnects.

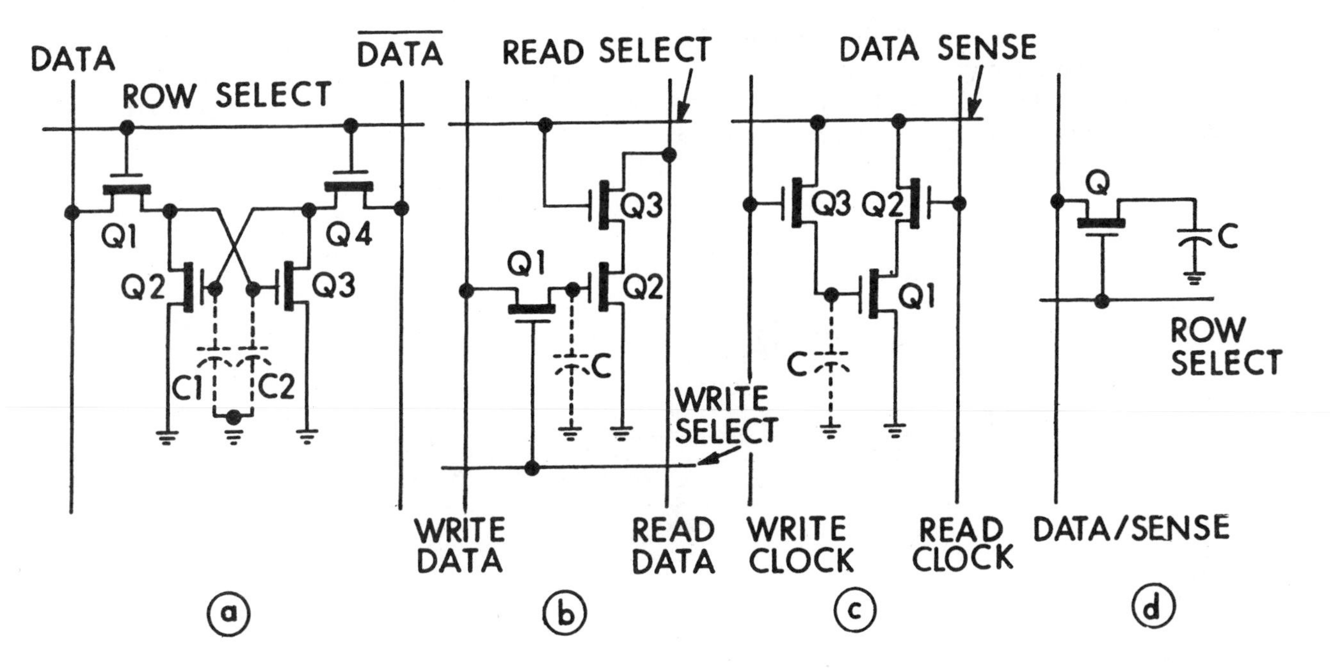

Fig. 11-38.4

Dynamic RAM cells differ in the number of transistors and the corresponding number of interconnects. Cells can have as many as four, and as few as one, transistor per cell (a through d, respectively). The four and three-transistor cells (a, b and c) use 3-1/2 and 4-1/2 interconnects.

data in the static state if the input clock signal stops in the active state. This constraint causes an upper limit to be placed on clock pulse width. Conversely, when ϕ_3 does latch up the cell, data are retained without refresh as long as power is applied.

MOS for ROMs

A typical MOS mask-programmed ROM cell appears in Fig. 11-38.7. To store a positive logic ONE in a particular cell, a hole is cut in the source-drain diffusion mask for transistor Q_1 at the location of the cell. Conversely, for a logic ZERO, no hole is cut.

Then to gain access to a given cell, the appropriate row and column are selected through address decoders. The bit-line voltage either pulls up to the substrate voltage (through Q_1) by means of a mask-programmed gate, or it pulls down to V_{GG} through load device Q_3. The resulting level is sensed, buffered and presented to the appropriate output.

The data in a ROM cell matrix are nonvolatile

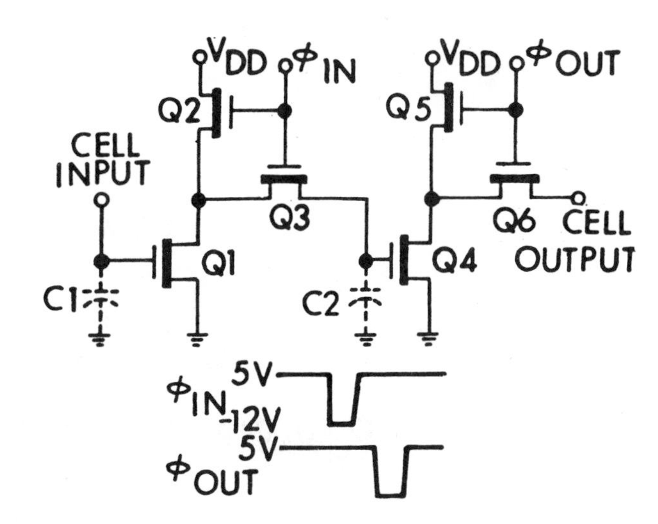

Fig. 11-38.5

A dynamic shift-register cell. It uses high-voltage clocks to minimize the size of both transfer devices and load devices.

—supply voltages don't affect the presence or absence of a device. However some of the peripheral logic on a ROM chip can be dynamic. For example, some large ROMs use dynamic address decoders to save power. These ROMs require a logic signal that is used as a clock. The IC is called a dynamic ROM, even though the stored data are not dynamic.

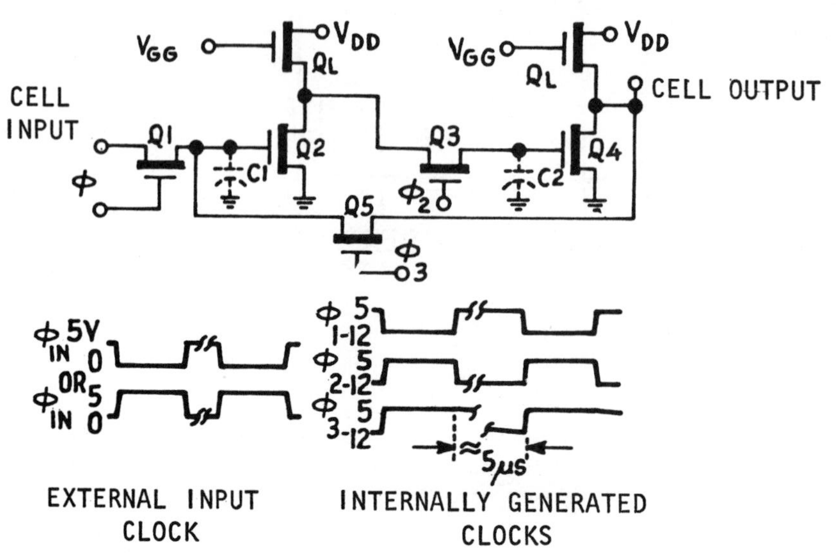

Fig. 11-38.6

Static shift registers require more complex cell structures than dynamic shift registers. The purpose of the additional circuitry is to provide level shift and multiphase high-voltage clock signals.

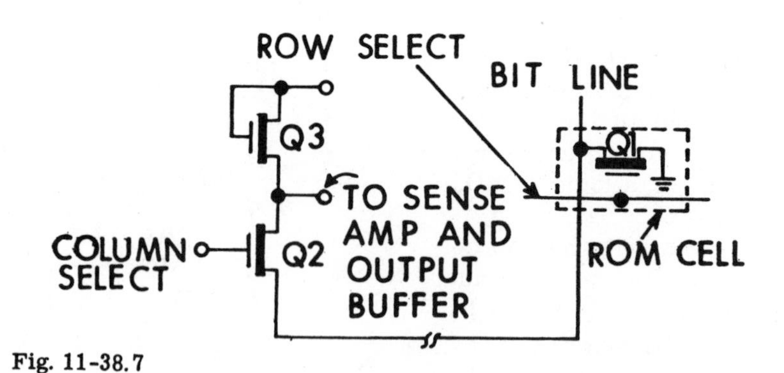

Fig. 11-38.7

A ROM is formed by mask programming a device where a ONE is required and omitting a device where a ZERO is called for. Row and column select lines are used to provide access to a cell.

11.5.11.10 Noise Immunity

An important characteristic of a logic circuit is its ability to be used in a variety of environments with a minimum of "hand trimming" and "de-bugging". This requires an ability to have a high degree of immunity to noise, in both the "on" and "off" states. This ability reduces, or eliminates, the need for attention to precise clocking and phasing, as well as the need for elaborate shielding and grounding.

There are essentially two types of "noise" that can cause malfunction in a digital circuit. The first of these, commonly referred to as d-c noise, is characterized by its presence in the form of pulses that are so wide that they present, in essence, a d-c input to the gates. Such pulses are commonly associated with power supply transients and resistively coupled paths. The second type of noise is known as pulsed noise. Far more frequently encountered, this noise consists of low duty cycle pulses, and is capable of being generated or picked-up whenever short rise-time signals are being transmitted or processed.

11.6 COMPUTER AIDED CIRCUIT LAYOUT

Once a circuit has been designed, its reduction to microelectronic form involves the layout and interconnection of components on a substrate. This step and the production of the corresponding art-work are lengthy and expensive processes. Thin-film circuits can take several man-weeks of effort and the more elaborate silicon integrated circuits may require many man-months of work.

The risk of human error during these processes rises rapidly as circuits become more complex. However, many of the stages involved can be described in mathematical terms which can be handled quickly and accurately by a digital computer.

Steps such as the calculation of the mechanical dimensions of components can be programmed easily, but it is the process of arranging these elements on the substrate, which is difficult. This is because it calls for both visual perception and imagination.

In one method outlined in Figure 11-39, a rough layout is automatically formed and is unconstrained in size. Modification is then made with the aid of a computer. Any attempt to constrain the size of the rough layout results in conflicts which, at present, need human intervention to resolve them. When modifying the rough layout, the designer communicates his experience and imagination to the computer, using a simple drawing language. A control typewriter or teleprinter forms a suitable input device, and the progress of the work is viewed using an on-line plotter. During this phase of the work, the designer does not have to worry about the underlying co-ordinate geometry of the situation, or the constraints imposed by the particular technology.

When the final layout is decided upon, the computer can verify that it is electrically correct before proceeding to produce detailed drawings. To eliminate any possible errors, it is preferable that the drawings be

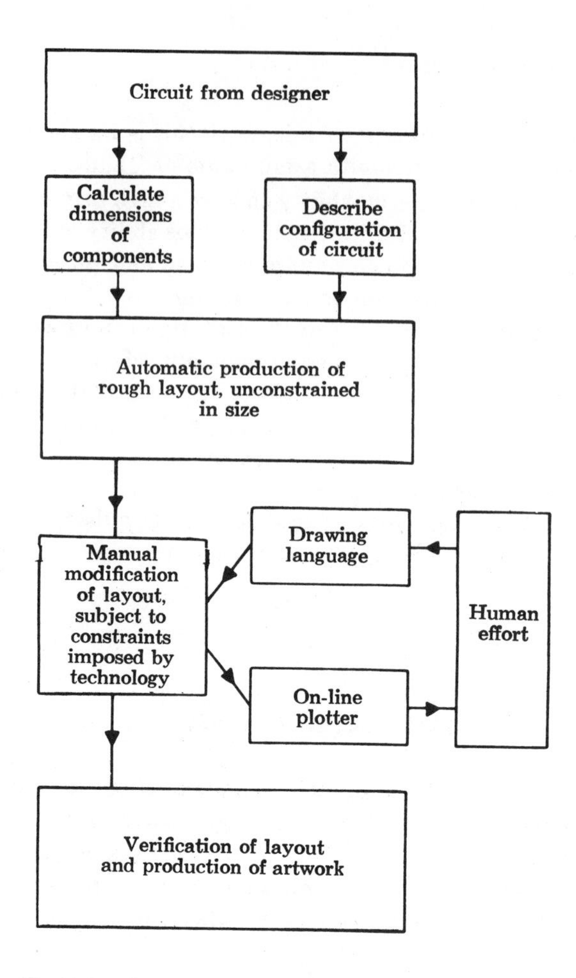

Fig. 11-39 Method for computer-aided layout of circuits.

produced on-line. One of the more difficult parts of the process outlined in Figure 11-39, is the creation of a drawing language. This is a means of communicating with the machine which enables the designer to modify the rough layout.

11.6.1 DECIDING ON THE ELECTRICAL STRUCTURE

The ideal microcircuit is one which is planar and can therefore be arranged on a substrate, without introducing any crossovers between conducting materials. Such networks are rare, and many practical microcircuits contain capacitors formed by one conductor bridging another, and separated from it by a dielectric. Part of the art of layout is to maneuver the capacitors which are specified in the designer's circuit, into positions where they can be employed as crossovers.

A suitable way of describing the structure, is to draw the circuit on a grid shown in Figure 11-40 so that it may be defined in simple numerical terms. In the example of Figure 11-40, the capacitor C3 is used as a cross-over, and once this decision has been made, it is not altered in the program. The positions of the solder pads for external connections are also defined in this drawing.

From this starting point, the subsequent steps in creating the layout involve the calculations of the physical sizes of components, and their relative positions on the substrate.

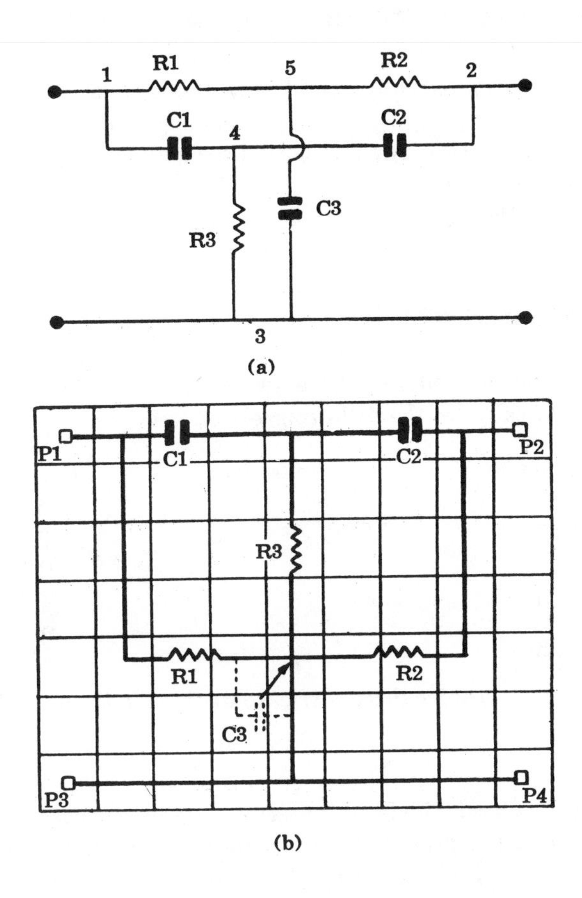

Fig. 11-40 Description of electrical structure.
(a) Conventional circuit.
(b) Corresponding microcircuit.

11.6.2 MAN/MACHINE COMMUNICATIONS

It is the interaction between man and machine which eases the task of laying out the circuit. The computer acts as a store for the parts of the puzzle, and provides the means for moving them to positions decided by eye. An optimum solution is difficult to obtain.

The input to the machine should be in a form which can be easily understood by the user. A drawing language consisting of familiar words such as MOVE, TURN and GROW fulfils this need. Using a graphical plotter connected on-line to the computer, communication from the machine to the designer can be readily achieved.

A cathode-ray tube and light-pen can be used to provide the required communication. However, the disadvantages of this system are that few installations have a suitable interface, and the amount of fast access store needed for this problem is very large.

11.6.3 NUMERICAL DESCRIPTION OF DESIGNER'S CIRCUIT

The circuit shown in Figure 11-40b has to be converted into a suitable form of input for the computer. Two possible methods are either to define the branches of the network by numbering the nodes, or to define the drawing directly using a matrix.

11.6.3.1 Numbering of Nodes

By numbering the nodes of the network of Figure 11-40a, the circuit is defined electrically by the following set of data:

Component	*Node A*	*Node B*
R1	1	5
R2	5	2
R3	4	3
C1	1	4
C2	4	2
C3	5	3

Transistors require three entries to fix them in the network.

This method is used widely for entering data into programs for circuit analysis, but does not provide an adequate description for the purposes of layout. The major problem is that the precise positions of the crossovers are not defined. As a result, it is difficult to determine an acceptable set of paths for the inter-connections.

11.6.3.2 Using a Matrix

It is far better if the network is defined by a matrix which shows the precise placing of the crossovers, and gives a close indication of the pattern of the conductors. Figure 11-41 is an example of such a matrix, and comparison with Figure 11-40b shows how this array of alphanumeric characters has the appearance of the original circuit. Experience has shown that this method is easy to check.

P1	–	C1	–	–	–	C2	–	P2
	I			I			I	
	I			R3			I	
	I			I			I	
	–	R1	–	J3	–	R2	–	
				I				
P3	–	–	–	–	–	–	–	P4

Fig. 11-41 Matrix of alphanumeric characters representing Fig. 11-40b

11.6.4 COMPUTER MODELS OF INTEGRATED CIRCUITS

Modeling procedures developed to describe the characteristics of integrated circuits for computer analysis and design, depend to a large extent upon how well components are isolated from each other in the same circuit. For example, when a transistor is fabricated on a substrate and is well isolated from the remainder of the circuit, the model is identical to the discrete transistor model. However, more commonly, there is considerable interaction between elements of an IC, either at boundaries or through the substrate. These interactions, usually distributed in nature, require the use of approximation techniques to predict the performance of the circuit. The problem of obtaining and using IC models is additionally complicated by non-linear and high frequency effects.

Development of analysis and design techniques including all of these factors is extremely difficult, even with the aid of a computer. The speed and storage capacity of a computer are used most efficiently when a suitable model has been selected based upon good engineering judgement. Procedures for modeling the geometrical and material structure of IC's must, therefore, be based upon a compromise between accurate representation of the physical processes, and simplification through approximation techniques which distinguish clearly between first order and higher order effects.

11.6.4.1 Planar Diffused IC Models

Integrated circuits fabricated by the planar diffusion process, result in components defined within a single crystalline substrate by regions of alternate doping, and are electrically isolated by either reverse-biased PN junction boundaries or dielectric regions. In the dielectrically-isolated planar-diffused IC, leakage currents between circuit elements are minimized, and these elements may therefore be represented by discrete equivalent circuit models with capacitors placed between appropriate terminals to represent displacement currents. However, when reverse-biased PN junctions are employed to isolate circuit components, the model must include the parasitic effects of these junctions.

A typical PN junction-isolated IC is shown in Figure 11-42. The NPN transistor fabricated in the planar diffused IC is modelled by two discrete transistors – an intrinsic NPN that represents the desired transistor and a PNP that represents the parasitic junction effects. The intrinsic transistor is drawn in solid lines, while the parasitic transistor, drawn in dashed lines, has its collector labeled substrate. Each of these devices is modelled according to its discrete equivalent circuit, depending on the signal amplitude and frequency of operation. The diode, resistor and capacitor are modelled in a similar fashion, with their parasitic elements. The equivalent circuit of an IC gate modelled in this manner is illustrated in Figure 11-43.

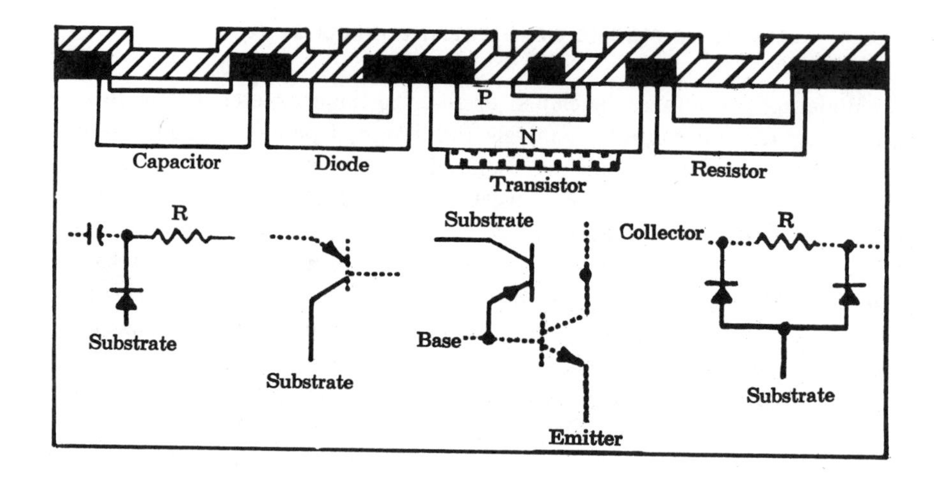

Fig. 11-42

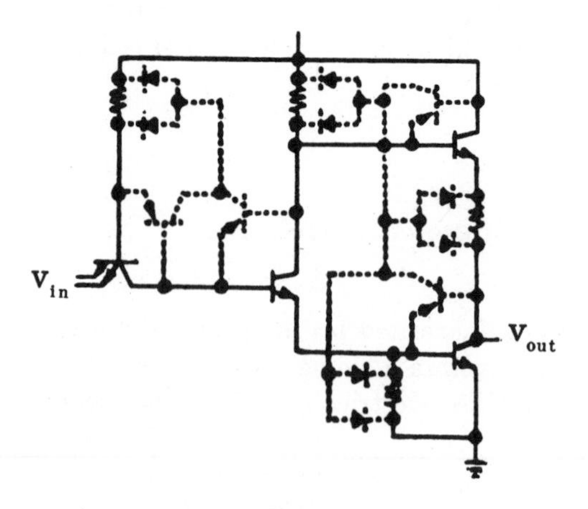

Fig. 11-43

For greater accuracy over a wider range of operation, a distributed model is useful, as illustrated by the diffused resistor of Figure 11-44, and corresponding model, Figure 11-45. The isolation junction between the P and N regions is represented by a parallel combination of ideal diodes with their associated capacitances, distributed along the junction. The parallel combination of capacitance C_S and ideal diode represents the isolation junction between the heavily doped N and substrates regions. The number of elements needed for the distributed model varies depending on the desired accuracy. In the final model, the distributed resistance is represented by a T-section as shown in the M-stage schematic representation of Figure 11-46.

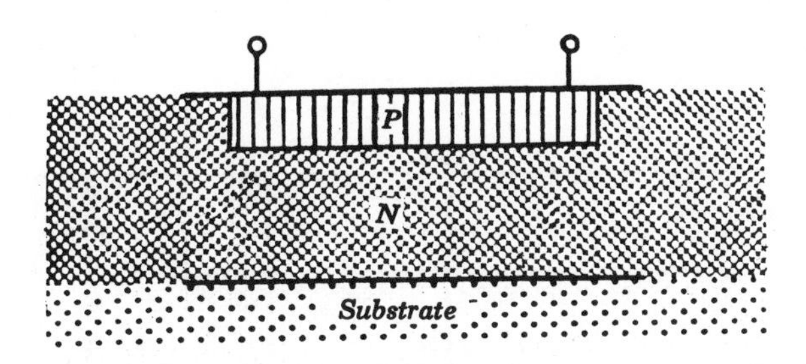

Fig. 11-44

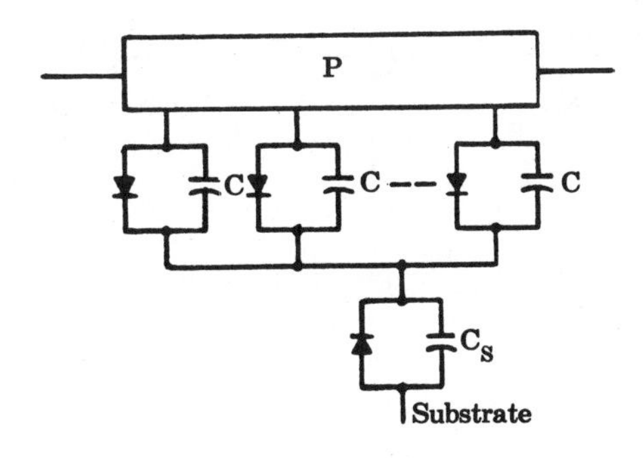

Fig. 11-45

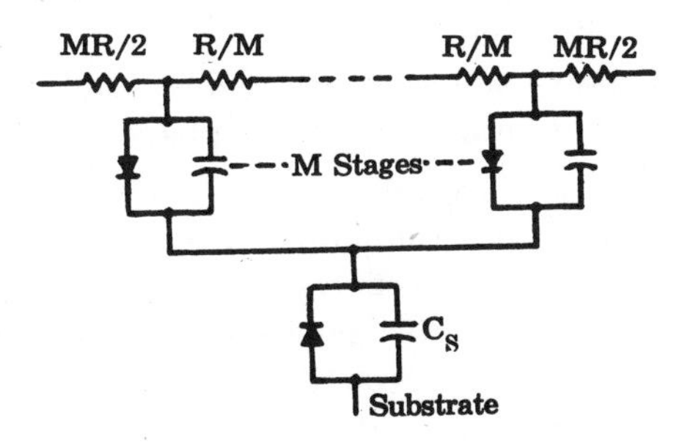

Fig. 11-46

An example of a distributed model of a bipolar NPN transistor is illustrated in Figure 11-47, together with the model schematic. Each region of the model corresponds to its physical counterpart.

A more complex approach, the lumped-parameter method, is required for circuits that operate over a wide range of environmental conditions. It is based directly on the physics of the structure, and may be employed to model thermal gradients, high minority carrier level and electro-magnetic, nuclear and cosmic radiation effects. The method differs from the distributed parameter technique in the way the physical structure is partitioned for modeling. Instead of dividing the structure into equivalent diodes and associated capacitances, the material is separated into non-uniform lumps, each representing a significant portion of the structure.

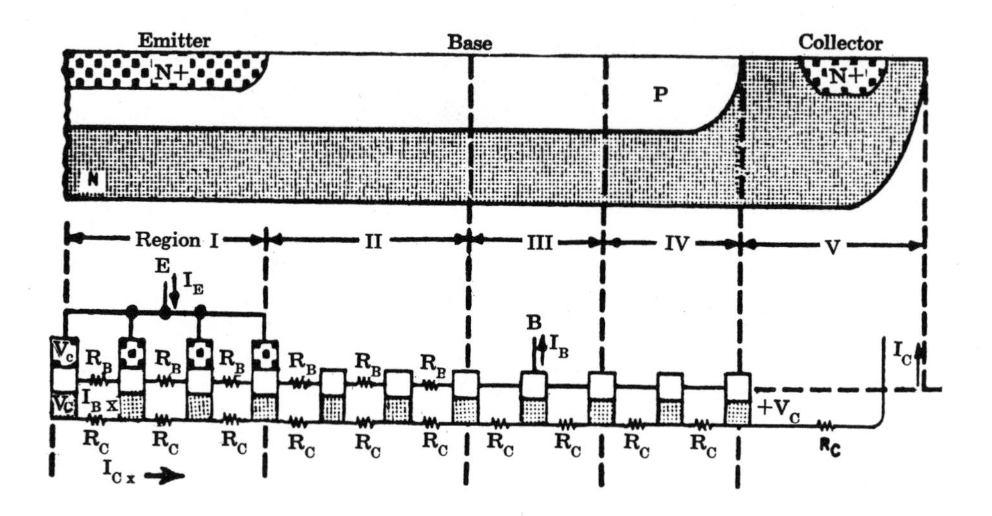

Fig. 11-47

11.6.4.2 Modeling Thin and Thick Film IC's

Film circuits may be in the form of films on passive substrates (ceramic or glass), and films on active substrates (silicon). Film components deposited on passive subtrates are well isolated from each other, and may be considered generally as closely placed discrete components with few interactions.

Films on active substrates containing active devices, are deposited there to reduce the interactions between components and the substrate. This increases the quality of the passive components compared with planar diffused elements. The interactions are here also small, and the components may be represented by discrete model representations. Thin-film transistors are modelled in a manner similar to the MOS-FET's described in the next section.

11.6.4.3 Modeling MOS-FET Integrated Circuits

MOS-FET integrated circuits are often used extensively in large arrays because of their area-per-function advantage over the equivalent

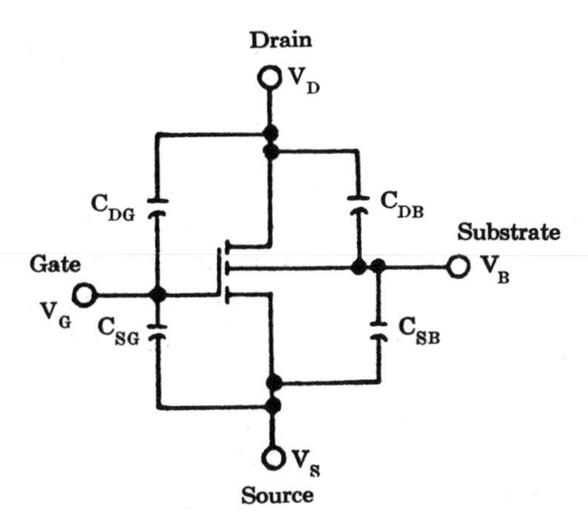

Fig. 11-48

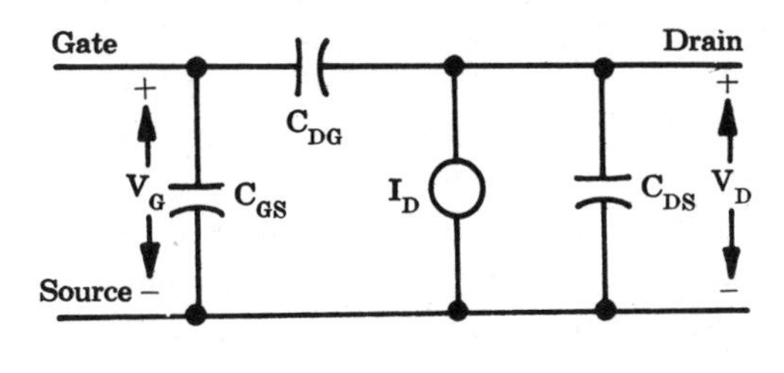

Fig. 11-49

bipolar IC's. The primary difference between discrete MOS-FET operation and their characteristics in large arrays of IC's, is their interaction through the common substrate between devices.

The model for an integrated MOS transistor is shown in Figure 11-48. This model may be represented in the form shown in Figure 11-49, where the controlled current source is a function of the gate and substrate voltages.

11.7 SYSTEM DESIGN

When microelectronic components are connected to perform an electrical function, their original individual characteristics are modified. Usually the effect of intraconnection and interaction is deleterious to component performance. However, different fabrication approaches (e.g., IC, Hybrid, or LSI) affect different constraints or the same constraints in different degrees. Accordingly it is possible to obtain better performance (both technical and economic) by selecting the approach whose trade-offs are most advantageous for a given application. The basic characteristics that affect system design are:

- Packaging
- Interconnection
- Intraconnection
- Operating frequency
- Thermal
- Logic choice
- Redundancy
- System partitioning
- Maintainability and logistics

Although the preceding topics are the essential ones, additional considerations may apply. Each system design should be considered unique, with some characteristics emphasized more than others.

IC packages have evolved from the transistor type unit (modified to include more leads), which are still used, to the widely used flat pack, and package configurations. Generally, these packages are limited in the number of leads that can be accommodated. Packages with a larger number of leads are being developed for LSI devices. Such techniques as the flip-chip Hybrid Microcircuit, require special package configurations.

Interconnections (connections made external to the device) are critical because they affect system performance, cost, and reliability. When a large number of IC's are mounted on a single board, it is necessary to have multiple layers of connections within the board. These multilayer boards are expensive and difficult to fabricate with high reliability.

The number of interconnections can be reduced if LSI techniques are employed. Such a reduction in interconnections improves frequency response and reliability.

The number of interconnections required with an LSI device, is primarily dependent upon the manner in which the system or subsystem is partitioned. Accordingly, partitioning is a major consideration for LSI design.

Interconnections are made external to the package. Intraconnections, on the other hand, which are often referred to as metalizations, are made on the silicon substrate. Once they are made, they cannot be altered.

In complex arrays, intraconnections may be substituted for interconnections. This substitution is advantageous if metalizations are less expensive and more reliable than the interconnections for which they are substituted. This is the case when a single level of metalization is required. However, LSI devices require two or more layers of metalization. In addition, it is necessary to intraconnect only the "good" circuits on a substrate. Thus, intraconnection techniques can become very complex.

Two intraconnection techniques have been proposed for LSI: fixed wiring and discretionary wiring. The fixed-wiring technique requires all circuits on some portion of the wafer to be good, since the intraconnection scheme is never varied. Discretionary wiring provides much more flexibility, since it makes it possible to cope with "bad" circuits in an array. With the discretionary approach, a different intraconnection pattern is generated for each LSI device. This method is much more costly than the fixed-wiring approach, but the effective yield may be sharply increased. This method is made thereby more practical than the fixed-wiring approach.

The intraconnection pattern can be fixed when the hybrid microcircuit approach is used, since each device can be tested before it is bonded to the substrate.

One of the most important contributions of complex arrays, is that they extend the operating-frequency range of monolithic circuits. Gate delays of less than a nanosecond are possible only with complex arrays. The number of leads, the conductor lengths, and the number of interconnections all introduce parasitics that limit operating frequency. Arrays provide close circuit proximity, reduce transmission-line lengths, and reduce the number of external connections.

Thermal effects and power dissipation are important in monolithic-circuit design. High-power circuits, greater than 1 watt, are usually not compatible with monolithic technology.

The question of which IC logic type to use, is of major importance to the designer. He must make a comprehensive evaluation of the logic forms that are available, and correctly choose the type that is best for his application. Trade-offs must be made between speed, power, and cost for each application.

Microelectronic devices, particuarly IC's, have characteristics that are highly useful in redundancy. The small size, weight, and power consumption of these devices, allow the addition of redundancy to a system without severe physical penalties. Also, the cost of integrated circuits does not increase linearly with increased complexity. Thus, a single package containing dual gates does not cost as much as two single gates in separate packages. Furthermore, as the complexity within a single package increases, reliability decreases slowly. Accordingly, dual gates in the same package are only slightly less reliable than a single gate, and much more reliable than two separate gates.

Once the equipment is designed and produced, it will be necessary to maintain it and provide spares. The repair level must be determined to the extent that a decision must be made whether to repair a module or discard it. In addition, the method of supplying spares must be chosen. These decisions should be made while the equipment is in the conception or design phase, rather than after production.

11.8 MASS DATA STORAGE

Laminated ferrite and integrated semiconductor are appealing technologies to combine for obtaining a mass random-access memory. The high bit-packing density and low-power operating capability of laminated ferrite arrays are a natural match for the high-density and relatively low-power handling characteristic of large-scale integrated semiconductors. The combination of MOS and bipolar devices, called BIMOS, is attractive for laminated ferrite electronics. The BIMOS integrated arrays are circuits with low power drain because of their complementary nature but with the high performance necessary for signal amplification and current driving.

A memory module which has been designed using laminated ferrite arrays is shown in Figure 11-50. The stack plane layout is illustrated in

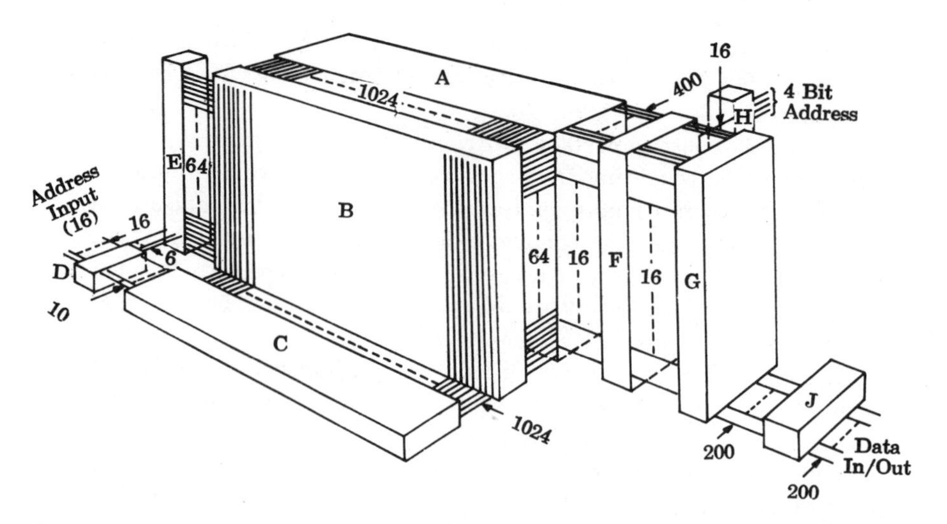

Fig. 11-50 Memory module – exploded view. A. ferrite stack, B. word drivers, C. word decoder (selects 1 word of 1024), D. address register, E. clocked supply decoder (selects 1 plane of 64), F. multiplex devices, G. amplifier/detector assembly, H. digit control (selects 1 segment of 16), and J. data register.

Figure 11-51. In this layout, each of the 64 planes containing eight ferrite arrays, 1024 word driver circuits, and a part of the word driver clock circuit, constitutes a replaceable plane which is mounted on a separate circuit board. This circuit board is tested independently of the others, and then stacked and connected to the rest of the system. Replacement of the board is possible by cutting its connections, removing it, sliding in a fresh board, and patching up the connections.

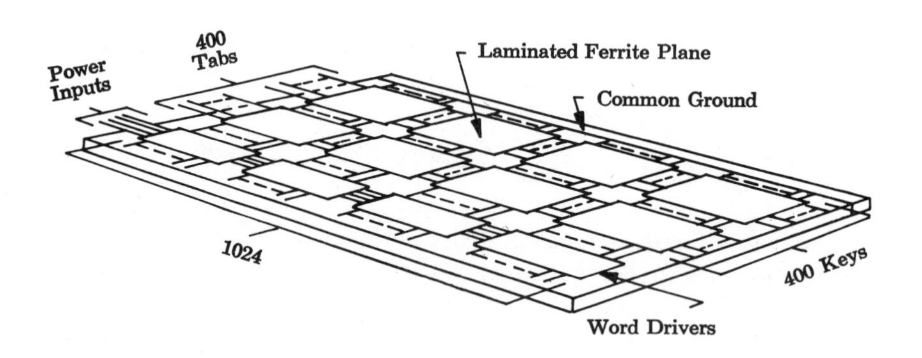

Fig. 11-51 Stack plane layout.

Each ferrite array, consisting of 256 x 200 crossovers, is encapsulated in a semihermetic metal-glass epoxy enclosure using solder and epoxy for sealing as shown in Figure 11-52. The contacts for interconnection to this package, are made using a "piano keyboard" construction along two edges, and a protruding tab construction at the other two edges. The connections could be on 10- or 15-mil centers, and are joined by a solder reflow technique using a point welder with two coplanar electrodes for supplying the heat pulse. Replacement of a faulty array may be made by cutting the leads, removing the serially connected arrays, placing new arrays in their place, and reconnecting.

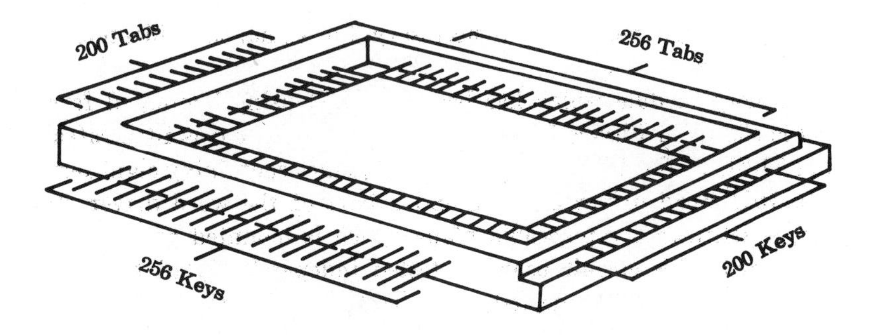

Fig. 11-52 Laminate assembly (cover removed).

The semiconductor devices are hermetically encapsulated using a beam-lead technique in which the semiconductor is covered by a layer of silicon nitride or aluminum oxide, with the leads of plated gold or nickel protruding from the strips on 10- or 15-mil centers. These tabs are directly connected to the ferrite array package keyboards, and to the vertical intraconnecting bars on the outside of the stack.

Connections from the word-driver strips to the word-decoder circuit are made via vertical intraconnection bars which join directly to the tabs of the word-driver strips by the solder reflow technique. The same is true for the word-driver clock lines.

On the opposite side of the stack, the word lines are joined by a ground connection which runs vertically between the planes.

The digit lines on the digit side of the stack are connected in series through four planes, resulting in 4096 bits per multiplex circuit. Thus, the terminations and the multiplex circuits repeat in 16 segments which are controlled by a digit decoder selecting 1 of 16 four-plane segments. Every fourth word plane, consequently, contains part of the digit multiplex circuit and terminations, in addition to the ferrite planes and word drivers. This multiplex circuit is interconnected by vertical bars leading to a digit-decoder circuit which connects the main sense amplifier and the digit-current source to the selected digit segment. From the multiplex circuit, 400 lines go to the sense/regeneration circuitry.

FET technology has become a prime contender for low-power, high-density memories and logic, whereas the bipolar transistor technology is considered to be best suitable for high- speed applications. The main reasons seem to be that an FET device has a simpler structure (self isolation!) and a higher impedance, but on the other hand a bipolar transistor allows a higher transconductance per unit area permitting a higher speed. However, by making use of integrated circuit design concepts very-low-power, high-density memory and logic chips can be realized in standard bipolar technologies.

To achieve a very low power dissipation in conventional flip-flop memory cells, extremely high ohmic load resistors, occupying a substantial amount of silicon area, are necessary. This drawback is overcome by replacing the usual ohmic resistors by constant current source load devices which can be advantageously implemented by common-base PNP transistors. They not only permit a static memory cell operation in the nanowatt power range, but also require little silicon real estate as illustrated by the all-transistor memory cell layout in Fig.11-52.1a. A further advantage of these PNP load transistors is that the cell current can be increased during addressing by orders of magnitude for fast reading and writing.

The fundamental layout in Fig.11-52.1b discloses another key measure for achieving a high density, namely the use of upside down operated NPN transistors (T_1, T_2). They considerably reduce the area necessary for device isolation and, in addition, provide a low-ohmic diffused address line. Thus, the memory array can be wired in single layer metallization without special cross-unders. As a third design feature, the decoupling devices of the bit lines (T_5, T_6) have been merged with the PNP load transistors (T_3, T_4) to further cut down the cell size. Compared to a conventional resistor load memory cell, the highly integrated memory device of Fig.11-52.1 significantly saves silicon area and reduces the power dissipation by orders of magnitude.

Yet, the cell components can be further merged as shown by the super integrated all-transistor memory device in Fig.11-52.2. Its equivalent cell circuit in Fig. 11-52.2a is identical with that of Fig.11-52.1a, except that the common base of the PNP load transistors and the common emitter of the cross-coupled NPN transistors are tied together. Hence, all memory cell components can be implemented in a common N-plane as illustrated by the schematical cross-section Fig.11-52.2b Furthermore, the P-regions comprising the bases of the NPN-transistors T_2 and T_5 and the collector of the PNP-transistor T_3 are connected together, and thus a single P-region can be used for these three transistors. As a result, the base currents of the NPN-transistors are supplied by direct injection of minority carriers into their common N-region.

With the injection-coupled memory cell in Fig.11-52.3 the principle of direct carrier injection is also utilized for the coupling to the bit lines. By this measure, a still higher degree of device integration is achieved. Fewer metal contacts and array lines are required, resulting in a smaller cell size.

The following table summarizes the main parameters of cell integration for the various approaches. The size is based on 5 µ minimum line dimensions in a standard buried layer technology. It is significantly reduced with a modern technology (e. g. oxide isolation).

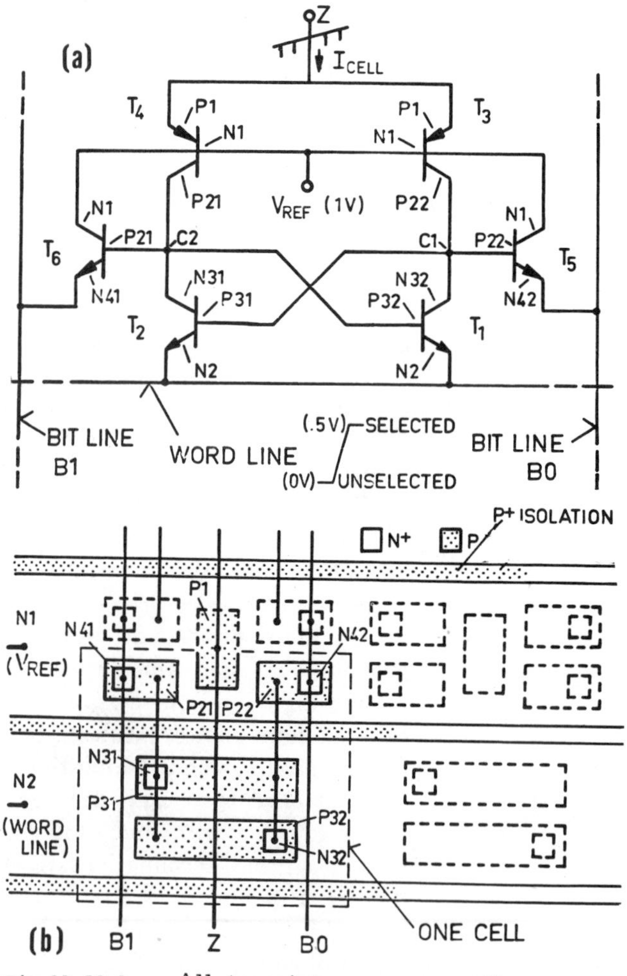

Fig.11-52.1 All-transistor memory cell

The above design concept of merging devices can also be utilized to achieve low-cost logic circuitry as shown by the schematical structure and equivalent circuit of the Merged Transistor Logic (MTL). The fabrication process becomes as simple as that of single transistors, requiring only four mask steps through metallization. The self-isolation of the devices and the complete omission of ohmic load resistors are the key for the high functional density.

	Size [mil^2]	Number of contacts	Number of array metal lines
Resistor load cell	25	13	4
All-transistor cell	7	8 1/2	3
Super-integrated all-transistor cell	4	7	4
Injection-coupled cell	3	5 1/2	1 1/2

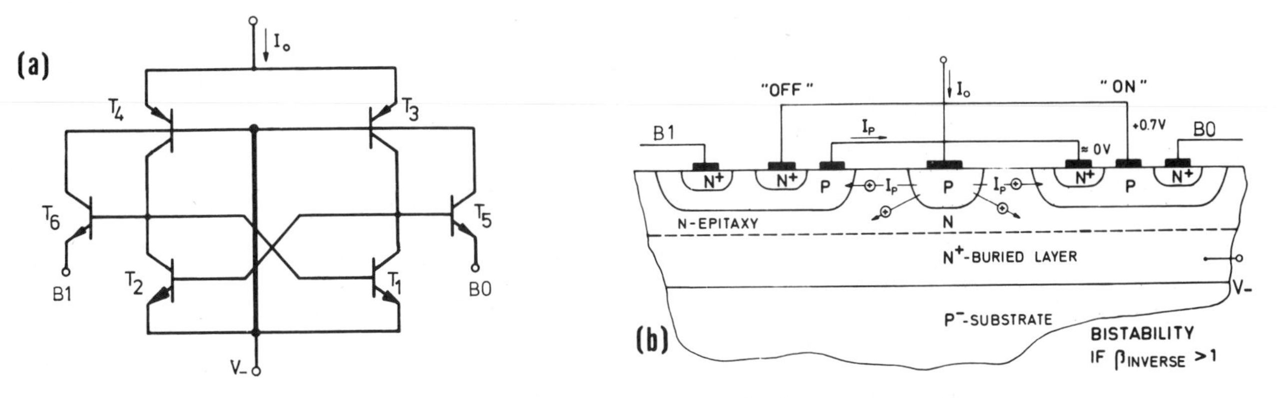

Fig.11-52.2 Super-integrated all-transistor memory cell

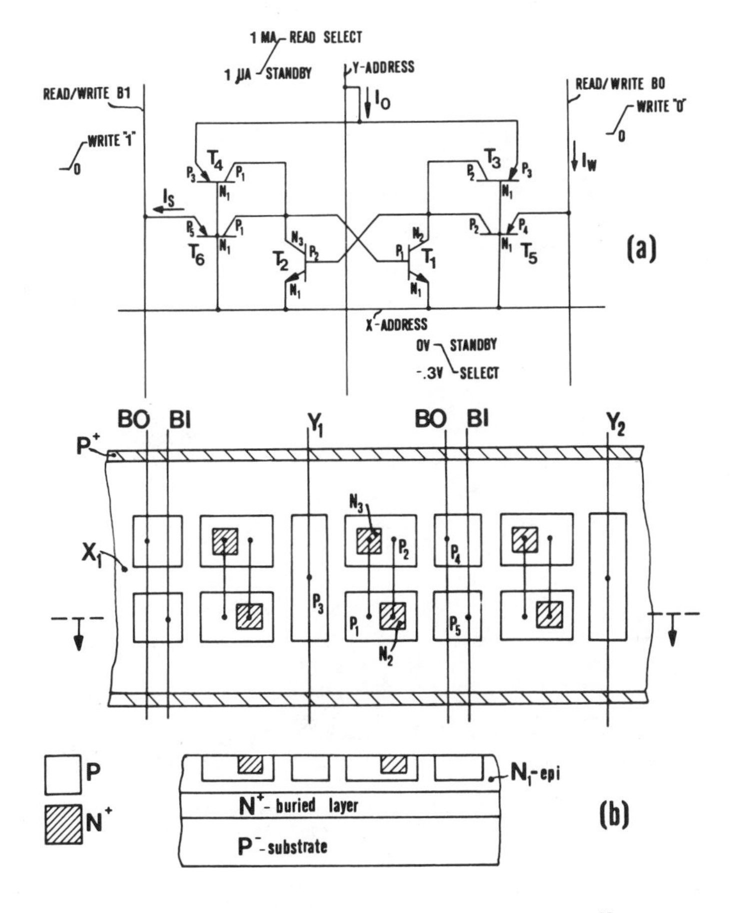

Fig.11-52.3 Injection-coupled memory cell

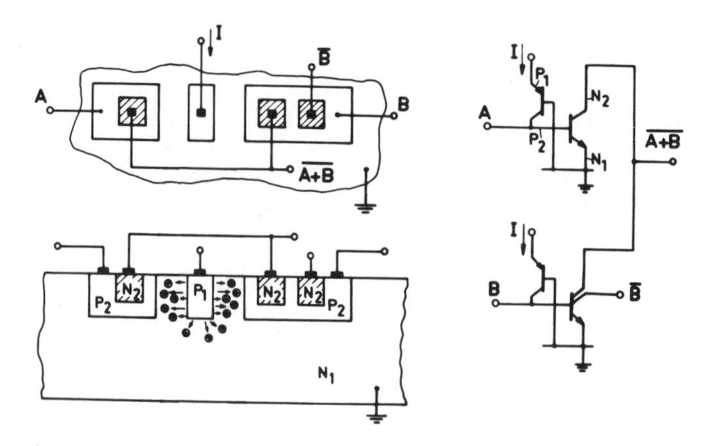

Fig.11-52.4 Merged-Transistor Logic (MTL) Circuit

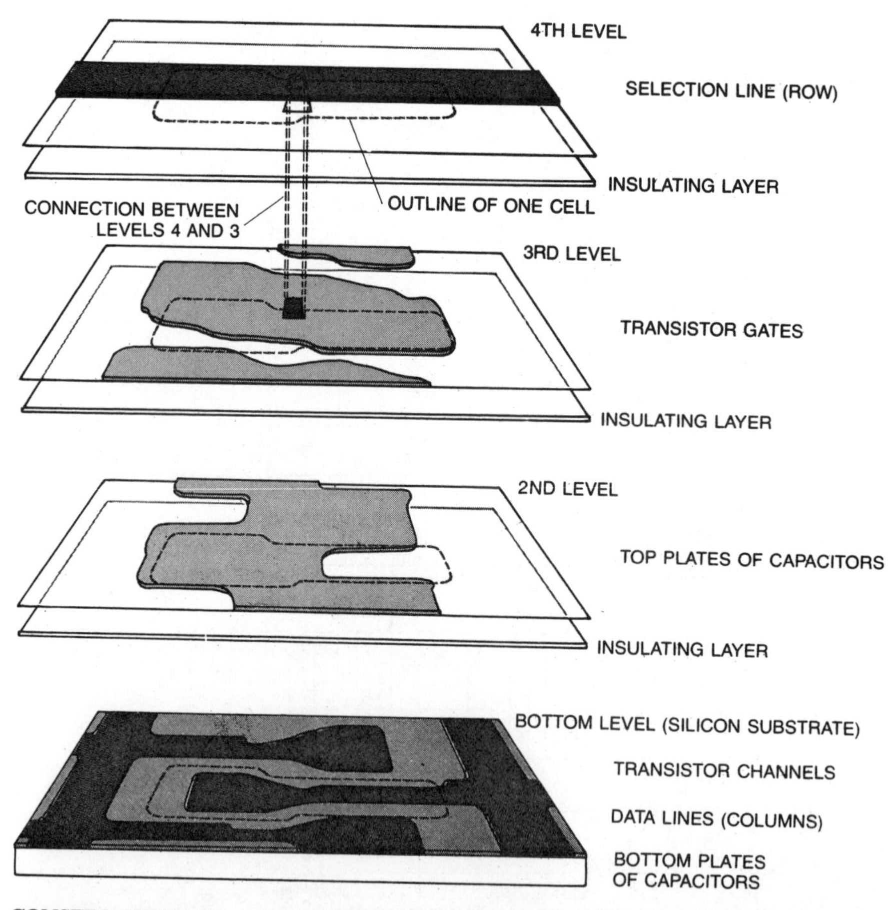

CONSTRUCTION OF A RAM CELL begins with the creation of multipurpose structures (*color*) in the bottom level by "doping" certain areas of the substrate with arsenic, which converts those areas into *n*-type silicon. The undoped area is chemically transformed into silicon nitride, an insulator (*gray*). The second and third levels of the cell are formed by the deposition of polycrystalline silicon. Fourth level is formed by deposition of aluminum. Thin insulating layers of silicon dioxide are deposited between levels. This cell measures 15 by 30 micrometers.

11.8.1 CHARGE-COUPLED DEVICE

A three-layer MOS structure (Fig. 11-53) is used in which charges are stored (coupled) in potential wells (spatially defined depletion regions) at the surface of the substrate. Minority carriers (holes in n-type silicon) are injected into a depletion region by surface avalanching at the semiconductor-oxide interface by light-induced hole-electron pairs, or by forward biasing a p-n junction.

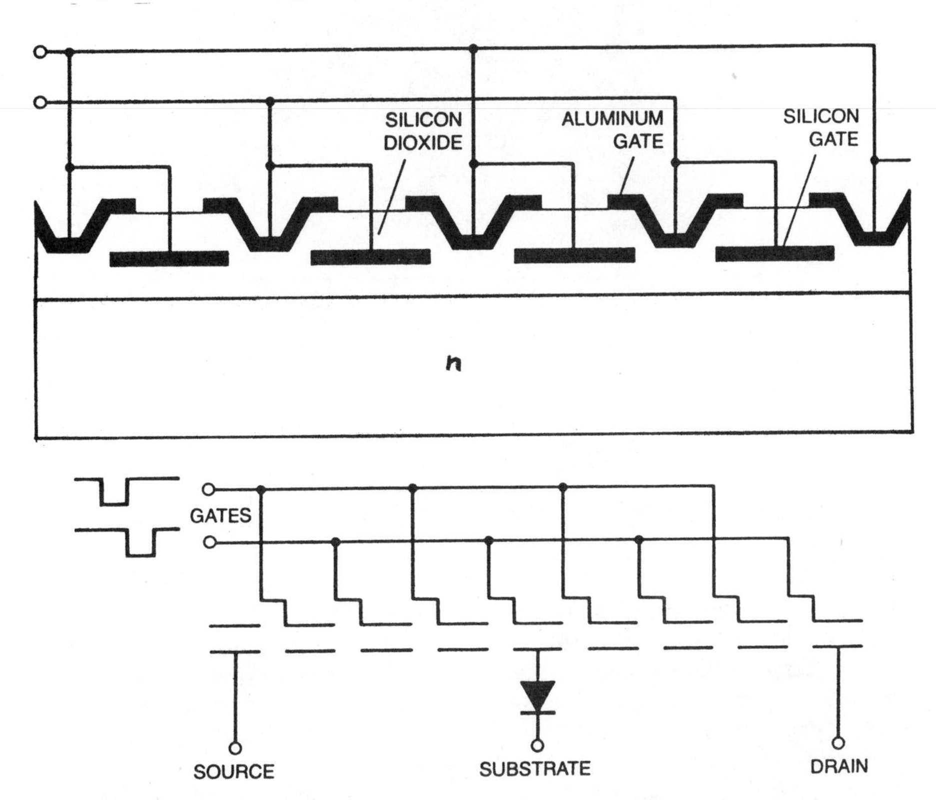

CHARGE-COUPLED DEVICE is a microelectronic circuit element whose function could not be duplicated by a practical assembly of discrete components. The device can be regarded as a stretched MOS transistor with a long string of gates (perhaps as many as 1,000) between the source and the drain. In the *p*-channel device shown here a charge packet, consisting of a concentration of holes, can be held in place for a short time by applying a steady negative voltage to one of the gates. If that voltage is then dropped and simultaneously the next gate in line is energized, the charge packet moves to a new position under the second gate. By applying pulses to alternate gates a sequence of charge packets can be transferred from source to drain.

Charges are moved about the surface by moving the well to which they are coupled. The charges do not leave the substrate. The metal pattern deposited on the insulating layer forms an array of conductor-insulator-substrate capacitors. Applied voltage controls well depth.

Above, charges flow from the well at V_2 into the deeper well at V_3 produced by a larger negative voltage. Proper voltage adjustments move the charges to the right or left. Detection can be done at any electrode by sensing capacitance changes due to charge presence by measuring the surface potential, or by transferring the charge into an output p-n junction. A digital system would detect the presence or absence of charge packets. An analog system could detect the amount of charge, which is variable controlled at the injector.

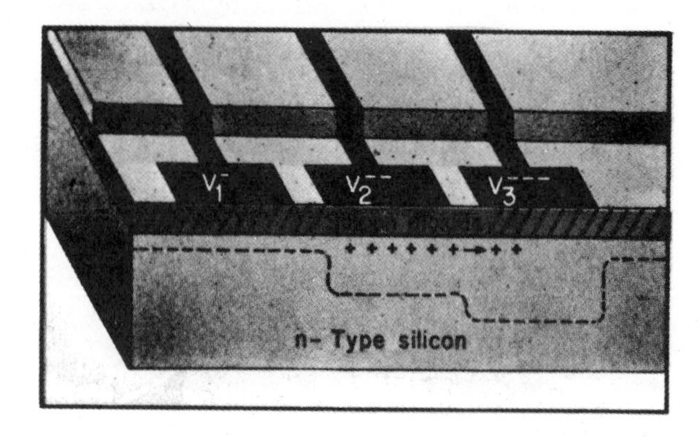

Fig. 11-53 A charge-coupled device for memory applications.

Non-volatile, Electrically Erasable Memory

EAROM's, often called electrically erasable ROM's (EEROM's) or non-volatile memories (NVM's), fit in the memory hierarchy between ROM's and RAM's (Fig. 11-53.2). They should not be confused with ultraviolet erasable PROM's (generally labeled EPROM's), which have an entirely different erase procedure and are used for different applications in many cases. EAROM's tend to exhibit the permanent storage capability of a ROM, for they can store data a long time including under power off conditions. They also exhibit some attributes of a RAM, since the stored data can be changed by applying control voltages.

EAROM's are built using two technologies: metal nitride oxide semiconductor (MNOS), and a variation of the floating-gate, avalanche-injection metal oxide semiconductor (FAMOS). process. Both techniques end up storing charge that must be removed or erased before new data can be inserted or rewritten. The stored charge stays in place when the power is turned off, retaining the stored data (Fig. 11-53.3).

The erase/write cycle is relatively slow, taking from 10 milliseconds to a couple of minutes depending on the type of device and how much must be rewritten. Some devices require that the entire device be erased and rewritten. Some are block or word erasable while others offer both whole chip and word erase. The amount of memory it is desirable to erase at one time depends on each application.

The MNOS types were the first EAROM's on the scene and improvements have progressed steadily. First generation EAROM's required a lot of external high voltage switching for erase/write. The second generation put control circuits on the chip, but the user still had to switch programming voltages, which are on the order of 30 V. High voltage switching was put on-chip in third generation devices, so the user only has to supply a steady high voltage plus TTL level erase/write signals. Since the high voltage current requirements are miniscule, this voltage can easily be supplied by bias supplies or modular dc converters. Some of the next generation devices will include on-chip converters so the user can forget the

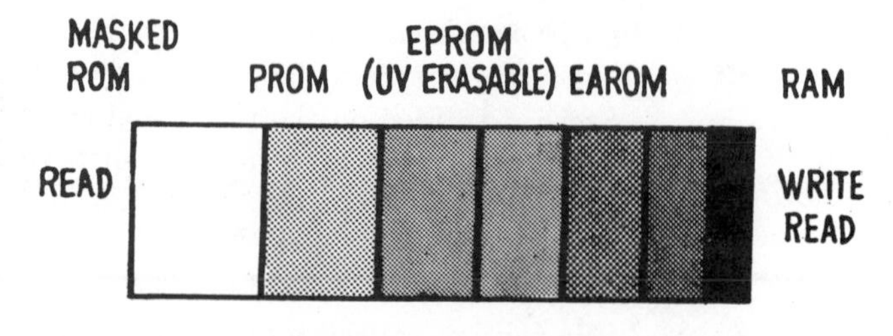

Fig. 11-53.2 Semiconductor memory hierarchy.

odd supply voltages. Others will only require that the high voltage be present during the erase/write cycle. Either way, the trend is to make the supply and control requirements simpler for the user.

Since EAROM's are based on stored charge, and the storage time of this charge is affected by temperature and the number of erase/write cycles, the potential user needs to be aware of two terms: *retention* and *endurance*. Retention is a measure of how long the device will hold stored data. It can be based on storage with the power off, with power on, or with the stress of many read cycles. The manufacturers have not standardized test conditions; therefore, while retention is commonly rated at 10 years, it is hard to make any data sheet comparison. Endurance is another longevity related term. Extensive erase/write cycling causes a permanent degradation in the memory characteristics; endurance is the number of such cycles that can be performed before the device shifts so it can't be used.

Endurance and retention are interrelated (Fig. 11-53.4) and the decrease in performance with cycling is one of the reasons these units are not advocated for RAM applications. However, when the memory is refreshed, it starts out on a new decay cycle, so there are applications where it may be desirable to deliberately introduce erase/rewrite cycles.

While endurance and retention are figures of merit, there are many applications in which the memory need only last for a period of hours, days, or weeks — such as during a power failure, weekend, or vacation. With these relaxed requirements, the user can evaluate any tradeoffs such as price, more convenient voltages, or shorter write cycles.

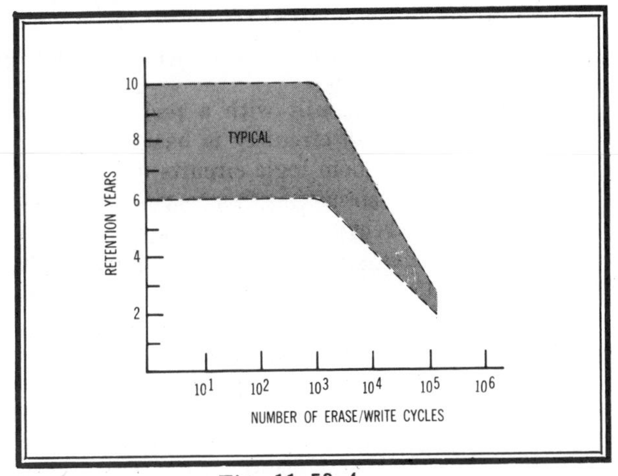

Fig. 11-53. 4

In EAROM's, data retention is a function of the number of times the memory has been stressed by erase/write cycles. This curve shows the general characteristics of the relationship, even though the exact breakpoints are not representative of any manufacturer's product line.

Fig. 11-53. 3

MNOS MEMORY OPERATION

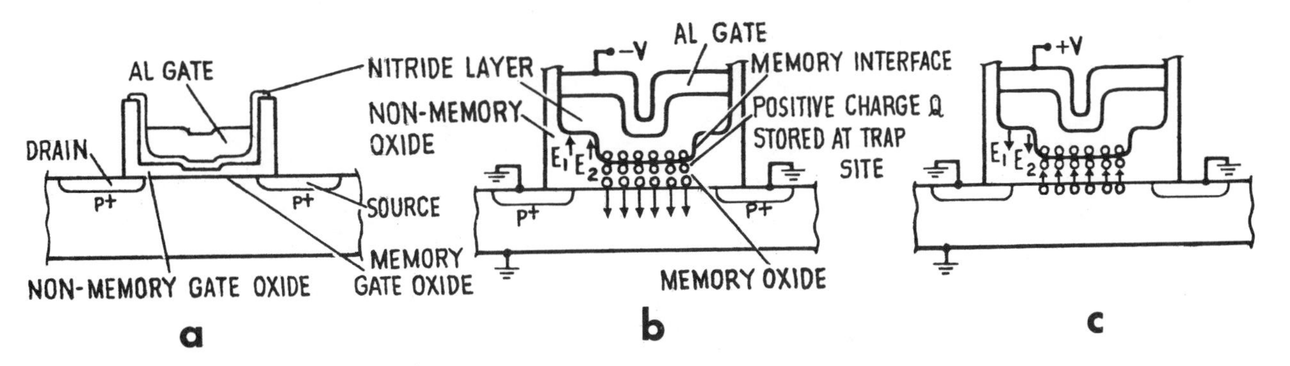

The MNOS memory cell consists of a nitride layer placed over a very thin gate oxide of a p-channel transistor (a). The device is erased by applying sufficiently high positive bias to the gate, so that it causes charges to tunnel from the substrate to the oxide-nitride interface (b). To write into the memory cell, a negative voltage is applied to the gate, repelling electrons from the nitride trap sites (c). This trapped charge shifts the threshold level of the transistor and the memory is read by applying intermediate level gate voltages and observing if the transistor turns on.

Magnetic bubbles

In its simplest form, a magnetic bubble is a cylindrical magnetic domain in a single-crystal layer of a ferro- or ferrimagnetic material. The magnetization of the domain is normal to the layer. The bubble is bounded by a domain wall which separates material with upward magnetization from material with down ward magnetization (Fig. a).

In magnetic bubble material, the upward and downward magnetic domains assume a minimum energy state which results in a serpentine pattern (Fig. b). When an external magnetic bias field is applied to the material in a downward direction, the total energy of the wafer changes in such a way that domains whose polarity is opposite to that of the external field shrink (Fig. c). At a critical value of field intensity, the single walled or island domains —the are two in Fig. c—become cylindrical, resulting in a magnetic bubble (Fig. d).

The movement of magnetic bubbles can be controlled to form useful devices by adding circuit elements of permalloy (Fig. e). In such a device, the bubble is stabilized by the vertical bias field and can be thought of as a vertical bar magnet. When a rotating magnetic drive field is applied, as illustrated, the bubble seeks an energy minimum under pole 1. By uniformly rotating the drive field in a clockwise direction through 360 degrees, the bubble can be made to move through one cycle, that is through positions 2 and 3 to position 4.

If the magnetic bias field is applied by some permanent-magnet configuration, the bubbles will remain permanently stable and thus a nonvolatile memory becomes possible.

To enter data into a bubble memory, electrical impulses representing binary bits are converted into bubbles—one bubble per pulse—by means of a bubble generator. In the generator, a source bubble is maintained under a circuit element and generation is achieved by splitting

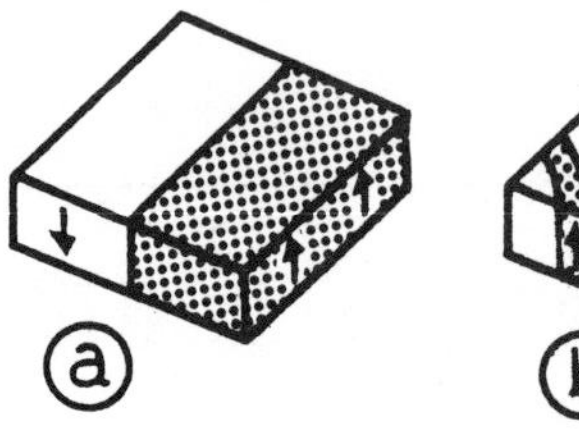

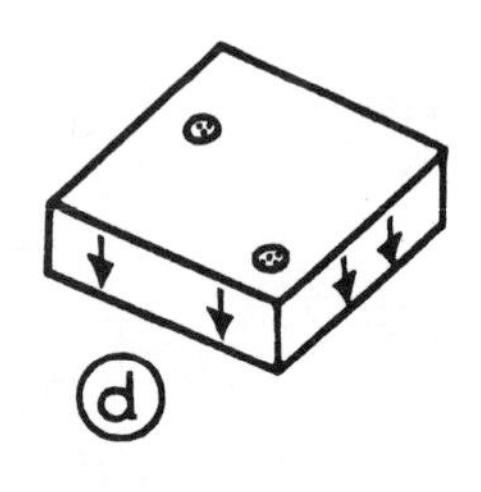

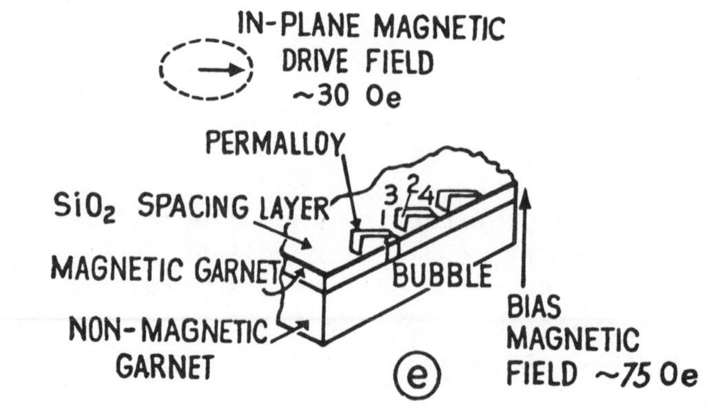

the source bubble in two. One of the bubbles is then propagated along the circuit, while the other remains as a source for further bubble generation.

To read information out of a bubble memory, the presence of a bubble must result in an electrical pulse. The pulse may be produced either by a direct magneto-electrical conversion—such as magneto-resistive change in a permalloy sensor element, or by a magneto-optical conversion —via the Faraday Effect—followed by an opto-electrical conversion. Either way the readout is nondestructive.

If, however, bubble destruction after readout is desired, an appropriately controlled bubble annihilator can be used.

A magnetic bubble is a physical phenomenon not unique to any one class of chemical compositions. [However, at present practically all bubble devices are made with single-crystal films of multicomponent magnetic rare earth–iron oxides having the garnet structure. Bubble physics considerations dominate device design and materials selection. The stability, size, and speed of bubbles are the key design parameters affecting device reliability, capacity, and data rate, respectively. Stability refers to how well bubbles resist destruction by environmental perturbations, such as the drive fields used to propagate domains. Smaller bubbles are packed more densely and thereby increase data capacity. Since small bubbles are more closely spaced, they move shorter distances and give better data rates than large bubbles with the same velocity. Also, of course, increased bubble velocity improves the data rate in a device. (In present-day devices, the coil drive circuitry limits the data rate.)

Stability of bubble domains is measurable by the in-plane field H_u (called the anisotropy field) required to tip the magnetization M from its easy axis by 90° into the film plane. The product $H_u M/2$ is the uniaxial anisotropy energy constant K_u. (In some bubble materials, this magnetic anisotropy can have three components: growth-induced, strain-induced, and intrinsic.) The stability factor is defined as the ratio of H_u to saturation magnetization, or in energy terms as

$$Q \equiv \frac{K_u}{2\pi M_s^2} \tag{1}$$

the ratio of the anisotropy energy to the magnetostatic energy. For stable isolated bubbles to exist, Q must be greater than unity. Furthermore, unless Q is appreciable, the in-plane drive field can strip out bubbles into stripes, a failure mode for devices. Stable bubbles exist in a range of bias fields between the (higher) collapse field H_0, where the entire film becomes a single domain, and the (lower) strip-out field (about 0.7 H_0), where bubbles change back into stripe domains. Depending on film material parameters and thickness, H_0 is 0.4 to 0.6 of $4\pi M_s$, the saturation magnetization.

The film thickness is generally made to be just less than the bubble diameter for energetic and stability reasons, and the bubble diameter is about eight times the characteristic length

$$\ell = \frac{\sqrt{AK_u}}{\pi M_s^2} \tag{2}$$

where A is the magnetic exchange stiffness constant, which is a measure of how strongly neighboring ions are magnetically coupled. The characteristic length is the ratio of the domain wall energy to the demagnetization energy, which is directly related to the magnetostatic energy. Equation 2 shows that bubble size is determined mainly by magnetization. The important thing to remember is that smaller bubbles require larger magnetization.

From a materials standpoint, the data rate of a device and the time required to access data are governed by the speed of bubbles. It will be seen that stability and speed have contrary materials requirements. Stability demands relatively high anisotropy, and speed needs low anisotropy. The speed of a bubble is determined by the product of the drive field across the bubble minus the threshold field for movement (coercivity) and the bubble mobility (velocity per drive field gradient)

$$\mu = \frac{\gamma}{\alpha}\sqrt{\frac{A}{K_u}} \qquad (3)$$

where γ is the gyromagnetic ratio, which has to do with the dynamics of atoms changing their directions of magnetization, and α is the Gilbert damping (magnetic viscosity) parameter. The quantity $\sqrt{A/K_u}$ times π is the domain wall width within which the magnetization reverses direction in any one or a combination of possible wall magnetic structures. For example, simple bubbles have a Bloch wall, wherein the magnetization vector rotates in the plane of the wall in either a left- or right-handed (chirality) direction, but in a Néel wall, which can partly form near the film surfaces, the magnetization vector rotates directly into the wall. Different structures within a given wall meet along Bloch lines, which can meet at Bloch points. These magnetic wall structures provide a mechanism for information coding of a higher order than binary.

Walls, whether simple or complex, define bubbles, and defects that lie in the plane of a wall are particularly effective in blocking bubble motion. Accordingly, films for devices must have a high degree of physical perfection and freedom from flaws. The growth techniques developed for making garnet films and the properties of garnets themselves provide products with the high quality required for devices.

The early devices had large bubbles and were made with platelets cut and polished from large single crystals. Now competition from other technologies creates the need for devices with smaller bubbles in thinner sheets, which are produced as thin films on flat substrates. Today, magnetic garnet films are grown as single-crystal layers, which are a crystallographic continuation of single-crystal substrates that are nonmagnetic platelets. This process for making films is called epitaxy.

Bubbles have been studied in single-crystal platelets of Fe_2O_3-based compounds with magnetoplumbite, orthoferrite, and garnet crystal structures and in single-crystal films of the latter. Also, films based on metallic $GdCo_5$ alloy compositions with a glasslike amorphous structure have been used. Devices have been fabricated with all but the magnetoplumbites.

Amorphous films were developed late and have bubble properties that are susceptible to temperature change in larger bubble films; hence they could not displace garnet films, which have satisfied design needs to date. Amorphous films could be revived for future devices using very small bubbles, but they would require an electrical insulation layer because they are conductors.

It is possible to slice and polish bulk crystal platelets, 1 cm^2 in area, down to a thickness of about 25 μm; this sets a similar minimum bubble diameter, which in turn limits data storage density. A preferable technology, which considerably relieves both the areal and thickness constraints, is to grow magnetic films epitaxially on thicker nonmagnetic crystal substrate wafers that have the same crystallographic structure and lattice spacing. An epitaxial film has chemically different, but similarly sized, ions in exact positional registration with ions in its supporting substrate lattice. Fortunately, the nonmagnetic gallates (Ga_2O_3-based compounds) have a crystal chemistry related to that of ferrites, and they function well as substrates in the case of garnets. Both Ga and Fe prefer to be trivalent and have similar ionic radii.

Magnetic Bubble Device Characteristics

High data-storage-density, low power operation, good production yield, nonvolatility, and radiation resistance are the outstanding desirable characteristics of the magnetic bubble device. Bit-storage-density of a future magnetic bubble memory system could become as high as one billion bits per square centimeter. Generally, it appears that bubble density is limited primarily by the fabrication techniques in deposition of small guide patterns and detectors rather than the actual size of the bubbles. The closest developing, competing technology, charge-coupled devices, will have great difficulty in approaching such densities at comparative yields.

Most of the power required to propagate bubbles is dissipated in the elements which produce the magnetic fields that move the bubbles along their guide pattern paths. Experiments show that small bubbles require less propagation power than the larger bubbles. Consequently, the smaller the bubble, the greater the density and the less power per bubble is required. In addition, some smaller bubbles are capable of moving at higher velocities. Experimental bubble memories are capable of operating at 10 nanowatts per bit; a figure which semiconductors will find hard to compete with.

Despite the attractive bubble device features, several drawbacks deter bubble device production:

Temperature sensitivity
Relative poor signal-to-noise ratio
Lack of standard material
Low voltage interface
Serial memory operation
Low speed data rate
Limited production technology

Slow data rate, serial data access, and temperature limitations are the most serious of these drawbacks. After ten years of bubble research, most laboratory memory models cannot exceed a data rate of one megahertz (1 MHz) under favorable temperature conditions. Other memory devices such as crossties, semiconductors, and optical memories are expected to be faster than the magnetic bubble memory and are intrinsically less sensitive to temperature.

Magnetic bubble materials, such as garnets, require doping by several elements to reduce undesirable characteristics such as temperature sensitivity and low speed operation. By comparison, silicon usually requires doping by only one element.

Fortunately, bubble device research facilities are able to make use of several fabrication and production processes which are already established in the silicon technology. Photo, E-beam, and X-ray lithography and ion implantation techniques can be applied to bubble hosts. However, the utilization of these existing fabrication techniques does not guarantee that bubble characteristics are exploited to their fullest extent. If bubble devices are to realize their fullest potential, fabrication techniques capable of producing micro-miniature circuits of one to two orders smaller than are presently available will be necessary. In the meantime, viability of the bubble device could very well depend on technical acceptance by the electronic market rather than on device performance.

Technical Obstacles in Magnetic Bubble Devices

The most severe obstacle in the development of magnetic bubble devices is the requirement for the use of relatively new garnet materials rather than the use of well-developed silicon materials.

The second obstacle is the reluctance of equipment manufacturers to accept an unproved memory device.

The third obstacle is the combination of several undesirable technical characteristics which are inherent to magnetic bubble devices: temperature sensitivity, low speed, and low-signal-level operation. It is anticipated that the temperature problem will eventually be solved. The low speed, low-signal-level operation problems of bubbles are considered the most difficult obstacles by scientists and keep the magnetic bubble memory from becoming a large, fast random access memory.

To overcome these obstacles, magnetic bubble device manufacturers must offer compensating device features such as super-high bit-storage-density, ultralow power operation, and low cost.

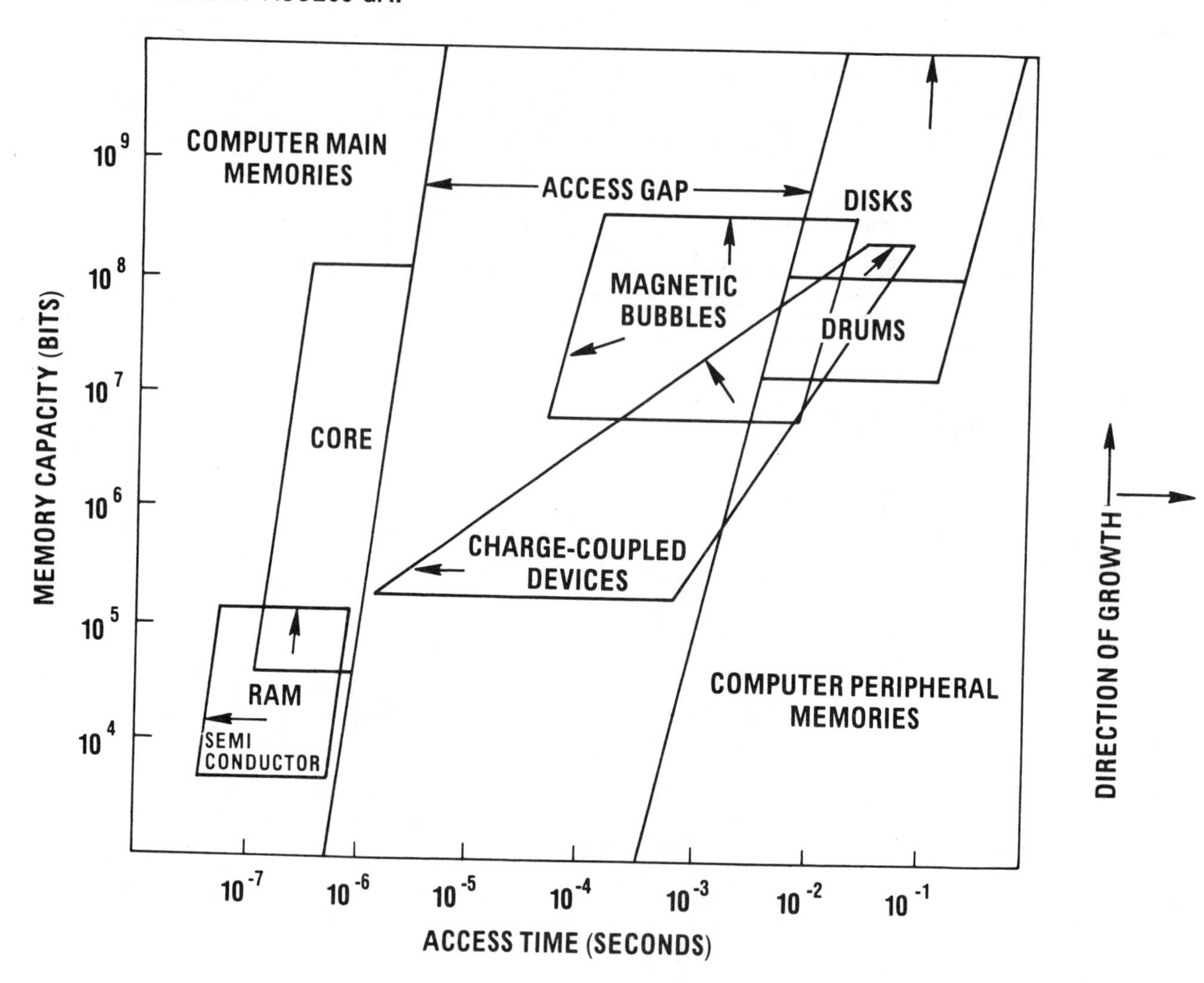

Magnetic Bubble Materials

Since the discovery of the magnetic bubble, researchers have been involved in the task of determining which magnetic material is best suited for use in producing devices. Most of the materials used for magnetic media contain rare earths. The inclusion of these elements produces magnetic media with the desirable magnetic properties.

Magnetic bubble materials should be low cost, readily available, and should exhibit the following characteristics:

a. Stable magnetic axes as a function of temperature, humidity, etc.

b. Easily workable crystal structures.

As the result of much research, three groups of materials have emerged as candidates for magnetic media:

a. Orthoferrites

b. Garnets

c. Amorphous film materials

Orthoferrites are characterized by the formula A Fe O_3, where A can be an element similar to any of those used in garnets. The general formula for garnets is A_3 B_5 O_{12}, where A and B are different types of elements with A being a large radius ion such as any of the rare earths and B a smaller radius ion such as any of the transition metals like iron (Fe). Amorphous films are unlike garnets and orthoferrites since they exhibit noncrystalline structures. The amorphous films are composed of metals such as cobalt and certain rare earths like gadolinium (Gd).

Magnetic Bubble Diameters

As previously mentioned, magnetic bubbles behave like magnetic domain cylinders. Experiments indicate that the smallest diameter of a cylinder with acceptable stability is obtained when it approaches the dimension of the film thickness. Thus, if bubbles, supported by small bubble guides, are very small and the magnetic film is very thin, the amount of data which is stored per given area can be high. Therefore, not only bubble device properties but also the fabrication techniques of the materials determine the storage densities.

The diameter of the domain can be made to vary from that of a strip domain to a magnetic bubble cylinder with a diameter of less than 1 micrometer by increasing the perpendicular magnetic field. If the magnetic field is increased to a critical value which depends upon magnetic characteristics of the bubble host material, the bubble collapses. A "safe" value for a stable bubble is one which is about twice the diameter of the magnetic bubble before collapsing.

Amorphous films have been produced with bubble diameters smaller (0.1 micrometer) than garnets.

Magnetic Bubble Memory Storage Density

A superior characteristic of magnetic bubbles over alternative memory technologies is their potentially high information-storage-density. Experts predict that bubble density will surpass 10^{11} bits per square centimeter (as compared to 10^8 bits for semiconductors). The primary limitation of bubble density is the fabrication process. The limits of photo, E-beam, and X-ray lithographic processes have been under constant study while the latest magnetic bubble research breakthroughs have occurred by the use of ion implantation processes.

It must be noted that high bit-storage-densities in materials have not always enabled the production of high bit density devices (such as memories). The additional area required by bubble guides and detectors almost always drastically reduce chip-bit densities. The speed of the bubble is usually a compromise between low material coercivity and high bubble stability. Bubble speed has increased steadily, but not drastically, from 1970 to the present. Some of this increase in speed has been achieved through the use of improved magnetic bubble materials. In general, garnets can have higher bubble speeds than either ferrites or orthoferrites.

The data rate of magnetic bubbles is measured in bits per second and relates to the rate at which data propagates through either the material or device,

Access time is related to data rate since bubble memories are serial devices. For example, the average access time for a 2048-bit serial bubble memory operating at a 500 KHz clock rate is determined by the following formula:

$$\text{T-access (average)} = \frac{1}{\text{(Bubble Propagation Rate)}} \times \frac{\text{Number of bubbles per device}}{2}$$

$$= \frac{1}{(5 \times 10^5 \text{ bits/sec})} \times \frac{2048 \text{ Bubbles}}{2} = 2048 \text{ microseconds}$$

The worst access time occurs when the required data have just left the detector and must travel the entire register again. The worst access time is twice the average access time. Obviously, the best access time occurs when the data are just about to arrive at the detector, i.e., one clock cycle away from the detector.

Since bubble memories are characteristically serially oriented, bubble devices will usually be comprised of several optimum length registers in one chip. In this approach, storage capacity can be great and the access time will be only slightly greater than the access time of one single register. The slight increase in time results from the additional time needed to choose which register is being addressed.

Magnetic bubble devices operate at very low power levels. For example, it is estimated that a bubble bit dissipates several hundred times less power than even the smallest transistor. According to one source, the power to perform 10^{12} binary operations a second in magnetic bubble devices would be only 0.04 watt as compared to 10 watts needed by semiconductor switches.

Information found in both non-U.S. and U.S. literature indicates that much of magnetic bubble LSI circuits are currently produced from garnet materials rather than from silicon. Silicon semiconductors are most often used in the manufacturing of contemporary LSI circuitry.

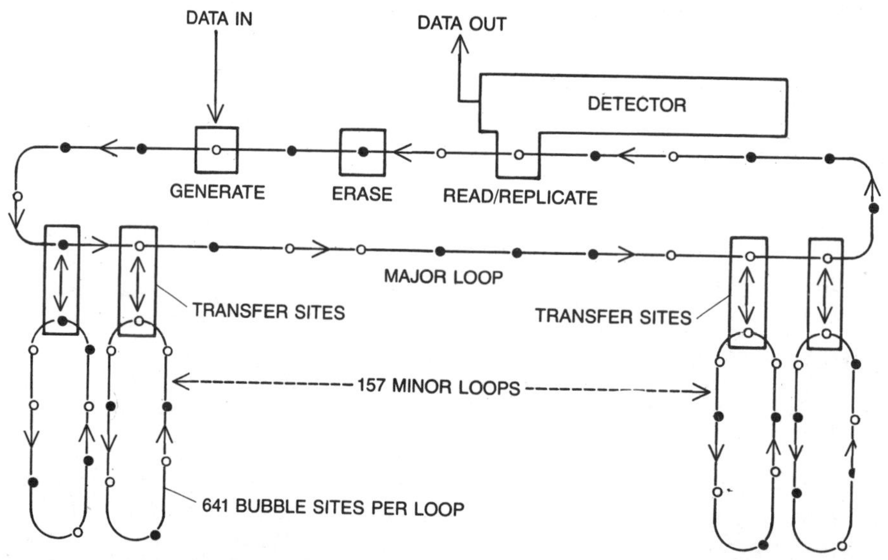

DIAGRAM OF MAGNETIC-BUBBLE MEMORY shows the pattern of circulation in the 100,637-bit memory. The major loop holds a single data block, consisting of 1's and 0's (bubbles or no bubbles) that are being written into the memory, read out, replicated or erased. In this particular device the data block contains 157 bits. In the writing cycle the 157 bits first enter the major loop, whence they are transferred simultaneously, at a signal, to the 157 minor loops, one bit per loop. Each minor loop in turn provides sites for 641 bubbles. Thus total capacity of device is 157 × 641, or 100,637, bits. In read cycle 157 bits are transferred simultaneously, at a signal, from minor loops into major loop, which carries them past read head.

Early magnetic bubble devices used orthoferrites initially as media for bubbles, but it is generally reported that these materials exhibit inferior memory characteristics when compared to garnet devices. Therefore, orthoferrites have been dropped by many of the advanced magnetic bubble device researchers. Available information reveals that virtually all major researchers and manufacturers of bubble devices in Japan, Great Britain, The Netherlands, the Federal Republic of Germany, and the United States use nonmagnetic garnet substrates and magnetic garnet films for their bubble devices.

Reportedly, amorphous materials are relative new-comers on the scene of magnetic bubble materials. In many instances, these materials are said to support smaller bubbles with potentially higher operating speeds than bubbles of the current garnet technology. Garnets, however, have already been widely accepted as a result of intensive material research that has been reported for over five years.

In one device, nickel-cobalt is the material used in the domain memory. These memories are called moving domain memories (MOD). MOD memories are said to present advantages over contemporary bubble memories in that less exotic substrate materials (glass) and readily available magnetic films (nickel-cobalt) can be used. In addition, nonvolatility of the stored MOD information is said to be assured without the use of permanent magnets that are necessary for magnetic bubble device data retentivity. However, the advertised access times and bit-storage-densities do not measure up to those of garnet device parameters although the MOD memories are potentially less expensive to produce.

Technical information published by magnetic bubble researchers stress the importance of high data-storage-density and fast data access times as goals for their devices. In addition, bubble researchers claim that the nonvolatility and low-power consumption of the magnetic bubble device packages make this product ideal for portable, battery operated, computing apparatus.

The application of the bubble device was found to occur first as memory in space and military digital data recorders. This memory application makes use of the high-reliability (no moving mechanical parts), high data-storage-density, low-power-operation, and high-radiation-resistance characteristics of the garnet bubble devices. Future applications that are found in bubble device evaluation reports include desk calculators, hand-held calculators, computer peripherals (disk and drum memories), microminiature bubble displays, and electronic test and measurement devices.

FLOPPY DISC MEMORY

The floppy disc itself consists of a flexible plastic storage medium similar to a 45 rpm record enclosed in a non-metallic jacket. When inserted into a floppy disc drive, a spindle attaches to the storage media through a hole on the center of the jacket. The disc revolves at 300 rpm within the jacket. A slot, cut in the jacket, allows a read/write head access to any of up to 77 tracks on the disc. Through a small circular hole in the jacket a sensor in the drive can detect a similar hole in the disc that indicates the rotational position of the medium.

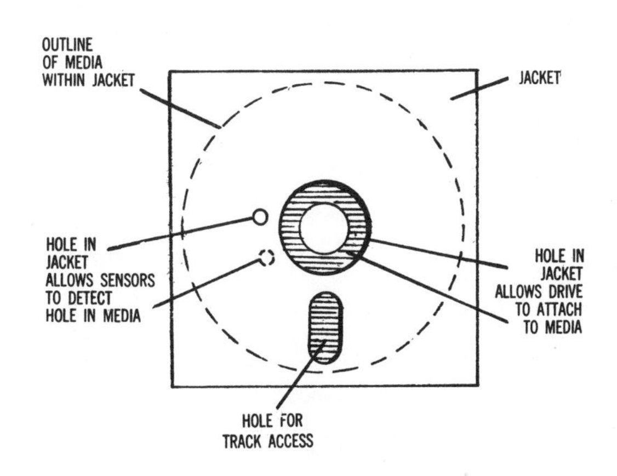

11.8.2 HOLOGRAPHIC MEMORY

Acousto-optic crystals deflect a laser beam (Fig. 11-54) in X and Y in proportion to the frequency of sound waves passing through them. This spacially selects a hololens (optical reference pattern) from a lens matrix.

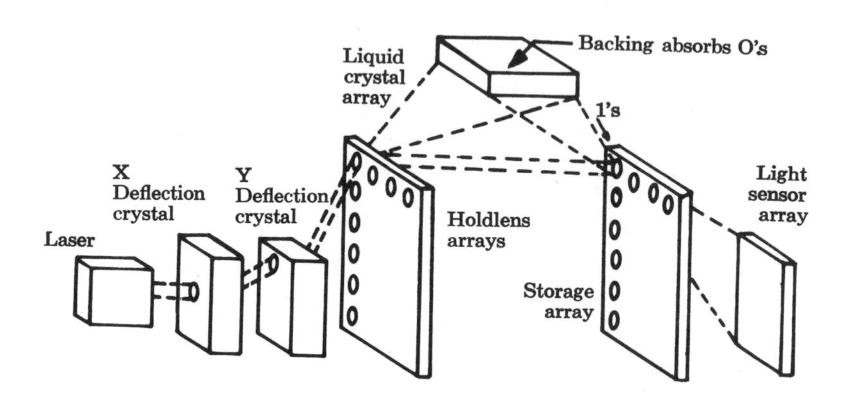

Fig. 11-54 Holographic memory system.

The hololens splits the beam, sending one portion to its corresponding point in the multi-element storage array. The storage medium is a unidirectionally magnetized film, usually manganese bismuth (MnBi). The other portion of the beam diffracts to an array of liquid crystal cells.

Each cell can be made transparent or reflective (a 1 or a 0) electronically. Transmitted light is absorbed behind the crystals. Reflected light refocuses on the same storage area as the undiffracted beam. Where the beams coincide, they locally heat the MnBi above its Curie point. When the light is removed, the film cools, and formerly hot areas reverse their direction of magnetization. This stores data in a magnetic hologram in the film.

To read, a reduced power laser shines on the film, but not on the liquid crystals. The film's transmission properties vary with magnetic state, so the magnetic hologram is converted to light and dark areas that are detected by a multi-bit sensor array.

Erasing is done by uniformly illuminating the storage medium. This heats an entire storage area above the Curie point. Applying a magnetic field during cooling returns the film to its original, uniformly magnetized state.

Holographic memory capacity is the product of storage areas and liquid crystal cells, so the number of discrete components increases roughly as twice the square root of capacity.

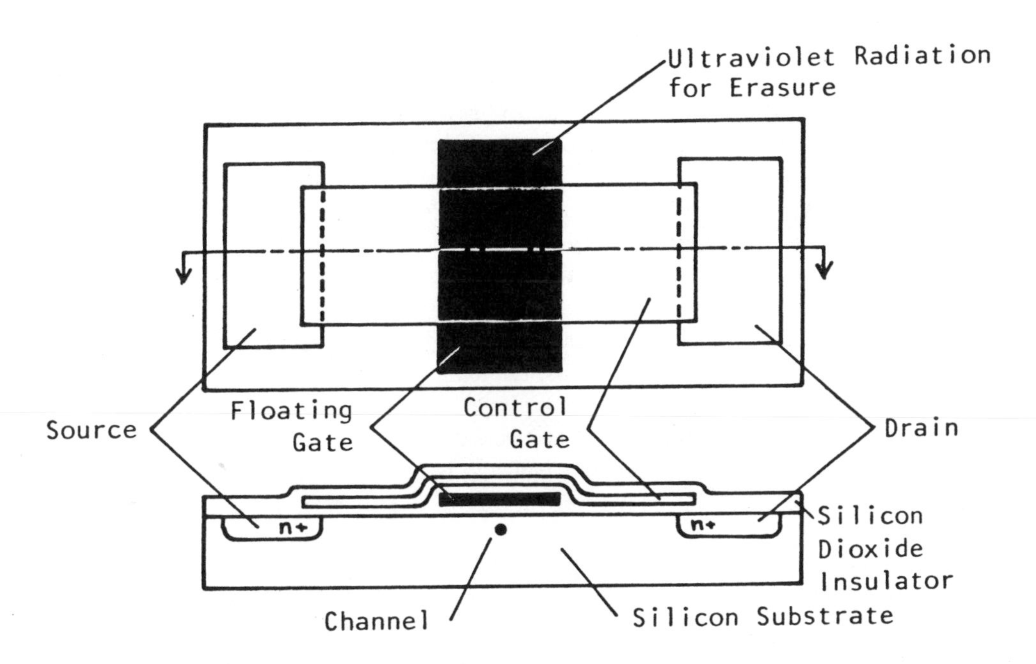

OPTICALLY ERASABLE READ-ONLY MEMORY CELL is shown in plan view and in cross section. The floating gate, which is not electrically connected to anything, holds a binary digit of information in the form of a stored charge that modifies the electrical characteristics of the device. Information is "written" (stored) in the floating gate by applying about 25 volts to the control gate and drain while the source and the substrate are grounded. The resulting high electric fields in the channel accelerate electrons to a considerable velocity. A small fraction of the electrons are able to cross the thin insulator and become trapped on the floating gate. The charge on the gate is not lost during "read" operations or when external power is removed. The stored charge can be erased, however, by exposing the cell to ultraviolet radiation, which temporarily makes the silicon dioxide sufficiently conductive for charge to leak away. All the cells in a memory are erased at one time, after which a new pattern can be written in.

11.9 UNIVERSAL LOGIC BLOCKS

In the past, logical design was based upon the optimum use of logic gates (NAND, AND, etc.) to realize given switching functions. It was desired to minimize the number of diodes (input leads) and/or transistors used. Using integrated circuits however, logical modules can be produced at a cost practically independent of the number of diodes and transistors contained in the module. Therefore, it seems reasonable to assume that the cost of a complex switching circuit composed of such modules, will be determined by the number of modules used, together with the number of interconnections between modules — and not by the complexity of circuits within a module. The problem, then, is two-fold. First, efficient design techniques must be established for the synthesis of digital circuits using a fairly complicated module as the logical connective. Second, logical modules must be found which simplify the synthesis procedure while decreasing the associated cost.

The rapidly advancing integrated circuit technology has also placed many new and often unforeseen demands on logic packaging techniques, and is severely affecting traditional computer design concepts. For instance, one of the most pertinent and immediate requirements is the optimum utilization of Input/Output (I/O) connections, since the package size is strongly dependent on such connections. The package efficiency is measured in part by the I/O pin-to-circuit ratio, assuming the circuits in a package are connected in a way to provide an optimum logic **function**. Another potential problem to be considered is power dissipation, since integrated circuits may be contained in extremely small areas.

Consequently, it is essential for a logic designer to interconnect many circuits with a package, thereby reducing I/O pins. At the same time, it is essential not to overburden a package with more heat than it can safely dissipate by the cooling provided. The above considerations are further affected by the fact that integrated circuit technology should inherently provide inexpensive circuits. Thus, the logic can be reasonably redundant to accomplish the minimum I/O pin-to-circuit objective as well as to minimize package types. The more versatile a package, the fewer types are required. The most versatile package is a simple logic connective as, for example, a NAND. However, the pin-to-circuit ratio is unacceptably high.

One possible solution to the problem, is to use Large Scale Integration (LSI), letting most packages be customized and still large enough to reduce the quantity of packages in any machine. The customization of each part leads to inventory problems at the place of manufacture. In addition, the complexity of each part may lead to many wiring layers which may be inaccessible for engineering changes. If an error is discovered during the fabrication process, the part may have to be scrapped. This delays delivery and increases cost of acceptable parts.

Another possible solution is the utilization of chips or cells having a certain defined logical complexity, flexibility, and acceptable pin-to-circuit ratio. This would enable engineering changes and modifications at the chip or cell level, thus making engineering change of the interconnections between cells possible.

The traditional procedure used in designing a computer, is first to define its major logical sections. The data flow paths are usually designed initially, since they are generally well defined. Controls and non-iterative sections are determined last, since they are heavily dependent on data flow organization. In recent years, the controls have become more organized through use of read-only memories (i.e., micro-programming). The read-only memory replaced much isolated and non-iterative hardware.

The various parts of the computer are still packaged in much the same way as they were first organized on paper. For example, the adders, registers, shifters, and other iterative networks are packaged to take advantage of their repetitive occurrence with the machine. The non-iterative sections are added without apparent organization.

To apply integrated circuits, the designer may package iterative logical networks in an attempt to make efficient use of the integrated circuits. The adders, registers, etc., have a high circuit density relative to the input/output connections to an integrated circuit element. Unit logic may be packaged on integrated circuit elements to satisfy the non-iterative logical sections of the machine. These elements have a *low circuit density* relative to the input/output connections of an integrated circuit element. Accordingly, this design philosophy will not permit efficient use of monolithic technology.

The problem, therefore, is to determine the logical function that complex cell elements should generate.

One investigation of the problem used the basic logic circuit of Figure 11-55 in an automated analysis and synthesis technique. The investigation was carried out through a computer using the flow chart of Figure 11-56. The conclusion of the investigation was that automated techniques can be effectively applied to logic analysis for selecting a universal cell set which can be used to efficiently implement machines of various types.

The trade-off to be made must consider that large, high performance machines generally have smaller markets than small low performance machines. Therefore, part number reduction may be more important in

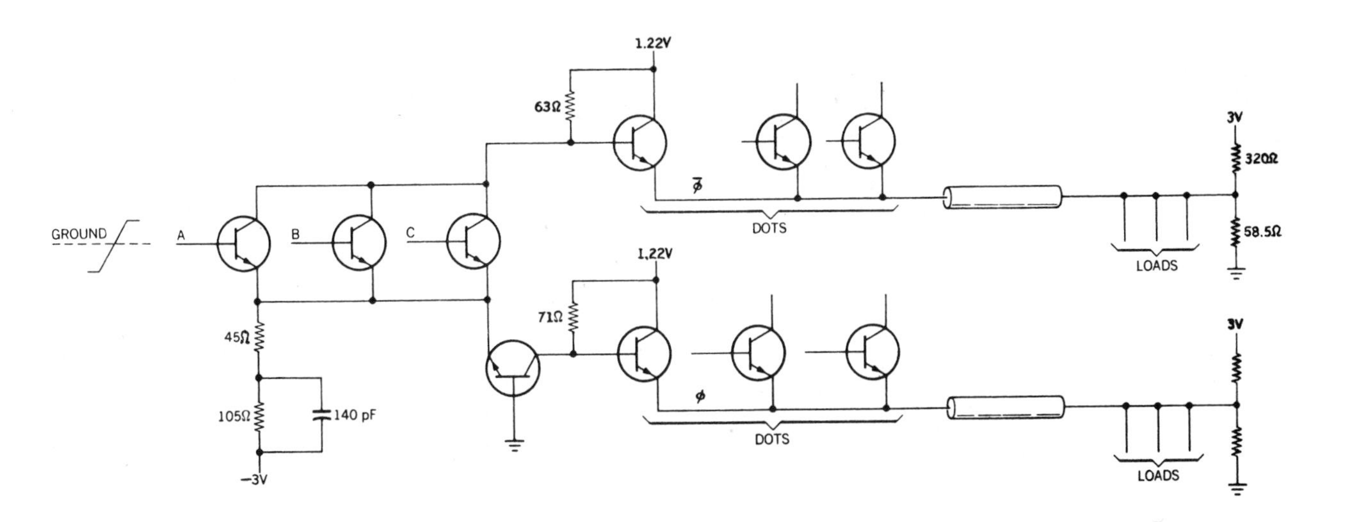

Fig. 11-55 **Basic logic circuit used to investigate universal cells.**

the high performance machines while total parts reduction may be more important in low performance machines. On the other hand, the high performance machine requires high density to get performance while the low performance machine can stand more delay. The density and the part number considerations oppose each other, thus emphasizing the need for automated logic analysis techniques.

If the functions used to reimplement a machine were built as structures diffused into the substrate rather than a combination of unit logic circuits and metallized connections, then power dissipation, delay, and reliability should be considerably improved. By eliminating the metallized connections and their associated impedance, the power dissipation and delay caused by them will be reduced. Also, since the metallized connections are a source of possible failure, reliability should be increased.

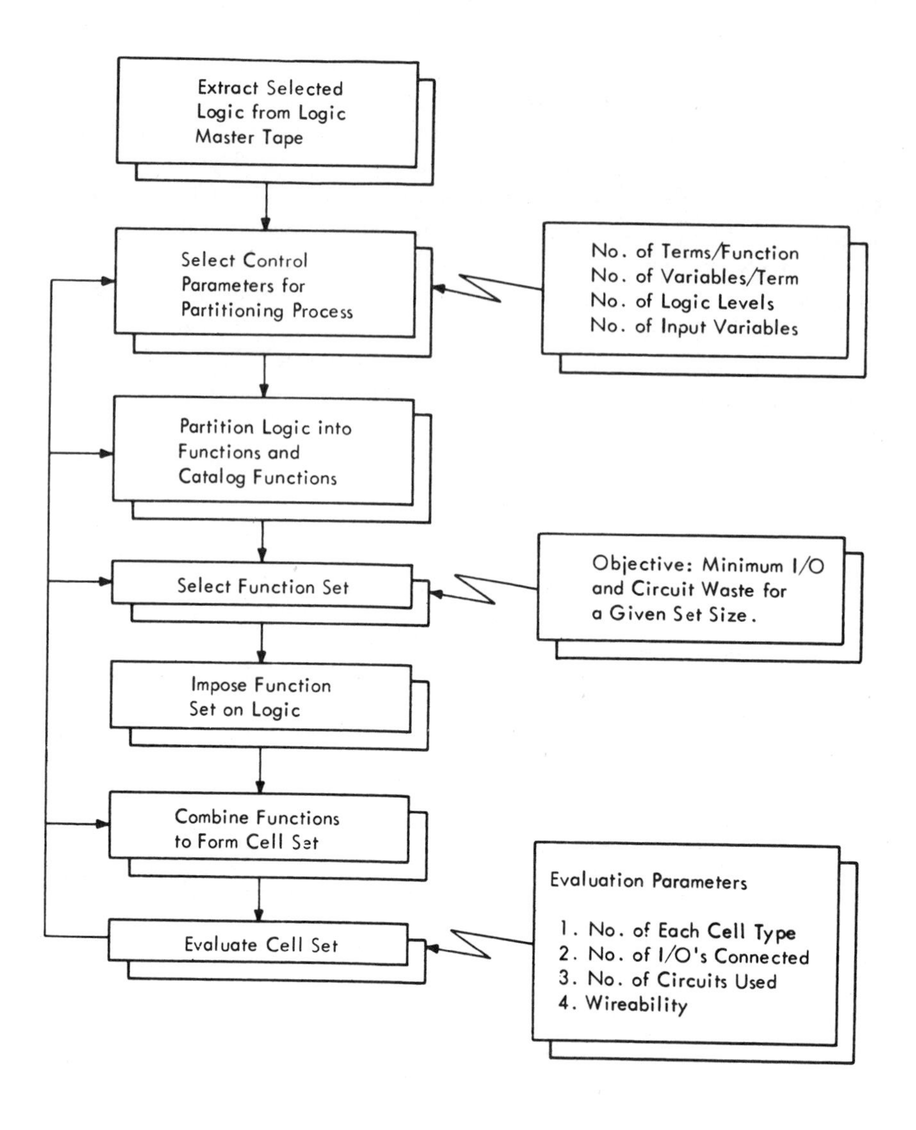

Fig. 11-56 Flow chart for universal cell analysis.

11.10 LARGE SCALE INTEGRATION DESIGN

Over the past decades computer circuitry has experienced the technological changes of first, the transistor, and secondly, the integrated circuit. In the forthcoming future, a change equally as profound, can be expected as the integrable complexity of components increases. However, in contrast with the revolutionary aspects of the transistor and the integrated circuit, the production of sophisticated circuit modules as exemplified by Large Scale Integration can be expected to evolve from existing semiconductor technology as manufacturing yields improve and automated design techniques are developed.

The first monolithic integrated circuit, assembled in 1958, was an electrically simple, rather crude-looking, hand-made affair with many of the connections between devices being accomplished by wire bonds. Yet the promise of mass production could be clearly seen, provided practical batch fabrication methods for interconnection could be developed. The planar process provided the necessary technology to achieve a plane surface with adherent thin film, batch fabricated, interconnections. The combination of planar technology with the original concept of the monolithic structure, resulted in the integrated circuit.

The first commercially available integrated circuits which appeared in 1960, were digital circuits because of the potential need for large numbers of similar type circuitry in digital equipment. These first circuits were simple gates similar to those manufactured with discrete components. It became apparent rather quickly, however, that the same sense of values which existed with regard to discrete devices, was not applicable to integrated structures. This led to a succession of circuit families, as manufacturers attempted to optimize performance and economics for monolithic construction.

At the same time that the basic circuit configurations were evolving, more complex circuits such as flip-flops were being developed to yield a series of compatible elements within each family. As manufacturing techniques and yields improved, it was found feasible to produce multiple devices on a single chip. Much of the discussion of LSI revolves around where, and under what conditions, the advantages of increased electronic complexity can offset the reduced flexibility of not having each input individually accessible. Not only does this internal connection reduce the number of package pins, since it is unnecessary to bring each input and output separately out of the package, but it eliminates the need for much external interconnection – thus simplifying the problems of the user.

LSI is a technology which can be defined in terms of device and circuit

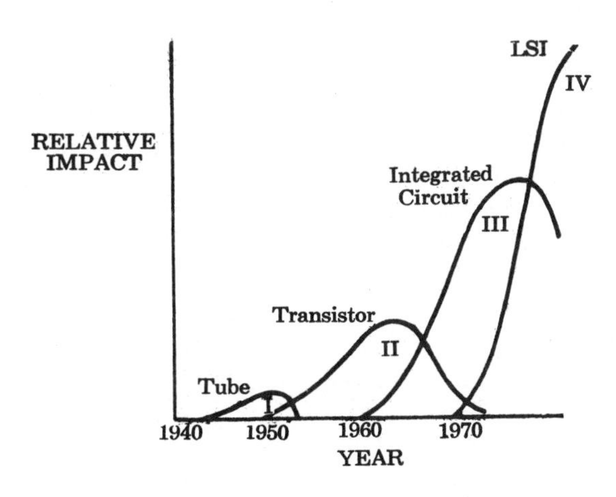

Fig. 11-57 Generations of digital equipment.

construction. Devices (resistors, transistors, etc.) are interconnected to form integrated circuits (gates, flip-flops, etc.). In a similar fashion these circuits may be interconnected through the technology of LSI to form what may be called Integrated Equipment Components or LSI components. The terminology of increasing complexity is, therefore, device, circuit, and component. Circuits are the building blocks of components in the same way that devices are the building blocks of circuits. Because digital circuitry is made from combinations of similar type circuits (in principle digital systems could be assembled from nothing more than gates, although this would not be an optimum arrangement), LSI is particularly well suited to digital configurations. It is generally conceded that to qualify as LSI, a component must have the equivalent of 100 or more gates. Analog components having greater than 100 analog circuits are rare, and for this reason LSI has dealt more with digital circuitry.

The various generations of digital equipment are shown in Figure 11-57. Each generation is built around an active device technology – tube, transistor, integrated circuit, and LSI. This qualitative plot of the "impact" of each technology was designed to illustrate that:

(1) Each succeeding component technology builds on previous work and consequently impacts digital systems more than the previous technology.

(2) Overlap occurs during the transition period between technologies resulting in hybrid mixture of technologies.

(3) The technology now emerging from the laboratories and which will have the greatest impact on digital systems in the next decade, will be LSI.

11.10.1 CHARACTERIZATION OF LSI BY COMPLEXITY

One method by which LSI can be characterized is by its complexity. When dealing with digital equipment, the usual method of doing this is to rate the component in terms of the number of equivalent gates required to perform the designed function. In most cases of logic, this can be accomplished in a straight-forward manner by adding up the composite circuits making up the structure, recognizing that circuits such as flip-flops may equal 4 equivalent gates, etc. In the case of shift registers or memory elements, the repetitive gate circuits tend to be very simple, and it may be reasonable to reduce the number of equivalent gates that each represents. Gate equivalency of a component is thus neither strictly in terms of area, or strictly in terms of circuits, but represents a combination of both in which a degree of judgement must be exercised. Within these limits, however, it is possible to characterize components according to the following categories:

Equivalent Gate Count	*Description*
1-10	Standard individually accessed, multifunction, integrated circuits (quads, duals, etc.) currently being produced in high volume.
10-100	General purpose, medium scale integration (MSI) components.

100-1000	LSI random function components.
> 1000	LSI regular function components.

MSI – A large number of designs have common, general purpose functions such as decoding, converting, and multiplexing which are primarily useful in input and output applications in any digital system. While no one system may be expected to use large numbers of these components, their wide appeal and relatively straightforward fabrication technology may result in MSI components quickly becoming a significant fraction of digital components.

LSI – Random Function Components – Percentage-wise, components in the 100 to 1000 gate range may not account for more than 10% of the total digital parts. In terms of equivalent gates, however, these components may make up the bulk of the logic circuitry. Since each system tends to be configured differently, there may be little commonality among components in this complexity range. The upper complexity of random type circuitry will probably be limited primarily by manufacturing yield, and may be estimated to be 1000 equivalent gates.

LSI – Regular Function Components – Above 1000 equivalent gates per package, circuit design must necessarily involve repetitious circuitry such as shift registers and memory elements because of size and yield limitations. Layout of these devices is straightforward, in this case, because the elements can be arranged in a regular fashion. This regularity also permits simplified test and design procedures to be used. Because of the relative ease with which regular functions can be designed, fabricated, and tested, they, more than other LSI components, may generate new applications instead of merely replacing existing circuitry. For this reason, the greater than 1000 gate component range represents the most rapidly expanding of all the parts catagories.

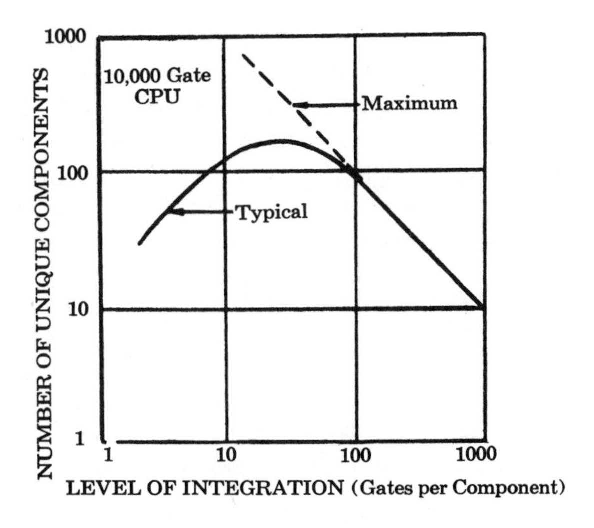

Fig. 11-58 Composition of a digital System as a function of the level of integration.

Characterization of LSI components as standard or custom design is fundamental to the nature of LSI. Consider, for example, the curve of Figure 11-58, where the number of unique components has been plotted as a function of the number of equivalent gates per component (level of integration) for a hypothetical 10,000 gate digital system. The curve illustrates that, in theory, the system may be composed of either a single component of great complexity, a large number of simple circuits, or any combination inbetween. The number of unique (electrically different) components has a maximum in the vicinity of 50 gates per component. Generally speaking both the systems fabricator and the components manufacturer would like to minimize the number of unique components. The systems fabricator because he will have fewer parts to inventory, and the components manufacturer because he can concentrate all his design and processing skill on a few components. However, the user and supplier may disagree on the best method of minimizing part count. The user would prefer to work to the right of the maximum of Figure 11-58, and ultimately to have only a single component. This minimizes his interconnection and handling problems. The manufacturer, on the other hand, would prefer to work to the left of the maximum where he can manufacture only a few types of circuits in high volume. This allows him to take full advantage of the batch fabrication processes of integrated circuits. In this illustration, custom designed circuits would lie to the right of the maximum, while standard circuits are to the left. The distinction in this case is volume. Standard components are characterized by their design costs being small compared with their manufacturing costs, while the reverse is true for custom components. This leads naturally into a consideration of LSI manufacturing and design costs, and what the economic benefits are to the user.

11.10.2 COMPUTER AIDS IN THE DESIGN CYCLE

The initial specification for the LSI system is in the form of a set of Boolean expressions. This can be translated into a logic system which then requires verification. As the systems become more complex, the cost and the time involved in achieving this through the traditional ''breadboarding'' become intolerable, and logic simulation programs are being extensively used to provide the engineer with a rapid verification of his design. The programs currently available provide a checking facility for both combinational and sequential logic. This indicates any logic or timing errors which may result in a redesign and a second run of the simulation program. The basic requirements for such a program are briefly described below:

The program should have a simple input language and have facilities such that if say a JK flip flop is used many times in one design, then it should only be assigned once, and thereafter just referred to. The program should be on-line and have on-line editing facilities so that the engineer can easily correct any typing errors. The output of the program should be easy to understand and as engineer-oriented as possible and should include—

(a) an indication of waveforms showing circuit operation.

(b) an analysis of logic hazards present in the system.

(c) a "timed" analysis utilizing realistic time delays for all the logic gates.

The first item allows the engineer to know whether the system is correct (or not); the second tells him which parts of the system should be carefully designed, to avoid race hazards causing the system to operate incorrectly; and the third item enables the engineer to see what effect production spreads of gate delays might have on his circuit.

Finally, the program should have an on-line "scope probe" facility which can be put on any point of the logic circuit, to monitor its operation at any time.

Once the logic has been established, the circuit design can be undertaken. This consists essentially of the interconnection of standard logic blocks, and the provision of any additional specialized circuit blocks if required.

The standard "blocks" of circuit layout are held under specified names in a computer library. The new, non-standard, blocks must be outlined and, if considered critical to the operation of the overall circuit (speed, gain etc.), they must be analyzed in detail without the need to fabricate them. This is accomplished by the use of specially developed d.c. and transient analysis programs. Once this is satisfactorily completed, the new blocks can be assigned a name and stored in the computer library, thus becoming further standard cells.

A further cycle of analysis may be undertaken when the layout has been finalized and more information about interblock dependence becomes available.

11.10.3 TRENDS IN LSI COMPUTERS

An analysis of current technical papers shows four trends in implementing computers with LSI arrays:

(1) new partitioning of a still existing third generation computer. This choice makes it possible to leave software unchanged, and therefore to reduce investments. A severe initial constraint, however, is imposed.

(2) design of new lines of computers in accordance with conventional architecture.

(3) computer design with an original structure (omnibus machine, multi-processors etc.). The unit consists of parts of arithmetic units with a control logic and a local memory on each. Data flows through the various units via a bus. Such a structure, although it minimizes the proportion of anarchical logic, increases the equipment required to process the data.

(4) adoption of DDA structures for airborne computers or industrial process-control. Integration, in one or two packages, of a complete digital integrator has interested component manufacturers for some time. The reason for this interest lies in the fact that any DDA can be completely obtained in MOS and bipolar technology.

However, the difficulty of programming a differential analyzer, the problem raised by their initialization, and the error propagation phenomenon, tend to limit their use. Despite these disadvantages, however, the use of DDA in inertial guidance systems (integration of accelerations, axis changes), in the solving of differential equations and transcendant functions has been successfully developed.

Microprocessors

Before the introduction of microprocessors, complex logic systems used discrete and random logic to perform the necessary functions. Integated circuit families, such as TTL and ECL, developed small and medium-scale integrated functions that simplified random-logic designs. General and special-purpose computer manufacturers used such devices to build their systems. Now third-generation computers use large numbers of these devices, coupled with various types of separate memory systems, to complete their architecture.

However, this computer architecture is too cumbersome and costly for most large digital systems, compared to microprocessors. Such systems include CRT terminals, point-of-sale and other table-top equipment, as well as lightweight airborne equipment.

A new direction was launched with the emergence of calculator chips. The present calculator can be defined as a small, highly specialized computer. The memory structure consists of both a fixed and a variable memory. The fixed portion, a read-only memory (ROM), provides a system control program called firmware—meaning nonchangeable instructions. This contrasts on the one hand with general-purpose computers programmed by software, and on the other hand with random-logic systems that use hard-wired circuitry. The tradeoffs for the three are indicated in Fig. 11-59

An extension of calculator design has been the development of larger word-length systems that are closer to true computer architecture. This evolution has resulted in the microprocessor, for computation and control applications besides calculators.

Fig.11-59

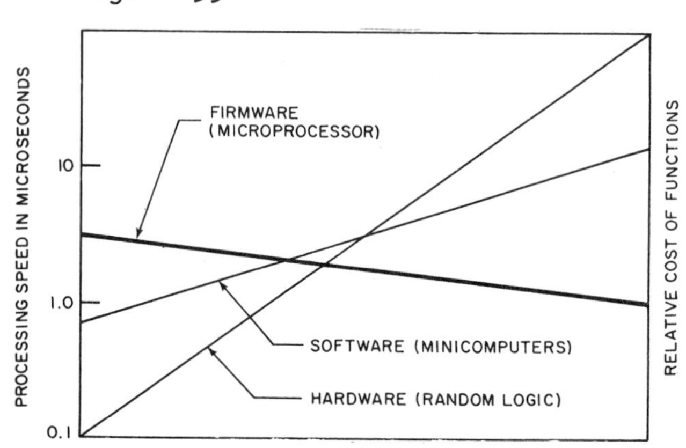

Microprocessors offer the lowest cost, but at the lowest speed, with system control provided by firmware. For the highest speed, but at the highest cost, random logic is the way to go. Minicomputers fall between the two approaches in this speed-cost tradeoff.

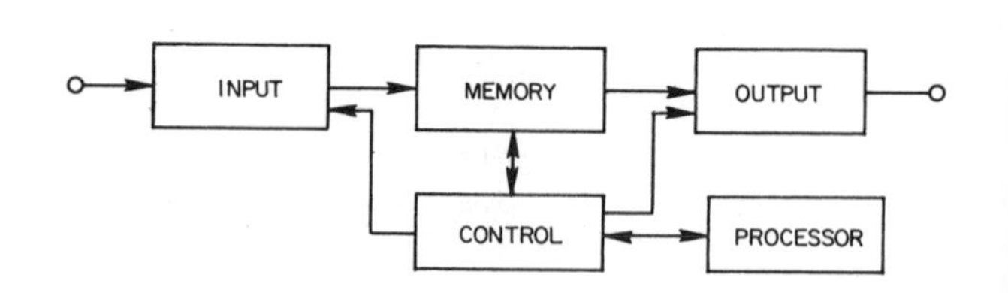

Fig.11-60

The basic architecture of a microprocessor. This functional diagram covers a wide range of industrial and process-control systems.

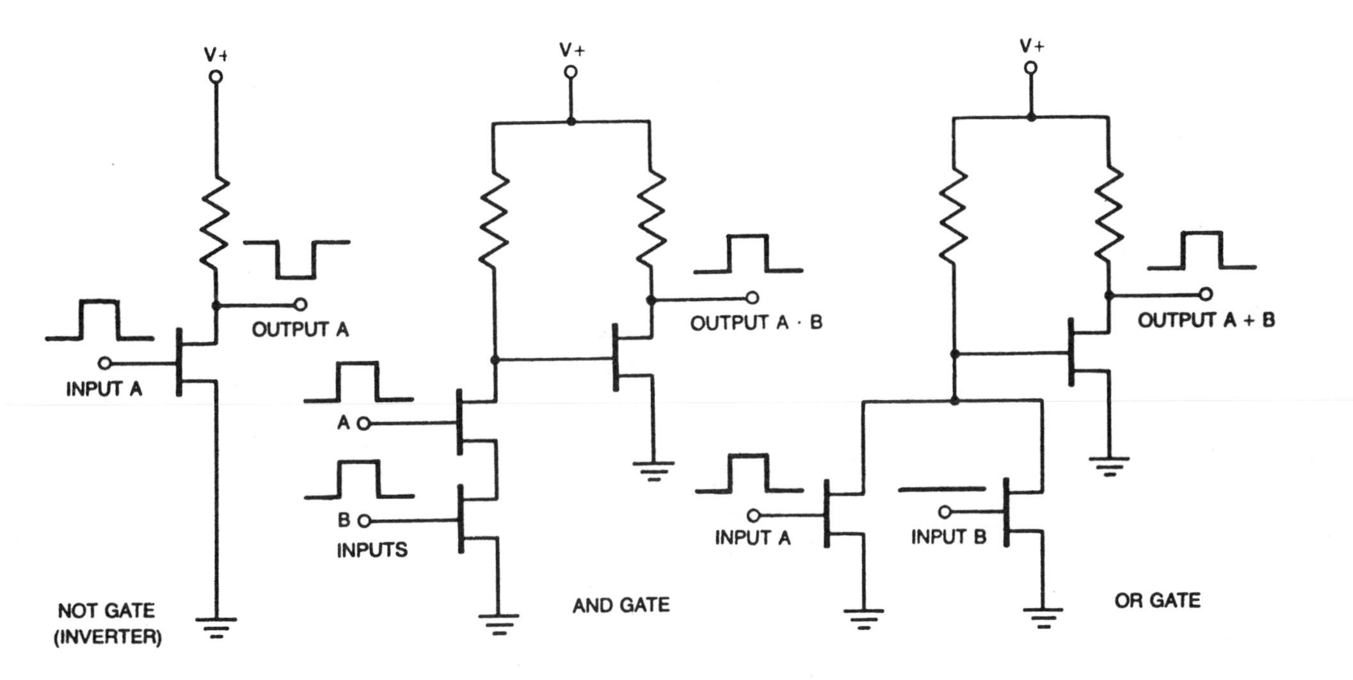

ELECTRONIC LOGIC GATES evaluate arithmetic and logic expressions in which binary values are represented by voltages. By convention a binary 1 is represented by a high voltage and a 0 by a low voltage. The gates shown here are constructed from metal-oxide-semiconductor transistors. The simplest is the "not" gate, or inverter. When the input to this gate is in the low state, the transistor does not conduct and only a negligible current flows from the supply voltage ($V+$) through the resistor and transistor to ground. As a result there is little voltage drop across the resistor and the output is in effect conput, the transistor conducts and the comparatively large current flowing through the circuit produces a considerable voltage drop across the resistor. The output voltage is now near ground and is therefore in the low state. The "and" gate has two input transistors connected in series; current flows through them only when both receive a high signal simultaneously. In order to restore the proper polarity of the signal the output of the two series transistors is followed by an inverter. In an "or" gate the input transistors are arranged in parallel, so that a high signal applied to either one of them results in conduction. Again, an inverter is required to change the polarity of the output.

Microprocessor basics

A microprocessor is a computer in IC form. A computer is a device capable of automatically carrying out a sequence of operations on data expressed in discrete (digital) or continuous (analog) form. Its purpose is to solve a problem or class of problems; it may be one of control, analysis, or a combination of the two. In digital computers, numbers are represented by the presence of voltage levels or pulses on given lines. A single line defines one bit (short for binary digit or a base-2 number). A group of lines considered together is called a "word"; a word may represent a computational quantity (operand) or it may be a directive specifying how the machine is to operate on computational quantities.

To accomplish automated computation or control, the computer must perform various internal functions. The most obvious is to do arithmetic type of operations (add, subtract, etc.) on two operands. The section performing this function is the Arithmetic Unit (AU).

Something must control the arithmetic unit to make it follow the specific sequence of operations necessary to solve a given problem. Accordingly, a sequencing mechanism is required; furthermore, if the computer is to be programmed, the sequencer must also be programmable. Some storage is necessary in which to hold the required sequence of operations before beginning a computation. The sequencer can be separated into two functional units: program storage and control.

The control unit can be viewed as sensing external conditions and issuing commands to other machine elements. For example, the control unit sends commands to the arithmetic unit to initiate arithmetic operations, or sends commands to the program storage, which causes it to look-up the next program directive. The control unit senses such conditions as the completion of an arithmetic operation, the sign of a result, and the presence of stop/start signals from the computer operators.

- **Memory** — A storage unit. Can be read-only (ROM), for program storage, or read/write random access (RAM) for program, operand or temporary storage. Data is usually stored in binary form.

- **Arithmetic unit** — Often referred to as the arithmetic and logic unit (ALU); it performs the arithmetic operations on operands or provides partial results within the computer.

- **Control unit** — Referred to as the brain of any computer because it coordinates all units of the computer in a timed, logical sequence. In fixed-instruction computers, this unit receives directives from the program memory (hereafter directives will be called "instructions" since they instruct the computer what actions to take and when to take them). These instructions are in sequences, called programs. They reside in the memory and are referred to as software. The control unit is closely synchronized to the memory cycle speed and the execution time of each fixed instruction is often a multiple of the memory speed.

- **Input/Output** — The means by which the computer communicates with a wide variety of devices, referred to as peripherals. They include switches, indicator lamps, teletypewriters, CRT's, magnetic or paper tape units, line printers, A/D or D/A converters, card readers and punches, communication modems, etc. The I/O lines can be connected to intermediate storage devices for use with mass memories, including magnetic discs, magnetic drums and large-scale RAM systems.

Microcomputer performance criteria

In microcomputers, the basic time interval is the microcycle. Since both the instruction fetch and instruction execute subcycles are each comprised of one or more microcycles, depending on the machine and instruction, the cycle time calculation becomes ambiguous and complex. To illustrate, consider a microcomputer that requires two microcycles to fetch an instruction, one microcycle to decode, and one microcycle to execute a "register add" instruction, two microcycles to execute a "jump to subroutine" instruction, etc. If we assume a 2 μsec microcycle, this machine would require a cycle time of 6 μsec to add two registers (3 microcycles) or 8 μsec to jump to subroutine (4 microcycles).

The point is: Cycle speed or cycle time alone is not a valid evaluation criterion for a computer, and especially not for a microcomputer. To provide a performance indicator, the efficiency of the instruction set must also be considered — what can an instruction really do and how long does it take to fetch it, execute it and be ready to fetch the next instruction?

The simplest ALU consists of an adder and an accumulator. The adder adds (or performs similar logical operations, e.g., OR) two inputs, A and B, and produces the output. The accumulator holds intermediate results of a computation or numbers for a pending computa-

The remainder of the CPU, the control portion, is implemented using an instruction register (IR), a control decoder and sequencer, and a program counter (PC). A machine instruction is transferred from program storage memory into the IR and is subsequently interpreted by the decoder/sequencer, which issues the appropriate control pulses to the other computer elements. The PC contains, at any given time, the address in memory of the next instruction. This counter is normally incremented by one immediately following the reading of a new instruction. The PC contents can be replaced by the contents of a specified memory location if the last instruction was of the jump class. This causes the next instruction to be read from a program-specified location instead of from the next sequential location.

Variations on a theme

Common improvements, additions and/or alternations to the classic architecture described above include multiple accumulators, sophisticated I/O structures, index registers, indirect addressing, interrupts, push-down stacks and microprogrammed control units. Such features can enhance a microprocessor's capabilities and are often the basis for comparisons between various machines, as well as providing a theme for competitive advertising and salesmanship. In view of these three facts, a discussion of basic computer variations follows:

Accumulators, multiple — By definition, an accumulator provides a temporary storage medium. Temporary storage allows programs to execute faster and more efficiently by obviating the need to store partial or intermediate results in main memory and subsequently to retrieve them for use in additional computations. Multiple accumulator registers allow several partial results to be maintained at the computer's fingertips, thereby eliminating the many program steps that would otherwise be required to store and then retrieve data (shorter programs cost less to write, less to store and execute faster than longer programs). Four accumulators are able to provide a great deal of programming and operational versatility, and it is often considered an optimum number.

I/O structures — A basic input/output system provides a single input port and a single output port. A port allows transfer of one word of data across the

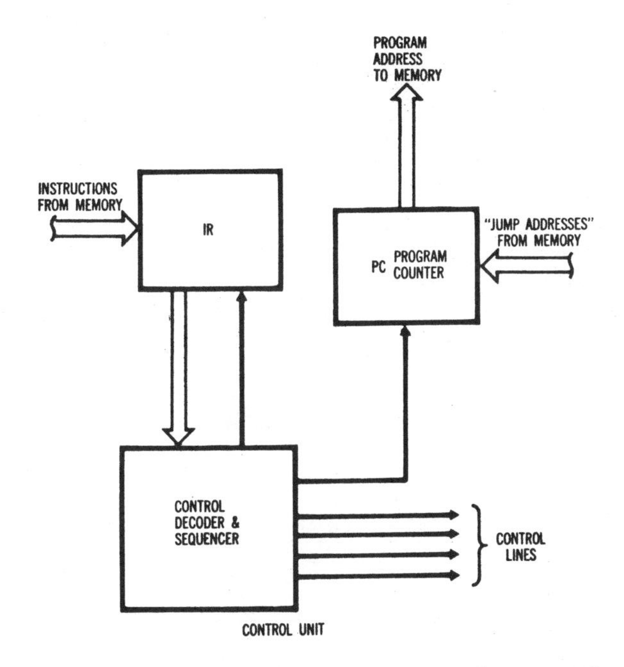

Elements of a memory unit. In the operation of a memory, the MAR contains the address where information is to be stored or read. The MDR holds the data to be stored or receives the data as it is read.

computer's boundary. More sophisticated I/O units facilitate the use of multiple input and output ports, allowing virtually simultaneous communication with many peripheral devices. Another powerful I/O technique, referred to as DMA (for direct memory access), allows peripheral devices to transfer data directly into and out of memory, independent of the control unit and the operating program. This contrasts to the more conventional programmed I/O, where an explicit program

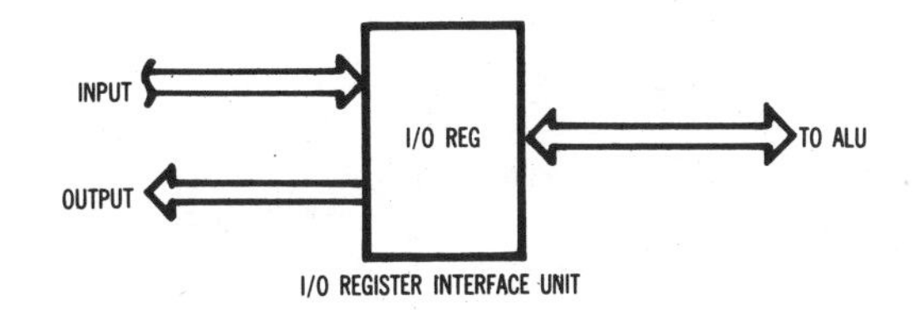

This simple ALU contains an adder and an accumulator. The accumulator provides temporary storage. For example, here it can hold one operand while another is obtained from memory in order to perform addition.

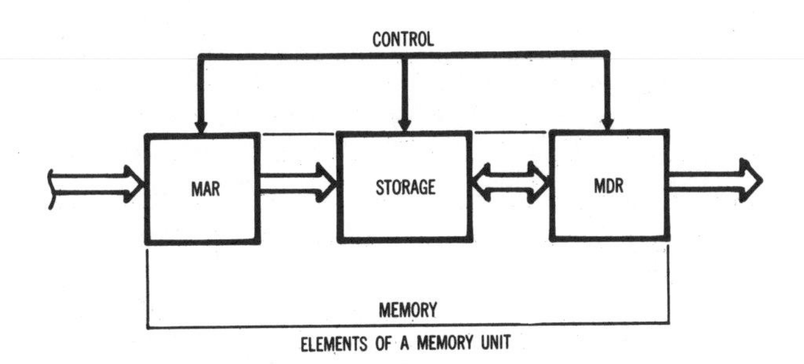

In the control unit, the instruction register receives the machine instructions from program storage. These are then interpreted by the decoder/sequencer, which controls the various microprocessor elements.

instruction is required for any data transfer. The DMA technique facilitates faster data exchanges with memory, with fewer program steps, and is considered most applicable to bulk storage device (disc) interfaces and computer-to-computer connections. Contrary to occasional misuse of the term, DMA is not a uniquely defined off-the-shelf circuit, but can be implemented in a variety of ways in any general purpose computer.

Index registers — This feature provides programming flexibility by providing the user with more memory addressing modes. As a rule, when a programmer wishes to retrieve an operand from memory, he specifies its address in the instruction that calls for work to be performed on that operand (e.g., add it to an accumulator). The presence of an index register(s) allows the programmer to modify or index the operand address with the number contained in the register. Typically, the operand address from the instruction would be added to the content of the index register. Such a feature greatly simplifies the transfer of an array or field of data into or out of memory. Machines with two index registers offer enhanced programming versatility over machines with a single register.

Indirect addressing — This is done when the address contained in the instruction specifies only the address of a memory word, which itself, specifies the operand address. An indirect address is an address in an instruction that indicates the location of the address of the referenced operand. Think of it as a computerized treasure hunt — the instruction does not tell the location of the "treasure" (operand), but tells where to go to find a clue that gives its location. Multi-level indirection is possible, although not considered necessary. Here the system jumps through two or more clues until the operand is found. Indirect addressing provides great programming flexibility by allowing operand address to be continuously modified by the program.

Interrupts — A machine operates under its own control but frequently it is desirable to have an external event cause the computer to shift its attention to another problem. This can be done in many ways:

- The computer program can include a section that causes it to look for possible external events each time it cycles. This may consume a lot of memory, make the computer operate its program more slowly and may not permit the computer to respond quickly to the external event.
- Interrupt signals may be forced into the computer. This requires extensive programming to insure that, when the external event has been serviced, the computer can return to its prior location.
- The computer can have interrupt capability built into its hardware, thus allowing the computer to service the interrupt quickly, with a minimum expenditure of program and memory storage space.

Push-down stack — Or "Push-down, pop-up" stack, LIFO (last-in, first-out) stack, etc. This is a useful feature for the "nesting" of interrupts and subroutines. Nesting refers to the entry into a second (or third) interrupt service program or subroutine prior to completion of service or execution of the first. The stack stores the current program execution address (contents of PC) each time the computer is directed to a new ancillary task, thereby allowing the computer to return and clean-up unfinished work in reverse order. The stack is also useful for storing partial results of computations. (Subroutine: A set of instructions necessary to direct the computer to carry out a well-defined mathematical, logical or analytical operation, usually arranged so that control can be transferred to it from the main program and so that, at the conclusion of the subroutine, control reverts to the main program.)

Microprogrammed control unit — In a computer with a microprogrammed control unit (MCU), three of the basic elements are nearly identical to our classic fixed-instruction computer; the significant difference is that the control unit has its own memory. This control memory contains the stored sequence of control functions that dictate the end-user architecture and the instruction repertoire of the microprogrammed computer. Thus, the instruction set can be modified or increased to adapt the microprocessor to system needs.

Instructions are machine directives and are the prime constituent of programs. They are fetched one-at-a-time by the control unit, which then carries out the operation(s) indicated in the instruction.

Instructions for most modern computers can be grouped into eight functional classes: load/store, arithmetic, logical, skip, shifts, transfer of control, register and I/O. A brief description and example of each class follows.

Load/store — This instruction class performs the function of exchanging data between main memory and temporary storage registers (accumulator, index, etc.). Load transfers contents of a selected memory location into a designated register. Store reverses the operation. Typical class members include load, load indirect, store and store indirect.

Arithmetic — Almost self-explanatory; these instructions perform an arithmetic operation upon two operands, one of which is in a register and the other in memory; the result usually replaces the operand in the register. Typical members include add, subtract, multiply and divide.

Logical — These perform a logical operation on two operands, one of which is in a register and the other in memory; the result usually replaces the operand in the register. Included are AND, OR, EXCLUSIVE-OR.

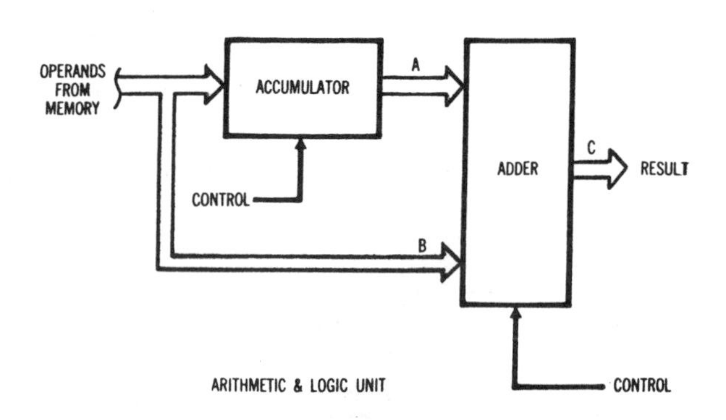

Input/output register interface unit. This component provides the data exchange link between the microprocessor and the outside world.

Example: Logical OR

Operand 1:	0 1 1 0 1 0 1 1	(Register)
Operand 2:	0 0 1 1 0 0 1 0	(Memory)
Result:	0 1 1 1 1 0 1 1	(Register)

Skip — These are usually 2-phase instructions; that is, an arithmetic or logical operation is performed on one or two operands and the result is tested for a specific condition (e.g., positive). If the condition is met, the next sequential instruction in the program is skipped. Class members include: increment and skip if zero, decrement and skip if zero, skip if greater, and skip if not equal.

Example: Decrement and skip if zero

A specified memory location has 1 subtracted from it; if the result is equal to zero, the next instruction is skipped.

Example: Skip if not equal

The contents of the specified register are compared to the contents of a specified memory location; if the two contents are not exactly identical, the next instruction is skipped.

Shifts — The contents of a designated register are shifted one bit to the left or right. The bit position that is vacated can be filled with a zero (shift) or the bit that "fell off" the other end (rotate). Rotate is merely a circular shift.

Example: Shift left

Before 1 1 0 0 0 1 1 0 1 1

After 1 0 0 0 1 1 0 1 1 0 ← 0

1

Example: Rotate right

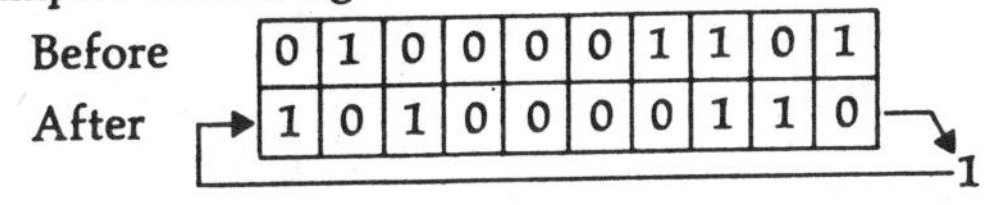

Transfer of control — This class of instruction causes the Program Counter (PC) to jump to an instruction — specified point in the program; that is, control of the computer is transferred to a new program element. Such transfers can be conditioned (based upon some operation and/or test) or unconditional. Conditional transfer includes Branch if Accumulator Positive, Branch if Condition, and Branch if Register = 0. Unconditional transfers include Jump, Jump to Subroutine, and Re-

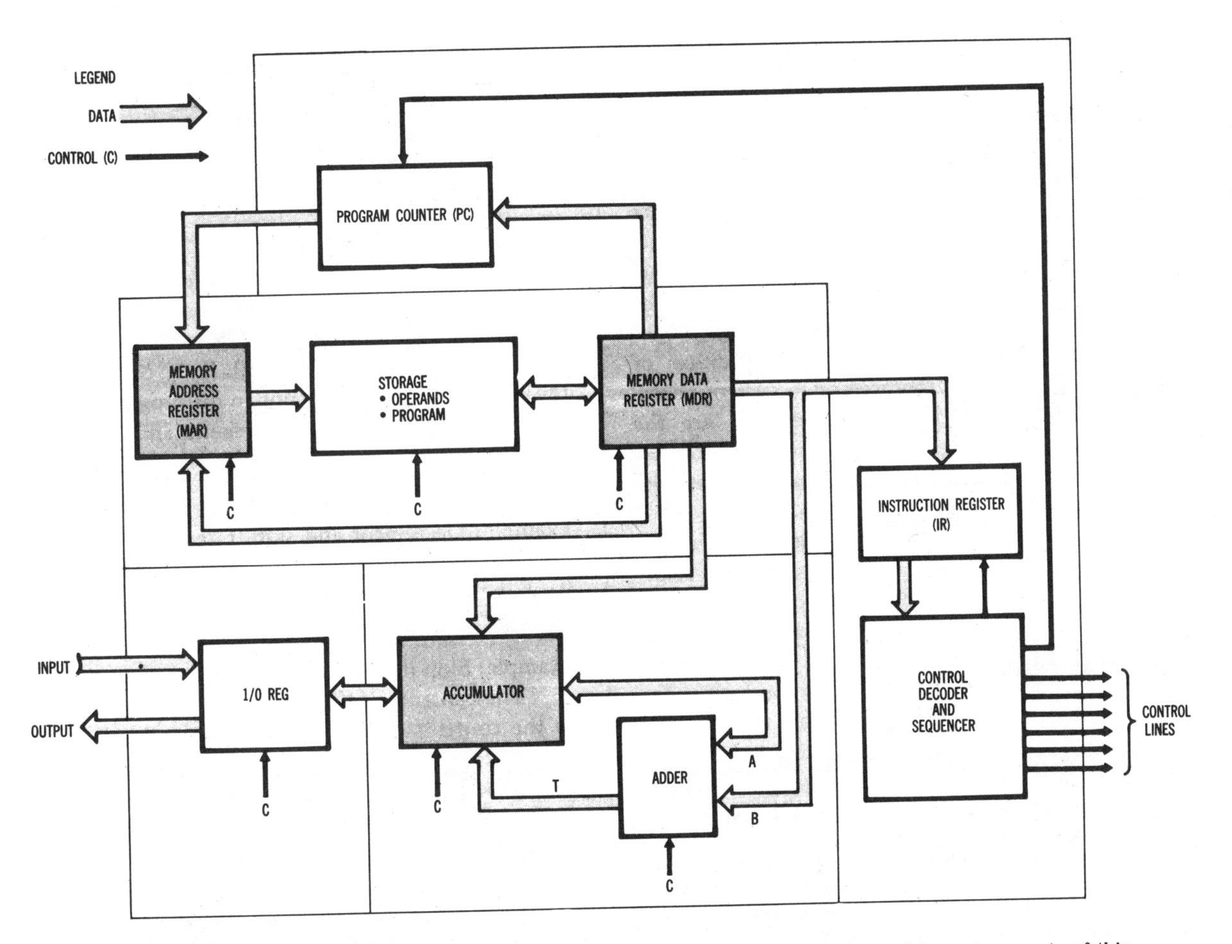

Combining the sub-elements, the whole computer emerges. Memory is performed in various parts of this microprocessor. The memory register contains the address where the information is to be read and stored, while the memory data register holds the data being exchanged with memory. The accumulator serves as a temporary storage unit.

turn From Interrupt. An immense amount of programming power is found or lost here.

Register — Included here are instructions that perform arithmetic or logical operations on the contents of two registers or after the content of a single register. Examples of two register instructions: Exchange Register and Stack, Register Add, Register Copy. Single Register instructions include Load Immediate, as well as Complement and Add Immediate. In an immediate instruction, the operand is inherently included.

Input/Output — An enormous variety of instructions are possible here; commoner instructions are those that transfer the content of a specific register to an output port (Register Out) or transfer the word appearing in an input port into a register (Register In).

There are as many instruction sets as there are computers, and it is quite difficult to say which are good and which are bad. The number of instructions is not a good indication of the power of a computer since each manufacturer counts differently. An instruction set that one manufacturer states has 43 instructions might be called 352, using another manufacturer's procedure (i.e., a register to register add might be counted as one instruction but in a machine with four registers it could be counted as sixteen).

One of the true measures of a machine is how many instructions it requires to execute a given "benchmark" program. It is important to note here that a computer with a microprogrammed control unit can be configured to execute any instruction; therefore the number of instructions required for a specific class of jobs can be minimized by tailoring the instruction set to the peculiar requirements of those jobs. Instruction efficiency is, in turn, related back to the architecture and word size of the individual computer.

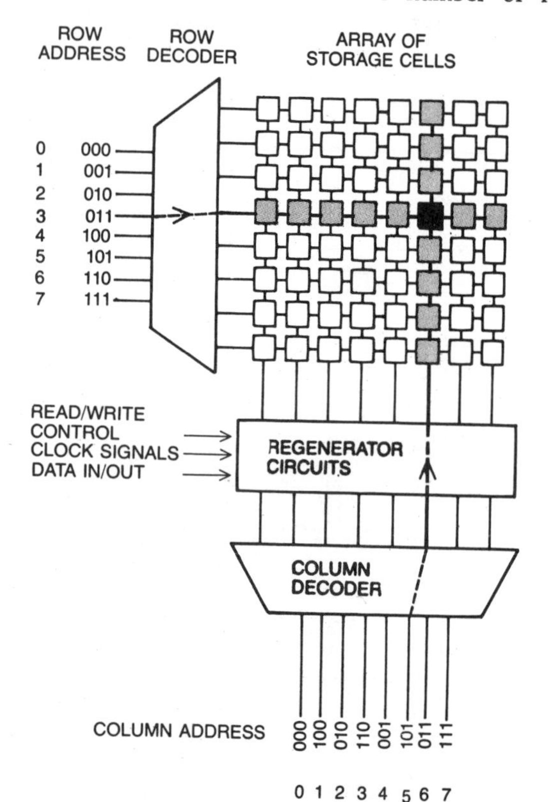

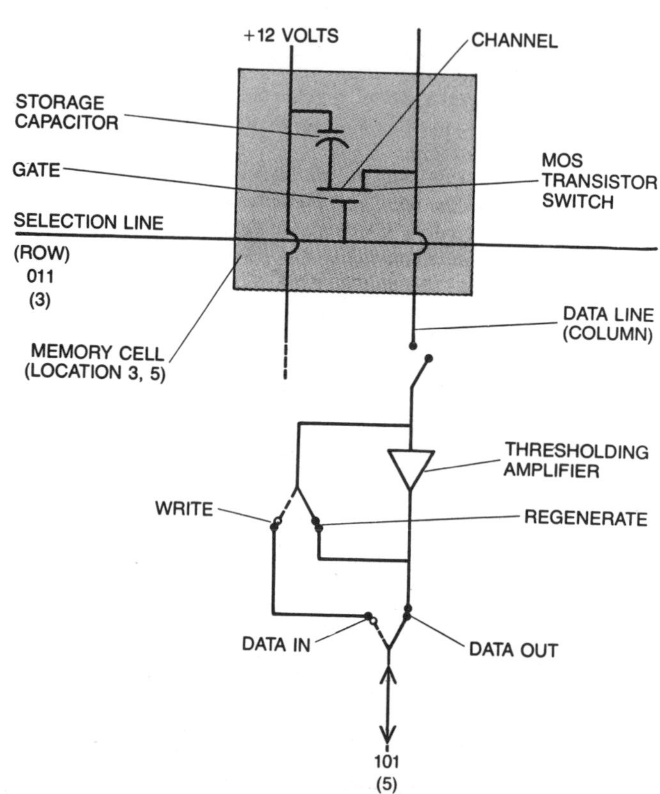

RANDOM-ACCESS MEMORIES are usually organized in rectangular arrays of rows and columns. The diagram at the left shows an eight-by-eight array for storage of 64 bits, one bit being stored in the cell at each intersection. To specify a particular memory location three binary digits are needed to indicate the row and another three to indicate the column. In this example row address 011 (binary for 3) and column address 101 (binary for 5) specify the memory location 3,5. (The locations start with 0,0 at the upper left and end with 7,7 at the lower right, specifying 64 locations in all.) The organization of a single one-transistor memory cell of the array is depicted at the right. Binary information is stored as a charge on a small capacitor. For example, zero charge might represent a binary 0 and a charge of 500 × 10^{-15} coulomb might represent a binary 1. When one of the selection lines, or rows, in an array is activated (here it is row 3), it turns on all the transistor switches connected to it. The transistor functions as an on-off switch to connect the storage capacitor to its particular data line, which corresponds to a column in the array. The simultaneous activation of a row and a column identifies the cell selected for reading or writing (here cell 3,5). Because the storage capacitor loses charge both by being read and by leakage it must be regenerated periodically, usually once every two milliseconds. The regenerated charge, which is supplied by the thresholding amplifier, is returned to the capacitor by the closing of the switch in the data line. All the switching in this type of memory device is accomplished by transistors.

Software

Software is a term used to describe the programs that make a computer do a specific task. In fact, when used in the context of computers, the word software can be interchanged with the word program. In general, a program is a series of sequential steps that accomplish an objective.

A computer is a device that can recognize and act on a predetermined set of instructions. Even though the specific set of instructions it can use is fixed by its design, a computer is general purpose because it can execute a list of these instructions (a program) to perform some functions, execute another list of instructions to perform some other function, and so on.

Since many applications for microcomputers can also accept a hardware solution, you should compare the design steps you would use for each. Since software is designed like hardware, it is interesting to note how similar the following steps are:

Software:

- Define the problem and what data, inputs, and outputs are available and/or required for its solution.
- Determine the best form of the solution.
- Outline the method of solution on a flow chart.
- Write the entire program, step by step, using the computer's instruction set. Assemble or compile the program (if necessary).
- Load the program into the computer memory and run it to test and debug.

Hardware:

- Define the program and what data, inputs and outputs are available and/or required for its solution.
- Determine the best form of the solution.
- Outline the method of solution on a flow chart. A static diagram can also be used.
- Draw up the detailed logic diagram using the available and compatible SSI, MSI and LSI functions.
- Make wire lists, etc.
- Wire circuit boards; operate to test and debug.

Defining the program is the most important and probably the most difficult part of either solution. Step 2 depends largely on what resources the designer has at his disposal. This is the point where a decision will be made to go hardware or software. Note that for some design problems the flow chart for hardware and software may look the same.

Writing the program, Step 4, determines the incremental cost of the system, since it defines the amount of memory required to store the program. Since the number of instructions required to perform a certain function may be different for each computer on which the function is programmed, the cost of performing a given function will depend on the instruction set of the computer used.

The speed at which a given function may be performed depends on the instruction set of the computer as well as the actual time it takes the machine to cycle through a given instruction. Because of this, a machine that is considered fast may take much longer to perform a given function than a machine that is considered slow. This paradox is part of the reason why "proper CPU selection is not easy." One almost has to write his program for several machines before making an accurate comparison of cost and performance. These tests are sometimes referred to as benchmark tests.

Software design is the analog of hardware logic design. The efficiency of the software design is measured primarily in the amount of memory used to store the software and the time required for execution of the program. Hardware design efficiency is measured in

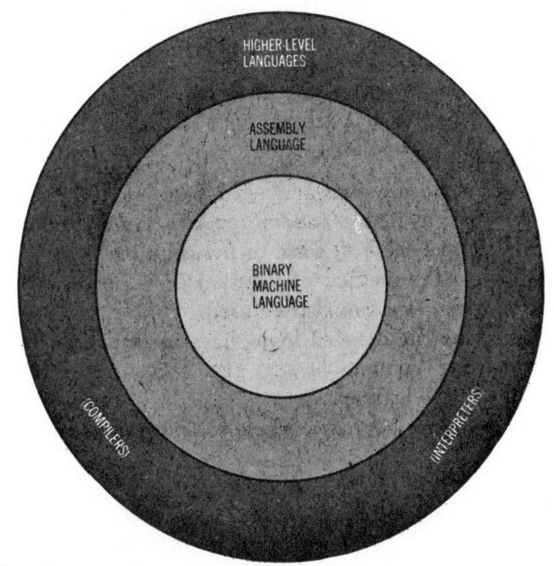

Fig. A.

In the hierarchy of programming, languages can be represented as a sphere. In the center is the machine code and each layer away from the center is closer to human language.

number of gates and functions used (packages x cost). It must be assumed that the efficient hardware design would perform the function as fast as required. There would be no advantage in greater speed, since a hardware based system would not lend itself to doing more functions in its spare time, while a software based system would.

Commanding the computer

We hear talk about software and programming — and most talk about programming in some language or another. This is because the way we command the machine is very much like the way we communicate by a written language. We have rules about how we start and end sentences and paragraphs and how we spell words. The way we communicate with a computer is through a programming language, which also has rules of spelling and punctuation, but these rules are much more strictly enforced. If you misspell a few words your reader will probably understand you anyway. A computer language is not that forgiving and will not produce the desired result if its rules are broken.

There are a number of levels of programming languages as shown in Fig. A. The innermost level is that of the actual machine language. Each instruction is uniquely defined by binary code (pattern) of ones and zeros. The central processing unit (CPU) examines each instruction code and performs the exact sequence of events to produce the operation defined by that instruction. Assume a 0011000100000000 code tells the computer to add register (accumulator) zero to register

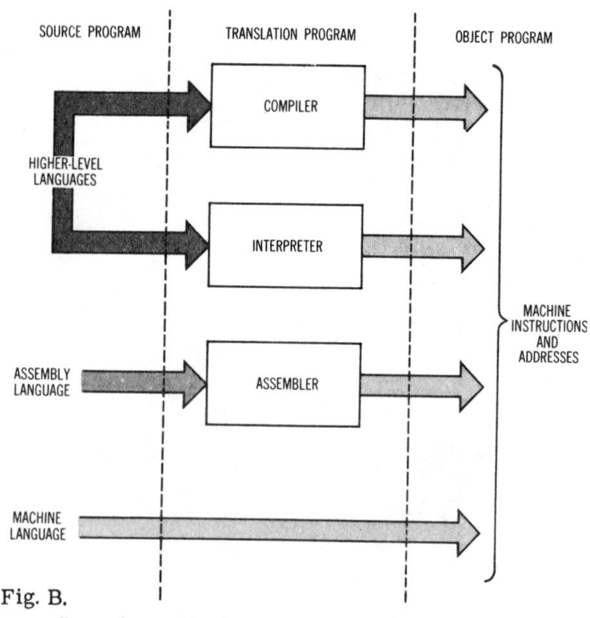

Fig. B.

Except for machine language itself, each user-written source program must be translated into a machine language object program before the computer can use it.

one and put the result in register one. When programming in machine language, the programmer must enter 0011 0001 0000 0000 to add register zero to register one. This can be awkward as well as quite slow; it isn't easy to remember all the codes. In spite of its disadvantages, the use of machine language is a perfectly reasonable way to program when the application is not too complex and the effort is on a low budget.

To make programming easier, assemblers have been developed; they are the next level on the software sphere. An assembler or assembler program is a computer program that accepts coded instructions or mnemonics that are more meaningful to use and translates them into a binary machine code the computer can execute. The mnemonics used for each instruction are much easier to remember, and they make a listing of the program much easier to read. The mnemonic for the register to register add mentioned above might be RADD 0, 1.

Keeping track of each instruction

The use of easy to remember and easy to work with symbolic codes in place of the ones and zeros of machine language is not the only improvement assemblers can provide. An assembler keeps track of the location of each instruction, which is important because it allows the programmer to use symbolic labels for important locations in the program. These labels allow references to be made to locations in a program without keeping track of the exact memory locations (which would change if instructions are inserted or deleted between the location and where it is referenced). This "bookkeeping" feature allows the assembler program to choose the best addressing modes (i.e., indirect, indexing, etc.) automatically if the instruction set has a variety of addressing modes.

In addition to allowing the use of mnemonics and labels, assemblers permit listings to include comments that help to document the programmer's work, macros that assign a mnemonic to groups of code, listings of labels and where they are found, and many other such refinements.

The outer layer of the software sphere is the area of the higher-level languages, which come the closest to natural or human languages. They are problem oriented and contain familiar words and expressions; however, they have very strictly defined structure and syntax. There are two types of higher level languages, compilers and interpreters. Both types are programs that take the higher-level language program the programmer writes and turn it into machine language the computer can use. The major difference between a compiler language and an interpretive language is how the language program converts to binary machine language. A compiler takes the whole program and converts (translates) it into binary machine language before it is ready to execute it, while an interpreter translates the program into executable binary machine code on a statement basis (and usually executes them at the time).

High level languages are often written for specific needs and special uses. Some that may be familiar are ALGOL and FORTRAN for scientific users, COBAL for large business systems, RPG for small business systems, BASIC and APL for time sharing, and PL-1 for large, general systems.

A higher level language (such as BASIC) might have a statement like the following:

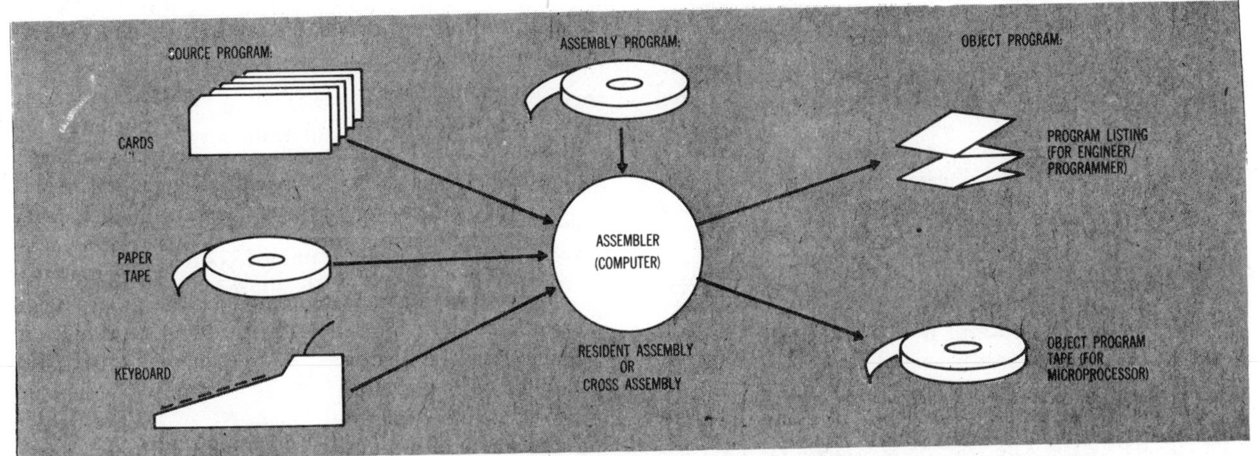

Fig. C. **By various means, the source program can be entered into a computer where it is processed to prepare an object program (machine language) for the microprocessor. If the computer is the same microprocessor that will execute the program, this step is called resident assembly; if it is another computer, it's called cross-assembly.**

LET ANS = A x B + C/D.

This statement computes the value ANS by multiplying the previously defined values of A and B and adding the result to C divided by D. The same statement written for another computer, but in the same language, would look essentially the same because the details of the instruction set, addressing modes, register assignments, etc., are taken care of by the compiler (or interpreter). Therefore, the compiler (or interpreter) program becomes, for all practical purposes, your computer, and it isn't necessary for you to know or care about the detailed operation of the host computer.

Defining the program

At this point, source programs and object programs must be defined. A source program is a program written by the programmer in any of the languages discussed. The object program is the list of binary machine instructions (and addresses) that is ready to be loaded into the computer and be executed. The object program is generally produced from the source program by one of the types of computer programs, i.e., compiler, interpreter, or assembler. These relationships are illustrated in Fig. B. Note that a machine language program requires no intermediate step.

It isn't necessary for the translation program to be run on the same type of computer that the object program is generated for. In fact, it is often not practical and sometimes, not even possible, because compilers are long programs and so take a lot of memory. Furthermore, some microcomputer instruction sets will not support a practical assembler. When the translation program is run on another machine, it is called a cross-assembler or cross-compiler (Fig. C).

The main advantages of machine language programming are that it can be completed without the aid of another program, and it allows the programmer to keep track of and control every detail of the machine operation. Assembly language programming allows the programmer to retain complete control over the important details of the computer operation, but takes care of all the drudgery of the binary coding, address calculations, and the like.

Compilers have an advantage in that programs can be written without regard to which machine they will run on. The higher-level instruction example given above might take 30 to 50 machine language instructions; this shows how much work a higher level language might save a programmer. This relative simplicity allows a person to be trained as a programmer in a fairly short time. With compilers, the programmer does not concern himself with the inner workings of the computer or even the details of how the compiler generates code, e.g., to produce a multiply. These very advantages, however, can be cited as disadvantages. Take the case of a multiply: The multiply function can be written in many ways; one uses very few instructions (not much memory) but is very slow; another uses a lot of instructions but is very fast. Which one should the compiler use?

The programmer has no control over these types of decisions and must accept all the constraints and compromises designed into the compiler. Other disadvantages of compilers include their often inefficient use of the machine instruction set in applications for which the compiler has not been specifically optimized, the problems involved in debugging the resultant object code on the actual machine, and the loss of control over things such as interrupts, register assignments and manipulations of individual bits (necessary in control applications). Compiler generated object programs generally take considerably more memory than the same program written in assembly language. Whether you consider this as an advantage or disadvantage depends on how many systems you will build and whether you are buying or selling the memory.

The following is a list of other software that is encountered while using microprocessors.

- **Simulators** — Software simulators are sometimes used to debug programs using another computer. They are especially useful if the actual computer is not available (or hasn't been built yet). If hardware is available, the use of a simulator is an unnecessary extra step, since the software must still be debugged on the hardware. The cost of the computer time to run the simulator effectively is often more than the cost of a prototyping system.
- **Debug programs** — Debug programs help the programmer to find errors in his programs while they are running on the computer, and allow him to replace or patch instructions into (or out of) his program.
- **Diagnostic programs** — These programs check the various hardware parts of a system for proper operation; CPU diagnostics check the CPU, memory diagnostics check the memory, and so forth.
- **Loaders** — The various applications (user written) programs must be placed in the proper locations of the system memory. The programs that do this job are called loaders. Loader programs range from simple ones that load absolute binary object code with no error detection, to sophisticated loaders that load relocatable binary object code, resolve global (between program) symbolic label linkages, perform error detection, and execute various commands, including starting the program just loaded.
- **Editor** — As an aid in preparing source programs, certain on-line programs have been developed that manipulate text material. These programs, called editors, text editors or paper tape editors make life easier for those who have system time to write source programs on-line.
- **I/O handlers** — Input/output handlers, sometimes called device drivers, are subroutines that service specific peripheral devices such as teletypewriters and card readers. They help prevent "reinvention of the wheel" every time a programmer wants to use a standard peripheral.

What is microprogramming?

Microprogramming has a number of different meanings. To some people microprogramming means the use of ROM for program storage instead of RAM. To others it means the combining of instruction codes such as can be done with a PDP-8., The preferred meaning refers to the programming of the control section of a computer. A macroinstruction is decoded by the control section of the computer; the control section then "pulls the proper strings" to do the operation specified by the instruction. With a microprogrammed controller, this string pulling is carried out by microinstructions. This is an alternative to the use of random logic to do the control section function. The greatest advantage of a microprogrammed controller is that the microinstruction set can be altered by changing the microprogram instead of rewiring a bunch of logic. (This procedure is much more difficult to execute on an LSI chip.)

Why use a microprocessor?

The main advantages of using a microprocessor approach to system design are:

- **Short design cycles** — The use of microprocessors allows rapid design once a basic set of boards and I/O interfaces have been developed. Because of the standardized nature of the logic, many aspects of the design can proceed in parallel. Logic design for special I/O and programming can proceed together once basic ground rules have been set

  And the ease with which the product can be modified allows earlier entry to the marketplace and faster resolution of any shortcomings.
- **Lower cost** — The use of fewer components can result in large cost savings for moderate-sized systems. The use of the same circuit boards for a variety of applications results in economies of scale.
- **Flexibility of the end product** — Allows redefinition of product without costly redesign. A wide variety of changes are possible by reprogramming.

For example, the company planning a product prepares a business specification, followed by a hardware spec based on what is practical and what is salable.

If the hard-wired system approach is chosen, the entire system must be designed logically. When complete or nearly complete, power requirements can be totaled and power supplies ordered.

The design must be breadboarded, which may point out logical design errors or may even force rewriting the equipment specification. Then the system is tested; if it doesn't meet specifications, partial or complete redesign is required. Next, the board layout is done, which may require two or three iterations, or even rebreadboarding. Finally, the mechanical design and system are tested. There is no guarantee of passing system test and more redesign may be required.

If, after the business specification is written, the equipment specifications include a microprocessor, the events change. First, the logic design of the interfaces is made. In parallel, after some basic design decisions

MICROPROCESSOR CHECKLIST

COST CONSIDERATIONS

Normally chip cost is not the largest factor, but it could be in a simple system. The major cost is usually in the circuitry needed to support the microprocessor. The engineering cost to put your system together, however, can be greatly influenced by the amount of quality of support the chip supplier has available.

Support circuitry

- ☐ Clock circuitry and clock drivers — how many phases?
- ☐ Power supplies — common or special?
- ☐ Buffers — MOS to TTL input/output?
- ☐ How much control logic (i.e., address and data latches, etc.)?
- ☐ Memory — standard or special?
- ☐ Power of the instruction set (a good instruction set can reduce memory requirements by 40%).
- ☐ Support requirements (the larger these are, the more expensive the PCB or the greater the number of PCB's required).
- ☐ Ease of checkout (test vs purchased card).
- ☐ Processor card availability (small production and preproduction will use off-the-shelf cards).

Support from supplier

- ☐ Hardware support
  - prototyping system
  - mechanical hardware
  - processor cards
  - memory cards
  - interface cards and cables
- ☐ Software support
  - high level languages
  - assemblers
  - utility programs — debug and edit
  - loaders — absolute and relocatable
  - peripheral drivers (TTY, card reader, line printer, floppy disc, etc.)
  - special subroutines (BCD to binary and binary to BCD, floating point math package, etc.)

Literature

How well written the hardware and software manuals are determines how much time is spent in learning the system.

- ☐ Technical manuals
- ☐ Software manuals
- ☐ Application notes
- ☐ Special interfaces — D/A, A/D peripherals

Technical support

This can reduce the engineering time and cost required to get the new product designed and into production.

- ☐ Area system specialist
- ☐ Field application engineer
- ☐ Plant applications and engineering groups

PERFORMANCE

Speed

- ☐ Efficiency of the instruction set — how many instructions are needed to solve a particular problem (a math problem, a process control problem, data handling, etc.)?
- ☐ Execution time of each instruction.
- ☐ Microprogrammability — can the instruction set be changed?

Interface

- ☐ Input/output flexibility and capability
  - How many peripherals can be handled?
  - How many commands to each peripheral?
  - How large is the subroutine to handle any of the peripherals?
  - How much logic is required?
- ☐ Interrupt flexibility and capability
  - Can vectored interrupts and/or polled interrupts be handled by the processor?
  - How many interrupts can be handled by the processor?
- ☐ Special control features
  - enable signals (single line control where fast response or ease of interface is important).
  - sense inputs (test a single input and respond accordingly).

are made, the software effort can begin and the interfaces breadboarded. Since the interfaces are usually fairly simple, the number of errors are reduced and rework is held to a minimum. The board is then laid out and, like the breadboard, the opportunity for errors is reduced because the hardware is reduced.

The program is tested, revised and retested. Finally, the mechanical design and systems test is performed. Any failure to meet specifications can probably be corrected by changes to the program. At this point the programmable system is ready to go to the field. Less time and less cost have been expended than in the hard-wired system, but even so, all the advantages have not been exploited.

When either system, hard-wired or programmable, is sold, the customer may ask for modifications. With the hard-wired system, the modification can be made if there are a couple of unused pins on the board and if an extra board is required, there is room in the card cage. But with the programmable microprocessor version, a new interface board can be assembled and the program modified to effect the desired change quickly and at a fraction of the cost of modifying the hardwired system.

Small quantities of systems are not economical because of the cost of developing software. However, when the quantity passes five or six or as the unit price passes $10,000, a microprocessor should be considered. But what size (number of bits) and which manufacturer's processor should be used are key questions that must be taken into consideration.

Choosing the right microprocessor

The choice of which processor size to use must realistically start with what performance is needed and how much reserve you want for future growth (i.e., modifications, options, greater performance, etc.) If the choice is based on bit size and the support is equal, then the choice is much easier. Once you know what the system must do and how much time it has to do it, you can better determine if you need a 4 bit, 8 bit or 16 bit system. There is no easy way to classify one application as an 8 bit problem, and another as a 16 bit problem. Some typical matches include:

4 bit systems

- Man/machine interface (BCD)
  - Accounting systems
  - Terminals (simple)
  - Instrumentation
  - Calculators
  - Store and forward
- Non BCD type
  - Games
  - Replace random logic designs

8 bit systems

- Traffic controllers
- Point-of-sale terminals
- Control systems
- Process control systems
- Smart terminals

16 bit systems

- Smart terminals
- Multiple intersection controllers
- Numerical control
- Process control
- Front end processor.

Applications of Microprocessors

The potential uses of microprocessors are virtually limitless. A few comments on their functions in major areas cover only a small part of what they can accomplish.

• Commercial building control systems perform the following functions: building automation (temperature control, lights turned on and off, etc.), building fire protection (when a fire is detected, the air flow contains the fire), building security (the system monitors windows, doors, etc.). In the past, this type of control system was implemented with hardwired processors or simple-to-complex logic systems. The current trend is to replace these hardwired processors with one or more microprocessors. Home control systems are very simple (thermostats), but will be a major user when very-low-cost microprocessors become available.

• Industrial control systems include process control and test instruments. Process control systems perform water treatment, waste treatment, metals processing/mining, ceramics, petroleum, petro-chemical refining and power plant regulation. These are now done by a hardwired controller or minicomputer, with future trends leaning in the direction of microcomputers. A general rule of thumb is that hardwired controllers and minicomputers work to only 15% to 25% of capacity in a process control environment; therefore, microcomputers can replace most hardwired controllers and minicomputers, even though the microcomputer may be slower.

• The primary use for information system computers is in the area of electronic data processing (EDP). Some traditional tasks of EDP computers are payroll, inventory control, management information and general accounting. Microcomputers are beginning to replace the traditional low-end EDP computers, but are not expected to compete with the medium or large EDP computers.

• Another area opening up to microcomputers in EDP is the replacement of hardwired logic in such devices as card readers, mag tapes, CRT's, front-end processors for telecommunications, plus time-sharing and point-of-sale terminals (intelligent cash registers). These new areas will be high volume users of microcomputers.

Emerging applications

There are several positions now being usurped by the "computer on a chip" concept that were previously the domain of other techniques. Airline ticketing functions are a prime example of the functional switchover from a man-oriented system to a computer-oriented one.

The changeover from analog to digital computers is becoming apparent in many process control applications where the condition of one stage of process effects a previous one. Until recently, digital methods have not been cost competitive with linear computational devices despite the labor overhead required to keep the analog computer on line. A primary consideration is the drift free operation of the microprocessor.

Evidence of the transition from the use of large computers to microcomputers can be seen in machine card control applications, where once one computer controlled several machine tools; dedication of each tool to a single microprocessor reduces line stoppages when there is a computer failure. Quite frequently the changeover from a large to a small computer can mean the life or death of an idea. A case in point is that of the satellite navigation system originally employed for spacecraft tracking, ships and sophisticated military vessels.

• Sometime in the next few years, you are going to yell at your television set and it will answer you back. With the advent of cable TV will come home high speed communication channels controlled by microprocessors. Time shared computer access, citywide town meetings in which instantaneous citizen response is available, and maybe even the ability to boo the visiting team, will be channeled through the cable TV set. High speed data channels exist now that will proliferate even more. Smart terminals, such as teaching machines, library researching units, and off-track betting machines, will perform portions of the job and refer tougher parts to central mainframes.

• A major automotive application will be the on-board car and truck computer. The computer will monitor and control such things as spark advance, carburetor gas flow, transmission shift, etc. It will also provide driver warnings of such things as alternator failure and what to do about it. It may even drive the car for you. The car controller, however, will become a one chip custom device used in very high volumes, so this may be considered custom LSI rather than a microprocessor.

• Automated gas stations are being tried using minicomputers, but a microprocessor will do this, too

• Microprocessors will end up in electric typewriters as controllers for self-justifying and executive spacing and as data communicators to CPU devices for such functions as editing, typesetting and translation.

• Specialized calculators, too low in volume potential for specialized chips, will appear. Private boating and aviation navigation aids are examples, along with hand carried mortar trajectory calculators for the Army and Marine Corps.

• Microprocessors will control machines involved in mail sorting, inventory pulling and stocking, and palletization of freight. Irrigating systems will sense crop needs for water and fertilizer, delivering the required amounts to whole fields or specific areas, depending on the microclimate.

• Very low cost processors will encourage the use of "throw-aways" in such areas as weather data collection, oceanographic monitoring, and weapons.

• Automatic controllers for: traffic lights, tools, stoves, drafting machines, looms, photography processing, paint mixers, asphalt makers, grape crushers, banana peelers, packaging machines, power switching, railroads, piano tuners, anti-skid braking systems, no-slip four wheel drive, fast food businesses, automated radio stations.

Software to carry out multiplication

The use of hardware multiplication to speed microcomputer computations also entails the writing of software. Though not a difficult task, digital multiplication could present problems to designers not familiar with the procedure.

A simple example (top right) will show how to develop a basic multiplication subroutine (bottom right).

Assume that two binary numbers, X = 13 and Y = 11, must be multiplied. X is called the multiplier and Y the multiplicand. The procedure requires that Y be added to the partial product Z_i whenever the least-significant multiplier bit X_i is 1. When $X_i = 0$, zero is added. After each addition, the partial product and the multiplier are shifted to the right by 1 bit. Thus after multiplication of two n-bit operands, a 2n-bit or double-precision product results.

Most microprocessors perform multiplication sequentially by software. A typical multiplication subroutine consists of three steps:

(1) Initialize,
(2) Loop and
(3) Finalize.

In Step 1, CPU registers AC0, AC2, AC3 are loaded with the operands X, Y and the index n; AC1 is reset to zero. In Step 2, we add Y into AC0 when the LSB of AC0 ≠ 0. We omit the addition when the LSB of AC0 = 0. The contents of AC0 and AC1 are shifted to the right 1 bit at a time, for each pass through the loop, while the index counter is decremented. In the last step we transfer the double-precision product from AC0 and AC1 to specified memory locations.

```
Y = 11            X = 13
 1011      x       1101

Z0 =   0000
      +1011
Z1 =   1011
        1011  ------> SHIFT RIGHT
     +00000
Z2 =   01011
         1011  -----> SHIFT RIGHT
     +101100
Z3 =   110111
        110111 -----> SHIFT RIGHT
     +1011000
Z4 =   1000111

143 = DOUBLE-PRECISION ANSWER

OPERATIONS: AC1, AC0 <-- AC0 x AC2

INITIALIZE: AC0 <-- X, AC1 <-- 0, AC2 <-- Y,
            AC3 <-- n

LP: JUMP +2 IF AC0 LSB = 0
    ADD AC1 <-- AC1 + AC2
    RIGHT SHIFT AC1       LSB --> L
    RIGHT SHIFT AC0       L --> MSB
    DECREMENT COUNTER, SKIP IF ZERO
    JUMP TO LP

FINALIZE: STORE AC0 AND AC1 IN MEMORY
```

A Microprocessor in a Computing System

To form a computer system, an 8-bit parallel microprocessor typically requires the external circuity shown in **Fig. 11-61**. The 8-bit bidirectional data bus connects the microprocessor to the external memory and external registers I_1 and I_2. During states T_1 and T_2 at phase ϕ_{22}, the control logic strobes data into the external registers. It also determines which way data travel on the bus to and from the microprocessor. In addition the control logic determines when the memory reads or writes, and it activates I/O channels.

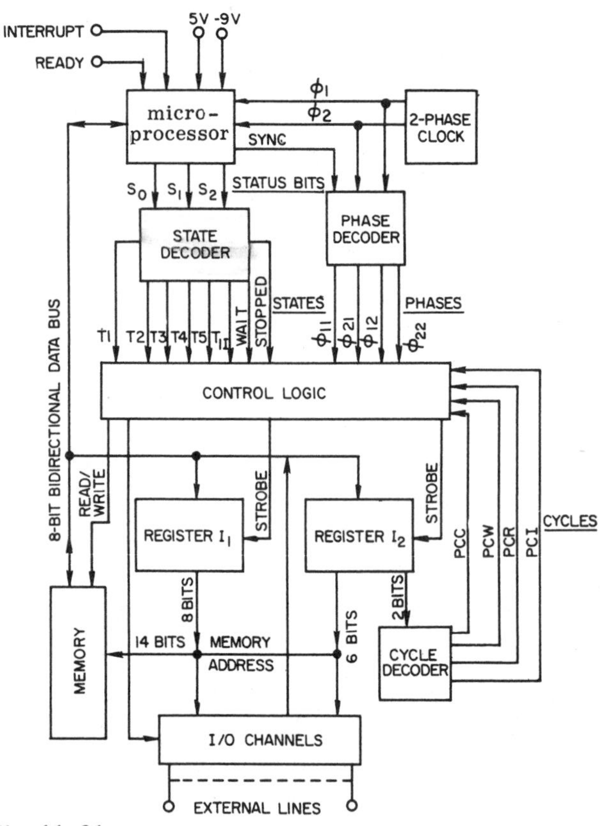

Fig. 11-61

A computing system using the microprocessor. The external components provide the interface logic for control of the microprocessor and transfer of data to and from memory and I/O devices.

External registers I_1 and I_2 supply addresses to memory as well as data bytes and pointers to the I/O channels. Data from the two most-significant-bit positions of register I_2 are trans ferred to the cycle decoder. The decoder produce four cycles—PCI, PCR, PCW and PCC—whic coordinate the internal operation of the micro processor with the external circuitry.

When power is applied to the processor chip 32 clock periods are needed to clear all register and to initiate the internal microprogram. The the processor goes into the STOPPED state—a indicated by status bits S_0 through S_2—until a interrupt occurs. When this happens—and durin states T_{11} and T_2—the address of memory loca tion 0 is referenced for the first instruction byte

If the first instruction has been loaded int location 0, the instruction byte will be fetche during state T_3. Otherwise the internal instru tion decoder attempts to execute the instructio represented by the random-bit pattern in mem ory location 0. Then the microprocessor sequen tially executes instructions or branches, as pro grammed.

Table 1 shows the execution of typical instruc tion LMI (Load Memory Immediate—with th next byte in memory following the instructio byte). The instruction occupies two bytes o memory, while its execution requires three cycles At the beginning of the instruction—and durin state T_1—the eight low-order bits of the progran counter are sent to register I_1, and the progran counter is incremented by one. The six high order bits of the program counter and the tw control bits transfer to register I_2 during stat T_2.

At the end of state T_2 the cycle decoder send a PCI signal to the timing logic, indicating tha this is an instruction-fetch cycle. During state T the memory location addressed by the content of register I_1 and I_2 is read into the instructio register of the microprocessor. The instructio decoder recognizes that this is an immediate typ of instruction and transfers control to the ap propriate microprogram. Since states T_4 and T_5 are not required, the cycle ends with state T_3.

Similarly the second-cycle and third-cycle operations are executed. Register b, in the second cycle, refers to one of two auxiliary registers (the other is called a). Both are used by the microprogram to transfer data internally. Registers H and L—in the third cycle—are two of seven 8-bit registers internal to the processor chip. The seven are labeled A through E, H and L. They

Table 1. Nine states execute LMI instruction

Cycle	State	Operation
PCI	T_1	Low-order bits of program counter to I_1. Increment program counter.
	T_2	High-order bits of program counter and control bits to I_2
	T_3	Fetch instruction byte from memory location addressed by I_1 and I_2 Put into instruction register.
		Skip states T_4 and T_5 and go to the next cycle
PCR	T_1	Low-order bits of program counter to I_1 Increment program counter.
	T_2	High-order bits of program counter and control bits to I_2
	T_3	Read byte (immediate data) from memory location addressed by I_1 and I_2 Put into auxiliary register b.
		Skip states T_4 and T_5 and go to the next cycle.
PCW	T_1	Register L out to I_1 (low-order address)
	T_2	Register H and control bits to I_2 (high-order address)
	T_3	Register b to memory location addressed by I_1 and I_2
		Skip states T_4 and T_5 and go to the next cycle.
	END OF EXECUTION OF INSTRUCTION	

Table 2. Subroutine links registers H and L

	Label	Instruction	Binary	Comment
INCREMENT REGISTERS	INCHL	INL RFZ INH RET	00110000 00001011 00101000 00xxx111	INCREMENT REGISTER L RETURN IF NOT ZERO INCREMENT REGISTER H RETURN
DECREMENT REGISTERS	DECHL	DCL CPI 377_8 RFZ DCH RET	00110001 00111100 11111111 00001011 00101001 00xxx111	DECREMENT REGISTER L COMPARE AGAINST THE NEXT BYTE ALL ONEs RETURN IF NOT MATCHED DECREMENT REGISTER H RETURN

Note: x = don't care

make up the accumulator (register A) and scratch-pad memory (the remaining registers).

To execute a memory-reference instruction, the logic takes the 14-bit memory address from the contents of register H (containing the 6 most significant bits) and register L (containing the least significant 8 bits). The instructions contain no field for the memory address.

Since registers H and L can be incremented and decremented, as well as operated logically and arithmetically with register A, it's possible to scan and index all memory locations. When register L is incremented or decremented through a count of zero, register H should be incremented or decremented to continue the scan.

The internal circuitry of the microprocessor

does not link registers H and L; so they must be linked by a software subroutine. Table 2 shows one subroutine for incrementing and one for decrementing through all 16k of memory. The contents of registers H and L are independent of the internal instruction address and the program counter.

The Ready line of the microprocessor may be activated by a logic ZERO at any time during an instruction. When it is, the microprogram proceeds normally until it reaches the end of a T_2 state. Then, instead of going into state T_3, the microprogram enters the WAIT state; the microprocessor just marks time by repeating the four clock phases. Deactivation of the Ready line by a logic ONE and after phase ϕ_{22} causes the microprocessor to resume normal operation by going into state T_3.

The Interrupt line may also be activated at any time during an instruction. When this occurs, the microprogram continues to complete execution

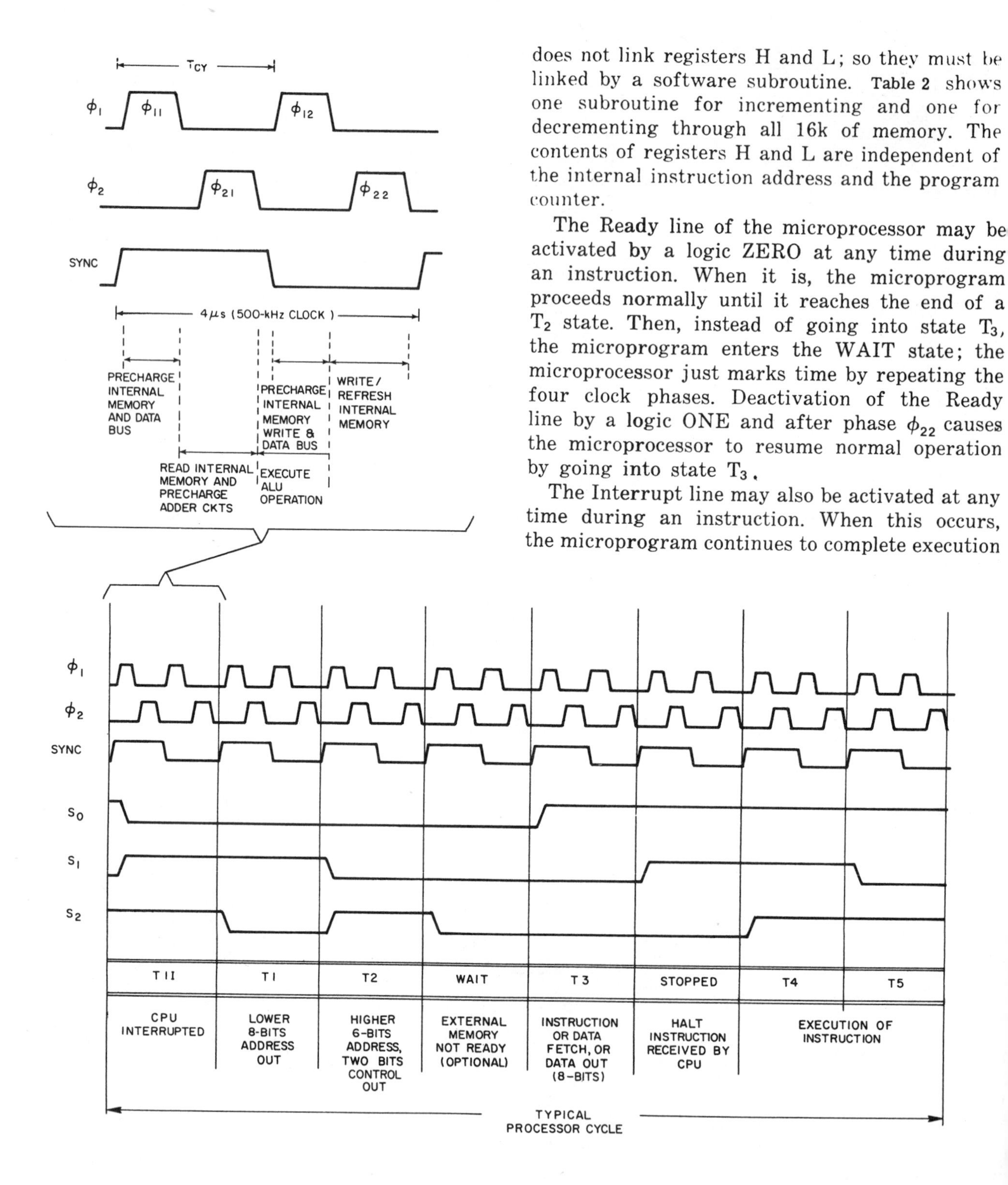

Fig. 11-62

Internal microprogram control uses cycles and states to convey relevant internal and expected external operations to the rest of the circuitry. A typical execution time for nonmemory instructions is 20 μs (with a 500-kHz clock), which covers states T_1, T_2, T_3, T_4 and T_5—or a typical processor cycle.

of the remaining cycles of the instruction. Then, instead of going to the T_1 state of the PCI cycle, the microprocessor goes to the T_{11} state. As a result, the external circuitry can jam an RST (restart) instruction onto the data lines during state T_3. The instruction calls one of the eight locations in low-order memory that contains the subroutine that services the interrupt.

On interrupt the contents of the program counter are not incremented but are pushed down in an internal stack. Hence a RETURN instruction should be programmed at the end of the interrupt service. This instruction causes the program-counter stack to pop up the original program counter, and normal operation is resumed.

By means of the cycles and states, the chip conveys information on what is happening internally and what should happen externally. Each instruction requires one, two or three cycles to complete its execution, and each cycle is composed of three, four or five states. In turn, each state is composed of four sequential pulses derived from the system clock.

The clock phases are called ϕ_1 and ϕ_2 (Fig. 11-62) The chip internally divides ϕ_2 by two to form a signal called SYNC, which is made available to the external circuitry. Each cycle of SYNC contains two pulses from ϕ_1 and two from ϕ_2—one each from the two phases when SYNC is high and one each when SYNC is low. A complete cycle of SYNC is called a state, and that is made up of four sequential pulses called phases—ϕ_{11}, ϕ_{21}, ϕ_{12} and ϕ_{22}.

Three parallel bits on the status lines define the present state from eight possible states for the chip (Table 3a). Each state begins with the completion of phase ϕ_{22} of the preceding state. The states indicate time slots for functions performed by internal and external operations of the microprocessors.

Normally only three to five of the eight states are used in a cycle. The remaining states can be used for interrupt (T_{11}), direct-memory-access conditions (WAIT) and processor halt (STOPPED).

Under normal operation 8 bits are outputted from the chip onto the external data lines during state T_1. Since memory addresses require 14 bits, two passes are needed to output the complete memory address. The low-order part of the address is the byte transferred during state T_1.

The external circuitry does not know at this time whether the contents of register I_1 is part of an I/O instruction, or the memory address of an instruction byte or a data byte.

In the case of an I/O instruction, the byte outputted during state T_1 is the contents of register A rather than part of a memory address. In any event, the byte outputted during state T· must be stored in external register I_1 until the external circuitry determines the type of cycle.

During state T_2 the microprocessor delivers to the external data lines the six high-order bits of the memory address (or the pointer to the I/O device) as well as two control bits. The control bits are the two most significant of the byte; they define which cycle the microprocessor is presently in. The byte outputted during state T_2 must also be stored in an external register, I_2, until the cycle has been decoded (Table 3b). The cycle defined by the two control bits determines what is to be done with the contents of the two registers.

In state T_3 the microprocessor either inputs a byte of data, fetches an instruction byte or outputs a byte of data. These transfers are performed via the bidirectional data bus, with the particular operation depending on the instruction being executed.

During states T_4 and T_5 the microprocessor executes the instruction and transfers data between its internal registers. Some instructions do not require these states. In these cases the states are either left idle or the cycle ends by skipping to state T_1.

Whenever a HALT instruction occurs, the microprocessor goes into the STOPPED state. The internal registers continue to refresh themselves periodically, maintaining the stored data. The microprocessor remains STOPPED until it receives an interrupt signal. It then goes to the T_{11} state at the beginning of the next instruction. When this is detected on the status lines, the INTERRUPT line should be put back to logic ZERO; the microprocessor then resumes normal operation.

A specific microprogram cycle is defined by the two control bits (most significant two bits) of external register I_2 at the end of state T_2. Each cycle performs a particular portion of an

Table 3. Status and control bits yield states and cycles

State	Status Bits		
	S_0	S_1	S_2
T_1	0	1	0
T_2	0	0	1
T_3	1	0	0
T_4	1	1	1
T_5	1	0	1
T_{1I}	0	1	1
WAIT	0	0	0
STOPPED	1	1	0

3a

Cycle	Control Bits		Function
	MSB	2SB	
PCI	0	0	INSTRUCTION FETCH CYCLE
PCR	1	0	MEMORY READ CYCLE
PCW	1	1	MEMORY WRITE CYCLE
PCC	0	1	I/O CYCLE

3b

instruction, which may require one, two or three cycles for completion.

Each instruction must begin with a PCI cycle, which fetches the next instruction. The instruction's location in memory is contained in the program counter and transferred to registers I_1 and I_2 during states T_1 and T_2, respectively. Fig. 11-63 shows the processor-state transitions for each cycle.

During the fourth phase, ϕ_{22}, of state T_2 it is convenient to strobe the two control bits through combinatorial logic into one of four cycle flip-flops. In this manner the external circuitry knows which cycle is being performed.

For the PCI cycle the contents of the addressed memory location are put onto the data bus during the next sequential state (T_3). The data byte enters the microprocessor and transfers via the internal data bus to the instruction register and decoder, which determines the operation to be performed. The program counter is incremented by one during the PCI cycle.

Some instructions—such as Register-Register, Register-Arithmetic-Logic, Rotate and Return—are executed internally during states T_4 and T_5 of the PCI cycle. For other instructions, the PCI cycle terminates with state T_3, and additional cycles are required to complete the execution.

When either the PCR (memory-read) or PCW (memory-write) cycles are indicated, the external register, I_1, holds the contents of register L. The six least significant bits of external register I_2 hold the contents of register H.

For a PCR cycle, the contents of the addressed memory location appear on the data bus during state T_3. The data byte is entered into the microprocessor, and transferred via the internal data bus to the designated register. Depending upon the particular instruction, states T_4 and T_5 may or may not be used.

For a PCW cycle, the contents of the appropriate register, as designated by the instruction, will appear on the data bus. The data byte is written into the memory location addressed by the contents of registers I_1 and I_2. States T_4 and T_5 are skipped during a PCW cycle, and the cycle terminates with state T_3.

The PCC cycle is used only for INPUT and OUTPUT instructions. These instructions are composed only of a three-state PCI cycle followed by a PCC cycle.

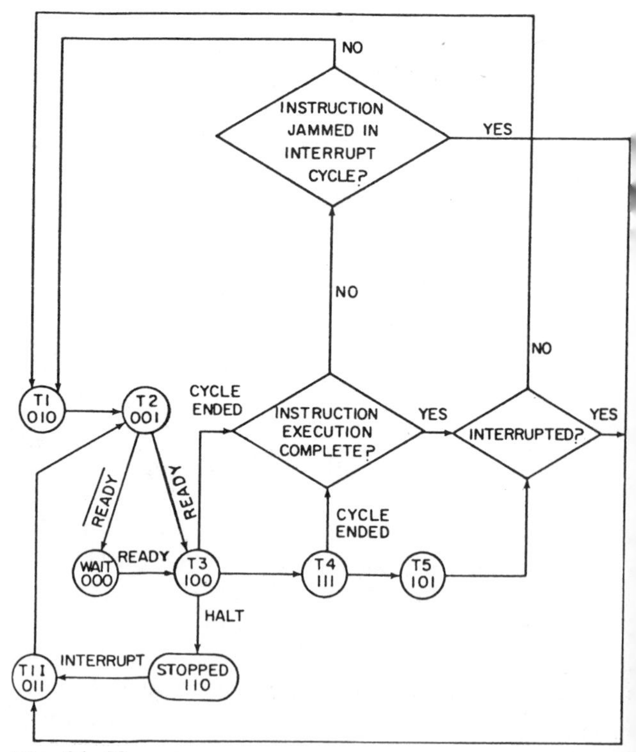

Fig. 11-63
Each cycle causes microprocessor state transitions. A cycle begins when status lines indicate state T_1.

At the end of state T_1 external register I_1 holds the contents of register A (accumulator). External register I_2 holds the least significant six bits of the instruction byte that was transferred during state T_2. The instruction byte contains a field of bits that points to the particular I/O device addressed. The instruction also indicates if it is an input or an output operation.

If an output operation is called, state T_3 is idle and the microprocessor merely marks time. Meanwhile the external circuitry must transfer the contents of register I_1 (contents of the accumulator) to one of 24 possible output devices, as indicated by the pointer field in register I_2. Register I_1 contains the data byte placed into register A before the OUTPUT instruction was executed. The cycle ends with state T_3.

To call an input operation, the input byte must appear on the data bus prior to state T_3. During state T_4 conditional flags will appear on the data bus. These may be examined for test purposes or to show status conditions. At the end of state T_5 the inputted data byte appears in register A.

The input device may be addressed in one of two ways, depending on the external circuitry. One technique uses the pointer field in register I_1—as was done during the output operation. This allows only one of eight possible input devices to be addressed. A different instruction is required to input from each device, because of the pointer field in the instruction byte.

The second method uses the entire 8-bit byte in register I_1 (instead of the pointer field of register I_2) to address one of 256 possible input devices. Since the contents of register A are placed into external register I_1 during T_1, this technique requires the device address to be placed into register A before execution of the INPUT instruction. Thus a single input instruction can address several input devices by changing the contents of register A.

MICROPROCESSOR GLOSSARY

Microprocessor/Microcomputer

These terms are often confused and used interchangeably. The current trend in terminology associates the component product with the term "microprocessor" and the board product with the term "microcomputer."

A microprocessor can be a single chip, a self-contained component, or an unassembled collection of processor-related components consisting of memories, peripheral and I/O (input/output) chips, a CPU (central processor unit) chip, clocks and interface chips. The composition varies depending on the application and user. Software is generally not provided in component form except in the case of high level language interpreters, which have appeared in ROM.

A microcomputer is an assembled kit of microprocessor components on a p-c board. The board contains all components required to make a working computer. It may have resident software and may be expandable into a larger system. Component kit variations are usually limited to a few standard types.

Very Low End Microprocessor

VLEM defines a class of processor products that generally meets high volume, very low cost requirements. The applications are generally intended for the consumer; they are typified by microwave ovens, washing machine controls, etc. The typical VLEM product is a one chip microprocessor with less than 2k of program memory, 64 words of data memory, clock, and less than 32 specialized I/O lines. The I/O requirements usually include a keyboard or an array of switches and often a glass touch panel keyboard. VLEM's may also include a numeric display using LED's, LCD's or fluorescent technology. Triac and relay drivers are also a popular output function. Sensor interrogation and A/D (analog to digital) conversion are often performed as input duties. VLEM microprocessor products lack expandability and are limited to one chip facilities. The processor speed is not critical and the instruction cycle can be greater than 10 microseconds.

Low End Microprocessor

The first microprocessor, introduced in November 1971, was a Low End microprocessor — the 4004 This product, and successive Low End microprocessor products, was aimed at the traditional electronics user. TTL and DTL users have been shifting their control and I/O intensive design application demands toward processor products in this category.

A Low End application typically has some kind of man/machine interface such as a keyboard, series of switches, displays, indicator, etc. Processor speeds for the Low End are not critical. Instruction cycle times between 2μs and 10 μs are generally acceptable.

The man/machine interface also places a requirement for BCD capability on the system, since it is more convenient to handle numerical data in BCD form than in binary, where human entry and display are concerned. A Low End application will have a program memory size of greater than 1k words, but less than 4k. More than half of these devices use greater than 1.5k of program memory. Data memory needs from 32 words to 256 words, with 50% of the units using less than 64 words. I/O requirements run between 16 to 64 I/O lines, with 40% of the Low End microprocessors having over 32 I/O lines. Other characteristics of the Low End are lots of I/O and bit manipulation, and instruction mixes that contain a large quantity of conditional branches and jumps. Word size can be 4 or 8 bits.

Mid Range Microprocessor

The mid range application segment has recently been dominated by the moderate performance 8 bit MOS microprocessor.

The byte orientations of these processors have much to do with where and how they are used. The terminal and business machine environments, which account for the majority of applications, are ASCII code (8 bit) based. Here, byte manipulation is important. Speed and throughput are also important, since many applications have real time components. Mid range microprocessors have typical cycle times between 2.5 μs and 1 μs. Program memory size ranges between 4k and 16k, with the majority falling between 4k and 8k. Data memory runs upwards of 256 words, with most falling below 1k in size. I/O tends to be of a more complex nature, requiring peripheral chips and special functions.

High End Microprocessor

High end microprocessors are characterized by the use of more memory, Interrupt and DMA (Direct Memory Access). Such performance can be obtained by using faster microprocessors, including microprocessors with large word widths and other improved facilities. Although this application segment is currently the smallest, it is expected to grow in the coming years to encompass 25% of the microprocessor component dollar volume. Many semiconductor suppliers are planning products in this area. More data and control can be incorporated in the 16 bit word. Two data bytes can be stored and retrieved from a single memory location. In applications such as process control, the advantage of the large word size is more significant. The High End microprocessors require combined program and data memory of between 8k and where the natural data word is greater than 8 bytes, 1M words; DMA and Interrupt capabilities are a must, either directly on the CPU chip or via peripheral chips.

Slice Architecture

Modular component building blocks, called bit slices, are used to fabricate microprocessors of varying word width such as 8 bit, 12 bit, 16 bit, 24 bit, etc. The completed products are generally customized and offer higher performance than conventional machines. Unlike conventional machines that have a fixed predefined instruction set, the slice architectures require an instruction set to be defined and implemented via a microprogram. This provides a system that is efficiently tailored to a particular application. Quite often, the high performance slice component products are used to emulate existing older technology computer architectures. In this case, a microprogram is used to make the slice machine execute the instruction of the emulated computer. For this application, there are two levels of programming: first, the microprogram firmware, and second, the software of the emulated machine. This two level complexity has discouraged universal acceptance of these devices compared to fixed instruction microprocessors. Applications that require high performance and efficient architectures are the domain of the slice approach.

Multiprocessor System

A multiprocessor system is a collection of two or more independent CPU's tied together via some kind of network and communication link. A **Tightly Coupled** system has all its CPU's on a common bus sharing common memory and I/O facilities; A **Loosely Coupled** system has the CPU's with their own independent memory and I/O facilities communicating to one another on a message basis via a parallel or serial bus link. The Tightly Coupled systems are on a parallel bus and interleave access to the memory and I/O, on a cycle by cycle basis. Subgrouping of multiprocessor systems also includes Master/Slave and Master/Master networks. In a Master/Slave network, one processor (the Master) controls the communication and activity assignments of the other processors (slaves). The slaves are generally not self-initiating. In a Master/Master scheme, each processor is independent and self-initiating. An arbitration algorithm and specific hardware control the communication flow. Each processor has a preassigned function.

Another multiprocessing subgroup defines the Symmetrical and the Asymmetrical system. In a Symmetrical system, every processor is performing the identical task as every other processor. The only difference is the use of different data variables. In an Asymmetrical System, each processor has a different task. These tasks are related and generally dependent on data from other processors in the network.

Central Processing Unit

A typical microprocessor incorporates a central process unit (CPU), a memory and Input/Output ports. The CPU reads instructions from the program memory in a logical determined sequence and uses them to initiate processor actions. To perform these functions, a CPU must incorporate registers, an arithmetic/logic unit (ALU) and central circuitry. Registers, temporary storage units inside the CPU, differ in their function capabilities. Some, like program counter and instruction registers, serve dedicated applications; Others, like the accumulator, serve general purpose functions.

An accumulator usually stores one of the operands manipulated by the ALU. A typical instruction could direct the ALU to add another register's contents to those of the accumulator and store the results in the accumulator. In general, the accumulator serves both as a source and destination register. In addition to the accumulator, the CPU may contain additional general purpose registers that programmers can use to store source or intermediate "scratchpad" data.

Word/Byte/Nibble

These terms are often misused in describing microprocessor data. For a specific microprocessor, a Word is the number of bits associated with the instruction or data length. This can be 4, 8, 16 bits etc, depending on the machine. A *Byte* specifically refers to an 8 bit word; a byte can be manipulated by a 4, 8, or 16 bit microprocessor. For example, instructions are often provided to deal with byte data in 4 or 16 bit processors. This is called Byte handling, and is independent of the natural word size of the machine.

A *Nibble* is 4 bits, and it is rather humorous to consider that it takes two nibbles to make a byte. Nibble (or 4 bit) control can be found on many 8 bit word machines as well as on some 16 bit machines. Four bit operations are usually associated with Hexadecimal (HEX) or Binary Coded Decimal (BCD) operations. Applications that have a man/machine interface, such as a control keyboard or a numeric display, are good candidates for nibble control.

I/O Lines/Ports

Microprocessor Input/Output (I/O) ranges from complex floppy disc control peripheral chips to simple 8 bit (I/O) latches. The Input and Output facilities are the devices by which the CPU communicates with the external environment. If the microprocessor has 8 bit internal structure, the I/O bus and each respective I/O port will also be organized for 8 bits. A port is a collection of individual I/O lines — their number is equal to the length of the basic microprocessor word. If a port is bidirectional, it can accept input or output, depending on the nature of the I/O command. Unidirectional structures are predetermined and can only accept one type of command. I/O ports that are "Quasi" bidirectional allow input and output lines to be mixed on one port. This is convenient when the number of inputs and outputs are not even multiples of the basic word size.

The majority of microprocessor applications have I/O requirements ranging between 32 and 64 I/O lines. This includes simple I/O as well as complex I/O peripherals, such as communication devices, CRT controllers, etc. Ninety-five percent of all microprocessor applications use less than 100 I/O lines while 40% use greater than 32 I/O lines.

machines have a separately addressed and generally smaller data memory than the Princeton machines. The Princeton machines can and do have data memory and program memory combined and mixed in one continuous address space.

Instruction Cycle

Instruction cycle is the periodic process of fetching an instruction from the microprocessor memory. This can consist of one access to memory or several accesses, depending on the type of instruction and the data required by the instruction. These are called memory cycles, which are a subset of instruction cycles. A literal type instruction generally requires one cycle since the instruction data is included in the instruction word. A conditional jump may require one cycle to fetch the direct or indirect address.

When a microprocessor manufacturer specifies a cycle time, the user should consider three quantities: minimum cycle time (the number often stated), typical cycle time, and maximum cycle time. Typical cycle time tells you what the overall periodic timing is for an instruction, considering a typical mix of instructions. This is the best measure of instruction cycle time. Maximum cycle time is the period of the largest instruction. A machine is efficient when the typical and minimum cycle times are close.

Program/Data Memory

Microprocessor architecture today falls into two categories: the first type, with separate program and data memory, is referred to as "Harvard" architecture; the second type, called "Princeton" architec-

ture, has program and data memory combined. Program memory is that memory which contains the instruction flow that, when taken as a whole, difines the program. In machines that have Harvard architecture, this program memory is usually ROM (Read Only Memory). For Princeton machines, this memory may be either ROM or RAM.

Data contains the modifiable system and manipulable data variable used in the computation. This is the alterable system input or output data. Harvard

Interrupt

To improve the efficiency response to real time demands, many CPU's have Interrupt facilities. In an Interrupt approach, an I/O device attracts the CPU's attention by activating an Interrupt line, to indicate that some external activity is either nearing completion or about to be initiated. The Interrupt, if enabled by the CPU, causes the CPU to suspend temporarily the operation it is performing and effect a jump to Interrupt subroutine. A system may have several sources of Interrupt, so the microprocessor must determine which of them is requesting service, since each requires different attention. This may be handled in two ways by the CPU. Some CPU's require that the Interrupt device provide an address at the time of Interrupt acknowledge. This address is used to compute the effective address of the Interrupt routine of interest. This approach is called an Interrupt vector. The second way is to have the CPU jump to a fixed Interrupt service location and have a small subroutine sort out which device initiated the Interrupt. When this sorting is done, a software jump to a subroutine to service the Interrupt will ensue. The former mode requires address generation logic; the latter method does not, but requires time before the desired interrupt routine may be started.

Software/Firmware

The microprocessor is generally a stored program computer, with its collection of programs and instructional procedures referred to as Software. Software, by directing the hardware, enables the microprocessor to perform a functional system related task. In a fixed instruction microprocessor, a set number of instructions or operations are defined with fixed word lengths, and these exercise the CPU independent of the data. Software is alterable and accessible by the user.

Firmware can be considered an extension to a computer's basic instruction repertoire that creates microprograms for a software instruction set. This extension to the basic instruction set is often permanently burned into read only memory, rather than being implemented in software. Firmware programs may be composed of instructions of variable width; the number of instructions in a Firmware program is generally smaller than in a Software program, although the instructions are usually much wider. A Firmware program can be used to implement a Software instruction set; this occurs in the emulation of larger minicomputers by slice microprocessors.

Simulate/Emulate

A simulator is a device used to imitate one system with another, using a software program written in assembly or high level language. The simulator accepts the same data, executes the programs and accomplishes the same results as the system imitated. The simulator is generally much slower than the machine being simulated and bears no physical resemblance to it. Simulators may be used to get a finer insight or control of workings of the imitated machine. Generally, larger minicomputers, such as software development systems, will be used to simulate smaller microprocessors for the purpose of developing and debugging software.

An emulator differs from a simulator, inasmuch as the latter uses software to imitate. The emulator uses a microprogram and specific hardware to imitate the desired system at the same speed as or faster than the imitated system's cycle time. The slice microprocessors are used to emulate large minicomputers with fewer components and faster cycle times. The emulator can be made to resemble physically the imitated machine.

GLOSSARY OF TERMS APPLICABLE TO MEMORY TECHNOLOGY

Amorphous film - a film with a noncrystalline structure.

Angelfish - a bubble guide pattern consisting of a series of permalloy arrowheads by which bubbles are propagated directly in response to the modulation of an external magnetic field which is applied normal to the chip surface.

Bubble - an isolated cylindrical domain which exhibits various magnetic properties.

Bubble device - an independent unit (such as memory or a logic processor) wherein the bubble is the active element.

Bubble host - a material such as a thin-film magnetic garnet epitaxially grown on a nonmagnetic garnet substrate. When exposed to the proper magnetic fields normal to the film surface, the garnet film can sustain small stable areas of reverse magnetization, referred to as bubble domains.

Bubble memory - a storage device in which magnetic bubbles are the active elements.

Coercivity - the characteristic of magnetic material which opposes the demagnetizing effect of an external magnetic field.

Conductor access - a method involving conductors in which currents in the conductors generate the required fields for magnetic bubble propagation.

Direct-optical sensing - a method in which the detector reacts to changes in the intensity of light caused by the passage of a bubble over a light source.

Domain wall - the boundary separating adjacent magnetic domains, which is the transition zone through which the magnetization reverses direction.

Electromagnetic-induction detection - a detection method in which the bubble serves as a tiny moving magnet dipole that induces a voltage in a pickup loop located in the reading head.

Epitaxy - a controlled growth or deposition of a crystalline material having the same crystal orientation as the substrate onto which the deposition is formed.

Evaporation - a process by which alloys are formed when beams of thermally agitated atoms intermingle and condense on the underside of a substrate that is suspended in their paths, each beam rising from a heated dish containing a liquid from one of the elements of which the alloy is composed.

Field-access - a method for propagating bubbles by the use of an external pulsating or rotating magnetic field.

Galvano-magnetism/Electromagnetism - magnetism developed by the flow of electrical current.

Garnet - any hard vitreous crystalline mineral from a group comprised of silicates of calcium, magnesium, iron, or manganese with aluminum or iron. For magnetic bubbles, any material having a garnet-like crystal structure with a formula of $(A)_3 Fe_5 O_{12}$, where A is a material such as yttrium or one of the rare earths.

Hall effect detection - a phenomenon in which a voltage appears across a semiconductor slab carrying a direct current when the magnetic field of the bubble acts perpendicularly to the slab (at right angles to the current).

Hexagonal ferrite - any phase containing Fe and having a crystal structure related to, but not necessarily the same as, magnetoplumbite.

In-plane rotation - a rotating magnetic field which is applied parallel to the surface of a magnetic platelet containing bubbles.

Liquid phase epitaxy (LPE) - a process of developing magnetic materials, such as garnets, by dipping a non-magnetic crystal into a solution so that a magnetic epitaxial film or solute crystal grows on the surface of the substrate.

Magnetic thin film - a layer of magnetic material on the order of one micron thick.

Magnetoplumbite - $Pb\ Fe_{12}\ O_{19}$ or any phase, containing Fe, having a similar crystal structure.

Magnetoresistance - the effect in which a change in electrical resistance of a material occurs which results from the presence of a magnetic field.

Magnetorestrictivity - the effect or phenomenon in which a magnetic material develops mechanical strains or stresses when magnetized.

Nonvolatility - a property of storage media indicating a capability of information retention independent of sustained power application.

Nucleation - the formulation of a nucleus or the process of clustering.

Oersted - the oersted (CGS system) is the intensity of a magnetic field in which a unit magnet pole experiences a force of one dyne.

Orthoferrite - generally, a mineral consisting of iron and other materials with an orthorhombic distortion of the perovskite crystal structure, whose chemical formula is $A\ Fe\ O_3$ where A can be any rare earth material such as gadolinium.

Permalloy - a bubble guide material, usually consisting of nickel-iron alloy (Ni Fe), which is deposited on a magnetic material.

Photolithography - a process of depositing images by using photographic techniques.

Sputtering - a process by which groups of atoms are chipped away by impact of material from the surface of a target material having the same composition as that of the desired alloy film, and fall onto a substrate, thus building up the layer of an alloy film.

T-bar - a bubble guide pattern which consists alternately of "T" or "I" shaped patterns of permalloy which are deposited on a magnetic material.

Uniaxial structure - a crystalline configuration wherein the magnetic moment vectors are all aligned in the same direction along an easy axis.

Wafer - a thin platelet of material such as a slice of silicon, germanium, or garnet.

Y-bar - a bubble guide pattern which consists of permalloy Y's and bars alternately deposited on a magnetic material.

Glossary of Memory Terms

Access time - the time between input memory addressing and output information availability.

Angstrom - a unit of measurement of wavelength equal to 10^{-4} micrometer or 10^{-8} centimeter.

Associative memory - a memory which produces addresses of words based on the contents of words, rather than the contents of words based on addresses of words; also referred to as content-addressable, catalog, or search memory.

B-H curve - a plot of flux density (B) versus magnetizing force of a magnetic material which shows the characteristic hysteresis loop.

Bipolar memory - a memory in which the storage element utilizes devices such as transistors whose conduction involves both minority and majority carriers.

Bit density - a measure of the number of bits stored per unit of length or area in the storage medium.

Bit plane - a screen or 2-dimensional surface each layer of which contains one bit associated with many words.

Block transfer - the transmission of a group of consecutive words considered as a unit from one plane to another.

Cache - a high-speed memory whose contents are continually updated in blocks from a larger, slower memory to make the effective memory access time approach that of the higher speed (cache) memory.

Charge-coupled device - a semiconductor device which propagates signals by the movement of charge packets.

CMOS (Complementary-Metal-Oxide-Semiconductor) - circuitry and logic employing both P-channel and N-channel MOS transistors with opposite, or complementary, switching characteristics.

Coincident current selection - the addressing, or selection, of a magnetic memory element for reading or writing by the simultaneous drive from two or more current sources.

Content addressable memory - (see associative memory).

Core - a magnetic toroid, usually a ferrite, capable of being magnetized and remaining in one of two conditions of magnetization, thus capable of providing information storage.

Destructive readout - a reading process that destroys the information that has been read out.

Direct access - a method which access to the next piece of information is not dependent upon the last accessed source of information.

Dynamic cell - a storage cell where stored information is affected by time.

Easy axis of magnetization - a preferred direction of magnetization in a magnetic material.

ECL (Emitter-Coupled Logic) - a nonsaturated structure logic where the multiple gate inputs are coupled together by their respective emitters.

Error rate - the total number of errors divided by the total number bits transmitted.

Flux density - the number of magnetic lines of force passing through a unit area.

Full-select current - the total current required in a selection winding or windings through a magnetic core to saturate the core.

Half-select current - approximately one-half the total current required in a winding or windings through a magnetic core to saturate the core.

Hard-axis of magnetization - a direction of magnetization perpendicular to the easy axis of magnetization of a material.

Head gap - the spacing in the magnetic circuit of a record/reproduce head from which the fringing fields emanate.

Hysteresis loop - a graph showing a plot of flux density in a magnetic material on the ordinate and the magnetizing force which produces the flux on the abscissa.

Latency - a delay time in accessing information from a storage device.

Logic speed - the typical logic gate-to-gate propagation speed or clock rate.

LSI (Large-Scale-Integration) - a silicon (or other material) chip containing many gates.

Mass memory - storage devices such as drums, disks, tapes, etc., which are capable of storing large quantities of information.

Memory bus - one or more conductors used as a path over which information may be transmitted from any of several memories to any of several computer system destinations.

Memory capacity - the number of bits, bytes, words, or other elementary quantity of data that can be stored in a memory device.

Memory port - an interface allowing access to the contents of a memory.

MOS (Metal-Oxide-Semiconductor) - a field-effect transistor where the gate is insulated from the channel between source and drain by a metal oxide.

MTBF - "Mean-Time-Between-Failures" (abbreviation).

MTTR - "Mean-Time-To-Repair" (abbreviation).

Nondestructive readout - the sensing of information contained in a storage device without significantly changing the physical representation of the stored information.

Paging - the grouping of storage segments by page and transferring these pages back and forth between main memory and the slower, larger backup memory.

Partial switching - the flux change in a core resulting from the application of partial read or partial write pulses.

Plated wire memory - memories which are obtained by depositing a material such as permalloy in the form of thin-film on a wire.

PROM (Programmable-Read-Only-Memory) - a memory having factory or user programmed contents which are only capable of being read.

RAM (Random-Access-Memory) - a memory capable of being addressed in any sequence for either reading or writing.

Remanent magnetization - the state of magnetization of a material after the external magnetic field has been removed.

ROM (Read-Only-Memory) - a memory having preprogrammed contents capable of being sensed or read only.

Saturation - the magnetic state of a material beyond which an increase in the magnetization is not possible.

Semiconductor memory - a memory whose basic storage elements are semiconductor devices.

SOS (Silicon-On-Saphire) - a metal-oxide-semiconductor circuit developed on an insulator (sapphire) base.

Thin-film memory - a memory that utilizes magnetic thin-films deposited on materials such as glass.

Transfer rate - the rate at which elements of information are transmitted in sequence from source to destination.

TTL (Transistor-Transistor Logic, or T^2L logic) - a standard integrated logic structure consisting of interconnected transistors.

Virtual storage - a combination of hardware and software which automatically allows programmers to program the combination main memory and mass memory as if they were one large main memory.

Volatile storage - a memory device which loses data when power is removed.

REFERENCES

1. Micro-Notes-Information on Microelectronics for Navy Equipments, Naval Ammunition Depot, Crane, Indiana, July 1968.

2. Wagner, S. and W. Doelp, ''Low Cost Integrated Circuit Techniques'', Final Report, Technical Report ECOM-01424-F, prepared by Philco-Ford Corporation, Microelectronics Division for U.S. Army Electronics Command, Fort Monmouth, N.J., Contract No. DA28-043-AMC-01424(E), May 1967.

3. Multipurpose Logic Cell Study, Ronald Waxman, et al, IBM, Hopewell Junction, N. Y., Contract No. F19628-67-C0302, Oct. 1968.

4. Microelectronic Device Data Handbook Vol. I, ARINC Research Corp. July 1968.

5. Integrated Logic Networks, James F. Allison, RCA Laboratories, Princeton, New Jersey, Oct. 1968.

6. R. Shahbender, ''Laminated Ferrite Memories – Review and Evaluation'', *RCA Review 29,* 180 (1968).

7. H. I. Moss and A. D. Robbi, ''High-Pressure Sintering of Ferrites'', Final Report, Contract DA36-039-AMC-03719(E), United States Army Electronics Command, Fort Monmouth, New Jersey, July 1967.

8. I. Gordon, R. L. Harvey, H. I. Moss, A. D. Robbi, J. W. Tuska, J. T. Wallmark, and C. Wentworth, ''An MOS-Transistor-Driven Laminated-Ferrite Memory'', *RCA Review 29,* 199 (1968).

9. A. D. Robbi and J. W. Tuska, ''Integrated MOS Transistor-Laminated Ferrite Memory'', (Abstract), *IEEE Trans.* Magnetics *Mag*-3 329 (1967).

10. R. L. Harvey, I. Gordon and A. D. Robbi, ''Laminated Ferrite Memory—Phase II'', Final Technical Report, Contract NASW-979, National Aeronautics and Space Administration, Washington, D. C., Aug. 1966 (''Low Drive, Temperature Stable Ferrite for Laminated Array'', *IEEE Trans.* Magnetics *Mag*-3 526 (1967).

11. Microelectronic Engineering Vol. I, Fabrication Technology, D. Abraham, Johns Hopkins University, 1966.

12. Computer Aided Layout of Microcircuits, W. J. Cullyer, etal, Signals Research and Development Establishment, Sept. 1968.

13. Perret, R., and David, R., ''Synthesis of Sequential Circuits Using Basic Cell Elements'', Proc. of Ninth *Joint Automatic Control Conference,* U. of Michigan, pp. 582-594, June 1968.

14. Elspac, B., Kautz, W. H., and Stone, H. S., ''Properties of Cellular Arrays for Logic and Storage'', Stanford Research Inst., Final Tech. Report No. AFCRL-68-0005, Nov. 1967.

15. King, W. F., and Giusti, A., ''The Design of a More Complex Building Block for Digital Systems'', Cambridge Research Laboratories AFCRL, 67-0516, *Phys. Sci. Res.* Paper No. 341, Sept. 1967.

16. Yau, S. S., and Tang, C. K., ''Universal Logic Circuits and Their Modular Realizations'', Proc. AFIPS Conference Proceedings of the 1968 *Spring Joint Computer Conference,* pp. 297-305, 1968.

17. Susskind, A. K., et al., ''Threshold Elements and the Design of Sequential Switching Networks'', *MIT Electronic Systems Laboratory Tech.* Report No. RADC-TR-67-255, July 1967.

18. Elspac, B., et al., ''Properties of Cellular Arrays for Logic and Storage'', Stanford Research Inst., SR3, Tech. Report No. AFCRL-67-0463, July 1967.

19. Patt, Y. N., ''A Complex Logic Module for the Synthesis of Combinational Switching Circuits'', AFIPS Conference *Proceedings of the 1967 Spring Joint Computer Conference,* pp. 699-705, 1967.

20. Universal Logic Blocks, Barry P. Shay, Naval Research Laboratory, Washington, D. C., Sept. 17, 1967.

21. Modeling Integrated Circuits for Computing, Gerald J. Herskowitz, Stevens Institute of Technology, Hoboken, New Jersey, Oct. 1967.

22. EIA. *Registration Data: Semiconductor Integrated Bistable Logic Circuits,* 1967.

23. Sanford, Robert, ''Understanding IC Logic'', *Electronics,* March 6, 1967.

24. Marshall, D. E., Jr., ''Large Scale Integration Problems and Possibilities'', *Computer Design,* (1967), pp. 18-21.

25. Hugle, William B., ''Integrated Logic Circuits: A Comparative Evaluation'', *Computer Design,* (1967), pp. 36-47.

26. Mays, H. C., ''Computer-Aided Design for Large-Scale Integration'', *International Solid State Circuits Conference,* (1967).

27. Boysel, B., ''Memory on A Chip: A Step Toward Large-Scale Integration'', *Electronics,* (1967), 93-97.

28. G. E. Moore, ''What Level of LSI?'', *Electronics,* February 16, 1970.

29. R. C. Foss, A. Richardson, J. S. Brothers, ''Design Considerations for Integrated Digital Systems'', *IEEE 1967 Microelectronics Conference Procedures Publication* #30.

30. Large Scale Integration in Microelectronics, J. W. Lathrop, *Lecture Series, Sponsored by the Avionics Panel of Agard,* London (U.K.) and Dusseldorf (Germany), July, 1970.

31. Study of Failure Modes of Multilevel Large Scale Integrated Circuits, Philco-Ford Corporation, Blue Bell, Pennsylvania, May 1968.

32. Wagner, S. R. and W. L. Doelp, Jr., ''Silane Deposited SiO_2 in LSI, *International Electron Devices Meeting,* Washington, D. C., October 18-20, 1967.

33. F. P. Heiman, *Integrated Logic Nets,* Final Report, Contract No. AF19 (628)-4208, Jan. 1967.

34. Research to Develop Mass Data Storage (U), A. D. Robbi, et al, *RCA Laboratories,* Aug. 1968.

PART IV

INTERCONNECTIONS

PACKAGING

CIRCUIT TESTING

CHAPTER 12

INTERCONNECTIONS

New generations of electronic systems will require increasingly dense and complex interconnections, and the trend toward higher operating frequencies will necessitate extremely careful design, even of single conductors. Conductor width, spacing, and dielectric material are electrically critical, and in many cases they are directly related to signal propagation, noise figure, pulse shape, and other system parameters. Therefore, it is mandatory that designers of systems and circuits become thoroughly familiar with new interconnection techniques.

The newly developed and improved joining processes must afford advantages in producibility, inspectability, repairability, and reliability. These four basic criteria are defined as:

Producibility: requires that intra/interconnection be made rapidly and economically, with optimum, uniform material combinations and configurations, eliminating human variables wherever possible.

Inspectability: requires that any joining process to be used for production work, be inspectable either visually or by a reliable process-monitoring system.

Repairability: requires that any assembly of greater value than the established maximum throw-away value be easily repairable without unusual equipment or procedures, and without damage to materials or components.

Reliability: requires that a minimum failure rate objective, dependent on projected reliability requirements above contractual commitments, be met or exceeded. (A typical missile guidance and control system requirement allows not more than one interconnection failure in 2×10^9 joint-hours of operation.)

In addition to the four basic criteria, consideration must also be given to current carrying capacity, voltage drop, contact resistance, connection noise level, size (connection density), thermal sensitivity of joined members, mechanical strength, and metallurgical characteristics of the connection.

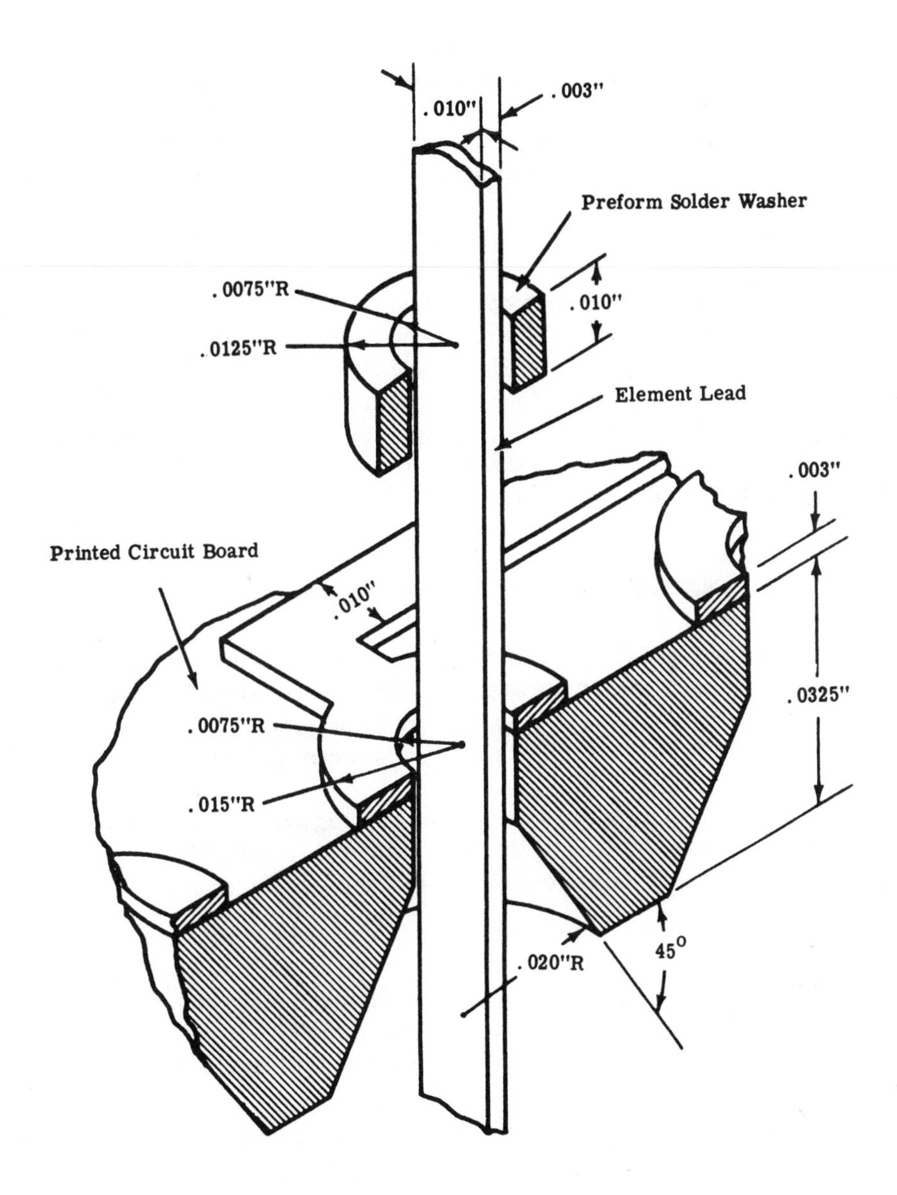

Fig. 12-1 Soldered joint assembly

12.1 LEAD ATTACHMENT AND BONDING METHODS

Joints to microelectronic circuits may be generally classified among three groups: a) permanent joints in which one or both lead ends must be destroyed to separate them as, for example, in welded joints; b) semi-permanent joints, such as solder joints, in which special tools or processes are required to separate the two leads, and once separated, they may be rejoined at the same point; and c) temporary or quick-disconnect joints, such as plug-in connectors.

Joints can be made by many techniques: soldering with hand irons, dip or flow machines, resistance or induction heating, ultrasonic or optical energy, and hot-gas jets; welding with opposed-electrode, parallel-gap or series resistance, or a percussive arc; welding with laser or electron beams; ultrasonic and thermocompression bonding; metal-to-metal diffusion; mechanically joining by wrapped-wire methods, spring clips, clamp screws and crimps; friction connection; films deposited by plating, vacuum evaporation, pyrolysis, sputtering and metal spraying; and conductive adhesives.

New variations and new ways are being tried continuously. There is little agreement, among the authorities in the field, as to which method to choose for a particular application. The pros and cons change continuously, as new materials and methods appear. In selecting the bonding method, to be used, it is important to note that close control of the process is required because a) nondestructive tests are difficult to make, b) excessive bonding pressure or temperature could damage semiconductor devices or thin-film conductors, and c) inadequate or excessive bonding pressure or temperature can result in weak bonds.

12.2 SOLDERING

Soldering is defined as a group of joining processes wherein coalescence is produced by heating to a suitable temperature below 800°F and by using a non-ferrous filler material, having a melting point below that of the base metals. The alloy must be distributed between, and molecularly "wet" the joining members.

Basic soldering processes consist of hand bit, dip, flow, and resistance techniques to produce the metallurgical connection between circuitry and components. The definition for each of the techniques is:

Hand Bit – a manual operation where the operator uses a miniature soldering iron with small diameter wire, or preform, and flux to produce the joint. This process is used for fabrication of printed circuit board, cordwood modules, wire and cables, thin-film, etc. The presence of the human variable requires close control of the process for high reliability production.

Dip – an automatic operation where circuit boards are flux coated, and then dipped into a molten solder bath of controlled temperature for a predetermined length of time to allow adequate hole penetration and wetting of component leads. The elimination of the human variable and the rapidity of soldering make this a desirable production process for fabrication of printed circuit boards with discrete components. Adaptations of the basic process have been made to allow the dip soldering of ceramic thin-film circuits.

Resistance – a semi or automatic operation where components are individually soldered to circuitry, using energy which is generated in the lead wire by I^2R heating between two electrodes. Flux is not always required. A specially developed automated version of the process has been

Fig. 12-2 Pulse-heated multi-lead soldering tip.

successfully used in fabrication of ceramic thin-film hybrid circuits.

Flow or Wave – an automatic operation where a prefluxed printed circuit board with discrete components is passed through a wave of solder. The associated conveyor system may be readily adapted to an automated production process. This process has also been successfully used to fabricate soldered cordwood modules.

The basic criteria of an adequate solder joint is sufficient and uniform ''wetting'' of the joining members with the solder. Visual criteria have been established which assure the integrity of the solder connection. Generally, flux is required to promote ''wetting'' by chemically reducing detrimental oxides. Care must be taken to use a noncorrosive flux or to completely remove the flux and flux residue after soldering.

Gold plating, tin and solder coatings are used extensively to protect component leads and circuit board materials, as well as to enhance their solderability.

Environmental study programs have been conducted to determine the effects that solder joint imperfections have on reliability. For this purpose, large numbers of solder joints, with and without various imperfections, were subjected to severe humidity, vibration, and aging conditions to establish correlations between the type and magnitude of imperfections and solder joint reliability. Electrical and metallographic evaluations were performed to establish failure criteria. The evaluations minimized needless rework of functionally acceptable solder joint imperfections. Rework is known to be a potential threat to high-reliability components.

The data from the studies comparing various lead material compositions and various aging environments, are shown graphically in Figure 12-3. Solder-coated leads and copper circuit board conductors show good solderability with little of the effect from accelerated aging that was noted with some other coatings. Solderability of components is primarily determined by the coatings. Thus, the base metal has little influence on solderability except in cases where diffusion of base lead materials, which subsequently oxidize, through the coating, tends to decrease wettability.

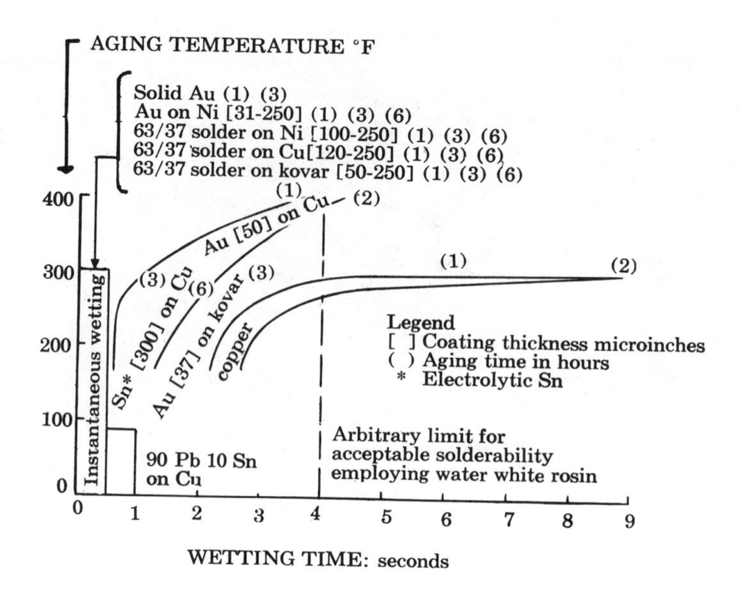

Fig. 12-3 The effect of aging temperature and time on the solderability of different component lead materials.

Soft soldering of leads to thin-film substrates is a technique used extensively throughout the industry. The soldering operation is dependent upon the formation of an alloy between the solder and the metal that is soldered. This alloying takes place at relatively low temperatures (361° F to 500° F). Most solders consist of tin and lead in various proportions, although various other materials, such as silver, antimony, indium, cadmium, etc., are sometimes used for special soldering operations. A good solder joint requires the presence of a soldering flux to remove surface oxides from the parts to be joined, and to reduce interfacial tensions.

Since the metallic films to be soldered are extremely thin, special precautions and techniques must be used to insure the formation of a reliable joint. Special small soldering irons are generally used, although several special soldering systems have been devised to produce automated solder joints. One system uses a specially shaped soldering tip made of nichrome. This tip is mounted in a holder whose vertical motion may be automatically controlled. Both the length of travel of the soldering head, and the pressure exerted by the head are controllable. Timing circuitry controls the heating time and the temperature cooling periods. In operation, the work piece is positioned under the soldering head and a foot switch is depressed. From this point, all operations are automatic – the soldering head makes contact with the work, the tip is heated, and the tip is lifted from the work.

This method of soldering is used on both ceramic and glass substrates using both copper ribbon and gold plated wire.

Figure 12-1 illustrates an assembly used to package microelectronic circuits in flat packages to form a system. The leads, which are flat ribbons, are inserted through holes in multilayer printed circuit boards. A solder preform is dropped over the flat lead, and a liquid, water soluble flux, is applied using a hypodermic needle.

Figure 12-4 shows the optical system in one soldering machine using infrared. The energy source is the crater of a carbon arc. Since carbon arcs are available in a variety of sizes and intensities, it is a good heat source for this application. The first lens is the condenser and it collimates the light from the arc crater. The second lens focuses the collimated light onto the work. The spot size may be determined from:

Crater diameter × focal length of focusing lens/focal length of condenser.

The rotating shutter is used to control the "on" time and is electrically

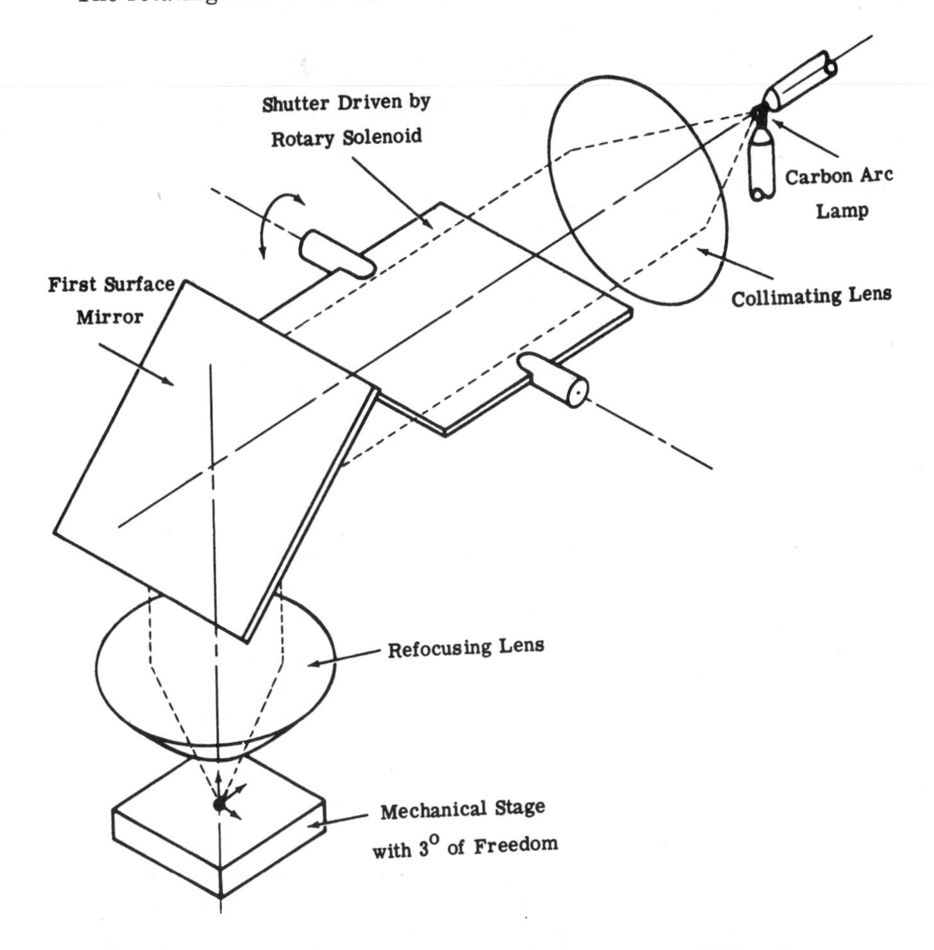

Fig. 12-4 Optical system used to produce a small intense light spot for heating joints in soft soldering

timed. A microscope is used to observe the melting of the preform and the formation of the bead. The required time is 2 to 3 seconds. Visual inspection of the joint can be used as an indication of the joint quality. Hot air for melting the solder may also be used.

The use of a flux with the soft solders is mandatory. The selection of an appropriate flux depends upon the materials used, but regardless of the flux which is finally selected, the residue should be removed completely by a washing procedure. Many of the proprietary fluxes, now available, are water soluble, and immersion for a few minutes in warm water should not affect a thin film circuit.

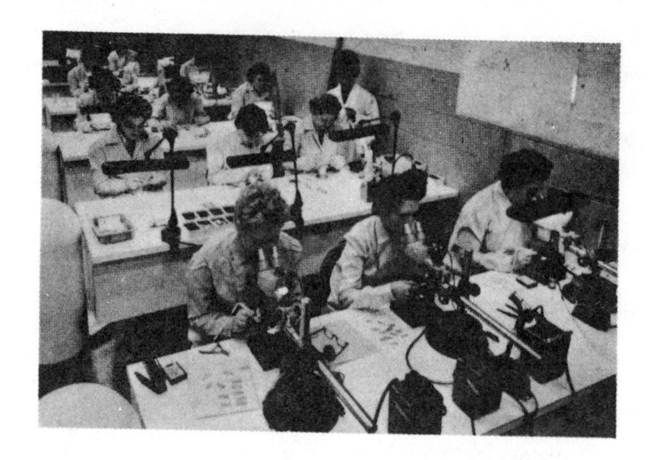

Fig. 12-5 Hand assembling and soldering discrete components to the patterned substrate.

12.3 RESISTANCE WELDING

Resistance welding of electronic parts has become a fairly common technique in recent years. The process involves clamping the materials to be welded between a pair of electrodes, and passing an electrical pulse through them. Heat is generated by resistance at the facing surface and a welded joint forms by fusion or solid state bonding, depending upon the materials involved. Although this method is useful for cordwood and foil-tab types of integrated-circuit modules, it is difficult to apply to lap joints on printed circuit boards, because both sides of the joint are not accessible.

This difficulty may be overcome through the use of series welders in which both welding electrodes are placed on the same side of the work piece at some fixed spacing as shown in Fig. 12-6a-b. One application of this welding technique is where one of the materials to be joined, is contacted by a welding electrode at some point remote from the point to be welded. The other electrode is positioned directly over the point to be welded. The advantage of this technique is that the thinner member of the joint system can be welded to the thicker material regardless of joint design. This process may be used for welding ribbons and wires to plated film as thin as one mil.

A refinement of the series welding technique is the single point parallel gap welder shown in Fig. 12-6c. This welder incorporates the two current-carrying probes necessary for resistance welding into one instrument. The two sides of the electrode are separated by an insulating spacer. In operation, the instrument is placed over the area to be welded, and welding heat is generated in the work beneath the welding probe at a point on the centerline of the gap. The width of the gap is directly proportional to the depth of penetration of the heat. Gaps range from 0.001 to 0.005 inches, being on the same order of size as the wire to be welded. In practice, the welding head is energized by a series of pulses whose shape, duration, and number can be controlled to adapt to different work pieces. Once the correct sequence and strength of the pulses have been determined for a specific job, the same type of joint may be made automatically, with no

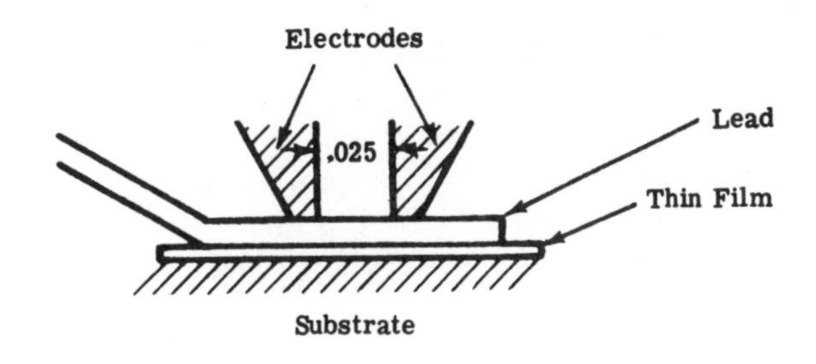

(a) Series Weld

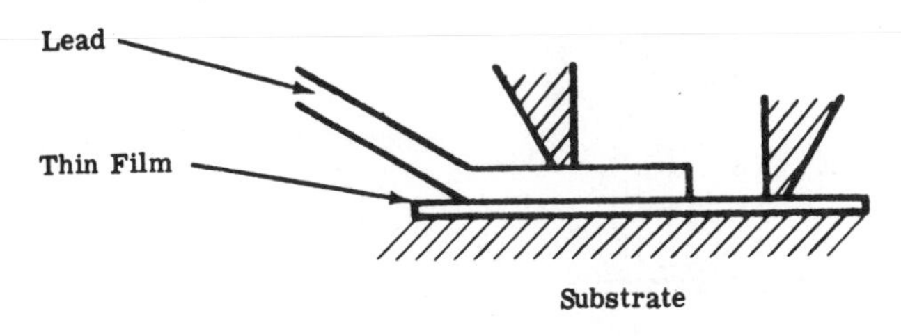

(b) Step Series Weld

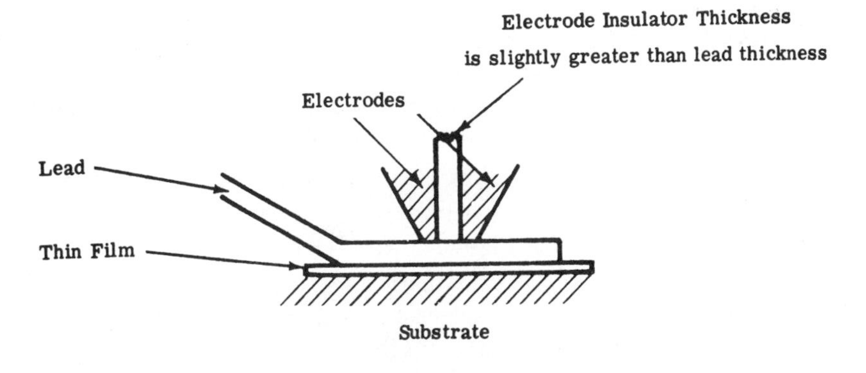

(c) Parallel Gap Weld

Fig. 12-6 Electrode arrangements used in welding leads to thin films

additional adjustment. This makes this type of welder extremely attractive for an assembly-line type of operation.

Microgap bonding is similar to parallel-gap welding. However, the series electrodes are smaller and closer together, and the bonds are made by diffusion rather than fusion. This bonding method is preferred for discrete microcomponents, where the leads (up to 10 mils diameter), are too large for capillary-type bonders. The shape of the electrodes is shown in Fig. 12-7. The gap is about the size of the wire diameter. When pressed on the ductile lead, the electrodes deform it under each contact. Passing an a-c current or a d-c pulse through the lead between the electrodes, heats the lead and causes diffusion at the interface between the lead and thin film. Although bonding times are relatively short, skilled operators are required to align the electrodes, and stray currents may damage circuit components.

It has been stated that leads can be welded to films as thin as 250 Å. Wires as small as 0.3 mil and ribbons as thin as 0.13 mil may be welded. Consistent production of a large number of good welds is not easily achieved. The power generated during a welding cycle should usually be varied for best results. This can best be done by using a shaped single dc pulse or a series of dc pulses of variable height. The use of ac power of

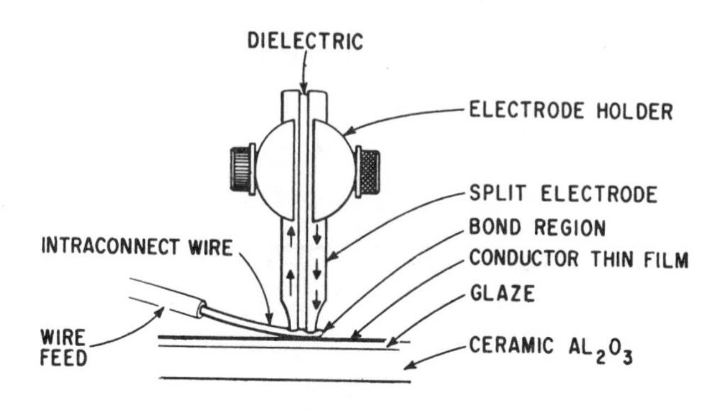

Fig. 12-7 Microgap bonding

variable amplitude is difficult, since the power "on" time may be short as a few milliseconds, and the load impedance is of the order of a milliohm.

While bonds can be made to films a few hundred angstroms thick, this is not recommended. They should be preferably from 0.3 to 1 micron thick. The leads should be 1 to 5 mils in diameter or thickness.

Pull tests or metallurgical examination of the weld nugget are destructive. Accordingly, they are not useful as a 100-percent testing procedure. However, they may be used in setting up a good weld schedule. X-ray radiography is a useful nondestructive weld examination procedure.

Fig. 12-8 Parallel gap weld head with adjustable electrodes.

12.4 PERCUSSIVE ARC WELDING

Percussive arc welding is a method introduced to make high-density welds of wires to flat surfaces, at high speed. The work pieces, which act as anode and cathode, are separated by a gap which is ionized by a burst of RF energy. A capacitor bank discharging across this gap, creates an arc which heats the work pieces. As the arc is struck, the work pieces are accelerated toward one another, and the resultant force percussively joins them together to make the weld. This technique requires low pressure, which permits the joining of materials with low columnar strength. High heat concentration permits the welding of a small wire to a large surface, resulting in good electrical properties, since a true weld is formed. Temperature sensitive components are safely welded using this technique, since the heat is at the weld interface and lasts for only a fraction of a millisecond. Tests indicate that the joints thus formed, in most cases, has a tensile strength greater than that of the parent materials. The process is highly repeatable and relatively insensitive to gap setting and capacitance.

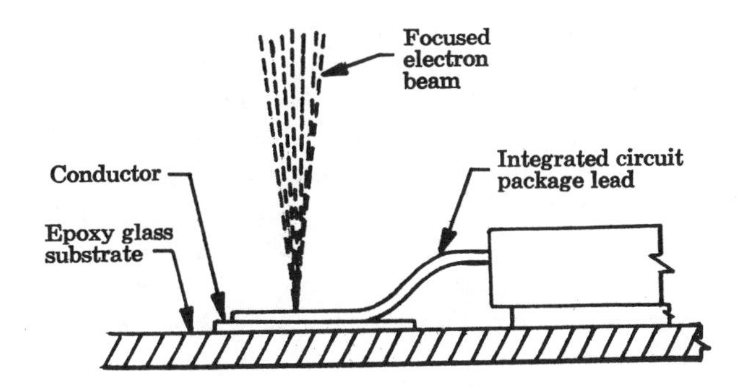

Fig. 12-9 Electron beam welding.

12.5 ELECTRON-BEAM WELDING

Electron beam welding shown in Fig. 12-9, offers advantages over other welding techniques for some applications. Higher packaging density may be obtained since the beam diameter may be extremely small – 0.0005 to 0.010 inch. Since electrode contact to the materials is not required, circuit conductors can be more closely spaced.

The welding energy is produced by the bombardment of a high-energy beam of electrons on the work pieces. The kinetic energy of the focused electrons is converted to heat at the surface of the pieces to be welded. Power densities of about 10^9 watts/sq. in. can be obtained, which is sufficient to melt or vaporize most known materials.

Advantages of the electron beam include: 1) precise control of heat; 2) since the entire operation is performed in vacuum, most contaminating elements are eliminated; 3) materials which are usually impossible to join due to high melting point may be welded; 4) the technique is not sensitive to surface irregularities, reflectivity, or minor impurities; 5) high pressure contact, and electrodes in close proximity to the weld joint are not required. Operation in a vacuum may be considered a disadvantage since this tends to complicate the process. Furthermore, the penetration of high-velocity electrons into the substrate causes spalling of the substrate.

The precision electron beam may also have other capabilities for use in fabricating microelectronics. Machining, drilling, and device processing may be carried out through the use of the electron beam. These additional electron beam applications appear to be capable of solving some of the critical production requirements for high reliability and densely-packaged microelectronic systems.

12.6 LASER WELDING

Laser welding is similar to electron beam welding in that it utilizes a focused, high-energy beam. The laser beam is, however, theoretically capable of higher power densities than an electron beam.

The laser process uses the heat produced by a focused beam of coherent light (electromagnetic radiation) to melt the joining members thereby forming a weld.

The laser beam is operated in the pulsed mode, between 1 to 4 milliseconds and at an energy level from 2 to 15 joules. The high power densities obtainable with the laser, present some difficulties when metals joining is to be accomplished. Unless careful control over the energy output is exercised, more energy may be put into the workpiece than can be conducted out of it. This excess intensity causes boiling at the surface of the metal, and hence vaporization of the metal. The reaction forces of the vaporized metal are sufficient to expel liquid, and this leaves craters on the surface of the metal. Therefore, the pulsed output of the laser must be carefully controlled.

Ruby or neodymium doped glass lasers are the most widely used for welding applications. The coherent light is produced by placing the laser crystal in an optical cavity and pulsing with a sufficiently high intensity light to produce a population inversion of the Cr^{+3} ion, photon cascading, and ultimately laser action. The magnitude of the pulsing light is such that the only reasonable source is the xenon flash tube. However, the operation of these tubes for more than a few milliseconds will quickly destroy them. Thus, the laser welder is thereby limited to pulse-type operation at a low repetition rate. Work is being done on continuous solid-state lasers for welding applications. Research is also being conducted to develop a multiple-pulse laser which would more closely resemble a continuous beam.

The utilization of the laser as a means for welding presents a number of problems. Present lasers have extremely low output efficiencies (1-2%) compared to electron beam systems (90%). Furthermore, since absorption of light is dependent upon the reflectivity of the surface of the joining members, light colored metals with polished surfaces can reflect as much as 90 percent of the laser beam.

Both the output of the ruby rod and the output of the xenon-flash tube are inversely affected by temperature. Steady-state temperatures of these components are required for constant energy output. However, this does not necessarily insure a uniform output energy. Deterioration of the optical cavity and flash tube is continuously occurring.

Particularly severe hazards are involved from reflected light. Secondary radiation is capable of causing permanent eye damage.

The most attractive feature of the laser welding technique, as opposed to electron beam welding, is that the work may be performed without a vacuum, and thereby allow direct access to the work pieces. Also, focusing of the beam may be accomplished with simple optical systems.

12.7 THERMOCOMPRESSION BONDING

Thermocompression Bonding is defined as a joining process wherein coalescence of two materials, at least one of which is ductile, is accomplished over a controlled area by diffusion between the members being joined. The members have been brought into intimate contact by a shaped bonding tool, at a temperature below that required for interface melting.

Plastic flow and diffusion occurs during a controlled time, temperature, and pressure cycle. At least one of the two materials to be joined, must be ductile to eliminate elastic ''spring-back'' which could degrade the joint when the bonding tool is removed. Heating the bonding tip and/or the substrate promotes intimate contact with a minimum of applied force, by lowering the compressive yield strength of the ductile material. Heating, also, accelerates the diffusion process. The basic requirements for developing thermocompression bonding procedures are:

1. Choose appropriate lead materials such as soft, fully, or partially annealed metals to minimize undesired elastic ''spring-back.''

2. Assure that adequate deformation of the wire occurs. The cross-section thickness of the bond region should be approximately one-half the original thickness of the wire. Under normal conditions, pressure should not exceed the compressive yield strength of the ductile bonding wire.

3. Radiused wedge bonding tools require the tip to have a radius of 1 to 4 times that of the wire, to prevent the formation of stress risers. The size and shape of the tool should produce bonds of adequate contact area and configuration when a moderate force (10 to 200 grams) is applied.

4. Choose an appropriate temperature within the annealing range of the ductile metals, but below the eutectic or alloying temperatures.

5. Optimize bonding force, time, and temperature by experimentation. The most desirable bonding will occur at minimum values of time, temperature, and pressure.

6. Precautions must be used to ensure that undesired films (oil, oxides, etc.,) do not exist on the bonding surfaces.

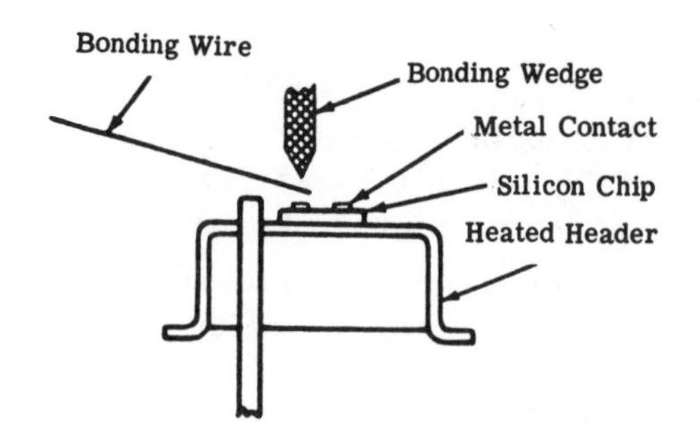

Fig. 12-10 Wedge Bonding

Small gold, copper, silver, aluminum, platinum, and tin lead wires and ribbons have been successfully thermocompression bonded directly to various semiconductor materials, or to an intermediate vapor deposited metallic conductor film (aluminum, gold, silver, etc.,). The gold wire to aluminum vapor deposited circuitry combination is the most predominantly used for bonding to surfaces of integrated circuits, and many transistor devices. The terminal end of the gold wire is then generally attached to gold-plated output leads of the protective package.

Several types of thermocompression bonding are presently in use. The three principal methods are wedge bonding, ball (nailhead) bonding, and stitch bonding, as shown in Figures 12-10 to 12-12, respectively.

Wedge bonding utilizes a very fine sapphire or silicon carbide wedge tool which is brought down upon the end of the wire to be bonded.

The wedge tool plastically deforms the lead into intimate contact with the bonding surface. After a controlled time period, the tool is raised completing the bonding operation.

In ball bonding, a fine wire is fed through a capillary tube, the exposed end of this wire is melted into a ball by means of a very small hydrogen flame, and the ball is brought down with pressure upon the area of contact.

The capillary is first lowered until the ball is brought into contact with the capillary tip. The capillary is then further lowered to the surface of the device where a predetermined amount of force is applied to deform the ball, thus bringing it into intimate contact with the surface to be bonded. Bonding temperatures are attained using a heated substrate and/or heated capillary. After a predetermined length of time, the capillary is raised from the substrate, leaving the "nailhead" bonded wire attached to the device. The hydrogen flame then severs the wire, producing another ball for the next bonding operation.

Stitch bonding also uses a capillary arrangement; here, however, the exposed end of the wire is bent at a 90° angle instead of being balled, and then bonded as in ball bonding above.

In all cases, bonding is done with the aid of fine micro-positioning equipment and observation microscopes. In addition, the silicon chip is heated to the appropriate temperature on a heat column while the bonding is being accomplished.

Thermocompression bonding methods, other than those above, are also in use. In resistance-heated thermocompression bonding, the major source of bonding heat is a tungsten-carbide capillary tool, as shown in Fig. 12-13. The tool is similar to those used in ball bonding. Heat generated

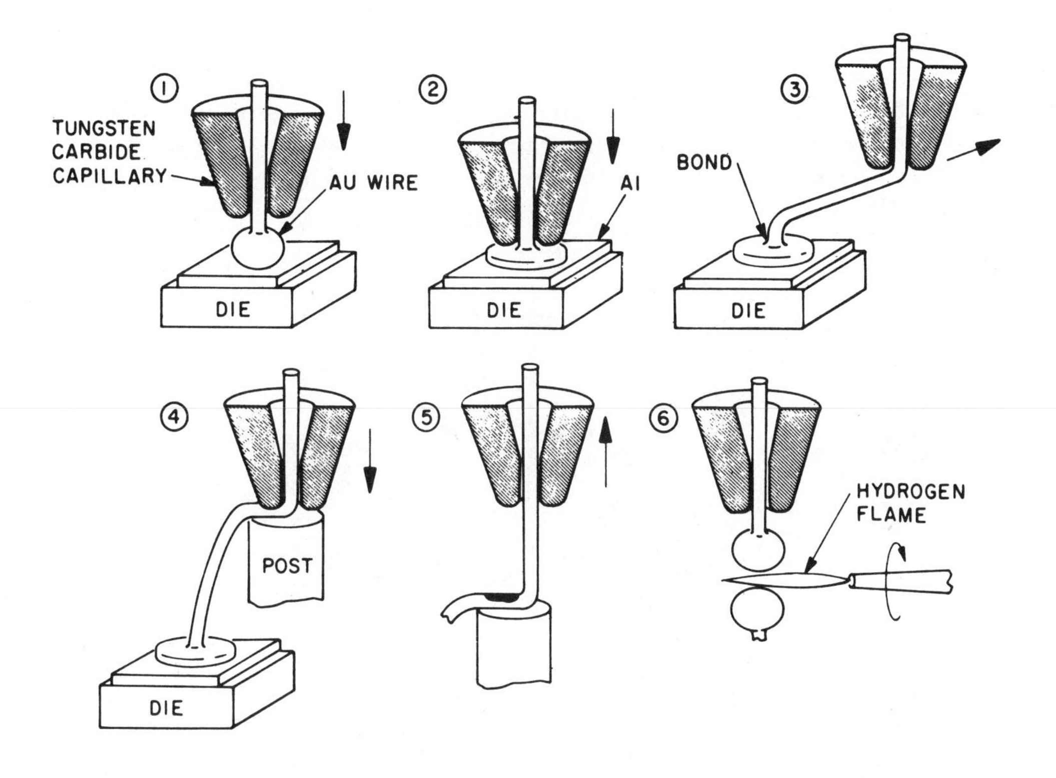

Fig. 12-11 Steps showing the ball bonding operation.

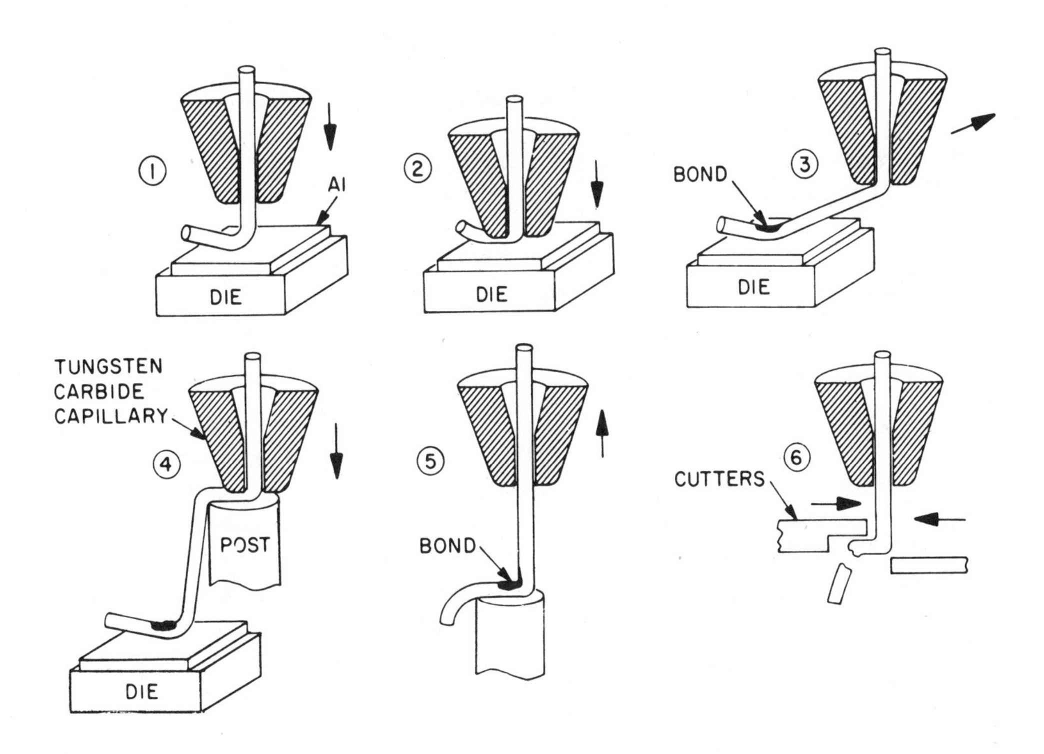

Fig. 12-12 Stitch bonding operation.

in its resistive element by 60-cycle current is conducted through the tool tip and the wire to the bond interface. Combining the wire feed with the bonding tool eliminates the need to align them separately as in the wedge bonding method. However, the bonds do not have equal pull strength in all directions. Bonds to the device and the thin-film circuit may require different heats.

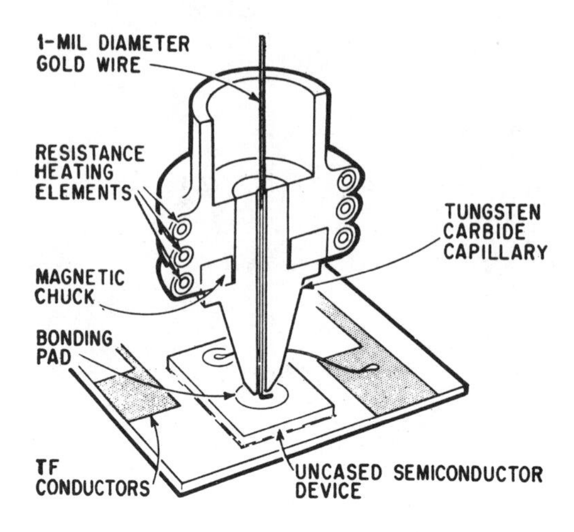

Fig. 12-13 Resistance - heated thermocompression bonding tool

Pulse-heated thermocompression bonding, shown in Fig. 12-14, is similar to the resistance-heated method, except that the tungsten carbide bonding tool is heated by a pulse of direct current. Slotting the capillary makes it hottest at its end, where bonding occurs. Bonding pressure, which is predetermined, is applied through the capillary.

Pulse-heated bonders are especially advantageous for temperature-sensitive components. Keeping the components at ambient temperature until the pulse is applied, minimizes heat damage, provided the pulse is controlled to avoid thermal shock.

12.8 ULTRASONIC BONDING

Ultrasonic bonding is a solid-state metals joining method which uses no fluxes or filler materials. Combined ultrasonic energy and induced pressure are applied to the joining members, overcoming the barriers to welding by plastically deforming the mating surfaces in such a way that oxide films or minute foreign materials are dispersed, and the irregular surfaces are made to conform to each other to ultimately produce a metallurgical bond.

Ultrasonic bonds may be produced by pressing a lead wire against the surface to which it is to be joined. The pressure is applied, generally, by the tip of a transducer that is vibrating at about 60 kilocycles per second. The mechanical force and scrubbing action cause molecular mingling of the two surfaces, in contact, and thereby form a bond.

The method is adaptable to the joining of dissimilar materials, as well as sections having different thicknesses. Ultrasonic bonding is a cold bonding technique in which no external heat is applied. Although local temperature rise may result due to the combined effects of such phenomena as elastic hysteresis, localized slip and plastic deformation, the method is often used when changes in component values, due to heating, is to be avoided. Surface cleaning is not exceptionally critical when preparing most materials for ultrasonic bonding. This is due to the condition that the vibratory displacements, occurring during the bonding process, tend to disrupt the oxide layers and other surface films on the mating surfaces. Furthermore, since fluxes, electrode shielding coatings, or adhesives are not required by this process, there is little contamination of the bond or its surroundings. The resulting joints are also claimed to be in the class of low-resistance and low-noise junctions, because they are relatively free of voids.

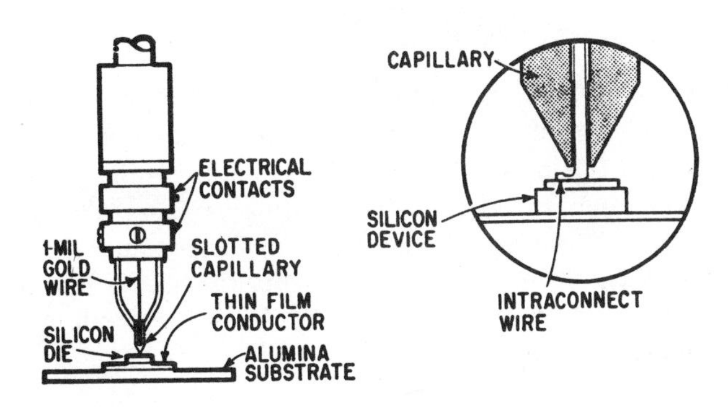

Fig. 12-14 Pulse - heated thermocompression bonding tool

An arrangement for carrying out ultrasonic bonding is shown in Fig. 12-15. The transducer, usually magnetostrictive, drives a metal tip. The wire lies in the groove in the tip, to avoid complete mashing of the wire. Ribbon may also be used. The oscillator which drives the transducer generates 100 watt at 40 kcs and has an "on" time and power control. The force between the tip and the work can be controlled pneumatically. While this technique has been used successfully, the weld schedules are critical. During vibration, dynamic stresses are induced which are sufficiently intense to cause plastic deformation of the interface material. Moisture and oxide films are broken up and dispersed, and irregularities in the surfaces are eliminated. This causes intimate nascent metal contact, and the result is a metallurgical bond, formed in the solid state, with no melting of the metals. It is possible to bond aluminum wire and ribbon to a wide range of materials including glass. A problem encountered in using this equipment, is the removal of the thin film around the weld so that the weld is surrounded by clear substrate. In this case the weld is of very high resistance. Careful adherence to a good weld schedule is necessary to overcome this condition.

Ultrasonic welding is best suited to ductile metals such as aluminum,

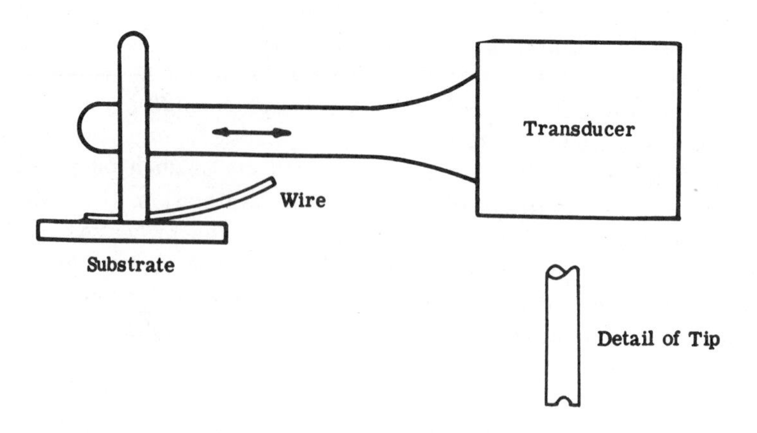

Fig. 12-15 Arrangement used in ultrasonic welding of small wires or ribbons to thin films

gold, silver, copper, etc. Since these metals have high electrical conductivity, they are especially suited for electrical interconnection.

Even though this process is termed a "cold" joining process, there is a substantial temperature rise of the bonding surfaces. The temperatures sometimes reach 30 to 50 percent of the melting point of the materials being joined. This phenomenon results from the elastic hysteresis of the highly stressed portion of the weld zone, during the welding operation. Other advantages of the process are:

1. Lack of surface preparation, other than degreasing.
2. No special environments or shielding gases are necessary.
3. Relatively simple tooling requirements.
4. Joint temperatures remain well below the melting points of the materials being joined under normal conditions. Contamination by thermal oxidation of the welding interface is, therefore, not a problem.
5. Metallurgically sound welds can be obtained between many dissimilar metals.
6. Intermetallic phases are seldom formed in the weld zone of dissimilar materials.
7. There are no electrical discharges to contaminate or damage the device.
8. Fine wires can be joined to thin foils or to thick sections.
9. Hermetic seam welds can be produced.

A disadvantage of the process is that precise alignment of the welding probe must be maintained. Close control of clamping pressure is essential when welding to thin films in order not to disrupt the film. Other disadvantages are:

1. One of the joining members must have a certain degree of ductility.
2. There is a definite upper limit to the thickness of materials which can be satisfactorily bonded with present state-of-the-art equipment.
3. Satisfactory butt welding techniques using ultrasonic bonding are difficult to obtain.

Some Mechanical Problems of Ultrasonic Bonding Machines

Low Frequency Motion and Bond Formation

Low-frequency differential-motion between the bonding tool and the bonding pad during ultrasonic bonding is a major cause of unreliable bonds. Such motion can originate from within the bonding machine due to the torque of its programming motor and cams, from various forms of operator movement, or from external motion such as building vibration. On assembly lines, often as many as 10 bonders are placed close together on a single long table. Vibrations originating from one machine or operator may interfere with bonds made on adjacent machines. Bonders are frequently installed on upper floors of buildings and/or close to heavy machinery or both. Vibrations can often be felt by merely touching a table or wall. This becomes meaningful when one realizes that any vibrations which can be sensed in this manner represent motions in the order of 1 mil (25 μm), and that wave cancellation and reinforcements can result in much of the vibration amplitude appearing as differential motion between the bonding tool and the device. This can seriously affect the quality of ultrasonic bonds made with small diameter wire (∿ 1 mil).

The conclusions to be drawn from the following work are: (1) Extraneous motion is the major cause of bond to bond deformation variations and a significant cause of unreliable bonds; (2) the operator should be cautioned about the detrimental effect that her own movements can have on bond formation; (3) each bonding machine should be placed on a separate table; (4) bonding assembly lines should be located in a vibration free environment, preferably on a ground-level cement-slab floor.

Capacitor-microphone displacement measurements revealed a lack of mechanical rigidity in some bonding machines (NBS Tech. Note 527, p. 39). This results in undesired movement of the bonding tool, held by the transducer, with respect to the transistor or integrated circuit, held by the work stage. Such movement can be transmitted to the machine in a number of ways, including motion of the bonding operator's arm as it rests on the machine, motion of the work stage as a result of loose mechanical tolerances due to poor design or wear, or operator movement if the work stage is rotated by hand. In the last case it has been observed that a slight twitch of the operator's hand can result in work stage movement on a typical machine of greater than 0.001 in. (25 μm). With respect to machine instability, the torque of the programming cam motor may produce detrimental vibrations. If any of these occur during the actual bonding period, a lift-off or low pull-strength bond may result.

Motion Perpendicular to the Wire

Experiments were conducted in which work-stage motion was intentionally introduced during the bonding period to characterize the effects of such motion. Sideway motion was introduced by mechanically driving the work stage with an electro-mechanical transducer at various frequencies and displacements. Bonds were observed visually and with the scanning electron microscope. As the degree of motion increased, the bond deformation began to vary widely from bond to bond and an increasingly greater number of lift-off bonds were produced. With motion in the order of 0.001 in. (25 μm) more than half of the bonds lifted off.

Examples of lift-off patterns for bonds made while the work stage had an intentional sideways motion of less than 0.0005 in. (13 μm) are shown in Fig.12-15.1. This amount of motion has been frequently observed in typical bonding machine installations on commercial production lines. While the effects of the motion are clearly apparent when the pattern is examined with a scanning electron microscope, they would be difficult to recognize from examination with an optical microscope.

Motion Parallel to the Wire

Motion was introduced parallel to the wire direction and ultrasonic tool motion at 20 Hz to simulate building vibrations and 60 Hz to simulate electrical equipment vibration.

The relative positions of the wire and tool during formation of the first bond are indicated in Fig.12-15.2. It is in this position that the normally sharp heel of the tool deforms the wire during bonding
shape. The greatest effect of the front-to-back motion is on the heel shape; increased motion results in more deformation and cracking. The extent of this depends on the radius of the tool heel. Three tools with different heel radii were used in the test. One had a sharp heel with no intentional radius; heel radii of the others were specified as 0.2 and

1.0 mil (5 and 13 μm). Motion amplitudes of approximately 0.2, 0.4 and one mil (5 μm, 10 μm, and 25 μm) were used at both the 20-Hz and 60-Hz driving frequencies. SEM photomicrographs of typical bonds made under these conditions are shown in Fig.12-15.3. A bond made with the sharp-heeled tool under ordinary conditions without motion is shown in Fig.12-15.3a. There is only a small crack in the heel; both other tools produced similar bonds under this condition. For 0.2 mil motion at 20 Hz, the crack in the heel was slightly enlarged, but the actual shape of the bond was not greatly different from the case of no motion. For 10-μm motion at 20 Hz, the sharp-heeled tool almost severed the wire from the bond as shown in Fig.12-15.3b. Increasing the motion to one mil often caused the wire to break. For the tool with a 0.2 mil heel radius less heel damage occurred. Although the heel was thinned, cracking was not severe. The deformation of the wire increased with motion, often by a factor of two in going from

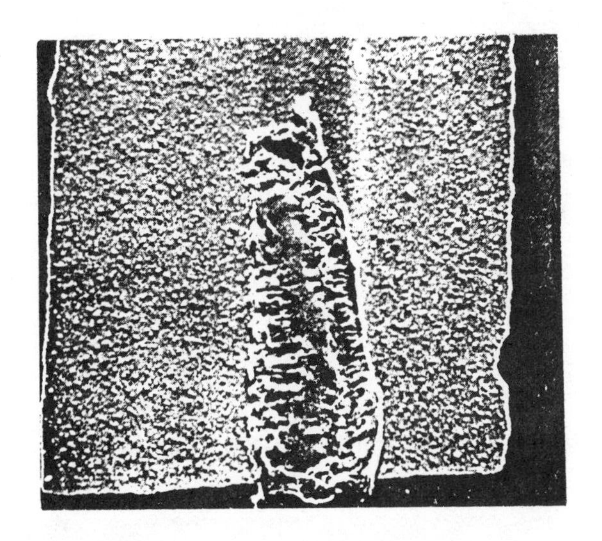

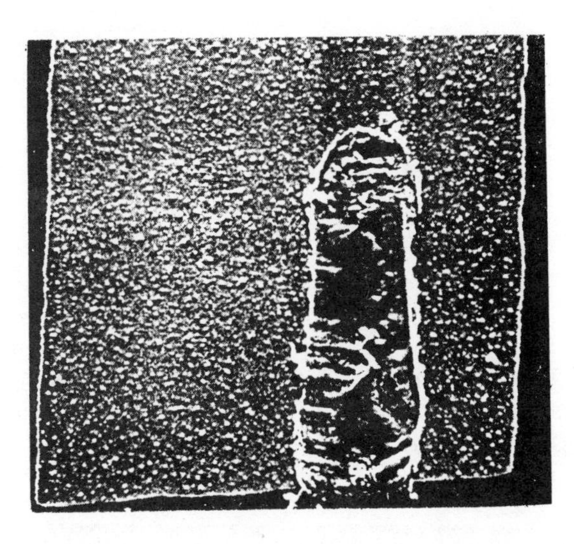

a: Magnification: 460 X. b: Magnification: 475 X.

Fig.12-15.1. SEM photomicrographs of lift-off patterns of bonds made with moderate (<0.0005 in.) sideways motion of the work stage during bonding. In both cases, some sideways tearing of the bond is evident. Additional evidence of motion in case a is the incomplete wire-to-metallization imprint on the left side of the pattern, and in case b, the smoothing of a portion of the normally rough weld surface.

no motion to one mil motion, but the wire was not cut. An example of a bond made with the 0.2 mil heel radius and one mil motion at 20 Hz is shown in Fig.12-15.3c.

Motion at 60Hz produced similar results but the deformation was increased. For example, 0.2 mil motion at 60 Hz caused the bond to have an appearance similar to that of one made with 0.4 mil motion at 20 Hz. An example of a bond made with the 0.5 mil heel radius and 0.4 mil motion at 60 Hz, shown in Fig.12-15.3d, illustrates the thinning at the heel that accompanies the increased

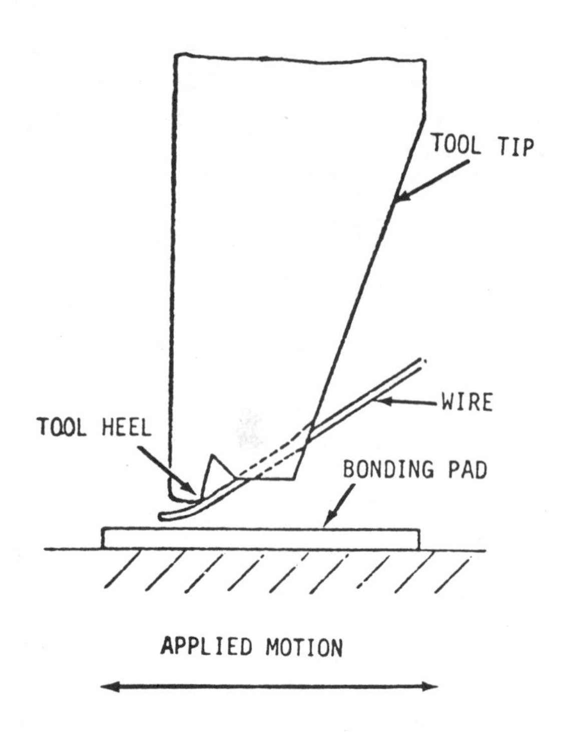

Fig.12-15.2. Relative positions of wire and bonding tool tip just prior to formation of the first bond.

deformation. As expected, typical bond shapes for bonds made with the tool with a 0.2 mil heel radius under the different motion conditions were between the two extremes just described.

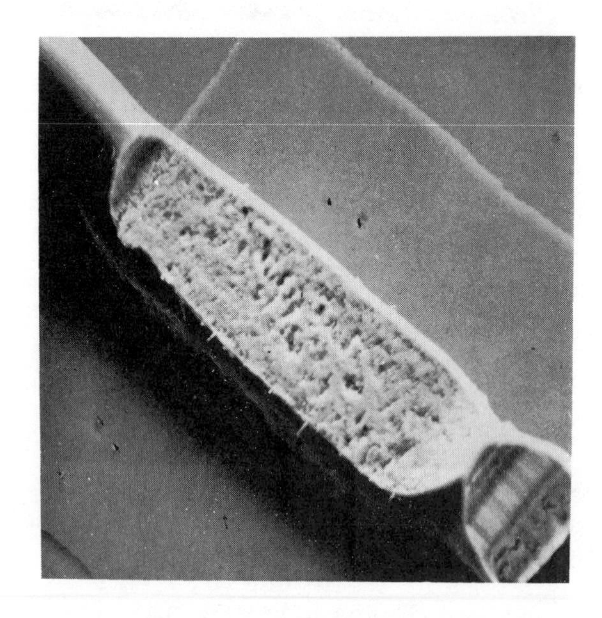

a: No motion; Magnification 650 x.

c: Amplitude: ∿ 25 μm; Frequency: 20 Hz; Magnification: 600 x.

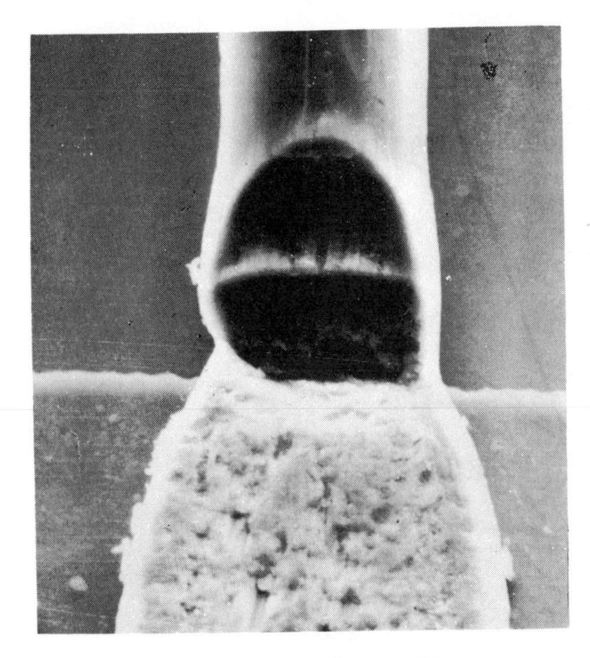

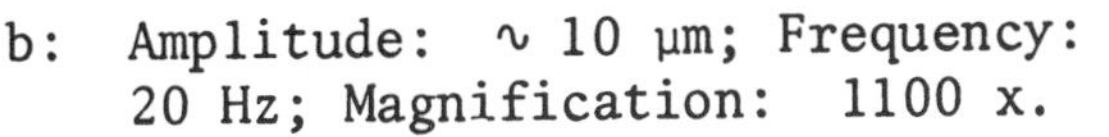

b: Amplitude: ∿ 10 μm; Frequency: 20 Hz; Magnification: 1100 x.

d: Amplitude: ∿ 10 μm; Frequency: 60 Hz; Magnification: 550 x.

Fig.12-15.3 SEM photomicrographs of typical first-bonds made with various relative front-to-back motion between bonding tool and work stage.

Two catastrophic failure modes due to parallel motion are illustrated in Fig.12-15.4. Actual severance of the wire due to cutting by the sharp heel is shown in Fig.12-15.4a. Lift-off which occurs when the motion has broken the weld at the wire-metallization interface is shown in Fig.12-15.4b. The type of failure mode depends on the portion of the bonding cycle in which the maximum motion occurs. Higher frequency motion tends to produce greater but more reproducible damage at lower displacements, since several cycles of motion may be integrated over the bonding period.

Self-Induced Motion in Bonding Machines

An electromagnetic displacement sensor(described in NBS Tech. Note 560, p.37) was used to seek the source of the extraneous self-induced motions which have been observed in bonding machines. In this application of the sensor, part of the detector is mounted on the machine housing and the other part is clamped to the work stage.

The most severe motions were found in the side-to-side and vertical directions. In the side-to-side direction, the upper part

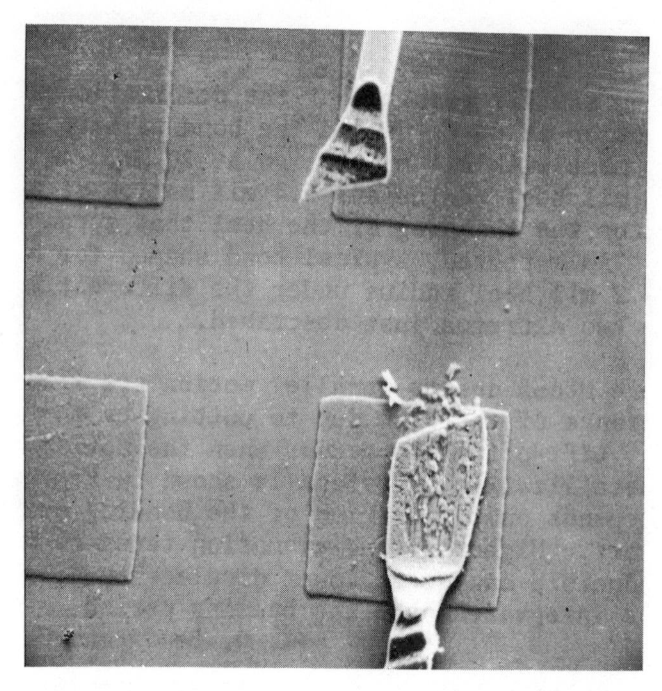

a. Severance of wire by heel of tool. Magnification: 200 X.

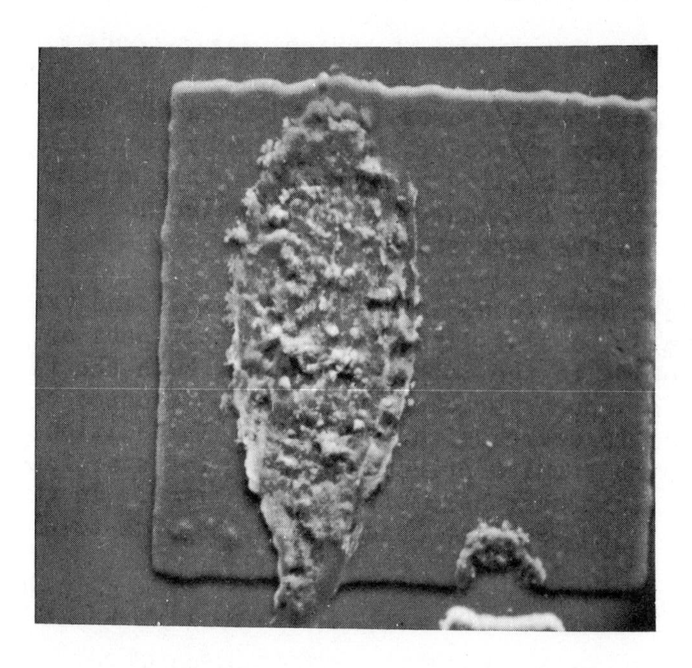

b. Lift-off at weakened weld. Magnification: 500 X.

Fig.12-15.4. SEM photomicrographs of catastrophic failure modes that result from front-to-back motion.

of the bonding machine moved in relation to the work stage during both the first and second bond cycles. A support bracket was made and mounted between the upper and lower bonder parts to decrease the movement. In addition to the movement, a vibration,

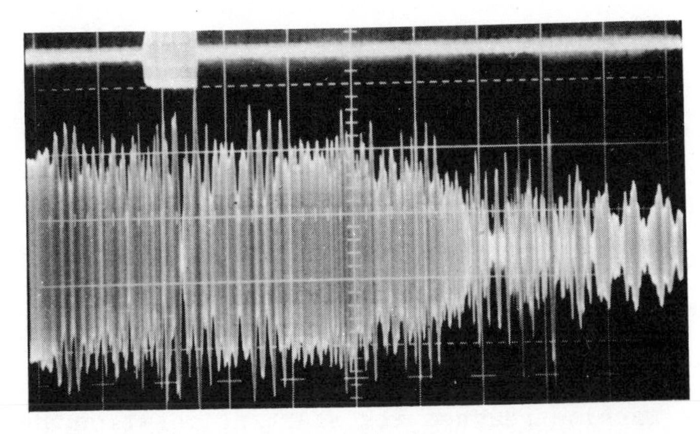

a: First search position through bonding and into loop position.

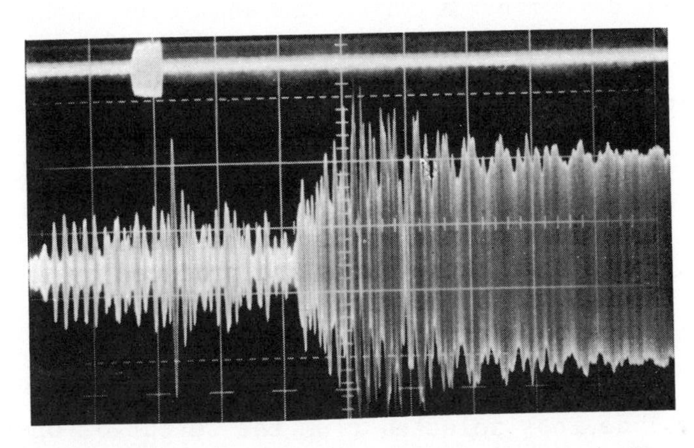

b: Second search position through bonding and wire cutoff to reset position.

Fig.12-15.5. Measurement of the self-induced vibrations of a bonding machine. The upper trace displays the ultrasonic bonding pulse. The lower trace gives the output of the displacement sensor. The vertical scale for the lower trace (displacement) is 2.5 μm/div. The horizontal scale for both traces is 100 ms/div.

which results principally from the starting torque of the stepping motor and camshaft assembly, was observed. In the vertical direction, no movement such as that observed in the side-to-side direction is seen.
A typical oscillogram of the motion between the machine housing, which supports the bonding tool, and the work stage with the motor and all the cams operating is shown in Fig.12-15.5. The indicated motion is in the side-to-side direction. The trace in Fig.12-15.5a covers the part of the bonding cycle from first search through bonding to loop formation. The upper trace displays the ultrasonic bonding pulse to indicate when the bonding occurs The lower curve, on the same time scale, shows the various self-

induced bonding machine vibrations. Maximum vibration amplitudes of about 0.5 mil (13μm) peak-to-peak that occur during and immediately after the bond is formed are indicated by fluctuations in the envelope amplitude.

The trace in Fig.12-15.5b covers the next part of the bonding cycle from second search, through second bonding and wire cutoff, to the start or reset position. A temporary displacement of the housing which occurred earlier in the cycle continued until wire cutoff, at which time the housing moved back to its original position accompanied by large vibrations with an initial amplitude approaching 0.5 mil. After this, the vibrations damped out as the machine reached its starting position and the cam motor stopped.

A vibration study was made of each individual cam, including the operating levers, in order to determine in which cam and lever combination the vibration originated. It was found that each cam and lever has its own vibration characteristics, but if all cams and levers are operating, some vibrations are amplified and others are cancelled out.

Wire Clamp Problems

Bonds from several devices selected from the output of operational production lines were examined with the scanning electron microscope (SEM).

The crack at the bond heel shown in Fig.12-15.6a is a feature of all first bonds made on these production lines. In addition, deep indentations were observed on the sides of the wire. It has been determined by additional SEM photographs that these marks are produced by the jaws of the clamping system through which the wire is fed from the spool to the bonding tool. In addition to marring the surface finish, these imprints could work-harden the wire and thus affect bonding conditions. Occasionally the wire twists after passing through the clamping system, and the indented portion then becomes a part of the actual bond interface as shown in Fig.12-15.6b. Of particular concern is the possibility of contamination from the clamp becoming embedded in the wire and directly interfering with the bonding process. Since the twists, such as those depicted in Fig.12-15.6b, occur only occasionaly and since not all of these need be significantly contaminated, the possibility should be considered that this may be a cause of occasional, otherwise unexplained weak bonds. One solution to this problem is to use large, 2 inch diameter, wire spools instead of the more usual 1/2 inch spools. Wire from the large spools shows little tendency to twist. Also the wire clamps should be polished and the clamping force should be minimized.

a: Normal clamp mark configuration.

b: Wire has twisted after being clamped and before bonding so that the clamp marks are incorporated in the weld interfacial region.

Fig.12-15.6 SEM photomicrographs (575X) of aluminum ultrasonic wire bonds showing heel crack and clamp marks. The clamp marks appear on both sides of the wire.

12.9 CONDUCTIVE ADHESIVES

Several conductive adhesives are available commercially. These are suspensions of a metal powder in a thermosetting plastic. The metal powders used are silver, gold, copper, and aluminum, while the adhesive is generally an epoxy-type resin. These adhesives have some disadvantages in use: the work must be clamped until the adhesive sets up; the setup time is long unless high temperatures are used, and even then it is long enough to be troublesome; joint resistances can be high enough (occasionally 5 ohms) to cause problems, and they are also variable on temperature cycling; and permissible operating temperatures are limited.

12.10 MULTILAYER INTERCONNECTIONS

One interconnection method of this class is based on plated-through holes for electrically connecting the layers of multilayer boards. Holes are etched or drilled through the laminated assembly and the interior of the holes plated to join the conductors on different layers. Reliability problems associated with the plated joints, at the narrow edges of the internal layers, can be minimized through etching and abrasion to enlarge the joint area.

In the method often referred to as the prefabrication method, buried layers are used for interconnection patterns that are normally fixed, such as ground, supply voltage, and sometimes clock and reset signal wiring. These are made in advance as core boards and stocked. The variable signal wiring is placed on external surfaces, or on layers readily added to the core boards.

In another concept, the additional signal layers are etched on thin, insulating stock. Etched fingers, which hang over the edge of the stock, are soldered to mating pads on a signal-voltage core.

Multilayer interconnections are particularly useful with high-density packaging. Multilayer boards are used (1) to save weight and space in interconnecting circuit modules, (2) to eliminate costly, complicated wiring harnesses, (3) to provide shielding of a large number of conductors, (4) to provide uniformity in conductor impedance in high-speed switching systems, and (5) to provide greater wiring density on boards.

Figure 12-16 illustrates the various individual boards that are mated to form a multilayer unit. The three proven methods for interconnecting the circuitry from layer to layer are described below:

12.10.1 CLEARANCE-HOLE METHOD

In the clearance-hole process, a hole is drilled in the copper island (terminating end) of the appropriate conductor on the top layer. This provides access to a conductor on the second layer. See Figure 12-17, hole A. The clearance hole is filled with solder, and the desired connection is completed. Usually the hole is extended through the entire assembly at the connection site. This small hole is necessary for the solder-flow process normally used with this interconnection method.

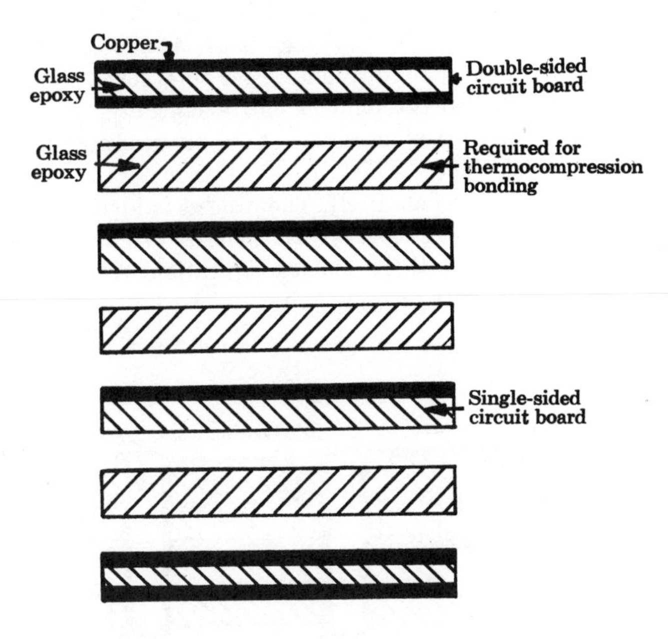

Fig. 12-16 Multilayer board construction.

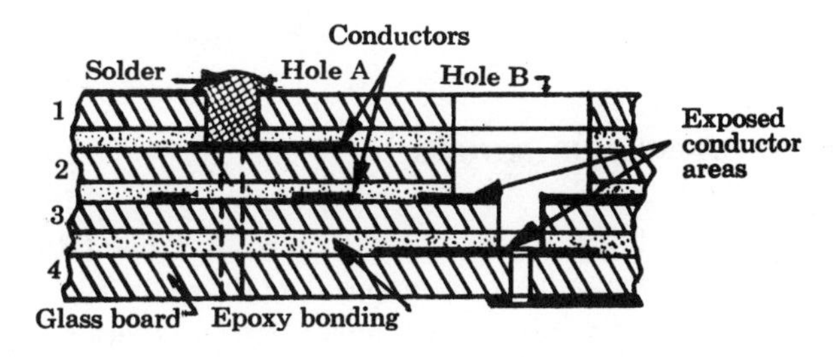

Fig. 12-17 Clearance-hole interconnection technique.

To interconnect conductors located several layers below the top, a stepped-down-hole process is employed as shown in Figure 12-17, hole B. Before assembly, a clearance hole is provided down to the first layer to be interconnected (in Figure 12-17, through layers 1 and 2 to layer 3). The first layer to be interconnected (3) is predrilled with a smaller hole than that of layers 1 and 2. Succeeding layers to be connected have progressively smaller clearance holes. After assembly, a portion of the island areas is exposed on the conductors that are to be interconnected. The stepped-down hole is then filled with solder, completing the connection. The larger the number of interconnections required at one point, the larger must be the diameter of the clearance hole on the top layer. Large clearance holes on the top layer, however, result in less space for components and therefore reduce packaging density.

12.10.2 PLATED-THROUGH-HOLE METHOD

The plated-through-hole method of interconnecting conductors is illustrated in Figure 12-18. The first step is to assemble, temporarily, all the layers into their final configurations. After holes corresponding to required connections are drilled through the entire assembly, the unit is disassembled. The internal walls of those holes, to be interconnected, are plated with metal (0.001 inch thick), which, in effect, transfers the conductor on the board surface into the hole itself. The process is identical to that used for standard printed-wiring boards. The boards are then reassembled and permanently fixed into their final configuration. Afterward all the holes are plated through with metal.

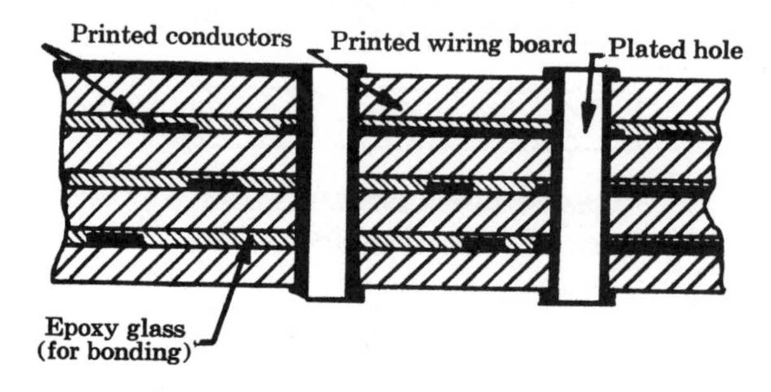

Fig. 12-18 Plated-through technique.

12.10.3 LAYER-BUILD-UP METHOD

In the layer-build-up method, conductors and insulation layers are alternately deposited on a backing material as shown in Figure 12-19. This method yields all-copper interconnections between layers and therefore is more reliable than the preceding two techniques.

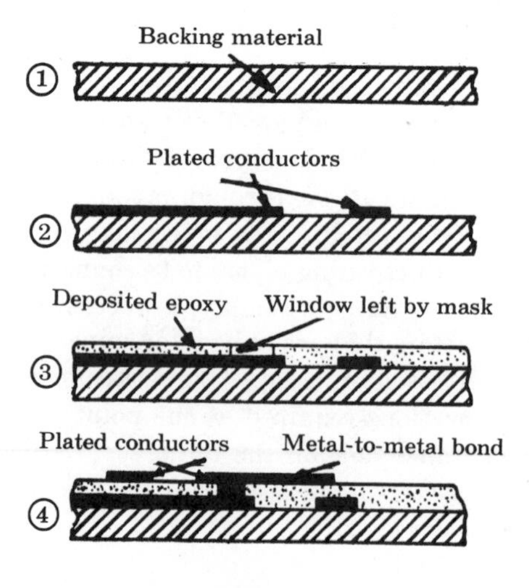

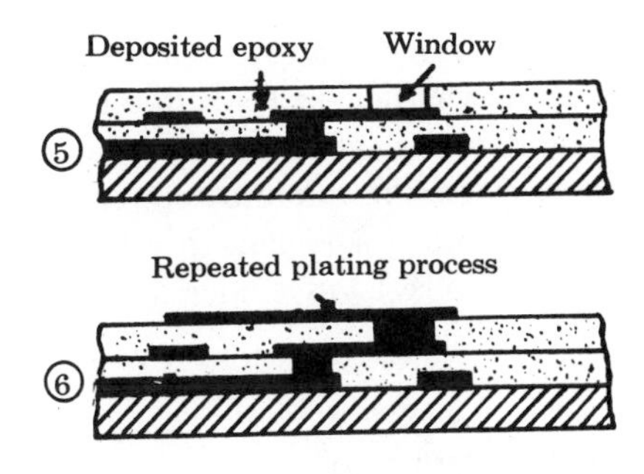

Fig. 12-19 Layer build-up technique.

12.10.4 COMPARISON OF METHODS

Each multilayer interconnection technique has the following advantages and limitations:

Cost:—For small quantities, the clearance-hole method is the cheapest, since it requires a minimum of equipment and engineering capability. However, the cost of applying this method goes up drastically for larger quantities — even though fabrication is simple, the operations are quite time-consuming. The plated-through-hole method is usually moderate in cost for both small and large quantities. Built-up layers are generally more expensive than the other two types, regardless of the quantities involved. However, various interconnection designs can have an effect on the multilayer-board price. For example, in the built-up layers, interlayer connections are made without bringing the connection to the surface (as is necessary in the other two methods), allowing greater flexibility in wiring layout. Thus, fewer layers are needed for the complete board, and a price lower than that for the plated-through-hole method may result.

Density:—The build-up method yields the highest packaging density, since its internal interconnections do not require corresponding holes in other layers. The clearance-hole technique offers the lowest packaging density of the three methods because of the oversized holes required.

Reliability:—As previously mentioned, the built-up layers form the most reliable interconnections, because of their solid-copper makeup. The clearance-hole method is the next most reliable, whereas the plated-through technique is the least reliable. In the latter case, uniform metalization has proven difficult to achieve.

Maintainability:—In general, interconnections made by the clearance-hole method are the least difficult to rework. Plated-through connections rank second, and built-up layers are third. In the built-up layers, it is usually difficult — if not impossible — to rework any of the internal connections.

12.10.5 PACKAGE TIE-DOWN TECHNIQUES

For connecting modules or components to multilayer boards, the conventional method is to solder their leads or pins into the holes of the boards.

Other methods of component attachment which have been tried successfully include resistance welding of component leads to tabs, pins, or eyelets protruding over the surface of the boards; and parallel-gap welding to tabs made of nickel, Kovar, special foils, or plain copper on the board surface. The common hand, dip, and wave soldering processes have been supplemented with techniques such as resistance microsoldering, controlled-heat-zone soldering, and hot-air soldering.

12.11 POINT-TO-POINT WIRING

Multilayer boards are difficult to use for Level II* wiring, in engineering models or in systems produced in small quantities. Point-to-point wiring, on the other hand, may be changed more readily. The type of wiring usually depends on the wire terminating methods which include those called wire wrap and poke-home connectors.

In the wire-wrap method, wire may be automatically wrapped on terminal pins by machine. Square pins 25 mils across flats and spaced 100 mils apart, may be wrapped by this method. The wiring routes can be programmed during design, and the wiring performed automatically with the wire-wrap machines. In some high-volume production systems, variable signal wires are wrapped to multilayer boards.

Poke-home connectors are miniature connectors whose pins lock into place when they are inserted in the connector body. The wire can be staked to the pin in advance, and two wires can be staked into each pin, so that relatively long wiring assemblies can be tied together in advance.

In another method, a small spring clip forces the lead wire into intimate contact with a post. The properties of the joint are similar to those realized with the wire-wrap method. An advantage of this method is that a clip at the bottom of the post can be snapped off and a new one added at the top of the post, forcing the higher pins down the post. Furthermore, the wire can be stranded, and the terminations can be made quickly. Changes in engineering models can be accommodated only by leaving a loop of wire at each post.

12.12 DIRECT INTERCONNECTIONS OF UNCASED CHIPS INTO A BEAM-LEADED MATRIX

A fabrication technique for direct interconnection of uncased integrated circuit chips in substantial arrays has been investigated. The technology consists of additive and subtractive chemical processes using fine-detail photo resist techniques for control of form factors. Considerable quantities of interconnections can be fashioned using a single piece of double-clad mylar laminate. The only physical terminations which need be made are to exit electrodes in the package enclosure and to the aluminum film electrodes on the planar surface of the integrated circuit chips. These connections are

*(Level I is a module of interconnected circuits, Level II a subsystem of interconnected Level I modules, and Level III a system of Level II modules.)

generally accomplished using ultrasonics and face-up bonding. Figures 12-20a and 12-20b illustrate the beam-lead matrix. Some advantages of this type of hybrid microassembly are:

1. Each I.C. chip is mechanically and thermally anchored to a ceramic substrate by means of the large-area base of the chip. Superior environmental performance can be achieved compared to "flip-chip" hybrid arrays where the mechanical support and thermal flow is by means of metal bumps.

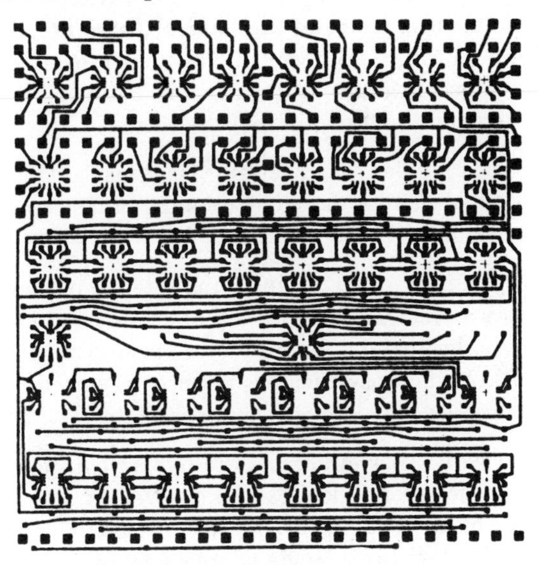

a) Top surface conductors.

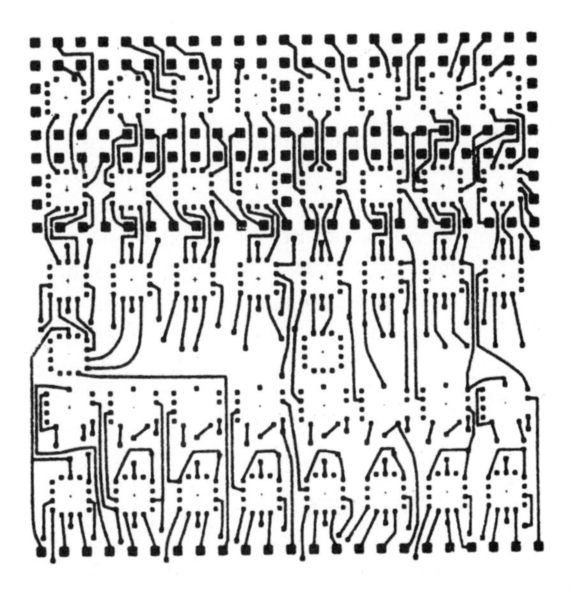

b) Bottom surface conductors.

Fig. 12-20 Beam lead matrix.

2. The quantity of man-made connections is reduced to a minimum consistent with hybrid construction. The quantity of bonds is the same as found in flip-chip construction and is approximately *one-half* that of a comparable hybrid assembly using aluminum or gold jumper wires. The quantity of man-made bonds is approximately *one-third* that found in a comparable microassembly using individually cased chips.
3. Thin, ribbon-like beam leads are formed integral to the laminate part which provide a measure of mechanical flexibility during thermal expansion and contraction cycles. The ability to withstand thermal shock is judged intermediate, part way between the wire jumper construction and the flip-chip construction.
4. As in the wire jumper construction, visual inspection can be made of each individual bond and malfunctioning chips can be replaced at time of assembly.
5. Where the quantity of I.C. chips being interconnected in the hybrid array causes a considerable quantity of crossovers in the topology of conductor pathways, the separate beam-lead matrix part permits a density of hybrid chips which has not been reported for multi-levels of conductors deposited on a hybrid substrate.
6. In common with other approaches to hybrid chip microassembly, a wide area of design freedom can be enjoyed by the engineer, since any mixture of uncased planar silicon chips available in the industry can be combined within a single microassembly array. This includes those chip devices which exhibit high impedance (F.E.T.), those which provide serial storage (M.O.S. shift registers), and those which offer high breakdown voltages, high cut-off frequencies, or high current-carrying capability.

12.13 ELECTRONIC JOINTS

Interconnections based on electronic phenomena in contrast to mechanical structures are being investigated. In one method, optoelectronic coupling is applied to link a photogenerating component, in one circuit, with a photoreceptor in an adjacent, electrically isolated circuit. Light pipes are used, in this method, to substitute for multiconductor cables and shielded transmission lines. Another method resides on the use of controlled forms of capacitive and inductive couplings. These types of couplings are usually the cause of the ''cross-talk'' which is difficult to eliminate in high-density wiring structures.

12.14 NONDESTRUCTIVE TESTS

Once a reliable joining process is developed, adequate nondestructive testing techniques are required to insure the highest level reliability. Where the joint is noninspectable (thermocompression, ultrasonic and micro-gap bonding) by visual means, a more elaborate testing technique is required. Ohmic contact measurements, and unique X-ray techniques are the most revealing methods presently used. Micro-electronic devices and assemblies are mounted in hermetic packages which are opaque. Radiographs of these devices will in some instances reveal internal discontinuities which affect the reliability. Many of the devices contain gold wires for intraconnects which is a good absorber of X-radiation and therefore reveals a definitive radiograph.

A recent development of an X-ray imaging system using television for instantaneous display of the radiographed part, shows promise as a non-destructive method of evaluating miniature bonds. A high resolution scanning electron beam appears to be also an effective means for nondestructive inspection of semiconductor and thin-film devices.

12.15 BOND FAILURE MODES

To obtain satisfactory results, each bonding process, which may be used, must be carefully controlled. Failure of a bond may stem from a number of different sources. The main sources of failure are as follows:

a) *Open Nail Head (or Ball) bonds Using Gold Wires Making Contact to Aluminum Pads.* This open bond type of failure was, in some instances, caused by the gold-aluminum eutectic formation known as "purple" plague". For quite some time, it was believed that the "purple plague", a formation of $AuAl_2$ in gold-aluminum bonds, resulted from high temperature (300°C) cycling after bonding, and was brittle and contributed toward bond failure. However, further investigation showed that the "purple plague" formation is not brittle and, indeed, was a good electrical conductor. It was proved that failure and brittleness of the bond was caused by a white powder, Au_2Al, which formed under the purple layer, and that this "white plague" occurred only when silicon, aluminum, and gold were present together in specific proportions. To avoid the problems with "plagues", aluminum wire is often substituted for the gold. When aluminum wire is used, it is usually wedge or stitch bonded to the chips, often with the use of ultrasonic agitation of the bonding tip.

Another cause of this type of failure is the poor adhesion of the aluminum film to the silicon dioxide. This failure is always triggered by mechanical stresses. This type of failure mode is a production process problem and can be cured by more careful process control.

Insufficient pressure or temperature (underbonding) is another cause of the ball-bond failures. This is usually caused by lack of adequate bond between the gold and the aluminum, so that no real metallurgical bond is formed between the gold and the aluminum. This type of failure is again the result of poor process control.

Overbonding (too much heat or pressure, or both) may also cause bond failures, usually resulting in the pulling up of the aluminum and silicon dioxide.

b) *Opens in the aluminum interconnects* are normally caused by the formation of hydrated alumina at the dissimilar metal contacts in the presence of excessive moisture. This failure mode, which can form at room temperature, may be accelerated by baking.

It should be noted that opens in the aluminum interconnects can also be caused by electrical overheating resulting from the reduction of the cross section of the interconnect, by scratches or by other defects from the deposition of the aluminum film and subsequent etching.

c) *Faulty oxide removal* at the windows may cause the build-up of a dielectric layer between the silicon and the aluminum film.

d) *Bulk shorts* due to secondary breakdowns are failure modes which can usually be eliminated through proper specification of the electrical characteristics.

e) *Aluminum interconnect* shorts to the silicon through the silicon dioxide. This is a voltage-dependent failure mode, and is attributed to the poor dielectric strength of the silicon dioxide. Pin holes, entrapped impurities, etc., cause weak SiO_2 dielectric.

f) *Poor chip layout.* Shorts may also occur when the bonds are too close to the oxide edge because of poor chip layout.

g) *Improper surface preparation* can produce surface problems which result in the long term changes in circuit characteristics such as the degradation of gain, and leakage currents under reverse bias.

h) *Internal lead wires.* The method of making contact from the aluminum pads on the chip surface to the posts of the package may give rise to shorts. Such shorts are always process problems which can be remedied by better chip and package layout, or by more careful process controls.

i) *Faulty handling, processing and bonding* can cause open breaks to occur in lead wires. Nicks and cuts can give rise to such breaks during mechanical stressing.

j) *Loose material and particles* within a package are always a potential cause of failures. The presence of free conducting material (extra length of leads which have been cut off, etc.) are a troublesome cause of failures. Non-conducting particles can cause mechanical damage during vibration and shock testing.

12.16 LARGE SCALE INTEGRATION

A major advantage of LSI is the reduction in interconnections. The total number of connections required for a system may be reasonably constant, regardless of how the system may be fabricated. The type of connection, however, and its frequency of occurrence can vary drastically from one approach to another. LSI substitutes intraconnections for interconnections, i.e., metalization within the package for soldered or welded connections external to the package.

The reduction in external leads afforded by the use of LSI devices instead of equivalent numbers of IC's, is difficult to predict because of the effect system partitioning has on external-pin requirements. It is only possible to indicate that the pin requirement for an LSI device is considerably lower.

One investigation has shown that the external-pin requirement for relatively simple circuits, depends on where the circuits are partitioned. The minimum pin requirement appears at a function level or a multiple of it. This is illustrated in Figure 12-21.

Figure 12-22 illustrates the same type of comparison as that of Figure 12-21 except on a higher function level. The minimum point represents some functional level.

The pin requirements are different for the two types of logic having the same circuit complexity. Control circuits will typically have a greater requirement for pins than data-processing circuits, because the control circuits usually require intermodule connections.

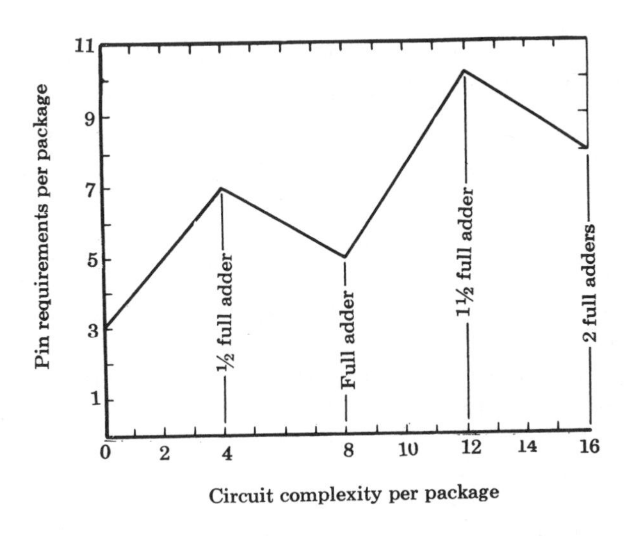

Fig. 12-21 Relationship of pin requirements to circuit complexity.

12.17 INTRACONNECTIONS

Intraconnections are metalizations which connect various components on the die. They may also be considered as metalizations within a package

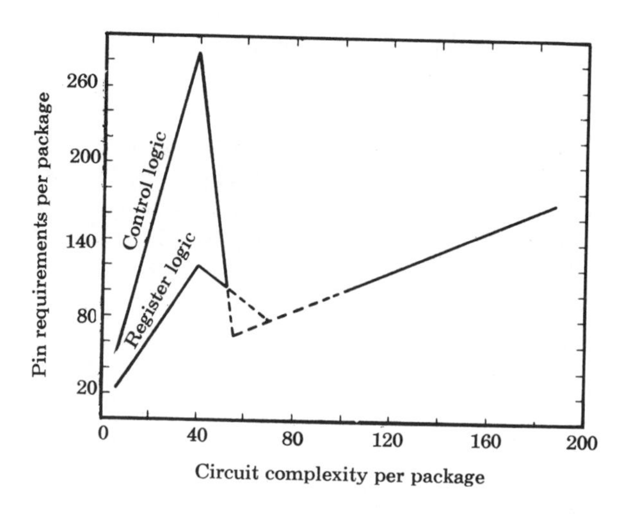

Fig. 12-22 Relationship of pin requirements to circuit complexity.

connecting various flip-chip dice. The difficulties associated with interconnections are also associated with intraconnections, primarily in multilevel metalizations required for high-density packaging.

LSI devices require a multilevel intraconnection system. The problems associated with intraconnecting the gates on a wafer are topological and electrical. Crossovers for bias and signal-voltage metalizations are required. The resistance of the metalizations and contact resistance at the crossover-oxide steps must be low. A multilevel crossover system is composed of the following:

a) The first level of metalization intraconnects the components that form each circuit. These circuits form a matrix array on the wafer.

b) A layer of insulation separates the first layer of metalization from the next layer. Holes are etched through the insulation to the first level of metalization.

c) A second level of metalization is deposited on the insulating layer and forms the second level of intraconnections. Connection is made between the two levels via the holes in the insulation layer. Additional layers of metalization may be fabricated by repeating the insulation and metalization steps. A typical two-layer metalization system is shown in Figure 12-23.

The intraconnection of a large number of gates on a wafer requires (1) a method of intraconnecting only good gates, and (2) multiple levels intraconnections. There are two basic approaches used for intraconnecting the good gates: fixed wiring and discretionary wiring.

The fixed-wiring approach uses a single set of metalization masks for identical LSI devices. Each circuit element in the wafer is tested before metalization. The bad circuits are automatically marked, and the wafer is examined to determine which portions of the wafer meet the topological requirement for the mask set. These are then separated from the wafer and metalized. The good portions of the wafer that cannot be used for the LSI device, are diced and packaged as individual IC's. This method provides low metalization costs, but is highly dependent upon the wafer yield. Many device manufacturers believe this is the best approach to the metalization problem.

The discretionary-wiring approach is not as sensitive to wafer yield,

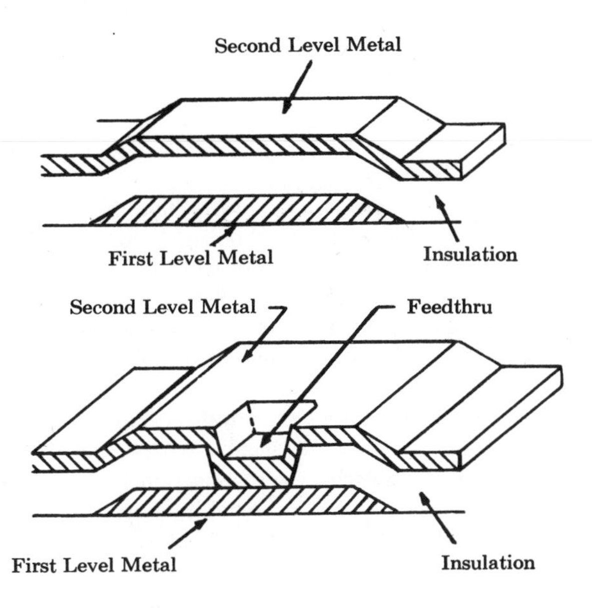

Fig. 12-23 Two-level lead crossover.

but is highly dependent upon costly computing equipment. In this process, the circuit elements on the wafer are tested, and the results are stored in a computer. The computer is programmed to provide an intraconnection pattern for a particular array. A mask is made by an automatic-exposure process under control of the computer.

This system uses a flying-spot scanner whose image is projected onto a high-resolution CRT. The CRT image is used to expose a high-resolution plate, to be used as the mask. The scanning routine starts with the shortest paths and proceeds to the longest. It may be necessary to perform the routine more than once.

The result is that each wafer has its own custom intraconnection pattern of good circuits. The bad circuits remain on the wafer, but are not intraconnected.

Either of these methods can be modified at the matrix-element level. The matrix elements are intraconnected by the first metalization, which can vary within certain constraints to provide different circuit arrangements within each matrix element.

Typical two-level metalizations are composed of Al-SiO_2-Al. A serious problem affecting this technique, is the incidence of pinholes that reduce the device yield. Pinholes are caused by perturbation or discontinuities in the metal or oxide. The result is a short or near-short through the oxide, which causes the device to fail.

Another important phenomenon, which manifests itself by frequent short (or near-short) circuits, is the inherent mechanical weakness of the Al-SiO_2-Al system. During testing, it is necessary to prove the wafer, but a light touch of the steel point will often cause a short. This situation is not peculiar to LSI, but may be complicated by the multiple layers of metalization.

A complex array may contain as many as 5000 crossovers. The failure of a single crossover may cause the entire array to fail. The cost to process a wafer to this point is high. Accordingly, the frequency of failure must be extremely low if reasonable prices for LSI devices are to be obtained.

The biggest user misconception about sockets is that they're all alike and will perform equally well. They aren't and they won't. Every socket is a compromise reflecting the designer's estimated importance of each tradeoff factor. Contacts must accept a variety of leads, receptacles must withstand heat and solvents, wrap pins must stay in (even when struck by a wrapping bit), physical size must be minimized, and the socket must sell at a competitive price.

The socket is built around its contacts, which account for more than half the manufacturing cost. There are two types, solder tail and wrap. Solder tail contacts terminate in a short tab and are made for soldering onto a p-c board. They need only be flat, solderable leads that fit through the holes in the p-c board. Some solder tails are round or channeled which makes them firmer and easier to solder. Many users, however, prefer the flat tabs, because they can be bent over before soldering to keep the socket in place during handling. Wrap contacts end in a longer "tail." They're usually 0.025" square in cross section and are for interconnection by point-to-point wiring techniques. For a reliable wrap connection, the corner radii of the tail should not exceed 0.003". Burrs or rounded edges on

Fig. 12-25.

All 14 contacts are loaded into the socket body from the top and locked in place by the snap-in cap, producing a rugged, yet repairable IC socket.

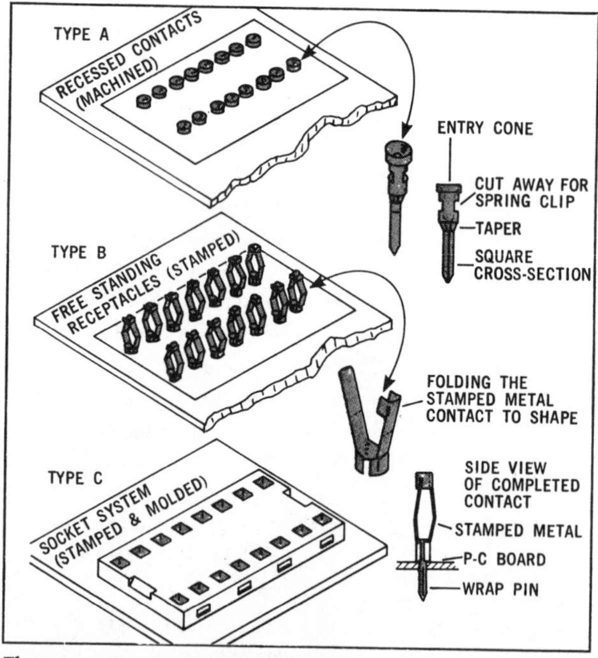

There are many ways to provide plug-in connection for an IC. Type A gives the lowest profile but its machined parts cost more than the stamped ones of type B. When the number of socketed items is low, thus not justifying the cost of labor saving automatic insertion tooling required by types A & B, then the complete assembly of pins in an insulated housing is the answer, type C.

the tail will cause a poor wrap. That's why the channel section isn't suitable for wrapping; it closes under pressure. Many users who try to stay with soldered channel tails, will occasionally try to wrap them, but contact reliability isn't high.

The considerations that go into the choice of each contact type are similar. The contact is evaluated by its configuration, material and finish. Designing a contact configuration that will mate reliably with a variety of device leads is a big problem in socket design. It would be convenient if the IC manufacturers would standardize on a single lead configuration with reasonable tolerances, but they haven't. JEDEC TO-116 is about the only controlling standard on lead sizes. It allows a wide ratio of maximum to minimum dimensions — an unreasonable situation. And not all leads fit this broadly drawn standard.

Although a well designed socket will accept nearly all lead shapes, the contact deflection range required is often excessive. Smaller, less expensive sockets would become feasible if standardized IC leads could be used.

One of the most successful solutions to the lead variation problem is the configuration shown in Fig. 12-26. Its two spring members grip the narrow faces of a TO-116 lead. By having them span the full lead width, a more favorable deflection range is obtained. This also produces higher contact pressures than can be obtained by allowing the contact to ride the wide lead faces. High contact pressure is desirable for reliability, but it increases abrasion during insertion. This is seldom a problem when only a few insertions are anticipated, but it is important when products must undergo many insertions, such as in test sockets. They're designed to contact the wide lead faces (instead of the narrow ones) and minimize abrasion.

Other configurations often use spring members that contact the wide faces of the TO-116 leads. They have either dual spring members or a single member in a bellows configuration. There are many variations of these general types.

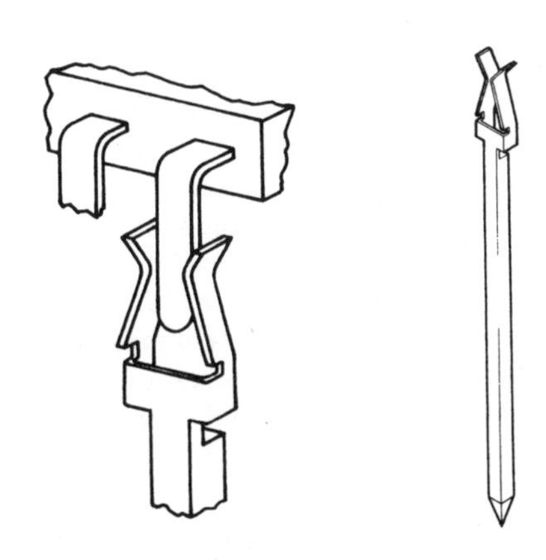

Fig. 12-26

Contact pins that grip the narrow edge exert high contact pressure without requiring high insertion force.

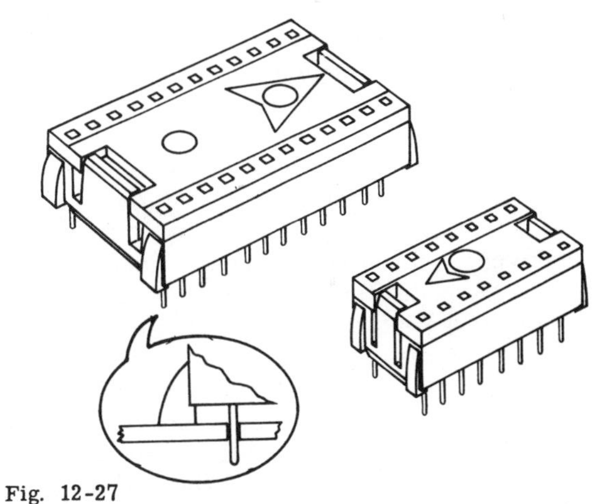

Fig. 12-27

In some sockets, solder relief, the space between socket and board that is used for cleaning out residues, is obtained by molding corner posts into the body.

BACKPANELS

The backpanel—an interconnection board into which logic cards or other cards can be plugged—provides a common electrical ground, all system voltage requirements, signal paths, input/output terminals and generally serves as a structural member of the packaging cabinetry. It also lends itself to automatic or semiautomatic wire wrapping.

Backpanels can take many forms, the simplest being discrete connectors mounted on rails or a metal frame. The majority of today's backpanels, however, are two sided or multilayer p-c boards, or metal panels. In either case, the panel provides a supporting structure for the p-c card edge connectors and the wire wrap pins.

With metal backplanes, power and ground planes are provided by stacking two aluminum plates together and separating them with thin sheets of an insulating dielectric material such as vinyl. One plate acts as the ground while the other acts as a voltage plane, in effect, creating a capacitor that filters out noise and smooths out momentary power

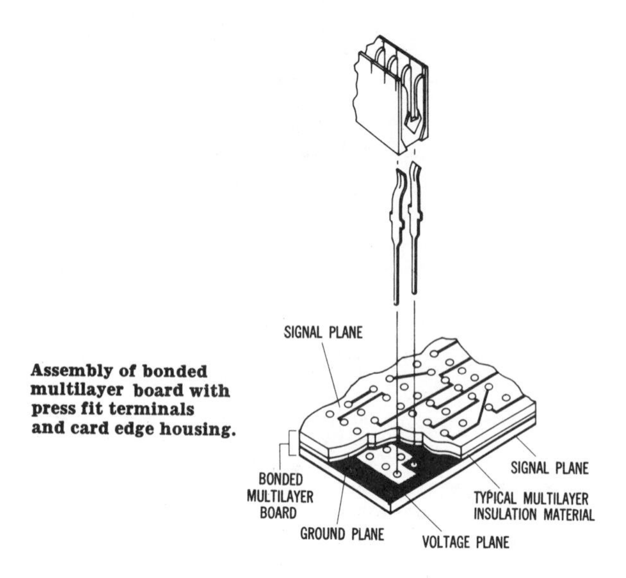

Assembly of bonded multilayer board with press fit terminals and card edge housing.

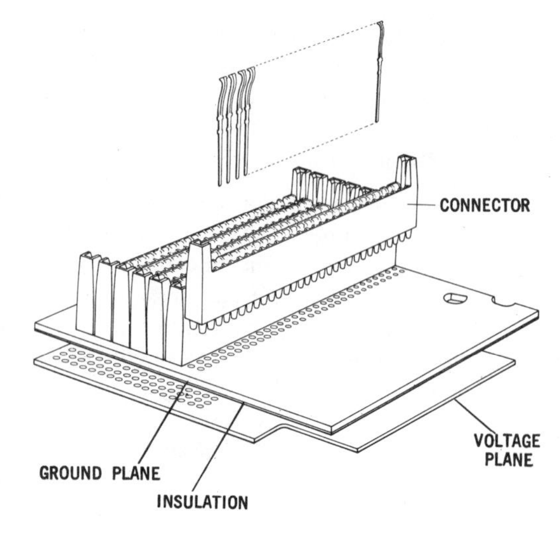

Exploded view of metal plate backpanel.

fluctuations. The holes are either punched or drilled, and the pins are pressed into nylon insulating bushings. Metal voltage/ground plane assemblies provide excellent structural rigidity and have high current carrying capacity.

Two sided p-c boards normally have power and ground planes on opposite sides, but they can be located adjacent to each other using wide traces. Wire wrapped pins are either press-fit or flow soldered into the plated through holes.

Multilayer p-c backpanels can reduce total packaging costs significantly. They allow the use of printed wiring for most of the circuitry and busing, thereby substantially reducing the number of wire wrap connections required. And the use of solderless or wire wrap pins in plated through holes retains the main advantage of wire wrapping — the ability to make circuit changes after the equipment is built. Multilayer boards use copper clad glass epoxy boards on which circuits are etched. Up to eight layers of voltage, ground and signal planes can be incorporated. Controlling impedance and crosstalk requires careful control of the distance between signal layers and power or ground planes. However, crosstalk can be minimized by shielding or by locating signal paths opposite the power levels.

The use of wire wrapped pins on backpanels stemmed from Bell Laboratories' efforts to improve the mechanical connection between the wire and terminal before soldering. The technique developed involves wrapping a wire tightly around a square post in a helical configuration. The sharp corners of the post bite into the wire to produce a "gas tight" corrosion-free connection that is so reliable, subsequent soldering is totally unnecessary. Because of solid state diffusion, the mechanical strength and reliability of the connection actually improves with time. With a life expectancy of over 40 years, wire wrap connections have no known failure rate.

CHAPTER 13

PACKAGING

To package microelectronic circuitry so that it will be useful and function properly under various environmental conditions, it is essential to select the correct packaging materials and techniques. The properly designed package protects the microelectronic circuit from environmental effects which tend to interfere with the reliable operation of the circuit. Depending on the intended use of the circuit, the packaging requirements may be relatively simple if, for example, the circuit is to be confined within a stationary household appliance. An extremely complex design may result, on the other hand, if the circuit is intended for space applications where it may be subjected to a variety of severe environmental conditions.

One of the major considerations in applying integrated circuits to a particular system, is selecting the proper package. In this regard, certain factors merit special attention by both the manufacturer and the user.

The device (integrated circuit plus its package) manufacturer has primary responsibility for the reliability of the IC package, whether he fabricates the package or purchases it. The device manufacturer, then, must consider such characteristics as mechanical design, hermetic seal, lead stresses, etc.

The user of the device is concerned with package reliability, space efficiency, and design flexibility.

General requirements of an IC package fall under the following headings:

a) Electrical
b) Thermal
c) Chemical
d) Mechanical
e) Radiation
f) Magnetic
g) Hermetic
h) Space efficiency
i) Design flexibility

13.1 FUNCTIONAL REQUIREMENTS FOR PACKAGING DESIGN

a) Electrical:—The package must carry current between external elements and specific regions of the interior, with suitable electrical isolation between these current-carrying members. The isolation may have to provide protection against inductive, capacitive, as well as resistive coupling. Electrical requirements of the IC package are as follows:

The number of leads projecting from the package, must be sufficient to accommodate a fairly complex circuit chip. Unused leads can be used for package tie-down or can be clipped off. In many applications, the package must be capable of providing electrical shielding. In that case, the package may have to be made of metal, with one of its leads electrically grounded to the case.

b) Thermal:—The package must provide a path for the flow of heat away from the device inside the package. One result of the operation of an electrical circuit is power dissipation which, in turn, can result in a build up of heat within the package. Without a good thermal path for heat dissipation, an active device within the package may rapidly exceed its maximum allowable operating temperature, and become inoperable. In some cases, if the power dissipation within the package exceeds the ability of the package to dissipate heat, more cooling area, such as cooling fins or a stud mounted heat sink, must be employed.

Of the three possible methods of heat transfer, shown in Figure 13-1, conduction can generally transfer the most heat. Accordingly, the IC package should be constructed of materials that have high thermal conductivity.

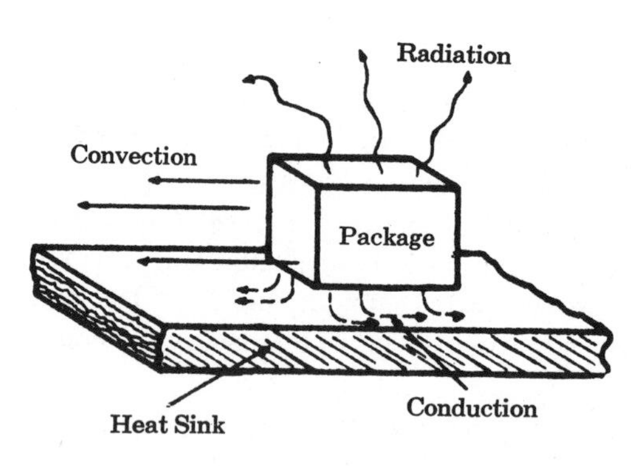

Fig. 13-1 Heat-transfer modes of an IC package.

c) Chemical:—The package must be resistant to changes due to reactions with external reactive agents in its operating environment. Under normal conditions, these external agents may be oxygen, moisture, etc. However, under exceptional conditions, the package may be required to operate in almost any conceivable environment. Electrochemical effects, such as electrolytic corrosion, must also be considered. In the presence of these external agents, the package must be capable of providing an interior environment compatible with device performance and reliability.

d) Mechanical:—The package must be strong enough to permit the entire structure to withstand the stresses occurring during assembly, after assembly, during connections to other packages in a system, and in actual use. These stresses may include, among other things, frequent bending of the leads, dropping, soldering, etc. The mechanical properties of the package may limit the extremes of temperature and pressure that the device can be subjected to. Mechanical construction affects the ease of handling and flexibility of the processes that can be employed in joining the package in larger arrays. It also determines the space requirements of the package itself, and of the final system.

For example, during fabrication and operation, the IC package must withstand exposure to certain levels of shock, acceleration, vibration, soldering or welding heat, temperature cycling, thermal shock, moisture, and lead stress. Various tests have been designed to ensure that the package will withstand the stresses encountered in device assembly (installing the chip in the package and performing the final seal), system fabrication (mounting and attaching the package to the mother board), and system operation.

e) Radiation:—The package may have to isolate the actual device or circuit, within it, from the effects of light or other external radiation. A similar requirement may be to prevent the device or circuit from radiating to the external circuitry.

f) Magnetic:—The package may have to provide a shield against external magnetic fields, or may be required to provide a path for magnetic flux.

g) Hermetic: — The internal environment of the IC package must remain constant, since variations can cause failure. Although an exact correlation between leakage rate and device life is extremely difficult to establish, packages with a leakage rate of less than 10^{-8} cubic centimeters per second have proven to be reliable from a hermetic standpoint.

h) Space Efficiency: — The package should conform closely to the shape of its contents. Standardized packages are often unable to meet this criterion, but the poor fit can often be offset by an increase in circuit functions (i.e., more or larger chips).

i) Design Flexibility: — The device user will require that the IC package have certain features that facilitate handling and assembly. These features, collectively called flexibility, include the following:

(1) The package must be easily handled during device manufacturing, testing, shipping, and fabrication into a particular equipment. Handling difficulties encountered during any of these phases will result in an increase in equipment costs and often a decrease in reliability.

(2) The package must be amenable to various mounting techniques (planar, stacked, on edge, etc.), to standard interconnection methods (flow-soldering, dip soldering, welding, or thermocompression bonding), and to various means of heat dissipation.

(3) The package should have high-density potential. Integrated circuits play a major role in microminiature systems, where high-density packaging is mandatory.

13.2 PACKAGING DENSITY

The method usually applied to determine packaging density is to note the maximum number of components that can be enclosed within a certain unit of volume. This is usually accomplished by scaling a typical circuit or assembly. Very often it excludes the volume required for lead wires, inter-connections, environmental protection, heat sinks, and the fact that in actuality, different size circuits are generally fitted together. As a result, packaging density numbers are not always meaningful.

For purposes of comparison, however, and some examples of package densities, most television sets and radios have a density of approximately 100 parts/ft^3. Pocket transistor radios have about 3000 parts/ft^3. The micromodule has a density of 5×10^5 parts, while miniature resistors are at the 5×10^6 level. A standard one-watt resistor can be packaged at a density of 10,000 elements/ft^3. Integrated circuits can be packed at the rate of 1×10^8 parts/ft^3. The human brain has a density of $3 \times 10^{11}/ft^3$. For uniformity, the preceding packaging density numbers are based on parts per unit volume for systems or components less leads, contact pins, and encapsulation.

When sub-miniature components having a density less than 5×10^6 parts/ft^3 are used, components, assemblies and systems can readily be handled without the application of special tooling. When the density increases to 5×10^9 parts/ft^3, components can only be handled with such tools while assemblies and systems made from these components can still be handled in the normal manner. When the package density exceeds 5×10^9 parts/ft^3, the components can no longer be handled through manual means. Assemblies made from these components can be handled if tools are used, and systems can still be handled in the usual manner.

13.3 CIRCUIT PACKAGING METHODS

A wafer which has proven to be satisfactory on testing must be cut up for individual circuit packaging. Each circuit is separated from the array, for this purpose, by "dicing." Dicing can be accomplished by scribing and breaking, ultrasonic cutting or etching. In laying out the original artwork, each circuit in the array is spaced 10 to 20 mils from its neighbor to allow room for this operation. The amount of spacing provided depends upon the dicing method chosen. For production purposes, scribing and dicing is usually favored. Compared to ultrasonic cutting and etching, scribing and dicing can be accomplished quicker and with the least amount of waste. Both ultrasonic cutting and etching require inherently greater spacings between circuits, and are wasteful, in this respect, of usable surface area. Scribing and dicing result in dice of more uniform dimensions. When dicing by etching, photolithographic techniques are used to confine the etch to the desired areas.

Scribing is accomplished within the spacings provided, by pulling a loaded diamond stylus across the surface of the wafer. The wafer is waxed down to a glass plate which, in turn, is positioned for scribing. The glass plate is held on a vacuum table capable of coplanar translational and rotational motion. After scribing, the wafer is diced. Dicing is accomplished by removing the wafer from the plate, laying it down on a plastic tape, and flexing it along the scribed lines by drawing the tape over a sharp edge. Commercial machines with varying degrees of automation are available for both the scribing and dicing operations.

Some losses can be expected from the scribing and dicing operations. Many times, the causes of the losses cannot be remedied. For example, crystal strains which may be present in the wafer may ''give'' in preference to the scribed line, and cause the wafer to break in an unpredictable manner. Similarly, if the scribing operation is not carefully performed, cracks arising from nonuniformities in the scribed line, may be propagated

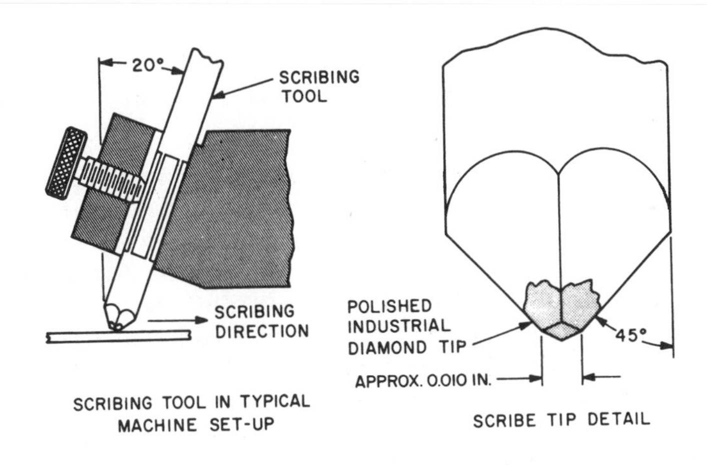

Fig. 13-2 Diamond tipped scribing tool produces cleavage channels around each circuit die for breaking them apart.

into the circuit dice during the breaking operation, thereby causing a decrease in yield.

Once a wafer has been diced, the dice or ''chips'' are cleaned in a solvent such as trichloroethylene. This solvent removes loose silicon particles, resulting from the scribing operation, and other organic contaminating films caused by handling. If the slices were mounted on a tape during scribing, the solvent also serves to dissolve the tape adhesive clinging to the back sides of the dice.

The packages and soldering or alloying preforms (if used) are cleaned in similar solvents, and then vacuum baked at 120° C or higher for periods of 3 hours or more to drive off gases absorbed on the surfaces of these parts. Cleaning of all parts is necessary to assure proper conditions for bonding of the silicon to the package. Component cleaning also provides protection against internal contamination which might result from the redeposition of contaminants on the surfaee of the chip. Such contaminants could arise from contaminants on the package itself which are vaporized during subsequent high-temperature processing.

The silicon chips or dice may be mounted in the package by alloying, soldering, or brazing, or by the use of cement. If electrical isolation is required, as in high-frequency circuits, the chip is mounted upon metallized lands on an insulator adapted to fit into the package. All package

Fig. 13-3 Manual dicing of alumina substrates.

metals and metallized lands are usually of metals which have temperature coefficients matching that of silicon. Kovar is often used for the packages and molybdenum-maganese for the lands. The insulating material is usually a ceramic, such as alumina, which provides good electrical isolation, but which is also a good thermal conductor. Eutectic alloying preforms are chosen so that the alloying temperature, necessary to secure the chip to the package or land, is lower than the eutectic aluminum-silicon temperature (approximately 576° C) used in making ohmic contact to the chip. This is done to prevent the realloying of the previously formed aluminum contacts. In most cases, gold doped with a trace of germanium or silicon is preferred for preform material, since it alloys readily with silicon and has the desired melting point range. The presence of the germanium or silicon in the gold slightly lowers its solubility in the silicon chip at the alloying temperature, and thus provides for shallow alloying of the gold

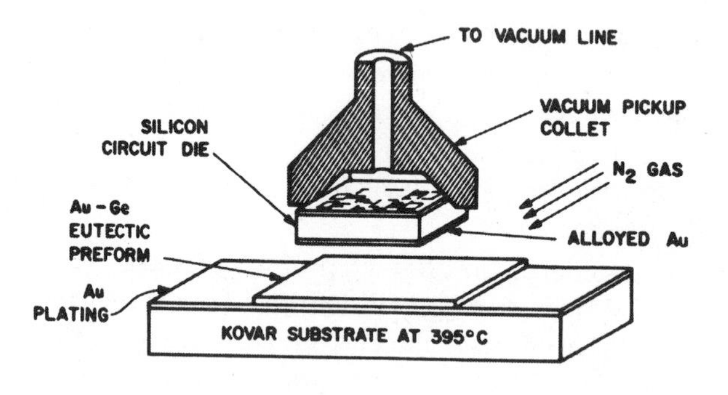

Fig. 13-4 The basic die bonding operation. Nitrogen gas minimizes oxidation during heating and bonding.

in the chip. Such preforms provide good bonding and, at the same time, form bonds which are not brittle. In general, solders have lower melting points but form brittle bonds which tend to fatigue on thermal cycling.

Alloying may be performed either in a furnace having a hydrogen or forming gas atmosphere, or on a special die-attachment machine provided with a forming gas atmosphere. The furnace process lends itself well to mass production, but it is not very satisfactory when using the relatively nonuniform dice that result from the scribing operation.

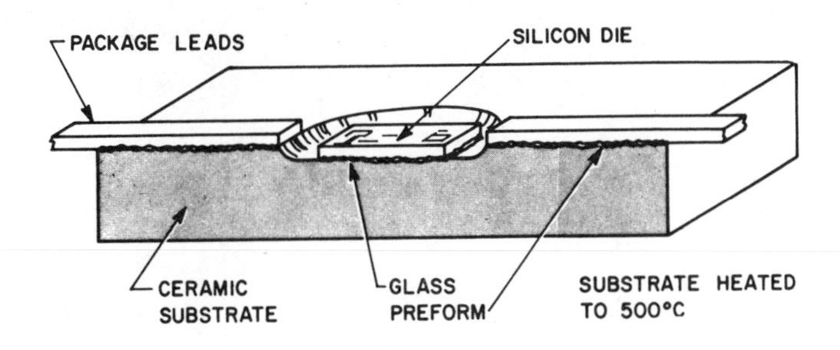

Fig. 13-5 With glass or ceramic packages, die attachment is done with a glass frit in an inert atmosphere.

13.3.1 TO-TYPE CANS

The advent of transistors brought about a major revision in electronic component packaging. Earlier active devices were situated in glass or metal-cased vacuum tubes. The small size and simplicity of the transistor necessitated the formulation of a new packaging concept compatible with the device. Since the semiconductor chip may be affected by its environment, the package had to be hermetically sealed against the elements. Whereas seven to twelve leads exited previously from the package, now only three leads had to be provided for. This simplified fixturing during the construction of the glass or metal envelope, as well as reduced somewhat glass to metal sealing problems.

One approach to transistor packaging was finally standardized – at least for low power transistors, through the TO series cans. These cans utilize a ''pie'' shape header through which the leads protrude and a ''top hat'' shaped cover. A glass-to-metal seal assures the hermeticity of the package where the leads protrude through the header. Special alloys were developed to match the thermal coefficient of expansion of the sealing glass. These alloys, notably Kovar, are used to form the metal parts of the transistor case. An enlarged view of a typical TO-series transistor is shown in Figure 13-6. The most popular sizes have been the TO-5 and the TO-18.

Using transistor cans to package the first microcircuits was a natural step, since the semiconductor chip is similar in size and shape to a transistor chip. However, as the state of the microelectronic art progressed, several shortcomings of this packaging method became apparent. Since complete circuits were packaged in one case, three leads were no longer sufficient to provide inputs and outputs to and from the circuit. The modified TO-5 can originated as a result of this shortcoming. The number of leads was increased to ten or twelve, and the diameter of the pin circle formed by the leads was increased. The overall height of the TO-5 case is not conducive to efficient packaging, since most of the internal volume is wasted. A reduction of the ''top hat'' height was brought about to reduce this problem.

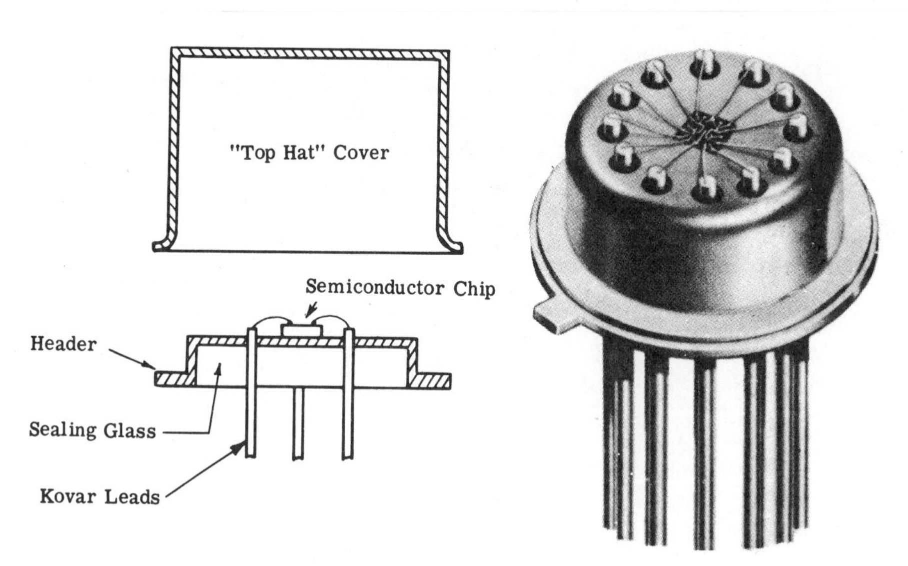

Fig. 13-6 Typical TO - type can

The TO container is made entirely of metal, with glass being used to insulate the leads from the package at the point of entry. The metal and glass materials are themselves sufficiently impervious to outside contaminants to adequately protect the chip. However, for the finished package to provide the required amount of protection, the various metal-to-metal and metal-to-glass seals must also be impervious to outside contamination. The metal-to-metal seal produces no particular problem and may be accomplished by many techniques such as soldering, brazing, or hot or cold welding. The primary problem resides essentially in producing an adequate glass-to-metal seal. A basic difficulty encountered in this area, is due to the condition that most metals have a significantly different thermal coefficient of expansion from most glasses. When two materials with different thermal coefficients of expansion are joined directly together, they tend to separate as the temperature of the seal is varied, and one material contracts or expands more than the other. This problem, as far as glass and metal are concerned, was partially solved by the development of a special metal and glass combination with similar coefficients of expansion. The metal and glass combination, most commonly used, is Kovar which has a thermal coefficient of expansion of 47×10^{-7}, and glass with a coefficient of expansion of 46×10^{-7}. While this close matching of coefficients may produce a reliable seal over a limited temperature range, it does not completely solve the problem if an extended temperature range is encountered. To insure a reliable seal over the complete temperature range to which the package may be subjected, either the two materials must be exactly matched or a buffer material between the glass and metal is required. The purpose of the buffer is to absorb the difference in expansion of the two materials, thus preventing any separations. On glass to metal seals, the buffer is an oxide layer grown on the metal. When the metal and glass are fused together, the outer surface of the oxide layer dissolves into the glass providing a continuous transition from metal-to-metal oxides, to metal oxides in glass, to glass.

In this way, the buffer of metal oxide provides the transition medium so that there is no discontinuity in composition from metal to glass.

Since the metal oxides are more porous to gases then the metal itself, the thickness of the oxide layer must be kept small. If made too thin, however, the entire oxide layer may dissolve in the glass. This would eliminate the buffer and produce a leaky seal.

In the general configuration of the TO container, the leads and eyelet are Kovar, and the can or top is nickel, silver, or Kovar. The portion of the container which includes the leads, eyelet, and glass seal is normally referred to as the header. The two header configurations most frequently encountered in TO containers, are illustrated in Figure 13-7. The header shown in Figure 13-7a has an all Kovar base, with sealing glass being used only in the holes surrounding the leads. The header shown in Figure 13-7b has a hollowed out Kovar base with sealing glass being used not only in holes surrounding the leads, but also in the cavity. The purpose of the

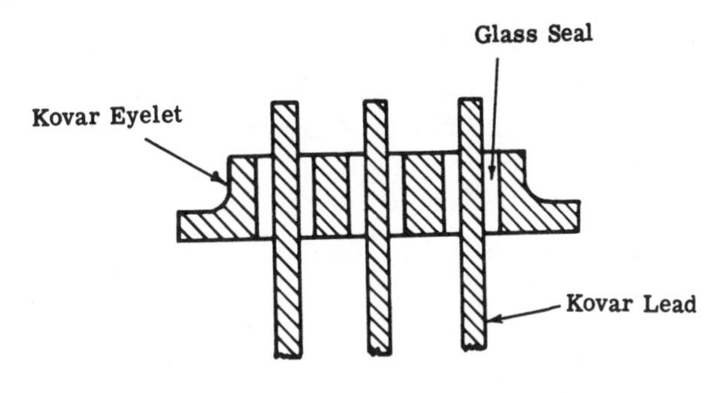

(a) All Kovar Header

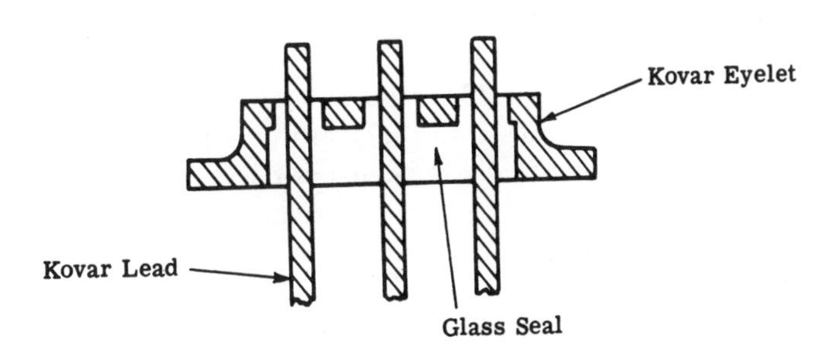

(b) Kovar-Glass Header

Fig. 13-7 Typical TO - type headers

flange around the circumference of the base piece, is to provide a platform to which they can may be welded.

Both the height and the diameter dimension of the various TO containers vary widely, depending upon the size of the chip inside the package, and upon the amount of heat sink required. The various size containers have been standardized, however, by the Joint Electron Device Engineering Council (JEDEC), into a series of TO containers of varying sizes. Each of the containers in the series, TO-1 through TO-51, is produced to a standard set of dimensions by all manufacturers.

The most common TO containers utilized for integrated circuits, are the TO-5, TO-18, and TO-47 packages. The only difference between the TO containers in the diagrams and those used as packages for integrated circuits, is that the integrated circuit containers have a greater number of leads and a larger pin diameter.

As indicated earlier, in an attempt to increase the volumetric efficiency of the integrated circuit containers, certain modifications in the typical TO configuration are sometimes made. The most frequently encountered modification is the reduction of the height of the container. These modifications are not standardized, and the height of a modified TO-5 can, for example, may be .15″, .175″, .2″, etc., depending upon the manufacturer and the requirements of the user. In general, no modification is made in the diameter of the container.

The manufacturing processes involved in the production of TO packages have been standardized to a great extent. The first step in the manufacturing process is to clean the metal and glass parts and then de-oxidize the metal parts. Next, eyelet and leads are oxidized to a controlled oxide thickness. As indicated above, the thickness of the oxide layer must be very closely controlled to assure that a good hermetic seal is obtained between the glass and metal parts. Normally the amount of oxide is controlled by temperature, time, and moisture content. After oxidation, the leads, eyelet, and glass preform are assembled in a graphite mold for fusing. Fusing, which is normally accomplished in a neutral atmosphere to preserve the oxide structure, takes place at a temperature of approximately 1000° C. At this temperature, the glass melts and flows into the mold cavity, metallurgically bonding to the oxide of the Kovar parts, and forming a hermetic seal. When the header (leads, glass, and eyelet) exits from the furnace, it is cleaned, clipped to the desired post height, and gold plated. The header is now ready for circuit die attachment. The die is bonded to the eyelet of the header, utilizing the gold-silicon eutectic which forms as the silicon of the circuit die alloys with the gold in the header. Fine wires are then attached from the circuit die to the leads on the header, and the circuit is tested electrically. Final sealing is accomplished by welding the can to the flange on the eyelet. The welding is usually done in a neutral atmosphere, to assure stability of the circuit under operating conditions.

Figure 13-8 is a cutaway view of a modified TO-5 package, with heat-flow paths indicated. Most of the heat is conducted from the chip to the header assembly. The header assembly then conducts the heat radially to the lip and sides of the package, where it is transferred to the atmosphere by convection and radiation.

The heat-dissipating capability of a modified TO package can be improved by either of two methods. A metal container (with or without fins) can be fitted over the top of the package, to increase its surface area, or forced-air cooling may be used.

The most signifivant thermal parameters in IC packages are the junction-to-case and junction-to-ambient thermal resistances (θ_{JC} and θ_{JA}, respectively). Values of these parameters vary among manufacturers, but are typically 80°C per watt for θ_{JC} and 180°C per watt for θ_{JA} for the modified TO-5 package. The θ_{JC} of a hybrid microcircuit mounted in the modified TO-5 package is approximately 10 percent higher than that of a standard circuit because of the ceramic disc between the chips and the header.

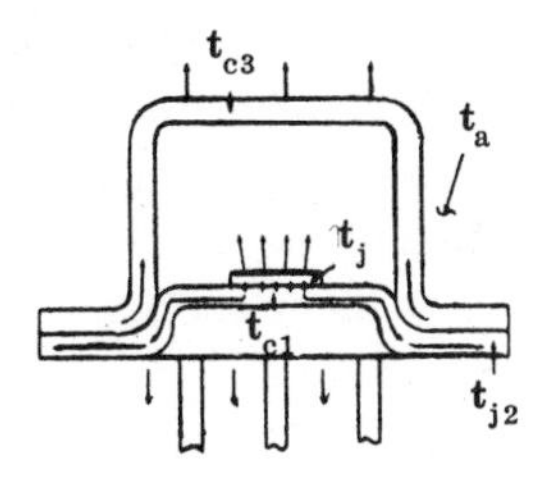

Fig. 13-8 Thermal paths in TO package.

13.3.2 GLASS AND METAL FLAT PACKAGES

Since the overall configuration of the TO-type case is not compatible with the shape of the semiconductor chip or thin-film substrate, a new package geometry has been generated. This package is a flat rectangular parallelepiped, which closely resembles the geometry of the chip or thin film substrate that it incloses. The package has several advantages over the TO-type cases: 1) greater volumetric efficiency, 2) Leads may be brought out from four sides of the package, greatly increasing thereby the flexibility and adaptibility of the package, and 3) the flat, rectangular shape of the package allows more compact system packaging.

There are two major approaches to flat pack construction – the glass and metal flat pack, and the encapsulated type flat pack. The glass and metal package utilizes many of the proven sealing techniques developed for TO type cans. These packages have proven reliability, both from the strength standpoint and the hermetical standpoint. Thin film substrates are generally not packaged in this type container, due to the larger size of the substrate.

The encapsulated type package is not actually a package in itself. Generally, the semiconductor chip or thin film substrate is embedded in an epoxy or some other special plastic type resin. The result is a circuit which is surrounded by some plastic material, effectively sealing the circuit against its environment. A variation of this method is to place the circuit wafer in an open plastic or epoxy box and actually potting the circuit with epoxy. Most thin film circuits are packaged using this encapsulated approach.

Flat Packages

Whereas the TO cans had been designed as a container for transistors and later adapted to microelectronic circuits, the flat pack was specifically designed as a container for microelectronic circuits. Primarily to minimize the wasted volume of the TO cans, the flat pack was designed to conform as closely as possible to the geometry of the "chip" which it was to contain. Another innovation of the flat pack was to pass the leads through the walls of the package rather than through the base, as in the case of the TO can. This resulted in several advantages. First, with the leads projecting from the walls of the package rather than the base, the packaged circuits are more conformable to planar system packaging techniques. Second, by utilizing the walls of the package, there is a capability for bringing out more leads from a package of a given size, and still retain a reasonable spacing between leads. This increase in leads permits increased circuit versatility, or an increase in the number of functions a single circuit can perform. On the other hand, if it is desired to have a given number of leads attached to a package, the spacing between leads can be greater. Third, passing the leads through the walls of the package, removes the necessity of drilling holes in the base.

In general, manufacturers utilize a matched glass-to-metal seal where the leads enter the package. This is to allow the proven techniques of TO can technology to be utilized as much as possible in flat pack manufacture. Although almost all flat packs utilize matched glass-metal seals, package characteristics such as material, configuration, etc., vary considerably from manufacturer to manufacturer. Most of the various packages belong, however, to one of three major groups: (1) packages which have a ceramic or metal base and lid, and an *all glass wall;* (2) packages which have a *ceramic wall,* base and lid; and (3) packages which have a *glass base and wall,* and a glass, ceramic, or metal lid.

The packages of the first group generally utilize a flat, ceramic or metal plate, which can be produced with a stamping operation, for the base and lid. A glass preform forms the wall, and a metal frame is generally used to provide a base for lid attachment. This type of package has the advantage of using the same Kovar or alumina which has been extensively used in TO can manufacture. By using the metal base, an excellent means for heat dissipation is obtained. The use of an all glass wall assures sufficient volume of glass around the leads for a reliable seal. A disadvantage of this type of package resides in the number of parts which must be assembled. A minimum of four pieces, exclusive of the leads, is required. Since a major cost factor of these packages is the labor required for assembly, one-piece construction can become a primary consideration.

A typical example of the group (1) package is shown in Figure 13-9. The base, leads, frame, and lid are Kovar. The walls of the package are composed of a Boro-silicate glass.

The base is a flat, rectangular Kovar plate, produced by a stamping operation. If device isolation is a requirement, a ceramic (alumina) base is used. The rectangular Kovar leads, typically .005″ × .015″, pass through the glass walls on .050″ centers. The Kovar frame covers the top of the glass walls to provide a metallic base for lid attachment. The lid is a flat, rectangular Kovar plate, identical to the base, with outside dimensions matching those of the metal frame. The lid is attached using either a gold-germanium or a gold-tin eutectic preform. All metal parts of the package are electrolytically gold plated to permit ease in device mounting and lead attachment.

Continued in Vol. II